Fundamentals
of
Plant
Physiology

Fundamentals
of
Plant
Physiology

Lincoln Taiz
Professor Emeritus, University of California, Santa Cruz

Eduardo Zeiger
Professor Emeritus, University of California, Los Angeles

Ian Max Møller
Professor Emeritus, Aarhus University, Denmark

Angus Murphy
Professor, University of Maryland

 SINAUER ASSOCIATES

NEW YORK   OXFORD
OXFORD UNIVERSITY PRESS

About the cover

The eight plant illustrations shown on the cover are (clockwise from upper left): Arabidopsis, the pitcher plant, wheat, eastern skunk cabbage, mangrove, orchid, fall foliage, and *Lithops*. Arabidopsis, whose genome was the first to be sequenced, is the preferred model system for studying the molecular basis of plant growth and development. The pitcher plant is a carnivorous plant that supplements its nitrogen supply from the soil by digesting and assimilating insect protein. Wheat is an important cereal crop that, together with rice, provides more food energy for the world's population than any other crop. Skunk cabbage is an unusual plant whose flower generates heat that helps it to emerge through the winter snow in early March; the heat also helps to volatilize the putrid smell the flower generates, that attracts fly pollinators. Mangrove trees, which grow with their roots in saline or brackish water, are adapted to salinity. Epiphytic orchids grow on the surfaces of trees and must obtain their water and mineral nutrients from the air, rain, and dust to which they are exposed. The fall foliage of deciduous trees exemplifies the process of seasonal leaf senescence. Finally, *Lithops*, or stone plant, is an example of a desert-adapted species that exhibits CAM-type photosynthesis.

Fundamentals of Plant Physiology

Oxford University Press is a department of the University of Oxford. It furthers the University's objective of excellence in research, scholarship, and education by publishing worldwide. Oxford is a registered trade mark of Oxford University Press in the UK and certain other countries.

Published in the United States of America by Oxford University Press
198 Madison Avenue, New York, NY 10016, United States of America

Address editorial correspondence to:
Sinauer Associates
23 Plumtree Road
Sunderland, MA 01375 U.S.A.
publish@sinauer.com

Address orders, sales, license, permissions, and translation inquiries to:
Oxford University Press U.S.A.
2001 Evans Road
Cary, NC 27513 U.S.A.
Orders: 1-800-445-9714

Library of Congress Cataloging-in-Publication Data

Names: Taiz, Lincoln, author. | Zeiger, Eduardo, author. |Møller, I. M. (Ian Max), 1950- author. | Murphy, Angus S., author.
Title: Fundamentals of plant physiology / Lincoln Taiz, Eduardo Zeiger, Ian Max Møller, Angus Murphy.
Description: First edition. | New York, NY : Published in the United States of America by Oxford University Press, 2018. | Includes bibliographical references and index.
Identifiers: LCCN 2018014331 | ISBN 9781605357904 (paperbound)
Subjects: LCSH: Plant physiology--Textbooks.
Classification: LCC QK711.2 .T34 2018 | DDC 571.2--dc23
LC record available at https://lccn.loc.gov/2018014331

9 8 7 6 5 4 3 2 1
Printed in the United States of America

Brief Contents

Table of Contents

CHAPTER 12
Signals and Signal Transduction   341

CHAPTER 18
Biotic Interactions 507

CHAPTER 19
Abiotic Stress   537

Preface

Over the course of publishing six editions of *Plant Physiology* (now *Plant Physiology and Development*), it has become abundantly clear to us that, as a consequence of the spectacular advances in plant developmental genetics, the field of plant physiology has broadened in scope and increased in depth to the point where it is no longer possible for a single textbook to serve the needs of all courses on plant physiology.

Based on extensive feedback from instructors affiliated with a wide range of universities and colleges, we designed *Fundamentals of Plant Physiology* for instructors seeking a shorter text that emphasizes concepts as opposed to detailed, comprehensive coverage. To accommodate instructors wishing to focus their lectures on canonical plant physiology topics, the treatment of plant development, including hormonal signaling and photoreceptors, has been greatly condensed and reorganized. In keeping with a conceptual approach, genetic pathways have been streamlined by prioritizing functional mechanisms and by substituting descriptive terms for three-letter gene names whenever possible.

Throughout the text, explanations have been revised for easier comprehension, and figures have been simplified and annotated for greater clarity. To further improve readability, we have switched to a single-column format that includes a running glossary in the margins. In working on *Fundamentals of Plant Physiology*, our primary goal has been to provide a more concise, accessible, and focused treatment of the field of plant physiology, while at the same time maintaining the high standard of scientific accuracy and pedagogical richness for which *Plant Physiology and Development* is known.

We wish to thank Massimo Maffei for his suggested revisions to the physiological chapters. This book could not have been produced without the extensive help and support of the outstanding team of professionals at Sinauer Associates, now an imprint of Oxford University Press. We are especially grateful to our project editor, Laura Green, whose considerable expertise in plant physiology and biochemistry, and thorough dissection and critique of each chapter, which sometimes entailed major reorganizing, helped us to optimize the logical flow of topics and avoid as many factual errors as is humanly possible. Our second line of defense against errors and stylistic problems was our tireless copyeditor, Liz Pierson, who did wonders to shape the newly designed chapters into a seamless, integrated, and coherent whole. As always, the talented artist Elizabeth Morales implemented all the changes we requested to existing figures and created several new, attractive figures as well.

We owe the new single-column format of the book, including the running glossary, to the creative talents of Chris Small and his production team. Ann Chiara designed the inside of the book and crafted the layout. Another new design feature was use of chapter openers featuring plants pertinent to each chapter's content. Mark Siddall, our photo researcher, is responsible for finding the many fine and appropriate photos for use as chapter openers, and Beth Roberge Friedrichs deserves credit for designing the distinctive new cover for *Fundamentals of Plant Physology*. We are also grateful to Johanna Walkowicz for laying the foundation for the book's conception by devising and compiling surveys obtained from a wide selection of instructors. Thanks to her detailed analysis of the data, we were able to identify those topics of plant physiology that were most often covered in courses.

Finally, we extend our deepest gratitude to our publisher, Andy Sinauer, whose creative and innovative contributions to *Fundamentals of Plant Physiology*, as well as to previous editions of *Plant Physiology and Development*, are too numerous to list. Andy is retiring this year, and we are extremely sorry to see him go. He leaves behind a magnificent legacy, Sinauer Associates, that is second to none among publishers of scientific textbooks. Although we will sorely miss Andy's wisdom and invaluable insights, we look forward to a productive ongoing relationship with our new publisher, Oxford University Press.

Lincoln Taiz   Eduardo Zeiger

Ian Max Moller Angus Murphy

April 2018

Editors

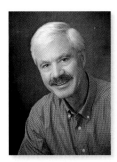 **Lincoln Taiz** is Professor Emeritus of Molecular, Cellular, and Developmental Biology at the University of California at Santa Cruz. He received his Ph.D. in Botany from the University of California at Berkeley in 1971. Dr. Taiz's main research focus has been on the structure, function, and evolution of vacuolar H⁺-ATPases. He has also worked on gibberellins, cell wall mechanical properties, metal tolerance, auxin transport, and stomatal opening.

 **Eduardo Zeiger** is Professor Emeritus of Biology at the University of California at Los Angeles. He received a Ph.D. in Plant Genetics at the University of California at Davis 1970. His research interests include stomatal function, the sensory transduction of blue-light responses, and the study of stomatal acclimations associated with increases in crop yields.

 **Ian Max Møller** is Professor Emeritus of Molecular Biology and Genetics at Aarhus University, Denmark. He received his Ph.D. in Plant Biochemistry from Imperial College, London, UK. He has worked at Lund University, Sweden and, more recently, at Risø National Laboratory and the Royal Veterinary and Agricultural University in Copenhagen, Denmark. Professor Møller has investigated plant respiration throughout his career. His current interests include turnover of reactive oxygen species and the role of protein oxidation in plant cells.

 **Angus Murphy** has been a Professor and Chair of the Department of Plant Science and Landscape Architecture at the University of Maryland since 2012. He earned his Ph.D. in Biology from the University of California at Santa Cruz in 1996 and moved to Purdue University as an assistant professor in 2001. Dr. Murphy studies ATP-Binding Cassette transporters, the regulation of auxin transport, and the mechanisms by which transport proteins are regulated in plastic plant growth.

Contributors

Sarah M. Assmann

Christine Beveridge

Robert E. Blankenship

Arnold J. Bloom

Eduardo Blumwald

John Browse

Bob B. Buchanan

Victor Busov

John Christie

Daniel J. Cosgrove

Susan Dunford

James Ehleringer

Jürgen Engelberth

Lawrence Griffing

N. Michele Holbrook

Andreas Madlung

Ron Mittler

Gabriele B. Monshausen

Wendy Peer

Allan G. Rasmusson

Darren R. Sandquist

Graham B. Seymour

Sally Smith

Joe H. Sullivan

Heven Sze

Bruce Veit

Philip A. Wigge

Ricardo A. Wolosiuk

Media and Supplements

to accompany *Fundamentals of Plant Physiology*

eBook

(ISBN 978-1-60535-791-1)

Fundamentals of Plant Physiology is available as an eBook, in several different formats, including RedShelf, VitalSource, and Chegg. All major mobile devices are supported.

For the Instructor

Ancillary Resource Center (oup-arc.com)

Includes the following resources to aid instructors in developing their courses and delivering their lectures:

- *Textbook Figures and Tables*: All the figures and tables from the textbook are provided in JPEG format, reformatted for optimal readability, with complex figures provided in both whole and split formats.

- *PowerPoint Resources*: A PowerPoint presentation for each chapter of the textbook includes all of the figures and tables from the chapter, with figure numbers and titles on each slide and complete captions in the Notes field.

To learn more about any of these resources, or to get access, please contact your local OUP representative.

1 Plant and Cell Architecture

Plant physiology is the study of plant *processes*—how plants grow, develop, and function as they interact with their physical (abiotic) and living (biotic) environments. Although this book will emphasize the physiological and biochemical functions of plants, it is important to recognize that, whether we are talking about gas exchange in the leaf, water conduction in the xylem, photosynthesis in the chloroplast, ion transport across membranes, signal transduction pathways involving light and hormones, or gene expression during development, all of these physiological and biochemical functions depend entirely on structures. Function derives from structures interacting at every level of scale, from molecules to organisms.

The cell is the fundamental organizational unit of plants and all other living organisms. The term *cell* is derived from the Latin *cella*, meaning "storeroom" or "chamber." It was first used in biology in 1665 by the English scientist Robert Hooke to describe the individual units of the honeycomb-like structure he observed in cork under a compound microscope. The cork "cells" Hooke observed were actually the empty lumens of dead cells surrounded by cell walls, but the term is an apt one, because cells are the basic building blocks that define plant structure.

Groups of specialized cells form specific tissues, and specific tissues arranged in particular patterns are the basis of three-dimensional organs. Plant anatomy is the study of the macroscopic arrangements of cells and tissues within organs, and plant cell biology is the study of the organelles and other small components that make up each cell. Improvements in both light and electron microscopy continue to reveal the astonishing variety and dynamics of cellular processes and their contributions to physiological functions.

This chapter provides an overview of the basic anatomy and cell biology of plants, from the macroscopic structure of organs and tissues to the microscopic ultrastructure of cellular organelles. Subsequent chapters will treat these structures in greater detail from the perspective of their physiological and developmental functions at different stages of the plant life cycle.

Plant Life Processes: Unifying Principles

The spectacular diversity of plant size and form is familiar to everyone. Plants range in height from less than 1 cm to more than 100 m. Plant morphology, or form, is also surprisingly diverse. At first glance, the tiny plant duckweed (*Lemna*) seems to have little in common with a giant saguaro cactus or a redwood tree. No single plant shows the entire spectrum of environmental adaptations that allow plants to occupy almost every niche on Earth, so plant physiologists often study model organisms that are representative of major plant functions and are easy to manipulate in research studies. These model systems are useful because all plants, regardless of their specific adaptations, carry out fundamentally similar processes and are based on the same architectural plan.

We can summarize the major unifying principles of plants as follows:

- As Earth's primary producers, plants and algae are the ultimate solar collectors. They harvest the energy of sunlight by converting light energy to chemical energy, which they store in bonds formed when they synthesize carbohydrates from carbon dioxide and water.

- Except for certain reproductive cells, plants do not move from place to place; they are sessile. As a substitute for motility, they have evolved the ability to grow toward essential resources, such as light, water, and mineral nutrients, throughout their life span.

- Plants are structurally reinforced to support their mass as they grow toward sunlight against the pull of gravity.

- Plants have mechanisms for moving water and minerals from the soil to the sites of photosynthesis and growth, as well as mechanisms for moving the products of photosynthesis to nonphotosynthetic organs and tissues.

- Plants lose water continuously by evaporation and have evolved mechanisms for avoiding desiccation.

- Plants develop from embryos that derive nutrients from the mother plant, and these additional food stores facilitate the production of large self-supporting structures on land.

Plant Classification and Life Cycles

Based on the principles listed above, we can define plants generally as sessile, multicellular organisms derived from embryos, adapted to land, and able to convert carbon dioxide into complex organic compounds through the process of photosynthesis. This broad definition includes a wide spectrum of organisms, from the mosses to the flowering plants (**Box 1.1**).

With the exception of the great coniferous forests of Canada, Alaska, and northern Eurasia, angiosperms dominate the landscape today. More than 250,000 species are known, with tens of thousands of additional undescribed species predicted by computer models. Many of the predicted species are imperiled because they occur primarily in regions of rich biodiversity where habitat destruction is common. The major anatomical innovation of the angiosperms is the flower;

Box 1.1   Evolutionary Relationships among Plants

Plants share with (mostly aquatic) green algae the primitive trait that is so important for photosynthesis in both clades: Their chloroplasts contain the pigments chlorophyll *a* and *b* and β-carotene. **Plants**, or **embryophytes**, share the evolutionarily derived traits for surviving on land that are absent in the algae (see figure). Plants include the **nonvascular plants**, referred to as **bryophytes** (mosses, hornworts, and liverworts), and the **vascular plants**, or **tracheophytes**. The vascular plants, in turn, consist of the **non-seed plants** (ferns and their relatives) and the **seed plants** (gymnosperms and angiosperms).

Because plants have many agricultural, industrial, timber, and medical uses, as well as an overwhelming dominance in terrestrial ecosystems, most research in plant biology has focused on the plants that have evolved in the last 300 million years, the seed plants (see figure). The **gymnosperms** (from the Greek for "naked seed") include the conifers, cycads, ginkgo, and gnetophytes (which include *Ephedra*, a popular medicinal plant). About 800

species of gymnosperms are known. The largest group of gymnosperms is the **conifers** ("cone-bearers"), which include such commercially important forest trees as pine, fir, spruce, and redwood. The evolution of the **angiosperms** (from the Greek for "vessel seed"), remains, in the words of Darwin, an "abominable mystery." The earliest known angiosperm fossil dates to the early Cretaceous, ~125 million years ago. However, based on analyses of the DNA sequences of extant angiosperms, the basal (i.e., primitive) angiosperms (Amborellales, Nyphaeles, and Austrobaileyales) may have appeared much earlier, in the Middle to Late Triassic period, ~224–240 million years ago. The same study suggests that the diversification of the angiosperms into the monocots, magnoliids, and eudicots probably occurred during the Jurassic period, ~154–191 million years ago, around the time when important pollinating insects, such as bees and butterflies, were also evolving. If these estimates are correct, fossils corresponding to these earlier dates may turn up in the future.

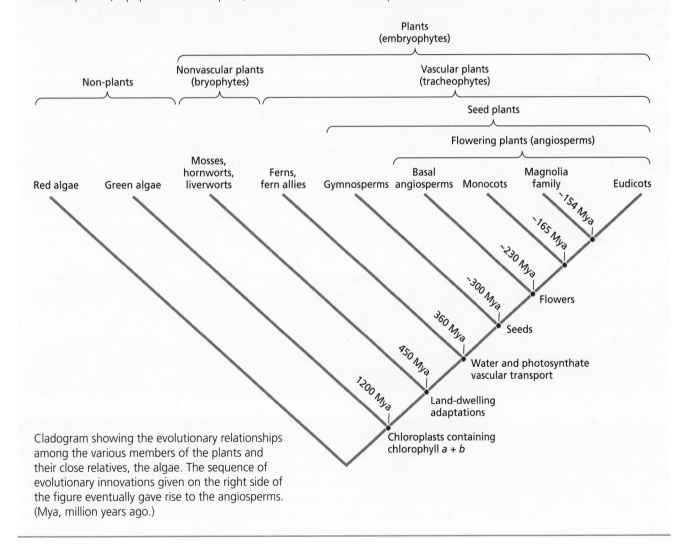

Cladogram showing the evolutionary relationships among the various members of the plants and their close relatives, the algae. The sequence of evolutionary innovations given on the right side of the figure eventually gave rise to the angiosperms. (Mya, million years ago.)

alternation of generations
The presence of two genetically distinct multicellular stages, one haploid and one diploid, in the plant life cycle. The haploid gametophyte generation begins with meiosis, while the diploid sporophyte generation begins with the fusion of sperm and egg.

gamete A haploid (1N) reproductive cell.

meiosis The "reduction division" whereby two successive cell divisions produce four haploid (1N) cells from one diploid (2N) cell. In plants with alternation of generations, spores are produced by meiosis. In animals, which don't have alternation of generations, gametes are produced by meiosis.

spores Reproductive cells formed in plants by meiosis in the sporophyte generation. They give rise by mitotic divisions to the gametophyte generation.

sporophyte The diploid (2N) multicellular structure that produces haploid spores by meiosis.

mitosis The ordered cellular process by which replicated chromosomes are distributed to daughter cells formed by cytokinesis.

gametophyte The haploid (1N) multicellular structure that produces haploid gametes by mitosis and differentiation.

pollen Small structures (microspores) produced by anthers of seed plants. Contain haploid male nuclei that will fertilize the egg in the ovule.

fertilization The formation of a diploid (2N) zygote from the cellular and nuclear fusion of two haploid (1N) gametes, the egg and the sperm. In angiosperms, fertilization also involves fusion of a second sperm nucleus with the haploid nuclei (usually two) of the central cell to form the endosperm (usually triploid).

megaspore The haploid (1N) spore that develops into the female gametophyte.

microspores The haploid (1N) cell that develops into the pollen tube or male gametophyte.

monoecious Refers to plants in which male and female flowers are found on the same individuals, such as cucumber (*Cucumis sativus*) and maize (corn; *Zea mays*).

hence they are referred to as flowering plants, distinguished from gymnosperms by the presence of a carpel that encloses the seeds.

Plant life cycles alternate between diploid and haploid generations

Plants, unlike animals, alternate between two distinct multicellular generations to complete their life cycle, a distinctive feature termed **alternation of generations**. One generation has diploid cells, cells with two copies of each chromosome and abbreviated as having 2N chromosomes, and the other generation has haploid cells, cells with only one copy of each chromosome, abbreviated as 1N. Each of these multicellular generations may be more or less physically dependent on the other, depending on their evolutionary grouping.

When diploid (2N) animals such as humans produce haploid **gametes**, egg (1N) and sperm (1N), they do so directly by the process of **meiosis**, cell division resulting in a reduction of the number of chromosomes from 2N to 1N. This cycle is depicted in the inner portion of **Figure 1.1**. In contrast, the products of meiosis in diploid plants are **spores**, and diploid plant forms are therefore called **sporophytes**. Each spore is capable of undergoing **mitosis**, cell division that does not change the number of chromosomes in the daughter cells, to form a new haploid multicellular individual, the **gametophyte**. These cycles are depicted in the outer portion of Figure 1.1. The haploid gametophytes produce gametes, egg and sperm, by simple mitosis, whereas haploid gametes in animals are produced by meiosis. This is a fundamental difference between plants and animals and gives the lie to some stories about "the birds and the bees"—bees do not carry around sperm to fertilize female flowers, they carry around the male gametophyte, the **pollen**, which is a multicellular structure that produces sperm cells. When placed on receptive sporophytic tissue, the pollen grain germinates to form a pollen tube that must grow through sporophytic tissue until it reaches the female gametophyte. The male gametophyte penetrates the female gametophyte and releases sperm to fertilize the egg.

Once the haploid gametes fuse and **fertilization** takes place to create the 2N zygote, the life cycles of animals and plants are similar (see Figure 1.1). The 2N zygote undergoes a series of mitotic divisions to produce the embryo, which eventually grows into the mature diploid adult.

Thus, all plant life cycles encompass two separate generations: the diploid, spore-producing sporophyte generation and the haploid, gamete-producing gametophyte generation. A line drawn between fertilization and meiosis divides these two separate stages of the generalized plant life cycle shown in Figure 1.1. Increasing the number of mitoses between fertilization and meiosis increases the size of the sporophyte generation and the number of spores that can be produced. Having more spores per fertilization event could compensate for low fertility when water becomes scarce on land. This could explain the marked tendency for the increase in size of the sporophyte generation, relative to the gametophyte generation, during the evolution of plants.

The sporophyte generation is dominant in the seed plants (gymnosperms and angiosperms) and gives rise to the **megaspores**, which develop into the female gametophyte, and the **microspores**, which develop into the male gametophyte (see Figure 1.1). Both megaspores and microspores produce gametophytes with relatively few cells compared with the sporophyte. In the vast majority of angiosperms, both male and female gametophytes occur in a single hermaphroditic ("perfect") flower, as in tulips. In some angiosperms, the male and female gametophytes occur on different flowers on the same plant, as in the male (staminate) and female (pistillate) flowers of maize (corn; *Zea mays*). Such species are referred to as **monoecious** (from the Greek for "one house"). Alternatively, if the male and female flowers occur on separate individuals, as in spinach plants or willow

SPOROPHYTE GENERATION

GAMETOPHYTE GENERATION

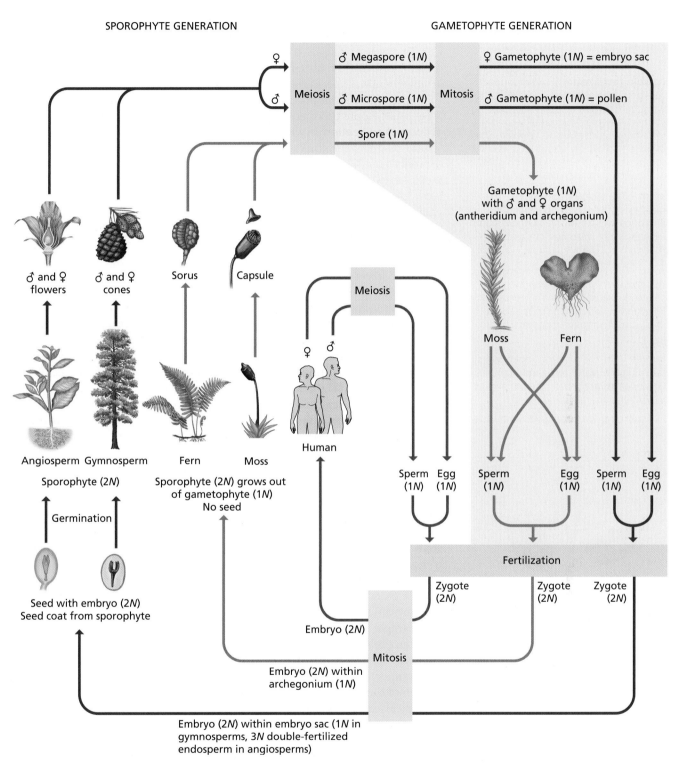

Figure 1.1   Diagram of the generalized life cycles of plants and animals. In contrast to animals, plants exhibit alternation of generations. Rather than producing gametes directly by meiosis as animals do, plants produce vegetative spores by meiosis. These 1*N* (haploid) spores divide to produce a second multicellular individual called the gametophyte. The gametophyte then produces gametes (sperm and egg) by mitosis. Following fertilization, the resulting 2*N* (diploid) zygote develops into the mature sporophyte generation, and the cycle begins again. In angiosperms, the process of double fertilization produces a 3*N* (triploid) or higher ploidy level (see Chapter 17) feeding tissue called the endosperm.

dioecious Refers to plants in which male and female flowers are found on different individuals, such as spinach (*Spinacia*) and hemp (*Cannabis sativa*).

double fertilization A unique feature of all angiosperms whereby, along with the fusion of a sperm with the egg to create a diploid zygote, a second male gamete fuses with the polar nuclei in the embryo sac to generate the endosperm tissue (with a triploid or higher number of chromosomes).

stem The typically above ground primary axis of the shoot that bears leaves and buds. May also occur underground in the form of rhizomes, corms, and tubers.

root The organ, usually underground, that serves to anchor the plant in the soil, and to absorb water and mineral ions, and conduct them to the shoot. In contrast to shoots, roots lack buds, leaves, or nodes.

leaves The main lateral appendages radiating out from stems and branches. Green leaves are usually the major photosynthetic organs of the plant.

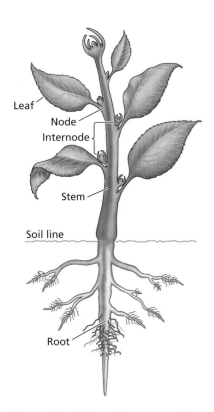

Figure 1.2 Schematic representation of the body of a typical eudicot.

trees, the species is termed **dioecious** (from the Greek for "two houses"). In gymnosperms, ginkos and cycads are dioecious, while conifers are monoecious. Conifers produce female cones, called megastrobili, which are usually higher up on the tree, and male cones, called microstrobili, on the lower branches, an arrangement that increases the chances of cross-pollination with other trees.

Sperm and egg production, as well as the dynamics of fertilization, differs among gametophytes of seed plants. Angiosperms carry out the unique process of **double fertilization**, in which two sperm cells are produced, only one of which fertilizes the egg. The other sperm fuses with two nuclei in the female gametophyte to produce the 3N (three sets of chromosomes) endosperm, the storage tissue for the angiosperm seed. (Some angiosperms produce endosperm of higher ploidy levels; see Chapter 17.) The storage tissue for the seed in gymnosperms is 1N gametophytic tissue because there is no double fertilization (see Figure 1.1). So the seed of seed plants is not at all a spore (defined as a cell that produces the gametophyte generation), but it does contain gametophytic (1N) storage tissue in gymnosperms and gametophyte-derived 3N storage tissue in angiosperms.

In the lower plants such as ferns and mosses, the sporophyte generation gives rise to spores that grow into adult gametophytes that then have regions that differentiate into male and female structures, the male antheridium and the female archegonium. In ferns the gametophyte is a small monoecious prothallus, which has antheridia and archegonia that divide mitotically to produce motile sperm and egg cells, respectively. The dominant leafy gametophyte generation in mosses contains antheridia and archegonia on the same (monoecious) or different (dioecious) individuals. The motile sperm then enters the archegonium and fertilizes the egg, to form the 2N zygote, which develops into an embryo enclosed in the gametophytic tissue, but no seed is formed. The embryo directly develops into the adult 2N sporophyte.

Overview of Plant Structure

Despite their apparent diversity, all seed plants have the same basic body plan (**Figure 1.2**). The vegetative body is composed of three organs—the stem, the root, and the leaves—each with a different direction, or polarity, of growth. The **stem** grows upward and supports the aboveground part of the plant. The **root**, which anchors the plant and absorbs nutrients and water, grows down below the ground. The **leaves**, whose primary function is photosynthesis, grow out laterally from the stem at the **nodes**. Variations in leaf arrangement can give rise to many different forms of **shoots**, the term for the leaves and stem together. For example, leaf nodes can spiral around the stem, rotating by a fixed angle between each **internode** (the region between two nodes). Alternatively, leaves can arise oppositely or alternating on either side of the stem.

Organ shape is defined by directional patterns of growth. The polarity of growth of the **primary plant axis** (the main stem and root) is vertical, whereas the typical leaf grows laterally at the margins to produce the flattened **leaf blade**. The growth polarities of these organs are adapted to their functions: Leaves function in light absorption, stems elongate to lift the leaves toward sunlight, and roots elongate in search of water and nutrients from the soil. The cellular component that directly determines growth polarity in plants is the cell wall.

Plant cells are surrounded by rigid cell walls

The outer boundary of the living cytoplasm of plant cells is the **plasma membrane** (also called the **plasmalemma**), similar to the situation in animals, fungi, and bacteria. The plasma membrane surrounds the **cytoplasm**, which consists of organelles and supporting elements suspended in an aqueous **cytosol**. (The nucleus, which is surrounded by a double membrane and contains the chromo-

The figure labels for the plant diagram: Leaf, Node, Internode, Stem, Soil line, Root.

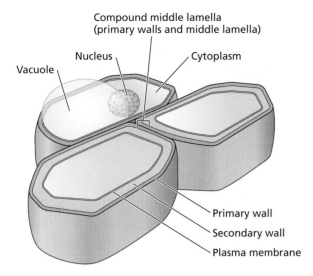

Compound middle lamella
(primary walls and middle lamella)

Nucleus   Cytoplasm

Vacuole

Primary wall
Secondary wall
Plasma membrane

Figure 1.3 Primary and secondary cell walls and their relationship to the rest of the cell. The two adjacent primary walls, along with the middle lamella, form a composite structure called the compound middle lamella.

somes, is considered separate from the cytoplasm.) Outside the plasma membrane of plant cells is a rigid **cell wall** composed of cellulose and other polymers that add rigidity and strength (**Figure 1.3**). During animal development, cells are able to migrate from one location to another, and tissues may thus contain cells that originated in different parts of the organism. In plants, however, cell walls limit cellular migration, and development depends solely on patterns of cell division and enlargement.

Plant cells have two types of walls: primary and secondary (see Figure 1.3). **Primary cell walls** are typically thin (less than 1 μm) and are characteristic of young, growing cells. **Secondary cell walls** are thicker and stronger than primary walls and are deposited on the inner surface of the primary wall after most cell enlargement has ended. The cell walls of neighboring cells are cemented by a **middle lamella** composed of the carbohydrate polymer pectin. Because it is sometimes difficult to distinguish the middle lamella from the primary wall, particularly when a secondary wall is present, the two adjacent primary walls and middle lamella are termed the compound middle lamella.

Primary and secondary cell walls differ in their components

Cell walls contain various types of polysaccharides that are named after the principal sugars they contain. For example, a **glucan** is a polymer of glucose units linked end to end. Polysaccharides may be linear unbranched chains of sugar residues, or they may contain side branches attached to the backbone. For branched polysaccharides, the backbone of the polysaccharide is usually indicated by the last part of the name. For example, xyloglucan has a glucan backbone (a linear chain of glucose residues) with xylose sugars attached as side chains.

The specific linkages between sugar rings, including the specific carbons that are linked together and the configuration of the linkage, are important for the properties of the polysaccharide. For instance, amylose (a component of starch in the plastid) is an α(1,4)-linked glucan (C-1 and C-4 carbons of adjacent glucose rings are linked through an O-glycosidic bond in an α configuration), whereas cellulose is a glucan made of β(1,4)-linkages (**Figure 1.4**). These differences in linkages make huge differences in the physical properties, enzymatic digestibility, and functional roles of these two polymers of glucose.

node   Position on the stem where leaves are attached.

shoots   The organ, usually aboveground, that includes the stem, leaves, buds, and reproductive structures. Function in photosynthesis and reproduction.

primary plant axis   The longitudinal axis of the plant defined by the positions of the shoot and root apical meristems.

leaf blade   The broad, expanded area of the leaf; also called the lamina.

plasma membrane (plasmalemma)   A bilayer of polar lipids (phospholipids or glycosylglycerides) and embedded proteins that together form a selectively permeable boundary around a cell.

cytoplasm   The cellular matter enclosed by the plasma membrane exclusive of the nucleus.

cytosol   The aqueous phase of the cytoplasm containing dissolved solutes but excluding supramolecular structures, such as ribosomes and components of the cytoskeleton.

cell wall   The rigid cell surface structure external to the plasma membrane that supports, binds, and protects the cell. Composed of cellulose and other polysaccharides and proteins.

primary cell walls   The thin (less than 1 μm), unspecialized cell walls that are characteristic of young, growing cells.

secondary cell wall   Cell wall synthesized by nongrowing cells. Often multilayered and containing lignin, it differs in composition and structure from the primary wall.

middle lamella   A thin layer of pectin-rich material at the junction where the primary walls of neighboring cells come into contact.

glucan   A polysaccharide made from glucose units.

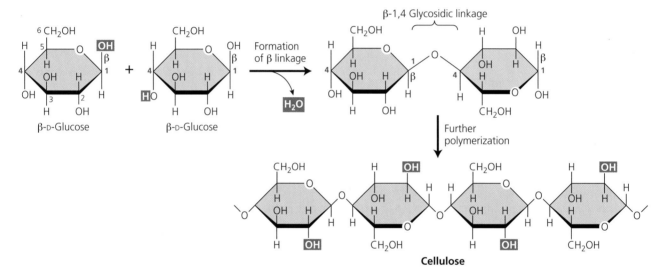

Figure 1.4 Structures of amylose and cellulose. The hydroxyl groups of amylose and cellulose are highlighted in blue to emphasize the structural differences between the polymers.

cellulose A linear chain of (1,4)-linked β-D-glucose. The repeating unit is cellobiose.

microfibril The major fibrillar component of the cell wall composed of layers of cellulose molecules packed tightly together by extensive hydrogen bonding.

hemicelluloses Heterogeneous group of polysaccharides that bind to the surface of cellulose, linking cellulose microfibrils together into a network.

pectins A heterogeneous group of complex cell wall polysaccharides that form a gel in which the cellulose–hemicellulose network is embedded.

Cell wall polysaccharides are classified into three groups: cellulose, hemicelluloses, and pectin. **Cellulose** is the major fibrillar component of the cell wall and is composed of an array of β(1,4)-linked glucans aggregated to form a **microfibril** with well-ordered and less-ordered regions. The simplest cellulose microfibrils are narrow structures, approximately 3 nm wide (1 nm = 10^{-9} m), which reinforce the cell wall like the steel rods in reinforced concrete. Each microfibril is composed of an estimated 18 parallel chains of (1,4)-linked β-D-glucose tightly packed together to form a crystalline core with extensive hydrogen bonding within and between glucan chains (**Figure 1.5**). Microfibrils are insoluble in water and have a high tensile strength, about half the tensile strength of steel.

Hemicelluloses are a heterogeneous group of polysaccharides with β(1,4)-linked backbones. Hemicelluloses typically require a strong reagent, such as 1–4 M NaOH, to be extracted from the cell wall. The third group of major cell wall components is **pectin**, a diverse group of hydrophilic, gel-forming polysaccha-

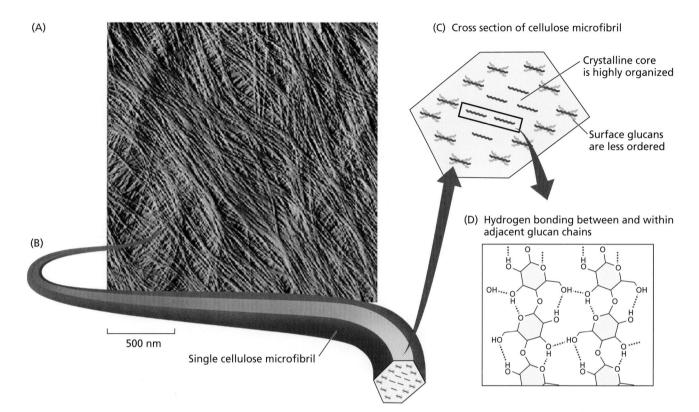

Figure 1.5 Structure of a cellulose microfibril. (A) Atomic force microscopy image of the primary cell wall from onion epidermis. Note its fibrillar texture, which arises from layers of cellulose microfibrils. (B) A single cellulose microfibril composed of (1,4)-β-D-glucan chains tightly bonded to each other to form a crystalline microfibril. (C) Cross section of a cellulose microfibril, illustrating one model of cellulose structure, with a crystalline core of highly ordered (1,4)-β-D-glucans surrounded by a less organized layer. (D) The crystalline regions of cellulose have precise alignment of glucans, with hydrogen bonding (indicated by dotted red lines) within, but not between, layers of (1,4)-β-D-glucans. (After Matthews et al. 2006; micrograph from Zhang et al. 2013.)

rides rich in acidic sugar residues, such as galacturonic acid. Many pectins are readily solubilized from the wall with hot water or with Ca^{2+} chelators. Together, hemicelluloses and pectins make up the **matrix polysaccharides**.

Typical primary cell walls of eudicots are rich in pectins, with smaller amounts of cellulose and hemicelluloses, while secondary cell walls are high in cellulose and a different form of hemicellulose. Instead of pectin, secondary walls have varying amounts of **lignin**, a highly cross-linked polymer of aromatic alcohols that confers rigidity to secondary walls. Whereas the high pectin content of primary cell walls enables them to expand during cell enlargement, the cellulose–hemicellulose–lignin complex of secondary cell walls forms a rigid matrix that is well suited for strength and compression resistance.

The evolution of lignified secondary cell walls provided plants with the structural reinforcement necessary to grow vertically above the soil and to colonize the land. Bryophytes, which lack lignified cell walls, are unable to grow more than a few centimeters above the ground.

The cellulose microfibrils and matrix polymers are synthesized via different mechanisms

Cellulose microfibrils are synthesized by large, ordered protein complexes, called cellulose synthase (CESA) complexes, which are embedded in the plas-

matrix polysaccharides   Polysaccharides comprising the matrix of plant cell walls. In primary cell walls they consist of pectins, hemicelluloses, and proteins.

lignin   Highly branched phenolic polymer made up of phenylpropanoid alcohols that is deposited in secondary cell walls.

Figure 1.6 Cellulose microfibrils are synthesized at the cell surface by plasma membrane–bound complexes containing cellulose synthase (CESA) proteins. Computational model of a CESA complex extruding glucan chains that coalesce to form a microfibril.

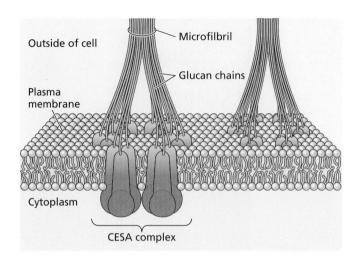

Outside of cell

Microfilbril

Glucan chains

Plasma membrane

Cytoplasm

CESA complex

cellulose synthase Enzyme that catalyzes the synthesis of individual (1,4)-linked β-D-glucans that make up cellulose microfibrils.

plasmodesmata (singular plasmodesma) Microscopic membrane-lined channel connecting adjacent cells through the cell wall and filled with cytoplasm and a central rod derived from the ER called the desmotubule.

symplast The continuous system of cell protoplasts interconnected by plasmodesmata.

symplastic transport The intercellular transport of water and solutes through plasmodesmata.

apoplastic transport Movement of molecules through the cell wall continuum that is called the apoplast. Molecules may move through the linked cell walls of adjacent cells, and in that way move throughout the plant without crossing a plasma membrane.

size exclusion limit (SEL) The restriction on the size of molecules that can be transported via the symplast. It is imposed by the width of the cytoplasmic sleeve that surrounds the desmotubule in the center of the plasmadesma.

ma membrane (**Figure 1.6**). These rosette-like structures are made up of six subunits, each of which is believed to contain three to six units of **cellulose synthase**, the enzyme that synthesizes the individual glucans that make up the microfibril. The catalytic domain of cellulose synthase, which is located on the cytoplasmic side of the plasma membrane, transfers a glucose residue from a sugar nucleotide donor, uridine diphosphate glucose (UDP-glucose), to the growing glucan chain.

In contrast to the cellulose microfibrils, the matrix polysaccharides are synthesized by membrane-bound enzymes that polymerize sugar molecules within the Golgi apparatus and are delivered to the cell wall via the exocytosis of tiny vesicles (**Figure 1.7**). Additional sugar residues may be added as branches to the polysaccharide backbone by other sets of enzymes. Unlike cellulose, which forms a crystalline microfibril, the matrix polysaccharides are much less ordered and are often described as amorphous. This amorphous character is a consequence of the structure of these polysaccharides—their branching and their nonlinear conformation. The predominant hemicellulose in primary cell walls of most land plants is xyloglucan, whereas the major hemicellulose in the primary cell wall of grasses (Poaceae) is arabinoxylan.

Plasmodesmata allow the free movement of molecules between cells

The cytoplasm of neighboring cells is usually connected by means of **plasmodesmata** (singular *plasmodesma*), tubular structures 40 to 50 nm in diameter and formed by the connected membranes of adjacent cells (**Figure 1.8**). Plasmodesmata facilitate intercellular movement of proteins, nucleic acids, and other macromolecular signals that coordinate developmental processes between plant cells. Plant cells interconnected in this way form a cytoplasmic continuum referred to as the **symplast**, and transport of small molecules through plasmodesmata is called **symplastic transport** (see Chapters 3 and 6). Transport through the permeable cell wall space outside the cells is called **apoplastic transport**. Both forms of transport are important in the vascular system of plants (see Chapter 6).

The symplast can transport water, solutes, and macromolecules between cells without crossing the plasma membrane. However, there is a restriction on the size of molecules that can be transported via the symplast; this restriction is called the **size exclusion limit**, and it varies with cell type, environment, and developmental stage. The transport can be followed by studying the movement

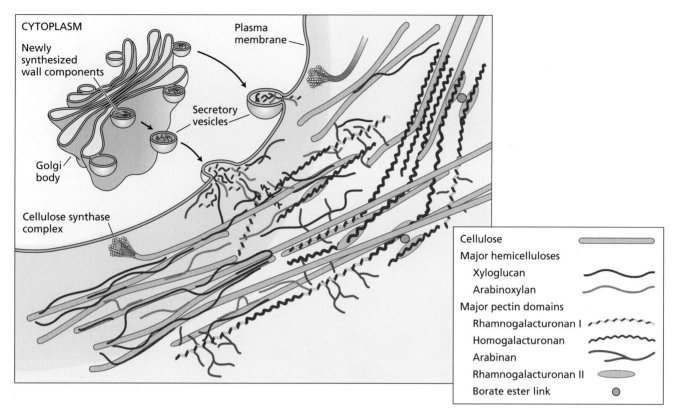

Figure 1.7 Schematic diagram of the major structural components of the primary cell wall and their possible arrangement. Cellulose microfibrils (gray rods) are synthesized at the cell surface and are partially coated with hemicelluloses (blue and purple strands), which may separate microfibrils from one another. Pectins (red, yellow, and green strands) form an interlocking matrix that controls microfibril spacing and wall porosity. Pectins and hemicelluloses are synthesized in the Golgi apparatus and delivered to the wall via vesicles that fuse with the plasma membrane and thus deposit these polymers to the cell surface. For clarity, the hemicellulose–cellulose network is emphasized on the left, and the pectin network is emphasized on the right. (After Cosgrove 2005.)

of fluorescently labeled proteins or dyes between cells. The movement through plasmodesmata can be regulated, or gated, by altering the dimensions of the surrounding structures as well as the lumen inside the internal structures (see Figure 1.8).

New cells originate in dividing tissues called meristems

Plant growth is concentrated in localized regions of cell division called **meristems**. Nearly all nuclear division (mitosis) and cell division (cytokinesis) occurs in these meristematic regions. In a young plant, the most active meristems are the **apical meristems**, located at the tips of the stem and the root (**Figure 1.9**). The phase of plant development that gives rise to new organs and to the basic plant form is called **primary growth**, which gives rise to the **primary plant body**. Primary growth results from the activity of apical meristems. Cell division in the meristem produces cuboidal cells about 10 μm on each side. Division is followed by progressive cell expansion, and cells typically elongate to become much longer than they are wide (30–100 μm long, 10–25 μm wide—about half the width of a baby's fine hair and about 50 times the width of a typical bacterium). The increase in length produced by primary growth amplifies the plant's axial (top-to-bottom) polarity, which is established in the embryo.

meristems Localized regions of ongoing cell division that enable growth during post-embryonic development.

apical meristems Localized regions made up of undifferentiated cells undergoing cell division without differentiation at the tips of shoots and roots.

primary growth The phase of plant development that gives rise to new organs and to the basic plant form.

primary plant body The part of the plant directly derived from the shoot and root apical meristems and primary meristems.

axillary buds Secondary meristems that are formed in the axils of leaves.

(A)

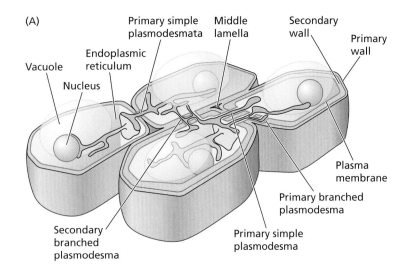

(B)

(C)

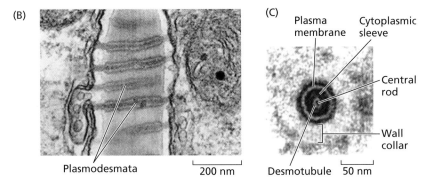

(D)

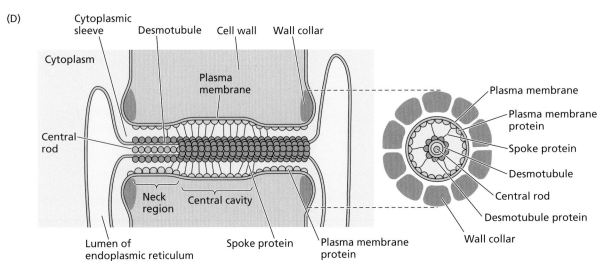

Figure 1.8 Plasmodesmata. (A) Diagrammatic representation of the cell walls surrounding four adjacent plant cells. Cells with only primary walls and with both primary and secondary walls are illustrated. The secondary walls form inside the primary walls. The cells are connected by plasmodesmata, which form during cell division. (B) Electron micrograph of a wall separating two adjacent cells, showing simple plasmodesmata in longitudinal view. (C) Tangential section through a cell wall showing a plasmodesma. (D) Schematic surface and cross-section views of a plasmodesma. The pore consists of a central cavity through which a strand of endoplasmic reticulum called the desmotubule runs, connecting the endoplasmic reticulum of the adjoining cells. The plasma membrane is connected to the desmotubule via spoke proteins. A wall collar, or sphincter, surrounds the neck of each plasmodesma. (B from Robinson-Beers and Evert 1991, courtesy of R. Evert; C from from Bell and Oparka 2011.)

lateral roots Arise from the pericycle in mature regions of the root through establishment of secondary meristems that grow out through the cortex and epidermis, establishing a new growth axis.

pericycle Meristematic cells forming the outermost layer of the vascular cylinder in the stem or root, interior to the endodermis.

Meristematic tissue is also found along the length of the root and shoot. **Axillary buds** are meristems that develop at the node in the leaf axil—the region between the leaf and the shoot. Axillary buds become the apical meristems of branches. The branches of roots, the **lateral roots**, arise from meristematic cells in the **pericycle**, or root branch meristem (see Figure 1.9D). This meristematic tissue then becomes the apical meristem of the lateral root.

The primary plant body can be further expanded through secondary growth (**Box 1.2**).

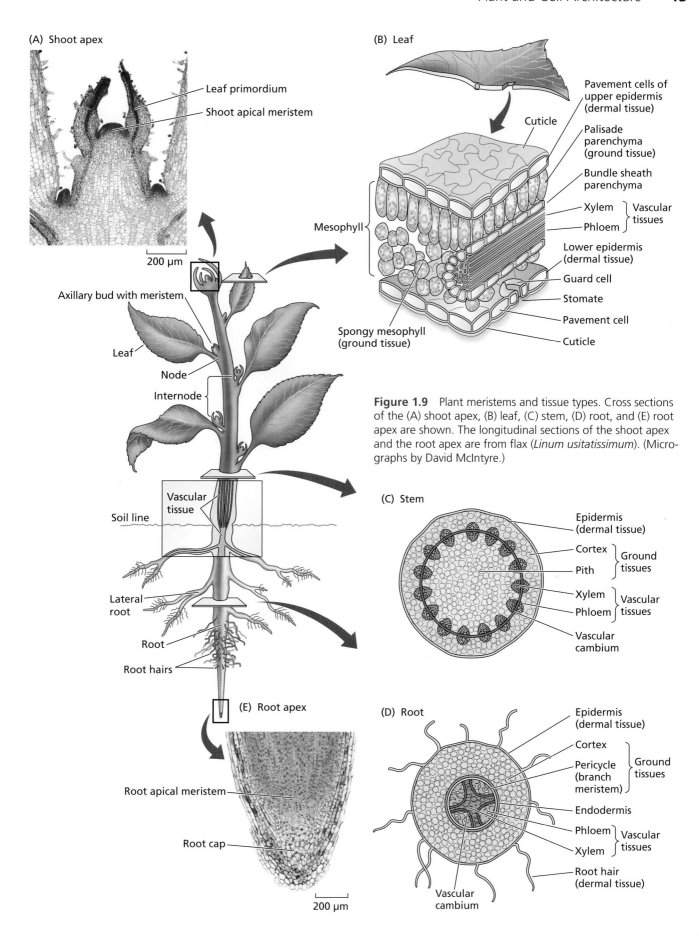

(A) Shoot apex

Leaf primordium

Shoot apical meristem

200 µm

Axillary bud with meristem

Leaf

Node

Internode

Vascular tissue

Soil line

Lateral root

Root

Root hairs

(E) Root apex

Root apical meristem

Root cap

200 µm

(B) Leaf

Pavement cells of upper epidermis (dermal tissue)

Cuticle

Palisade parenchyma (ground tissue)

Bundle sheath parenchyma

Mesophyll

Xylem } Vascular tissues
Phloem

Lower epidermis (dermal tissue)

Guard cell

Stomate

Pavement cell

Cuticle

Spongy mesophyll (ground tissue)

Figure 1.9 Plant meristems and tissue types. Cross sections of the (A) shoot apex, (B) leaf, (C) stem, (D) root, and (E) root apex are shown. The longitudinal sections of the shoot apex and the root apex are from flax (*Linum usitatissimum*). (Micrographs by David McIntyre.)

(C) Stem

Epidermis (dermal tissue)

Cortex } Ground tissues
Pith

Xylem } Vascular tissues
Phloem

Vascular cambium

(D) Root

Epidermis (dermal tissue)

Cortex

Pericycle (branch meristem) } Ground tissues

Endodermis

Phloem } Vascular tissues
Xylem

Root hair (dermal tissue)

Vascular cambium

Box 1.2 The Secondary Plant Body

The apical meristems of a plant add primarily to the length of the plant's organs. Another type of meristematic tissue, the **cambium**, gives rise to **secondary growth**, which produces an increase in the width or diameter of shoots and roots (see figure). The cambial layer that produces wood is called the **vascular cambium**. This meristem arises in the vascular system, between the xylem and the phloem of the primary plant body. The cells of the vascular cambium divide longitudinally to produce derivatives toward the inside and the outside of the vascular cambium in both stems and roots. The cells on the inside of the vascular cambium differentiate into **secondary xylem**, which conducts water and nutrients from the soil upward to other plant organs and is characterized by thick secondary walls. In temperate climates, summer wood is darker and denser than spring wood; alternating layers of summer and spring wood form **annual rings**. The vascular cambium derivatives displaced toward the outside of the secondary stem or root give rise to **secondary phloem**, which, like primary phloem, conducts

the products of photosynthesis either downward from the leaves to other organs of the plant, or upward from the leaves to reproductive structures. The cells of the vascular cambium can also divide transversely to produce ray cells that form the **rays**. The function of rays is to transport water, minerals and organic material from the xylem radially to the outer tissues of the stem.

The **cork cambium**, or **phellogen**, is the cambial layer that produces the protective **periderm** (see figure) on the outside of woody plants. The cork cambium typically arises each year within the secondary phloem. The production of layers of water-resistant cork cells isolates the outer phloem tissues of the stem or root from their water supply, the xylem, causing them to shrivel and die. The **bark** of a woody plant is the collective term for several tissues—the secondary phloem, cortex (in stems), pericycle (in roots), and periderm—that can be peeled off as a unit at the soft layer of vascular cambium.

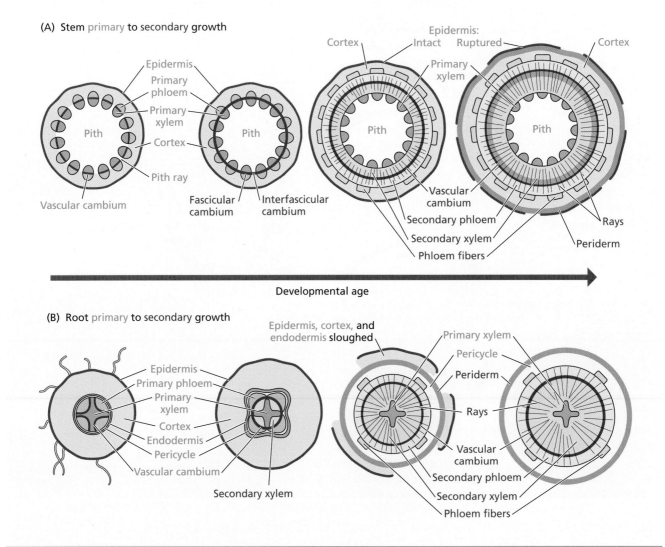

(A) Stem primary to secondary growth

Developmental age

(B) Root primary to secondary growth

Box 1.2 (*continued*)

Figure caption Secondary growth in stems and roots. (A) Stem primary to secondary growth. Primary growth is labeled in green, secondary growth in brown. The vascular cambium starts as separated growth regions in the vascular bundles, or fascia, of primary xylem and phloem. As the plant grows, the bundled, fascicular cambium becomes connected by interfascicular cambium between the bundles. Once the vascular cambium forms a continuous ring, it divides inward to generate secondary xylem and it divides outward to generate the secondary phloem. Regions in the cortex develop into phloem fibers and the periderm, which contains the phellogen, or cork cambium, and the outer phelloderm. With growth, the epidermis ruptures, and rays connect the inner and outer vasculature. (B) Root primary to secondary growth. The central vascular cylinder contains the primary phloem and primary xylem. As in the stem, the vascular cambium becomes connected and grows outward, generating secondary phloem and rays. As roots increase in girth, the pericycle generates the root periderm, while the outer epidermis, cortex, and endodermis are sloughed off. The pericycle produces the phloem fibers and rays as well as lateral roots (not shown). The vascular cambium produces secondary phloem and rings of secondary xylem.

Plant Cell Types

Three major tissue systems are present in all plant organs: dermal tissue, ground tissue, and vascular tissue (see Figure 1.9B–D). **Dermal tissue** forms the outer protective layer of the plant and is called the **epidermis** in the primary plant body; **ground tissue** fills out the three-dimensional bulk of the plant and includes the **pith** and **cortex** of primary stems and roots, and the **mesophyll** in leaves. **Vascular tissue**, consists of two types of tissues: **xylem** and **phloem**, each of which consists of conducting cells, generalized parenchyma cells, and thick-walled fibers.

Dermal tissue covers the surfaces of plants

The epidermis of plants comprises several cell types, including pavement cells, which are shaped either like puzzle pieces or the rounded bricks used in decorative paving; stomatal guard cells, which regulate gas exchange in the leaf; and hairlike trichomes found on leaves, stems, and roots. The pavement cells of the epidermis of onion are shown in **Figure 1.10A**. In roots, root hairs differentiate from the epidermis. Many plants have relatively few chloroplasts in leaf epidermal tissue, perhaps because chloroplast division is shut down. Stomatal guard cells are typically the only epidermal cells with chloroplasts.

Ground tissue forms the bodies of plants

The ground tissue makes up the bulk of the primary plant body. Ground tissue consists mainly of three cell types: parenchyma, collenchyma, and sclerenchyma. **Parenchyma** cells are living cells with thin primary walls (**Figure 1.10B**). These cells make up much of the bulk of the primary plant body, including stems, leaves, and roots. Parenchyma cells also have the capacity to continue division and can differentiate into a variety of other ground tissues and vascular tissues, after being produced by meristems.

Parenchyma cells can differentiate into another type of ground tissue, called **collenchyma**, which has thickened primary walls but can nevertheless continue to elongate (**Figure 1.10C**). A familiar example is the collenchyma in the ribs of celery stalks, which have thick, pectin-rich cell walls. Parenchyma can also differentiate into **sclerenchyma**, the cells of which (sclereids and fibers) have thick secondary walls (**Figure 1.10D**). Sclereids derive from parenchyma cells in leaves (e.g., water lily), flowers (e.g., camellia), and fruits (e.g., pear).

Fibers develop from parenchyma and form elongated support structures with thickened secondary walls both in ground tissue and in vascular tissue. Fibers can elongate into the longest cells of higher plants. For example, individual phloem fiber cells of the *Ramie* plant can be 25 cm long! Because the walls are hardened

dermal tissue The tissue system that covers the outside of the plant body; the epidermis or periderm.

epidermis The outermost layer of plant cells, typically one cell thick.

ground tissue The internal tissues of the plant, other than the vascular tissues.

pith The ground tissue in the center of the stem or root.

cortex Ground tissue in the region of the primary stem or root located between the vascular tissue and the epidermis, mainly consisting of parenchyma.

mesophyll Leaf tissue found between the upper and lower epidermal layers.

vascular tissues Plant tissues specialized for the transport of water (xylem) and photosynthetic products (phloem).

xylem The vascular tissue that transports water and ions from the root to the other parts of the plant.

phloem The tissue that transports the products of photosynthesis from mature leaves (or storage organs) to areas of growth and storage, including the roots.

parenchyma Metabolically active ground tissue consisting of thin-walled cells.

collenchyma A specialized parenchyma with irregularly thickened, pectin-rich, primary cell walls.

sclerenchyma Plant tissue composed of cells (sclereids and fibers), often dead at maturity, with thick, lignified secondary cell walls.

fiber An elongated, tapered sclerenchyma cell that provides support in vascular plants.

(A) Dermal tissue: epidermal cells

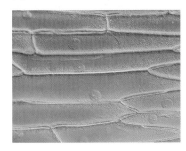

(B) Ground tissue: parenchyma cells

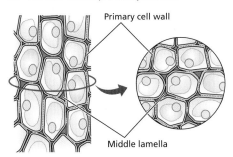

Primary cell wall

Middle lamella

Figure 1.10 Illustrations of basic plant cell types. (A) The epidermis of a leaf. Ground tissue cell types include (B) parenchyma, (C) collenchyma, and (D) sclerenchyma. The vascular tissue is made up of conducting cells of the (E) xylem and (F) phloem. (A ©Heiti Paves/Alamy Stock Photo; B–F after Esau 1977.)

(C) Ground tissue: collenchyma cells

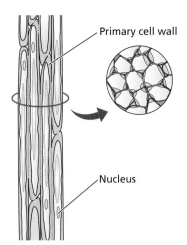

Primary cell wall

Nucleus

(D) Ground tissue: sclerenchyma cells

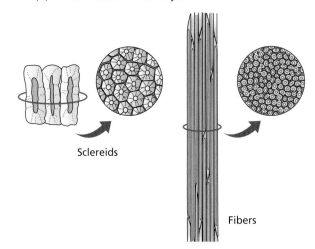

Sclereids

Fibers

(E) Vascular tisssue: xylem

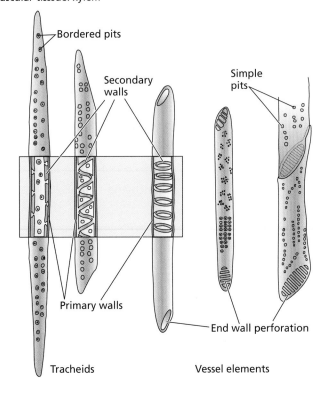

Bordered pits

Secondary walls

Simple pits

Primary walls

End wall perforation

Tracheids

Vessel elements

(F) Vascular tisssue: phloem

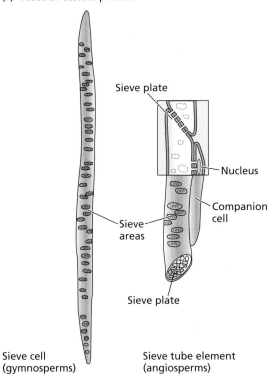

Sieve plate

Nucleus

Companion cell

Sieve areas

Sieve plate

Sieve cell (gymnosperms)

Sieve tube element (angiosperms)

with lignin after elongation, the cells have very high tensile strength, and it is no wonder that such fibers, called bast fibers, are used to make textiles, rope, paper, and other materials. Specialized holes in the secondary wall, called **pits**, connect the living fiber cells with each other. The pits of adjoining fibers align with each other, forming **pit pairs**. Pit pairs in living cells are often grouped together to form fields of plasmodesmata.

Vascular tissue forms transport networks between different parts of the plant

The xylem cells that conduct water and minerals from the root, called **tracheary elements**, are nonliving at maturity and consist of **tracheids** (in all vascular plants) and shorter, wider **vessel elements** (mostly in angiosperms) (**Figure 1.10E**). Vessel elements stack end-to-end to form columns of vessel elements called **vessels**. Protoxylem cells with primary walls begin to differentiate into mature tracheary elements by laying down secondary cell walls with spiral bands of cellulose and strengthened with lignin. Once elongation has stopped, the top and bottom end walls get large perforations in them. On the side walls, secondary walls continue to thicken, except in spots containing pits, which begin as plasmodesmata fields and ultimately become channels in the walls shared between adjacent cells. The cells of the tracheids and vessels undergo a process called **programmed cell death** and die, leaving behind the hardened bundle of tubes made by the secondary walls and connected side-to-side by pits. The pits are important because flow through these narrow tubes depends on having a continuous liquid stream (see Chapters 3 and 6).

Phloem cells, which conduct the products of photosynthesis from the leaves to the root, and upward to flowers and seeds (see Chapter 10), are living at maturity and have nonlignified cell walls. They include **sieve tube elements**—which stack to form **sieve tubes**—in angiosperms, and **sieve cells**—which overlap one another—in gymnosperms (**Figure 1.10F**). Living sieve tube elements are connected to specialized **companion cells** via fields of plasmodesmata called **sieve areas**. Sieve tube elements are metabolically dependent on their companion cells, and an injury to the companion cell results in the death of the sieve tube element

Plant Cell Organelles

All plant cells have the same basic organization: They contain cytosol, a nucleus, and other subcellular organelles enclosed by the plasma membrane and cell wall (**Figure 1.11**). All plant cells *begin* with a similar complement of organelles. These organelles fall into two main categories based on how they arise and whether they can divide independently:

1. *The endomembrane system*: the endoplasmic reticulum, nuclear envelope, Golgi apparatus, vacuole, endosomes, and plasma membrane. Other organelles derived from the endomembrane system include oil bodies, peroxisomes, and glyoxysomes, which function in lipid storage and carbon metabolism in seeds and leaves. The endomembrane system plays a central role in secretory processes, membrane recycling, and the cell cycle. The plasma membrane regulates transport into and out of the cell. Endosomes arise from vesicles derived from the plasma membrane and process or recycle the vesicle contents.

2. *Semiautonomous (independently dividing) organelles*: plastids and mitochondria, which function in energy metabolism and storage, and synthesize a wide range of metabolites and structural molecules.

pit A microscopic region where the secondary wall of a tracheary element is absent and the primary wall is thin and porous.

pit pair Two pits occurring opposite one another in the walls of adjacent tracheids or vessel elements. Pit pairs constitute a low resistance path for water movement between the conducting cells of the xylem.

tracheary elements Water-transporting cells of the xylem.

tracheids Spindle-shaped, water-conducting cells with tapered ends and pitted walls without perforations found in the xylem of both angiosperms and gymnosperms.

vessel elements Nonliving water-conducting cells with perforated end walls found only in angiosperms and a small group of gymnosperms.

vessel A stack of two or more vessel elements in the xylem.

programmed cell death (PCD) The process whereby individual cells activate an intrinsic senescence program accompanied by a distinct set of morphological and biochemical changes similar to mammalian apoptosis.

sieve tube elements The highly differentiated sieve elements typical of the angiosperms.

sieve tube Tube formed by the joining together of individual sieve tube elements at their end walls.

sieve cells The relatively unspecialized sieve elements of gymnosperms.

sieve area A depression in the cell wall of a sieve tube element that contains a field of plasmodesmata.

companion cells In angiosperms, metabolically active cells that are connected to their associated sieve element by large, branched plasmodesmata and take over many of the metabolic activities of the sieve element. In source leaves, they function in the transport of photosynthate into the sieve elements.

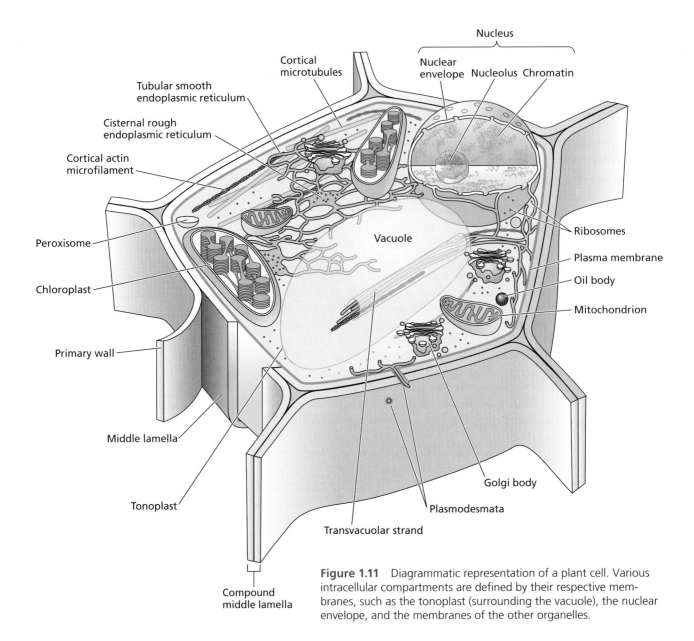

Figure 1.11 Diagrammatic representation of a plant cell. Various intracellular compartments are defined by their respective membranes, such as the tonoplast (surrounding the vacuole), the nuclear envelope, and the membranes of the other organelles.

Because all of these cellular organelles are membranous compartments, we will begin with a description of membrane structure and function.

Biological membranes are bilayers that contain proteins

All cells are formed enclosed in a membrane that serves as their outer boundary, separating the cytoplasm from the external environment. This plasma membrane allows the cell to take up and retain certain substances while excluding others. Various transport proteins embedded in the plasma membrane are responsible for this selective traffic of solutes—water-soluble ions and small, uncharged molecules—across the membrane. The accumulation of ions or molecules in the cytosol through the action of transport proteins consumes metabolic energy. In eukaryotic cells, membranes enshroud the genetic material, delimit the boundaries of other specialized internal organelles of the cell, and regulate the fluxes of ions and metabolites into and out of these compartments.

According to the **fluid-mosaic model**, all biological membranes have the same basic molecular organization. They consist of a double layer (*bilayer*) of lipids in which proteins are embedded (**Figure 1.12**). Each layer is called a *leaflet* of the bilayer. In most membranes, proteins make up about half of the membrane's mass. However, the composition of the lipid components and the properties of the proteins vary from membrane to membrane, conferring on each membrane its unique functional characteristics.

LIPIDS The most prominent membrane lipids found in plants are phospholipids, a class of lipids in which two fatty acids are covalently linked to glycerol, which is covalently linked to a phosphate group. Attached to the phosphate group in the phospholipid is a variable component, called the *head group*, such as choline, glycerol, or inositol (see Figure 1.12B). The nonpolar hydrocarbon chains of the fatty acids form a region that is hydrophobic—that is, that excludes water. In contrast to the fatty acids, the head groups are highly polar; consequently, phospholipid molecules display both hydrophilic and hydrophobic properties (i.e., they are *amphipathic*). Various phospholipids are distributed asymmetrically across the plasma membrane, giving the membrane sidedness; in terms of phospholipid composition, the outside leaflet of the plasma membrane that faces the outside of the cell is different from the inside leaflet that faces the cytosol. The inner membranes of **plastids**, the group of

fluid-mosaic model The common molecular lipid–protein structure for all biological membranes. A double layer (bilayer) of polar lipids (phospholipids or, in chloroplasts, glycosylglycerides) has a hydrophobic, fluid-like interior. Membrane proteins are embedded in the bilayer and may move laterally due to its fluid-like properties.

plastids Cellular organelles found in eukaryotes, bounded by a double membrane, and sometimes containing extensive membrane systems. They perform many different functions: photosynthesis, starch storage, pigment storage, and energy transformations.

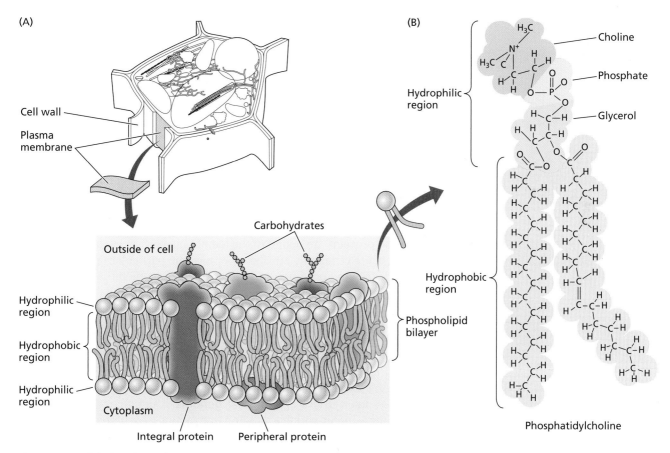

Figure 1.12 (A) The plasma membrane, endoplasmic reticulum, and other endomembranes of plant cells consist of proteins embedded in a phospholipid bilayer. (B) Chemical structure of phosphatidylcholine, a typical plant membrane phospholipid. The two fatty acid chains differ in the number of double bonds (see Chapter 11). (B after Buchanan et al. 2000.)

integral membrane proteins
Proteins that are embedded in the lipid bilayer of a membrane via at least one transmembrane domain.

peripheral membrane proteins
Proteins that are bound to the membrane surface by noncovalent bonds, such as ionic bonds or hydrogen bonds.

anchored proteins   Proteins that are bound to the membrane surface via lipid molecules, to which they are covalently attached.

nucleus   (plural *nuclei*) The organelle that contains the genetic information primarily responsible for regulating cellular metabolism, growth, and differentiation.

nuclear genome   The entire complement of DNA found in the nucleus.

nucleoplasm   The soluble matrix of the nucleus in which the chromosomes and nucleolus are suspended.

nuclear envelope   The double membrane surrounding the nucleus.

membrane-bound organelles to which chloroplasts belong, are unique in that their lipid component consists almost entirely of glycosylglycerides, the glycosyl polar head groups of which are galactose derivatives.

The fatty acid chains of phospholipids and glycosylglycerides vary in length but usually consist of 16 to 24 carbons. If the carbons are linked by single bonds, the fatty acid chain is *saturated* (with hydrogen atoms), but if the chain includes one or more double bonds, it is *unsaturated*.

Double bonds in a fatty acid chain create a kink in the chain that prevents tight packing of the phospholipids in the bilayer (i.e., the bonds adopt a kinked *cis* configuration, as opposed to an unkinked *trans* configuration). The kinks promote membrane fluidity, which is critical for many membrane functions. Membrane fluidity is also strongly influenced by temperature. Because plants generally cannot regulate their body temperature, they are often faced with the problem of maintaining membrane fluidity under conditions of low temperature, which tends to decrease membrane fluidity. To maintain membrane fluidity at cold temperatures, plants can produce membrane phospholipids with a higher percentage of unsaturated fatty acids, such as *oleic acid* (one double bond), *linoleic acid* (two double bonds), and *linolenic acid* (three double bonds).

PROTEINS   The proteins associated with the lipid bilayer are of three main types: integral, peripheral, and anchored.

Integral membrane proteins are embedded in the lipid bilayer (see Figure 1.12A). Most integral proteins span the entire width of the phospholipid bilayer, so one part of the protein interacts with the outside of the cell, another part interacts with the hydrophobic core of the membrane, and a third part interacts with the interior of the cell, the cytosol. Proteins that serve as ion channels (see Chapter 6) are always integral membrane proteins, as are certain receptors that participate in signal transduction pathways (see Chapter 12). Some integral membrane proteins on the outer surface of the plasma membrane recognize and bind tightly to cell wall constituents, effectively cross-linking the membrane to the cell wall.

Peripheral membrane proteins are bound to the membrane surface (see Figure 1.12A) by noncovalent bonds, such as ionic bonds or hydrogen bonds, and can be dissociated from the membrane with high-salt solutions or chaotropic agents, which break ionic and hydrogen bonds, respectively. Peripheral proteins serve a variety of functions in the cell. For example, some are involved in interactions between the membranes and the major elements of the cytoskeleton, the microtubules and actin microfilaments.

Anchored proteins are bound to the membrane surface via lipid molecules, to which they are covalently attached. Because the particular lipids are asymmetrically distributed on different sides of the plasma membrane, they make the two sides of the membrane even more distinct.

The Nucleus

The **nucleus** (plural *nuclei*) is the membrane-bound organelle that contains the genetic information primarily responsible for regulating the metabolism, growth, and differentiation of the cell. Collectively, these genes and their intervening sequences are referred to as the **nuclear genome**. The size of the nuclear genome in plants is highly variable, ranging from about 1.2×10^8 base pairs for the mustard relative *Arabidopsis thaliana* to 1×10^{11} base pairs for the lily *Fritillaria assyriaca*. The remainder of the genetic information of the cell is contained in the two semiautonomous organelles—the plastid and the mitochondrion—which we will discuss later in this chapter.

The nucleus consists of a complex matrix, the **nucleoplasm**, surrounded by a double membrane called the **nuclear envelope** (**Figure 1.13A**), which is a

(A)

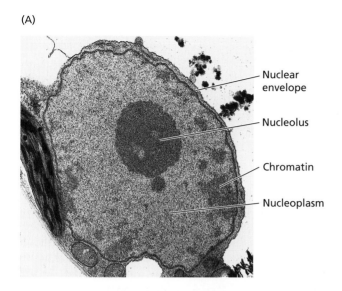

Nuclear envelope

Nucleolus

Chromatin

Nucleoplasm

Figure 1.13   (A) Transmission electron micrograph of a plant nucleus, showing the nucleolus and the nuclear envelope. (B) Organization of the nuclear pore complexes (NPCs) on the nuclear surface in tobacco culture cells. The NPCs that touch each other are colored brown; the rest are colored blue. The first inset (top right) shows that most of the NPCs are closely associated, forming rows of 5–30 NPCs. The second inset (bottom right) shows the tight associations of the NPCs. (A courtesy of R. Evert; B from Fiserova et al. 2009.)

(B)

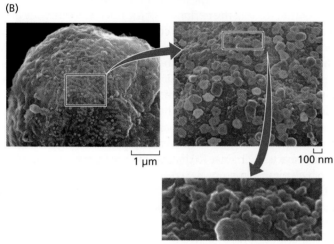

1 μm   100 nm

subdomain of the endoplasmic reticulum (ER; see below). **Nuclear pores** form selective channels across both membranes, connecting the nucleoplasm with the cytoplasm (**Figure 1.13B**). There can be very few to many thousands of nuclear pores on an individual nuclear envelope.

The nuclear "pore" is actually an elaborate structure composed of more than 100 different nucleoporin proteins arranged octagonally to form a 105-nm nuclear pore complex (NPC). The nucleoporins lining the 40-nm channel of the NPC form a meshwork that acts as a supramolecular sieve. Several proteins required for nuclear import and export have been identified. A specific amino acid sequence called the nuclear localization signal is required for a protein to gain entry into the nucleus.

The nucleus is the site of storage and replication of the **chromosomes**, composed of DNA and its associated proteins (**Figure 1.14**). Collectively, this DNA–protein complex is known as **chromatin**. The linear length of the entire DNA in any plant genome is usually millions of times greater than the diameter of the nucleus in which it is found. To solve the problem of packaging this chromosomal DNA within the nucleus, segments of the linear double helix of DNA are coiled twice around a solid cylinder of eight **histone** protein molecules, forming a **nucleosome**. Nucleosome assembly is also assisted by large protein complexes called condensins. Nucleosomes are arranged like beads on a string along the length

nuclear pores   Sites where the two membranes of the nuclear envelope join, forming a partial opening between the interior of nucleus and the cytosol.

chromosome   A threadlike or rod-shaped structure of DNA and protein found in the nucleus of most living cells, encoding genetic information in the form of genes.

chromatin   The DNA–protein complex found in the interphase nucleus. Condensation of chromatin occurs during cell division to form the rod-shaped mitotic and meiotic chromosomes.

histones   A family of proteins that interact with DNA and around which DNA is wound to form a nucleosome.

nucleosome   A structure consisting of eight histone proteins around which DNA is coiled.

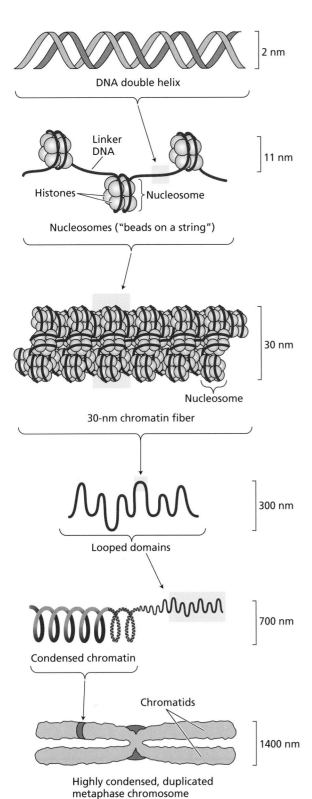

2 nm

DNA double helix

Linker DNA

11 nm

Histones — Nucleosome

Nucleosomes ("beads on a string")

30 nm

Nucleosome

30-nm chromatin fiber

300 nm

Looped domains

700 nm

Condensed chromatin

Chromatids

1400 nm

Highly condensed, duplicated metaphase chromosome of a dividing cell

Figure 1.14   Packaging of DNA in a metaphase chromosome. The DNA is first aggregated into nucleosomes and then wound to form the 30-nm chromatin fibers. Further coiling leads to the condensed metaphase chromosome. (After Alberts et al. 2007.)

of each chromosome. When the nucleus is not dividing, the chromosomes maintain their spatial independence; they do not "tangle," and instead remain quite discrete.

During mitosis, the chromatin condenses, first by coiling tightly into a 30-nm chromatin fiber, with six nucleosomes per turn, followed by further folding and packing processes that depend on interactions between proteins and nucleic acids (see Figure 1.14). At interphase (discussed later), two types of chromatin are distinguishable, based on their degree of condensation: heterochromatin and euchromatin. **Heterochromatin** is a highly compact and transcriptionally inactive form of chromatin and accounts for about 10% of the DNA. Most of the heterochromatin is concentrated along the periphery of the nuclear membrane and is associated with regions of the chromosomes that contain few genes, such as telomeres and centromeres. The rest of the DNA consists of **euchromatin**, the dispersed, transcriptionally active form. Only about 10% of the euchromatin is transcriptionally active at any given time. The remainder exists in a state of condensation intermediate between that of the transcriptionally active euchromatin and that of heterochromatin. The chromosomes reside in specific regions of the nucleoplasm, each in its own separate space, giving rise to the possibility of separate regulation of each chromosome.

During the cell cycle, chromatin undergoes dynamic structural changes. In addition to transient local changes that are required for transcription, heterochromatic regions can be converted to euchromatic regions, and vice versa, by cytosine methylation or by the addition or removal of functional groups on the histone proteins. Such global changes in the genome can give rise to stable changes in gene expression. In general, stable changes in gene expression that occur without changes in the DNA sequence are referred to as *epigenetic regulation*.

Nuclei contain a densely granular region called the **nucleolus** (plural *nucleoli*), which is the site of **ribosome** synthesis (see Figure 1.13A). Typical cells have one nucleolus per nucleus; some cells have more. The nucleolus includes portions of one or more chromosomes where ribosomal RNA (rRNA) genes are clustered to form a structure called the **nucleolar organizer region** (**NOR**). Even though chromosomes remain largely separate within the nucleus, parts of several may come together in their middles to help form the nucleolus. The nucleolus assembles the proteins and rRNA of the ribosome into a large and a small subunit, each exiting the nucleus separately through the nuclear pores. The two subunits unite in the cytoplasm to form a complete ribosome (see Figure 1.15A, number 1). Assembled ribosomes are protein-synthesizing nanomachines. Those produced by the nucleus for cytoplasmic, "eukaryotic" protein synthesis, the 80S ribosomes, are larger than the ribosomes assembled in and remaining in mitochondria and plastids for their separate program of "prokaryotic" protein synthesis, the 70S ribosomes.

Gene expression involves both transcription and translation

The nucleus is the site of read-out, or **transcription**, of the cell's DNA (**Figure 1.15A**). Some of the DNA is transcribed as messenger RNA (mRNA), which codes for proteins. Ribosomes read mRNA in one direction, from the 5′ to the 3′ end (**Figure 1.15B**). Other regions of the DNA are transcribed into transfer RNA (tRNA) and rRNA to be used in **translation** (see Figure 1.15A, numbers 1–3). The RNA moves through nuclear pores to the cytoplasm, where the polyribosomes (groups of ribosomes translating a single RNA strand) that are "free" in the cytosol (not membrane-bound) translate the RNA for proteins destined for the cytosol and organelles that receive proteins independently of the endomembrane pathway. Endomembrane and secreted proteins enter the endomembrane system during the process of translation (co-translationally) on polyribosomes bound to the ER. The mechanism of co-translational insertion of proteins into the ER is complex, involving the ribosomes, the mRNA that codes for the secretory protein, and a special protein-translocating pore, the **translocon**, in the ER membrane (see Figure 1.15A, numbers 4–7).

The proteins synthesized on cytosolic ribosomes that are targeted to membrane organelles after translation employ posttranslational insertion to cross the organelle membrane. The process of translation on either cytosolic or membrane-bound polysomes produces the primary protein sequence of the protein, which includes not only the sequence involved in protein function, but also sequence information required to "target" the protein to different destinations within the cell.

Posttranslational regulation determines the life span of proteins

Protein stability plays an important role in regulating a protein's longevity, and thus gene expression. Once synthesized, a protein has a finite life span in the cell, ranging from a few minutes to hours or days. Thus, steady-state levels of cellular proteins reflect equilibrium between the synthesis and the degradation of those proteins, referred to as **turnover**. In both plant and animal cells, there are two distinct pathways of protein turnover, one in specialized lytic vacuoles (called lysosomes in animal cells; see next section) and the other in both the cytosol and nucleus.

The pathway of protein turnover that occurs in the cytosol and nucleoplasm involves the ATP-dependent formation of a covalent bond between the protein that is to be degraded and a small, 76–amino acid polypeptide called **ubiquitin**. Addition of one or many molecules of ubiquitin to a protein is called *ubiquitination*. Ubiquitination marks a protein for destruction by a large, ATP-dependent proteolytic complex called the **26S proteasome**, which specifically recognizes such "tagged" molecules (**Figure 1.16**). More than 90% of the short-lived proteins in eukaryotic cells are degraded via the ubiquitin pathway.

Ubiquitination is initiated when the ubiquitin-activating enzyme (E1) catalyzes the ATP-dependent adenylation of the C terminus of ubiquitin. The adenylated ubiquitin is then transferred to a cysteine residue on a second enzyme, the ubiquitin-conjugating enzyme (E2). Proteins destined for degradation are bound by a third type of protein, a ubiquitin ligase (E3). The E2–ubiquitin complex then transfers its ubiquitin to a lysine residue of the protein bound to E3. This process can occur multiple times to form a polymer of ubiquitin. The ubiquitinated protein is then targeted to the proteasome for degradation.

As we now know, there are a multitude of protein-specific ubiquitin ligases that regulate the turnover of specific target proteins. In Chapter 15 we will discuss examples of this pathway in more detail in relation to plant hormone action.

heterochromatin   Chromatin that is densely packed and transcriptionally modified or suppressed.

euchromatin   The dispersed, transcriptionally active form of chromatin.

nucleolus   (plural *nucleoli*) A densely granular region in the nucleus that is the site of ribosome synthesis.

ribosome   The site of cellular protein synthesis; consists of RNA and protein.

nucleolar organizer region (NOR)   Site in the nucleolus where chromosomal regions that code for ribosomal RNA are clustered and are transcribed.

transcription   The process by which the base sequence information in DNA is copied into an RNA molecule.

translation   The process whereby a specific protein is synthesized on ribosomes according to the sequence information encoded by the mRNA.

translocons   Pores in the rough endoplasmic reticulum that enable proteins synthesized on ribosomes to enter the ER lumen.

turnover   The balance between the rate of synthesis and the rate of degradation, usually applied to protein or RNA. An increase in turnover typically refers to an increase in degradation.

ubiquitin   A small polypeptide that is covalently attached to proteins, and that serves as a recognition site for a large proteolytic complex, the proteasome.

26S proteasome   A large proteolytic complex that degrades intracellular proteins marked for destruction by the attachment of one or more copies of the small protein ubiquitin.

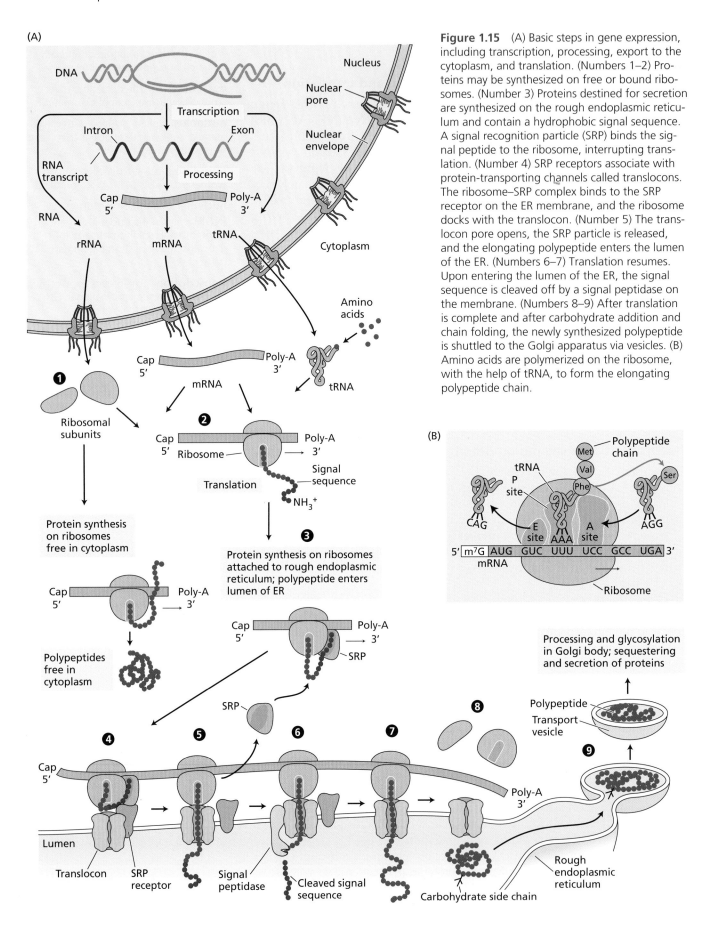

(A)

Figure 1.15 (A) Basic steps in gene expression, including transcription, processing, export to the cytoplasm, and translation. (Numbers 1–2) Proteins may be synthesized on free or bound ribosomes. (Number 3) Proteins destined for secretion are synthesized on the rough endoplasmic reticulum and contain a hydrophobic signal sequence. A signal recognition particle (SRP) binds the signal peptide to the ribosome, interrupting translation. (Number 4) SRP receptors associate with protein-transporting channels called translocons. The ribosome–SRP complex binds to the SRP receptor on the ER membrane, and the ribosome docks with the translocon. (Number 5) The translocon pore opens, the SRP particle is released, and the elongating polypeptide enters the lumen of the ER. (Numbers 6–7) Translation resumes. Upon entering the lumen of the ER, the signal sequence is cleaved off by a signal peptidase on the membrane. (Numbers 8–9) After translation is complete and after carbohydrate addition and chain folding, the newly synthesized polypeptide is shuttled to the Golgi apparatus via vesicles. (B) Amino acids are polymerized on the ribosome, with the help of tRNA, to form the elongating polypeptide chain.

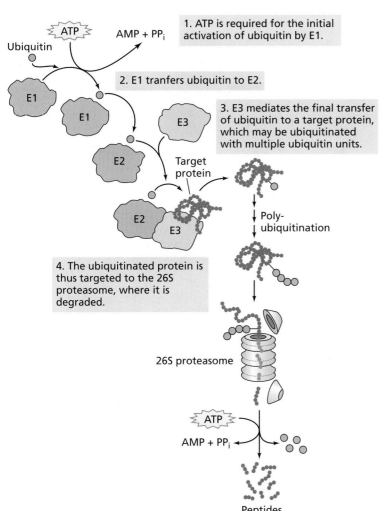

Figure 1.16 Generalized diagram of the cytoplasmic pathway of protein degradation.

1. ATP is required for the initial activation of ubiquitin by E1.

2. E1 tranfers ubiquitin to E2.

3. E3 mediates the final transfer of ubiquitin to a target protein, which may be ubiquitinated with multiple ubiquitin units.

4. The ubiquitinated protein is thus targeted to the 26S proteasome, where it is degraded.

The Endomembrane System

The endomembrane system of eukaryotic cells is the collection of related internal membranes that subdivides the cell into functional and structural compartments and that distributes membranes and proteins via vesicular traffic among cellular organelles. Having already described the nucleus, here we describe the other major endomembrane compartments.

The endoplasmic reticulum is a network of internal membranes

The **endoplasmic reticulum** (**ER**) is composed of an extensive network of tubules that is continuous with the nuclear envelope. The tubules join together to form a network of flattened saccules called **cisternae** (singular *cisterna*) (**Figure 1.17**). The tubules spread throughout the cell, forming close associations with other organelles. The ER network may therefore be a communication network between organelles within a cell, while also serving as a synthesis and delivery system for proteins and lipids. The ER that lies just under, and is probably attached to, the plasma membrane resides in the outer layer of cytoplasm called the cell cortex. This portion of the ER is therefore called the **cortical ER**.

The region of the ER that has many membrane-bound ribosomes is called **rough ER** because the bound ribosomes give the ER a rough appearance in electron micrographs (see Figure 1.17A and B). ER without bound ribosomes is called **smooth ER** and is the site of lipid biosynthesis (see Figure 1.17C).

endoplasmic reticulum (ER)
A continuous membrane system within the cytoplasm of eukaryotic cells that serves multiple functions, including the synthesis, modification, and intracellular transport of proteins.

cisternae (singular *cisterna*) A network of flattened saccules and tubules that compose the endoplasmic reticulum.

cortical ER The network of endoplasmic reticulum that lies just under the plasma membrane and is associated with the plasma membrane at specific contact points.

rough ER The endoplasmic reticulum to which ribosomes are attached. Rough ER synthesizes proteins that are transported by vesicles either to internal organelles or the plasma membrane.

smooth ER The endoplasmic reticulum lacking attached ribosomes and usually consisting of tubules. Functions in lipid synthesis.

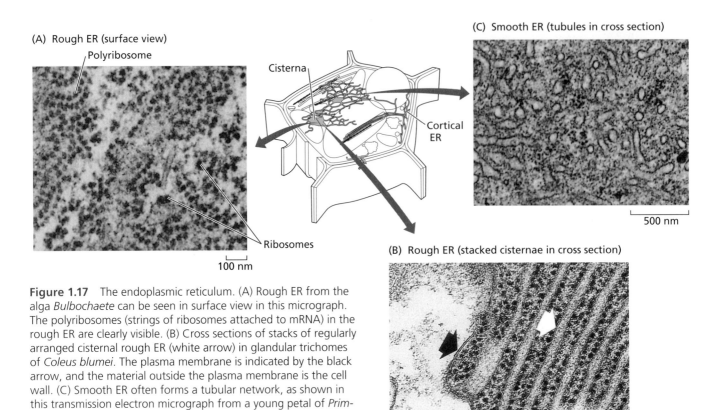

(A) Rough ER (surface view)

Polyribosome

Ribosomes

100 nm

Cisterna

Cortical ER

(C) Smooth ER (tubules in cross section)

500 nm

(B) Rough ER (stacked cisternae in cross section)

100 nm

Figure 1.17 The endoplasmic reticulum. (A) Rough ER from the alga *Bulbochaete* can be seen in surface view in this micrograph. The polyribosomes (strings of ribosomes attached to mRNA) in the rough ER are clearly visible. (B) Cross sections of stacks of regularly arranged cisternal rough ER (white arrow) in glandular trichomes of *Coleus blumei*. The plasma membrane is indicated by the black arrow, and the material outside the plasma membrane is the cell wall. (C) Smooth ER often forms a tubular network, as shown in this transmission electron micrograph from a young petal of *Primula kewensis*. (Micrographs from Gunning and Steer 1996.)

The ER is the major source of membrane phospholipids and provides membrane proteins and protein cargo for the other compartments in the endomembrane pathway: the nuclear envelope, the Golgi apparatus (see Figure 1.11), vacuoles, the plasma membrane, and the endosomal system. It even transports some proteins to the chloroplast. Most of this transport occurs via specialized vesicles moving between the endomembrane organelles.

Vacuoles have diverse functions in plant cells

The plant vacuole was originally defined by its appearance in the microscope—a membrane-enclosed compartment without cytoplasm. Instead of cytosol, it contains **vacuolar sap** composed of water and solutes. The increase in volume of plant cells during growth takes place primarily through an increase in the volume of vacuolar sap. A large central vacuole occupies up to 95% of the total cell volume in many mature plant cells. The fact that vacuoles can differ in size, appearance, and contents suggests how diverse in form and function the plant vacuolar compartment can be. Meristematic cells have no large central vacuole, but rather many small vacuoles. Some of these probably fuse together to form the large central vacuole as the cell matures.

The vacuolar membrane, or **tonoplast**, contains proteins and lipids that are synthesized initially in the ER. In addition to its role in cell expansion, the vacuole can also serve as a storage compartment for secondary metabolites involved in plant defense against herbivores (see Chapter 22). Inorganic ions, sugars, organic acids, and pigments are just some of the solutes that can accumulate in vacuoles, because of the presence of a variety of specific membrane transporters (see Chapter 6). Protein-storing vacuoles, called **protein bodies**, are abundant

vacuolar sap The fluid contents of a vacuole, which may include water, inorganic ions, sugars, organic acids, and pigments.

tonoplast The vacuolar membrane.

protein bodies Protein storage organelles enclosed by a single membrane; found mainly in seed tissues.

in seeds. Protein bodies and other vacuolar compartments that can also serve as sites of proteolysis are termed **lytic vacuoles**.

Oil bodies are lipid-storing organelles

Many plants synthesize and store large quantities of oil during seed development. These oils accumulate in organelles called **oil bodies** (also known as **oleosomes**). Oil bodies are unique among the organelles in that they are surrounded by a "half–unit membrane"—that is, a phospholipid monolayer—derived from the ER. The phospholipids in the half–unit membrane are oriented with their polar head groups toward the aqueous phase of the cytosol and their hydrophobic fatty acid tails facing the lumen, dissolved in the stored lipid. Oil bodies initially form as regions of differentiation within the ER. The nature of the storage product, **triglycerides** (three fatty acids covalently linked to a glycerol backbone), dictates that this storage organelle will have a hydrophobic lumen. Consequently, as triglyceride is stored, it appears to be initially deposited in the hydrophobic region between the outer and inner leaflets of the ER membrane.

Microbodies play specialized metabolic roles in leaves and seeds

Microbodies are a class of spherical organelles surrounded by a single membrane and specialized for one of several metabolic functions. **Peroxisomes** and **glyoxysomes** are microbodies specialized for the β-oxidation of fatty acids and the metabolism of glyoxylate, a two-carbon acid aldehyde (see Chapter 11). The glyoxysome is associated with mitochondria and oil bodies, while the peroxisome is associated with mitochondria and chloroplasts (**Figure 1.18**).

lytic vacuoles   Analogous to lysosomes in animal cells, plant lytic vacuoles contain hydrolytic enzymes that break down cellular macromolecules during senescence.

oil bodies   Organelles that accumulate and store triacylglycerols. They are bounded by a single phospholipid leaflet ("half–unit membrane" or "phospholipid monolayer") derived from the endoplasmic reticulum. Also known as oleosomes or spherosomes.

triglycerides   Three fatty acyl groups in ester linkage to three hydroxyl groups of glycerol. Fats and oils.

microbodies   A class of spherical organelles surrounded by a single membrane and specialized for one of several metabolic functions.

peroxisome   A type of microbody in which organic substrates are oxidized by O_2. These reactions generate H_2O_2 that is broken down to water by the peroxisomal enzyme catalase.

glyoxysome   A type of microbody found in the oil-rich storage tissues of seeds in which fatty acids are oxidized.

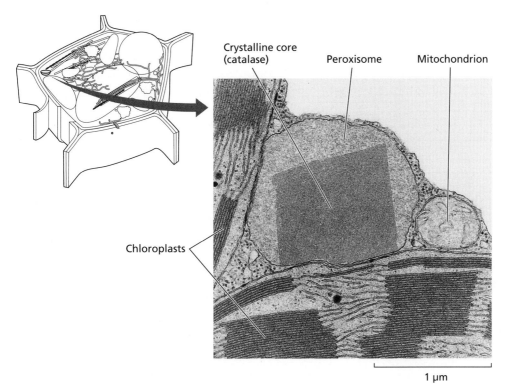

Figure 1.18   Catalase crystal in a peroxisome of a mature leaf of tobacco. Note the close association of the peroxisome with two chloroplasts and a mitochondrion, organelles that exchange metabolites with peroxisomes, especially during photorespiration (see Chapter 8). (Micrograph by S. E. Frederick, courtesy of E. H. Newcomb.)

Independently Dividing Semiautonomous Organelles

mitochondrion (plural *mitochondria*) The organelle that is the site for most reactions in the respiratory process in eukaryotes.

cristae Folds in the inner mitochondrial membrane that project into the mitochondrial matrix.

matrix The aqueous, gel-like phase of a mitochondrion that occupies the internal space into which the cristae extend. Contains the DNA, ribosomes, and soluble enzymes required for the tricarboxylic acid cycle, oxidative phosphorylation, and other metabolic reactions.

A typical plant cell has two types of energy-producing organelles: mitochondria and chloroplasts. Both types are separated from the cytosol by a double membrane (an outer and an inner membrane) and contain their own DNA and ribosomes.

Mitochondria (singular *mitochondrion*) are the cellular sites of respiration, a process in which the energy released from sugar metabolism is used for the synthesis of ATP (adenosine triphosphate) from ADP (adenosine diphosphate) and inorganic phosphate (P_i) (see Chapter 11).

Mitochondria are highly dynamic structures that can undergo both fission and fusion. Mitochondrial fusion can result in long, tubelike structures that may branch to form mitochondrial networks. Regardless of shape, all mitochondria have a smooth outer membrane and a highly convoluted inner membrane (**Figure 1.19**). The inner membrane contains an ATP synthase that uses a proton gradient to synthesize ATP for the cell. The proton gradient is generated through the cooperation of electron transporters called the electron transport chain, which is embedded in, and peripheral to, the inner membrane (see Chapter 11). The infoldings of the inner membrane are called **cristae** (singular *crista*). The compartment enclosed by the inner membrane, the mitochondrial **matrix**, contains the enzymes of the pathway of intermediary metabolism called the tricarboxylic acid cycle.

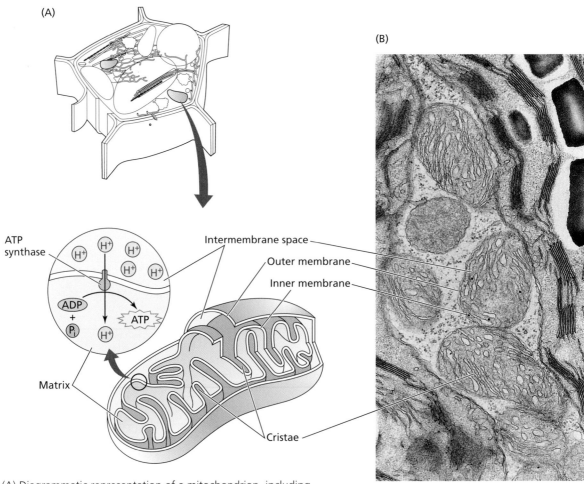

Figure 1.19 (A) Diagrammatic representation of a mitochondrion, including the location of the ATP synthases involved in ATP synthesis on the inner membrane. (B) Electron micrograph of mitochondria from a leaf cell of Bermuda grass (*Cynodon dactylon*). (Micrograph by S. E. Frederick, courtesy of E. H. Newcomb.)

Chloroplasts (**Figure 1.20A**) belong to another group of double-membrane–enclosed organelles called plastids, and are the sites of photosynthesis. In addition to their inner and outer envelope membranes, chloroplasts possess a third system of membranes called **thylakoids**. A stack of thylakoids forms

chloroplast The organelle that is the site of photosynthesis in photosynthetic eukaryotic organisms.

thylakoids The specialized, internal, chlorophyll-containing membranes of the chloroplast where light absorption and the chemical reactions of photosynthesis take place.

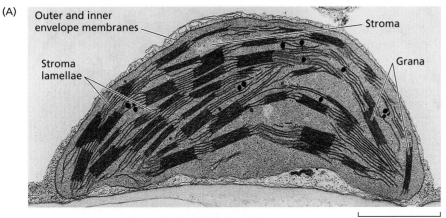

(A) Outer and inner envelope membranes
Stroma lamellae
Stroma
Grana
2 μm

(B)
Thylakoid
Granum
Stroma
Stroma lamellae
0.5 μm

(C)
Outer envelope membrane
Inner envelope membrane
Thylakoids
Stroma
Granum (stack of thylakoids)
Thylakoid lumen
Thylakoid membrane

(D)
Stroma
ATP synthase
H⁺
ADP + Pᵢ
ATP
H⁺

Figure 1.20 (A) Electron micrograph of a chloroplast from a leaf of timothy grass (*Phleum pratense*). (B) The same preparation at higher magnification. (C) Three-dimensional view of grana stacks and stroma lamellae, showing the complexity of the organization. (D) Diagrammatic representation of a thylakoid, showing the location of the ATP synthases in the thylakoid membrane. (Micrographs by W. P. Wergin, courtesy of E. H. Newcomb.)

granum (plural *grana*) In the chloroplast, a stack of thylakoids.

stroma lamellae Unstacked thylakoid membranes within the chloroplast.

stroma The fluid component surrounding the thylakoid membranes of a chloroplast.

chromoplasts Plastids that contain high concentrations of carotenoid pigments, rather than chlorophyll.

leucoplasts Nonpigmented plastids, the most important of which is the amyloplast.

amyloplast A starch-storing plastid found abundantly in storage tissues of shoots and roots, and in seeds. Specialized amyloplasts in the root cap also serve as gravity sensors.

a **granum** (plural *grana*) (**Figure 1.20B**). Proteins and pigments (chlorophylls and carotenoids) that function in the photochemical events of photosynthesis are embedded in the thylakoid membrane. Adjacent grana are connected by unstacked membranes called **stroma lamellae** (singular *lamella*). The fluid compartment surrounding the thylakoids, called the **stroma**, is analogous to the matrix of the mitochondrion and contains what may be Earth's most abundant protein, Rubisco, the protein involved in converting the carbon from carbon dioxide into organic acids during photosynthesis (see Chapter 8). The large subunit of Rubisco is encoded by the chloroplast genome, while the small subunit is encoded by the nuclear genome. Coordinated expression of each subunit (and other proteins) by each genome is necessary as chloroplasts grow and divide.

The various components of the photosynthetic apparatus are localized in different areas of the grana and the stroma lamellae. The ATP synthases of the chloroplast are located on the thylakoid membrane (**Figure 1.20C**). During photosynthesis, light-driven electron transfer reactions result in a proton gradient across the thylakoid membrane (**Figure 1.20D**) (see Chapter 7). As in the mitochondria, ATP is synthesized when the proton gradient is dissipated via ATP synthase. In the chloroplast, however, the ATP is not exported to the cytosol, but is used for many stromal reactions, including the fixation of carbon from carbon dioxide in the atmosphere, as described in Chapter 8.

Plastids that contain high concentrations of carotenoid pigments, rather than chlorophyll, are called **chromoplasts**. Chromoplasts are responsible for the yellow, orange, and red colors of many fruits and flowers, as well as of autumn leaves (**Figure 1.21**).

Nonpigmented plastids are called **leucoplasts**. Leucoplasts in specialized secretory tissues, such as the nectary, make monoterpenoids, volatile molecules (in essential oils) that often have a strong smell. The most important type of leucoplast is the **amyloplast**, a starch-storing plastid. Amyloplasts are abundant in the storage tissues of shoots and roots, and in seeds. Specialized amyloplasts in the root cap also serve as gravity sensors that direct root growth downward into the soil (see Chapter 15).

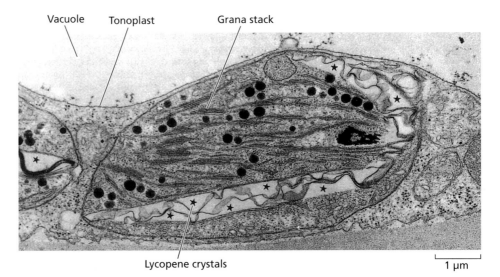

Vacuole   Tonoplast   Grana stack

Lycopene crystals   1 μm

Figure 1.21 Electron micrograph of a chromoplast from tomato (*Solanum esculentum*) fruit at an early stage in the transition from chloroplast to chromoplast. Small grana stacks are still visible. Stars indicate crystals of lycopene, a type of carotenoid pigment. (From Gunning and Steer 1996.)

Proplastids mature into specialized plastids in different plant tissues

Meristem cells contain **proplastids**, which have few or no internal membranes, no chlorophyll, and an incomplete complement of the enzymes necessary to carry out photosynthesis (**Figure 1.22A**). In angiosperms and some gymnosperms, chloroplast development from proplastids is triggered by light. Upon illumination, enzymes are formed inside the proplastid or imported from the cytosol; light-absorbing pigments are produced; and membranes proliferate rapidly, giving rise to stroma lamellae and grana stacks (**Figure 1.22B**).

Seeds usually germinate in the soil in the dark, and their proplastids mature into chloroplasts only when the young shoot is exposed to light. If, instead, germinated seedlings are kept in the dark, the proplastids differentiate into **etioplasts**, which contain semicrystalline tubular arrays of membrane known as **prolamellar bodies** (**Figure 1.22C**). Instead of chlorophyll, etioplasts contain a pale yellow-green precursor pigment, protochlorophyllide. Within minutes after exposure to light, an etioplast differentiates, converting the prolamellar body into thylakoids and stroma lamellae, and the protochlorophyllide into chlorophyll. The maintenance of chloroplast structure depends on the presence of light; mature chloroplasts can revert to etioplasts during extended periods of darkness. Likewise, under different environmental conditions, chloroplasts can be converted to chromoplasts (see Figure 1.21), as in autumn leaves and ripening fruit.

Chloroplast and mitochondrial division are independent of nuclear division

Like mitochondria, plastids divide by the process of fission, consistent with their prokaryotic origins. Fission and organellar DNA replication are regulated independently of nuclear division. For example, the number of chloroplasts per cell volume depends on the cell's developmental history and its local environment.

proplastid Type of immature, undeveloped plastid found in meristematic tissue.

etioplast Photosynthetically inactive form of chloroplast found in etiolated seedlings.

prolamellar bodies Elaborate semicrystalline lattices of membrane tubules that develop in plastids that have not been exposed to light (etioplasts).

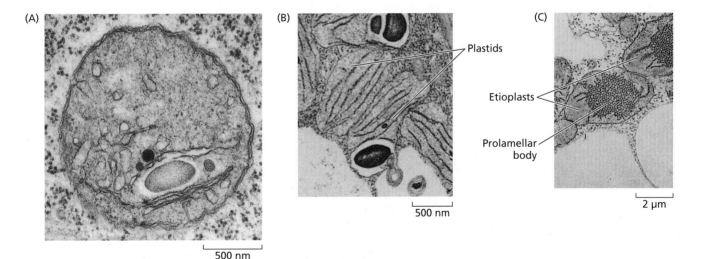

Figure 1.22 Electron micrographs illustrating several stages of plastid development. (A) Proplastid from the root apical meristem of the broad bean (*Vicia faba*). The internal membrane system is rudimentary, and grana are absent. (B) Mesophyll cell of a young oat (*Avena sativa*) leaf grown in the light, at an early stage of differentiation. The plastids are developing grana stacks. (C) Cell of a young oat leaf from a seedling grown in the dark. The plastids have developed as etioplasts, with elaborate semicrystalline lattices of membrane tubules called prolamellar bodies. When exposed to light, the etioplast can convert to a chloroplast by the disassembly of the prolamellar body and the formation of grana stacks. (From Gunning and Steer 1996.)

cytoskeleton Composed of polarized microfilaments of actin or microtubules of tubulin, the cytoskeleton helps control the organization and polarity of organelles and cells during growth.

microtubule A component of the cell cytoskeleton and the mitotic spindle, and a player in the orientation of cellulose microfibrils in the cell wall. Made of tubulin.

tubulin A family of cytoskeletal GTP-binding proteins with three members, α-, β-, and γ-tubulin. α-tubulin forms heterodimers with β-tubulin, which polymerize to form microtubules.

protofilament A column of polymerized tubulin monomers (α- and β-tubulin heterodimers) or a chain of polymerized actin subunits.

microfilament A component of the cell cytoskeleton made of actin; it is involved in organelle motility within cells.

actin A major ATP-binding cytoskeletal protein. The monomeric, globular form of actin is called G-actin; the polymerized form in microfilaments is F-actin.

Accordingly, there are many more chloroplasts in the mesophyll cells in the interior of a leaf than in the epidermal cells forming the outer layer of the leaf.

Although the timing of the fission of chloroplasts and mitochondria is independent of the timing of cell division, these organelles require nuclear-encoded proteins to divide. In both bacteria and the semiautonomous organelles, fission is facilitated by proteins that form rings on the inner membrane at the site of the future division plane. In plant cells, the genes that encode these proteins are located in the nucleus. The proteins may be delivered to the site through associated ER, which forms a ring around the dividing organelle. Mitochondria and chloroplasts can also increase in size without dividing, in order to meet energy or photosynthetic demand. If, for example, the proteins involved in mitochondrial division are experimentally inactivated, the fewer mitochondria become larger, allowing the cell to meet its energy needs.

Both plastids and mitochondria can move around plant cells. In some plant cells the chloroplasts are anchored in the outer, cortical cytoplasm of the cell, but in others they are mobile. Plastids and mitochondria are motorized by plant myosins that move along actin microfilaments. Actin microfilament networks are among the main components of the plant cytoskeleton, which we will describe next.

The Plant Cytoskeleton

The cytoplasm is organized into a three-dimensional network with filamentous proteins, called the **cytoskeleton**. This network provides the spatial organization for the organelles and serves as scaffolding for the movements of organelles and other cytoskeletal components. It also plays fundamental roles in mitosis, meiosis, cytokinesis, wall deposition, the maintenance of cell shape, and cell differentiation.

The plant cytoskeleton consists of microtubules and microfilaments

Two major types of cytoskeletal elements have been demonstrated in plant cells: microtubules and microfilaments. Each type is filamentous, having a fixed diameter and a variable length, up to many micrometers.

Microtubules are hollow cylinders with an outer diameter of 25 nm; they are composed of polymers of the protein **tubulin**. The tubulin monomer is a heterodimer composed of two similar polypeptide chains (α- and β-tubulin) (**Figure 1.23A**). A single microtubule consists of hundreds of thousands of tubulin monomers arranged in columns called **protofilaments**.

Microfilaments are solid, with a diameter of 7 nm. They are composed of the monomeric form of the protein **actin**, called globular actin, or G-actin.

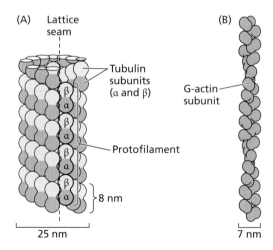

Figure 1.23 (A) Drawing of a microtubule in longitudinal view. Each microtubule is typically composed of 13 protofilaments (but varies with species and cell type). The organization of the α and β subunits is shown. (B) Diagrammatic representation of a microfilament, showing an F-actin strand (protofilament) with a helical pitch based on the asymmetry of the monomers, the G-actin subunits.

Monomers of G-actin polymerize to form a single chain of actin subunits, also called a protofilament. The actin in the polymerized protofilament is referred to as filamentous actin, or F-actin. A microfilament is helical, a shape resulting from the polarity of association of the G-actin monomers (**Figure 1.23B**).

Actin, tubulin, and their polymers are in constant flux in the living cell

In the cell, actin and tubulin subunits exist as pools of free proteins that are in dynamic equilibrium with their polymerized forms. The cycle of polymerization–depolymerization is essential for cell life; drugs that halt this cycle will eventually kill the cell. Each of the monomers contains a bound nucleotide: ATP or ADP in the case of actin, GTP or GDP (guanosine tri- or diphosphate) in the case of tubulin. Both microtubules and microfilaments are polarized; that is, the two ends are different. The polarity is displayed by the different rates of growth of the two ends, with the more active end being the plus end and the less active end being the minus end.

In microfilaments, the polarity arises from the polarity of the actin monomer itself; the ATP/ADP nucleotide-binding cleft of the monomer orients toward the slow-growing negative end of the microfilament, while the side opposite the ATP/ADP nucleotide-binding cleft orients toward the fast-growing positive end (**Figure 1.24A**). In microtubules, the polarity arises from the polarity of the α- and β-tubulin heterodimer (see Figure 1.23). The α-tubulin monomer exists only in the GTP form and is exposed on the minus end, and the β-tubulin can bind either GTP or GDP and is exposed on the plus end (see Figure 1.25A). Microfilaments and microtubules have half-lives, usually counted in minutes, determined by accessory proteins that regulate the dynamics of microfilaments and microtubules.

In the presence of ATP, Mg^{2+}, and K^+, and without accessory proteins, G-actin polymerizes in a concentration-dependent manner (**Figure 1.24B**). During nucleation, free G-actin binds to ATP and associates to form trimers. The monomers assemble at both the positive and negative ends of the growing F-actin microfilament, although net growth is faster at the positive end. ATP hydrolysis is not

(A)

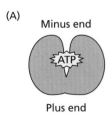

(B)

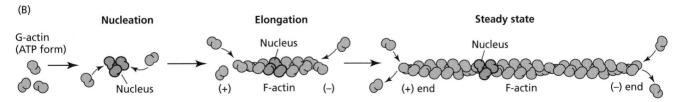

Figure 1.24 Models for the assembly of actin microfilaments. (A) Structure of G-actin. Each G-actin molecule has an ATP/ADP nucleotide-binding cleft at the minus end of the monomer. The side opposite the nucleotide-binding cleft is termed the plus end. (B) F-actin assembly. Individual G-actin monomers bind ATP and aggregate to form a trimer, which nucleates formation of a new filament. Additional G-actin then polymerizes at the plus and minus ends of the growing F-actin filament. ATP hydrolysis to ADP occurs after the ATP-charged units are polymerized, and the G-actin that comes off has ADP. (After Lodish et al. 2007.)

treadmilling During interphase, a process by which microfilaments or microtubules in the cortical cytoplasm appear to migrate around the cell periphery due to addition of G-actin or tubulin heterodimers, respectively, to the plus end at the same rate as their removal from the minus end.

cytoplasmic streaming The coordinated movement of particles and organelles through the cytosol.

directed organelle movement Movement of an organelle in a particular direction, which can be driven by the interaction with molecular motors associated with the cytoskeleton.

required for polymerization, but is required for depolymerization at the negative end of the microfilament. At steady state, the rate of assembly of the microfilament at the positive end is at dynamic equilibrium with the rate of depolymerization at the negative end, a condition referred to as **treadmilling**. Treadmilling allows microfilaments to appear to migrate around the cell cortex in the direction of the positive pole. However, what appears to be microfilament motility is actually caused by differential growth at the positive and negative ends.

Microtubule protofilaments first assemble into flat sheets before curling into cylinders

Microtubule nucleation and initiation of growth occur at *microtubule organizing centers (MTOCs)*, also called *initiation complexes*, but the precise nature of the initiation complex is still unresolved. One type of initiation complex contains a much less abundant type of tubulin called *γ-tubulin*, which forms ring complexes in the cortical cytoplasm from which microtubules can grow (**Figure 1.25A–C**).

Each tubulin heterodimer contains two GTP molecules, one on the α-tubulin monomer and the other on the β-tubulin monomer. GTP on the α-tubulin monomer is tightly bound and nonhydrolyzable, while the GTP on the β-tubulin site is hydrolyzed to GDP after the heterodimer assembles onto the plus end of a microtubule. If the rate of GTP hydrolysis "catches up" with the rate of addition of new heterodimers, the GTP-charged cap of tubulin vanishes and the protofilaments come apart from each other, initiating a catastrophic depolymerization (see Figure 1.25C). Such catastrophes can be rescued (depolymerization stopped and polymerization resumed) if the increase in the local free tubulin concentration (with GTP) caused by the catastrophe once again favors polymerization. The slowly growing minus end does not depolymerize if it is capped by γ-tubulin. However, plant microtubules can be released from γ-tubulin ring complexes by an ATPase, katanin (from the Japanese word *katana*, "samurai sword"), which severs the microtubule at the point where the growing microtubule branches off another microtubule (**Figure 1.25D**). Once the microtubules are released by katanin, they travel, snakelike, through the cell cortex by treadmilling.

During treadmilling, tubulin heterodimers are added to the growing plus end at about the same rate that they are removed from the shrinking minus end (see Figure 1.25D). As we will discuss in the section *Cell Cycle Regulation* below, the transverse orientation of the cortical microtubules determines the orientation of the newly synthesized cellulose microfibrils in the cell wall. The presence of transverse cellulose microfibrils in the cell wall reinforces the wall in the transverse direction, promoting growth along the longitudinal axis. In this way, microtubules play an important role in the polarity of plant growth.

Cytoskeletal motor proteins mediate cytoplasmic streaming and directed organelle movement

Mitochondria, peroxisomes, and Golgi bodies are extremely dynamic in plant cells. These approximately 1-μm particles move at rates of about 1–10 μm s^{-1} in seed plants. This movement is quite fast; it is equivalent to a 1-m object moving at 10 m sec^{-1}, approximately the speed of the world's fastest human. Actin and its motor protein, myosin, act together in the plant endoplasm (the more fluid inner layer of the cytoplasm) to generate this movement and are therefore frequently referred to as the actomyosin cytoskeleton.

Cytoplasmic streaming refers to the coordinated bulk flow of the cytosol and cytoplasmic organelles inside the cell. The movement of individual organelles can be part of cytoplasmic streaming but is perhaps better called **directed organelle movement** because organelles can frequently move past each other in opposite directions. If directed organelle movement exerts sufficient viscous drag on

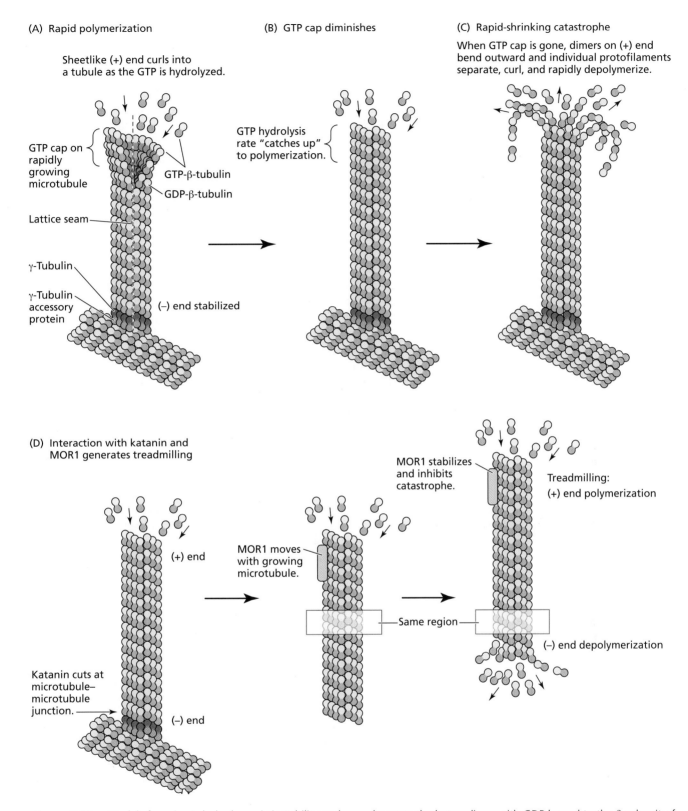

(A) Rapid polymerization

Sheetlike (+) end curls into
a tubule as the GTP is hydrolyzed.

GTP cap on
rapidly
growing
microtubule

GTP-β-tubulin

GDP-β-tubulin

Lattice seam

γ-Tubulin

γ-Tubulin
accessory
protein

(−) end stabilized

(B) GTP cap diminishes

GTP hydrolysis
rate "catches up"
to polymerization.

(C) Rapid-shrinking catastrophe

When GTP cap is gone, dimers on (+) end
bend outward and individual protofilaments
separate, curl, and rapidly depolymerize.

**(D) Interaction with katanin and
MOR1 generates treadmilling**

(+) end

MOR1 moves
with growing
microtubule.

Katanin cuts at
microtubule–
microtubule
junction.

(−) end

Same region

MOR1 stabilizes
and inhibits
catastrophe.

Treadmilling:
(+) end polymerization

(−) end depolymerization

Figure 1.25 Models for microtubule dynamic instability and treadmilling. (A) The minus ends of microtubules at new initiation sites can be stabilized by γ-tubulin ring complexes, some of which are found along the sides of previously existing microtubules. Microtubule plus ends grow rapidly, producing a "cap" of tubulin that has GTP bound to the β subunit. The newly added cap has a sheetlike structure that curls into a tubule as the GTP is hydrolyzed. (B) As the rate of growth slows or GTP hydrolysis increases, the GTP cap is diminished. (C) When the GTP cap is gone, the microtubule protofilaments peel apart because the heterodimer with GDP bound to the β subunit of tubulin is slightly curved. The protofilaments are unstable, and rapid, catastrophic depolymerization ensues. (D) If the microtubule is severed at a branch point by the ATPase katanin, the minus end becomes unstable and can depolymerize. If the microtubule is stabilized at the plus end by MOR1 (microtubule organizing 1 protein), a microtubule-associated protein, then the rate of addition on the plus end can match the depolymerization at the minus end and treadmilling ensues.

the surrounding cytosol and the organelles in it, then it will drive cytoplasmic streaming. As cells enlarge, movement rates tend to increase. In the giant cells of the green algae *Chara* and *Nitella*, cytoplasmic streaming occurs in a helical path down one side of a cell and up the other side, occurring at speeds of up to 75 μm s^{-1}. Directed organelle movement or stabilization (tethering) involves interactions via attachment sites to other organelles, endomembranes, the cytoskeleton, or the plasma membrane. In particular, the ER interacts with the plasma membrane and other organelles via these attachment sites.

Molecular motors participate in both directed organelle movements and tethering. Plants have two types of motors, the myosins and the kinesins. The myosins are actin-binding proteins. The kinesins are microtubule-associated proteins. When myosins and kinesins move along the cytoskeleton, they do so in a particular direction along the polar cytoskeletal polymers. Myosins generally move toward the plus end of the actin filaments. Although members of the kinesin family interact with some organelle membranes, they tend to tether the organelles rather than mediate movement along microtubules. Kinesins can also bind to chromatin or to other microtubules, and help organize the spindle apparatus during mitosis (see below).

How is it that motors can both move and tether organelles? All of the above motors have separate head, neck, and tail domains, as does myosin XI (**Figure 1.26**). The globular head domain binds to the cytoskeleton reversibly, depending on the phosphorylation state of the nucleotide in the domain's ATPase active site. The neck

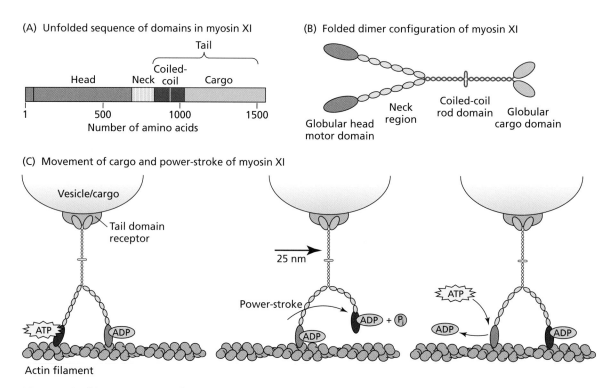

Figure 1.26 Myosin-driven movement of organelles. (A) Extended amino acid domains of a class XI myosin motor. The tail domain includes a coiled-coil region for dimerization and a cargo domain for interacting with membranes. (B) The head domain folds to become globular. Near the extended neck domain, ATP/ADP binds to the head domain. The neck consists of regions with a particular amino acid composition that can interact with modulating proteins. (C) Movement and power-stroke of myosin XI. The tail binds to the organelle via its cargo domains and a receptor complex on the membrane. The two heads, shown in red and pink, have ATPase and motor activity such that a change in the configuration of the neck region adjacent to the head produces a "walking," processive movement along the actin filament during the power-stroke of the motor, when ATP is hydrolyzed to ADP and inorganic phosphate (P$_i$). The cargo moves about 25 nm with each step. The back domain in pink releases its ADP and binds ATP, thus allowing the process to repeat.

region changes angle upon hydrolysis of ATP, flexing the head relative to the tail. The tail domain usually contains regions for dimerization, and the end of the globular tail domain binds to specific organelles or "cargo" and is called the *cargo domain* (see Figure 1.26B). In order for a motor to tether an organelle via the cytoskeleton to the plasma membrane, the motor head binds to the cytoskeleton, which is bound to the organelle, while the cargo domain binds to a protein in the plasma membrane. Often tethering motors are monomeric, so when the myosin-bound ATP is hydrolysed, the head domain releases from the cytoskeleton and the organelle bound to the cytoskeleton via the motor is released. In order for a motor to move an organelle, the motor dimerizes; the two molecules interact and bind to the organelle at the cargo domain. The two heads of the dimer then alternately bind to the cytoskeleton and "walk" forward while the neck flexes as ATP is hydrolyzed (see Figure 1.26C). In this way, the organelle is moved along the cytoskeleton.

Cell Cycle Regulation

The cell division cycle, or cell cycle, is the process by which cells reproduce themselves and their genetic material, the nuclear DNA (**Figure 1.27**). The cell cycle consists of four phases: **gap 1 (G$_1$)**, **DNA synthesis phase (S**, or **S phase)**, **gap 2 (G$_2$)**, and **mitotic phase (M**, or **M phase)**. G$_1$ is the phase when a newly formed daughter cell has not yet replicated its DNA. DNA is replicated during S phase. G$_2$ is the phase when a cell with replicated DNA has not yet proceeded to mitosis, which occurs during M phase. Collectively, the G$_1$, S, and G$_2$ phases are referred to as **interphase**. M phase includes all the stages of mitosis, from metaphase to telophase (discussed below). In vacuolated cells, the vacuole enlarges throughout interphase, and the plane of cell division bisects the vacuole during mitosis (see Figure 1.27).

Each phase of the cell cycle has a specific set of biochemical and cellular activities

Nuclear DNA is prepared for replication in G$_1$ by the assembly of a prereplication complex at the origins of replication along the chromatin. DNA is replicated during S phase, and G$_2$ cells prepare for mitosis.

The whole architecture of the cell is altered as it enters mitosis. If the cell has a large central vacuole, this vacuole must first be divided in two by a coalescence of cytoplasmic transvacuolar strands that contain the nucleus; this becomes the region where nuclear division will occur. (Compare Figure 1.27, division in a vacuolated cell, with Figure 1.29, division in a nonvacuolated cell.) Golgi bodies and other organelles divide and partition themselves equally between the two halves of the cell. As described below, the endomembrane system and cytoskeleton are extensively rearranged.

As a cell enters mitosis, the chromosomes change from their interphase state of organization within the nucleus and begin to condense to form the metaphase chromosomes (**Figure 1.28**; see also Figure 1.14). The metaphase chromosomes are held together by special proteins called *cohesins*, which reside in the centromeric region of each chromosome pair. In order for the chromosomes to separate, these proteins must be cleaved by the enzyme *separase*. This occurs when the kinetochore attaches to spindle microtubules (described below).

At a key regulatory point, or **checkpoint**, late in G$_1$ of the cell cycle, the cell becomes committed to the initiation of DNA synthesis. In mammalian cells, DNA replication and mitosis are linked—the cell division cycle, once begun, is not interrupted until all phases of mitosis have been completed. In contrast, plant cells can leave the cell division cycle either before or after replicating their DNA (i.e., during G$_1$ or G$_2$). As a consequence, whereas most animal cells are diploid (having two sets of chromosomes), plant cells are frequently tetraploid (having

G$_1$ Gap 1. The phase of the cell cycle preceding the synthesis of DNA.

S phase DNA synthesis phase. The phase in the cell cycle during which DNA is replicated; it follows G$_1$ and precedes G$_2$.

G$_2$ Gap 2. The phase of the cell cycle following the synthesis of DNA.

M phase Mitotic phase. The phase of the cell cycle in which the replicated chromosomes condense, move to opposite poles, and come to reside in the nuclei of two identical cells.

interphase Collectively the G$_1$, S, and G$_2$ phases of the cell cycle.

checkpoint One of several key regulatory points in the cell cycle. One checkpoint, late in G$_1$, determines if the cell is committed to the initiation of DNA synthesis.

endoreduplication Cycles of nuclear DNA replication without mitosis resulting in polyploidization.

cyclin-dependent kinases (CDKs) Protein kinases that regulate the transitions from G_1 to S, from S to G_2, and from G_2 to mitosis, during the cell cycle.

four sets of chromosomes) or sometimes even polyploid (having many sets of chromosomes) after going through additional cycles of nuclear DNA replication without mitosis, a process called **endoreduplication**.

The cell cycle is regulated by cyclins and cyclin-dependent kinases

The biochemical reactions governing the cell cycle are evolutionarily highly conserved in eukaryotes, and plants have retained the basic components of this mechanism. Progression through the cycle is regulated mainly at three checkpoints: during late G_1 phase (as mentioned above), late S phase, and at the G_2/M boundary.

The key enzymes that control the transitions between the different phases of the cell cycle, and the entry of nondividing cells into the cell cycle, are the **cyclin-dependent kinases**, or **CDKs**. Protein kinases are enzymes that phosphorylate proteins using ATP. Most multicellular eukaryotes have several protein kinases that are active in different phases of the

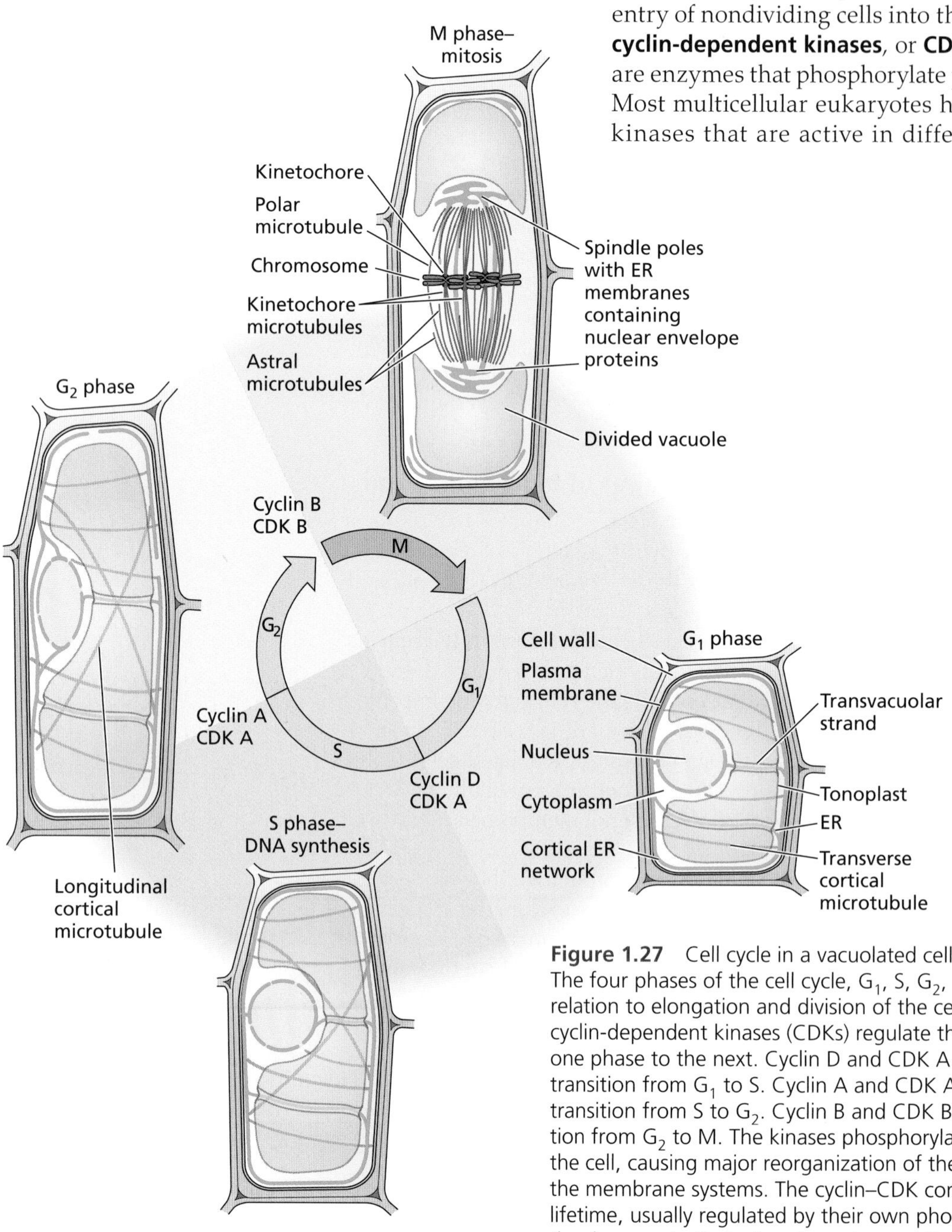

Figure 1.27 Cell cycle in a vacuolated cell type (a tobacco cell). The four phases of the cell cycle, G_1, S, G_2, and M, are shown in relation to elongation and division of the cell. Various cyclins and cyclin-dependent kinases (CDKs) regulate the transitions from one phase to the next. Cyclin D and CDK A are involved in the transition from G_1 to S. Cyclin A and CDK A are involved in the transition from S to G_2. Cyclin B and CDK B regulate the transition from G_2 to M. The kinases phosphorylate other proteins in the cell, causing major reorganization of the cytoskeleton and the membrane systems. The cyclin–CDK complexes have a finite lifetime, usually regulated by their own phosphorylation state; the decrease in their abundance toward the end of each phase allows progression to the next stage of the cell cycle.

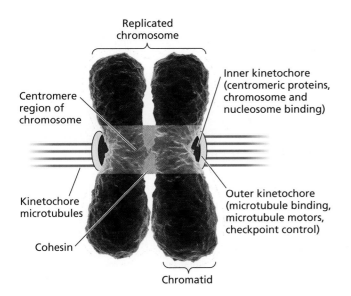

Replicated chromosome

Centromere region of chromosome

Inner kinetochore (centromeric proteins, chromosome and nucleosome binding)

Kinetochore microtubules

Outer kinetochore (microtubule binding, microtubule motors, checkpoint control)

Cohesin

Chromatid

Figure 1.28 Structure of a metaphase chromosome. The centromeric DNA is highlighted, and the region where cohesion molecules bind the two chromosomes together is shown in orange. The kinetochore is a layered structure (inner layer is shown in purple, outer layer in yellow) that contains microtubule-binding proteins, including kinesins that help depolymerize the microtubules during shortening of kinetochore microtubules in anaphase. (© Sebastian Kaulitzki/Shutterstock.)

cell cycle. All depend on regulatory subunits called **cyclins** for their activities. Several classes of cyclins have been identified in plants, animals, and yeast. Three cyclins have been shown to regulate the tobacco cell cycle, as shown in Figure 1.27:

1. G_1/S cyclins, cyclin D, active late in G_1

2. S-type cyclins, cyclin A, active in late S phase

3. M-type cyclins, cyclin B, active just prior to the mitotic phase

The critical restriction point late in G_1, which commits the cell to another round of DNA replication, is regulated primarily by the D-type cyclins and CDKs.

Mitosis and cytokinesis involve both microtubules and the endomembrane system

Mitosis is the process by which previously replicated chromosomes are aligned, separated, and distributed in an orderly fashion to daughter cells (**Figure 1.29**). Microtubules are an integral part of mitosis. The period immediately prior to prophase is called **preprophase**. During preprophase, the G_2 microtubules are completely reorganized into a **preprophase band** of microtubules and microfilaments around the nucleus at the site of the future cell plate—the precursor of the cross-wall (see Figure 1.29A). The position of the preprophase band, the underlying *cortical division site*, and the partition of cytoplasm that divides the central vacuoles determine the plane of cell division in plants, and thus play a crucial role in development (see Chapters 14–17).

At the start of **prophase**, microtubules, polymerizing on the surface of the nuclear envelope, begin to gather at two foci on opposite sides of the nucleus, initiating spindle formation (see Figure 1.29B). Although not associated with the centrosomes (which plants, unlike animal cells, lack), these foci serve the same function in organizing microtubules. During prophase, the nuclear envelope remains intact, but it breaks down as the cell enters **metaphase** in a process that involves reorganization and reassimilation of the nuclear envelope into the ER (see Figure 1.29C). Throughout the division cycle, cell division kinases interact with microtubules to help reorganize the spindle by phosphorylating microtubule-associated proteins and kinesins. The nucleolus completely disappears during mitosis and gradually reassembles after mitosis, as the chromosomes decondense and reestablish their positions in the daughter nuclei.

cyclins Regulatory proteins associated with cyclin-dependent kinases that play a crucial role in regulating the cell cycle.

preprophase In mitosis, the stage just before prophase during which the G_2 microtubules are completely reorganized into a preprophase band.

preprophase band A circular array of microtubules and microfilaments formed in the cortical cytoplasm just prior to cell division that encircles the nucleus and predicts the plane of cytokinesis following mitosis.

prophase The first stage of mitosis (and meiosis) prior to disassembly of the nuclear envelope, during which the chromatin condenses to form distinct chromosomes.

metaphase A stage of mitosis during which the nuclear envelope breaks down and the condensed chromosomes align in the middle of the cell.

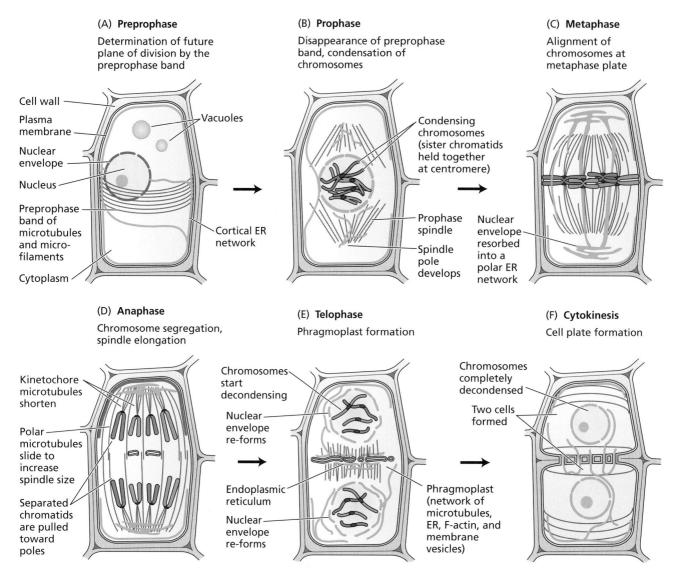

(A) Preprophase

Determination of future plane of division by the preprophase band

Cell wall

Plasma membrane

Nuclear envelope

Nucleus

Preprophase band of microtubules and micro-filaments

Cytoplasm

Vacuoles

Cortical ER network

(B) Prophase

Disappearance of preprophase band, condensation of chromosomes

Condensing chromosomes (sister chromatids held together at centromere)

Prophase spindle

Spindle pole develops

(C) Metaphase

Alignment of chromosomes at metaphase plate

Nuclear envelope resorbed into a polar ER network

(D) Anaphase

Chromosome segregation, spindle elongation

Kinetochore microtubules shorten

Polar microtubules slide to increase spindle size

Separated chromatids are pulled toward poles

(E) Telophase

Phragmoplast formation

Chromosomes start decondensing

Nuclear envelope re-forms

Endoplasmic reticulum

Nuclear envelope re-forms

(F) Cytokinesis

Cell plate formation

Chromosomes completely decondensed

Two cells formed

Phragmoplast (network of microtubules, ER, F-actin, and membrane vesicles)

Figure 1.29   Changes in cellular organization that accompany mitosis in a meristem-atic (nonvacuolated) plant cell.

mitotic spindle   The mitotic structure involved in chromosome movement. Polymerized from α- and β-tubulin monomers formed by the disassembly of the preprophase band in early metaphase.

centromere   The constricted region on the mitotic chromosome where the kinetochore forms and to which spindle fibers attach.

telomeres   Regions of repetitive DNA that form the chromosome ends and protect them from degradation.

kinetochore   The site of spindle fiber attachment to the chromosome in anaphase.

In early metaphase, the preprophase band disappears and new microtubules polymerize to complete the **mitotic spindle**. The mitotic spindles of plant cells are more boxlike in shape than those in animal cells. The spindle microtubules in a plant cell arise from a diffuse zone consisting of multiple foci at opposite ends of the cell and extend toward the middle in nearly parallel arrays.

Metaphase chromosomes are completely condensed through close packing of the histones into nucleosomes, which are further wound into condensed fibers (see Figures 1.14 and 1.28). The **centromere**, the region where the two chromatids are attached near the center of the chromosome, contains repetitive DNA, as does the **telomere**, which forms the chromosome ends that protect it from degradation. Some of the spindle microtubules bind to the chromosomes in the special region of the centromere called the **kinetochore**, and the condensed chromosomes align at the metaphase plate (see Figures 1.27 and 1.29). Some of the unattached microtubules overlap with microtubules from the opposite polar region in the spindle midzone.

The mechanism of chromosome separation during **anaphase** has two components (see Figure 1.29D). In early anaphase, the sister chromatids separate and begin to move toward their poles. During late anaphase, the polar microtubules slide relative to each other and elongate to push the spindle poles farther apart. At the same time, the sister chromosomes are pushed to their respective poles. In plants, the spindle microtubules are apparently not anchored to the cell cortex at the poles, and so the chromosomes cannot be pulled apart. Instead they are probably pushed apart by kinesins in the overlapping polar microtubules of the spindle.

At **telophase**, a new network of microtubules and F-actins called the **phragmoplast** appears (see Figure 1.29E). The phragmoplast organizes the region of the cytoplasm where cytokinesis takes place. The microtubules have now lost their spindle shape but retain polarity, with their minus ends still pointed toward the separated, now decondensing, chromosomes, where the nuclear envelope is in the process of re-forming. The plus ends of the microtubules point toward the midzone of the phragmoplast, where small vesicles accumulate, partly derived from endocytotic vesicles from the parent cell plasma membrane. These vesicles have long tethers that may aid in the formation of the cell plate in the next cell cycle stage: cytokinesis.

Cytokinesis is the process that establishes the **cell plate**, a precursor of the cross-wall that will separate the two daughter cells (see Figure 1.29F). This cell plate, with its enclosing plasma membrane, forms as an island in the center of the cell that grows outward toward the parent wall by vesicle fusion. The site at which the forming cell plate fuses with the parental plasma membrane is determined by the location of the preprophase band (which disappeared earlier) and specific microtubule-associated proteins. As the cell plate assembles, it traps ER tubules in plasmalemma-lined membrane channels spanning the plate, thus connecting the two daughter cells (**Figure 1.30**). The cell plate–spanning ER tubules establish the sites for the primary plasmodesmata (see Figure 1.8). After cytokinesis, microtubules re-form in the cell cortex. The new cortical microtubules have a transverse orientation relative to the cell axis, and this orientation determines the polarity of future cell extension.

anaphase The stage of mitosis during which the two chromatids of each replicated chromosome are separated and move toward opposite poles.

telophase Prior to cytokinesis, the final stage of mitosis (or meiosis) during which the chromatin decondenses, the nuclear envelope re-forms, and the cell plate extends.

phragmoplast An assembly of microtubules, membranes, and vesicles that forms during late anaphase or early telophase and precedes fusion of vesicles to form the cell plate.

cytokinesis In plant cells, following nuclear division, the separation of daughter nuclei by the formation of new cell wall.

cell plate A wall-like structure that separates newly divided cells. Formed by the phragmoplast and later becomes the cell wall.

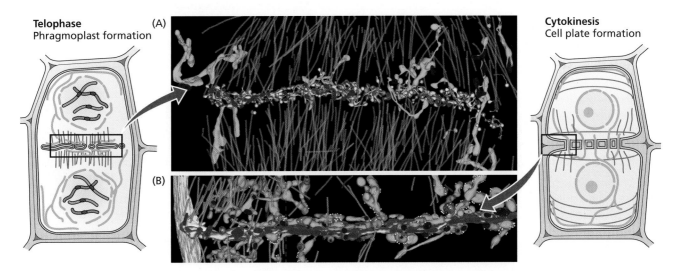

Telophase
Phragmoplast formation   (A)

(B)

Cytokinesis
Cell plate formation

Figure 1.30 Changes in the organization of the phragmoplast and ER during cell plate formation. (A) The forming cell plate (yellow, seen from the side) in early telophase has only a few sites where it interacts with the ER (blue). The solid phragmoplast microtubules (purple) also have few ER cisternae among them. (B) Side view of the forming peripheral cell plate (yellow), showing that although many cytoplasmic ER tubules (blue) intermingle with microtubules (purple) in the peripheral growth zone, there is little direct contact between ER tubules and cell plate membranes. The small white dots are ER-bound ribosomes. (Three-dimensional tomographic reconstruction of electron microscopy of the phragmoplast from Seguí-Simarro et al. 2004.)

Summary

Despite their great diversity in form and size, all plants carry out similar physiological and biochemical functions. These functions depend entirely on structures, from the molecular to the anatomical level.

Plant Life Processes: Unifying Principles

- All plants convert solar energy to chemical energy. They use growth instead of motility to obtain resources, and have vascular systems, rigid structures, and mechanisms to avoid desiccation on land. Plants develop from embryos sustained and protected by tissues from the mother plant.

Plant Classification and Life Cycles

- Plant classification is based on evolutionary relationships (**Box 1.1**).

- Plant life cycles alternate between diploid and haploid generations (**Figure 1.1**).

Overview of Plant Structure

- All plants have the same basic body plan, consisting of stem, root, and leaves (**Figure 1.2**).

- Plant cells are surrounded by rigid cell walls (**Figure 1.3**).

- Primary cell walls are synthesized in actively growing cells, whereas secondary cell walls are deposited on the inner surface of the primary walls, after most cell expansion has ended (**Figure 1.3**).

- Cell walls contain various types of polysaccharides that are named after the principal sugars they contain (**Figure 1.4**). These cell wall polysaccharides are classified into three groups: cellulose, hemicelluloses, and pectin.

- Cellulose is composed of highly ordered arrays of glucan chains, called microfibrils, which are synthesized at the surface of the cell by protein complexes called cellulose synthase (CESA) complexes. These rosette-like structures contain three to six units of cellulose synthase, the enzyme that synthesizes the individual glucans that make up the microfibril (**Figures 1.5, 1.6**).

- Matrix polysaccharides are synthesized in the Golgi apparatus and delivered to the cell wall via the exocytosis of vesicles (**Figure 1.7**).

- Because of plants' rigid cell walls, plant development depends solely on patterns of cell division and cell enlargement (**Figures 1.3, 1.8**).

- The cytoplasm of neighboring cells is connected by plasmodesmata, forming the symplast, which allows water and small molecules to move between cells without crossing the plasma membrane (**Figure 1.8**).

- New cells are produced in meristems (**Figure 1.9**).

- Secondary growth results in the increase in girth of roots and shoots through the action of the specialized meristems, the vascular and cork cambium (**Box 1.2**).

Plant Cell Types

- Dermal tissue, ground tissue, and vascular tissue are the three major tissue systems present in all plant organs (**Figure 1.9**).

- Dermal tissue includes the epidermis, which has multiple cell types, including pavement cells, stomatal guard cells, and trichomes (**Figure 1.10**).

- Ground tissue consists of three main cell types: parenchyma, collenchyma, and sclerenchyma. Parenchyma cells have the capacity to continue division and can differentiate into a variety of other ground tissues and vascular tissues (**Figure 1.10**).

- Vascular tissue has thickened secondary cell walls, perforated end walls, and pits that develop into channels in the walls shared between adjacent cells (**Figure 1.10**).

Plant Cell Organelles

- All plant cells contain cytosol, a nucleus, and other subcellular organelles enclosed by the plasma membrane and cell wall (**Figure 1.11**).

- All plants begin with a similar complement of organelles, which fall into two main categories based on how they arise: from the endomembrane system or semiautonomously.

- The endomembrane system plays a central role in secretory processes, membrane recycling, and the cell cycle (**Figure 1.11**).

- Plastids and mitochondria are semiautonomous, independently dividing organelles that are not derived from the endomembrane system.

- Biological membranes are composed of lipids and three main types of proteins: peripheral proteins, integral proteins, and anchored proteins (**Figure 1.12**).

- The composition and fluid-mosaic structure of all plasma membranes permit regulation of transport into and out of the cell and between subcellular compartments (**Figure 1.12**).

The Nucleus

- The nucleus is the site of storage, replication, and transcription of the DNA in the chromatin, as well as the site for the synthesis of ribosomes and mRNA (**Figures 1.13–1.15**).

(Continued)

Summary (continued)

- The specialized membranes of the nuclear envelope are derived from the endoplasmic reticulum (ER), a component of the endomembrane system (**Figure 1.13**).

- Translation of mRNA into proteins occurs in the cytoplasm (**Figure 1.15**).

- Posttranslational regulation of gene expression involves protein turnover (**Figure 1.16**).

The Endomembrane System

- The ER is a network of membrane-bound tubules that form a complex and dynamic structure (**Figure 1.17**).

- The rough ER is involved in synthesis of proteins that enter the lumen of the ER; the smooth ER is the site of lipid biosynthesis (**Figure 1.17**).

- Secretion of proteins from cells begins with the rough ER (**Figure 1.15**).

- Glycoproteins and polysaccharides destined for secretion are processed in the Golgi apparatus (**Figures 1.7 and 1.15**).

- Vacuoles serve multiple functions in plant cells; in addition to being involved in the uptake of water and solutes, they play a role in cell expansion and in the storage of secondary metabolites that protect plants against herbivores (**Figure 1.11**).

- Oil bodies, peroxisomes, and glyoxysomes perform specialized metabolic functions in the cell (**Figure 1.18**).

Independently Dividing Semiautonomous Organelles

- Mitochondria and chloroplasts each have an inner and an outer membrane (**Figures 1.19, 1.20**).

- Chloroplasts have an additional internal thylakoid membrane system that contains chlorophylls and carotenoids (**Figure 1.20**).

- Plastids may contain high concentrations of pigments or starch (**Figure 1.21**).

- Proplastids pass through distinct developmental stages to form specialized plastids (**Figure 1.22**).

- In plastids and mitochondria, fission and organellar DNA replication are regulated independently of nuclear division.

The Plant Cytoskeleton

- A three-dimensional network of polymerizing and depolymerizing filamentous proteins—tubulin in microtubules, and actin in microfilaments—organizes the cytosol and is required for life (**Figure 1.23, 1.24**).

- Microtubules show dynamic instability, but can stabilize and treadmill through the cell with accessory proteins (**Figure 1.25**).

- Molecular motors reversibly bind the cytoskeleton and can tether organelles or direct organelle movement (**Figure 1.26**).

- During cytoplasmic streaming, bulk flow of the cytosol is driven by the viscous drag in the wake of motor-driven organelles.

Cell Cycle Regulation

- The cell cycle, during which cells replicate their DNA and reproduce themselves, consists of four phases (**Figure 1.27**).

- Cyclins and cyclin-dependent kinases (CDKs) regulate the cell cycle, including the separation of paired metaphase chromosomes (**Figures 1.27, 1.28**).

- Successful mitosis (**Figure 1.29**) and cytokinesis (**Figure 1.30**) require the participation of microtubules and the endomembrane system.

Suggested Reading

Albersheim, P., Darvill, A., Roberts, K., Sederoff, R., and Staehelin, A. (2011) *Plant Cell Walls: From Chemistry to Biology*. Garland Science, Taylor and Francis Group, New York.

Bell, K., and Oparka, K. (2011) Imaging plasmodesmata. *Protoplasma* 248: 9–25.

Burch-Smith, T. M., Stonebloom, S., Xu, M., and Zambryski, P. C. (2011) Plasmodesmata during development: Re-examination of the importance of primary, secondary, and branched plasmodesmata structure versus function. *Protoplasma* 248: 61–74.

Burgess, J. (1985) *An Introduction to Plant Cell Development*. Cambridge University Press, Cambridge.

Carrie, C., Murcha, M. W., Giraud, E., Ng, S., Zhang, M. F., Narsai, R., and Whelan, J. (2013) How do plants make mitochondria? *Planta* 237: 429–439.

Chapman, K. D., Dyer, J. M., and Mullen, R. T. (2012) Biogenesis and functions of lipid droplets in plants: Thematic Review Series: Lipid droplet synthesis and metabolism: from yeast to man. *J. Lipid Res.* 53: 215–226.

Griffing, L. R. (2010) Networking in the endoplasmic reticulum. *Biochem. Soc. Trans.* 38: 747–753.

Gunning, B. E. S. (2009) *Plant Cell Biology on DVD*. Springer, New York, Heidelberg.

Henty-Ridilla, J. L., Li, J., Blanchoin, L., and Staiger, C. J. (2013) Actin dynamics in the cortical array of plant cells. *Curr. Opinion Plant Biol.* 16: 678–687.

Hu, J., Baker, A., Bartel, B., Linka, N., Mullen, R. T., Reumann, S., and Zolman, B. K. (2012) Plant peroxisomes: biogenesis and function. *Plant Cell* 24: 2279–2303.

Jones, R., Ougham, H., Thomas, H., and Waaland, S. (2013) *The Molecular Life of Plants*. Wiley-Blackwell, Oxford.

Joppa, L. N., Roberts, D. L., and Pimm, S. L. (2011) How many species of flowering plants are there? *Proc. R. Soc. B* 278: 554–559.

Leroux, O. (2012) Collenchyma: a versatile mechanical tissue with dynamic cell walls. *Ann. Bot.* 110: 1083–1098.

McMichael, C. M., and Bednarek, S. Y. (2013) Cytoskeletal and membrane dynamics during higher plant cytokinesis. *New Phytol.* 197: 1039–1057.

Müller, S., Wright, A. J., and Smith, L. G. (2009) Division plane control in plants: new players in the band. *Trends Cell Biol.* 19: 180–188.

Wasteneys, G. O., and Ambrose, J. C. (2009) Spatial organization of plant cortical microtubules: Close encounters of the 2D kind. *Trends Cell Biol.* 19: 62–71.

Williams, M. E. (July 16, 2013). How to be a plant. Teaching tools in plant biology: Lecture notes. *Plant Cell* 25(7): DOI 10.1105/tpc.113.tt0713.

2 Water and Plant Cells

Water plays a crucial role in the life of the plant. Photosynthesis requires that plants draw carbon dioxide from the atmosphere, and at the same time exposes them to water loss and the threat of dehydration. To prevent leaf desiccation, water must be absorbed by the roots and transported through the plant body. Even slight imbalances between the uptake and transport of water and the loss of water to the atmosphere can cause water deficits and severe malfunctioning of many cellular processes. Thus, balancing the uptake, transport, and loss of water represents an important challenge for land plants.

A major difference between plant and animal cells, which has a large impact on their respective water relations, is that plant cells have cell walls and animal cells do not. Cell walls allow plant cells to build up large internal hydrostatic pressures, called **turgor pressure**. Turgor pressure is essential for many physiological processes, including cell enlargement, stomatal opening, transport in the phloem, and various transport processes across membranes. Turgor pressure also contributes to the rigidity and mechanical stability of nonlignified plant tissues. In this chapter we consider how water moves into and out of plant cells, emphasizing the molecular properties of water and the physical forces that influence water movement at the cell level.

turgor pressure Force per unit of area in a liquid. In a plant cell, turgor pressure pushes the plasma membrane against the rigid cell wall and provides a force for cell expansion.

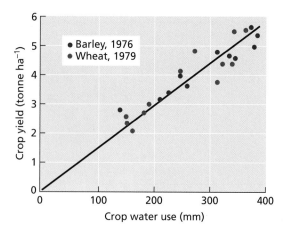

Figure 2.1   Grain yield as a function of water used under a range of irrigation treatments for barley in 1976 and wheat in 1979 in southeastern England. (After Jones 1992; data from Day et al. 1978 and Innes and Blackwell 1981.)

Water in Plant Life

Of all the resources that plants need to grow and function, water is the most abundant and yet often the most limiting. The practice of crop irrigation reflects the fact that water is a key resource limiting agricultural productivity (**Figure 2.1**). Water availability likewise limits the productivity of natural ecosystems (**Figure 2.2**), leading to marked differences in vegetation type along precipitation gradients.

The reason that water is frequently a limiting resource for plants, but much less so for animals, is that plants use water in huge amounts. Most (~97%) of the water absorbed by a plant's roots is carried through the plant and evaporates from leaf surfaces. Such water loss is called **transpiration**. In contrast, only a small amount of the water absorbed by roots actually remains in the plant to supply growth (~2%) or to be consumed in the biochemical reactions of photosynthesis and other metabolic processes (~1%).

Water loss to the atmosphere appears to be an inevitable consequence of carrying out photosynthesis on land. The uptake of CO_2 is coupled to the loss of water through a common diffusional pathway: As CO_2 diffuses into leaves, water vapor diffuses out. Because the driving gradient for water loss from leaves is much larger than that for CO_2 uptake, as many as 400 water molecules are lost for every CO_2 molecule gained. This unfavorable exchange has had a major influence on the evolution of plant form and function and explains why water plays such a key role in the physiology of plants.

We will begin our study of water by considering how its structure gives rise to some of its unique physical properties. We will then examine the physical basis for water movement, the concept of water potential, and the application of this concept to cell–water relations.

The Structure and Properties of Water

Water has special properties that enable it to act as a wide-ranging solvent and to be readily transported through the body of the plant. These properties derive primarily from the hydrogen bonding ability and polar structure of the water molecule. In this section we examine how the formation of hydro-

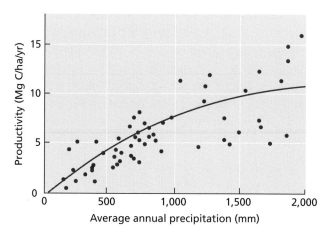

Figure 2.2   Productivity of various ecosystems worldwide as a function of annual precipitation. Above 2,000 to 3,000 mm yearly precipitation productivity actually starts to decrease. (Mg = 10^6 g) (After Schuur 2003.)

transpiration   The evaporation of water from the surface of leaves and stems.

gen bonds contributes to the high specific heat, surface tension, and tensile strength of water.

Water is a polar molecule that forms hydrogen bonds

The water molecule consists of an oxygen atom covalently bonded to two hydrogen atoms (**Figure 2.3A**). Because the oxygen atom is more **electronegative** than hydrogen, it tends to attract the electrons of the covalent bond. This attraction results in a partial negative charge at the oxygen end of the molecule and a partial positive charge at each hydrogen, making water a **polar** molecule. These partial charges are equal, so the water molecule carries no *net* charge.

Water molecules are tetrahedral in shape. At two points of the tetrahedron are the hydrogen atoms, each with a partial positive charge. The other two points of the tetrahedron contain lone pairs of electrons, each with a partial negative charge. Thus, each water molecule has two positive poles and two negative poles. These opposite partial charges create electrostatic attractions between water molecules, known as **hydrogen bonds** (**Figure 2.3B**).

Hydrogen bonds take their name from the fact that effective electrostatic bonds are formed only when highly electronegative atoms such as oxygen are covalently bonded to hydrogen. The reason for this is that the small size of the hydrogen atom allows the partial positive charges to be more concentrated, and thus more effective in bonding electrostatically.

Hydrogen bonds are responsible for many of the unusual physical properties of water. Water can form up to four hydrogen bonds with adjacent water molecules, resulting in very strong intermolecular interactions. Hydrogen bonds can also form between water and other molecules that contain electronegative atoms (O or N), especially when these are covalently bonded to H.

Water is an excellent solvent

Water dissolves greater amounts of a wider variety of substances than do other related solvents. Its versatility as a solvent is due in part to the small size of the water molecule. However, it is the hydrogen-bonding ability of water and its polar structure that make it a particularly good solvent for ionic substances and for molecules such as sugars and proteins that contain polar —OH or —NH$_2$ groups.

electronegative Having the capacity to attract electrons and thus producing a slightly negative electric charge.

polarity Property of some molecules, such as water, in which differences in the electronegativity of some atoms result in a partial negative charge at one end of the molecule and a partial positive charge at the other end.

hydrogen bonds Weak chemical bonds formed between a hydrogen atom and an oxygen or nitrogen atom.

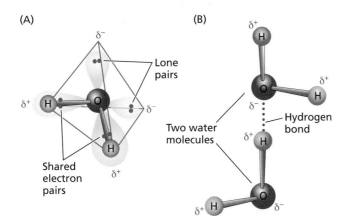

Figure 2.3 Structure of the water molecule. (A) The strong electronegativity of the oxygen atom means that the two electrons that form the covalent bond with hydrogen are unequally shared, such that each hydrogen atom has a partial positive charge. Each of the two lone pairs of electrons of the oxygen atom produces a partial negative charge. (B) The opposite partial charges (δ^- and δ^+) on the water molecule lead to the formation of intermolecular hydrogen bonds with other water molecules. Oxygen has six electrons in the outer orbitals; each hydrogen has one.

specific heat   Ratio of the heat capacity of a substance to the heat capacity of a reference substance, usually water. Heat capacity is the amount of heat needed to change the temperature of a unit of mass by 1 degree Celsius. The heat capacity of water is 1 calorie (4.184 Joule) per gram per degree Celsius.

latent heat of vaporization
The energy needed to separate molecules from the liquid phase and move them into the gas phase at constant temperature.

surface tension   A force exerted by water molecules at the air–water interface, resulting from the cohesion and adhesion properties of water molecules. This force minimizes the surface area of the air–water interface.

Hydrogen bonding between water molecules and ions, and between water and polar solutes, effectively decreases the electrostatic interaction between the charged substances and thereby increases their solubility. Similarly, hydrogen bonding between water and macromolecules such as proteins and nucleic acids reduces interactions between macromolecules, thus helping draw them into solution.

Water has distinctive thermal properties relative to its size

The extensive hydrogen bonding between water molecules results in water having both a high specific heat capacity and a high latent heat of vaporization.

Specific heat capacity is the heat energy required to raise the temperature of a substance by a set amount. Temperature is a measure of molecular kinetic energy (energy of motion). When the temperature of water is raised, the molecules vibrate faster and with greater amplitude. Hydrogen bonds act like rubber bands that absorb some of the energy from applied heat, leaving less energy available to increase motion. Thus, compared with other liquids, water requires a relatively large heat input to raise its temperature. This is important for plants, because it helps buffer temperature fluctuations.

Latent heat of vaporization is the energy needed to separate molecules from the liquid phase and move them into the gas phase—a process that occurs during transpiration. The latent heat of vaporization decreases as temperature increases, reaching a minimum at the boiling point (100°C). For water at 25°C, the heat of vaporization is 44 kJ mol^{-1}—the highest value known for any liquid. Most of this energy is used to break hydrogen bonds between water molecules.

Latent heat does not change the temperature of water molecules that have evaporated, but it does cool the surface from which the water has evaporated. Thus, the high latent heat of vaporization of water serves to moderate the temperature of transpiring leaves, which would otherwise increase due to the input of radiant energy from the sun.

Water molecules are highly cohesive

Water molecules at an air–water interface are attracted to neighboring water molecules by hydrogen bonds, and this interaction is much stronger than any interaction with the adjacent gas phase. As a consequence, the lowest-energy (i.e., most stable) configuration is one that minimizes the surface area of the air–water interface. To increase the area of an air–water interface, hydrogen bonds must be broken, which requires an input of energy. The energy required to increase the surface area of a gas–liquid interface is known as **surface tension**.

Surface tension can be expressed in units of energy per area (J m^{-2}) but is generally expressed in the equivalent, but less intuitive, units of force per length (J m^{-2} = N m^{-1}). A joule (J) is the SI unit of energy, with units of force × distance (N m); a newton (N) is the SI unit of force, with units of mass × acceleration (kg m s^{-2}). If the air–water interface is curved, surface tension produces a net force perpendicular to the interface (**Figure 2.4**). As we will see later, surface tension

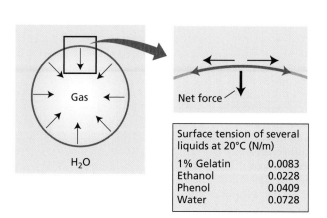

Figure 2.4   A gas bubble suspended within a liquid assumes a spherical shape such that its surface area is minimized. Because surface tension acts along the tangent to the gas–liquid interface, the resultant (net) force is inward, leading to compression of the bubble. The magnitude of the pressure (force/area) exerted by the interface is equal to $2T/r$, where T is the surface tension of the liquid (N/m) and r is the radius of the bubble (m). Water has an extremely high surface tension compared with other liquids at the same temperature.

Surface tension of several liquids at 20°C (N/m)

1% Gelatin	0.0083
Ethanol	0.0228
Phenol	0.0409
Water	0.0728

and adhesion (defined below) at the evaporative surfaces in leaves generate the physical forces that pull water through the plant's vascular system.

The extensive hydrogen bonding in water also gives rise to the property known as **cohesion**, the mutual attraction between molecules. A related property, called **adhesion**, is the attraction of water to a solid phase such as a cell wall or glass surface, which again is due primarily to the formation of hydrogen bonds. The degree to which water is attracted to the solid phase versus to itself can be quantified by measuring the **contact angle** (**Figure 2.5A**). The contact angle describes the shape of the air–water interface and thus the effect that the surface tension has on the pressure in the liquid.

Cohesion, adhesion, and surface tension give rise to a phenomenon known as **capillarity** (**Figure 2.5B**). Consider a vertically oriented glass capillary tube with wettable walls (contact angle < 90°). At equilibrium, the water level in the capillary tube will be higher than that of the water supply at its base. Water is drawn into the capillary tube because of (1) the attraction of water to the polar surface of the glass tube (adhesion) and (2) the surface tension of water. Together, adhesion and surface tension pull on the water molecules, causing them to move up the tube until this upward force is balanced by the weight of the water column. The narrower the tube, the higher the equilibrium water level.

Water has a high tensile strength

Hydrogen bonding gives water a high **tensile strength**, defined as the maximum force per unit of area that a continuous column of water can withstand before

cohesion The mutual attraction between water molecules due to extensive hydrogen bonding.

adhesion The attraction of water to a solid phase such as a cell wall or glass surface, due primarily to the formation of hydrogen bonds.

contact angle A quantitative measure of the degree to which a water molecule is attracted to a solid phase versus to itself.

capillarity The movement of water for small distances up a glass capillary tube or within the cell wall, due to water's cohesion, adhesion, and surface tension.

tensile strength The ability to resist a pulling force. Water has a high tensile strength.

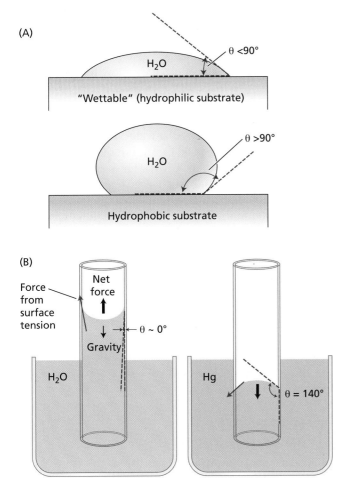

Figure 2.5 (A) The shape of a droplet placed on a solid surface reflects the relative attraction of the liquid to the solid versus to itself. The contact angle (θ), defined as the angle from the solid surface through the liquid to the gas–liquid interface, is used to describe this interaction. "Wettable" surfaces have contact angles of less than 90°; a highly wettable (i.e., hydrophilic) surface (such as water on clean glass or primary cell walls) has a contact angle close to 0°. Water spreads out to form a thin film on highly wettable surfaces. In contrast, nonwettable (i.e., hydrophobic) surfaces have contact angles greater than 90°. Water "beads" up on such surfaces. (B) Capillarity can be observed when a liquid is supplied to the bottom of a vertically oriented capillary tube. If the walls are highly wettable (e.g., water on clean glass), the net force is upward. The water column rises until this upward force is balanced by the weight of the water column. In contrast, if the liquid does not "wet" the walls (e.g., Hg on clean glass has a contact angle of approximately 140°), the meniscus curves downward, and the force resulting from surface tension lowers the level of the liquid in the tube.

Figure 2.6 A sealed syringe can be used to create positive and negative pressures in fluids such as water. Pushing on the plunger causes the fluid to develop a positive, hydrostatic pressure (white arrows) that acts in the same direction as the inward force resulting from the surface tension of the gas–liquid interface (black arrows). Thus, a small air bubble trapped within the syringe will shrink as the pressure increases. Pulling on the plunger causes the fluid to develop a tension, or negative pressure. Air bubbles in the syringe will expand if the outward force on the bubble exerted by the fluid (white arrows) exceeds the inward force resulting from the surface tension of the gas–liquid interface (black arrows).

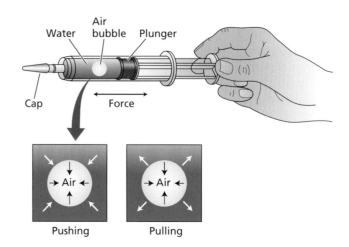

breaking. We do not usually think of liquids as having tensile strength; however, such a property is evident in the rise of a water column in a capillary tube.

We can demonstrate the tensile strength of water by placing it in a clean glass syringe (**Figure 2.6**). When we *push* on the plunger, the water is compressed, and a positive **hydrostatic pressure** builds up. Pressure is measured in units called *pascals* (Pa) or, more conveniently, *megapascals* (MPa). One MPa equals approximately 9.9 atmospheres. Pressure is equivalent to a force per unit of area ($1\ \text{Pa} = 1\ \text{N m}^{-2}$) and to an energy per unit of volume ($1\ \text{Pa} = 1\ \text{J m}^{-3}$). **Table 2.1** compares units of pressure.

If instead of pushing on the plunger we *pull* on it, a tension, or *negative hydrostatic pressure*, develops as the water molecules resist being pulled apart. Negative pressures develop only when molecules are able to pull on one another. Strong hydrogen bonds between water molecules allow tensions to be transmitted through water, even though it is a liquid. In contrast, gases cannot develop negative pressures because the interactions between gas molecules are limited to elastic collisions.

How hard must we pull on the plunger before the water molecules are torn away from each other and the water column breaks? Careful studies have demonstrated that water can resist tensions greater than 20 MPa. The water column in a syringe (see Figure 2.6), however, cannot sustain such large tensions because of the presence of microscopic gas bubbles. Because gas bubbles can expand, they interfere with the ability of the water in the syringe to resist the pull exerted by the plunger. The formation and expansion of gas bubbles due to tension in the surrounding liquid is known as **cavitation**. As we will see in Chapter 3, cavitation can have a devastating effect on water transport through the xylem.

hydrostatic pressure Pressure generated by compression of water into a confined space. Measured in units called pascals (Pa) or, more conveniently, megapascals (MPa).

cavitation The collapse of tension in a column of water resulting from the formation and expansion of tiny gas bubbles.

Table 2.1 Comparison of units of pressure

1 atmosphere = 14.7 pounds per square inch
= 760 mm Hg (at sea level, 45° latitude)
= 1.013 bar
= 0.1013 MPa
= 1.013×10^5 Pa

A car tire is typically inflated to about 0.2 MPa.

The water pressure in home plumbing is typically 0.2–0.3 MPa.

The water pressure under 30 feet (10 m) of water is about 0.1 MPa.

Diffusion and Osmosis

Cellular processes depend on the transport of molecules both to the cell and away from it. **Diffusion** is the spontaneous movement of substances from regions of higher to lower concentration. At the scale of a cell, diffusion is the dominant mode of transport. The diffusion of water across a selectively permeable barrier is referred to as **osmosis**.

In this section we examine how the processes of diffusion and osmosis lead to the net movement of both water and solutes.

Diffusion is the net movement of molecules by random thermal agitation

The molecules in a solution are not static; they are in continuous motion, colliding with one another and exchanging kinetic energy. The trajectory of any particular molecule after a collision is considered to be a random variable. Yet these random movements can result in the net movement of molecules.

Consider an imaginary plane dividing a solution into two equal volumes, A and B. As all the molecules undergo random motion, at each time step there is some probability that any particular solute molecule will cross our imaginary plane. The number we expect to cross from A to B in any particular time step will be proportional to the number at the beginning of the time step on side A, and the number crossing from B to A will be proportional to the number starting on side B.

If the initial concentration on side A is higher than that on side B, more solute molecules will be expected to cross from A to B than from B to A, and we will observe a net movement of solutes from A to B. Thus, diffusion results in the net movement of molecules from regions of high concentration to regions of low concentration, even though each molecule is moving in a random direction. The independent motion of each molecule explains why the system will evolve toward an equal number of A and B molecules on each side (**Figure 2.7**).

This tendency for a system to evolve toward an even distribution of molecules can be understood as a consequence of the second law of thermodynamics, which tells us that spontaneous processes evolve in the direction of increasing entropy, or disorder. Increasing entropy is synonymous with a decrease in free energy. Thus, diffusion represents the natural tendency of systems to move toward the lowest possible energy state.

It was Adolf Fick, who first noticed in the 1850s that the rate of diffusion is directly proportional to the concentration gradient ($\Delta c_s/\Delta x$)—that is, to the

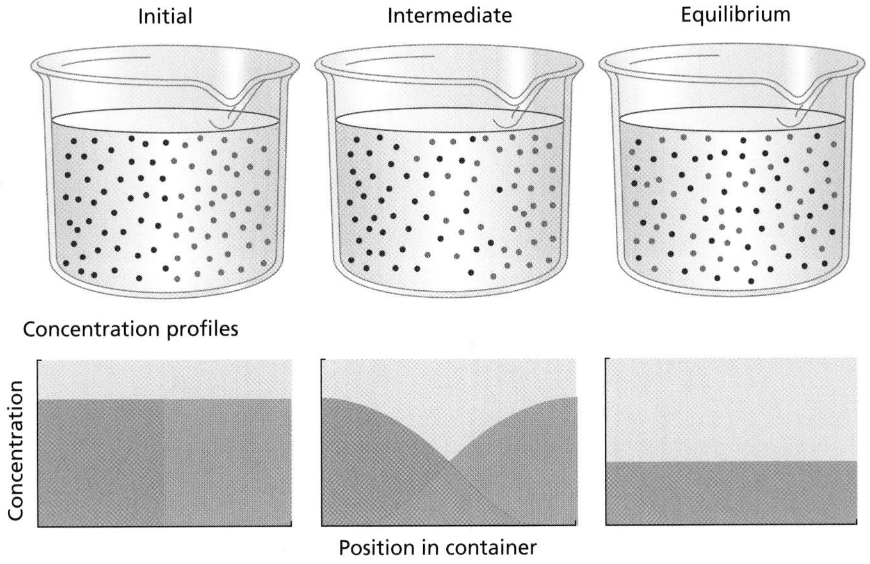

diffusion The movement of substances due to random thermal agitation from regions of high free energy to regions of low free energy (e.g., from high to low concentration).

osmosis The net movement of water across a selectively permeable membrane toward the region of more negative water potential, Ψ (lower concentration of water).

Initial   **Intermediate**   **Equilibrium**

Concentration profiles

Figure 2.7 Thermal motion of molecules leads to diffusion—the gradual mixing of molecules and eventual dissipation of concentration differences. Initially, two materials containing different molecules are brought into contact. The materials may be gas, liquid, or solid. Diffusion is fastest in gases, slower in liquids, and slowest in solids. The initial separation of the molecules is depicted graphically in the upper panels, and the corresponding concentration profiles are shown in the lower panels as a function of position. The purple color indicates an overlap in the concentration profiles of red and blue solutes. With time, the mixing and randomization of the molecules diminish net movement. At equilibrium the two types of molecules are randomly (evenly) distributed. Note that at all points and times the *total* concentration of solutes (i.e., both red and blue solutes) remains constant.

flux density (J_s) The rate of transport of a substance s across a unit of area per unit of time. J_s may have units of moles per square meter per second (mol m^{-2} s^{-1}).

diffusion coefficient (D_s) The proportionality constant that measures how easily a specific substance s moves through a particular medium. The diffusion coefficient is a characteristic of the substance and depends on both the medium and the temperature.

difference in concentration of substance s (Δc_s) between two points separated by a very small distance Δx. In symbols, we write this relation as Fick's first law:

$$J_s = -D_s \frac{\Delta c_s}{\Delta x} \qquad (2.1)$$

The rate of transport, expressed as the **flux density** (J_s), is the amount of substance s crossing a unit of cross-sectional area per unit of time (e.g., J_s may have the unit moles per square meter per second [mol m^{-2} s^{-1}]). The **diffusion coefficient** (D_s) is a proportionality constant that measures how easily substance s moves through a particular medium. The diffusion coefficient is a characteristic of the substance (larger molecules have smaller diffusion coefficients) and depends on both the medium (diffusion in air is typically 10,000 times faster than diffusion in a liquid, for example) and the temperature (substances diffuse faster at higher temperatures). The negative sign in Equation 2.1 indicates that the flux moves down a concentration gradient.

Fick's first law says that a substance will diffuse faster when the concentration gradient becomes steeper (Δc_s is large) or when the diffusion coefficient is increased. Note that this equation accounts only for movement in response to a concentration gradient, and not for movement in response to other forces (e.g., pressure, electric fields, and so on).

Diffusion is most effective over short distances

Consider a mass of solute molecules initially concentrated around a position $x = 0$ (**Figure 2.8A**). As the molecules undergo random motion, the concentration front moves away from the starting position, as shown for a later time point in **Figure 2.8B**.

Comparing the distribution of the solutes at the two times, we see that as the substance diffuses away from the starting point, the concentration gradient becomes less steep (Δc_s decreases); that is, the number of solute molecules that happen to step "backward" (i.e., toward $x = 0$) relative to those that step "forward" (away from $x = 0$) increases, and thus net movement becomes slower. Note that the average position of the solute molecules remains at $x = 0$ for all time, but that the distribution slowly flattens out.

As a direct consequence of the fact that each molecule is undergoing its own random walk, and thus is as likely to step toward some point of interest as away from it, the average time needed for a particle to diffuse a distance L grows

Figure 2.8 Graphical representation of the concentration gradient of a solute that is diffusing according to Fick's first law. The solute molecules were initially located in the plane indicated on the x-axis ("0"). (A) The distribution of solute molecules shortly after placement at the plane of origin. Note how sharply the concentration drops off as the distance, x, from the origin increases. (B) The solute distribution at a later time point. The average distance of the diffusing molecules from the origin has increased, and the slope of the gradient has flattened out. (After Nobel 1991.)

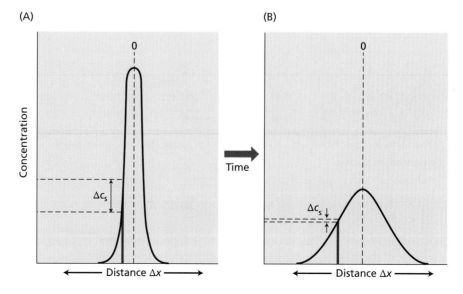

as L^2/D_s. In other words, the average time required for a substance to diffuse a given distance increases as the *square* of that distance.

The diffusion coefficient for glucose in water is about 10^{-9} m^2 s^{-1}. Thus, the average time required for a glucose molecule to diffuse across a cell with a diameter of 50 μm is 2.5 s. However, the average time needed for the same glucose molecule to diffuse a distance of 1 m in water is approximately 32 years. These values show that diffusion in solutions can be effective within cellular dimensions but is far too slow to be effective over long distances.

Osmosis describes the net movement of water across a selectively permeable barrier

Membranes of plant cells are **selectively permeable**; that is, they allow water and other small, uncharged substances to move across them more readily than larger solutes and charged substances. If the concentration of solutes is greater in the cell than in the solution surrounding it, water diffuses into the cell, but the solutes are unable to diffuse out of the cell. The net movement of water across a selectively permeable barrier is referred to as *osmosis*.

We saw earlier that the tendency of all systems to evolve toward increasing entropy results in solutes spreading out through the entire available volume. In osmosis, the volume available for solute movement is restricted by the membrane, and thus entropy maximization is realized by the volume of solvent diffusing through the membrane to dilute the solutes. Indeed, in the absence of any countervailing force, *all* the available water will flow to the solute side of the membrane.

Let's imagine what happens when we place a living cell in a beaker of pure water. The presence of a selectively permeable membrane means that the net movement of water continues until one of two things happens: (1) the cell expands until the selectively permeable membrane ruptures, allowing solutes to diffuse freely, or (2) the expansion of the cell volume is mechanically constrained by the presence of a cell wall such that the driving force for water to enter the cell is balanced by a pressure exerted by the cell wall.

The first scenario describes what would happen to an animal cell, which lacks a cell wall. The second scenario is relevant to plant cells. The plant cell wall is very strong. The resistance of cells walls to deformation creates an inward force that raises the hydrostatic pressure within the cell. The word *osmosis* derives from the Greek word for "pushing"; it is an expression of the positive pressure generated when solutes are confined.

We will soon see how osmosis drives the movement of water into and out of plant cells. First, however, let's discuss the concept of a composite or total driving force, representing the free-energy gradient of water.

Water Potential

All living things, including plants, require a continuous input of free energy to maintain and repair their highly organized structures, as well as to grow and reproduce. Processes such as biochemical reactions, solute accumulation, and long-distance transport are all driven by an input of free energy into the plant. In this section we examine how concentration, pressure, and gravity influence free energy.

The chemical potential of water represents the free-energy status of water

Chemical potential is a quantitative expression of the free energy associated with a substance. In thermodynamics, free energy represents the potential for performing work, force × distance. The unit of chemical potential is energy per

selective permeability The membrane property that allows diffusion of some molecules across the membrane to a different extent than other molecules.

chemical potential The free energy associated with a substance that is available to perform work.

water potential (Ψ) A measure of the free energy associated with water per unit of volume (J m^{-3}). These units are equivalent to pressure units such as pascals. Ψ is a function of the solute potential, the pressure potential, and the gravitational potential: $\Psi = \Psi_s + \Psi_p + \Psi_g$. The term Ψ_g is often ignored because it is negligible for heights under 5 m.

osmotic potential (Ψ_s) The effect of dissolved solutes on water potential. Also called solute potential.

osmolarity A unit of concentration expressed as moles of total dissolved solutes per liter of solution (mol L^{-1}). In biology, the solvent is usually water.

pressure potential (Ψ_p) The hydrostatic pressure of a solution in excess of ambient atmospheric pressure.

mole of substance (J mol^{-1}). Note that chemical potential is a relative quantity: It represents the difference between the potential of a substance in a given state and the potential of the same substance in a standard state.

The chemical potential of water represents the free energy associated with water. Water moves spontaneously, that is, without an input of energy, from regions of higher chemical potential to ones of lower chemical potential.

Historically, plant physiologists have most often used a related parameter called **water potential**, defined as the chemical potential of water divided by the partial molal volume of water (the volume of 1 mol of water): 18×10^{-6} m^3 mol^{-1}. Water potential is thus a measure of the free energy of water per unit of volume (J m^{-3}). These units are equivalent to pressure units such as the pascal, which is the common measurement unit for water potential. Let's look more closely at the important concept of water potential.

Three major factors contribute to cell water potential

The major factors influencing the water potential in plants are *concentration*, *pressure*, and *gravity*. Water potential is symbolized by Ψ (the Greek letter psi), and the water potential of solutions may be dissected into individual components, usually written as the following sum:

$$\Psi = \Psi_s + \Psi_p + \Psi_g \tag{2.2}$$

The terms Ψ_s and Ψ_p and Ψ_g denote the effects of solutes, pressure, and gravity, respectively, on the free energy of water. Energy levels must be defined in relation to a reference, analogous to how the contour lines on a map specify the distance above sea level. The reference state most often used to define water potential is pure water at ambient temperature and standard atmospheric pressure. The reference height is generally set either at the base of the plant (for whole-plant studies) or at the level of the tissue under examination (for studies of water movement at the cellular level). Let's consider each of the terms on the right-hand side of Equation 2.2.

SOLUTES The term Ψ_s, called the **solute potential** or the **osmotic potential**, represents the effect of dissolved solutes on water potential. Solutes reduce the free energy of water by diluting the water. This is primarily an entropy effect; that is, the mixing of solutes and water increases the disorder or entropy of the system and thereby lowers the free energy. This means that *the osmotic potential is independent of the specific nature of the solute*. For dilute solutions of nondissociating substances such as sucrose, the osmotic potential may be approximated by:

$$\Psi_s = -RTc_s \tag{2.3}$$

where R is the gas constant (8.32 J mol^{-1} K^{-1}), T is the absolute temperature (in degrees Kelvin, or K), and c_s is the solute concentration of the solution, expressed as **osmolarity** (moles of total dissolved solutes per volume of water [mol L^{-1}]). The minus sign in the equation indicates that dissolved solutes reduce the water potential of a solution relative to the reference state of pure water.

Equation 2.3 is valid for "ideal" solutions. Real solutions frequently deviate from the ideal, especially at high concentrations—for example, greater than 0.1 mol L^{-1}. Temperature also affects water potential. In our treatment of water potential, we will assume that we are dealing with ideal solutions.

PRESSURE The term Ψ_p, called the **pressure potential**, represents the effect of hydrostatic pressure on the free energy of water. Positive pressures raise the water potential; negative pressures reduce it. Both positive and negative pressures occur within plants. The positive hydrostatic pressure within cells is referred to as *turgor pressure*. Negative hydrostatic pressures, which frequently develop in xylem conduits, are referred to as **tension**. As you will see, tension is important in moving water long distances through the plant.

Hydrostatic pressure is often measured as the deviation from atmospheric pressure. Remember that water in the reference state is at atmospheric pressure, so by this definition $\Psi_p = 0$ MPa for water in the standard state. Thus, the value of Ψ_p for pure water in an open beaker is 0 MPa, even though its absolute pressure is approximately 0.1 MPa (1 atmosphere).

GRAVITY   Gravity causes water to move downward unless the force of gravity is opposed by an equal and opposite force. The **gravitational potential** (Ψ_g) depends on the height (h) of the water above the reference-state water, the density of water (ρ_w), and the acceleration due to gravity (g). In symbols, we write the following:

$$\Psi_g = \rho_w g h \tag{2.4}$$

where $\rho_w g$ has a value of 0.01 MPa m^{-1}. Thus, raising water a distance of 10 m translates into a 0.1 MPa increase in water potential.

The gravitational component (Ψ_g) is generally omitted in considerations of water transport at the cell level, because differences in this component among neighboring cells are negligible compared with differences in the osmotic potential and the pressure potential. Thus, in these cases Equation 2.2 can be simplified as follows:

$$\Psi = \Psi_s + \Psi_p \tag{2.5}$$

Water potentials can be measured

Cell growth, photosynthesis, and crop productivity are all strongly influenced by water potential and its components. Plant scientists have thus expended considerable effort in devising accurate and reliable methods for evaluating the water status of plants.

The principal approaches for determining Ψ use psychrometers, of which there are two types, or the pressure chamber. Psychrometers take advantage of water's large latent heat of vaporization, which allows accurate measurements of (1) the vapor pressure of water in equilibrium with the sample or (2) the transfer of water vapor between the sample and a solution of known Ψ_s. The pressure chamber measures Ψ by applying external gas pressure to an excised leaf until water is forced out of the living cells.

In some cells, it is possible to measure Ψ_p directly by inserting a liquid-filled microcapillary that is connected to a pressure sensor into the cell. In other cases, Ψ_p is estimated as the difference between Ψ and Ψ_s. Solute concentrations (Ψ_s) can be determined using a variety of methods, including psychrometers and instruments that measure freezing point depression.

In discussions of water in dry soils and plant tissues with very low water contents, such as seeds, one often finds reference to the **matric potential**, Ψ_m. Under these conditions, water exists as a very thin layer, perhaps one or two molecules deep, bound to solid surfaces by electrostatic interactions. These interactions are not easily separated into their effects on Ψ_s and Ψ_p, and are thus sometimes combined into a single term, Ψ_m.

Water Potential of Plant Cells

Plant cells typically have water potentials of 0 MPa or less. A negative value indicates that the free energy of water within the cell is less than that of pure water at ambient temperature, atmospheric pressure, and equal height. As the water potential of the solution surrounding the cell changes, water enters or

tension   Negative hydrostatic pressure.

gravitational potential (Ψ_g)   The part of the water potential caused by gravity. It is only of a significant size when considering water transport into trees and drainage in soils.

matric potential (Ψ_m)   The sum of osmotic potential (Ψ_s) + hydrostatic pressure (Ψ_p). Useful in situations (dry soils, seeds, and cell walls) where the separate measurement of Ψ_s and Ψ_p is difficult or impossible to obtain.

leaves the cell via osmosis. In this section we illustrate the osmotic behavior of water in plant cells with some numerical examples.

Water enters the cell along a water potential gradient

First imagine an open beaker full of pure water at 20°C (**Figure 2.9A**). Because the water is open to the atmosphere, the pressure potential of the water is the same as atmospheric pressure ($\Psi_p = 0$ MPa). There are no solutes in the water, so $\Psi_s = 0$ MPa. Finally, because we focus here on transport processes that take place within the beaker, we define the reference height as equal to the level of the beaker, and thus $\Psi_g = 0$ MPa. Therefore, the water potential is 0 MPa ($\Psi = \Psi_s + \Psi_p$).

Now imagine dissolving sucrose in the water to a concentration of 0.1 M (**Figure 2.9B**). This addition lowers the osmotic potential (Ψ_s) to –0.244 MPa and decreases the water potential (Ψ) to –0.244 MPa.

Next consider a flaccid plant cell (i.e., a cell with no turgor pressure) that has a total internal solute concentration of 0.3 M (**Figure 2.9C**). This solute concentration gives an osmotic potential (Ψ_s) of –0.732 MPa. Because the cell is flaccid, the internal pressure is the same as atmospheric pressure, so the pressure potential (Ψ_p) is 0 MPa and the water potential of the cell is –0.732 MPa.

What happens if this cell is placed in the beaker containing 0.1 M sucrose (see Figure 2.9C)? Because the water potential of the sucrose solution ($\Psi = -0.244$ MPa; see Figure 2.9B) is greater (less negative) than the water potential of the cell ($\Psi = -0.732$ MPa), water moves from the sucrose solution to the cell (from high to low water potential).

As water enters the cell, the plasma membrane begins to press against the cell wall. The wall stretches a little but also resists deformation by pushing back on the cell. This increases the pressure potential (Ψ_p) of the cell. Consequently, the cell water potential (Ψ) increases, and the difference between inside and outside water potentials ($\Delta\Psi$) is reduced.

Eventually, cell Ψ_p increases enough to raise the cell Ψ to the same value as the Ψ of the sucrose solution. At this point, equilibrium is reached ($\Delta\Psi = 0$ MPa), and net water transport ceases.

At equilibrium, water potential is equal everywhere: $\Psi_{(cell)} = \Psi_{(solution)}$. Because the volume of the beaker is much larger than that of the cell, the tiny amount of water taken up by the cell does not significantly affect the solute concentration of the sucrose solution. Hence, Ψ_s, Ψ_p, and Ψ of the sucrose solution are not altered. Therefore, at equilibrium, $\Psi_{(cell)} = \Psi_{(solution)} = -0.244$ MPa.

Calculation of cell Ψ_p and Ψ_s requires knowledge of the change in cell volume. In this example, let's assume that we know that the cell volume increased by 15%, such that the volume of the turgid cell is 1.15 times that of the flaccid cell. If we assume that the number of solutes within the cell remains constant as the cell hydrates, the final concentration of solutes will be diluted by 15%. The new Ψ_s can be calculated by dividing the initial Ψ_s by the relative increase in size of the hydrated cell: $\Psi_s = -0.732/1.15 = -0.636$ MPa. We can then calculate the pressure potential of the cell by rearranging Equation 2.5 as follows: $\Psi_p = \Psi - \Psi_s = (-0.244) - (-0.636) = 0.392$ MPa (see Figure 2.9C).

Water can also leave the cell in response to a water potential gradient

Water can also leave the cell by osmosis. If we now remove our plant cell from the 0.1 M sucrose solution and place it in a 0.3 M sucrose solution (**Figure 2.10A**), $\Psi_{(solution)}$ (–0.732 MPa) is more negative than $\Psi_{(cell)}$ (–0.244 MPa), and water moves from the turgid cell to the solution.

(A) Pure water

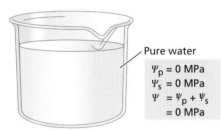

Pure water

$\Psi_p = 0$ MPa
$\Psi_s = 0$ MPa
$\Psi\ = \Psi_p + \Psi_s$
$\quad = 0$ MPa

(B) Solution containing 0.1 M sucrose

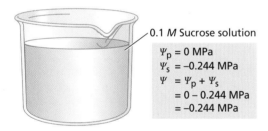

0.1 M Sucrose solution

$\Psi_p = 0$ MPa
$\Psi_s = -0.244$ MPa
$\Psi\ = \Psi_p + \Psi_s$
$\quad = 0 - 0.244$ MPa
$\quad = -0.244$ MPa

(C) Flaccid cell dropped into sucrose solution

Flaccid cell

$\Psi_p = 0$ MPa
$\Psi_s = -0.732$ MPa
$\Psi\ = -0.732$ MPa

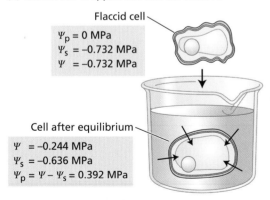

Cell after equilibrium

$\Psi\ = -0.244$ MPa
$\Psi_s = -0.636$ MPa
$\Psi_p = \Psi - \Psi_s = 0.392$ MPa

Figure 2.9 Water potential gradients can cause water to enter a cell. (A) Pure water. (B) A solution containing 0.1 M sucrose. (C) A flaccid cell (in air) is dropped into a 0.1 M sucrose solution. Because the starting water potential of the cell is less than the water potential of the solution, the cell takes up water. After equilibration, the water potential of the cell equals the water potential of the solution, and the result is a cell with a positive turgor pressure.

As water leaves the cell, the cell volume decreases. As the cell volume decreases, cell Ψ_p and Ψ decrease until $\Psi_{(cell)} = \Psi_{(solution)} = -0.732$ MPa. As before, we assume that the number of solutes within the cell remains constant as water flows from the cell. If we know that the cell volume decreases by 15%, the concentration of solutes will increase by 15%. Thus, we can calculate the new Ψ_s by multiplying the initial Ψ_s by the relative amount the cell volume has decreased: $\Psi_s = -0.636 \times 1.15 = -0.732$ MPa. This allows us to calculate that $\Psi_p = 0$ MPa using Equation 2.5.

If, instead of placing the turgid cell in the 0.3 M sucrose solution, we leave it in the 0.1 M solution and slowly squeeze it by pressing the cell between two plates (**Figure 2.10B**), we effectively raise the cell Ψ_p, consequently raising the cell Ψ and creating a $\Delta\Psi$ such that water now flows *out* of the cell. This is analogous to the industrial process of reverse osmosis in which externally applied pressure is used to separate water from dissolved solutes by forcing it across a semipermeable barrier. If we continue squeezing until half the cell's water is removed and then hold the cell in this condition, the cell will reach a new equilibrium. As in the previous example, at equilibrium, $\Delta\Psi = 0$ MPa, and the amount of water added to the external solution is so small that it can be ignored. The cell will thus return to the Ψ value that it had before the squeezing procedure. However, the components of the cell Ψ will be quite different.

Because half of the water was squeezed out of the cell while the solutes remained inside the cell (the plasma membrane is selectively permeable), the cell solution is concentrated twofold, and thus Ψ_s is lower (-0.636 MPa $\times 2 = -1.272$ MPa). Knowing the final values for Ψ and Ψ_s, we can calculate the pressure potential, using Equation 2.5, as $\Psi_p = \Psi - \Psi_s = (-0.244$ MPa$) - (-1.272$ MPa$) = 1.028$ MPa.

In our example we used an external force to change cell volume without a change in water potential. In nature, it is typically the water potential of the cell's

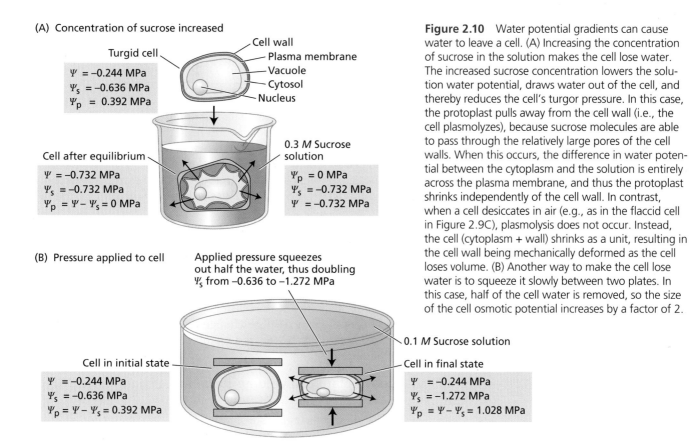

(A) Concentration of sucrose increased

Turgid cell
Cell wall
Plasma membrane
Vacuole
Cytosol
Nucleus

$\Psi = -0.244$ MPa
$\Psi_s = -0.636$ MPa
$\Psi_p = 0.392$ MPa

Cell after equilibrium

$\Psi = -0.732$ MPa
$\Psi_s = -0.732$ MPa
$\Psi_p = \Psi - \Psi_s = 0$ MPa

0.3 M Sucrose solution

$\Psi_p = 0$ MPa
$\Psi_s = -0.732$ MPa
$\Psi = -0.732$ MPa

(B) Pressure applied to cell

Applied pressure squeezes out half the water, thus doubling Ψ_s from -0.636 to -1.272 MPa

0.1 M Sucrose solution

Cell in initial state

$\Psi = -0.244$ MPa
$\Psi_s = -0.636$ MPa
$\Psi_p = \Psi - \Psi_s = 0.392$ MPa

Cell in final state

$\Psi = -0.244$ MPa
$\Psi_s = -1.272$ MPa
$\Psi_p = \Psi - \Psi_s = 1.028$ MPa

Figure 2.10 Water potential gradients can cause water to leave a cell. (A) Increasing the concentration of sucrose in the solution makes the cell lose water. The increased sucrose concentration lowers the solution water potential, draws water out of the cell, and thereby reduces the cell's turgor pressure. In this case, the protoplast pulls away from the cell wall (i.e., the cell plasmolyzes), because sucrose molecules are able to pass through the relatively large pores of the cell walls. When this occurs, the difference in water potential between the cytoplasm and the solution is entirely across the plasma membrane, and thus the protoplast shrinks independently of the cell wall. In contrast, when a cell desiccates in air (e.g., as in the flaccid cell in Figure 2.9C), plasmolysis does not occur. Instead, the cell (cytoplasm + wall) shrinks as a unit, resulting in the cell wall being mechanically deformed as the cell loses volume. (B) Another way to make the cell lose water is to squeeze it slowly between two plates. In this case, half of the cell water is removed, so the size of the cell osmotic potential increases by a factor of 2.

wilting   Loss of rigidity, leading to a flaccid state, due to turgor pressure falling to zero.

environment that changes, and the cell gains or loses water until its Ψ matches that of its surroundings.

One point common to all these examples deserves emphasis: Water flow across membranes is a passive process. That is, water moves in response to physical forces, toward regions of low water potential or low free energy. There are no known metabolic "pumps" (e.g., reactions driven by ATP hydrolysis) that can be used to drive water across a semipermeable membrane against its free-energy gradient.

The only situation in which water can be said to move across a semipermeable membrane against its water potential gradient is when it is coupled to the movement of solutes. The transport of sugars, amino acids, or other small molecules by various membrane proteins can "drag" up to 260 water molecules across the membrane per molecule of solute transported.

Such transport of water can occur even when the movement is against the usual water potential gradient (i.e., toward a higher water potential), because the loss of free energy by the solute more than compensates for the gain of free energy by the water. The net change in free energy remains negative. The amount of water transported in this way is generally quite small compared with the passive movement of water down its water potential gradient.

Water potential and its components vary with growth conditions and location within the plant

In leaves of well-watered plants, Ψ ranges from -0.2 to about -1.0 MPa in herbaceous plants and to -2.5 MPa in trees and shrubs. Leaves of plants in arid climates can have much lower Ψ, down to below -10 MPa under the most extreme conditions.

Just as Ψ values depend on the growing conditions and the type of plant, so too, the values of Ψ_s can vary considerably. Within cells of well-watered garden plants (examples include lettuce, cucumber seedlings, and bean leaves), Ψ_s may be as high as -0.5 MPa (low cell solute concentration), although values of -0.8 to -1.2 MPa are more typical. In woody plants, Ψ_s tends to be lower (higher cell solute concentration), allowing the more negative midday Ψ typical of these plants to occur without a loss in turgor pressure.

Although Ψ_s *within* cells may be quite negative, the apoplastic solution surrounding the cells—that is, in the cell walls and in the xylem—is generally quite dilute. The Ψ_s of the apoplast is typically -0.1 to 0 MPa, although in certain tissues (e.g., developing fruits) and habitats (e.g., high-salinity environments) the concentration of solutes in the apoplast can be high.

Values for Ψ_p within cells of well-watered plants may range from 0.1 to as much as 3 MPa, depending on the value of Ψ_s inside the cell. A plant **wilts** when the turgor pressure inside the cells of such tissues falls toward zero. As more water is lost from the cell, the walls become mechanically deformed, and the cell may be damaged as a result.

Cell Wall and Membrane Properties

Structural elements make important contributions to the water relations of plant cells. Cell wall elasticity defines the relation between turgor pressure and cell volume, while the permeability of the plasma membrane and the tonoplast to water influences the rate at which cells exchange water with their surroundings. In this section we examine how wall and membrane properties influence the water status of plant cells.

Small changes in plant cell volume cause large changes in turgor pressure

Cell walls provide plant cells with a substantial degree of volume homeostasis relative to the large changes in water potential that they experience every day as a consequence of the transpirational water losses associated with photosynthesis (see Chapter 3). Because plant cells have fairly rigid walls, a change in cell Ψ is generally accompanied by a large change in Ψ_p, with relatively little change in cell (protoplast) volume, as long as Ψ_p is greater than 0.

This phenomenon is illustrated by the *pressure–volume curve* shown in **Figure 2.11**. As Ψ decreases from 0 to –1.2 MPa, the relative or percent water content is reduced by only slightly more than 5%. Most of this decrease is due to a reduction in Ψ_p (by about 1.0 MPa); Ψ_s decreases by less than 0.2 MPa as a result of increased concentration of cell solutes.

Measurements of cell water potential and cell volume can be used to quantify how wall properties influence the water status of plant cells. Turgor pressure in most cells approaches zero as the relative cell volume decreases by 10 to 15%. However, for cells with very rigid cell walls, the volume change associated with turgor loss can be much smaller. In cells with extremely elastic walls, such as the water-storing cells in the stems of many cacti, this volume change may be substantially larger.

The volumetric elastic modulus, symbolized by ε (the Greek letter epsilon), can be determined by examining the relationship between Ψ_p and cell volume: ε is the change in Ψ_p for a given change in relative volume ($\varepsilon = \Delta\Psi_p/\Delta$[relative volume]). Cells with a large ε have stiff cell walls and thus experience larger changes in turgor pressure for the same change in cell volume than cells with a smaller ε and more elastic walls. The mechanical properties of cell walls vary among species and cell types, resulting in significant differences in the extent to which water deficits affect cell volume.

A comparison of the cell water relations within stems of cacti illustrates the important role of cell wall properties. Cacti are stem succulent plants, typically found in arid regions. Their stems consist of an outer, photosynthetic layer that surrounds nonphotosynthetic tissues that serve as a water storage reservoir (**Figure 2.12**). During drought, water is lost preferentially from these inner cells, despite the fact that the water potential of the two cell types remains in equilibrium (or very close to equilibrium). How does this happen?

Detailed studies of *Opuntia ficus-indica* demonstrate that the water storage cells are larger and have thinner walls than the photosynthetic cells, and are thus more flexible (have lower ε). For a given decrease in water potential, a water storage cell loses a greater fraction of its water content than a photosynthetic cell. In addition, the solute concentration of the water storage cells decreases during drought, in part due to the polymerization of soluble sugars into insoluble starch granules. A more typical plant response to drought is to accumulate solutes, in part to prevent water loss from cells.

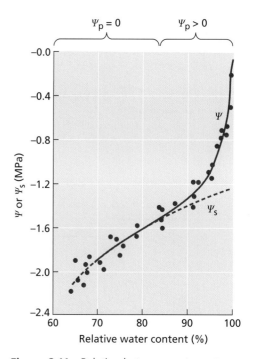

Figure 2.11 Relation between water potential (Ψ), solute potential (Ψ_s), and relative water content ($\Delta V/V$) in cotton (*Gossypium hirsutum*) leaves. Note that water potential (Ψ) decreases steeply with the initial decrease in relative water content. In comparison, osmotic potential (Ψ_s) changes little. As cell volume decreases below 90% in this example, the situation reverses: Most of the change in water potential is due to a drop in cell Ψ_s, accompanied by relatively little change in turgor pressure. (After Hsiao and Xu 2000.)

Figure 2.12 Cross section of a cactus stem, showing an outer, photosynthetic layer, and an inner, nonphotosynthetic tissue that functions in water storage. During drought, water is lost preferentially from nonphotosynthetic cells, so the water status of the photosynthetic tissue is maintained. (Photo by David McIntyre.)

(A)

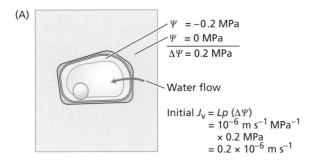

$\Psi = -0.2$ MPa
$\Psi = 0$ MPa
$\Delta\Psi = 0.2$ MPa

Water flow

Initial $J_v = Lp\,(\Delta\Psi)$
$= 10^{-6}$ m s^{-1} MPa^{-1}
$\times\ 0.2$ MPa
$= 0.2 \times 10^{-6}$ m s^{-1}

Figure 2.13 The rate of water transport into a cell depends on the magnitude of the water potential difference ($\Delta\Psi$) and the hydraulic conductivity of the plasma membranes (Lp). (A) In this example, the magnitude of the initial water potential difference is 0.2 MPa and Lp is 10^{-6} m s^{-1} MPa^{-1}. These values give an initial transport rate (J_v) of 0.2×10^{-6} m s^{-1}. (B) As water is taken up by the cell, the water potential difference decreases with time, leading to a slowing in the rate of water uptake. This effect follows an exponentially decaying time course with a half-time ($t_{1/2}$) that depends on the following cell parameters: volume (V), surface area (A), conductivity (Lp), volumetric elastic modulus (ε), and cell osmotic potential (Ψ_s).

(B)

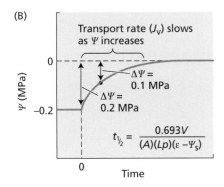

Transport rate (J_v) slows as Ψ increases

$\Delta\Psi = 0.1$ MPa

$\Delta\Psi = 0.2$ MPa

$t_{1/2} = \dfrac{0.693V}{(A)(Lp)(\varepsilon - \Psi_s)}$

Ψ (MPa)

Time

However, in the case of cacti, the combination of more flexible cell walls and a decrease in solute concentration during drought allows water to be withdrawn preferentially from the water storage cells, thus helping maintain the hydration of the photosynthetic tissues.

The rate at which cells gain or lose water is influenced by plasma membrane hydraulic conductivity

So far, we have seen that water moves into and out of cells in response to a water potential gradient. The direction of flow is determined by the direction of the Ψ gradient, and the rate of water movement is proportional to the magnitude of the driving gradient. However, for a cell that experiences a change in the water potential of its surroundings (e.g., see Figures 2.9 and 2.10), the movement of water across the plasma membrane decreases with time as the internal and external water potentials converge (**Figure 2.13**). The rate approaches zero in an exponential manner. The time it takes for the rate to decline by half—its half-time, or $t_{1/2}$—is given by the following equation:

$$t_{1/2} = \left(\frac{0.693}{(A)(Lp)}\right)\left(\frac{V}{\varepsilon - \Psi_s}\right)$$

(2.6)

where V and A are, respectively, the volume and surface area of the cell, and Lp is the **hydraulic conductivity** of the plasma membrane. Hydraulic conductivity describes how readily water can move across a membrane; it is expressed in terms of volume of water per unit of area of membrane per unit of time per unit of driving force (i.e., m^3 m^{-2} s^{-1} MPa^{-1}).

A short half-time means fast equilibration. Thus, cells with large surface-to-volume ratios, high membrane hydraulic conductivity, and stiff cell walls (large ε) come rapidly into equilibrium with their surroundings. Cell half-times typically range from 1 to 10 s, although some are much shorter. Because of their short half-times, single cells come to water potential equilibrium with their surroundings in less than 1 min. For multicellular tissues, the half-times may be much longer.

Aquaporins facilitate the movement of water across plasma membranes

For many years, plant physiologists were uncertain about how water moves across plant membranes. Specifically, it was unclear whether water movement into plant cells was limited to the diffusion of water molecules across the plasma membrane's lipid bilayer or if it also involved diffusion through protein-lined pores (**Figure 2.14**). Some studies suggested that diffusion directly across the lipid bilayer was not sufficient to account for observed rates of water movement across membranes, but the evidence in support of microscopic pores was not compelling.

This uncertainty was put to rest in 1991 with the discovery of **aquaporins** (see Figure 2.14). Aquaporins are integral membrane proteins that form water-selective

hydraulic conductivity   A measure of how readily water can move across a membrane; it is expressed in terms of volume of water per unit of area of membrane per unit of time per unit of driving force (i.e., m^3 m^{-2} s^{-1} MPa^{-1}).

aquaporins   Integral membrane proteins that form channels across a membrane, many of which are selective for water (hence the name). Such channels facilitate water movement across a membrane.

channels across the membrane. Because water diffuses much faster through such channels than through a lipid bilayer, aquaporins facilitate water movement into plant cells.

Although aquaporins may alter the *rate* of water movement across the membrane, they do not change the *direction of transport* or the *driving force* for water movement. However, aquaporins can be reversibly "gated" (i.e., transferred between an open and a closed state) in response to physiological parameters such as intercellular pH levels and Ca^{2+} concentrations As a result, plants have the ability to regulate the permeability of their plasma membranes to water.

Plant Water Status

The concept of water potential has two principal uses: First, water potential governs transport across plasma membranes, as we have described. Second, water potential is often used as a measure of the *water status* of a plant. In this section we discuss how the concept of water potential helps us evaluate the water status of a plant.

Physiological processes are affected by plant water status

Because of transpirational water loss to the atmosphere, plants are seldom fully hydrated. During periods of drought, they suffer from water deficits that lead to inhibition of plant growth and photosynthesis. **Figure 2.15** lists some of the physiological changes that occur as plants experience increasingly drier conditions.

The sensitivity of any particular physiological process to water deficits is, to a large extent, a reflection of that plant's strategy for dealing with the range of water availability that it experiences in its environment. According to Figure 2.15, the process that is most affected by water deficit is cell expansion. In many plants, reductions in water supply inhibit shoot growth and leaf expansion but *stimulate* root elongation. A relative increase in roots relative to leaves is an appropriate response to reductions in water availability, and thus the sensitivity of shoot growth to decreases in water availability can be seen as an adaptation to drought rather than a physiological constraint.

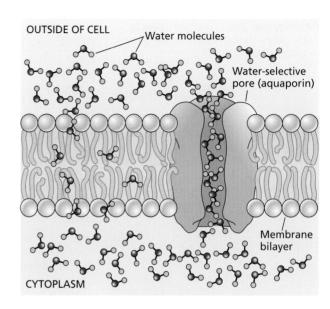

Figure 2.14 Water can cross plant membranes by diffusion of individual water molecules through the membrane bilayer, as shown on the left, and by the linear diffusion of water molecules through water-selective pores formed by integral membrane proteins such as aquaporins.

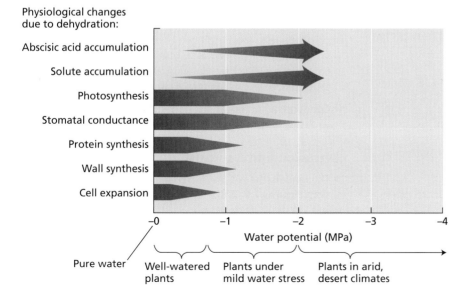

Figure 2.15 Sensitivity of various physiological processes to changes in water potential under various growing conditions. The thickness of the arrows corresponds to the magnitude of the process. For example, cell expansion decreases as water potential falls (becomes more negative). Abscisic acid is a hormone that induces stomatal closure during water stress (see Chapter 19). (After Hsiao 1973.)

halophytes Plants that are native to saline soils and complete their life cycles in that environment.

However, what plants cannot do is to alter the availability of water in the soil. (Figure 2.15 shows representative values for Ψ at various stages of water stress.) Thus, drought does impose some absolute limitations on physiological processes, although the actual water potentials at which such limitations occur vary with species.

Solute accumulation helps cells maintain turgor and volume

The ability to maintain physiological activity as water becomes less available typically incurs some costs. The plant may spend energy to accumulate solutes to maintain turgor pressure, invest in the growth of nonphotosynthetic organs such as roots to increase water uptake capacity, or build xylem conduits capable of withstanding large tensions. Thus, physiological responses to water availability reflect a trade-off between the benefits accrued by being able to carry out physiological processes (e.g., growth) over a wider range of environmental conditions and the costs associated with such capability.

Plants that grow in saline environments, called **halophytes**, typically have very low values of Ψ_s. A low Ψ_s lowers cell Ψ enough to allow root cells to extract water from saline water without allowing excessive levels of salts to enter at the same time. Plants may also exhibit quite negative Ψ_s under drought conditions. Water stress typically leads to an accumulation of solutes in the cytoplasm and vacuole of plant cells, thus allowing the cells to maintain turgor pressure despite low water potentials.

A positive turgor pressure ($\Psi_p > 0$) is important for several reasons. First, growth of plant cells requires turgor pressure to stretch the cell walls. The loss of turgor under water deficits can explain in part why cell growth is so sensitive to water stress, as well as why this sensitivity can be modified by varying the cell's osmotic potential (see Chapter 19). The second reason positive turgor is important is that turgor pressure increases the mechanical rigidity of cells and tissues.

Finally, although some physiological processes may be influenced directly by turgor pressure, it is likely that many more are affected by changes in cell volume. The existence of stretch-activated signaling molecules in the plasma membrane suggests that plant cells may sense changes in their water status via changes in volume, rather than by responding directly to turgor pressure.

Summary

Photosynthesis exposes plants to water loss and the threat of dehydration. To prevent desiccation, water must be absorbed by the roots and transported through the plant body.

Water in Plant Life

• Cell walls allow plant cells to build up large internal hydrostatic pressures (turgor pressure). Turgor pressure is essential for many plant processes.

• Water limits both agricultural and natural ecosystem productivity (**Figures 2.1, 2.2**).

• About 97% of the water absorbed by roots is carried through the plant and is lost by transpiration from the leaf surfaces.

• The uptake of CO_2 is coupled to the loss of water through a common diffusional pathway.

The Structure and Properties of Water

• The polarity and tetrahedral shape of water molecules permit them to form hydrogen bonds that give water its unusual physical properties: It is an excellent solvent and has a high specific heat, an unusually high latent heat of vaporization, and a high tensile strength (**Figures 2.3, 2.6**).

• Cohesion, adhesion, and surface tension give rise to capillarity (**Figures 2.4, 2.5**).

Diffusion and Osmosis

• The random thermal motion of molecules results in diffusion (**Figures 2.7, 2.8**).

• Diffusion is important over short distances (cellular level). The average time for a substance to diffuse a given distance increases as the square of that distance.

(Continued)

Summary (*continued*)

- Osmosis is the net movement of water across a selectively permeable barrier.

Water Potential

- Water's chemical potential measures the free energy of water in a given state.

- Concentration, pressure, and gravity contribute to water potential (Ψ) in plants.

- Ψ_s, the solute potential or osmotic potential, represents the reduction of the free energy of water caused by the dissolved solutes.

- Ψ_p, the pressure potential, represents the effect of hydrostatic pressure on the free energy of water. Positive pressure (turgor pressure) raises the water potential; negative pressure (tension) reduces it.

- Ψ_g, the gravitational potential, is generally omitted when calculating cell water potential. Thus, $\Psi = \Psi_s + \Psi_p$.

Water Potential of Plant Cells

- Plant cells typically have negative water potentials.

- Water enters or leaves a cell according to the water potential gradient.

- When a flaccid cell is placed in a solution that has a water potential greater (less negative) than the cell's water potential, water moves from the solution into the cell (from high to low water potential) (**Figure 2.9**).

- As water enters, the cell wall resists being stretched, increasing the turgor pressure (Ψ_p) of the cell.

- At equilibrium [$\Psi_{(cell)} = \Psi_{(solution)}$; $\Delta\Psi = 0$], the cell Ψ_p has increased sufficiently to raise the cell Ψ to the same value as the Ψ of the solution, and net water movement ceases.

- Water can also leave the cell by osmosis. When a turgid plant cell is placed in a sucrose solution that has a water potential more negative than the water potential of the cell, water moves from the turgid cell to the solution (**Figure 2.10**).

- If a cell is squeezed, its Ψ_p is raised, as is cell Ψ, resulting in a $\Delta\Psi$ such that water flows out of the cell (**Figure 2.10**).

Cell Wall and Membrane Properties

- Cell wall elasticity defines the relation between turgor pressure and cell volume, while the water permeability of the plasma membrane and tonoplast determines how fast cells exchange water with their surroundings.

- Because plant cells have fairly rigid walls, small changes in plant cell volume cause large changes in turgor pressure (**Figure 2.11**).

- For any non-zero initial $\Delta\Psi$, the net movement of water across the membrane decreases with time as the internal and external water potentials converge (**Figure 2.13**).

- Aquaporins are integral membrane proteins that form water-selective membrane channels (**Figure 2.14**).

Plant Water Status

- During drought, photosynthesis and growth are inhibited, while concentrations of abscisic acid and solutes increase (**Figure 2.15**).

- During drought, plants must use energy to maintain turgor pressure by accumulating solutes, as well as to support root and vascular growth.

- Stretch-activated signaling molecules in the plasma membrane may permit plant cells to sense changes in their water status via changes in volume.

Suggested Reading

Bartlett, M. K., Scoffoni, C., and Sack, L. (2012) The determinants of leaf turgor loss point and prediction of drought tolerance of species and biomes: A global meta-analysis. *Ecol. Lett.* 15: 393–405.

Chaumont, F., and Tyerman, S. D. (2014) Aquaporins: Highly regulated channels controlling plant water relations. *Plant Physiol.* 164: 1600–1618.

Goldstein, G., Ortega, J. K. E., Nerd, A., and Nobel, P. S. (1991) Diel patterns of water potential components for the crassulacean acid metabolism plant *Opuntia ficus-indica* when well-watered or droughted. *Plant Physiol.* 95: 274–280.

Kramer, P. J., and Boyer, J. S. (1995) *Water Relations of Plants and Soils.* Academic Press, San Diego.

Maurel, C., Verdoucq, L., Luu, D.-T., and Santoni, V. (2008) Plant aquaporins: Membrane channels with multiple integrated functions. *Annu. Rev. Plant Biol.* 59: 595–624.

Munns, R. (2002) Comparative physiology of salt and water stress. *Plant Cell Environ.* 25: 239–250.

Nobel, P. S. (1999) *Physicochemical and Environmental Plant Physiology.* 2nd ed. Academic Press, San Diego.

Tardieu, F., Parent, B., Caldeira, C. F., and Welcker, C. (2014) Genetic and physiological controls of growth under water deficit. *Plant Physiol.* 164: 1628–1635.

Wheeler, T. D., and Stroock, A. D. (2008) The transpiration of water at negative pressures in a synthetic tree. *Nature* 455: 208–212.

3 Water Balance of Plants

Life in Earth's atmosphere presents a formidable challenge to land plants. On the one hand, the atmosphere is the source of carbon dioxide, which is needed for photosynthesis. On the other hand, the atmosphere is usually quite dry, leading to a net loss of water due to evaporation. Because plants lack surfaces that can allow the inward diffusion of CO_2 while preventing water loss, CO_2 uptake exposes plants to the risk of dehydration. This problem is compounded because the concentration gradient for CO_2 uptake is much smaller than the concentration gradient that drives water loss. To meet the contradictory demands of maximizing carbon dioxide uptake while limiting water loss, plants have evolved adaptations to control water loss from leaves, and to replace the water lost to the atmosphere with water drawn from the soil.

In this chapter we examine the mechanisms and driving forces operating on water transport within the plant and between the plant and its environment. We will begin our examination of water transport by focusing on water in the soil. We will then consider how water moves from the soil into the roots and from the roots up through specialized transport cells to the leaves from which water is lost to the atmosphere. We will end the chapter by considering the ways in which the leaf can control the loss of water, as well as the entry of CO_2, by regulating the opening and closing of stomata, the small openings through which the major water loss occurs.

Water in the Soil

The water content and the rate of water movement in soils depend to a large extent on soil type and soil structure. At one extreme is sand, in which the soil particles may be 1 mm or more in diameter. Sandy soils have a relatively low surface area per gram of soil and have large spaces or channels between particles.

At the other extreme is clay, in which particles are smaller than 2 μm in diameter. Clay soils have much greater surface areas and smaller channels between particles. With the aid of organic substances such as humus (decomposing organic matter), clay particles may aggregate into "crumbs," allowing large channels to form that help improve soil aeration and infiltration of water.

When a soil is heavily watered by rain or by irrigation, the water percolates downward by gravity through the spaces between soil particles, partly displacing, and in some cases trapping, air in these channels. Because water is pulled into the spaces between soil particles by capillarity, the smaller channels become filled first. Depending on the amount of water available, water in the soil may exist as a film adhering to the surface of soil particles, it may fill the smaller but not the larger channels, or it may fill all of the spaces between particles.

In sandy soils, the spaces between particles are so large that water tends to drain from them and remain only on the particle surfaces and in the spaces where particles come into contact. In clay soils, the spaces between particles are so small that much water is retained against the force of gravity. A few days after a soaking rainfall, a clay soil might retain 40% water by volume. In contrast, sandy soils typically retain only about 15% water by volume after thorough wetting.

A negative hydrostatic pressure in soil water lowers soil water potential

Like the water potential of plant cells, the water potential of soils may be dissected into three components: the osmotic potential, the pressure potential, and the gravitational potential. The **osmotic potential** (Ψ_s; see Chapter 2) of soil water is generally negligible, because except in saline soils, solute concentrations are low; a typical value might be −0.02 MPa. In soils that contain a substantial concentration of salts, however, Ψ_s can be significant, perhaps −0.2 MPa or lower.

The second component of soil water potential is the **pressure potential** (Ψ_p) (**Figure 3.1**). For wet soils, Ψ_p is very close to zero. As soil dries out, Ψ_p decreases and can become quite negative. Where does the negative pressure potential in soil water come from?

Recall from our discussion of capillarity in Chapter 2 that water has a high surface tension that tends to minimize

osmotic potential (Ψ_s) The effect of dissolved solutes on water potential. Also called solute potential.

pressure potential (Ψ_p) The hydrostatic pressure of a solution in excess of ambient atmospheric pressure.

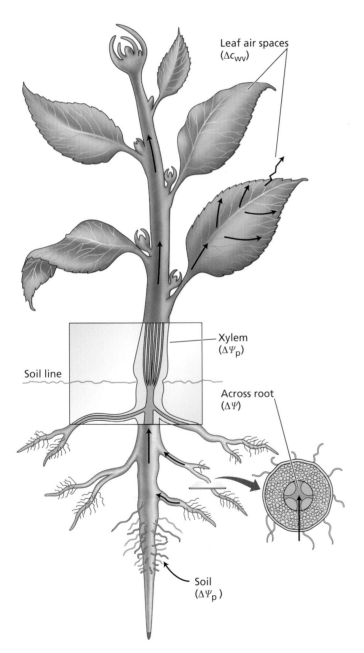

Leaf air spaces
(Δc_{wv})

Xylem
($\Delta\Psi_p$)

Soil line

Across root
($\Delta\Psi$)

Soil
($\Delta\Psi_p$)

Figure 3.1 Main driving forces for water flow from the soil through the plant to the atmosphere. Differences in water vapor concentration (Δc_{wv}) between leaf and air are responsible for the diffusion of water vapor from the leaf to the air; differences in pressure potential ($\Delta\Psi_p$) drive the bulk flow of water through xylem conduits; and differences in water potential ($\Delta\Psi$) are responsible for the movement of water across the living cells in the root.

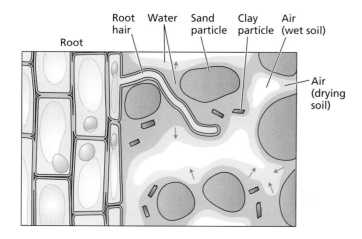

Root hair | Root | Water | Sand particle | Clay particle | Air (wet soil) | Air (drying soil)

Figure 3.2 Root hairs make intimate contact with soil particles and greatly amplify the surface area used for water absorption by the plant. The soil is a mixture of particles (sand, clay, silt, and organic material), water, dissolved solutes, and air. Water is adsorbed to the surface of the soil particles. As water is absorbed by the plant, the soil solution recedes into smaller pockets, channels, and crevices between the soil particles. At the air–water interfaces, this recession causes the surface of the soil solution to develop concave menisci (curved interfaces between air and water, marked in the figure by arrows), and brings the solution into tension (negative pressure) by surface tension. As more water is removed from the soil, the curvature of the air–water menisci increases, resulting in greater tensions (more negative pressures).

air–water interfaces. However, because of adhesive forces, water also tends to cling to the surfaces of soil particles (**Figure 3.2**).

As the water content of the soil decreases, the water recedes into the channels between soil particles, forming air–water surfaces whose curvature represents the balance between the tendency to minimize the surface area of the air–water interface and the attraction of the water for the soil particles. Water under a curved surface develops a negative pressure that may be estimated by the following formula:

$$\Psi_p = \frac{-2T}{r} \tag{3.1}$$

where T is the surface tension of water (7.28×10^{-8} MPa m) and r is the radius of curvature of the air–water interface. Note that here the soil particles are assumed to be fully wettable (see Figure 2.5, contact angle $\theta = 0$).

As soil dries out, water is first removed from the largest spaces between soil particles and subsequently from successively smaller spaces between and within soil particles. In this process, the value of Ψ_p in soil water can become quite negative due to the increasing curvature of air–water surfaces in pores of successively smaller diameter. For instance, a curvature of $r = 1$ μm (about the size of the largest clay particles) corresponds to a Ψ_p value of −0.15 MPa. The value of Ψ_p may easily reach −1 to −2 MPa as the air–water interface recedes into the smaller spaces between clay particles.

The third component of soil water potential is **gravitational potential** (Ψ_g). Gravity plays an important role in drainage. The downward movement of water is due to the fact that Ψ_g is proportional to elevation: higher at higher elevations, and lower at lower elevations.

Water moves through the soil by bulk flow

Bulk or mass flow is the concerted movement of molecules en masse, most often in response to a pressure gradient. Common examples of bulk flow are water moving through a garden hose or down a river. The movement of water through soils is predominantly by bulk flow.

Because the pressure in soil water is due to the existence of curved air–water interfaces, water flows from regions of higher soil water content, where the water-filled spaces are larger and thus Ψ_p is less negative, to regions of lower soil water content, where the smaller size of the water-filled spaces is associated with more curved air–water interfaces and a more negative Ψ_p. Diffusion of water vapor also accounts for some water movement, which can be important in dry soils.

As plants absorb water from the soil, they deplete the soil of water near the surface of the roots. This depletion reduces Ψ_p near the root surface and establishes

gravitational potential (Ψ_g) The part of the water potential caused by gravity. It is only of a significant size when considering water transport into trees and drainage in soils.

soil hydraulic conductivity
A measure of the ease with which water
moves through a soil.

root hairs   Microscopic extensions
of root epidermal cells that greatly
increase the surface area of the root for
absorption.

a pressure gradient with respect to neighboring regions of soil that have higher Ψ_p values. Because the water-filled pore spaces in the soil are interconnected, water moves down the pressure gradient to the root surface by bulk flow through these channels.

The rate of water flow in soils depends on two factors: the size of the pressure gradient through the soil, and the hydraulic conductivity of the soil. **Soil hydraulic conductivity** is a measure of the ease with which water moves through the soil, and it varies with the type of soil and its water content. Sandy soils, which have large spaces between particles, have a large hydraulic conductivity when saturated, whereas clay soils, with only minute spaces between their particles, have an appreciably smaller hydraulic conductivity.

As the water content (and hence the water potential) of a soil decreases, the hydraulic conductivity decreases dramatically. This decrease in soil hydraulic conductivity is due primarily to the replacement of water in the soil by air. When air moves into a soil channel previously filled with water, water movement through that channel is restricted to the periphery of the channel. As more of the soil spaces become filled with air, water flow is limited to fewer and narrower channels, and the hydraulic conductivity falls.

Water Absorption by Roots

Contact between the surface of the root and the soil is essential for effective water absorption by the root. This contact provides the surface area needed for water uptake and is maximized by the growth of the root and of root hairs into the soil. **Root hairs** are filamentous outgrowths of root epidermal cells that greatly increase the surface area of the root, thus providing greater capacity for absorption of ions and water from the soil. When 3-month-old wheat plants were examined, their root hairs were found to constitute more than 60% of the surface area of the roots (see Figure 4.7).

Water enters the root most readily near the root tip. Mature regions of the root are less permeable to water because they have developed a modified epidermal layer that contains hydrophobic materials in its walls. Although it might at first seem counterintuitive that any portion of the root system should be impermeable to water, the older regions of the root must be sealed off if there is to be water uptake (and thus bulk flow of nutrients) from the regions of the root system that are actively exploring new areas in the soil (**Figure 3.3**).

The contact between the soil and the root surface is easily ruptured when the soil is disturbed. It is for this reason that

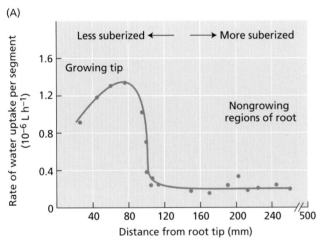

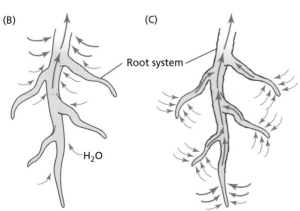

Figure 3.3   (A) Rate of water uptake by short segments (3–5 mm) at various positions along an intact pumpkin (*Cucurbita pepo*) root. (B and C) Diagrams of water uptake in which the entire root surface is equally permeable (B) or is impermeable in older regions due to the deposition of suberin, a hydrophobic polymer (C). When root surfaces are equally permeable, most of the water enters near the top of the root system, with more distal regions being hydraulically isolated as the suction in the xylem is relieved due to the inflow of water. Decreasing the permeability of older regions of the root allows xylem tensions to extend further into the root system, enabling water uptake from distal regions of the root system. (A from Kramer 1983, after Agricultural Research Council Letcombe Laboratory Annual Report [1973, p.10].)

newly transplanted seedlings and plants need to be protected from water loss for the first few days after transplantation. Thereafter, new root growth into the soil reestablishes soil–root contact, and the plant can better withstand water stress.

Water moves in the root via the apoplast, symplast, and transmembrane pathways

In the soil, water flows between soil particles. However, from the epidermis to the endodermis of the root, there are three pathways through which water can flow (**Figure 3.4**): the apoplast, the symplast, and the transmembrane pathway.

1. The apoplast is the continuous system of cell walls, intercellular air spaces, and the lumens of nonliving cells (e.g., xylem conduits and fibers). In this pathway, water moves through cell walls and extracellular spaces without crossing any membranes as it travels across the root cortex.

2. The symplast consists of the entire network of cell cytoplasm interconnected by plasmodesmata (see Chapter 1). In this pathway, water travels across the root cortex via the plasmodesmata.

3. The transmembrane pathway is the route by which water enters a cell on one side, exits the cell on the other side, enters the next in the series, and so on. In this pathway, water crosses the plasma membrane of each cell in its path twice (once on entering and once on exiting). Transport across the tonoplast may also be involved.

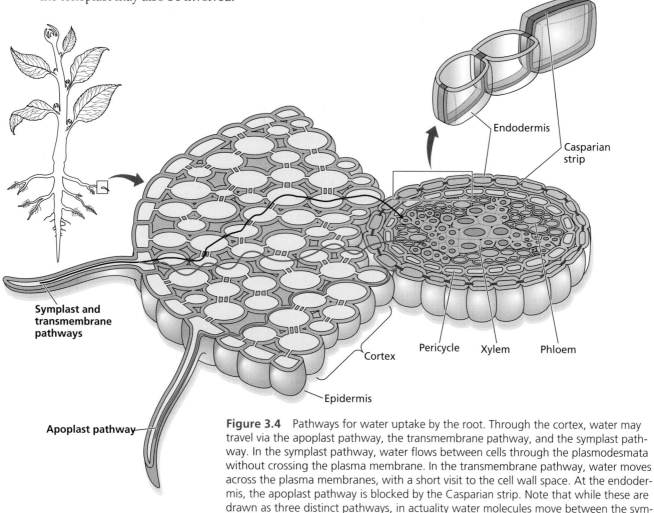

Figure 3.4 Pathways for water uptake by the root. Through the cortex, water may travel via the apoplast pathway, the transmembrane pathway, and the symplast pathway. In the symplast pathway, water flows between cells through the plasmodesmata without crossing the plasma membrane. In the transmembrane pathway, water moves across the plasma membranes, with a short visit to the cell wall space. At the endodermis, the apoplast pathway is blocked by the Casparian strip. Note that while these are drawn as three distinct pathways, in actuality water molecules move between the symplast and apoplast as directed by gradients in water potential and hydraulic resistances.

Casparian strip   A band in the cell walls of the endodermis that is impregnated with lignin. Prevents the apoplastic movement of water and solutes into the stele.

aquaporins   Integral membrane proteins that form channels across a membrane, many of which are selective for water (hence the name). Such channels facilitate water movement across a membrane.

root pressure   A positive hydrostatic pressure in the xylem of roots that typically occurs at night in the absence of transcription.

guttation   An exudation of liquid from the leaves due to root pressure.

Although the relative importance of the apoplast, symplast, and transmembrane pathways has not yet been fully established, experiments with the pressure probe technique indicate an important role for plasma membranes, and thus the transmembrane pathway, in the movement of water across the root cortex. And though we can define three pathways, it is important to remember that water moves not according to a single chosen path, but wherever the gradients and resistances direct it. A particular water molecule moving in the symplast may cross the membrane and move in the apoplast for a moment, and then move back into the symplast again.

At the endodermis, water movement through the apoplast pathway is obstructed by the Casparian strip (see Figure 3.4). The **Casparian strip** is a band within the radial cell walls of the endodermis that is impregnated with suberin and/or lignin, two hydrophobic polymers. The Casparian strip forms in the nongrowing part of the root, several millimeters to several centimeters behind the root tip (see Figure 3.3A), at about the same time that the first xylem elements mature. The Casparian strip breaks the continuity of the apoplast pathway, forcing water and solutes to pass through the plasma membrane in order to cross the endodermis.

The requirement that water move symplastically across the endodermis helps explain why the permeability of roots to water depends strongly on the presence of **aquaporins**, protein pores that facilitate movement of water across membranes. Down-regulating the expression of aquaporin genes markedly reduces the hydraulic conductivity of roots and can result in plants that wilt easily or that compensate by producing larger root systems.

Water uptake decreases when roots are subjected to low temperature or anaerobic conditions, or treated with respiratory inhibitors. Until recently, there was no explanation for the connection between root respiration and water uptake, or for the enigmatic wilting of flooded plants. We now know that the permeability of aquaporins can be regulated in response to intracellular pH. Decreased rates of respiration, in response to low temperature or anaerobic conditions, can lead to increases in intracellular pH. This increase in cytosolic pH alters the conductance of aquaporins in root cells, resulting in roots that are markedly less permeable to water. Thus, maintaining membrane permeability to water requires energy expenditure by root cells that is supplied by respiration.

Solute accumulation in the xylem can generate "root pressure"

Plants sometimes exhibit a phenomenon referred to as **root pressure**. For example, if the stem of a young seedling is cut off just above the soil, the stump will often exude sap from the cut xylem for many hours. If a manometer is sealed over the stump, positive pressures as high as 0.2 MPa (and sometimes even higher) can be measured.

When transpiration is low or absent, positive hydrostatic pressure builds up in the xylem because roots continue to absorb ions from the soil and transport them into the xylem. The buildup of solutes in the xylem sap leads to a decrease in the xylem osmotic potential (Ψ_s) and thus a decrease in the xylem water potential (Ψ). This lowering of the xylem Ψ provides a driving force for water absorption, which in turn leads to a positive hydrostatic pressure in the xylem. In effect, the multicellular root tissue behaves as an osmotic membrane does, building up a positive hydrostatic pressure in the xylem in response to the accumulation of solutes.

Root pressure is most likely to occur when soil water potentials are high and transpiration rates are low. As transpiration rates increase, water is transported through the plant and lost to the atmosphere so rapidly that a positive pressure resulting from ion uptake never develops in the xylem.

Plants that develop root pressure frequently produce liquid droplets on the edges of their leaves, a phenomenon known as **guttation** (**Figure 3.5**). Positive xylem pressure causes exudation of xylem sap through specialized pores called *hydathodes* that are associated with vein endings at the leaf margin. The "dewdrops" that can be seen on the tips of grass leaves in the morning are actually guttation droplets exuded from hydathodes. Guttation is most noticeable when transpiration is suppressed and the relative humidity is high, such as at night. It is possible that root pressure reflects an unavoidable consequence of high rates of ion accumulation. However, the existence of positive pressures within the xylem at night can help dissolve gas bubbles, and thus play a role in reversing the deleterious effects of cavitation described in the next section.

Water Transport through the Xylem

In most plants, the xylem constitutes the longest part of the pathway of water transport. In a plant 1 m tall, more than 99.5% of the water transport pathway through the plant is within the xylem, and in tall trees the xylem represents an even higher percentage of the pathway. Compared with the movement of water through layers of living cells, the xylem is a simple pathway of low resistivity. In the following sections we examine how the structure of the xylem contributes to the movement of water from the roots to the leaves, and how negative pressures generated by transpiration pull water through the xylem.

Figure 3.5 Guttation in a leaf from lady's mantle (*Alchemilla vulgaris*). In the early morning, leaves secrete water droplets through the hydathodes, located at the margins of the leaves. (Photo by David McInytre.)

The xylem consists of two types of transport cells

The conducting cells in the xylem have a specialized anatomy that enables them to transport large quantities of water with great efficiency. There are two main types of water-transporting cells in the xylem: tracheids and vessel elements (**Figure 3.6**). Vessel elements are found in angiosperms, a small group of gymnosperms called the Gnetales, and some ferns. Tracheids are present in both angiosperms and gymnosperms, as well as in ferns and other groups of vascular plants.

The maturation of both tracheids and vessel elements involves the production of secondary cell walls and the subsequent death of the cell—the loss of the cytoplasm and all of its contents. What remain are the thick, lignified cell walls, which form hollow tubes through which water can flow with relatively little resistance.

Tracheids are elongated, spindle-shaped cells (see Figure 3.6A) that are arranged in overlapping vertical files (see Figure 3.6B). Water flows between tracheids by means of the numerous **pits** in their lateral walls. Pits are microscopic regions where the secondary wall is absent and only the primary wall is present. Pits of one tracheid are typically located opposite pits of an adjoining tracheid, forming **pit pairs** (**Figure 3.7**). Pit pairs constitute a low-resistance path for water movement between tracheids. The water-permeable layer between pit pairs, consisting of two primary walls and a middle lamella, is called the **pit membrane** (not to be confused with the lipid membranes of a living cell).

Pit membranes in tracheids of conifers have a central thickening, called a **torus** (plural *tori*), surrounded by a porous and relatively flexible region known as the **margo** (see Figure 3.7A). The torus acts like a valve: When it is centered in the pit cavity, the pit remains open; when it is lodged in the circular or oval wall thickenings bordering the pit, the pit is closed. Such lodging of the torus effectively prevents gas bubbles from spreading into neighboring tracheids (we will discuss this formation of bubbles, a process called cavitation, shortly). With very few exceptions, the pit membranes in all other plants, whether in tracheids or vessel elements, lack tori. But because the water-filled pores in the pit membranes of nonconifers are very small, they also serve as an effective barrier against the

tracheids Spindle-shaped, water-conducting cells with tapered ends and pitted walls without perforations, found in the xylem of both angiosperms and gymnosperms.

pit A microscopic region where the secondary wall of a tracheary element is absent and the primary wall is thin and porous.

pit pair Two pits occurring opposite one another in the walls of adjacent tracheids or vessel elements. Pit pairs constitute a low-resistance path for water movement between the conducting cells of the xylem.

pit membrane The porous layer in the xylem between pit pairs, consisting of two thinned primary walls and a middle lamella.

torus A central thickening found in the pit membranes of tracheids in the xylem of most gymnosperms.

margo A porous and relatively flexible region of the pit membranes in tracheids of conifer xylem, surrounding a central thickening, the torus.

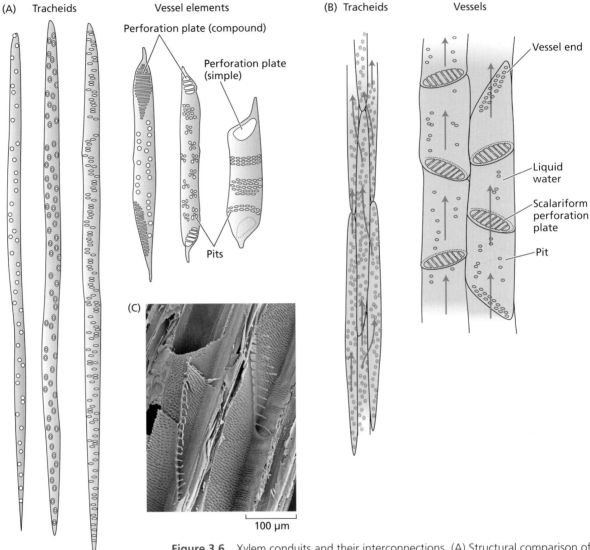

(A) Tracheids   Vessel elements

Perforation plate (compound)

Perforation plate (simple)

Pits

(B) Tracheids   Vessels

Vessel end

Liquid water

Scalariform perforation plate

Pit

(C)

100 μm

Figure 3.6 Xylem conduits and their interconnections. (A) Structural comparison of tracheids and vessel elements. Tracheids are elongated, hollow, dead cells with highly lignified walls. The walls contain numerous pits—regions where secondary wall is absent but primary wall remains. The shapes of pits and the patterns of wall pitting vary with species and organ type. Tracheids are present in all vascular plants. Vessels consist of a stack of two or more vessel elements. Like tracheids, vessel elements are dead cells and are connected to one another by perforation plates—regions of the wall where pores or holes have developed. Vessels are connected to other vessels and to tracheids through pits. Vessels are found in most angiosperms and are lacking in most gymnosperms. (B) Tracheids (left) and vessels (right) form a series of parallel, interconnected pathways for water movement. (C) Scanning electron micrograph showing two vessels (running diagonally from lower left to upper right). Pits are visible on the side walls, as are the scalariform end walls between vessel elements. (C © Steve Gschmeissner/Science Source.)

vessel elements Nonliving water-conducting cells with perforated end walls, found only in angiosperms and a small group of gymnosperms.

perforation plate The perforated end wall of a vessel element in the xylem.

vessel A stack of two or more vessel elements in the xylem.

movement of gas bubbles. Thus, pit membranes of both types play an important role in preventing the spread of gas bubbles, called *emboli*, within the xylem.

Vessel elements tend to be shorter and wider than tracheids and have perforated end walls that form a **perforation plate** at each end of the cell (See Figure 3.6A). Like tracheids, vessel elements have pits on their lateral walls (see Figure 3.6C). Unlike in tracheids, the perforated end walls allow vessel elements to be stacked end to end to form a much longer conduit called a **vessel** (see Figure

(A) Conifers

(B) Other vascular plants

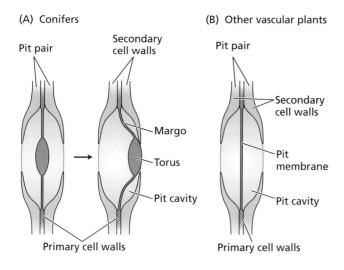

Figure 3.7 Pit pairs. (A) Diagram of a coniferous bordered pit with the torus centered in the pit cavity (left) or lodged to one side of the cavity (right). When the pressure difference between two tracheids is small, the pit membrane lies close to the center of the bordered pit, allowing water to flow through the porous margo region of the pit membrane; when the pressure difference between two tracheids is large, such as when one has cavitated and the other remains filled with water under tension, the pit membrane is displaced such that the torus becomes lodged against the overarching walls, thereby preventing the embolism from propagating between tracheids. (B) In contrast, the pit membranes of angiosperms and other nonconiferous vascular plants are relatively homogeneous in their structure. These pit membranes have very small pores compared with those of conifers, which prevents the spread of embolism but also imparts a significant hydraulic resistance. (A after Zimmermann 1983.)

3.6B). Vessels are multicellular conduits that vary in length both within and among species. Vessels range from a few centimeters in length to many meters. The vessel elements found at the extreme ends of a vessel lack perforations in their end walls and are connected to neighboring vessels via pits.

Water moves through the xylem by pressure-driven bulk flow

Pressure-driven bulk flow of water is responsible for long-distance transport of water in the xylem. It also accounts for much of the water flow through the soil and through the cell walls of plant tissues. In contrast to the diffusion of water across semipermeable membranes, pressure-driven bulk flow is independent of solute concentration gradients, as long as viscosity changes are negligible.

If we consider bulk flow through a tube, the rate of flow depends on the radius (r) of the tube, the viscosity (η) of the liquid, and the pressure gradient ($\Delta\Psi_p/\Delta x$) that drives the flow. Jean Léonard Marie Poiseuille (1797–1869) was a French physician and physiologist, and the relation just described is given by one form of Poiseuille's equation:

$$\text{Volume flow rate} = \left(\frac{\pi r^4}{8\eta}\right)\left(\frac{\Delta\Psi_p}{\Delta x}\right) \tag{3.2}$$

expressed in cubic meters per second ($m^3\ s^{-1}$). This equation tells us that pressure-driven bulk flow is extremely sensitive to the radius of the tube. If the radius is doubled, the volume flow rate increases by a factor of 16 (2^4). Vessel elements up to 500 μm in diameter, nearly an order of magnitude greater than the largest tracheids, occur in the stems of climbing species. These large-diameter vessels permit vines to transport large amounts of water despite the slenderness of their stems.

Equation 3.2 describes water flow through a cylindrical tube and thus does not take into account the fact that xylem conduits are of finite length, such that water must cross many pit membranes as it flows from the soil to the leaves. All else being equal, pit membranes should impede water flow through single-celled (and thus shorter) tracheids to a greater extent than through multicellular (and thus longer) vessels. However, the pit membranes of conifers are much more permeable to water than are those found in other plants, allowing conifers to grow into large trees despite producing only tracheids.

Water movement through the xylem requires a smaller pressure gradient than movement through living cells

The xylem provides a pathway of low resistivity for water movement. Some numerical values will help you appreciate the extraordinary efficiency of the xylem.

We will calculate the driving force required to move water through the xylem at a typical velocity and compare it with the driving force that would be needed to move water through a pathway made up of living cells at the same rate.

For the purposes of this comparison, we will use a value of 4 mm s^{-1} for the xylem transport velocity and 40 μm as the vessel radius. This is a high velocity for such a narrow vessel, so it will tend to exaggerate the pressure gradient required to support water flow in the xylem. Using a version of Poiseuille's equation (see Equation 3.2), we can calculate the pressure gradient needed to move water at a velocity of 4 mm s^{-1} through an *ideal* tube with a uniform inner radius of 40 μm. The calculation gives a value of 0.02 MPa m^{-1}.

Of course, *real* xylem conduits have irregular inner wall surfaces, and water flow through perforation plates and pits adds resistance to water transport. Such deviations from the ideal increase the frictional drag: Measurements show that the actual resistance is greater by approximately a factor of 2.

Let's now compare this value with the driving force that would be necessary to move water at the same velocity from cell to cell, crossing the plasma membrane each time. The driving force needed to move water through a layer of cells at 4 mm s^{-1} is 2×10^8 MPa m^{-1}. This is ten orders of magnitude greater than the driving force needed to move water through our 40-μm-radius xylem vessel. Our calculation clearly shows that water flow through the xylem is vastly more efficient than water flow across living cells. Nevertheless, the xylem can make a significant contribution to the total resistance to water flow through the plant.

What pressure difference is needed to lift water 100 meters to a treetop?

With the foregoing example in mind, let's see what pressure gradient is needed to move water up to the top of a very tall tree. The tallest trees in the world are the coast redwoods (*Sequoia sempervirens*) of North America and the mountain ash (*Eucalyptus regnans*) of Australia. Individuals of both species can exceed 100 m.

If we think of the stem of a tree as a long pipe, we can estimate the pressure difference that is needed to overcome the frictional drag of moving water from the soil to the top of the tree by multiplying the pressure gradient needed to move the water by the height of the tree. The pressure gradients needed to move water through the xylem of very tall trees are on the order of 0.01 MPa m^{-1}, smaller than in our previous example. If we multiply this pressure gradient by the height of the tree (0.01 MPa m^{-1} × 100 m), we find that the total pressure difference needed to overcome the frictional resistance to water movement through the stem is equal to 1 MPa.

In addition to frictional resistance, we must consider gravity. As described by Equation 2.4, for a height difference of 100 m, the difference in Ψ_g is approximately 1 MPa. That is, Ψ_g is 1 MPa higher at the top of the tree than at the ground level. So the other components of water potential must be 1 MPa more negative at the top of the tree to counter the effects of gravity.

To allow transpiration to occur, the pressure gradient due to gravity must be added to that required to cause water movement through the xylem. Thus, we calculate that a pressure difference of roughly 2 MPa, from the base to the top branches, is needed to carry water up the tallest trees.

The cohesion–tension theory explains water transport in the xylem

In theory, the pressure gradients needed to move water through the xylem could result from the generation of positive pressures at the base of the plant or negative pressures at the top of the plant. We mentioned previously that some roots can develop positive hydrostatic pressure in their xylem. However, root pressure is typically less than 0.1 MPa and disappears with transpiration or when soils are dry, so it is clearly inadequate to move water up a tall tree. Furthermore, because root pressure is generated by the accumulation of ions in the xylem, relying on this for transporting water would require a mechanism for dealing with these solutes once the water evaporates from the leaves.

Instead, the water at the top of a tree develops a large tension (a negative hydrostatic pressure), and this tension *pulls* water through the xylem. This mechanism, first proposed toward the end of the nineteenth century, is called the *cohesion–tension theory of sap ascent* because it requires the cohesive properties of water to sustain large tensions in the xylem water columns. One can readily demonstrate xylem tension by puncturing intact xylem through a drop of ink on the surface of a stem from a transpiring plant. When the tension in the xylem is relieved, the ink is drawn instantly into the xylem, resulting in visible streaks along the stem.

The xylem tensions needed to pull water from the soil develop in leaves as a consequence of transpiration. How does the loss of water vapor through open stomata result in the flow of water from the soil? When leaves open their stomata to obtain CO_2 for photosynthesis, water vapor diffuses out of the leaves. This causes water to evaporate from the surface of cell walls inside the leaves. In turn, the loss of water from the cell walls causes the water potential in the walls to decrease (**Figure 3.8**). This creates a gradient in water potential that causes water to flow toward the sites of evaporation.

One hypothesis for how a loss of water from cell walls results in a decrease in water potential is that as water evaporates, the surface of the remaining water is

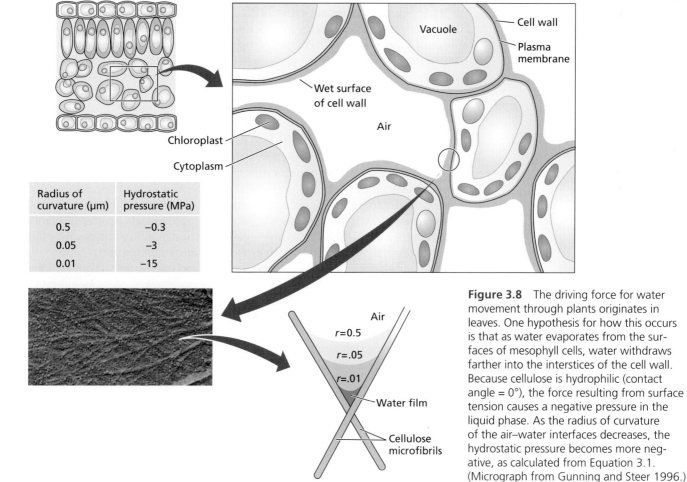

Radius of curvature (μm)	Hydrostatic pressure (MPa)
0.5	−0.3
0.05	−3
0.01	−15

Figure 3.8 The driving force for water movement through plants originates in leaves. One hypothesis for how this occurs is that as water evaporates from the surfaces of mesophyll cells, water withdraws farther into the interstices of the cell wall. Because cellulose is hydrophilic (contact angle = 0°), the force resulting from surface tension causes a negative pressure in the liquid phase. As the radius of curvature of the air–water interfaces decreases, the hydrostatic pressure becomes more negative, as calculated from Equation 3.1. (Micrograph from Gunning and Steer 1996.)

drawn into the interstices of the cell wall (see Figure 3.8) where it forms curved air–water interfaces. Because water adheres to the cellulose microfibrils and other hydrophilic components of the cell wall, the curvature of these interfaces induces a negative pressure in the water. As more water is removed from the wall, the curvature of these air–water interfaces increases and the pressure of the water becomes more negative (see Equation 3.1), a situation analogous to what occurs in the soil.

Some of the water that flows toward the sites of evaporation comes from the protoplasts of adjacent cells. However, because leaves are connected to the soil via a low-resistance pathway—the xylem—most of what replaces water lost from the leaves due to transpiration comes from the soil. Water will flow from the soil when the water potential of the leaves is low enough to overcome the Ψ_p of the soil, as well as the resistance associated with moving water through the plant. Note that for water to be pulled from the soil requires that there be a continuous liquid-filled pathway extending from the sites of evaporation, down through the plant, and out into the soil.

The cohesion–tension theory explains how the substantial movement of water through plants can occur without the direct expenditure of metabolic energy: The energy that powers the movement of water through plants comes from the sun, which, by increasing the temperature of both the leaf and the surrounding air, drives the evaporation of water. However, water transport through the xylem is not "free." The plant must build xylem conduits capable of withstanding the large tensions needed to pull water from the soil. In addition, plants must accumulate enough solutes in their living cells that they are able to remain turgid even as water potentials decrease due to transpiration.

The cohesion–tension theory has been a controversial subject for more than a century and continues to generate lively debate. The main controversy surrounds the question of whether water columns in the xylem can sustain the large tensions (negative pressures) necessary to pull water up tall trees. Recently, water transport through a microfluidic device designed to function as a synthetic "tree" demonstrated the stable flow of liquid water at pressures lower (more negative) than –7.0 MPa.

Xylem transport of water in trees faces physical challenges

The large tensions that develop in the xylem of trees and other plants present significant physical challenges. First, the water under tension transmits an inward force to the walls of the xylem. If the cell walls were weak or pliant, they would collapse under this tension. The secondary wall thickenings and lignification of tracheids and vessels are adaptations that offset this tendency to collapse. Plants that experience large xylem tensions tend to have dense wood, reflecting the mechanical stresses imposed on the wood by water under tension.

A second challenge is that water under such tensions is in a *physically metastable state*. Water is stable as a liquid when its hydrostatic pressure exceeds its saturated vapor pressure. When the hydrostatic pressure in liquid water becomes equal to its saturated vapor pressure, the water undergoes a phase change. We are all familiar with the idea of vaporizing water by increasing its temperature (raising its saturated vapor pressure). Less familiar, but still easily observed, is the fact that water can be made to boil at room temperature by placing it in a vacuum chamber (lowering the hydrostatic pressure of the liquid phase by reducing the pressure of the atmosphere).

In our earlier example, we estimated that a pressure gradient of 2 MPa would be needed to supply water to leaves at the top of a 100-m-high tree. If we assume that the soil surrounding this tree is fully hydrated and lacks significant concentrations of solutes (i.e., $\Psi = 0$), the cohesion–tension theory predicts that the hydrostatic pressure of water in the xylem at the top of the tree will be

–2 MPa. This value is substantially below the saturated vapor pressure (absolute pressure of ~0.002 MPa at 20°C), raising the question of what maintains the water column in its liquid state.

Water in the xylem is described as being in a metastable state because despite the existence of a thermodynamically lower energy state—the vapor phase—it remains a liquid. This situation occurs because (1) the cohesion and adhesion of water make the free-energy barrier for the liquid-to-vapor phase change very high, and (2) the structure of the xylem minimizes the presence of *nucleation sites*—sites that lower the energy barrier separating the liquid from the vapor phase.

The most important nucleation sites are gas bubbles. When a gas bubble grows to a sufficient size that the inward force resulting from surface tension is less than the outward force due to the negative pressure in the liquid phase, the bubble expands. Furthermore, once a bubble starts to expand, the inward force due to surface tension decreases, because the air–water interface has less curvature. Thus, a bubble that exceeds the critical size for expansion will expand until it fills the entire conduit (**Figure 3.9**).

The absence of gas bubbles of sufficient size to destabilize the water column when under tension is, in part, due to the fact that in the roots, water must flow across the endodermis to enter the xylem. The endodermis serves as a filter, preventing gas bubbles from entering the xylem. Pit membranes also function as filters as water flows from one xylem conduit to another. However, when pit membranes are exposed to air on one side—due to injury, leaf abscission, or the existence of a neighboring gas-filled conduit—pit membranes can serve as sites of entry for air. Air enters when the pressure difference across the pit membrane is sufficient either to allow air to penetrate the cellulose microfibriller matrix of structurally homogeneous pit membranes (see Figure 3.7B), or to dislodge the torus of a coniferous pit membrane (see Figure 3.7A). This phenomenon is called *air seeding*.

A second mode by which bubbles can form in xylem conduits is freezing of the xylem tissues. Because water in the xylem contains dissolved gases and the solubility of gases in ice is very low, freezing of xylem conduits can lead to bubble formation.

The phenomenon of bubble expansion is known as *cavitation*, and the resulting gas-filled void is referred to as an *embolism* (see Figure 3.9). Its effect is similar to that of a vapor lock in the fuel line of an automobile or an embolism in a blood vessel. Cavitation breaks the continuity of the water column and prevents the transport of water under tension.

Such breaks in the water columns in plants are not unusual. When plants are deprived of water, sound pulses or clicks can be detected. The formation and rapid ex-

Figure 3.9 Cavitation blocks water movement because of the formation of gas-filled (embolized) conduits. Because xylem conduits are interconnected through pits in their thick secondary walls, water can detour around the blocked vessel by moving through adjacent conduits. The very small pores in the pit membranes help prevent embolisms from spreading between xylem conduits. Thus, in the diagram on the right, the gas is contained within a single cavitated tracheid. In the diagram on the left, gas has filled the entire cavitated vessel, shown here as being made up of three vessel elements, each separated by scalariform (resembling the rungs of a ladder) perforation plates. In nature, vessels can be very long (up to several meters in length) and thus made up of many vessel elements.

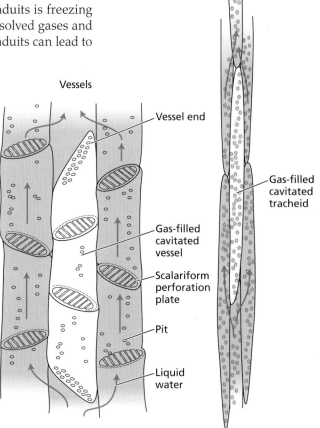

Tracheids

Vessels

Vessel end

Gas-filled cavitated vessel

Scalariform perforation plate

Pit

Liquid water

Gas-filled cavitated tracheid

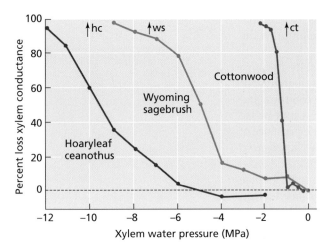

Figure 3.10 Xylem vulnerability curves represent the percentage loss of hydraulic conductance in stem xylem versus xylem water pressure, here in three species of contrasting drought tolerance. Data were obtained from excised branches subjected experimentally to increasing levels of xylem tension using a centrifugal force technique. Arrows on the upper axis indicate the minimum xylem pressure measured in the field for each species. (After Sperry 2000.)

pansion of air bubbles in the xylem, such that the pressure in the water is suddenly increased by perhaps 1 MPa or more, result in high-frequency acoustic shock waves through the rest of the plant. These breaks in xylem water continuity, if not repaired, would be disastrous to the plant. By blocking the main transport pathway of water, such embolisms would increase flow resistance and ultimately cause the dehydration and death of the leaves and other organs.

Vulnerability curves (**Figure 3.10**) provide a way of quantifying a species' susceptibility to cavitation and the impact of cavitation on flow through the xylem. A vulnerability curve plots the measured hydraulic conductivity (usually as a percent of maximum) of a branch, stem, or root segment versus the experimentally imposed level of xylem tension. Due to cavitation, xylem hydraulic conductivity decreases with increasing tensions until flow ceases entirely. However, the decrease in xylem hydraulic conductivity occurs at much lower tensions in species in moist habitats, such as birch, than in species from more arid regions, such as sagebrush.

Plants minimize the consequences of xylem cavitation

The impact of xylem cavitation on the plant can be minimized by several means. Because the water-transporting conduits in the xylem are interconnected, one gas bubble might, in principle, expand to fill the whole network. In practice, gas bubbles do not spread far, because an expanding gas bubble cannot easily pass through the small pores of the pit membranes. Because the capillaries in the xylem are interconnected, one gas bubble does not completely stop water flow. Instead, water can detour around the embolized conduit by traveling through neighboring, water-filled conduits (see Figure 3.9). Thus, the finite length of the tracheid and vessel conduits of the xylem, while resulting in an increased resistance to water flow, also provides a way to restrict the impact of cavitation.

Gas bubbles can also be eliminated from the xylem. As we have seen, some plants develop positive pressures (root pressures) in the xylem. Such pressures shrink bubbles and cause the gases to dissolve. Recent studies suggest that cavitation may be repaired even when the water in the xylem is under tension. A mechanism for such repair is not yet known and remains the subject of active research.

Finally, many plants have secondary growth in which new xylem forms each year. The production of new xylem conduits allows plants to replace losses in water-transport capacity due to cavitation.

Water Movement from the Leaf to the Atmosphere

On its way from the leaf to the atmosphere, water is pulled from the xylem into the cell walls of the mesophyll, where it evaporates into the air spaces of the leaf (**Figure 3.11**). The water vapor then exits the leaf through stomatal pores. The movement of liquid water through the living tissues of the leaf is controlled by gradients in water potential. However, transport in the vapor phase is by diffusion, so the final part of the transpiration stream is controlled by the *concentration gradient of water vapor.*

The waxy cuticle that covers the leaf surface is an effective barrier to water movement. It has been estimated that only about 5% of the water lost from leaves escapes through the cuticle. Almost all of the water lost from leaves is lost by diffusion of water vapor through the tiny stomatal pores. In most herbaceous

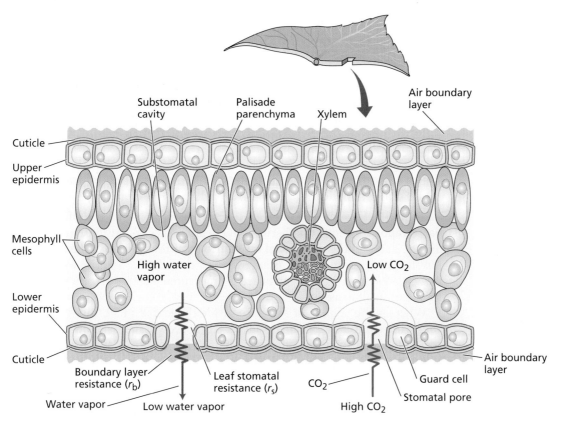

Figure 3.11 Water pathway through the leaf. Water is pulled from the xylem into the cell walls of the mesophyll, where it evaporates into the air spaces within the leaf. Water vapor then diffuses through the leaf air space, through the stomatal pore, and across the boundary layer of still air found next to the leaf surface. CO_2 diffuses in the opposite direction along its concentration gradient (low inside, higher outside).

species, stomata are present on both the upper and lower surfaces of the leaf, usually more abundant on the lower surface. In many tree species, stomata are located only on the lower surface of the leaf.

Leaves have a large hydraulic resistance

Although the distances that water must traverse within leaves are small relative to the entire soil-to-atmosphere pathway, the contribution of the leaf to the total hydraulic resistance is large. On average, leaves constitute 30% of the total liquid-phase resistance, and in some plants their contribution is much larger. This combination of short path length and large hydraulic resistance also occurs in roots, reflecting the fact that in both organs, water transport takes place across highly resistive living tissues as well as through the xylem.

Water enters into leaves and is distributed across the leaf lamina in xylem conduits. Water must exit through the xylem walls and pass through multiple layers of living cells before it evaporates. Leaf hydraulic resistance thus reflects the number, distribution, and size of xylem conduits, as well as the hydraulic properties of leaf mesophyll cells. The hydraulic resistance of leaves of diverse vein architectures varies as much as 40-fold. A large part of this variation appears to be due to the density of veins within the leaf and their distance from the evaporative leaf surface. Leaves with closely spaced veins tend to have lower hydraulic resistance and higher rates of photosynthesis, suggesting that the proximity of leaf veins to sites of evaporation exerts a significant impact on the rates of leaf gas exchange.

difference in water vapor concentration (Δc_{wv})   The difference between the water vapor concentration of the air spaces inside the leaf and that of the air outside the leaf.

diffusional resistance (r)
The restriction posed by the boundary layer and the stomata to the free diffusion of gases from and into the leaf.

The hydraulic resistance of leaves varies in response to growth conditions and exposure to low leaf water potentials. For example, leaves of plants growing in shaded conditions exhibit greater resistance to water flow than do leaves of plants grown in higher light. Leaf hydraulic resistance also typically increases with leaf age. Over shorter time scales, decreases in leaf water potential lead to marked increases in leaf hydraulic resistance. The increase in leaf hydraulic resistance may result from decreases in the membrane permeability of mesophyll cells, cavitation of xylem conduits in leaf veins, or in some cases, the physical collapse of xylem conduits under tension.

The driving force for transpiration is the difference in water vapor concentration

Transpiration from the leaf depends on two major factors: (1) the **difference in water vapor concentration** between the leaf air spaces and the external bulk air (Δc_{wv}) and (2) the **diffusional resistance** (r) of this pathway. The difference in water vapor concentration is expressed as $c_{wv(leaf)} - c_{wv(air)}$. The water vapor concentration of air ($c_{wv[air]}$) can be readily measured, but that of the leaf ($c_{wv[leaf]}$) is more difficult to assess.

Whereas the volume of air space inside the leaf is small, the wet surface from which water evaporates is large. Air space volume is about 5% of the total leaf volume in pine needles, 10% in maize (corn; *Zea mays*) leaves, 30% in barley, and 40% in tobacco leaves. In contrast to the volume of the air space, the internal surface area from which water evaporates may be from 7 to 30 times the external leaf area. This high surface-to-volume ratio makes for rapid vapor equilibration inside the leaf. Thus, we can assume that the air space in the leaf is close to water potential equilibrium with the cell wall surfaces from which liquid water is evaporating.

Within the range of water potentials experienced by transpiring leaves (generally greater than –2.0 MPa), the equilibrium water vapor concentration is within 2 percentage points of the saturation water vapor concentration. This allows one to estimate the water vapor concentration within a leaf from its temperature, which is easy to measure. Because the saturated water vapor content of air increases exponentially with temperature, leaf temperature has a marked impact on transpiration rates.

The concentration of water vapor, c_{wv}, changes at various points along the transpiration pathway. We see from **Table 3.1** that c_{wv} decreases at each step

---

Table 3.1   Representative values for relative humidity, absolute water vapor concentration, and water potential for four points in the pathway of water loss from a leaf

Location	Relative humidity	Water vapor Concentration (mol m⁻³)	Water vapor Potential (MPa)[a]
Inner air spaces (25°C)	0.99	1.27	–1.38
Just inside stomatal pore (25°C)	0.97	1.21	–7.04
Just outside stomatal pore (25°C)	0.47	0.60	–103.7
Bulk air (20°C)	0.50	0.50	–93.6

Source: Adapted from Nobel 1999.
Note: See Figure 3.11.
[a]Calculated with values for $RT/\bar{V}_w$ of 135 MPa at 20°C and 137.3 MPa at 25°C.

of the pathway from the cell wall surface to the bulk air outside the leaf. The important points to remember are that (1) the driving force for water loss from the leaf is the *absolute* concentration difference (difference in c_{wv}, in mol m^{-3}), and (2) this difference is markedly influenced by leaf temperature. The numbers in Table 3.1 also show that as long as the bulk air is less than 97% saturated, the water potential difference between the soil solution and the air is easily sufficient to overcome the frictional resistance and gravitational drag (total of 2 MPa) and carry water up the tallest trees.

Water loss is also regulated by the pathway resistances

The second important factor governing water loss from the leaf is the diffusional resistance of the transpiration pathway, which consists of two varying components (see Figure 3.11):

1. The resistance associated with diffusion through the stomatal pore, the **stomatal resistance** (r_s).

2. The resistance due to the layer of unstirred air next to the leaf surface through which water vapor must diffuse to reach the turbulent air of the atmosphere. This second resistance, r_b, is called the leaf **boundary layer resistance**. We will discuss this type of resistance before considering stomatal resistance.

The boundary layer contributes to diffusional resistance

The thickness of the boundary layer is determined primarily by wind speed and leaf size. When the air surrounding the leaf is very still, the layer of unstirred air on the surface of the leaf may be so thick that it is the primary deterrent to water vapor loss from the leaf. Increases in stomatal apertures under such conditions have little effect on transpiration rate (**Figure 3.12**), although closing the stomata completely will still reduce transpiration.

When wind velocity is high, the moving air reduces the thickness of the boundary layer at the leaf surface, reducing the resistance of this layer. Under such conditions, stomatal resistance largely controls water loss from the leaf.

Various anatomical and morphological aspects of the leaf can influence the thickness of the boundary layer. Hairs on the surface of leaves can serve as microscopic windbreaks. Some plants have sunken stomata that provide a sheltered region outside the stomatal pore. The size and shape of leaves and their orientation relative to the wind direction also influence the way the wind sweeps across the leaf surface. Most of these factors, however, cannot be altered on an hour-to-hour or even a day-to-day basis. For short-term regulation of transpiration, control of stomatal apertures by the guard cells plays a crucial role in the regulation of leaf transpiration.

Some species are able to change the orientation of their leaves and thereby influence their transpiration rates. For example, when plants orient their leaves parallel to the sun's rays, leaf temperature is reduced and with it the driving force for transpiration, Δc_{wv}. Many grass leaves roll up as they experience water deficits, in this way increasing their boundary layer resistance. Even wilting can help ameliorate high transpiration rates by reducing the amount of radiation intercepted, resulting in lower leaf temperatures and a decrease in Δc_{wv}.

leaf stomatal resistance (r_s)
The resistance to CO_2 diffusion imposed by the stomatal pores.

boundary layer resistance (r_b)
The resistance to the diffusion of water vapor, CO_2, and heat due to the layer of unstirred air next to the leaf surface. A component of diffusional resistance.

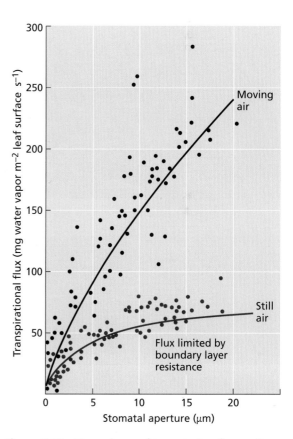

Figure 3.12 Dependence of transpiration flux on the stomatal aperture of zebra plant (*Tradescantia zebrina*) in still air and in moving air. The boundary layer is thicker and more rate-limiting in still air than in moving air. As a result, the stomatal aperture has less control over transpiration in still air. (After Bange 1953.)

guard cells A pair of specialized epidermal cells that surround the stomatal pore and regulate its opening and closing.

Stomatal resistance is another major component of diffusional resistance

Because the cuticle covering the leaf is nearly impermeable to water, most leaf transpiration results from the diffusion of water vapor through the stomatal pores (see Figure 3.11). When open, the microscopic stomatal pores provide a *low-resistance pathway* for diffusional movement of gases across the epidermis and cuticle. Changes in stomatal resistance are important for the regulation of water loss by the plant and for controlling the rate of carbon dioxide uptake necessary for sustained CO_2 fixation during photosynthesis.

The leaf cannot control $c_{wv(air)}$ or r_b (see Figure 3.11). However, it can regulate its stomatal resistance (r_s) by opening and closing of the stomatal pore. Stomatal resistance is governed by the degree to which the stomata are open. The biological control over this process is exerted by a pair of specialized epidermal cells, the **guard cells**, which surround the stomatal pore (**Figure 3.13**). Stomatal opening is dependent on the special features of the guard cell walls as well as on leaf water status and light conditions. Let us first consider the guard cell walls.

The cell walls of guard cells have specialized features

Guard cells are found in leaves of all vascular plants, and they are also present in some nonvascular plants, such as hornworts and mosses. Guard cells show considerable morphological diversity, but we can distinguish two main types: One is typical of grasses, while the other is found in most other flowering plants, as well as in mosses, ferns, and gymnosperms.

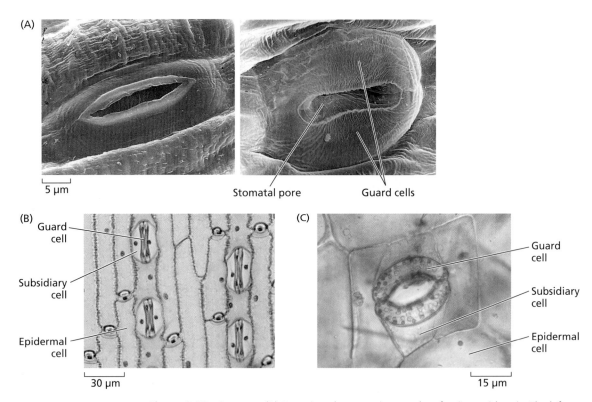

Figure 3.13 Stomata. (A) Scanning electron micrographs of onion epidermis. The left panel shows the outside surface of the leaf, with a stomatal pore inserted in the cuticle. The right panel shows a pair of guard cells facing the stomatal cavity, toward the inside of the leaf. (B) Stomata of maize (*Zea mays*), showing the dumbbell-shaped guard cells typical of grasses. (C) Most other plants have kidney-shaped guard cells, as seen in this open stoma of *Tradescantia zebrina*. (A from Zeiger and Hepler 1976 [left] and E. Zeiger and N. Burnstein [right]; B © age fotostock Spain, S.L./Alamy; C © Ray Simons/Science Source.)

In grasses (see Figure 3.13B), guard cells have a characteristic dumbbell shape, with bulbous ends. The pore proper is a long slit located between the two "handles" of the dumbbells. These guard cells are always flanked by a pair of differentiated epidermal cells called **subsidiary cells**, which help the guard cells control the stomatal pore. The guard cells, subsidiary cells, and pore are collectively called the **stomatal complex**.

In most other plants, guard cells have an elliptical contour (often called "kidney-shaped") with the pore at their center (see Figure 3.13C). Although subsidiary cells are common in species with kidney-shaped stomata, they can be absent, in which case the guard cells are surrounded by ordinary epidermal cells.

A distinctive feature of guard cells is the specialized structure of their walls. Portions of these walls are substantially thickened (**Figure 3.14**) and may be up to 5 µm across, in contrast to the 1 to 2 µm typical of epidermal cells. In kidney-shaped guard cells, a differential thickening pattern results in very thick inner and outer (lateral) walls, a thin dorsal wall (the wall in contact with epidermal cells), and a somewhat thickened ventral (pore) wall. The portions of the wall that face the atmosphere often extend into well-developed ledges, which form the pore proper.

The alignment of cellulose microfibrils, which reinforce all plant cell walls and are an important determinant of cell shape (see Chapter 1), plays an essential role in the opening and closing of the stomatal pore. In ordinary, cylindrically shaped cells, cellulose microfibrils are oriented transversely to the long axis of the cell. As a result, the cell expands in the direction of its long axis, because the cellulose reinforcement offers the least resistance at right angles to its orientation.

In guard cells the microfibril organization is different. Kidney-shaped guard cells have cellulose microfibrils fanning out radially from the pore (see Figure 3.14C). As a result, the inner wall (facing the pore) is much stronger than the outer wall. Thus, as a guard cell increases in volume, the outer wall expands more than the inner wall. This causes the guard cells to bow apart and the pore to open. In grasses, the dumbbell-shaped guard cells function like beams with inflatable ends. The orientation of the cellulose microfibrils is such that as the bulbous ends of the cells increase in volume, the beams are separated from each other and the slit between them widens (see Figure 3.14C).

An increase in guard cell turgor pressure opens the stomata

Guard cells function as multisensory hydraulic valves. Environmental factors such as light intensity and quality, temperature, leaf water status, and intracellular CO_2 concentrations are sensed by guard cells, and these signals are integrated into well-defined stomatal responses. If leaves kept in the dark are illuminated, the light stimulus is perceived by the guard cells as an opening signal, triggering a series of responses that result in opening of the stomatal pore.

The early aspects of this process are ion uptake and other metabolic changes in the guard cells, which we will discuss in detail in Chapter 6. Here we note the effect of decreases in osmotic potential (Ψ_s) resulting from ion uptake and from biosynthesis of organic molecules in the guard cells. Water relations in guard cells follow the same rules as in other cells. As Ψ_s decreases, the water potential decreases, and water consequently moves into the guard cells. As water enters the cell, turgor pressure increases and the stomata open.

In angiosperms, stomatal opening and closing involves changes in the volume and turgor pressure of both the guard cells and the subsidiary (or adjacent epidermal) cells (see Figure 3.14A). At the same time that the uptake of solutes into guard cells causes them to increase in volume and turgor pressure, the subsidiary (or adjacent epidermal) cells release solutes into the apoplast. The transfer of solutes out of subsidiary cells and into the guard cells causes the former to decrease in both turgor pressure and size, facilitating the expansion of guard cells in the

subsidiary cells Specialized epidermal cells that flank the guard cells and work with the guard cells in the control of stomatal apertures.

stomatal complex The guard cells, subsidiary cells, and stomatal pore, which together regulate leaf transpiration.

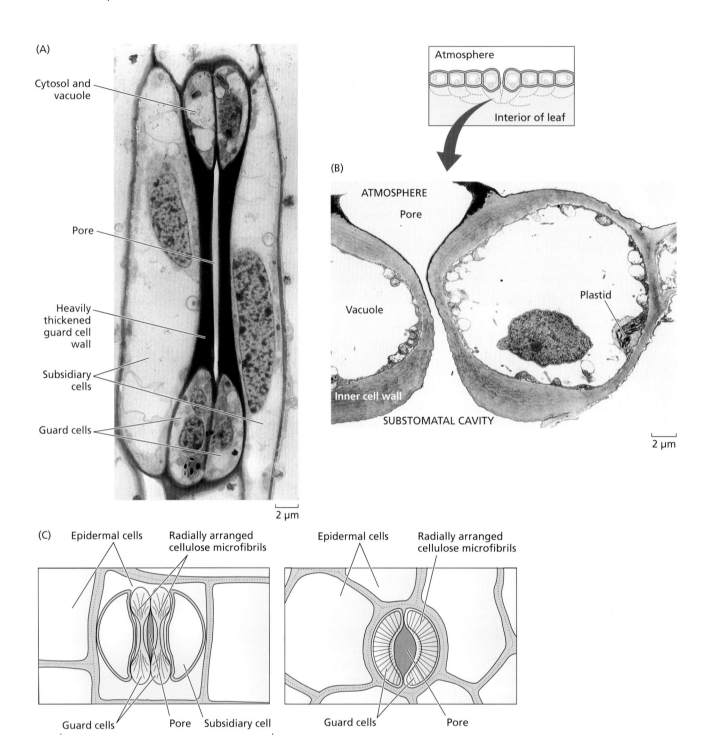

(A)

Cytosol and vacuole

Pore

Heavily thickened guard cell wall

Subsidiary cells

Guard cells

2 µm

Atmosphere

Interior of leaf

(B)

ATMOSPHERE

Pore

Vacuole

Plastid

Inner cell wall

SUBSTOMATAL CAVITY

2 µm

(C)

Epidermal cells

Radially arranged cellulose microfibrils

Guard cells   Pore   Subsidiary cell

Stomatal complex

Epidermal cells

Radially arranged cellulose microfibrils

Guard cells   Pore

Figure 3.14 Guard cell wall structure. (A) Electron micrograph of a stoma from a grass (*Phleum pratense*). The bulbous ends of each guard cell show their cytosolic content and are joined by the heavily thickened walls. The stomatal pore separates the two mid-portions of the guard cells. (B) Electron micrograph showing a pair of guard cells from tobacco (*Nicotiana tabacum*). The section was made perpendicular to the main surface of the leaf. The pore faces the atmosphere; the bottom faces the substomatal cavity inside the leaf. Note the uneven thickening pattern of the walls, which determines the asymmetric deformation of the guard cells when their volume increases during stomatal opening. (C) Radial alignment of the cellulose microfibrils in guard cells and epidermal cells of a grasslike stoma (left) and a kidney-shaped stoma (right). (A from Palevitz 1981, courtesy of B. Palevitz; B from Sack 1987, courtesy of F. Sack; C after Meidner and Mansfield 1968.)

direction away from the stomatal pore. Conversely, the transfer of solutes from guard cells to the subsidiary cells increases the size and turgor pressure of the latter, thus pushing the guard cells together and causing the stomata to close.

Subsidiary cells appear to play an important role in allowing angiosperm stomata to open quickly and to achieve large apertures. One consequence of these interactions is that decreases in leaf water potential are not passively linked to stomatal closure. The subsidiary cells must increase in volume and turgor pressure for the stomata to close. In Chapter 19 we will see how chemical signals play an important role in controlling stomatal aperture during drought.

Coupling Leaf Transpiration and Photosynthesis: Light-dependent Stomatal Opening

Under temperate conditions, light is the dominant stimulus causing stomatal opening (**Figure 3.15**). The two major factors involved in light-dependent opening are (1) photosynthesis in the guard cell chloroplast and (2) a specific response to blue light. In addition, increases in mesophyll photosynthesis lower intercellular CO_2 concentrations, which stimulates opening of stomata.

Stomatal opening is regulated by light

When water is abundant, the functional solution to the leaf's need to limit water loss while taking in CO_2 is the temporal regulation of stomatal apertures—open during the day, closed at night (**Figure 3.16**). At night, when there is no photosynthesis and thus no demand for CO_2 inside the leaf, stomatal apertures are kept small or closed, preventing unnecessary loss of water. On a sunny morning when the supply of water is abundant and the solar radiation incident on the leaf favors high photosynthetic activity, the demand for CO_2 inside the leaf is large, and the stomatal pores open wide, decreasing the stomatal resistance to CO_2 diffusion. Water loss by transpiration is substantial under these conditions, but since the water supply is plentiful, it is advantageous for the plant to trade water for the products of photosynthesis, which are essential for growth and reproduction.

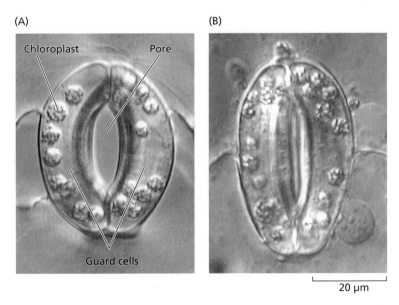

(A)   (B)

Chloroplast   Pore

Guard cells

20 µm

Figure 3.15   Light-stimulated stomatal opening in detached epidermis of *Vicia faba*. An open, light-treated stoma (A) is shown in the dark-treated, closed state in (B). Stomatal opening is quantified by microscopic measurement of the width of the stomatal pore. Unlike typical epidermal cells, guard cells contain chloroplasts. (Courtesy of E. Raveh.)

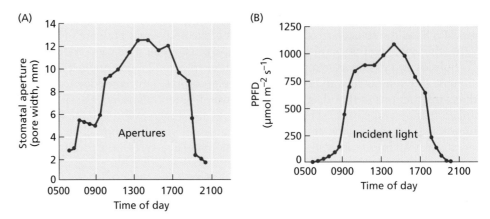

Figure 3.16 Stomatal opening tracks photosynthetically active radiation at the leaf surface. Stomatal opening in the lower surface of leaves of *Vicia faba* grown in a greenhouse, measured as the width of the stomatal pore (A), closely follows the levels of photosynthetically active radiation (400–700 nm) at the leaf surface (B), indicating that the response to light is the dominant response regulating stomatal opening. (PPFD, photosynthetic photon flux density.) (After Srivastava and Zeiger 1995.)

However, when soil water is less abundant, the stomata will open less or even remain closed on a sunny morning. By keeping its stomata closed in dry conditions, the plant avoids dehydration.

Stomatal opening is specifically regulated by blue light

Studies of the stomatal response to light have shown that dichlorophenyldimethylurea (DCMU), an inhibitor of photosynthetic electron transport (see Figure 7.27), causes a partial inhibition of light-stimulated stomatal opening. This indicates that photosynthesis in the guard cell chloroplast plays a role in light-dependent stomatal opening, but why is the response only partial? This partial response to DCMU points to the involvement of a DCMU-insensitive, nonphotosynthetic component of the stomatal response to light. Detailed studies carried out under colored light have shown that light activates two distinct responses of guard cells: photosynthesis in the guard cell chloroplast and a specific blue-light response.

Since blue light stimulates both the specific blue-light response of stomata and guard cell photosynthesis (see the action spectrum for photosynthesis in Figure 7.8), blue light alone cannot be used to study the specific stomatal response to blue light. To achieve a clear-cut separation between the two light responses, researchers use dual-beam experiments. First, high fluence rates of red light are used to *saturate* the photosynthetic response; such saturation prevents any further stomatal opening mediated by photosynthesis in response to further increases in red light. Then, low photon fluxes of blue light are added after the response to the saturating red light has been established (**Figure 3.17**). The addition of blue light causes substantial additional stomatal

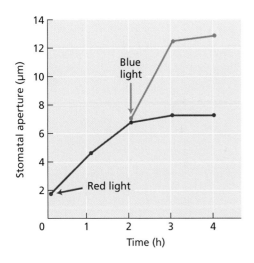

Figure 3.17 Response of stomata to blue light under a red-light background. Stomata from detached epidermis of common dayflower (*Commelina communis*) were treated with saturating photon fluxes of red light (red trace). In a parallel treatment, stomata illuminated with red light were also illuminated with blue light, as indicated by the arrow (blue trace). The increase in stomatal opening above the level reached in the presence of saturating red light indicates that a different photoreceptor system, stimulated by blue light, is mediating the additional increases in opening. Experiments carried out with detached epidermis eliminate CO_2 effects from the mesophyll. (From Schwartz and Zeiger 1984.)

opening that, as just explained, cannot be due to further stimulation of guard cell photosynthesis, because the background red light has saturated photosynthesis.

Light-stimulated stomatal movements are driven by changes in the osmoregulation of guard cells. These processes will be described at the end of Chapter 6. The signal transduction process that links the perception of blue light with the opening of stomata will be described in Chapter 13.

Water-use Efficiency

The effectiveness of plants in moderating water loss while allowing sufficient CO_2 uptake for photosynthesis can be assessed by a parameter called the **transpiration ratio**. This value is defined as the amount of water transpired by the plant divided by the amount of carbon dioxide assimilated by photosynthesis.

For plants in which the first stable product of carbon fixation is a three-carbon compound (C_3 plants; see Chapter 8), as many as 400 molecules of water are lost for every molecule of CO_2 fixed by photosynthesis, giving a transpiration ratio of 400. (Sometimes the reciprocal of the transpiration ratio, called the *water use efficiency*, is cited. Plants with a transpiration ratio of 400 have a water use efficiency of 1/400, or 0.0025.)

The large ratio of H_2O efflux to CO_2 influx results from three factors:

1. The concentration gradient driving water loss is about 50 times larger than that driving the influx of CO_2. In large part, this difference is due to the low concentration of CO_2 in air (about 0.04%) and the relatively high concentration of water vapor within the leaf.

2. CO_2 diffuses about 1.6 times more slowly through air than water does (the CO_2 molecule is larger than H_2O and has a smaller diffusion coefficient).

3. CO_2 must cross the plasma membrane, the cytoplasm, and the chloroplast envelope before it is assimilated in the chloroplast. These membranes add to the resistance of the CO_2 diffusion pathway.

Some plants use variations in the usual photosynthetic pathway for fixation of carbon dioxide that substantially reduce their transpiration ratio. Plants in which a four-carbon compound is the first stable product of photosynthesis (C_4 plants; see Chapter 8) generally transpire less water per molecule of CO_2 fixed than C_3 plants do; a typical transpiration ratio for C_4 plants is about 150. This is largely because C_4 photosynthesis results in a lower CO_2 concentration in the intercellular air space (see Chapter 8), thus creating a larger driving force for the uptake of CO_2 and allowing these plants to operate with smaller stomatal apertures and thus lower transpiration rates.

Desert-adapted plants with crassulacean acid metabolism (CAM) photosynthesis (Chapter 8), in which CO_2 is initially fixed into four-carbon organic acids at night, have even lower transpiration ratios; values of about 50 are not unusual. This is possible because their stomata have an inverted diurnal rhythm, opening at night and closing during the day. Transpiration is much lower at night, because the cool leaf temperature gives rise to only a very small Δc_{wv}.

Overview: The Soil–Plant–Atmosphere Continuum

You have seen that movement of water from the soil through the plant to the atmosphere involves different mechanisms of transport:

- In the soil and the xylem, liquid water moves by bulk flow in response to a pressure gradient ($\Delta \Psi_p$).

transpiration ratio The ratio of water loss to photosynthetic carbon gain. Measures the effectiveness of plants in moderating water loss while allowing sufficient CO_2 uptake for photosynthesis.

- When liquid water is transported across membranes, the driving force is the water potential difference across the membrane. Such osmotic flow occurs when cells absorb water and when roots transport water from the soil to the xylem.

- In the vapor phase, water moves primarily by diffusion, at least until it reaches the outside air, where convection (a form of bulk flow) becomes dominant.

However, the key element in the transport of water from the soil to the leaves is the generation of negative pressures within the xylem due to the capillary forces within the cell walls of transpiring leaves. At the other end of the plant, soil water is also held by capillary forces. This results in a "tug-of-war" on a rope of water by capillary forces at both ends. As a leaf loses water due to transpiration, water moves up the plant and out of the soil driven by physical forces, without the involvement of any metabolic pump. The energy for the movement of water is ultimately supplied by the sun.

This simple mechanism makes for tremendous efficiency energetically—which is critical when as many as 400 molecules of water are being transported for every CO_2 molecule being taken up in exchange. Crucial elements that allow this transport system to function are a low-resistivity xylem flow path that is protected from cavitation and a high-surface-area root system for extracting water from the soil.

Summary

There is an inherent conflict between a plant's need for CO_2 uptake and its need to conserve water, resulting from water being lost through the same pores that let CO_2 in. To manage this conflict, plants have evolved adaptations to control water loss from leaves, and to replace the water that is lost.

Water in the Soil

- The water content and rate of movement in soils depend on soil type and structure, which influence the pressure gradient in the soil and its hydraulic conductivity.

- In soil, water may exist as a surface film on soil particles, or it may partially or completely fill the spaces between particles.

- Osmotic potential, pressure potential, and gravitational potential influence the movement of water from the soil through the plant to the atmosphere (**Figure 3.1**).

- The intimate contact between root hairs and soil particles greatly increases the surface area for water absorption (**Figure 3.2**).

Water Absorption by Roots

- Water uptake is mostly confined to regions near root tips (**Figure 3.3**).

- In the root, water may move via the apoplast, the symplast,

or the transmembrane pathway (**Figure 3.4**).

- Water movement through the apoplast is obstructed by the Casparian strip in the endodermis, which forces water to move symplasmically before it enters the xylem (**Figure 3.4**).

- When transpiration is low or absent, the continued transport of solutes into the xylem fluid leads to a decrease in Ψ_s and a decrease in Ψ, providing the force for water absorption and a positive Ψ_p, which yields a positive hydrostatic pressure in the xylem (**Figure 3.5**).

Water Transport through the Xylem

- Xylem conduits, which can be either single-celled tracheids or multicellular vessels, provide a low-resistance pathway for the transport of water (**Figure 3.6**).

- Elongated, spindle-shaped tracheids and stacked vessel elements have pits in lateral walls (**Figures 3.6, 3.7**).

- Pressure-driven bulk flow moves water long distances through the xylem.

- The ascent of water through plants results from the decrease in water potential at the sites of evaporation within leaves (**Figure 3.8**).

- Cavitation breaks the continuity of the water column and prevents the transport of water under tension (**Figure 3.9**).

(Continued)

Summary *(continued)*

Water Movement from the Leaf to the Atmosphere

• Water is pulled from the xylem into the cell walls of leaf mesophyll before evaporating into the leaf's air spaces (**Figure 3.11**).

• The hydraulic resistance of leaves is large and varies in response to growth conditions and exposure to low leaf water potentials.

• Transpiration depends on the difference in water vapor concentration between the leaf air spaces and the external air and on the diffusional resistance of this pathway, which consists of leaf stomatal resistance and boundary layer resistance (**Figures 3.11, 3.12**).

• Opening and closing of the stomatal pore is accomplished and controlled by guard cells (**Figures 3.13, 3.14**).

Coupling Leaf Transpiration and Photosynthesis: Light-dependent Stomatal Opening

• Light is the dominant stimulus causing stomatal opening. The two major driving forces for light-dependent stomatal

opening are photosynthesis in the guard cell chloroplast and a specific response to blue light (**Figures 3.15, 3.16**).

• An inhibitor of photosynthetic electron transport, DCMU, inhibits stomatal opening, indicating that the photosynthetic process plays a role in stomatal opening. The inhibition is only partial, however, meaning that other opening mechanisms must be active. Another major mechanism is a specific stomatal response to blue light (**Figure 3.17**).

Water-use Efficiency

• The effectiveness of plants in limiting water loss while allowing CO_2 uptake is given by the transpiration ratio.

Overview: The Soil–Plant–Atmosphere Continuum

• Physical forces, without the involvement of any metabolic pump, drive the movement of water from soil to plant to atmosphere, with the sun being the ultimate source for the energy.

Suggested Reading

Assmann, S. M. (2010) Hope for Humpty Dumpty. Systems biology of cellular signaling. *Plant Physiol.* 152: 470–479.

Bramley, H., Turner, N. C., Turner, D. W., and Tyerman, S. D. (2009) Roles of morphology, anatomy and aquaporins in determining contrasting hydraulic behavior of roots. *Plant Physiol.* 150: 348–364.

Brodribb, T. J., and McAdam, S. A. M. (2011) Passive origins of stomatal control in vascular plants. *Science* 331: 582–585.

Franks, P. J., and Farquhar, G. D. (2007) The mechanical diversity of stomata and its significance in gas-exchange control. *Plant Physiol.* 143: 78–87.

Hacke, U. G., Sperry, J. S., Pockman, W. T., Davis, S. D., and McCulloh, K. (2001) Trends in wood density and structure are linked to prevention of xylem implosion by negative pressure. *Oecologia* 126: 457–461.

Lawson, T. (2009) Guard cell photosynthesis and stomatal function. *New Phytol.* 181: 13–34.

Pittermann, J., Sperry, J. S., Hacke, U. G., Wheeler, J. K., and Sikkema, E. H. (2005) Torus-margo pits help conifers compete with angiosperms. *Science* 310: 1924.

Roelfsema, M. R. G., and Kollist, H. (2013) Tiny pores with a global impact. *New Phytol.* 197: 11–15.

Zwieniecki, M. A., Thompson, M. V., and Holbrook, N. M. (2002) Understanding the hydraulics of porous pipes: Tradeoffs between water uptake and root length utilization. *J. Plant Growth Regul.* 21: 315–323.

std/Moment/Getty Images

4 Mineral Nutrition

Mineral nutrients are elements such as nitrogen, phosphorus, and potassium that plants acquire primarily in the form of inorganic ions from the soil. Although mineral nutrients continually cycle through all organisms, they enter the **biosphere** predominantly through the root systems of plants, so in a sense plants act as the "miners" of Earth's crust. The large surface area of roots and their ability to absorb inorganic ions at low concentrations from the soil solution increase the effectiveness of mineral acquisition by plants. After being absorbed by roots, the mineral elements are translocated to the different parts of the plant, where they serve in numerous biological functions. Other organisms, such as mycorrhizal fungi and nitrogen-fixing bacteria, often participate with roots in the acquisition of mineral nutrients.

The study of how plants obtain and use mineral nutrients is called **mineral nutrition**. This area of research is central for improving modern agricultural practices and environmental protection, as well as for understanding plant ecological interactions in natural ecosystems. High agricultural yields depend on fertilization with mineral nutrients. In fact, yields of most crop plants increase linearly with the amount of fertilizer they absorb. To meet increased demand for food, annual world consumption of the primary mineral elements used in fertilizers—nitrogen, phosphorus, and potassium—rose steadily from 30 million metric tons in 1960 to 143 million metric tons in 1990. For a decade after that, consumption remained relatively constant as fertilizers were used more judiciously in an attempt to balance rising costs. During the past 10 to 15 years, however, annual consumption has climbed past 180 million metric tons (**Figure 4.1**).

biosphere   The parts of the surface and atmosphere of Earth that support life as well as the organisms living there.

mineral nutrition   The study of how plants obtain and use mineral nutrients.

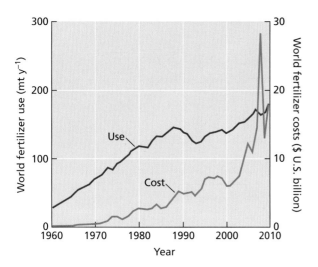

Figure 4.1 Worldwide fertilizer consumption and costs over the past 5 decades. (After http://faostat3.fao.org/faostat-gateway/go/to/download/R/*/E.)

Over half of the energy used in agriculture is expended on the production, distribution, and application of nitrogen fertilizers. Moreover, production of phosphorus fertilizers depends on nonrenewable resources that are likely to reach peak production during this century. Crop plants, however, typically use less than half of the fertilizer applied to the soils around them. The remaining minerals may leach into surface waters or groundwater, become associated with soil particles, or contribute to air pollution or climate change.

As a consequence of fertilizer leaching, many water wells in the United States now exceed federal standards for nitrate (NO_3^-) concentrations in drinking water, and the same problem occurs in many agricultural areas in the rest of the world. Enhanced nitrogen availability through nitrate (NO_3^-) and ammonium (NH_4^+) released to the environment from human activities and deposited in the soil by rainwater, a process known as atmospheric nitrogen deposition, is altering ecosystems throughout the world.

On a brighter note, plants are the traditional means of recycling animal wastes and are proving useful for removing deleterious materials, including heavy metals, from toxic-waste dumps. Because of the complex nature of plant–soil–atmosphere relationships, studies of mineral nutrition involve atmospheric chemists, soil scientists, hydrologists, microbiologists, and ecologists as well as plant physiologists.

In this chapter we discuss the nutritional needs of plants, the symptoms of specific nutritional deficiencies, and the use of fertilizers to ensure proper plant nutrition. Then we examine how soil structure (the arrangement of solid, liquid, and gaseous components) and root morphology influence the transfer of inorganic nutrients from the environment into a plant. Finally, we introduce the topic of symbiotic mycorrhizal associations, which play key roles in nutrient acquisition in the majority of plants. Chapters 5 and 6 will address additional aspects of nutrient assimilation and solute transport, respectively.

Essential Nutrients, Deficiencies, and Plant Disorders

Only certain elements have been determined to be essential for plants. An **essential element** is defined as one that is an intrinsic component in the structure or metabolism of a plant or whose absence causes severe abnormalities in plant growth, development, or reproduction and may prevent a plant from completing its life cycle. If plants are given these essential elements, as well as water and energy from sunlight, they can synthesize all the compounds they need for normal growth. **Table 4.1** lists the elements that are considered to be essential for most, if not all, higher plants. The first three elements—hydrogen, carbon, and oxygen—are not considered mineral nutrients because they are obtained primarily from water or carbon dioxide.

Essential mineral elements are usually classified as *macronutrients* or *micronutrients* according to their relative concentrations in plant tissue. In some cases the differences in tissue concentration between macronutrients and micronutrients are not as large as those indicated in Table 4.1. For example, some plant tissues, such as the leaf mesophyll, contain almost as much iron or manganese as they do sulfur or magnesium. Often elements are present in concentrations greater than the plant's minimum requirements.

Some researchers have argued that a classification into macronutrients and micronutrients is difficult to justify physiologically. Konrad Mengel and Ernest Kirkby have proposed that the essential elements be classified instead according

essential element A chemical element that is an intrinsic component of the structure or metabolism of a plant. When the element is in limited supply, a plant suffers abnormal growth, development, or reproduction.

Table 4.1 Tissue concentrations of essential elements required by most plants

Element	Chemical symbol	Concentration in dry matter (% or ppm)[a]	Relative number of atoms with respect to molybdenum
Obtained from water or carbon dioxide			
Hydrogen	H	6	60,000,000
Carbon	C	45	40,000,000
Oxygen	O	45	30,000,000
Obtained from the soil			
Macronutrients			
Nitrogen	N	1.5	1,000,000
Potassium	K	1.0	250,000
Calcium	Ca	0.5	125,000
Magnesium	Mg	0.2	80,000
Phosphorus	P	0.2	60,000
Sulfur	S	0.1	30,000
Silicon	Si	0.1	30,000
Micronutrients			
Chlorine	Cl	100	3,000
Iron	Fe	100	2,000
Boron	B	20	2,000
Manganese	Mn	50	1,000
Sodium	Na	10	400
Zinc	Zn	20	300
Copper	Cu	6	100
Nickel	Ni	0.1	2
Molybdenum	Mo	0.1	1

Source: Epstein 1972, 1999.

[a]The values for the nonmineral elements (H, C, O) and the macronutrients are percentages. The values for micronutrients are expressed in parts per million.

to their biochemical role and physiological function. **Table 4.2** shows such a classification, in which plant nutrients have been divided into four basic groups:

1. Nitrogen and sulfur constitute the first group of essential elements. Plants assimilate these nutrients via biochemical reactions involving oxidation and reduction to form covalent bonds with carbon and create organic compounds (e.g., amino acids, nucleic acids, and proteins).

2. The second group is important in energy storage reactions or in maintaining structural integrity. Elements in this group are often present in plant tissues as phosphate, borate, and silicate esters in which the elemental group is covalently bound to an organic molecule (e.g., sugar phosphate).

3. The third group is present in plant tissue as either free ions dissolved in the plant water or ions electrostatically bound to substances such as the pectic acids present in the plant cell wall. Elements in this group have

Table 4.2   Classification of plant mineral nutrients according to biochemical function

Mineral nutrient	Functions
Group 1	**Nutrients that are part of carbon compounds**
N	Constituent of amino acids, amides, proteins, nucleic acids, nucleotides, coenzymes, hexosamines, etc.
S	Component of cysteine, cystine, methionine. Constituent of lipoic acid, coenzyme A, thiamine pyrophosphate, glutathione, and biotin.
Group 2	**Nutrients that are important in energy storage or structural integrity**
P	Component of sugar phosphates, nucleic acids, nucleotides, coenzymes, phospholipids, phytic acid, etc. Has a key role in reactions that involve ATP.
Si	Deposited as amorphous silica in cell walls. Contributes to cell wall mechanical properties, including rigidity and elasticity.
B	Complexes with mannitol, mannan, polymannuronic acid, and other constituents of cell walls. Involved in cell elongation and nucleic acid metabolism.
Group 3	**Nutrients that remain in ionic form**
K	Required as a cofactor for more than 40 enzymes. Principal cation in establishing cell turgor and maintaining cell electroneutrality.
Ca	Constituent of the middle lamella of cell walls. Required as a cofactor by some enzymes involved in the hydrolysis of ATP and phospholipids. Acts as a second messenger in metabolic regulation.
Mg	Required by many enzymes involved in phosphate transfer. Constituent of the chlorophyll molecule.
Cl	Required for the photosynthetic reactions involved in O_2 evolution.
Zn	Constituent of alcohol dehydrogenase, glutamic dehydrogenase, carbonic anhydrase, etc.
Na	Involved with the regeneration of phospho*enol*pyruvate in C_4 and CAM plants. Substitutes for potassium in some functions.
Group 4	**Nutrients that are involved in redox reactions**
Fe	Constituent of cytochromes and nonheme iron proteins involved in photosynthesis, N_2 fixation, and respiration.
Mn	Required for activity of some dehydrogenases, decarboxylases, kinases, oxidases, and peroxidases. Involved with other cation-activated enzymes and photosynthetic O_2 evolution.
Cu	Component of ascorbic acid oxidase, tyrosinase, monoamine oxidase, uricase, cytochrome oxidase, phenolase, laccase, and plastocyanin.
Ni	Constituent of urease. In N_2-fixing bacteria, constituent of hydrogenases.
Mo	Constituent of nitrogenase, nitrate reductase, and xanthine dehydrogenase.

Source: After Evans and Sorger 1966 and Mengel and Kirkby 2001.

important roles as enzyme cofactors, in regulating osmotic potentials, and in controlling membrane permeability.

4. The fourth group, comprising metals such as iron, has important roles in reactions involving electron transfer.

Please keep in mind that this classification is somewhat arbitrary because many elements serve several functional roles. For example, manganese is listed in group 4 as a metal involved in several key electron transfer reactions, yet it is a mineral element that remains in ionic form, which would place it in group 3.

Some naturally occurring elements, such as aluminum, selenium, and cobalt, are not essential elements yet can also accumulate in plant tissues. Aluminum, for example, is not considered to be an essential element, although plants commonly contain from 0.1 to 500 μg aluminum per gram dry matter, and the addition of low amounts of aluminum to a nutrient solution may stimulate plant growth. Many species in the genera *Astragalus*, *Xylorhiza*, and *Stanleya* accumulate selenium, although plants have not been shown to have a specific requirement for this element. Cobalt is part of cobalamin (vitamin B_{12} and its derivatives), a component of several enzymes in nitrogen-fixing microorganisms; thus, cobalt deficiency blocks the development and function of nitrogen-fixing nodules, but plants that are not fixing nitrogen do not require cobalt. Crop plants normally contain only relatively small amounts of such nonessential elements.

The following sections describe the methods used to examine the roles of nutrient elements in plants.

Special techniques are used in nutritional studies

To demonstrate that an element is essential requires that plants be grown under experimental conditions in which only the element under investigation is absent. Such conditions are extremely difficult to achieve with plants grown in a complex medium such as soil. In the nineteenth century, several researchers, including Nicolas-Théodore de Saussure, Julius von Sachs, Jean-Baptiste-Joseph-Dieudonné Boussingault, and Wilhelm Knop, approached this problem by growing plants with their roots immersed in a **nutrient solution** containing only inorganic salts. Their demonstration that plants could grow normally with no soil or organic matter proved unequivocally that plants can fulfill all their needs from only mineral nutrient elements, water, air (CO_2), and sunlight.

The technique of growing plants with their roots immersed in a nutrient solution without soil is called **solution culture** or **hydroponics**. Successful hydroponic culture (**Figure 4.2A**) requires a large volume of nutrient solution or frequent adjustment of the nutrient solution to prevent nutrient uptake by roots from producing large changes in the mineral nutrient concentrations and pH of the solution. A sufficient supply of oxygen to the root system is also critical and may be achieved by vigorous bubbling of air through the solution.

Hydroponics is used in the commercial production of many greenhouse and indoor crops, such as tomato (*Solanum lycopersicum*), cucumber (*Cucumis sativus*), and hemp (*Cannabis sativa*). In one form of commercial hydroponic culture, plants are grown in a supporting material such as sand, gravel, vermiculite, rockwool, polyurethane foams, or expanded clay (i.e., kitty litter). Nutrient solutions are then flushed through the supporting material, and old solutions are removed by leaching. In another form of hydroponic culture, plant roots lie on the surface of a trough, and nutrient solutions flow in a thin layer along the trough over the roots. This **nutrient film growth system** ensures that the roots receive an ample supply of oxygen (**Figure 4.2B**).

Another technique, which has sometimes been heralded as the medium of the future for scientific investigations, is to grow the plants in **aeroponics**. In this technique plants are grown with their roots suspended in air while being sprayed continuously with a nutrient solution (**Figure 4.2C**). This approach provides easy manipulation of the gaseous environment around the roots, but it requires higher concentrations of nutrients than hydroponic culture does to sustain rapid plant growth. For this reason and other technical difficulties, the use of aeroponics is not widespread.

An ebb-and-flow system (**Figure 4.2D**) is yet another approach to solution culture. In such systems, the nutrient solution periodically rises to immerse plant roots and then recedes, exposing the roots to a moist atmosphere. Like aeroponics, ebb-and-flow systems require higher concentrations of nutrients than do other hydroponic or nutrient film systems.

Nutrient solutions can sustain rapid plant growth

Over the years, many formulations have been used for nutrient solutions. Early formulations developed by Knop in Germany included only KNO_3, $Ca(NO_3)_2$, KH_2PO_4, $MgSO_4$, and an iron salt. At the time, this nutrient solution was be-

nutrient solution   A solution containing only inorganic salts that supports the growth of plants in sunlight without soil or organic matter.

solution culture   A technique for growing plants with their roots immersed in nutrient solution without soil. Also called hydroponics.

nutrient film growth system A form of hydroponic culture in which the plant roots lie on the surface of a trough, and the nutrient solution flows over the roots in a thin layer along the trough.

aeroponics   The technique by which plants are grown with their roots suspended in air while being sprayed continuously with a nutrient solution.

(A) Hydroponic growth system

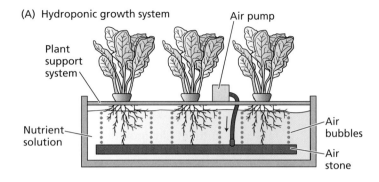

(B) Nutrient film growth system

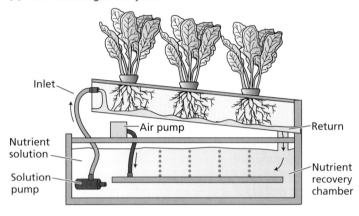

(C) Aeroponic growth system

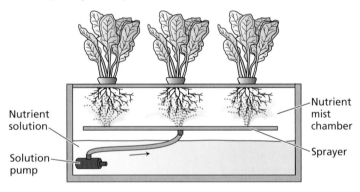

(D) Ebb-and-flow system

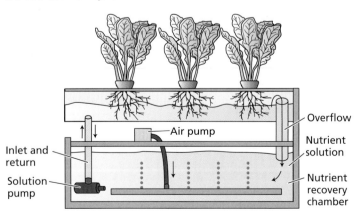

Figure 4.2 Various types of solution culture systems. (A) In a standard hydroponic culture, plants are suspended by the base of the stem over a tank containing a nutrient solution. The pumping of air through an air stone, a porous solid that generates a stream of small bubbles, keeps the solution fully saturated with oxygen. (B) In the nutrient film technique, a pump drives nutrient solution from a main reservoir along the bottom of a tilted tank, and down a return tube back to the reservoir. (C) In one type of aeroponics, a high-pressure pump sprays nutrient solution on roots enclosed in a tank. (D) In an ebb-and-flow system, a pump periodically fills an upper chamber containing the plant roots with nutrient solution. When the pump is turned off, the solution drains back through the pump into a main reservoir. (After Epstein and Bloom 2005.)

lieved to contain all the minerals required by plants, but these experiments were carried out with chemicals that were contaminated with other elements that are now known to be essential (such as boron or molybdenum). **Table 4.3** shows a more modern formulation for a nutrient solution. This formulation is called a modified **Hoagland solution**, named after Dennis R. Hoagland, a researcher who was prominent in the development of modern mineral nutrition research in the United States.

A modified Hoagland solution contains all known mineral elements needed for rapid plant growth. The concentrations of these elements are set at the highest possible concentrations without producing toxicity symptoms or salinity stress, and thus may be several orders of magnitude higher than those found in the soil around plant roots. For example, whereas phosphorus is present in the soil solution at concentrations normally less than 0.06 µg g^{-1} or 2 µM, here it is offered at 62 µg g^{-1} or 2 mM. Such high initial concentrations permit plants to be grown in a medium for extended periods without replenishment of the nutrient, but they may injure young plants. Therefore, many researchers dilute their nutrient solutions severalfold and replenish them frequently to minimize fluctuations of nutrient concentration in the medium and in plant tissue.

Another important property of this modified Hoagland formulation is that nitrogen is supplied as both ammonium (NH_4^+) and nitrate (NO_3^-). Supplying nitrogen in a balanced mixture of cations (positively charged ions) and anions (negatively charged ions) tends to decrease the rapid rise in the pH of the medium that is commonly observed when the nitrogen is supplied solely as nitrate anion. Even when the pH of the medium is kept neutral, most plants grow better if they have access to both NH_4^+ and NO_3^- because absorption and assimilation of the two nitrogen forms promote inorganic cation–anion balance in the plant.

Table 4.3 Composition of a modified Hoagland nutrient solution for growing plants

Compound	Molecular weight	Concentration of stock solution	Concentration of stock solution	Volume of stock solution per liter of final solution	Element	Final concentration of element	
	g mol^{-1}	mM	g L^{-1}	mL		μM	ppm
Macronutrients							
KNO$_3$	101.10	1,000	101.10	6.0	N	16,000	224
Ca(NO$_3$)$_2$·4H$_2$O	236.16	1,000	236.16	4.0	K	6,000	235
NH$_4$H$_2$PO$_4$	115.08	1,000	115.08	2.0	Ca	4,000	160
MgSO$_4$·7H$_2$O	246.48	1,000	246.49	1.0	P	2,000	62
					S	1,000	32
					Mg	1,000	24
Micronutrients							
KCl	74.55	25	1.864	⎤	Cl	50	1.77
H$_3$BO$_3$	61.83	12.5	0.773	⎟	B	25	0.27
MnSO$_4$·H$_2$O	169.01	1.0	0.169	⎬ 2.0	Mn	2.0	0.11
ZnSO$_4$·7H$_2$O	287.54	1.0	0.288	⎟	Zn	2.0	0.13
CuSO$_4$·5H$_2$O	249.68	0.25	0.062	⎟	Cu	0.5	0.03
H$_2$MoO$_4$ (85% MoO$_3$)	161.97	0.25	0.040	⎦	Mo	0.5	0.05
NaFeDTPA	468.20	64	30.0	0.3–1.0	Fe	16.1– 53.7	1.00– 3.00
Optional[a]							
NiSO$_4$·6H$_2$O	262.86	0.25	0.066	2.0	Ni	0.5	0.03
Na$_2$SiO$_3$·9H$_2$O	284.20	1,000	284.20	1.0	Si	1,000	28

Source: After Epstein and Bloom 2005.

Note: The macronutrients are added separately from stock solutions to prevent precipitation during preparation of the nutrient solution. A combined stock solution is made up containing all micronutrients except iron. Iron is added as sodium ferric diethylenetriaminepentaacetate (NaFeDTPA, trade name Ciba-Geigy Sequestrene 330 Fe; see Figure 4.3); some plants, such as maize, require the higher concentration of iron shown in the table.

[a]Nickel is usually present as a contaminant of the other chemicals, so it may not need to be added explicitly. Silicon, if included, should be added first and the pH adjusted with HCl to prevent precipitation of the other nutrients.

A significant problem with nutrient solutions is maintaining the availability of iron. When supplied as an inorganic salt such as FeSO$_4$ or Fe(NO$_3$)$_2$, iron can precipitate out of solution as iron hydroxide, particularly under alkaline conditions. If phosphate salts are present, insoluble iron phosphate will also form. Precipitation of the iron out of solution makes it physically unavailable to plants, unless iron salts are added at frequent intervals. Earlier researchers solved this problem by adding iron together with citric acid or tartaric acid. Compounds such as these are called **chelators** because they form soluble complexes with cations, such as iron and calcium ions, in which the cation is held by ionic forces rather than by covalent bonds. Chelated cations thus remain physically available to plants.

More modern nutrient solutions use the chemicals ethylenediaminetetraacetic acid (EDTA), diethylenetriaminepentaacetic acid (DTPA, or pentetic acid), or ethylenediamine-N,N'-bis(*o*-hydroxyphenylacetic) acid (*o,o*EDDHA) as chelating agents. **Figure 4.3** shows the structure of DTPA. The fate of the chelation complex during iron uptake by the root cells is not clear; iron may be released from the

Hoagland solution A type of nutrient solution for plant growth, originally formulated by Dennis R. Hoagland.

chelator A carbon compound (e.g., malic acid or citric acid) that can form a soluble noncovalent complex with certain cations, thereby facilitating their uptake.

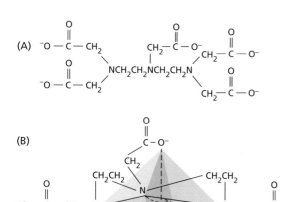

(A)

(B)

Figure 4.3   Chelator and chelated cation. Chemical structure of
the chelator diethylenetriaminepentaacetic acid (DTPA) by itself
(A) and chelated to Fe^{3+} (B). Iron binds to DTPA through interactions
with three nitrogen atoms and the three ionized oxygen atoms
of the carboxylate groups. The resulting ring structure clamps the
metallic ion and effectively neutralizes its reactivity in solution.
During the uptake of iron at the root surface, Fe^{3+} appears to be
reduced to Fe^{2+}, which is released from the DTPA–iron complex. The
chelator can then associate with other available Fe^{3+}. (After Sievers
and Bailar 1962.)

chelator when it is reduced from ferric iron (Fe^{3+}) to ferrous iron
(Fe^{2+}) at the root surface. The chelator may then diffuse back
into the nutrient (or soil) solution and associate with another
Fe^{3+} or other metal ion.

After uptake into the root, iron is kept soluble by chelation with
organic compounds present in plant cells. Citric acid may play
a major role as such an organic iron chelator, and long-distance
transport of iron in the xylem appears to involve an iron–citric
acid complex.

Mineral deficiencies disrupt plant metabolism and function

An inadequate supply of an essential element results in a nutritional disorder manifested by characteristic deficiency symptoms. In hydroponic culture,
withholding of an essential element can be readily correlated with a given set
of symptoms. For example, a particular deficiency might elicit a specific pattern
of leaf discoloration. Diagnosis of soil-grown plants can be more complex for
the following reasons:

- Deficiencies of several elements may occur simultaneously in different
  plant tissues.

- Deficiencies or excessive amounts of one element may induce
  deficiencies or excessive accumulations of another element.

- Some virus-induced plant diseases may produce symptoms similar to
  those of nutrient deficiencies.

Nutrient deficiency symptoms in a plant are the expression of metabolic disorders
resulting from the insufficient supply of an essential element. These disorders
are related to the roles played by essential elements in normal plant metabolism
and function (listed in Table 4.2).

Although each essential element participates in many different metabolic
reactions, some general statements about the functions of essential elements
in plant metabolism are possible. In general, the essential elements function in
plant structure, metabolism, and cellular osmoregulation. More specific roles
may be related to the ability of divalent cations such as Ca^{2+} or Mg^{2+} to modify
the permeability of plant membranes. In addition, research continues to reveal
specific roles for these elements in plant metabolism; for example, calcium ions
act as a signal to regulate key enzymes in the cytosol. Thus, most essential
elements have multiple roles in plant metabolism.

An important clue in relating an acute deficiency symptom to a particular
essential element is the extent to which an element can be recycled from
older to younger leaves. Some elements, such as nitrogen, phosphorus, and
potassium, can readily move from leaf to leaf; others, such as boron, iron, and
calcium, are relatively immobile in most plant species (**Table 4.4**). If an essen-

tial element is mobile, deficiency symptoms tend to appear first in older leaves. Conversely, deficiencies of immobile essential elements become evident first in younger leaves. Although the precise mechanisms of nutrient mobilization are not well understood, plant hormones such as cytokinins appear to be involved (see Chapter 12). Below we describe the particular deficiency symptoms and functional roles of the essential elements as they are grouped in Table 4.2. Please keep in mind that many symptoms are highly dependent on plant species.

GROUP 1: DEFICIENCIES IN MINERAL NUTRIENTS THAT ARE PART OF CARBON COMPOUNDS This group consists of nitrogen and sulfur. Nitrogen availability in soils limits plant productivity in most natural and agricultural ecosystems. By contrast, soils generally contain sulfur in excess. Despite this difference, nitrogen and sulfur are similar chemically in that their oxidation–reduction states range widely (see Chapter 5). Some of the most energy-intensive reactions in life convert highly oxidized inorganic forms, such as nitrate and sulfate, that roots absorb from the soil into highly reduced organic compounds, such as amino acids, within plants.

NITROGEN Nitrogen is the mineral element that plants require in the greatest amounts (see Table 4.1). It serves as a constituent of many plant cell components, including chlorophyll, amino acids, and nucleic acids. Therefore, nitrogen deficiency rapidly inhibits plant growth. If such a deficiency persists, most species show leaf **chlorosis** (yellowing of the leaves), especially in the older leaves near the base of the plant. Under severe nitrogen deficiency, these leaves become completely yellow (or tan) and fall off the plant. Younger leaves may not show these symptoms initially because nitrogen can be mobilized from older leaves. Thus, a nitrogen-deficient plant may have light green upper leaves and yellow or tan lower leaves.

When nitrogen deficiency develops slowly, plants may have markedly slender and often woody stems. This woodiness may be due to a buildup of excess carbohydrates that cannot be used in the synthesis of amino acids or other nitrogen-containing compounds. Carbohydrates not used in nitrogen metabolism may also be used in anthocyanin synthesis, leading to accumulation of that pigment. This condition is revealed as a purple coloration in leaves, petioles, and stems of nitrogen-deficient plants of some species, such as tomato and certain varieties of maize (corn; *Zea mays*).

SULFUR Sulfur is found in certain amino acids (i.e., cysteine and methionine) and is a constituent of several coenzymes and vitamins, such as coenzyme A, *S*-adenosylmethionine, biotin, vitamin B1, and pantothenic acid, which are essential for metabolism.

Many of the symptoms of sulfur deficiency are similar to those of nitrogen deficiency, including leaf chlorosis, stunting of growth, and anthocyanin accumulation. This similarity is not surprising, since sulfur and nitrogen are both constituents of proteins. The chlorosis caused by sulfur deficiency, however, generally arises initially in young and mature leaves, rather than in old leaves as in nitrogen deficiency, because sulfur, unlike nitrogen, is not easily remobilized to the younger leaves in most species. Nonetheless, in some plant species sulfur chlorosis may occur simultaneously in all leaves, or even initially in older leaves.

Table 4.4 Mineral elements classified on the basis of their mobility within a plant and their tendency to retranslocate during deficiencies

Mobile	Immobile
Nitrogen	Calcium
Potassium	Sulfur
Magnesium	Iron
Phosphorus	Boron
Chlorine	Copper
Sodium	
Zinc	
Molybdenum	

Note: Elements are listed in the order of their abundance in the plant.

chlorosis The yellowing of plant leaves such as occurs as a result of mineral deficiency. The leaves affected and the parts of the leaves that yellow can be diagnostic for the type of deficiency.

GROUP 2: DEFICIENCIES IN MINERAL NUTRIENTS THAT ARE IMPORTANT IN ENERGY STORAGE OR STRUCTURAL INTEGRITY This group consists of phosphorus, silicon, and boron. Phosphorus and silicon are found at concentrations in plant tissue that warrant their classification as macronutrients, whereas boron is much less abundant and is considered a micronutrient. These elements are usually present in plants as ester linkages between an inorganic acid group such a phosphate (PO_4^{3-}) and a carbon alcohol (i.e., X–O–C–R, where the element X is attached to a carbon-containing molecule C–R via an oxygen atom O).

PHOSPHORUS Phosphorus (as phosphate, PO_4^{3-}) is an integral component of important compounds in plant cells, including the sugar–phosphate intermediates of respiration and photosynthesis as well as the phospholipids that make up plant membranes. It is also a component of nucleotides used in plant energy metabolism (such as ATP) and in DNA and RNA. Characteristic symptoms of phosphorus deficiency include stunted growth of the entire plant and a dark green coloration of the leaves, which may be malformed and contain small areas of dead tissue called **necrotic spots**.

As in nitrogen deficiency, some species may produce excess anthocyanins under phosphorus deficiency, giving the leaves a slight purple coloration. Unlike in nitrogen deficiency, the purple coloration of phosphorus deficiency is not associated with chlorosis. In fact, the leaves may be a dark greenish purple. Additional symptoms of phosphorus deficiency include the production of slender (but not woody) stems and the death of older leaves. Maturation of the plant may also be delayed.

SILICON Only members of the family Equisetaceae—called *scouring rushes* because at one time their ash, rich in gritty silica, was used to scour pots—require silicon to complete their life cycle. Nonetheless, many other species accumulate substantial amounts of silicon in their tissues and show enhanced growth, fertility, and stress resistance when supplied with adequate amounts of silicon.

Plants deficient in silicon are more susceptible to lodging (falling over) and fungal infection. Silicon is deposited primarily in the endoplasmic reticulum, cell walls, and intercellular spaces as hydrated, amorphous silica ($SiO_2 \cdot nH_2O$). It also forms complexes with polyphenols and thus serves as a complement to lignin in the reinforcement of cell walls. In addition, silicon can lessen the toxicity of many metals, including aluminum and manganese.

BORON Although many of the functions of boron in plant metabolism are still unclear, evidence shows that it cross-links RG II (rhamnogalacturonan II, a small pectic polysaccharide) in the cell wall and suggests that it plays roles in cell elongation, nucleic acid synthesis, hormone responses, membrane function, and cell cycle regulation. Boron-deficient plants may exhibit a wide variety of symptoms, depending on the species and the age of the plant.

A characteristic symptom is black necrosis of young leaves and terminal buds. The necrosis of the young leaves occurs primarily at the base of the leaf blade. Stems may be unusually stiff and brittle. Apical dominance may also be lost, causing the plant to become highly branched; however, the terminal apices of the branches soon become necrotic because of inhibition of cell differentiation. Structures such as the fruits, fleshy roots, and tubers may exhibit necrosis or abnormalities related to the breakdown of internal tissues.

GROUP 3: DEFICIENCIES IN MINERAL NUTRIENTS THAT REMAIN IN IONIC FORM This group includes some of the most familiar mineral elements: the macronutrients potassium, calcium, and magnesium and the micronutrients chlorine, zinc, and sodium. These elements may be found as ions in solution in the cytosol or vacuoles, or they may be bound electrostatically or as ligands to larger, carbon-containing compounds.

POTASSIUM   Potassium, present in plants as the cation K^+, plays an important role in regulating the osmotic potential of plant cells (see Chapters 2, 3, and 6). It also activates many enzymes involved in respiration and photosynthesis.

The first observable symptom of potassium deficiency is mottled or marginal chlorosis, which then develops into necrosis primarily at the leaf tips, at the margins, and between veins. In many monocots, these necrotic lesions may initially form at the leaf tips and margins and then extend toward the leaf base. Because potassium can be mobilized to the younger leaves, these symptoms appear initially on the more mature leaves toward the base of the plant. The leaves may also curl and crinkle. The stems of potassium-deficient plants may be slender and weak, with abnormally short internodal regions. In potassium-deficient maize, the roots may have an increased susceptibility to root-rotting fungi present in the soil, and this susceptibility, together with effects on the stem, results in an increased tendency for the plant to be easily lodged.

CALCIUM   Calcium ions (Ca^{2+}) have two distinct roles in plants: (1) a structural/apoplastic role whereby Ca^{2+} binds to acidic groups of membrane lipids (phospho- and sulfolipids) and cross-links pectins, particularly in the middle lamellae that separate newly divided cells; and (2) a signaling role whereby Ca^{2+} acts as a second messenger that initiates plant responses to environmental stimuli. In its function as a second messenger, Ca^{2+} may bind to **calmodulin**, a protein found in the cytosol of plant cells. The calmodulin–Ca^{2+} complex then binds to several different types of proteins, including kinases, phosphatases, second messenger signaling proteins, and cytoskeletal proteins, and thereby regulates many cellular processes ranging from transcription control and cell survival to release of chemical signals (see Chapter 12).

Characteristic symptoms of calcium deficiency include necrosis of young meristematic regions such as the tips of roots or young leaves, where cell division and cell wall formation are most rapid. Necrosis in slowly growing plants may be preceded by a general chlorosis and downward hooking of young leaves. Young leaves may also appear deformed. The root system of a calcium-deficient plant may appear brownish, short, and highly branched. Severe stunting may result if the meristematic regions of the plant die prematurely.

MAGNESIUM   In plant cells, magnesium ions (Mg^{2+}) have a specific role in the activation of enzymes involved in respiration, photosynthesis, and the synthesis of DNA and RNA. Mg^{2+} is also part of the ring structure of the chlorophyll molecule (see Figure 7.6A). A characteristic symptom of magnesium deficiency is chlorosis between the leaf veins, occurring first in older leaves because of the high mobility of this cation. This pattern of chlorosis occurs because the chlorophyll in the vascular bundles remains unaffected longer than that in the cells between the bundles. If the deficiency is extensive, the leaves may become yellow or white. An additional symptom of magnesium deficiency may be senescence and premature leaf abscission.

CHLORINE   The element chlorine is found in plants as the chloride ion (Cl^-). It is required for the water-splitting reaction of photosynthesis through which oxygen is produced (see Chapter 7). In addition, chlorine may be required for cell division in leaves and roots. Plants deficient in chlorine develop wilting of the leaf tips followed by general leaf chlorosis and necrosis. The leaves may also exhibit reduced growth. Eventually, the leaves may take on a bronzelike

calmodulin   A Ca^{2+}-binding protein that regulates many cellular processes in a Ca^{2+}-dependent manner.

color ("bronzing"). Roots of chlorine-deficient plants may appear stunted and thickened near the root tips.

Chloride ions are highly soluble and are generally available in soils because seawater is swept into the air by wind and delivered to soil when it rains. Therefore, chlorine deficiency is only rarely observed in plants grown in native or agricultural habitats. Most plants absorb chlorine at concentrations much higher than those required for normal functioning.

ZINC   Many enzymes require zinc ions (Zn^{2+}) for their activity, and zinc may be required for chlorophyll biosynthesis in some plants. Zinc deficiency is characterized by a reduction in internodal growth, and as a result plants display a rosette habit of growth in which the leaves form a circular cluster radiating at or close to the ground. The leaves may also be small and distorted, with leaf margins having a puckered appearance. These symptoms may result from loss of the ability to produce sufficient amounts of the auxin indole-3-acetic acid (IAA) (see Chapter 12). In some species (e.g., maize, sorghum, and beans), older leaves may show chlorosis between the leaf veins and then develop white necrotic spots. This chlorosis may be an expression of a zinc requirement for chlorophyll biosynthesis.

SODIUM   Species using the C_4 and crassulacean acid metabolism (CAM) pathways of carbon fixation (see Chapter 8) may require sodium ions (Na^+). In these plants, Na^+ appears vital for regenerating phospho*enol*pyruvate, the substrate for the first carboxylation in the C_4 and CAM pathways. Under sodium deficiency, these plants exhibit chlorosis and necrosis, or even fail to form flowers. Many C_3 species also benefit from exposure to low concentrations of Na^+. Sodium ions stimulate growth through enhanced cell expansion and can partly substitute for potassium ions as an osmotically active solute.

GROUP 4: DEFICIENCIES IN MINERAL NUTRIENTS THAT ARE INVOLVED IN REDOX REACTIONS   This group of five micronutrients consists of the metals iron, manganese, copper, nickel, and molybdenum. All of these can undergo reversible oxidations and reductions (e.g., $Fe^{2+} \leftrightarrow Fe^{3+}$) and have important roles in electron transfer and energy transformation. They are usually found in association with larger molecules such as cytochromes, chlorophyll, and proteins (usually enzymes).

IRON   Iron has an important role as a component of enzymes involved in the transfer of electrons (redox reactions), such as cytochromes. In this role, it is reversibly oxidized from Fe^{2+} to Fe^{3+} during electron transfer.

As in magnesium deficiency, a characteristic symptom of iron deficiency is intervenous chlorosis. This symptom, however, appears initially on younger leaves because iron, unlike magnesium, cannot be readily mobilized from older leaves. Under conditions of extreme or prolonged deficiency, the veins may also become chlorotic, causing the whole leaf to turn white. The leaves become chlorotic because iron is required for the synthesis of some of the chlorophyll–protein complexes in the chloroplast. The low mobility of iron is probably due to its precipitation in the older leaves as insoluble oxides or phosphates. The precipitation of iron diminishes subsequent mobilization of the metal into the phloem for long-distance translocation.

MANGANESE   Manganese ions (Mn^{2+}) activate several enzymes in plant cells. In particular, decarboxylases and dehydrogenases involved in the tricarboxylic acid (Krebs) cycle are specifically activated by manganese ions. The best-defined function of Mn^{2+} is in the photosynthetic reaction through which oxygen (O_2)

is produced from water (see Chapter 7). The major symptom of manganese deficiency is intervenous chlorosis associated with the development of small necrotic spots. This chlorosis may occur on younger or older leaves, depending on plant species and growth rate.

COPPER Like iron, copper is associated with enzymes involved in redox reactions, through which it is reversibly oxidized from Cu^+ to Cu^{2+}. An example of such an enzyme is plastocyanin, which is involved in electron transfer during the light reactions of photosynthesis. The initial symptom of copper deficiency in many plant species is the production of dark green leaves, which may contain necrotic spots. The necrotic spots appear first at the tips of young leaves and then extend toward the leaf base along the margins. The leaves may also be twisted or malformed. Cereal plants exhibit a white leaf chlorosis and necrosis with rolled tips. Under extreme copper deficiency, leaves may drop prematurely and flowers may be sterile.

NICKEL Urease is the only known nickel-containing (Ni^{2+}) enzyme in higher plants, although nitrogen-fixing microorganisms require nickel (Ni^+ through Ni^{4+}) for the enzyme that reprocesses some of the hydrogen gas generated during fixation (hydrogen uptake hydrogenase) (see Chapter 5). Nickel-deficient plants accumulate urea in their leaves and consequently show leaf tip necrosis. Nickel deficiency in the field has been found in only one crop, pecan trees in the southeastern United States, because plants require only minuscule amounts of nickel (see Table 4.1).

MOLYBDENUM Molybdenum ions (Mo^{4+} through Mo^{6+}) are components of several enzymes, including nitrate reductase, nitrogenase, xanthine dehydrogenase, aldehyde oxidase, and sulphite oxidase. Nitrate reductase catalyzes the reduction of nitrate to nitrite during its assimilation by the plant cell; nitrogenase converts nitrogen gas to ammonia in nitrogen-fixing microorganisms (see Chapter 5). The first indication of a molybdenum deficiency is general chlorosis between veins and necrosis of older leaves. In some plants, such as cauliflower or broccoli, the leaves may not become necrotic but instead may appear twisted and subsequently die (whiptail disease). Flower formation may be prevented, or the flowers may abscise prematurely.

Because molybdenum is involved with both nitrate assimilation and nitrogen fixation, a molybdenum deficiency may bring about a nitrogen deficiency if the nitrogen source is primarily nitrate or if the plant depends on symbiotic nitrogen fixation. Although plants require only tiny amounts of molybdenum (see Table 4.1), some soils (for example, acidic soils in Australia) supply inadequate concentrations. Small additions of molybdenum to such soils can greatly enhance crop or forage growth at negligible cost.

Analysis of plant tissues reveals mineral deficiencies

Requirements for mineral elements change as a plant grows and develops. In crop plants, nutrient concentrations at certain stages of growth influence the yield of the economically important tissues (tuber, grain, etc.). To optimize yields, farmers use analyses of nutrient concentrations in soil and in plant tissue to determine fertilizer schedules.

Soil analysis is the chemical determination of the nutrient content in a soil sample from the root zone. As we will discuss later in the chapter, both the chemistry and the biology of soils are complex, and the results of soil analyses vary with sampling methods, storage conditions for the samples, and nutrient extraction techniques. Perhaps more important is that a particular soil analysis reflects the amount of nutrients *potentially* available to the plant roots from the

soil analysis The chemical determination of the nutrient content in a soil sample, typically from the root zone.

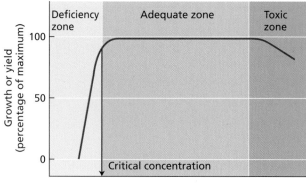

Figure 4.4 Relationship between yield (or growth) and the nutrient concentration of the plant tissue defines zones of deficiency, adequacy, and toxicity. Yield or growth may be expressed in terms of shoot dry weight or height. To obtain data of this type, plants are grown under conditions in which the concentration of one essential nutrient is varied while all others are in adequate supply. The effect of varying the concentration of this nutrient during plant growth is reflected in the growth or yield. The critical tissue concentration for that nutrient is the concentration below which yield or growth is reduced.

plant tissue analysis   In the context of mineral nutrition, the analysis of the concentrations of mineral nutrients in a plant sample.

deficiency zone   In plant tissue, the range of concentrations of a mineral nutrient below the critical concentration, the highest concentration where reduced plant growth or yield is observed

adequate zone   In plant tissue, the range of concentrations of a mineral nutrient beyond which further addition of the nutrient no longer increases growth or yield.

critical concentration (of a nutrient)   In plant tissue, the minimum concentration of a mineral nutrient that is correlated with maximum growth or yield.

toxic zone   In plant tissue, the range of concentrations of a mineral nutrient in excess of the adequate zone and where growth or yield declines.

soil, but soil analysis does not tell us how much of a particular mineral nutrient the plant actually needs or is able to absorb. This additional information is best determined by plant tissue analysis.

Proper use of **plant tissue analysis** requires an understanding of the relationship between plant growth (or yield) and the concentration of a nutrient in plant tissue samples. Bear in mind that the tissue concentration of a nutrient depends on the balance between nutrient absorption and dilution of the amount of nutrient through growth. **Figure 4.4** identifies three zones (deficiency, adequate, and toxic) in the response of growth to increasing tissue concentrations of a nutrient. When the nutrient concentration in a tissue sample is low, growth is reduced. In this **deficiency zone** of the curve, an increase in nutrient availability and absorption is directly related to an increase in growth or yield. As nutrient availability and absorption continue to increase, a point is reached at which further addition of the nutrient is no longer related to increases in growth or yield, but is reflected only in increased tissue concentrations. This region of the curve is called the **adequate zone**.

The point of transition between the deficiency and adequate zones of the curve reveals the **critical concentration** of the nutrient (see Figure 4.4), which may be defined as the minimum tissue concentration of the nutrient that is correlated with maximum growth or yield. As the nutrient concentration of the tissue increases beyond the adequate zone, growth or yield declines because of toxicity. This region of the curve is the **toxic zone**.

To evaluate the relationship between growth and tissue nutrient concentration, researchers grow plants in soil or a nutrient solution in which all the nutrients are present in adequate amounts except the nutrient under consideration. At the start of the experiment, the limiting nutrient is added in increasing concentrations to different sets of plants, and the concentrations of the nutrient in specific tissues are correlated with a particular measure of growth or yield. Several curves are established for each element, one for each tissue and tissue age.

Because agricultural soils are often limited in the elements nitrogen, phosphorus, and potassium (NPK), many farmers routinely take into account, at a minimum, growth or yield responses for these elements. If a nutrient deficiency is suspected, steps are taken to correct the deficiency before it reduces growth or yield. Plant analysis has proved useful in establishing fertilizer schedules that sustain yields and ensure the food quality of many crops.

Treating Nutritional Deficiencies

Many traditional and subsistence farming practices promote the recycling of mineral elements. Crop plants absorb nutrients from the soil, humans and animals consume locally grown crops, and crop residues and manure from humans and animals return the nutrients to the soil. The main losses of nutrients from such agricultural systems ensue from leaching that carries dissolved ions, especially nitrate, away with drainage water. In acidic soils, leaching of nutrients other than nitrate may be decreased by the addition of lime—a mix of CaO, $CaCO_3$, and $Ca(OH)_2$—to make the soil more alkaline because many mineral elements form less soluble compounds when the pH is higher than 6 (**Figure 4.5**). This decrease in leaching, however, may be gained at the expense of decreased availability of some nutrients, especially iron.

Figure 4.5 Influence of soil pH on the availability of nutrient elements in organic soils. The thickness of the horizontal bars indicates the degree of nutrient availability to plant roots. All of these nutrients are available in the pH range of 5.5 to 6.5. (After Lucas and Davis 1961.)

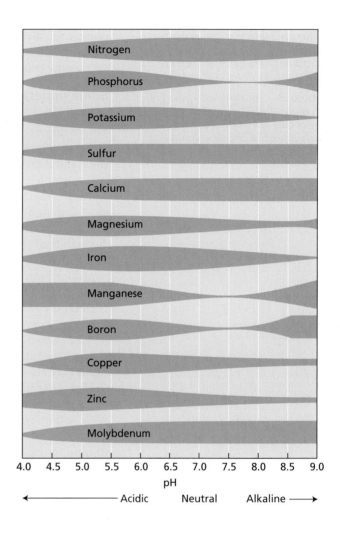

In the high-production agricultural systems of industrialized countries, a large proportion of crop biomass leaves the area of cultivation, and returning crop residues to the land where the crop was produced becomes difficult at best. This unidirectional removal of nutrients from agricultural soils makes it important to restore the lost nutrients to these soil through the addition of fertilizers.

Crop yields can be improved by the addition of fertilizers

Most **chemical fertilizers** contain inorganic salts of the macronutrients nitrogen, phosphorus, and potassium (see Table 4.1). Fertilizers that contain only one of these three nutrients are termed *straight fertilizers*. Some examples of straight fertilizers are superphosphate, ammonium nitrate, and muriate of potash (potassium chloride). Fertilizers that contain two or more mineral nutrients are called *compound fertilizers* or *mixed fertilizers*, and the numbers on the package label, such as "10-14-10," refer to the percentages of N, P, and K, respectively, in the fertilizer.

With long-term agricultural production, consumption of micronutrients by crops can reach a point at which they too must be added to the soil as fertilizers. Adding micronutrients to the soil may also be necessary to correct a preexisting deficiency. For example, many acidic, sandy soils in humid regions are deficient in boron, copper, zinc, manganese, molybdenum, or iron and can benefit from nutrient supplementation.

Chemicals may also be applied to the soil to modify soil pH. As Figure 4.5 shows, soil pH affects the availability of all mineral nutrients. Addition of lime, as mentioned previously, can raise the pH of acidic soils; addition of elemental sulfur can lower the pH of alkaline soils. In the latter case, microorganisms absorb the sulfur and subsequently release sulfate and hydrogen ions that acidify the soil.

Organic fertilizers are those approved for organic agricultural practices. In contrast to chemical fertilizers, they originate from natural rock deposits such as sodium nitrate and rock phosphate (phosphorite) or from the residues of plant or animal life. Natural rock deposits are chemically inorganic, but are acceptable for use in organic agriculture. Plant and animal residues contain many nutrient elements in the form of organic compounds. Before crop plants can acquire the nutrient elements from these residues, the organic compounds must be broken down, usually by the action of soil microorganisms through a process called **mineralization**. Mineralization depends on many factors, including temperature, water and oxygen availability, pH, and the type and number of microorganisms present in the soil. As a consequence, rates of mineralization are highly variable, and nutrients from organic residues become available to plants over periods that range from days to months to years. This slow rate of mineralization hinders efficient organic fertilizer use, so farms that rely solely on organic fertilizers may require the addition of substantially more nitrogen or phosphorus and may suffer even higher nutrient losses than farms that use chemical fertilizers. Residues from

chemical fertilizers Fertilizers that provide nutrients in inorganic forms.

organic fertilizer A fertilizer that contains nutrient elements derived from natural sources without any synthetic additions.

mineralization The process of breaking down organic compounds, usually by soil microorganisms, and thereby releasing mineral nutrients in forms that can be assimilated by plants.

foliar application The application, and subsequent absorption, of some mineral nutrients or other chemical compounds to leaves as sprays.

symbiosis The close association of two organisms in a relationship that may or may not be mutually beneficial. Often applied to beneficial (mutualistic) relationships.

organic fertilizers do improve the physical structure of most soils, enhancing water retention during drought and increasing drainage in wet weather. In some developing countries, organic fertilizers are all that is available or affordable.

Some mineral nutrients can be absorbed by leaves

In addition to absorbing nutrients added to the soil as fertilizers, most plants can absorb mineral nutrients applied to their leaves as a spray, a process known as **foliar application**. In some cases this method has agronomic advantages over the application of nutrients to the soil. Foliar application can reduce the lag time between application and uptake by the plant, which could be important during a phase of rapid growth. It can also circumvent the problem of restricted uptake of a nutrient from the soil. For example, foliar application of mineral nutrients such as iron, manganese, and copper may be more efficient than application through the soil, where these ions are adsorbed on soil particles and hence are less available to the root system.

Nutrient uptake by leaves is most effective when the nutrient solution is applied to the leaf as a thin film. Production of a thin film often requires that the nutrient solutions be supplemented with surfactant chemicals, such as the detergent Tween 80 or newly developed organosilicon surfactants, that reduce surface tension. Nutrient movement into the plant seems to involve diffusion through the cuticle and uptake by leaf cells, although uptake through the stomatal pores may also occur.

For foliar nutrient application to be successful, damage to the leaves must be minimized. If foliar sprays are applied on a hot day, when evaporation is high, salts may accumulate on the leaf surface and cause burning or scorching. Spraying on cool days or in the evening helps alleviate this problem. Addition of lime to the spray diminishes the solubility of many nutrients and limits toxicity. Foliar application has proved economically successful mainly with tree crops and vines such as grapes, but it is also used with cereals. Nutrients applied to the leaves can save an orchard or vineyard when soil-applied nutrients would be too slow to correct a deficiency. In wheat (*Triticum aestivum*), nitrogen applied to the leaves during the later stages of growth enhances the protein content of seeds.

Soil, Roots, and Microbes

Soil is complex physically, chemically, and biologically. It is a heterogeneous mixture of substances distributed in solid, liquid, and gaseous phases (see Chapter 3). All of these phases interact with mineral elements. The inorganic particles of the solid phase provide a reservoir of potassium, phosphorus, calcium, magnesium, and iron. Also associated with this solid phase are organic compounds containing nitrogen, phosphorus, and sulfur, among other elements. The liquid phase of soil constitutes the soil solution that is held in pores between the soil particles. It contains dissolved mineral ions and serves as the medium for ion movement to the root surface. Gases such as oxygen, carbon dioxide, and nitrogen are dissolved in the soil solution, but roots exchange gases with soils predominantly through the air-filled pores between soil particles.

From a biological perspective, soil constitutes a diverse ecosystem in which plant roots and microorganisms interact. Many microorganisms play key roles in releasing (mineralizing) nutrients from organic sources, some of which are then made directly available to plants. Under some soil conditions, free-living microbes compete with the plants for these mineral nutrients. In contrast, some specialized microorganisms, including mycorrhizal fungi and nitrogen-fixing bacteria, can form alliances with plants for their mutual benefit (**symbioses**, singular *symbiosis*). In this section we discuss the importance of soil properties, root structure, and symbiotic mycorrhizal relationships to plant mineral nutrition. Chapter 5 will address the symbiotic relationships of plants with nitrogen-fixing bacteria.

Negatively charged soil particles affect the adsorption of mineral nutrients

Soil particles, both inorganic and organic, have predominantly negative charges on their surfaces. Many inorganic soil particles are crystal lattices that are tetrahedral arrangements of the cationic forms of aluminum (Al^{3+}) and silicon (Si^{4+}) bound to oxygen atoms, thus forming aluminates and silicates. When cations of lesser charge replace Al^{3+} and Si^{4+} in the crystal lattice, these inorganic soil particles become negatively charged.

Organic soil particles originate from dead plants, animals, and microorganisms that soil microorganisms have decomposed to various degrees. The negative surface charges of organic particles result from the dissociation of hydrogen ions from the carboxylic and phenolic acid groups present in this component of the soil. Most of the world's soils are composed of aggregates formed from organic and inorganic particles.

Soils are categorized by particle size:

- Gravel consists of particles larger than 2 mm.
- Coarse sand consists of particles between 0.2 and 2 mm.
- Fine sand consists of particles between 0.02 and 0.2 mm.
- Silt consists of particles between 0.002 and 0.02 mm.
- Clay consists of particles smaller than 0.002 mm (2 μm).

The silicate-containing clay materials are further divided into three major groups—kaolinite, illite, and montmorillonite—based on differences in their structure and physical properties (**Table 4.5**). The kaolinite group is generally found in well-weathered soils; the montmorillonite and illite groups are found in less-weathered soils.

Mineral cations such as ammonium (NH_4^+) and potassium (K^+) are adsorbed to the negative surface charges of inorganic and organic soil particles or adsorbed within the lattices formed by soil particles. This cation adsorption is an important factor in soil fertility. Mineral cations adsorbed on the surface of soil particles are not readily leached when the soil is infiltrated by water and thus provide a nutrient reserve available to plant roots. Mineral nutrients adsorbed in this way can be replaced by other cations in a process known as **cation exchange** (**Figure 4.6**). The degree to which a soil can adsorb and exchange ions is termed its *cation exchange capacity* (CEC) and is highly dependent on the soil type. A

Table 4.5 Comparison of properties of three major types of silicate clays found in the soil

Property	Type of clay		
	Montmorillonite	Illite	Kaolinite
Size (μm)	0.01–1.0	0.1–2.0	0.1–5.0
Shape	Irregular flakes	Irregular flakes	Hexagonal crystals
Cohesion	High	Medium	Low
Water-swelling capacity	High	Medium	Low
Cation exchange capacity (milliequivalents 100 g^{-1})	80–100	15–40	3–15

Source: After Brady 1974.

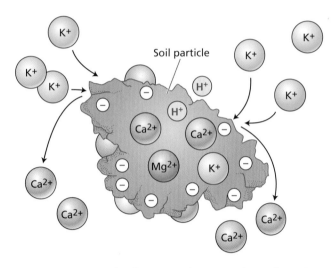

Figure 4.6 Principle of cation exchange on the surface of a soil particle. Cations are adsorbed on the surface of a soil particle because that surface is negatively charged. Addition of one cation, such as potassium (K^+), to the soil can displace other cations, such as calcium (Ca^{2+}), from the surface of the soil particle and make it available for uptake by roots.

soil with higher cation exchange capacity generally has a larger reserve of mineral nutrients.

Mineral anions such as nitrate (NO_3^-) and chloride (Cl^-) tend to be repelled by the negative charge on the surface of soil particles and remain dissolved in the soil solution. Thus, the anion exchange capacity of most agricultural soils is small compared with their cation exchange capacity. Nitrate in particular remains mobile in the soil solution, where it is susceptible to leaching by water moving through the soil.

Phosphate ions ($H_2PO_2^-$) may bind to soil particles containing aluminum or iron because the positively charged iron and aluminum ions (Fe^{2+}, Fe^{3+}, and Al^{3+}) are associated with hydroxyl ion (HO^-) groups that are exchanged for phosphate. Phosphate ions also react strongly with Ca^{2+}, Fe^{3+}, and Al^{3+} to form insoluble inorganic compounds. As a result, phosphate is frequently tightly bound at both low and high pH (see Figure 4.5), and its lack of mobility and availability in soil can limit plant growth. Formation of mycorrhizal symbioses (which we discuss later in this section) helps overcome this lack of mobility. Additionally, the roots of some plants, such as those of lupine (*Lupinus albus*) and members of the Proteaceae (e.g., *Macadamia, Banksia, Protea*), secrete large amounts of organic anions or protons into the soil that release phosphate from iron, aluminum, and calcium phosphates.

Sulfate (SO_4^{2-}) in the presence of Ca^{2+} forms gypsum ($CaSO_4$). Gypsum is only slightly soluble, but it releases sufficient sulfate to support plant growth. Most nonacidic soils contain substantial amounts of Ca^{2+}; consequently, sulfate mobility in these soils is low, and sulfate is not highly susceptible to leaching.

Soil pH affects nutrient availability, soil microbes, and root growth

Hydrogen ion concentration (pH) is an important property of soils because it affects the growth of plant roots and soil microorganisms. Root growth is generally favored in slightly acidic soils, at pH values between 5.5 and 6.5. Fungi generally predominate in acidic (pH below 7) soils; bacteria become more prevalent in alkaline (pH above 7) soils. Soil pH determines the availability of soil nutrients (see Figure 4.5). Acidity promotes the weathering of rocks that releases K^+, Mg^{2+}, Ca^{2+}, and Mn^{2+} and increases the solubility of carbonates, sulfates, and some phosphates. Increasing the solubility of nutrients enhances their availability to roots as concentrations increase in the soil solution.

Major factors that lower the soil pH are the decomposition of organic matter, ammonium assimilation by plants and microbes, and the amount of rainfall. Carbon dioxide is produced as a result of the decomposition of organic matter and equilibrates with soil water in the following reaction:

$$CO_2 + H_2O \leftrightarrow H^+ + HCO_3^-$$

This releases hydrogen ions (H^+), lowering the pH of the soil. Microbial decomposition of organic matter also produces ammonia/ammonium (NH_3/NH_4^+) and hydrogen sulfide (H_2S) that can be oxidized in the soil to form the strong acids nitric acid (HNO_3) and sulfuric acid (H_2SO_4), respectively. As plant roots absorb ammonium ions from the soil and assimilate them into amino acids, the roots generate hydrogen ions that they excrete into the surrounding soil (see Chapter 5). Hydrogen ions also displace K^+, Mg^{2+}, Ca^{2+}, and Mn^{2+} from the surfaces of soil particles. Leaching may then remove these ions from the upper soil layers, leaving a more acidic soil. By contrast, the weathering of rock in arid regions releases K^+, Mg^{2+}, Ca^{2+}, and Mn^{2+} into the soil, but because of the low rainfall, these ions do not leach from the upper soil layers, and the soil remains alkaline.

Excess mineral ions in the soil limit plant growth

When excess mineral ions are present in soil, the soil is said to be *saline,* and such soils may inhibit plant growth if the mineral ions reach concentrations that limit water availability or exceed the adequate concentration for a particular nutrient (see Chapter 19). Sodium chloride and sodium sulfate are the most common salts in saline soils. Excess mineral ions in soils can be a major problem in arid and semiarid regions because rainfall is insufficient to leach them from the soil layers near the surface.

Irrigated agriculture fosters soil salinization if the amount of water applied is insufficient to leach the salt below the root zone. Irrigation water can contain 100 to 1000 g of mineral ions per cubic meter. An average crop requires about 10,000 m^3 of water per hectare. Consequently, 1000 to 10,000 kg of mineral ions per hectare may be added to the soil per crop, and over several growing seasons, high concentrations of mineral ions may accumulate in the soil.

In saline soils, plants encounter **salt stress**. Whereas many plants are affected adversely by the presence of relatively low concentrations of salt, other plants can survive (**salt-tolerant plants**) or even thrive (**halophytes**) at high salt concentrations. The mechanisms by which plants tolerate high salinity are complex (see Chapter 19), involving biochemical synthesis, enzyme induction, and membrane transport. In some plant species, excess mineral ions are not taken up, being excluded by the roots; in others, they are taken up but excreted from the plant by salt glands associated with the leaves. To prevent toxic buildup of mineral ions in the cytosol, many plants sequester them in the vacuole. Efforts are under way to bestow salt tolerance on salt-sensitive crop species using both classic plant breeding and biotechnology, as detailed in Chapter 19.

Another important problem with excess mineral ions is the accumulation of heavy metals in the soil, which can cause severe toxicity in plants as well as humans. These heavy metals include zinc, copper, cobalt, nickel, mercury, lead, cadmium, silver, and chromium.

Some plants develop extensive root systems

The ability of plants to obtain both water and mineral nutrients from the soil is related to their ability to develop an extensive root system and to various other traits such as ability to secrete organic anions or develop mycorrhizal symbioses. In the late 1930s, H. J. Dittmer examined the root system of a single winter rye plant after 16 weeks of growth. He estimated that the plant had 13 million primary and lateral root axes, extending more than 500 km in length and providing 200 m^2 of surface area. This plant also had more than 10^{10} root hairs, providing another 300 m^2 of surface area. The total surface area of roots from a single rye plant equaled that of a professional basketball court. Other plant species may not develop such extensive root systems, which may limit their nutrient absorption capacity and increase their reliance on mycorrhizal symbioses (discussed below).

In the desert, the roots of mesquite (genus *Prosopis*) may extend downward more than 50 m to reach groundwater. Annual crop plants have roots that usually grow between 0.1 and 2.0 m in depth and extend laterally to distances of 0.3 to 1.0 m. In orchards, the major root systems of trees planted 1 m apart reach a total length of 12 to 18 km per tree. The annual production of roots in natural ecosystems may easily surpass that of shoots, so in many respects the aboveground portions of a plant represent only "the tip of the iceberg." Nonetheless, making observations on root systems is difficult and usually requires special techniques.

Plant roots may grow continuously throughout the year if conditions are favorable. Their proliferation, however, depends on the availability of water

salt stress   The adverse effects of excess minerals on plants.

salt-tolerant plants   Plants that can survive or even thrive in high-salt soils.

halophytes   Plants that are native to saline soils and complete their life cycles in that environment.

rhizosphere The immediate microenvironment surrounding the root.

primary root In monocots, a root generated directly by growth of the embryonic root or radicle.

nodal roots In monocots, adventitious roots that form after the emergence of primary roots.

taproot In eudicots, the main single root axis from which lateral roots develop.

meristematic zone The region at the tip of the root containing the meristem that generates the body of the root. Located between the root cap and the elongation zone.

root cap Cells at the root apex that cover and protect the meristematic cells from mechanical injury as the root moves through the soil. Site for the perception of gravity and signaling for the gravitropic response in roots.

gravitropic response The growth initiated by the root cap's perception of gravity and the signal that directs the roots to grow downward.

quiescent center The central region of the root meristem where cells divide more slowly than surrounding cells, or do not divide at all.

and minerals in the immediate microenvironment surrounding the root, the so-called **rhizosphere**. If the rhizosphere is poor in nutrients or too dry, root growth is slow. As rhizosphere conditions improve, root growth increases. If fertilization and irrigation provide abundant nutrients and water, root growth may not keep pace with shoot growth. Plant growth under such conditions becomes carbohydrate-limited, and a relatively small root system meets the nutrient needs of the whole plant. In crops for which we harvest aboveground parts, fertilization and irrigation cause greater allocation of resources to the shoot and reproductive structures than to roots, and this shift in allocation patterns often results in higher yields.

Root systems differ in form but are based on common structures

The *form* of the root system differs greatly among plant species. In monocots, root development starts with the emergence of three to six **primary** (or *seminal*) **root** axes from the germinating seed. With further growth, the plant extends new adventitious roots, called **nodal roots** or *brace roots*. Over time, the primary and nodal root axes grow and branch extensively to form a complex *fibrous root system* (**Figure 4.7**). In fibrous root systems, all the roots generally have similar diameters (except where environmental conditions or pathogenic interactions modify the root structure), so it is impossible to distinguish a main root axis.

In contrast to monocots, eudicots develop root systems with a main single root axis, called a **taproot**, which may thicken as a result of secondary cambial activity. From this main root axis, *lateral roots* develop to form an extensively branched root system (**Figure 4.8**).

The development of the root system in both monocots and eudicots depends on the activity of the root apical meristem and the production of lateral root meristems. **Figure 4.9** is a generalized diagram of the apical region of a plant root and identifies three zones of activity: the meristematic, elongation, and maturation zones.

In the **meristematic zone**, cells divide both in the direction of the root base to form cells that will differentiate into the tissues of the functional root and in the direction of the root apex to form the **root cap**. The root cap protects the delicate meristematic cells as the root expands into the soil. It commonly secretes a gelatinous material called *mucigel*, which surrounds the root tip. The precise function of mucigel is uncertain, but it may provide lubrication that eases the root's penetration of the soil, protect the root apex from desiccation, promote the transfer of nutrients to the root, and affect interactions between the root and soil microorganisms. The root cap is central to the perception of gravity, the signal that directs the growth of roots downward. This process is termed the **gravitropic response** (see Chapter 15).

Cell division in the root apex proper is relatively slow; thus this region is called the **quiescent center**. After a few generations of slow cell divisions, root cells displaced from the apex by about 0.1 mm begin to divide more rapidly. Cell division again tapers off at about 0.4 mm from the apex, and the cells expand equally in all directions.

(A) Dry soil   (B) Irrigated soil

30 cm

Figure 4.7 Fibrous root systems of wheat (a monocot). (A) The root system of a mature (3-month-old) wheat plant growing in dry soil. (B) The root system of a mature wheat plant growing in irrigated soil. It is apparent that the morphology of the root system is affected by the amount of water present in the soil. In a mature fibrous root system, the primary root axes are indistinguishable. (After Weaver 1926.)

(A) Sugar beet   (B) Alfalfa

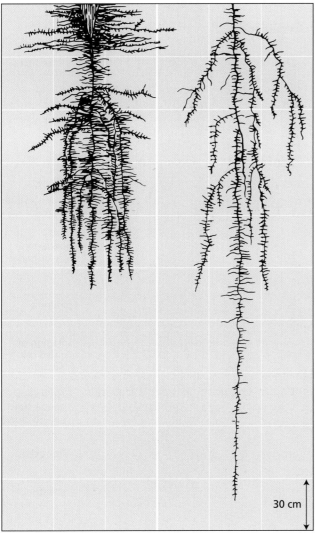

30 cm

Figure 4.8 Taproot system of two adequately watered eudicots: (A) sugar beet and (B) alfalfa. The sugar beet root system is typical of 5 months of growth; the alfalfa root system is typical of 2 years of growth. In both eudicots, the root system shows a major vertical root axis. In the case of sugar beet, the upper portion of the taproot system is thickened because of its function as storage tissue. (After Weaver 1926.)

elongation zone The region of rapid and extensive root cell elongation showing few, if any, cell divisions.

endodermis A specialized layer of cells surrounding the vascular tissue in roots and some stems.

Casparian strip A band in the cell walls of the endodermis that is impregnated with lignin. Prevents apoplastic movement of water and solutes into the stele.

cortex Ground tissue in the region of the primary stem or root located between the vascular tissue and the epidermis, mainly consisting of parenchyma.

stele In the root, the tissues located interior to the endodermis. The stele contains the vascular elements of the root: the phloem and the xylem.

phloem The tissue that transports the products of photosynthesis from mature leaves (or storage organs) to areas of growth and storage, including the roots.

xylem The vascular tissue that transports water and ions from the root to the other parts of the plant.

The **elongation zone** begins approximately 0.7 to 1.5 mm from the apex (see Figure 4.9). In this zone, cells elongate rapidly and undergo a final round of divisions to produce a central ring of cells called the **endodermis**. The walls of this endodermal cell layer become thickened, and suberin is deposited on the radial walls and forms the **Casparian strip**, a hydrophobic structure that prevents apoplastic movement of water or solutes across the root (see Figure 3.4).

The endodermis divides the root into two regions: the **cortex** toward the outside and the **stele** toward the inside. The stele contains the vascular elements of the root: the **phloem**, which transports metabolites from the shoot to the root and to fruits and seeds, and the **xylem**, which transports water and solutes to the shoot (see Chapter 3).

Phloem develops more rapidly than xylem, attesting to the fact that phloem function is critical near the root apex. Large quantities of carbohydrates must flow through the phloem to the growing apical zones in order to support cell division and elongation (see Chapter 10). Carbohydrates provide rapidly growing cells with an energy source and with the carbon skeletons required to synthesize organic compounds. Six-carbon sugars (hexoses) also function as osmotically active solutes in the root tissue. At the root apex, where the phloem is not yet

Figure 4.9 Diagrammatic longitudinal section of the apical region of the root. The meristematic cells are located near the tip of the root. These cells generate the root cap and the upper tissues of the root. In the elongation zone, cells differentiate to produce xylem, phloem, and cortex. Root hairs, formed by epidermal cells, first appear in the maturation zone.

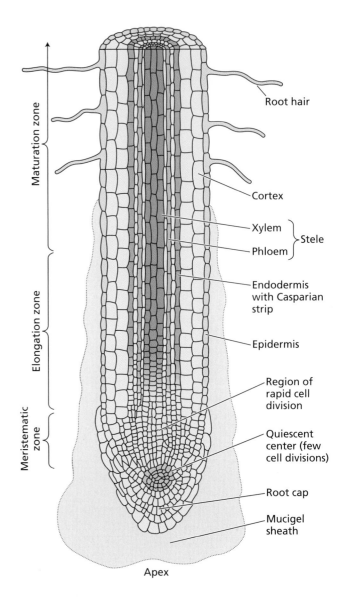

developed, carbohydrate movement depends on symplastic transport and is relatively slow. The low rates of cell division in the quiescent center may result from the fact that insufficient carbohydrates reach this centrally located region or that this area is kept in an oxidized state.

Root hairs, with their large surface area for absorption of water and solutes and for anchoring the root to the soil, first appear in the **maturation zone** (see Figure 4.9), and here the xylem develops the capacity to translocate substantial quantities of water and solutes to the shoot.

Different areas of the root absorb different mineral ions

The precise point of entry of minerals into the root system has been a topic of considerable interest. Some researchers have claimed that nutrients are absorbed only at the apical regions of the root axes or branches; others claim that nutrients are absorbed over the entire root surface. Experimental evidence supports both possibilities, depending on the plant species and the nutrient being investigated:

• Root absorption of calcium ions in barley (*Hordeum vulgare*) appears to be restricted to the apical region.

maturation zone   The region of the root where differentiation occurs, including the production of root hairs and functional vascular tissue.

- Iron may be taken up either at the apical region, as in barley and other species, or over the entire root surface, as in maize.

- Potassium ions, nitrate, ammonium, and phosphate can be absorbed freely at all locations of the root surface, but in maize the elongation zone has the maximum rates of potassium ion accumulation and nitrate absorption.

- In maize and rice and in wetland species, the root apex absorbs ammonium more rapidly than the elongation zone does. Ammonium and nitrate uptake by conifer roots varies significantly across different regions of the root and may be influenced by rates of root growth and maturation.

- In several species, the root apex and root hairs are the most active in phosphate absorption. For species with poorly developed root hairs, hyphae of arbuscular mycorrhizal fungi may play a significant role in uptake of phosphate and other nutrients, and the development of this symbiosis can change the regions of the root involved in uptake.

The high rates of nutrient absorption in the apical root zones result from the strong demand for nutrients in these tissues and the relatively high nutrient availability in the soil surrounding them. For example, cell elongation depends on the accumulation of solutes such as potassium, chloride, and nitrate ions to increase the osmotic pressure within the cell. Ammonium is the preferred nitrogen source to support cell division in the meristem because meristematic tissues are often carbohydrate-limited and because assimilation of ammonium into organic nitrogen compounds consumes less energy than assimilation of nitrate (see Chapter 5). The root apex and root hairs grow into fresh soil, where nutrients have not yet been depleted.

Within the soil, nutrients can move to the root surface both by bulk flow and by diffusion (see Chapter 2). In bulk flow, nutrients are carried by water moving through the soil toward the root. The amounts of nutrients provided to the root by bulk flow depend on the rate of water flow through the soil toward the plant, which itself depends on transpiration rates and on nutrient concentrations in the soil solution. When both the rate of water flow and the concentrations of nutrients in the soil solution are high, bulk flow can play an important role in nutrient supply. Hence, highly soluble nutrients such as nitrate are largely carried by bulk flow, but this process is less important for nutrients with low solubility, such as phosphate and zinc ions.

In diffusion, mineral nutrients move from a region of higher concentration to a region of lower concentration. Nutrient uptake by roots lowers the concentrations of nutrients at the root surface, generating concentration gradients in the soil solution surrounding the root. Diffusion of nutrients down their concentration gradients, along with bulk flow resulting from transpiration, can increase nutrient availability at the root surface.

When the rate of absorption of a nutrient by roots is high and the nutrient concentration in the soil solution is low, bulk flow can supply only a small fraction of the total nutrient requirement. Under these conditions, plant nutrient absorption becomes independent of plant transpiration rates, and diffusion rates limit the movement of the nutrient to the root surface. When diffusion is too slow to maintain high nutrient concentrations near the root, a **nutrient depletion zone** forms adjacent to the root surface (**Figure 4.10**). This zone may extend from about 0.2 to 2.0 mm from the root surface, depending on the mobility of the nutrient in the soil. The nutrient depletion zone is particularly important for phosphate.

nutrient depletion zone   The region surrounding the root surface showing diminished nutrient concentrations due to uptake into the roots and slow replacement by diffusion.

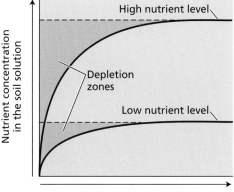

Figure 4.10   Formation of a nutrient depletion zone in the region of the soil adjacent to the plant root. A nutrient depletion zone forms when the rate of nutrient uptake by the cells of the root exceeds the rate of replacement of the nutrient by bulk flow and diffusion in the soil solution. This depletion causes a localized decrease in the nutrient concentration in the area adjacent to the root surface. (After Mengel and Kirkby 2001.)

mycorrhizal fungi Fungi that can form mycorrhizal symbioses with plants.

mycorrhiza (plural *mycorrhizae*) The symbiotic (mutualistic) association of certain fungi and plant roots. Facilitates the uptake of mineral nutrients by roots.

The formation of a depletion zone tells us something important about mineral nutrition. Because roots deplete the mineral supply in the rhizosphere, their effectiveness in mining minerals from the soil is determined not only by the rate at which they can remove nutrients from the soil solution, but by their continuous growth into undepleted soil. Without continuous growth, roots would rapidly deplete the soil adjacent to their surfaces. Optimal nutrient acquisition therefore depends both on the root system's capacity for nutrient uptake and on its ability to grow into fresh soil. The ability of the plant to form a mycorrhizal symbiosis is also critical in overcoming the effects of depletion zones, because the hyphae of the fungal symbionts grow beyond the depletion zone. These fungal structures take up nutrients far from the root (up to 25 cm in the case of arbuscular mycorrhizae) and translocate them rapidly to the roots, overcoming the slow diffusion in soil.

Nutrient availability influences root growth

Plants, which have limited mobility for most of their lives, must deal with changes in their local environment because they cannot move away from unfavorable conditions. Above the ground, light intensity, temperature, and humidity may fluctuate substantially during the day and across the canopy, but CO_2 and O_2 concentrations remain relatively uniform. In contrast, soil buffers the roots from temperature extremes, but the belowground concentrations of CO_2 and O_2, water, and nutrients are extremely heterogeneous, both spatially and temporally. For example, inorganic nitrogen concentrations in soil may range a thousandfold over a distance of centimeters or the course of hours. Given such heterogeneity, plants seek the most favorable conditions within their reach.

Roots sense the belowground environment—through gravitropism, thigmotropism, chemotropism, and hydrotropism—to guide their growth toward soil resources. Some of these responses involve auxin (see Chapter 15). The extent to which roots proliferate within a soil patch varies with nutrient concentrations (**Figure 4.11**). Root growth is minimal in poor soils because the roots become nutrient-limited. As soil nutrient availability increases, roots proliferate.

Where soil nutrients exceed an optimal concentration, root growth may become carbohydrate-limited and eventually cease. With high soil nutrient concentrations, a few roots—3.5% of the root system in spring wheat and 12% in lettuce—are sufficient to supply all the nutrients required, so the plant may diminish the allocation of its resources to roots while increasing its allocation to the shoot and reproductive structures. This resource shifting is one mechanism through which fertilization stimulates crop yields.

Mycorrhizal symbioses facilitate nutrient uptake by roots

Our discussion so far has centered on the direct acquisition of mineral elements by roots, but this process is usually modified by the association of **mycorrhizal fungi** with the root system to form a **mycorrhiza** (from

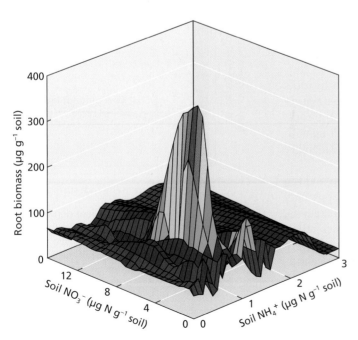

Figure 4.11 Root biomass as a function of extractable soil NH_4^+ and NO_3^-. The root biomass is shown (µg root dry weight g^{-1} soil) plotted against extractable soil NH_4^+ and NO_3^- (µg extractable N g^{-1} soil) for tomato (*Solanum lycopersicum* cv T-5) growing in an irrigated field that had been fallow for the previous 2 years. The colors emphasize the differences among biomasses, ranging from low (purple) to high (red). (After Bloom et al. 1993.)

the Greek words for "fungus" and "root"; plural *mycorrhizae*). The host plant supplies associated mycorrhizal fungi with carbohydrates and in return receives nutrients from the fungi. There are indications that drought and disease tolerance may also be improved in the host plant.

Mycorrhizal symbioses of two main types—**arbuscular mycorrhizae** and **ectomycorrhizae**—are widespread in nature, occurring in about 90% of terrestrial plant species, including most major crops. The majority, perhaps 80%, are arbuscular mycorrhizae, which are symbioses between a newly described phylum of fungi, the Glomeromycota, and a broad range of angiosperms, gymnosperms, ferns, and liverworts. Their importance in herbaceous species and in fruit trees of many types makes arbuscular mycorrhizae vital to agricultural production, particularly in nutrient-poor soils. This is the most ancient type of mycorrhiza, occurring in fossils of the earliest land plants. This symbiosis was probably important in facilitating plant establishment on land more than 450 million years ago, because early land plants had poorly developed underground organs.

By contrast, ectomycorrhizal symbioses evolved more recently. They are formed by far fewer plant species, notably trees in the families Pinaceae (pines, larches, Douglas fir), Fagaceae (beech, oak, chestnut), Salicaceae (poplar, aspen), Betulaceae (birch), and Myrtaceae (*Eucalyptus*). The fungal partners belong to either Basidomycota or, less frequently, Ascomycota. These symbioses play major roles in the nutrition of trees and therefore in the productivity of vast areas of boreal forest.

Some plant species, particularly those in the families Salicaceae (willow [*Salix*] and populus and aspen [*Populus*]) and Myrtaceae (*Eucalyptus*), can form both arbuscular and ectomycorrhizal symbioses. Other plants prove unable to form any kind of mycorrhiza. These include members of the families Brassicaceae, such as cabbage (*Brassica oleracea*) and the model plant *Arabidopsis thaliana*; Chenopodiaceae, such as spinach (*Spinacia oleracea*); and Proteaceae, such as macadamia nut (*Macadamia integrifolia*).

Certain agricultural practices may reduce or eliminate mycorrhiza formation in plants that normally form them. These practices include flooding (paddy rice does not form mycorrhizae, whereas upland rice does), extensive soil disturbance caused by plowing, application of high concentrations of fertilizer, and of course soil fumigation and application of some fungicides. Such practices may decrease yields in crops such as maize that are very dependent on mycorrhizae for nutrient uptake. Mycorrhizae also do not form in solution culture or in hydroponic cultivation. Nonetheless, for the majority of plants, mycorrhiza formation is the normal situation and the non-mycorrhizal state is essentially an artifact, brought about by particular agricultural practices.

Mycorrhizae modify the plant root system and influence plant mineral nutrient acquisition, but the way they do so varies between types. Arbuscular mycorrhizal fungi develop, outside the root of their host, a highly branched system (mycelium) of hyphae (fine filamentous structures 2 to 10 μm in diameter) that explore the soil (**Figure 4.12**). Different arbuscular mycorrhizal fungi vary considerably in their distance and intensity of soil exploration, but phosphate transfer from as far as 25 cm away from the root has been measured. The mycelium in soil also helps stabilize aggregates of soil particles, promoting good soil structure. The hyphae extend into soil well beyond the zone of depletion

arbuscular mycorrhizae Symbioses between fungi in the phylum Glomeromycota and the roots of a broad range of angiosperms, gymnosperms, ferns, and liverworts. The hyphae of arbuscular mycorrhizae penetrate the cortical cells of the root.

ectomycorrhizae Symbioses in which the fungus typically forms a thick sheath, or mantle, around the roots. The root cells themselves are not penetrated by the fungal hyphae, and instead are surrounded by a network of hyphae called the Hartig net.

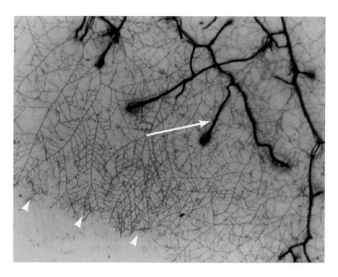

Figure 4.12 Visualization of the intact extraradical mycelium of *Glomus mosseae* spreading from colonized roots of cherry plum (*Prunus cerasifera*). The advancing front of the extraradical mycelium is indicated by arrowheads, and the plant roots by an arrow. Note the differences in lengths and diameters of roots and hyphae. (From Giovannetti et al. 2006.)

arbuscules Branched structures formed by mycorrhizal fungi inside cortical cells of the host plant root; the sites of nutrient transfer between the fungus and the plant.

hyphal coils Coiled structures formed by mycorrhizal fungi within the root cortical cells of its host plant; the sites of nutrient transfer between the fungus and the plant.

that develops around a root and thus can absorb an immobile nutrient such as phosphate from beyond the depletion zone. Hyphae also penetrate soil pores that are much narrower than those available to roots.

The root of the arbuscular mycorrhizal host plant looks almost the same as a non-mycorrhizal root, and the presence of the fungi can only be detected by staining and microscopy. Hyphae of arbuscular mycorrhizal fungi, growing from spores in the soil or roots of another plant, penetrate the root epidermis and colonize the root cortex, extending through intercellular spaces and invading the cortical cells to form either highly branched structures called **arbuscules** (Arum-type colonization; **Figure 4.13A**) or complex **hyphal coils** (Paris-type colonization; **Figure 4.13B**). The fungi are restricted to the cortex and never penetrate the endodermis or colonize the stele of the root. These structures increase the area of contact between the symbionts and remain surrounded by a plant membrane that is involved in transferring nutrients from fungal to plant cells. The penetration process is genetically controlled by a pathway that millions and millions of years later was partially co-opted for the colonization of legume roots by nitrogen-fixing bacteria (see Chapter 5).

Phosphate is delivered by the fungi directly to the root cortex. After export from the fungal arbuscules or coils, this phosphate is taken up by the plant cells. Some of the suite of plant phosphate transporters (see Chapter 6) are specifically or preferentially expressed only in the plant membrane surrounding the arbuscules and coils in the root cortex and are not expressed in non-mycorrhizal roots. The transporters play a key role in the transfer of phosphate from fungus to plant.

The hyphae of arbuscular mycorrhizal fungi have the capacity for fast growth, highly efficient absorption, and rapid translocation and transfer of nutrients such as phosphate to the root cells. This means they can explore soil much more effectively and with fewer resources than non-mycorrhizal roots. In a large number

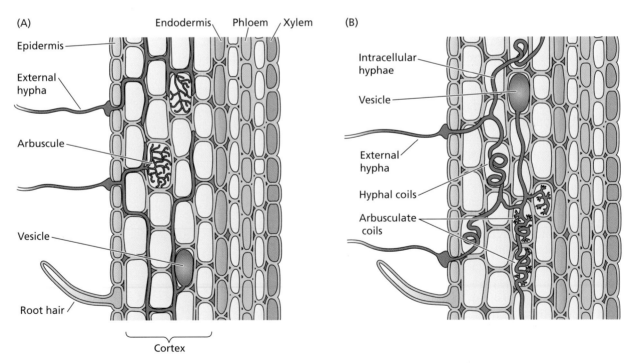

Figure 4.13 Diagrammatic representation of the two major forms of arbuscular mycorrhizal colonization of the root cortex. (A) Arum-type colonization, characterized by the formation of intracellular, highly branched arbuscules in root cortical cells. (B) Paris-type colonization, characterized by the formation of intracellular hyphal coils in root cortical cells, some of which (called arbuscatula coils) bear small arbuscule-like branches.

of plant species, the response to arbuscular mycorrhizal fungi colonization is increased phosphate uptake and hence growth, especially when soil phosphorus is poorly available. A wide variety of responses has been observed, however, ranging from large positive responses to zero or even negative responses. The conventional explanation for the negative responses is that the fungi consume excessive carbohydrate and fail to deliver adequate amounts of nutrients to the plant. Nonetheless, the fungi remain active in phosphate delivery, while at the same time decreasing the amount of phosphate that is absorbed directly through the root epidermis. The lack of positive responses may therefore derive from "cross talk" between the plant and fungal symbionts that interferes with the way the roots absorb nutrients. High phosphate availability in soil tends to decrease the stimulatory effect that arbuscular mycorrhiza formation has on plant phosphorus uptake, growth, and yield, but substantive evidence for specific plant control of fungal colonization and activity by phosphate is still lacking.

Harnessing arbuscular mycorrhizal symbiosis to optimize crop nutrition as fertilizers become increasingly expensive will depend on understanding how the symbiotic partners interact to influence nutrient acquisition. At present, arbuscular mycorrhizal fungi are known to be important in uptake of immobile nutrients such as phosphate and zinc. Their role in increasing nitrogen uptake remains to be established.

Roots colonized by ectomycorrhizal fungi can be clearly distinguished from non-mycorrhizal roots; they grow more slowly and often appear thicker and highly branched. The fungi typically form a thick sheath, or *mantle*, of mycelium around roots, and some of the hyphae penetrate between the epidermal and sometimes (in the case of conifers) the cortical cells (**Figure 4.14**). The root cells themselves are not penetrated by the fungal hyphae, but instead are surrounded by a network of hyphae called the **Hartig net**, which provides a large area of contact between the symbionts that is involved in nutrient transfers. The fungal mycelium also extends into the soil, away from the compact sheath, where it is present as individual hyphae, mycelial fans (**Figure 4.15**), or mycelial strands. The fans in particular play important roles in obtaining nutrients from the soil, especially soil organic matter.

Ectomycorrhizal fungi produce many of the toadstools, puffballs, and truffles found in forests. Often the amount of fungal mycelium is so extensive that its total mass is much greater than that of the roots themselves. The arrangement

Hartig net A network of fungal hyphae that surround, but do not penetrate, the cortical cells of roots in ectomycorrhizal symbioses.

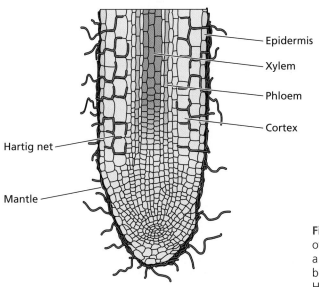

100 μm

Figure 4.14 Diagrammatic representation of a longitudinal section of an ectomycorrhizal root. Fungal hyphae (shown in brown) form a dense fungal mantle over the surface of the root and penetrate between epidermal cells, or epidermal and cortical cells, to form the Hartig net. Hyphae also grow extensively in soil, forming dense mycelium and/or mycelial strands. (After Rovira et al. 1983.)

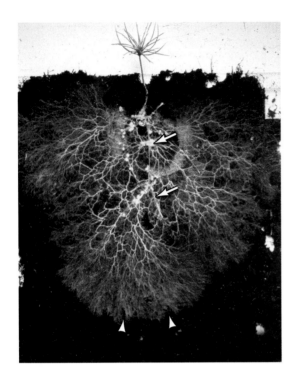

Figure 4.15 Seedling of pine (*Pinus*) showing mycorrhizal rootlets (upper arrow) colonized by an ectomycorrhizal fungus and grown in an observation chamber in forest soil. Note the differences between the mycelial front of dense hyphae advancing into the soil (arrowheads) and aggregated mycelial strands (lower arrow). (Courtesy of D. J. Read.)

and biochemical activities of the fungal structures in relation to the root tissues determine important aspects of nutrient acquisition by ectomycorrhizal roots and the form in which the nutrients pass from fungus to plant. In addition, all nutrients from the soil must pass through the fungal mantle covering the root epidermis before reaching the root cells themselves, giving the fungus a major role in uptake of all nutrients from the soil solution, including phosphate and inorganic forms of nitrogen (nitrate and ammonium). To what extent the fungi are actually involved in uptake of inorganic nitrogen and to what extent they may compete with the roots when nitrogen is in short supply are matters of active research. The fungal mycelium that develops in soil proliferates extensively in soil organic matter patches (see Figure 4.15). The hyphae have a marked ability to convert insoluble organic nitrogen and phosphorus to soluble forms and to pass these nutrients to the plants. In this way, ectomycorrhizal fungi enable their host plants to access organic sources of nutrients, avoid competition with free-living mineralizing organisms, and grow in highly organic forest soils that contain very low amounts of inorganic nutrients.

Nutrients move between mycorrhizal fungi and root cells

Movement of nutrients from soil via a mycorrhizal fungus to root cells involves complex integration of structure and function in both fungus and plant symbionts. The interfaces where fungus and plant are juxtaposed are critical zones for transport and are composed of the plasma membranes of both organisms, plus variable amounts of cell wall material. Therefore, nutrient movements from fungus to plant are potentially under the control of these two membrane types and subject to regulatory transport processes as described in Chapter 6. Movement of nutrients from soil to the plant via a mycorrhizal fungus requires (at least) uptake of a nutrient from soil by the fungus, long-distance translocation of the nutrient through fungal hyphae (and mycelial strands when present), release (or efflux) from the fungus to the apoplastic zone between the two membranes of the interface, and uptake by the plant plasma membrane. Important issues to be resolved include the form of nutrient that is transferred and the mechanism and amounts of transfers. Mechanisms promoting efflux from the fungus to the interfacial apoplastic zone are poorly understood, but uptake into the plant has received more attention. In the case of phosphate, the plant uptake step is an active process requiring energy and the presence of phosphate transporters, which are specifically or preferentially expressed in the plant membrane surrounding the intracellular fungal structures when the roots are mycorrhizal.

Transfer of nitrogen is more complex and more controversial. In ectomycorrhizae, for which a major role in plant nitrogen nutrition has long been accepted, organic nitrogen may move from fungus to plant, with the form (glutamine, glutamine and alanine, or glutamate) varying with the distribution of enzymes involved in inorganic nitrogen assimilation and the identity of the plant and fungal symbionts. Some transfer of nitrogen as ammonium or ammonia may also occur. As mentioned above, the involvement of arbuscular mycorrhizae in enhancing nitrogen uptake and transfer to host plants is not well established.

Summary

Plants are autotrophic organisms capable of using the energy from sunlight to synthesize all their components from carbon dioxide, water, and mineral elements. Although mineral nutrients continually cycle through all organisms, they enter the biosphere predominantly through the root systems of plants. After being absorbed by the roots, the mineral elements are translocated to the various parts of the plant, where they serve in numerous biological functions.

Essential Nutrients, Deficiencies, and Plant Disorders

- Studies of plant nutrition have shown that specific mineral elements are essential for plant life (**Tables 4.1, 4.2**).

- These elements are classified as macronutrients or micronutrients, depending on the relative amounts found in plant tissue (**Table 4.1**).

- Certain visual symptoms are diagnostic for deficiencies in specific nutrients in higher plants. Nutritional disorders occur because nutrients have key roles in plants. They serve as components of organic compounds, in energy storage, in plant structures, as enzyme cofactors, and in electron transfer reactions.

- Mineral nutrition can be studied through the use of solution culture, which allows the characterization of specific nutrient requirements (**Figure 4.2; Table 4.3**).

- Soil and plant tissue analysis can provide information on the nutritional status of the plant–soil system and can suggest corrective actions to avoid deficiencies or toxicities (**Figure 4.4**).

Treating Nutritional Deficiencies

- When crop plants are grown under modern high-production conditions, substantial amounts of nutrients are removed from the soil.

- To prevent the development of deficiencies, nutrients can be added back to the soil in the form of fertilizers, particularly nitrogen, phosphorus, and potassium.

- Fertilizers that provide nutrients in inorganic forms are called chemical fertilizers; those that derive from plant or animal residues or from natural rock deposits are considered organic fertilizers. In both cases, plants absorb the nutrients primarily as inorganic ions. Most fertilizers are applied to the soil, but some are sprayed on leaves.

Soil, Roots, and Microbes

- Soil is a complex substrate—physically, chemically, and biologically. The size of soil particles and the cation exchange capacity of the soil determine the extent to which a soil provides a reservoir for water and nutrients (**Table 4.5; Figure 4.6**).

- Soil pH also has a large influence on the availability of mineral elements to plants (**Figure 4.5**).

- If mineral elements, especially sodium or heavy metals, are present in excess in the soil, plant growth may be adversely affected. Certain plants are able to tolerate excess mineral elements, and a few species—for example, halophytes in the case of sodium—may thrive under these extreme conditions.

- To obtain nutrients from the soil, plants develop extensive root systems (**Figures 4.7, 4.8**), form symbioses with mycorrhizal fungi, and produce and secrete protons or organic anions into the soil.

- Roots continually deplete the nutrients from the immediate soil around them (**Figure 4.10**).

- The majority of plants have the ability to form symbioses with mycorrhizal fungi.

- The fine hyphae of mycorrhizal fungi extend the reach of roots into the surrounding soil and facilitate the acquisition of nutrients (**Figures 4.12, 4.14, 4.15**). Arbuscular mycorrhizae increase uptake of mineral nutrients, particularly phosphorus, whereas ectomycorrhizae play a significant role in obtaining nitrogen from organic sources.

- In return, plants provide carbohydrates to the mycorrhizal fungi.

Suggested Reading

Armstrong, F. A. (2008) Why did nature choose manganese to make oxygen? *Philos. Trans. R. Soc. Lond., B, Biol. Sci.* 363: 1263–1270.

Bucher, M. (2007) Functional biology of plant phosphate uptake at root and mycorrhiza interfaces. *New Phytol.* 173: 11–26.

Connor, D. J., Loomis, R. S., and Cassman, K. G. (2011) *Crop Ecology: Productivity and Management in Agricultural Systems*, 2nd ed. Cambridge University Press, Cambridge.

Cordell, D., Drangerta, J.-O., and White, S. (2009) The story of phosphorus: Global food security and food for thought. *Glob. Environ. Change* 19: 292–305.

Epstein, E., and Bloom, A. J. (2005) *Mineral Nutrition of Plants: Principles and Perspectives*, 2nd ed. Sinauer Associates, Sunderland, MA.

Fageria, N., Filho, M. B., Moreira, A., and Guimaraes, C. (2009) Foliar fertilization of crop plants. *J. Plant Nutr.* 32: 1044–1064.

Feldman, L. J. (1998) Not so quiet quiescent centers. *Trends Plant Sci.* 3: 80–81.

Fox, T. C., and Guerinot, M. L. (1998) Molecular biology of cation transport plants. *Annu. Rev. Plant Physiol. Plant Mol. Biol.* 49: 669–696.

Jeong, J., and Guerinot, M. L. (2009) Homing in on iron homeostasis in plants. *Trends Plant Sci.* 14: 280–285.

Jones, M. D., and Smith, S. E. (2004) Exploring functional definitions of mycorrhizas: Are mycorrhizas always mutualisms? *Can. J. Bot.* 82: 1089–1109.

Kochian, L. V. (2000) Molecular physiology of mineral nutrient acquisition, transport and utilization. In *Biochemistry and Molecular Biology of Plants*, B. Buchanan, W. Gruissem, and R. Jones, eds., American Society of Plant Physiologists, Rockville, MD, pp. 1204–1249.

Larsen, M. C., Hamilton, P. A., and Werkheiser, W. H. (2013) Water quality status and trends in the United States. In *Monitoring Water Quality: Pollution Assessment, Analysis, and Remediation*, S. Ahuja, ed., Elsevier, Amsterdam, pp. 19–57.

Mengel, K., and Kirkby, E. A. (2001) *Principles of Plant Nutrition*, 5th ed. Kluwer Academic Publishers, Dordrecht, Netherlands.

Sattelmacher, B. (2001) The apoplast and its significance for plant mineral. *New Phytol.* 149: 167–192.

Smith, F. A., Smith, S. E., and Timonen, S. (2003) Mycorrhizas. In *Root Ecology*, H. de Kroon and E. J. W. Visser, eds., Springer, Berlin, pp. 257–295.

Smith, S. E., and Read, D. J. (2008) *Mycorrhizal Symbiosis*, 3rd ed. Academic Press and Elsevier, Oxford.

Zegada-Lizarazu, W., Matteucci, D. and Monti, A. (2010) Critical review on energy balance of agricultural systems. *Biofuels, Bioprod. Biorefin.* 4: 423–446.

5 Assimilation of Inorganic Nutrients

Higher plants are autotrophic organisms that can synthesize all of their organic molecular components out of inorganic nutrients obtained from their surroundings. For many inorganic nutrients, this process involves absorption from the soil by the roots (see Chapter 4) and incorporation into the organic compounds that are essential for growth and development. This incorporation of inorganic nutrients into organic substances such as pigments, enzyme cofactors, lipids, nucleic acids, and amino acids is termed **nutrient assimilation**.

Assimilation of some nutrients—particularly nitrogen and sulfur—involves a complex series of biochemical reactions that are among the most energy-consuming reactions in living organisms. Assimilation of other nutrients, especially the macronutrient and micronutrient cations (see Chapter 4), involves the formation of complexes with organic compounds. For example, Mg^{2+} associates with chlorophyll pigments, Ca^{2+} associates with pectates in the cell wall, and Mo^{6+} associates with enzymes such as nitrate reductase and nitrogenase. These complexes are highly stable, and removal of the nutrient from the complex may result in total loss of function.

This chapter outlines the primary reactions through which the major nutrients (nitrogen, sulfur, phosphate, and iron) are assimilated and discusses the organic products of these reactions. We emphasize the physiological implications of the required energy expenditures and introduce the topic of symbiotic nitrogen fixation. Plants serve as the major conduit through which nutrients pass from slower geophysical domains into faster biological ones; this chapter thus highlights the vital role of plant nutrient assimilation in the human diet.

nutrient assimilation   The incorporation of mineral nutrients into carbon compounds such as pigments, enzyme cofactors, lipids, nucleic acids, or amino acids.

nitrogen fixation The natural or industrial processes by which atmospheric nitrogen N_2 is converted to ammonia (NH_3) or nitrate (NO_3^-).

Nitrogen in the Environment

Many prominent biochemical compounds in plant cells contain nitrogen (see Chapter 4). For example, nitrogen is found in the nucleotides and amino acids that form the building blocks of nucleic acids and proteins, respectively. Only the elements oxygen, carbon, and hydrogen are more abundant in plants than nitrogen. Most natural and agricultural ecosystems show dramatic gains in productivity after fertilization with inorganic nitrogen, attesting to the importance of this element and to the fact that it is present in suboptimal amounts.

Nitrogen passes through several forms in a biogeochemical cycle

Nitrogen is present in many forms in the biosphere. The atmosphere contains vast quantities (about 78% by volume) of molecular nitrogen (N_2). For the most part, this large reservoir of nitrogen is not directly available to living organisms. Acquisition of nitrogen from the atmosphere requires the breaking of an exceptionally stable triple covalent bond between two nitrogen atoms ($N \equiv N$) to produce ammonia (NH_3) or nitrate (NO_3^-). These reactions, known as **nitrogen fixation**, occur through both industrial and natural processes.

N_2 combines with hydrogen to form ammonia under elevated temperature (about 200°C) and high pressure (about 200 atmospheres) and in the presence of a metal catalyst (usually iron). The extreme conditions are required to overcome the high activation energy of the reaction. This nitrogen fixation reaction, called the *Haber–Bosch process*, is a starting point for the manufacture of many

---

Table 5.1 Major processes of the biogeochemical nitrogen cycle

Process	Definition	Rate (10^{13} g y^{-1})[a]
Industrial fixation	Industrial conversion of molecular nitrogen to ammonia	10
Atmospheric fixation	Lightning and photochemical conversion of molecular nitrogen to nitrate	1.9
Biological fixation	Prokaryotic conversion of molecular nitrogen to ammonia	17
Plant acquisition	Plant absorption and assimilation of ammonium or nitrate	120
Immobilization	Microbial absorption and assimilation of ammonium or nitrate	N/C
Ammonification	Bacterial and fungal catabolism of soil organic matter to ammonium	N/C
Anammox	Anaerobic ammonium oxidation: bacterial conversion of ammonium and nitrite to molecular nitrogen	N/C
Nitrification	Bacterial (*Nitrosomonas* spp.) oxidation of ammonium to nitrite and subsequent bacterial (*Nitrobacter* spp.) oxidation of nitrite to nitrate	N/C
Mineralization	Bacterial and fungal catabolism of soil organic matter to mineral nitrogen through ammonification or nitrification	N/C
Volatilization	Physical loss of gaseous ammonia to the atmosphere	10
Ammonium fixation	Physical embedding of ammonium into soil particles	1
Denitrification	Bacterial conversion of nitrate to nitrous oxide and molecular nitrogen	21
Nitrate leaching	Physical flow of nitrate dissolved in groundwater out of the topsoil and eventually into the oceans	3.6

Note: Terrestrial organisms, the soil, and the oceans contain about 5.2×10^{15} g, 95×10^{15} g, and 6.5×10^{15} g, respectively, of organic nitrogen that is active in the cycle. Assuming that the amount of atmospheric N_2 remains constant (inputs = outputs), the *mean residence time* (average time that a nitrogen molecule remains in organic forms) is about 350 years [(pool size)/(fixation input) = (5.2×10^{15} g + 95×10^{15} g)/(10×10^{13} g y^{-1} + 1.9×10^{13} g y^{-1} + 17×10^{13} g y^{-1})].

[a]N/C, not calculated.

industrial and agricultural products, including nitrogen fertilizers. Worldwide industrial production of nitrogen fertilizers amounts to more than 100 million metric tons per year (10×10^{13} g y^{-1}).

The following natural processes fix about 190 million metric tons per year of nitrogen (**Table 5.1**):

- *Lightning.* Lightning is responsible for about 8% of the nitrogen fixed by natural processes. Lightning converts water vapor and oxygen into highly reactive hydroxyl free radicals, free hydrogen atoms, and free oxygen atoms that attack molecular nitrogen (N_2) to form nitric acid (HNO_3). This nitric acid subsequently falls to Earth with rain.

- *Photochemical reactions.* Approximately 2% of the nitrogen fixed derives from photochemical reactions between gaseous nitric oxide (NO) and ozone (O_3) that produce nitric acid (HNO_3).

- *Biological nitrogen fixation.* The remaining 90% results from biological nitrogen fixation, in which bacteria or cyanobacteria (formerly called blue-green algae) fix N_2 into ammonia (NH_3). This ammonia dissolves in water to form ammonium (NH_4^+):

$$NH_3 + H_2O \rightarrow NH_4^+ + HO^- \qquad (5.1)$$

From an agricultural standpoint, biological nitrogen fixation is critical, because industrially produced nitrogen fertilizers are economically and environmentally costly and not affordable for many poor farmers.

Once fixed into ammonia or nitrate, nitrogen enters a biogeochemical cycle and passes through several organic or inorganic forms before it eventually returns to molecular nitrogen (**Figure 5.1**; see also Table 5.1). The ammonium

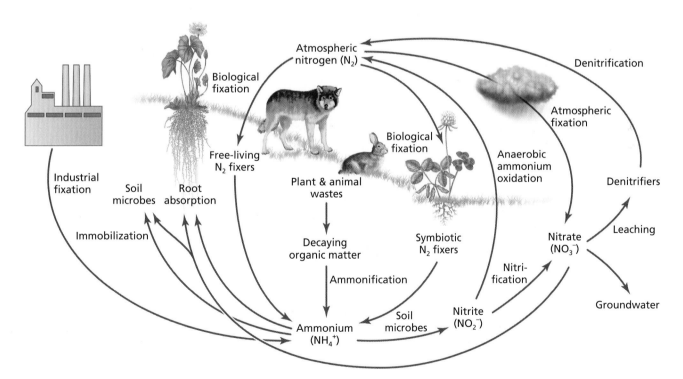

Figure 5.1 Nitrogen cycles through the atmosphere, changing from a gas to soluble ions before it is incorporated into organic compounds in living organisms. Some of the steps involved in the nitrogen cycle are shown.

(NH_4^+) and nitrate (NO_3^-) ions in the soil solution that are generated through fixation or released through decomposition of soil organic matter become the object of intense competition among plants and microorganisms. To be competitive, plants have evolved mechanisms for scavenging these ions rapidly from the soil solution (see Chapter 4). Under the elevated soil concentrations that occur after fertilization, the absorption of ammonium and nitrate by the roots may exceed the capacity of a plant to assimilate these ions, leading to their accumulation in the plant's tissues.

Unassimilated ammonium or nitrate may be dangerous

Ammonium is toxic to both plants and animals if it accumulates to high concentrations in the tissues. Ammonium dissipates transmembrane proton gradients (**Figure 5.2**) that are created by photosynthetic and respiratory electron transport (see Chapters 7 and 11) and that are used for sequestering metabolites in organelles such as the vacuole and for transporting nutrients across biological membranes (see Chapter 6). Probably because high levels of ammonium are dangerous, animals have developed a strong aversion to its smell. The active ingredient in smelling salts, a medicinal vapor released under the nose to revive a person who has fainted, is ammonium carbonate. Plants assimilate ammonium near the site of absorption or generation and rapidly store any excess in their vacuoles, thus minimizing toxic effects on membranes and the cytosol.

Unlike the case with ammonium, plants can store high levels of nitrate, and they can translocate it from tissue to tissue without deleterious effect. Yet if livestock or humans consume plant material that is high in nitrate, they may suffer methemoglobinemia, a disease in which the liver reduces nitrate to nitrite, which combines with hemoglobin and renders hemoglobin unable to bind oxygen. Humans and other animals may also convert nitrate into nitrosamines, which are potent carcinogens, or into nitric oxide, a potent signaling molecule involved in many physiological processes such as widening of blood vessels. Some countries limit the nitrate content in plant materials sold for human consumption.

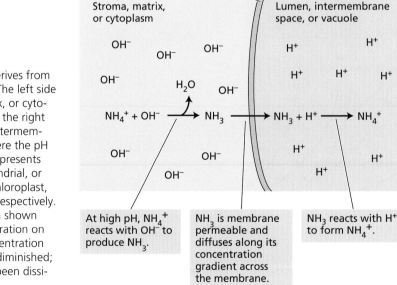

Figure 5.2 NH_4^+ toxicity derives from dissipation of pH gradients. The left side represents the stroma, matrix, or cytoplasm, where the pH is high; the right side represents the lumen, intermembrane space, or vacuole, where the pH is low; and the membrane represents the thylakoid, inner mitochondrial, or tonoplast membrane for a chloroplast, mitochondrion, or root cell, respectively. The net result of the reaction shown is that both the OH^- concentration on the left side and the H^+ concentration on the right side have been diminished; that is, the pH gradient has been dissipated. (After Bloom 1997.)

Nitrate Assimilation

Plant roots actively absorb nitrate from the soil solution via several low- and high-affinity nitrate–proton cotransporters (see Chapter 6). Plants eventually assimilate most of this nitrate into organic nitrogen compounds. The first step of this process is the conversion of nitrate to nitrite in the cytosol, a reduction reaction (for redox properties, see Chapter 11) that involves the transfer of two electrons. The enzyme **nitrate reductase** catalyzes this reaction:

$$NO_3^- + NAD(P)H + H^+ \rightarrow NO_2^- + NAD(P)^+ + H_2O \qquad (5.2)$$

where NAD(P)H indicates either NADH or NADPH. The most common form of nitrate reductase uses only NADH as an electron donor; another form of the enzyme that is found predominantly in nongreen tissues such as roots can use either NADH or NADPH.

The nitrate reductases of higher plants are composed of two identical subunits, each containing three prosthetic groups: flavin adenine dinucleotide (FAD), heme, and a molybdenum ion complexed to an organic molecule called a *pterin*.

A pterin (fully oxidized)

Nitrate reductase is the main molybdenum-containing protein in vegetative tissues; one symptom of molybdenum deficiency is the accumulation of nitrate that results from diminished nitrate reductase activity.

X-ray crystallography and comparison of the amino acid sequences for nitrate reductase from several species with the sequences of other well-characterized proteins that bind FAD, heme, or molybdenum ions have led to a multiple-domain model for nitrate reductase; a simplified three-domain model is shown in **Figure 5.3**. The FAD-binding domain accepts two electrons from NADH or NADPH. The electrons then pass through the heme domain to the molybdenum complex, where they are transferred to nitrate.

Many factors regulate nitrate reductase

Nitrate, light, and carbohydrates influence nitrate reductase at the transcription and translation levels. In barley seedlings, nitrate reductase mRNA was detected approximately 40 min after addition of nitrate, and maximum levels were

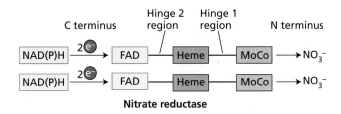

Nitrate reductase

Figure 5.3 Model of the nitrate reductase dimer, illustrating the three binding domains whose polypeptide sequences are similar in eukaryotes: molybdenum complex (MoCo), heme, and FAD. The NAD(P)H binds at the FAD-binding region of each subunit and initiates a two-electron transfer from the carboxyl (C) terminus, through each of the electron transfer components, to the amino (N) terminus. Nitrate is reduced at the molybdenum complex near the amino terminus. The polypeptide sequences of the hinge regions are highly variable among species.

nitrate reductase An enzyme located in the cytosol that reduces nitrate (NO_3^-) to nitrite (NO_2^-). It catalyzes the first step by which nitrate absorbed by roots is assimilated into organic form.

Figure 5.4 Stimulation of nitrate reductase activity follows the induction of nitrate reductase mRNA in shoots and roots of barley (g_{fw}, grams fresh weight). (After Kleinhofs et al. 1989.)

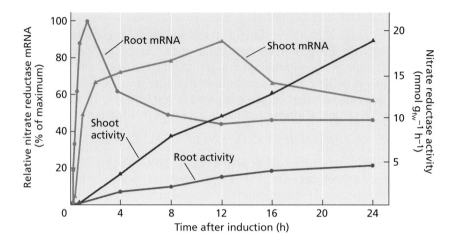

attained within 2 h in roots and in about 12 h in shoots (**Figure 5.4**). In contrast to the rapid mRNA accumulation, there was a gradual linear increase in nitrate reductase activity, reflecting the fact that each mRNA molecule is translated many times over quite a long period of time.

In addition to this regulation at the transcription level, the nitrate reductase protein is subject to posttranslational modification involving a reversible phosphorylation. Light, carbohydrate levels, and other environmental factors stimulate a protein phosphatase that dephosphorylates a key serine residue in the hinge 1 region of nitrate reductase (between the molybdenum complex and heme-binding domains; see Figure 5.3) and thereby activates the enzyme.

Operating in the reverse direction, darkness and Mg^{2+} stimulate a protein kinase that phosphorylates the same serine residue, which then interacts with a 14-3-3 inhibitor protein (one of a class of proteins that is present in all eukaryotic cells), and thereby inactivates nitrate reductase. *Regulation of nitrate reductase activity through phosphorylation and dephosphorylation provides more rapid control than can be achieved through synthesis or degradation of the enzyme (minutes versus hours).*

Nitrite reductase converts nitrite to ammonium

Nitrite (NO_2^-) is a highly reactive, potentially toxic ion. Plant cells immediately transport the nitrite generated by nitrate reduction (see Equation 5.2) from the cytosol into chloroplasts in leaves and plastids in roots. In these organelles, the enzyme nitrite reductase reduces nitrite to ammonium, a reaction that involves the transfer of six electrons, according to the following overall reaction:

$$NO_2^- + 6\ Fd_{red} + 8\ H^+ \rightarrow NH_4^+ + 6\ Fd_{ox} + 2\ H_2O \tag{5.3}$$

where Fd is ferredoxin and the subscripts *red* and *ox* stand for *reduced* and *oxidized*, respectively. Reduced ferredoxin is derived from photosynthetic electron transport in the chloroplasts (see Chapter 7) and from NADPH generated by the oxidative pentose phosphate pathway in nongreen tissues (see Chapter 11).

Chloroplasts and root plastids contain different forms of nitrite reductase, but both forms consist of a single polypeptide containing two prosthetic groups: an iron–sulfur cluster (Fe_4S_4) and a specialized heme. These groups act together to bind nitrite and reduce it to ammonium. Although no nitrogen compounds of intermediate redox states accumulate, a small percentage (0.02–0.2%) of the nitrite reduced is released as nitrous oxide (N_2O), a greenhouse gas. The electron flow through ferredoxin, Fe_4S_4, and heme can be represented as in **Figure 5.5**.

Nitrite reductase is encoded in the nucleus and synthesized in the cytoplasm with an N-terminal transit peptide that targets it to the plastids. Elevated concen-

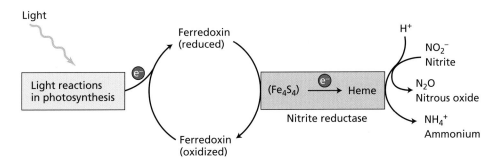

Figure 5.5 Model for coupling of photosynthetic electron flow, via ferredoxin, to the reduction of nitrite by nitrite reductase. The enzyme contains two prosthetic groups, Fe_4S_4 and heme, which participate in the reduction of nitrite to ammonium.

trations of NO_3^- or exposure to light induce the transcription of nitrite reductase mRNA. Accumulation of the end products of nitrate assimilation—the amino acids asparagine and glutamine—represses this induction.

Both roots and shoots assimilate nitrate

In many plants, when the roots receive small amounts of nitrate, nitrate is reduced primarily in the roots. As the supply of nitrate increases, a greater proportion of the absorbed nitrate is translocated to the shoot and assimilated there. Even under similar conditions of nitrate supply, the balance between root and shoot nitrate metabolism—as indicated by the proportion of nitrate reductase activity in each of the two organs or by the relative concentrations of nitrate and reduced nitrogen in the xylem sap—varies from species to species.

In plants such as cocklebur (*Xanthium strumarium*), nitrate metabolism is restricted to the shoot; in other plants, such as white lupine (*Lupinus albus*), most nitrate is metabolized in the roots (**Figure 5.6**). Generally, species native to temperate regions rely more heavily on nitrate assimilation by the roots than do species of tropical or subtropical origins.

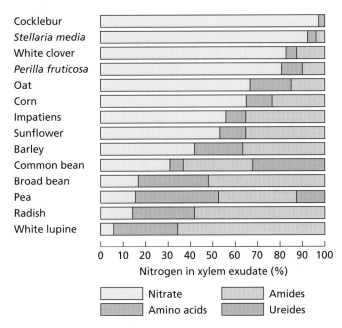

Figure 5.6 Relative amounts of nitrate and other nitrogen compounds in the xylem sap of various plant species. The plants were grown with their roots exposed to nitrate solutions, and xylem sap was collected by severing the stem. Note the presence of ureides in common bean and pea; only legumes of tropical origin export nitrogen from roots to shoots in such compounds. (After Pate 1973.)

glutamine synthetase (GS)
An enzyme that catalyzes the condensation of ammonium and glutamate to form glutamine. The reaction is critical for the assimilation of ammonium into essential amino acids. Two forms of GS exist—one in the cytosol and one in chloroplasts.

glutamate synthase (GOGAT)
An enzyme that transfers the amide group of glutamine to 2-oxoglutarate, yielding two molecules of glutamate. Also known as glutamine:2-oxoglutarate aminotransferase.

glutamate dehydrogenase (GDH)
An enzyme that catalyzes a reversible reaction that synthesizes or deaminates glutamate as part of the nitrogen assimilation process.

Ammonium Assimilation

Plant cells avoid ammonium toxicity by rapidly converting the ammonium generated from nitrate assimilation or photorespiration (see Chapter 8) into amino acids. The primary pathway for this conversion involves the sequential actions of glutamine synthetase and glutamate synthase. In this section we discuss the enzymatic processes that mediate the assimilation of ammonium into essential amino acids, and the role of amides in the regulation of nitrogen and carbon metabolism.

Converting ammonium to amino acids requires two enzymes

Glutamine synthetase (**GS**) combines ammonium with glutamate to form glutamine (**Figure 5.7A**):

$$\text{Glutamate} + NH_4^+ + ATP \rightarrow \text{glutamine} + ADP + P_i \qquad (5.4)$$

This reaction requires the hydrolysis of one ATP and involves a divalent cation such as Mg^{2+}, Mn^{2+}, or Co^{2+} as a cofactor. Plants contain two classes of GS, one in the cytosol and the other in root plastids or shoot chloroplasts. The cytosolic forms are expressed in germinating seeds or in the vascular bundles of roots and shoots and produce glutamine for intercellular nitrogen transport. The GS in root plastids generates amide nitrogen for local consumption; the GS in shoot chloroplasts reassimilates photorespiratory NH_4^+ (see Chapter 8). Light and carbohydrate levels alter the expression of the plastid forms of the enzyme, but they have little effect on the cytosolic forms.

Elevated plastid levels of glutamine stimulate the activity of **glutamate synthase** (also known as *glutamine:2-oxoglutarate aminotransferase*, or **GOGAT**). This enzyme transfers the amide group of glutamine to 2-oxoglutarate, yielding two molecules of glutamate (see Figure 5.7A). Plants contain two types of GOGAT; one accepts electrons from NADH, and the other accepts electrons from ferredoxin (Fd):

$$\text{Glutamine} + \text{2-oxoglutarate} + NADH + H^+ \rightarrow 2 \text{ glutamate} + NAD^+ \qquad (5.5)$$

$$\text{Glutamine} + \text{2-oxoglutarate} + 2 \text{ Fd}_{red} \rightarrow 2 \text{ glutamate} + 2 \text{ Fd}_{ox} \qquad (5.6)$$

The NADH type of the enzyme (NADH-GOGAT) is located in plastids of nonphotosynthetic tissues such as roots or the vascular bundles of developing leaves. In roots, NADH-GOGAT is involved in the assimilation of NH_4^+ absorbed from the rhizosphere (the soil near the surface of the roots); in vascular bundles of developing leaves, NADH-GOGAT assimilates glutamine translocated from roots or senescing leaves.

The ferredoxin-dependent type of glutamate synthase (Fd-GOGAT) is found in chloroplasts and serves in photorespiratory nitrogen metabolism. Both the amount of protein and its activity increase with light levels. Roots, particularly those of plants using nitrate as a nitrogen source, have Fd-GOGAT in plastids. Fd-GOGAT in the roots presumably functions to incorporate the glutamine generated during nitrate assimilation. Electrons to reduce Fd in roots are generated by the oxidative pentose phosphate pathway (see Chapter 11).

Ammonium can be assimilated via an alternative pathway

Glutamate dehydrogenase (**GDH**) catalyzes a reversible reaction that synthesizes or deaminates glutamate (**Figure 5.7B**):

$$\text{2-Oxoglutarate} + NH_4^+ + NAD(P)H \leftrightarrow \text{glutamate} + H_2O + NAD(P)^+ \qquad (5.7)$$

An NADH-dependent form of GDH is found in mitochondria, and an NADPH-dependent form is localized in the chloroplasts of photosynthetic organs. Although both forms are relatively abundant, they cannot substitute for the GS–GOGAT pathway for assimilation of ammonium, and their primary function is in deaminating glutamate during the reallocation of nitrogen (see Figure 5.7B).

(A)

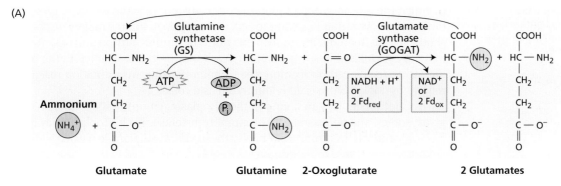

Glutamate   **Glutamine**   **2-Oxoglutarate**   **2 Glutamates**

(B)

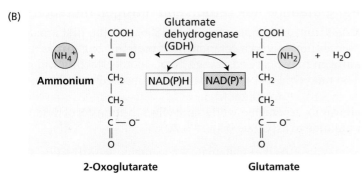

2-Oxoglutarate   **Glutamate**

Figure 5.7 Structure and pathways of compounds involved in ammonium metabolism. Ammonium can be assimilated by one of several processes. (A) The GS-GOGAT pathway that forms glutamine and glutamate. A reduced cofactor is required for the reaction: ferredoxin (Fd) in green leaves and NADH in nonphotosynthetic tissue. (B) The GDH pathway that forms glutamate using NADH or NADPH as a reductant. (C) Transfer of the amino group from glutamate to oxaloacetate to form aspartate (catalyzed by aspartate aminotransferase). (D) Synthesis of asparagine by transfer of an amino acid group from glutamine to aspartate (catalyzed by asparagine synthetase).

(C)

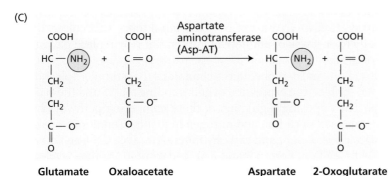

Glutamate   Oxaloacetate   Aspartate   2-Oxoglutarate

(D)

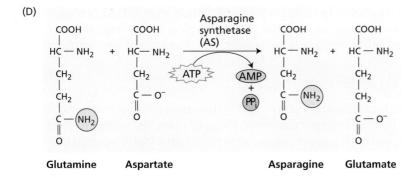

Glutamine   Aspartate   Asparagine   Glutamate

Transamination reactions transfer nitrogen

Once assimilated into glutamine and glutamate, nitrogen is incorporated into other amino acids via transamination reactions. The enzymes that catalyze these reactions are known as aminotransferases. An example is **aspartate aminotransferase (Asp-AT)**, which catalyzes the following reaction (**Figure 5.7C**):

$$\text{Glutamate + oxaloacetate} \rightarrow \text{2-oxoglutarate + aspartate} \qquad (5.8)$$

aspartate aminotransferase (Asp-AT)   An aminotransferase that transfers the amino group from glutamate to the carboxyl atom of oxaloacetate to form aspartate.

asparagine synthetase (AS)
An enzyme that transfers nitrogen as an amino group from glutamine to aspartate, forming asparagine.

in which the amino group of glutamate is transferred to the carboxyl group of oxaloacetate. Aspartate is an amino acid that participates in the malate–aspartate shuttle to transfer reducing equivalents from the mitochondrion and chloroplast into the cytosol and in the transport of carbon from the mesophyll to the bundle sheath for C_4 carbon fixation (see Chapter 8). All transamination reactions require pyridoxal phosphate (vitamin B6) as a cofactor.

Aminotransferases are found in the cytoplasm, chloroplasts, mitochondria, glyoxysomes, and peroxisomes. The aminotransferases in chloroplasts may have a significant role in amino acid biosynthesis, because plant leaves or isolated chloroplasts exposed to radioactively labeled carbon dioxide rapidly incorporate the label into glutamate, aspartate, alanine, serine, and glycine.

Asparagine and glutamine link carbon and nitrogen metabolism

Asparagine, isolated from asparagus as early as 1806, was the first amide to be identified. It serves not only as a component of proteins, but also as a key compound for nitrogen transport and storage because of its stability and high nitrogen-to-carbon ratio (2 N to 4 C for asparagine compared with 2 N to 5 C for glutamine and 1 N to 5 C for glutamate).

The major pathway for asparagine synthesis involves the transfer of the amide nitrogen from glutamine to aspartate (**Figure 5.7D**):

$$\text{Glutamine + aspartate + ATP} \rightarrow \text{glutamate + asparagine + AMP + PP}_i \quad (5.9)$$

Asparagine synthetase (AS), the enzyme that catalyzes this reaction, is found in the cytosol of leaves and roots and in nitrogen-fixing nodules (see the section *Biological Nitrogen Fixation*). In maize (corn; *Zea mays*) roots, particularly those under potentially toxic levels of ammonia, ammonium may replace glutamine as the source of the amide group.

High levels of light and carbohydrate—conditions that stimulate plastid GS and Fd-GOGAT—inhibit the expression of genes coding for AS and the activity of the enzyme. The opposing regulation of these competing pathways helps balance the metabolism of carbon and nitrogen in plants. Conditions of ample energy (i.e., high levels of light and carbohydrates) stimulate GS (see Equation 5.4) and GOGAT (see Equations 5.5 and 5.6), and inhibit AS; thus they favor nitrogen assimilation into glutamine and glutamate, compounds that are rich in carbon and participate in the synthesis of new plant materials.

In contrast, energy-limited conditions inhibit GS and GOGAT, stimulate AS, and thus favor nitrogen assimilation into asparagine, a compound that is rich in nitrogen and sufficiently stable for long-distance transport or long-term storage.

Amino Acid Biosynthesis

Humans and most animals cannot synthesize certain amino acids—histidine, isoleucine, leucine, lysine, methionine, phenylalanine, threonine, tryptophan, and valine, and in the case of young humans, arginine (adult humans can synthesize arginine)—and thus must obtain these so-called essential amino acids from their diet. In contrast, plants synthesize all of the 20 amino acids that are common in proteins. The nitrogen-containing amino group, as discussed in the previous section, derives from transamination reactions with glutamine or glutamate. The carbon skeletons for amino acids derive from 3-phosphoglycerate, phospho*enol*pyruvate, and pyruvate generated during glycolysis, and from 2-oxoglutarate and oxaloacetate generated in the tricarboxylic acid cycle (**Figure 5.8**; see also Chapter 11). Parts of these pathways required for synthesis of the essential amino acids are appropriate targets for herbicides (such as Roundup), because they are missing from animals, so substances that block these pathways are lethal to plants, but at low concentrations do not injure animals.

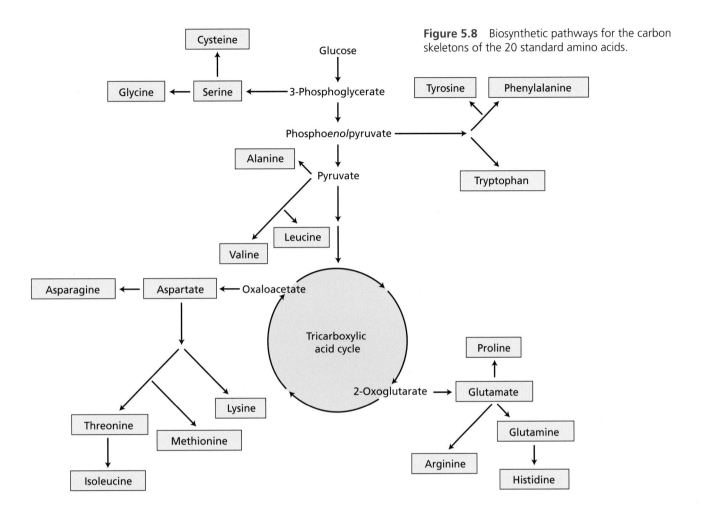

Figure 5.8 Biosynthetic pathways for the carbon skeletons of the 20 standard amino acids.

Biological Nitrogen Fixation

Biological nitrogen fixation accounts for most of the conversion of atmospheric N_2 into ammonium, and thus serves as the key entry point of molecular nitrogen into the biogeochemical cycle of nitrogen (see Figure 5.1). In this section we describe the symbiotic relationship between nitrogen-fixing organisms and higher plants; nodules, the specialized structures that form in roots when infected by nitrogen-fixing bacteria; the genetic and signaling interactions that regulate nitrogen fixation by symbiotic prokaryotes and their hosts; and the properties of the nitrogenase enzymes that fix nitrogen.

Free-living and symbiotic bacteria fix nitrogen

Some bacteria, as stated earlier, can convert atmospheric nitrogen into ammonia (**Table 5.2**). Most of these nitrogen-fixing prokaryotes live in the soil, generally independent of other organisms. Several form symbiotic associations with higher plants in which the prokaryote directly provides the host plant with fixed nitrogen in exchange for other nutrients and carbohydrates (see Table 5.2, top). Such symbioses occur in nodules that form on the roots of the plant and contain the nitrogen-fixing bacteria.

The most common type of symbiosis occurs between members of the plant family Fabaceae (Leguminosae) and soil bacteria of the genera *Azorhizobium*, *Bradyrhizobium*, *Mesorhizobium*, *Rhizobium*, and *Sinorhizobium* (collectively called **rhizobia**; **Table 5.3** and **Figure 5.9**). Another common type of symbiosis occurs between several woody plant species, such as alder trees, and soil bacteria of the

rhizobia A collective term for the genera of soil bacteria that form symbiotic (mutualistic) relationships with members of the plant family Fabaceae (Leguminosae).

Table 5.2   Examples of organisms that can carry out nitrogen fixation

SYMBIOTIC NITROGEN FIXATION

Host plant	N-fixing symbionts
Leguminous: legumes, *Parasponia*	*Azorhizobium, Bradyrhizobium, Mesorhizobium, Rhizobium, Sinorhizobium*
Actinorhizal: alder (tree), *Ceanothus* (shrub), *Casuarina* (tree), *Datisca* (shrub)	*Frankia*
Gunnera	*Nostoc*
Azolla (water fern)	*Anabaena*
Sugarcane	*Acetobacter*
Miscanthus	*Azospirillum*

FREE-LIVING NITROGEN FIXATION

Type	N-fixing genera
Cyanobacteria (formerly called blue-green algae)	*Anabaena, Calothrix, Nostoc*
Other bacteria	
Aerobic	*Azospirillum, Azotobacter, Beijerinckia, Derxia*
Facultative	*Bacillus, Klebsiella*
Anaerobic	
Nonphotosynthetic	*Clostridium, Methanococcus* (archaebacterium)
Photosynthetic	*Chromatium, Rhodospirillum*

genus *Frankia*; these plants are known as **actinorhizal** plants. Still other types of nitrogen-fixing symbioses involve the South American herb *Gunnera* and the tiny water fern *Azolla*, which form associations with the cyanobacteria *Nostoc* and *Anabaena*, respectively (**Figure 5.10**; also see Table 5.2). Finally, several types of nitrogen-fixing bacteria are associated with C_4 grasses such as sugarcane and *Miscanthus*.

Nitrogen fixation requires microaerobic or anaerobic conditions

Because nitrogen fixation involves the expenditure of large amounts of energy, the nitrogenase enzymes that catalyze these reactions have sites that facilitate the high-energy exchange of electrons. Oxygen, being a strong electron acceptor, can damage these sites and irreversibly inactivate nitrogenase, so nitrogen must be fixed under anaerobic conditions. Each of the nitrogen-fixing organisms listed in Table 5.2 either functions under natural anaerobic conditions or creates an internal, local anaerobic environment (microaerobic conditions) separated from the oxygen in the atmosphere that surrounds it.

In cyanobacteria, anaerobic conditions are created in specialized cells called *heterocysts* (see Figure 5.10). Heterocysts are thick-walled cells that differentiate when filamentous cyanobacteria are deprived of NH_4^+. These cells lack photosystem II, the oxygen-producing photosystem of chloroplasts (see Chapter 7), so they do not generate

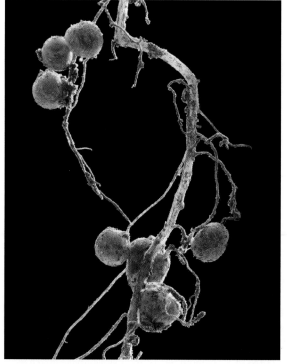

1 mm

Figure 5.9   Root nodules on a common bean (*Phaseolus vulgaris*). The nodules, the spherical structures, are a result of infection by *Rhizobium* spp. (Photo by David McIntyre.)

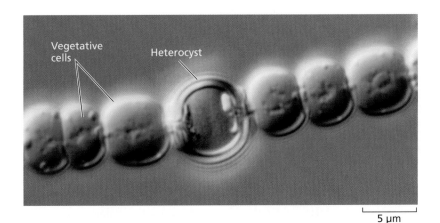

actinorhizal Pertaining to several woody plant species, such as alder trees, in which symbiosis occurs with soil bacteria of the nitrogen-fixing genus *Frankia*.

Figure 5.10 Heterocyst in a filament of the nitrogen-fixing cyanobacterium *Anabaena*, which forms associations with the water fern *Azolla*. The thick-walled heterocysts, interspersed among vegetative cells, have an anaerobic inner environment that allows cyanobacteria to fix nitrogen in aerobic conditions. (© Dr. Peter Siver/Visuals Unlimited.)

oxygen. Heterocysts appear to represent an adaptation for nitrogen fixation, in that they are widespread among aerobic cyanobacteria that fix nitrogen.

Cyanobacteria can fix nitrogen under anaerobic conditions such as those that occur in flooded fields. In Asian countries, nitrogen-fixing cyanobacteria of both the heterocyst and nonheterocyst types are a major means for maintaining an adequate nitrogen supply in the soil of rice fields. These microorganisms fix nitrogen when the fields are flooded and die as the fields dry, releasing the fixed nitrogen to the soil. Another important source of available nitrogen in flooded rice fields is *Azolla*, which associates with the cyanobacterium *Anabaena*. The *Azolla–Anabaena* association can fix as much as 0.5 kg of atmospheric nitrogen per hectare per day, a rate of fertilization that is sufficient to attain moderate rice yields.

Free-living bacteria that are capable of fixing nitrogen are aerobic, facultative, or anaerobic (see Table 5.2, bottom):

- *Aerobic* nitrogen-fixing bacteria such as *Azotobacter* are thought to maintain a low oxygen concentration (microaerobic conditions) through their high levels of respiration. Others, such as *Gloeothece*, evolve O_2 photosynthetically during the day and fix nitrogen during the night when respiration lowers oxygen levels.

Table 5.3 Associations between host plants and rhizobia

Plant host	Rhizobial symbiont
Parasponia (a nonlegume, formerly called *Trema*)	*Bradyrhizobium* spp.
Soybean (*Glycine max*)	*Bradyrhizobium japonicum* (slow-growing type); *Sinorhizobium fredii* (fast-growing type)
Alfalfa (*Medicago sativa*)	*Sinorhizobium meliloti*
Sesbania (aquatic)	*Azorhizobium* (forms both root and stem nodules; the stems have adventitious roots)
Bean (*Phaseolus*)	*Rhizobium leguminosarum* bv. *phaseoli; R. tropici; R. etli*
Clover (*Trifolium*)	*Rhizobium leguminosarum* bv. *trifolii*
Pea (*Pisum sativum*)	*Rhizobium leguminosarum* bv. *viciae*
Aeschynomene (aquatic)	Photosynthetic *Bradyrhizobium* clade (photosynthetically active rhizobia that form stem nodules, probably associated with adventitious roots)

nodules   Specialized organs of a plant host that contain symbiotic nitrogen-fixing bacteria.

leghemoglobin   An oxygen-binding heme protein produced by legumes during nitrogen-fixing symbioses with rhizobia. Found in the cytoplasm of infected nodule cells, it facilitates the diffusion of oxygen to the respiring symbiotic bacteria.

nodulin genes   Plant genes specific to nodule formation.

nodulation (*nod*) genes   Rhizobial genes, the products of which participate in nodule formation.

- *Facultative* organisms, which are able to grow under both aerobic and anaerobic conditions, generally fix nitrogen only under anaerobic conditions.

- Obligate *anaerobic* nitrogen-fixing bacteria that grow in environments devoid of oxygen can be either photosynthetic (e.g., *Rhodospirillum*) or nonphotosynthetic (e.g., *Clostridium*).

Symbiotic nitrogen fixation occurs in specialized structures

Some symbiotic nitrogen-fixing prokaryotes dwell within **nodules**, the special organs of the plant host that enclose the nitrogen-fixing bacteria (see Figure 5.9). In the case of *Gunnera*, these organs are preexisting stem glands that develop independently of the symbiont. In the case of legumes and actinorhizal plants, the nitrogen-fixing bacteria induce the plant to form root nodules.

Grasses can also develop symbiotic relationships with nitrogen-fixing organisms, but in these associations root nodules are not produced. Instead, the nitrogen-fixing bacteria anchor to the root surfaces, mainly around the elongation zone and the root hairs, or live as endophytes, colonizing plant tissues without causing disease. For example, the nitrogen-fixing bacteria *Acetobacter diazotrophicus* and *Herbaspirillum* spp. live in the apoplast of stem tissues in sugarcane and may provide their host with about 30% of its nitrogen, which lessens the need for fertilization. The potential for associative and endophytic nitrogen-fixing bacteria to supplement the nitrogen nutrition of maize, rice, and other grains has been explored, but the diversity of bacterial species found on roots and in tissues, and the variety of plant responses to these bacteria, have impeded progress.

Legumes and actinorhizal plants regulate gas permeability in their nodules, maintaining oxygen concentrations of 20 to 40 nanomolar (nM) within the nodule (about 10,000 times lower than equilibrium concentrations in water). These levels can support respiration but are sufficiently low to avoid inactivation of the nitrogenase. Gas permeability increases in the light and decreases under drought or upon exposure to nitrate. The mechanism for regulating gas permeability is not yet known, but it may involve potassium ion fluxes into and out of infected cells.

Nodules contain oxygen-binding heme proteins called **leghemoglobins**. Leghemoglobins are the most abundant proteins in nodules, giving them a heme-pink color, and are crucial for symbiotic nitrogen fixation. Leghemoglobins have a high affinity for oxygen (a K_m of about 10 nM), about ten times higher than the β chain of human hemoglobin.

Although leghemoglobins were once thought to provide a buffer for nodule oxygen, more recent studies indicate that they store only enough oxygen to support nodule respiration for a few seconds. Their function is to increase the rate of oxygen transport to the respiring symbiotic bacterial cells, which decrease substantially the steady-state level of oxygen in infected cells. To continue aerobic respiration under such conditions, the rhizobia use a specialized electron transport chain (see Chapter 11) in which the terminal oxidase has an affinity for oxygen even higher than that of leghemoglobins, a K_m of about 7 nM.

Establishing symbiosis requires an exchange of signals

The symbiosis between legumes and rhizobia is not obligatory. Legume seedlings germinate without any association with rhizobia, and they may remain unassociated throughout their life cycle. Rhizobia also occur as free-living organisms in the soil. Under nitrogen-limited conditions, however, the symbionts seek each other out through an elaborate exchange of signals. This signaling, the subsequent infection process, and the development of nitrogen-fixing nodules involve specific genes in both the host and the symbionts.

Plant genes specific to nodules are called **nodulin genes**; rhizobial genes that participate in nodule formation are called **nodulation (*nod*) genes**. The *nod* genes are classified as common *nod* genes or host-specific *nod* genes. The common *nod* genes—*nodA*, *nodB*, and *nodC*—are found in all rhizobial strains; the host-specific *nod* genes—such as *nodP*, *nodQ*, and *nodH*; or *nodF*, *nodE*, and *nodL*—differ among rhizobial species and determine the host range (the plants that can be infected). Only one of the *nod* genes, the regulatory *nodD*, is constitutively expressed, and as we will explain in detail, its protein product (NodD) regulates the transcription of the other *nod* genes.

The first stage in the formation of the symbiotic relationship between the nitrogen-fixing bacteria and their host is migration of the bacteria toward the roots of the host plant. This migration is a chemotactic response mediated by chemical attractants, especially (iso)flavonoids and betaines, secreted by the roots. These attractants activate the rhizobial NodD protein, which then induces transcription of the other *nod* genes. The promoter region of all *nod* operons, except that of *nodD*, contains a highly conserved sequence called the *nod* box. Binding of the activated NodD to the *nod* box induces transcription of the other *nod* genes.

Nod factors produced by bacteria act as signals for symbiosis

The *nod* genes, which NodD activates, code for nodulation proteins, most of which are involved in the biosynthesis of Nod factors. **Nod factors** are lipochitin oligosaccharide signal molecules, all of which have a chitin β-1,4-linked *N*-acetyl-D-glucosamine backbone (varying in length from three to six sugar units) and a fatty acid chain on the C-2 position of the nonreducing sugar (**Figure 5.11**).

Three of the *nod* genes (*nodA*, *nodB*, and *nodC*) encode enzymes (NodA, NodB, and NodC, respectively) that are required for synthesizing this basic structure:

1. NodA is an *N*-acyltransferase that catalyzes the addition of a fatty acyl chain.

2. NodB is a chitin-oligosaccharide deacetylase that removes the acetyl group from the terminal nonreducing sugar.

3. NodC is a chitin-oligosaccharide synthase that links *N*-acetyl-D-glucosamine monomers.

Host-specific NOD factors that vary among rhizobial species are involved in the modification of the fatty acyl chain or the addition of groups important in determining host specificity:

• NodE and NodF determine the length and degree of saturation of the fatty acyl chain; those of *Rhizobium leguminosarum* bv. *viciae* and *Sinorhizobium meliloti* result in the synthesis of an 18:4 and a 16:2 fatty acyl group, respectively. (The number before the colon gives the total number of carbons in the fatty acyl chain, and the number after the colon gives the number of double bonds; see Chapter 11.)

• Other enzymes, such as NodL, influence the host specificity of Nod factors through the addition of specific substitutions at the reducing or nonreducing sugar moieties of the chitin backbone.

A particular legume host responds to a specific Nod factor. The legume receptors for Nod factors are protein kinases with extracellular sugar-binding LysM domains (for lysin motif, a widespread protein module originally identified in enzymes that degrade bacterial cell walls, but also present in many other proteins) in the root hairs. Nod factors activate these domains, inducing oscillations in the concentrations of free calcium ions in the nuclear regions of

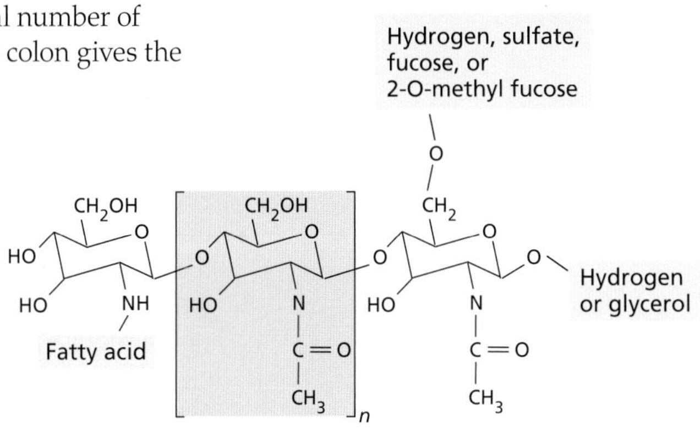

Nod factors Lipochitin oligosaccharide signal molecules active in regulating gene expression during nitrogen-fixing nodule formation. All Nod factors have a chitin β-1,4-linked *N*-acetyl-D-glucosamine backbone (varying in length from three to six sugar units) and a fatty acid chain on the C-2 position of the nonreducing sugar.

Figure 5.11 Nod factors are lipochitin oligosaccharides. The fatty acid chain typically has 16 to 18 carbons. The number of repeated middle sections (*n*) is usually two or three. (After Stokkermans et al. 1995.)

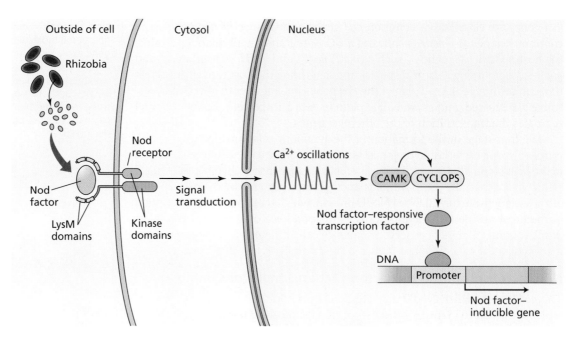

Figure 5.12 Nod factors induce expression of specific genes in the host plant.

root epidermal cells (**Figure 5.12**). Recognition of the Ca^{2+} oscillations requires a Ca^{2+}/calmodulin-dependent protein kinase (CaMK) that is associated with a protein of unknown function named CYCLOPS. Once the plant epidermal cell recognizes ongoing Ca^{2+} oscillations, Nod factor–responsive transcriptional regulators directly associate with the promoters of Nod factor–inducible genes. The overall process links Nod factor perception at the plasma membrane to gene expression changes in the nucleus and is called the symbiotic pathway because it shares elements with the process through which arbuscular mycorrhizal fungi initially interact with their hosts (see Chapter 4).

Nodule formation involves phytohormones

Two processes—infection and nodule organogenesis—occur simultaneously during root nodule formation. Rhizobia usually infect root hairs by first releasing Nod factors that induce a pronounced curling of the root hair cells (**Figure 5.13A and B**). The rhizobia become enclosed in the small compartment formed by the curling. The cell wall of the root hair degrades in these regions, also in response to Nod factors, allowing the bacterial cells direct access to the outer surface of the plant plasma membrane.

The next step is formation of the **infection thread** (**Figure 5.13C**), an internal tubular extension of the plasma membrane that is produced by the fusion of Golgi-derived membrane vesicles at the site of infection. The thread grows at its tip by the fusion of secretory vesicles to the end of the tube. Deeper into the root cortex, near the xylem, cortical cells dedifferentiate and start dividing, forming a distinct area within the cortex, called a *nodule primordium*, from which the nodule will develop. The nodule primordia form opposite the protoxylem poles of the root vascular bundle.

Different signaling compounds, acting either positively or negatively, control the development of nodule primordia. Nod factors activate localized cytokinin signaling in the root cortex and pericycle, leading to the localized suppression of polar auxin transport, which in turn induces nodule morphogenesis and stimulates cell division. Ethylene is synthesized in the region of the pericycle, diffuses into the cortex, and blocks cell division opposite the phloem poles of the root.

infection thread An internal tubular extension of the plasma membrane of root hairs through which rhizobia enter root cortical cells.

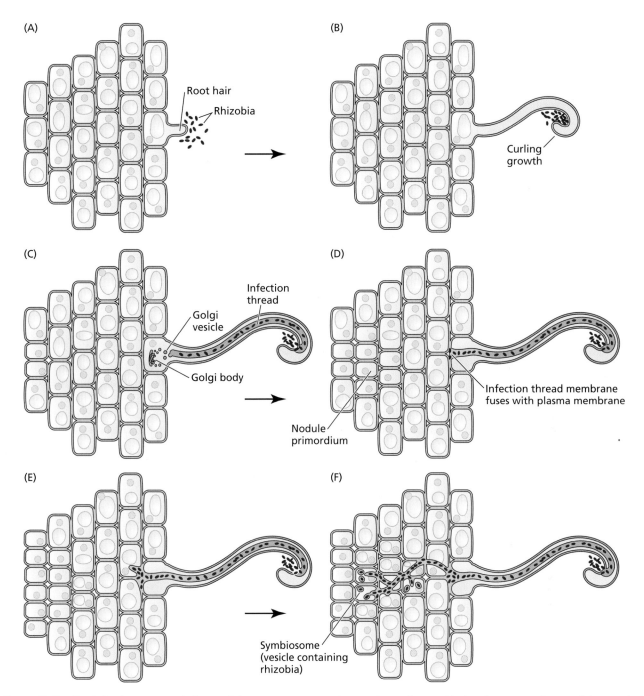

Figure 5.13 The infection process during nodule organogenesis. (A) Rhizobia bind to an emerging root hair in response to chemical attractants sent by the plant. (B) In response to factors produced by the bacteria, the root hair exhibits abnormal curling growth, and rhizobia cells proliferate in the coils. (C) Localized degradation of the root hair wall leads to infection and formation of the infection thread from Golgi secretory vesicles of root cells. (D) The infection thread reaches the end of the cell, and its membrane fuses with the plasma membrane of the root hair cell. (E) Rhizobia are released into the apoplast and penetrate the compound middle lamella to the subepidermal cell plasma membrane, leading to the initiation of a new infection thread, which forms an open channel with the first. (F) The infection thread extends and branches until it reaches target cells, where symbiosomes composed of plant membrane enclosing bacterial cells are released into the cytosol.

The infection thread filled with proliferating rhizobia elongates through the root hair and cortical cell layers, in the direction of the nodule primordium (**Figure 5.13D and E**). When the infection thread reaches the nodule primordium, its tip fuses with the plasma membrane of a host cell and penetrates into

bacteroids Endosymbiotic bacteria that have differentiated into a nondividing nitrogen-fixing state.

nitrogenase enzyme complex
The two-component protein complex that catalyzes the biological nitrogen fixation reaction in which ammonia is produced from molecular nitrogen.

the cytoplasm (**Figure 5.13F**). Bacterial cells are subsequently released into the cytoplasm, surrounded by the host plasma membrane, resulting in the formation of an organelle called the *symbiosome*. Branching of the infection thread inside the nodule enables the bacteria to infect many cells.

At first the bacteria in symbiosomes continue to divide, and the surrounding symbiosome membrane (also called the *peribacteroid membrane*) increases in surface area to accommodate this growth by fusing with smaller vesicles. Soon thereafter, upon an undetermined signal from the plant, the bacteria stop dividing and begin to differentiate into nitrogen-fixing **bacteroids**.

The nodule as a whole develops such features as a vascular system (which facilitates the exchange of fixed nitrogen produced by the bacteroids for nutrients contributed by the plant) and a layer of cells to exclude O_2 from the root nodule interior. In some temperate legumes (e.g., pea), the nodules are elongated and cylindrical because of the presence of a *nodule meristem*. The nodules of tropical legumes, such as soybean and peanut, lack a persistent meristem and are spherical.

The nitrogenase enzyme complex fixes N$_2$

Biological nitrogen fixation, like industrial nitrogen fixation, produces ammonia from molecular nitrogen. The overall reaction is

$$N_2 + 8\ e^- + 8\ H^+ + 16\ ATP \rightarrow 2\ NH_3 + H_2 + 16\ ADP + 16\ P_i \qquad (5.10)$$

Note that the reduction of N_2 to two NH_3, a six-electron transfer, is coupled to the reduction of two protons to evolve H_2. The **nitrogenase enzyme complex** catalyzes this reaction.

The nitrogenase enzyme complex can be separated into two components—the Fe protein and the MoFe protein—neither of which has catalytic activity by itself (**Figure 5.14**):

- The Fe protein is the smaller of the two components and has two identical subunits that vary in mass from 30 to 72 kDa each, depending on the bacterial species. Each subunit contains an iron–sulfur cluster (four Fe^{2+}/Fe^{3+} and four S^{2-}), which participates in the redox reactions that convert N_2 to NH_3. The Fe protein is irreversibly inactivated by O_2 with typical half-decay times of 30 to 45 s.

- The MoFe protein has four subunits, with a total molecular mass of 180 to 235 kDa, depending on the bacterial species. Each subunit has two Mo–Fe–S clusters. The MoFe protein is also inactivated by O_2, with a half-decay time in air of 10 min.

Figure 5.14 The reaction catalyzed by nitrogenase. Ferredoxin reduces the Fe protein. Binding and hydrolysis of ATP to the Fe protein are thought to cause a conformational change of the Fe protein that facilitates the redox reactions. The Fe protein reduces the MoFe protein, and the MoFe protein reduces the N_2. (After Dixon and Wheeler 1986; Buchanan et al. 2000.)

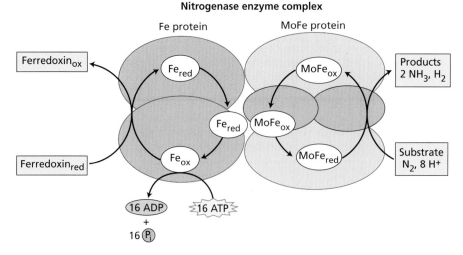

In the overall nitrogen reduction reaction (see Figure 5.14), ferredoxin serves as an electron donor to the Fe protein, which in turn hydrolyzes ATP and reduces the MoFe protein. The MoFe protein can then reduce numerous substrates (**Table 5.4**), although under natural conditions it reacts only with N_2 and H^+. One of the reactions catalyzed by nitrogenase, the reduction of acetylene to ethylene, is used in estimating nitrogenase activity.

The energetics of nitrogen fixation are complex. The production of NH_3 from N_2 and H_2 is an exergonic reaction, with a $\Delta G^{0'}$ (change in free energy) of -27 kJ mol^{-1}. However, industrial production of NH_3 from N_2 and H_2 is *endergonic*, requiring a very large energy input because of the activation energy needed to break the triple bond in N_2. For the same reason, the enzymatic reduction of N_2 by nitrogenase also requires a large investment of energy (see Equation 5.10), although the exact changes in free energy are not yet known.

Calculations based on the carbohydrate metabolism of legumes show that a plant respires 9.3 moles of CO_2 per mole of N_2 fixed. On the basis of Equation 5.10, the $\Delta G^{0'}$ for the overall reaction of biological nitrogen fixation is about -200 kJ mol^{-1}. Because the overall reaction is highly exergonic, ammonium production is limited by the slow operation (the number of N_2 molecules reduced per unit of time is about 5 s^{-1}) of the nitrogenase complex. To compensate for this slow turnover rate, the bacteroid synthesizes large amounts of nitrogenase (up to 20% of the total protein in the cell).

Under natural conditions, substantial amounts of H^+ are reduced to H_2 gas, and this process can compete with N_2 reduction for electrons from nitrogenase. In rhizobia, 30 to 60% of the energy supplied to nitrogenase may be lost as H_2, diminishing the efficiency of nitrogen fixation. Some rhizobia, however, contain hydrogenase, an enzyme that can split the H_2 formed and generate electrons for N_2 reduction, thus improving the efficiency of nitrogen fixation.

Table 5.4	Reactions catalyzed by nitrogenase
$N_2 \rightarrow NH_3$	Molecular nitrogen fixation
$N_2O \rightarrow N_2 + H_2O$	Nitrous oxide reduction
$N_3^- \rightarrow N_2 + NH_3$	Azide reduction
$C_2H_2 \rightarrow C_2H_4$	Acetylene reduction
$2 H^+ \rightarrow H_2$	H_2 production
$ATP \rightarrow ADP + P_i$	ATP hydrolytic activity

Amides and ureides are the transported forms of nitrogen

The symbiotic nitrogen-fixing prokaryotes release ammonia that, to avoid toxicity, must be rapidly converted into organic forms in the root nodules before being transported to the shoot via the xylem. Nitrogen-fixing legumes can be classified as amide exporters or ureide exporters, depending on the composition of the xylem sap. Amides (principally the amino acids asparagine or glutamine) are exported by temperate-region legumes, such as pea (*Pisum*), clover (*Trifolium*), broad bean (*Vicia*), and lentil (*Lens*).

Ureides are exported by legumes of tropical origin, such as soybean (*Glycine*), common bean (*Phaseolus*), peanut (*Arachis*), and southern pea (*Vigna*). The three major ureides are allantoin, allantoic acid, and citrulline (**Figure 5.15**). Allantoin is synthesized in peroxisomes from uric acid, and allantoic acid is synthesized

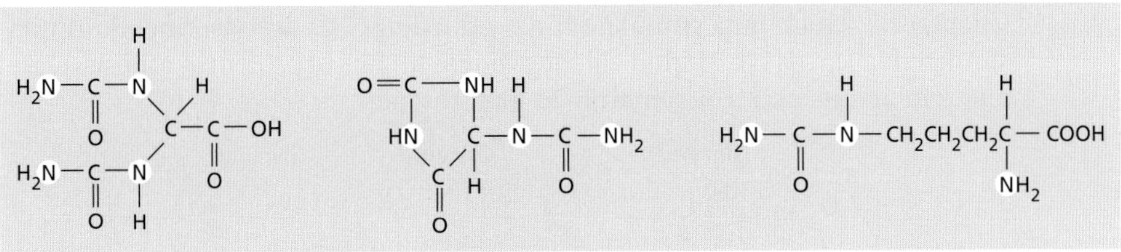

Allantoic acid   **Allantoin**   **Citrulline**

Figure 5.15   The major ureide compounds used to transport nitrogen from sites of fixation to sites where their deamination will provide nitrogen for amino acid and nucleoside synthesis.

from allantoin in the endoplasmic reticulum. The site of citrulline synthesis from the amino acid ornithine has not yet been determined. All three compounds are ultimately released into the xylem and transported to the shoot, where they are rapidly catabolized to ammonium. This ammonium enters the assimilation pathway described earlier.

Sulfur Assimilation

Sulfur is among the most versatile elements in living organisms. Disulfide bridges in proteins play structural and regulatory roles (see Chapter 8). Sulfur participates in electron transport through iron–sulfur clusters (see Chapters 7 and 11). The catalytic sites for several enzymes and coenzymes, such as urease and coenzyme A, contain sulfur. Secondary metabolites (compounds that are not involved in primary pathways of growth and development) that contain sulfur range from the rhizobial Nod factors discussed in the previous section to the antiseptic alliin in garlic and the anticarcinogen sulforaphane in broccoli.

The versatility of sulfur derives in part from the property that it shares with nitrogen: *multiple stable oxidation states*. In this section we discuss sulfur assimilation into the two sulfur-containing amino acids, cysteine and methionine.

Sulfate is the form of sulfur transported into plants

Most of the sulfur in higher-plant cells derives from sulfate (SO_4^{2-}) transported via an H^+–SO_4^{2-} symporter (see Chapter 6) from the soil solution. Sulfate in the soil comes predominantly from the weathering of parent rock material. Industrialization, however, adds an additional source of sulfate: atmospheric pollution. The burning of fossil fuels releases several gaseous forms of sulfur, including sulfur dioxide (SO_2) and hydrogen sulfide (H_2S), which find their way to the soil in rain.

In the gaseous phase, sulfur dioxide reacts with a hydroxyl radical and oxygen to form sulfur trioxide (SO_3). SO_3 dissolves in water to become sulfuric acid (H_2SO_4), a strong acid, which is the major source of acid rain. Plants can metabolize sulfur dioxide taken up in the gaseous form through their stomata. Nonetheless, prolonged exposure (more than 8 h) to high atmospheric concentrations (greater than 0.3 ppm) of SO_2 causes extensive tissue damage because of the formation of sulfuric acid.

Sulfate assimilation occurs mostly in leaves

The first steps in the synthesis of sulfur-containing organic compounds involve the reduction of sulfate and synthesis of the amino acid cysteine. The reduction of sulfate to cysteine changes the oxidation number of sulfur from +6 to –2, thus involving the transfer of eight electrons. Glutathione, ferredoxin, NAD(P)H, or O-acetylserine may serve as electron donors at various steps of the pathway, which we do not present in detail here. Leaves are generally much more active than roots in sulfur assimilation, presumably because photosynthesis provides reduced ferredoxin (see Chapter 7), and photorespiration generates serine (see Chapter 8), which may stimulate the production of O-acetylserine. Sulfur assimilated in leaves is exported via the phloem to sites of protein synthesis (shoot and root apices, and fruits) mainly as glutathione:

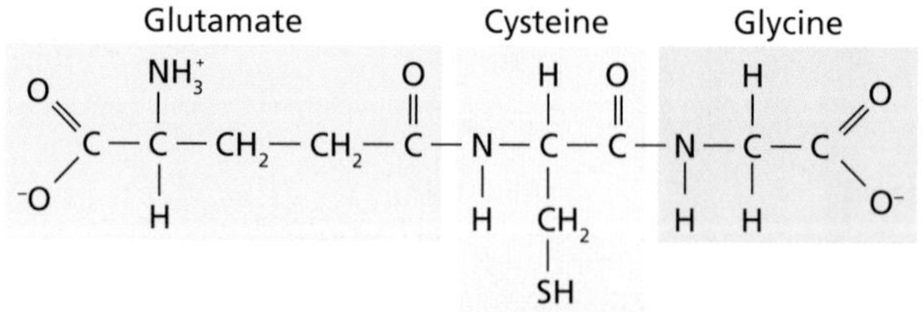

Reduced glutathione

Glutathione also acts as a signal that coordinates the transport of sulfate into roots and the assimilation of sulfate by the shoot.

Methionine is synthesized from cysteine

Methionine, the other sulfur-containing amino acid found in proteins, is synthesized in plastids from cysteine. After cysteine and methionine are synthesized, sulfur can be incorporated into proteins and a variety of other compounds, such as acetyl-CoA and S-adenosylmethionine. The former compound is used for the transfer of acetyl groups—that is, in intermediary metabolism such as the tricarboxylic acid cycle and the glyoxylate cycle (see Chapter 11)—while the latter is important in the synthesis of ethylene (see Chapter 12).

Phosphate Assimilation

Phosphate (HPO_4^{2-}) in the soil solution is readily transported into plant roots via an H^+–HPO_4^{2-} symporter (see Chapter 6) and incorporated into a variety of organic compounds, including sugar phosphates, phospholipids, and nucleotides. The main entry point of phosphate into assimilatory pathways occurs during the formation of ATP, the energy "currency" of the cell. In the overall reaction for this process, inorganic phosphate is added to the second phosphate group in adenosine diphosphate to form a phosphate ester bond.

In mitochondria, the energy for ATP synthesis derives from the oxidation of NADH or succinate by oxidative phosphorylation (see Chapter 11). ATP synthesis is also driven by light-dependent photophosphorylation in the chloroplasts (see Chapter 7). In addition to these reactions in mitochondria and chloroplasts, reactions in the cytosol such as glycolysis also assimilate phosphate.

Glycolysis incorporates inorganic phosphate into 1,3-bisphosphoglyceric acid, forming a high-energy acyl phosphate group. This phosphate can be donated to ADP to form ATP in a substrate-level phosphorylation reaction (see Chapter 11). Once incorporated into ATP, the phosphate group may be transferred via many different reactions to form the various phosphorylated compounds found in higher-plant cells.

Iron Assimilation

Cations taken up by plant cells form complexes with organic compounds in which the cation becomes bound to the complex by noncovalent bonds. Plants assimilate macronutrient cations such as potassium, magnesium, and calcium ions, as well as micronutrient cations such as copper, iron, manganese, cobalt, sodium, and zinc ions, in this manner.

Roots modify the rhizosphere to acquire iron

Iron is important in iron–sulfur proteins (see Chapter 7) and as a catalyst in enzyme-mediated redox reactions (see Chapter 4), such as those of nitrogen metabolism discussed earlier. Plants obtain iron from the soil, where it is present primarily as ferric iron (Fe^{3+}) in oxides such as $Fe(OH)^{2+}$, $Fe(OH)_3$, and $Fe(OH)_4^-$. At neutral pH, ferric iron is highly insoluble. To absorb sufficient amounts of iron from the soil solution, roots have developed several mechanisms that increase iron solubility and thus its availability (**Figure 5.16**). These mechanisms include:

- Soil acidification, which increases the solubility of ferric iron, followed by the reduction of ferric iron to the more soluble ferrous form (Fe^{2+}).

- Release of compounds that form stable, soluble complexes with iron. Recall from Chapter 4 that such compounds are called iron chelators (see Figure 4.3).

Figure 5.16 Two processes through which plant roots absorb iron. (A) A process common to eudicots such as pea, tomato, and soybean, which includes, from top to bottom, proton export, chelation and reduction to the ferrous form, and uptake of ferrous iron. The chelators include organic compounds such as malic acid, citric acid, phenolics, and piscidic acid. (B) A process common to grasses such as barley, maize, and oat. After the grass excretes the siderophore and it removes iron from soil particles, the complex may degrade and release the iron to the soil, exchange iron for another ligand, or be transported into the root. (After Guerinot and Yi 1994.)

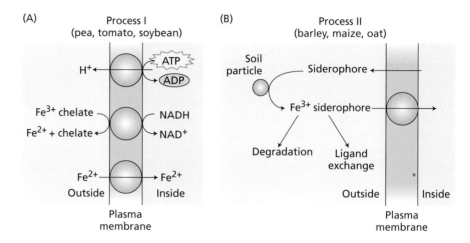

Roots generally acidify the soil around them. They export protons during the import and assimilation of cations, particularly ammonium, and release organic acids, such as malic acid and citric acid, which enhance iron and phosphate availability (see Figure 4.5). Iron deficiencies stimulate the export of protons by roots. In addition, plasma membranes in roots contain an enzyme, called *iron-chelate reductase*, that reduces ferric iron (Fe^{3+}) to the ferrous (Fe^{2+}) form, with cytosolic NADH or NADPH serving as the electron donor (see Figure 5.16A). The activity of this enzyme increases under iron deprivation.

Several compounds secreted by roots form stable chelates with iron. Examples include malic acid, citric acid, phenolics, and piscidic acid. Grasses produce a special class of iron chelators called *siderophores*. Siderophores are made of amino acids that are not found in proteins, such as mugineic acid, and form highly stable complexes with Fe^{3+}. Root cells of grasses have Fe^{3+}–siderophore transport systems in their plasma membranes that bring the chelate into the cytoplasm. Under iron deficiency, grass roots release more siderophores into the soil and increase the capacity of their Fe^{3+}–siderophore transport system (see Figure 5.16B).

Iron cations form complexes with carbon and phosphate

After the roots absorb iron cations or an iron chelate, they oxidize it to a ferric form and translocate much of it to the leaves as an electrostatic complex with citrate or nicotianamine.

Once in the leaves, the iron cation is reduced to Fe^{2+}, after which it undergoes an important assimilatory reaction catalyzed by the enzyme ferrochelatase through which it is inserted into the porphyrin precursor of heme groups found in the cytochromes located in chloroplasts and mitochondria (**Figure 5.17**). Most of the iron in the plant is found in heme groups. In addition, iron–sulfur proteins of the electron transport chain contain nonheme iron covalently bound to the sulfur atoms of cysteine residues in the apoprotein.

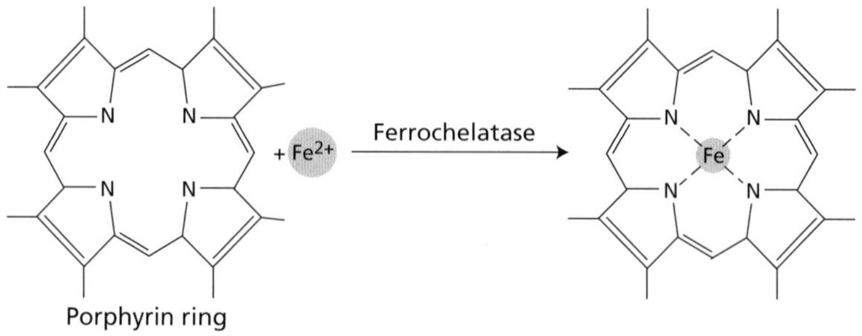

Figure 5.17 The ferrochelatase reaction. The enzyme ferrochelatase catalyzes the insertion of iron into the porphyrin ring to form a coordination complex.

Porphyrin ring $+ Fe^{2+}$ Ferrochelatase →

Free iron (iron ions that are not complexed with carbon compounds) may interact with oxygen to form highly damaging hydroxyl radicals, HO•. Plant cells may limit such damage by storing surplus iron ions in an iron–protein complex called **ferritin**. Mutants of Arabidopsis show that, although ferritins are essential for protection against oxidative damage, they do not serve as a major iron pool for either seedling development or proper functioning of the photosynthetic apparatus. Ferritin consists of a protein shell with 24 identical subunits forming a hollow sphere that has a molecular mass of about 480 kDa. Within this sphere is a core of 5400 to 6200 iron atoms present as a ferric oxide–phosphate complex.

How iron is released from ferritin is uncertain, but breakdown of the protein shell appears to be involved. The level of free iron in plant cells regulates the de novo biosynthesis of ferritin. Interest in ferritin is high because iron in this protein-bound form may be highly available to humans, and foods rich in ferritin, such as soybean, may address dietary anemia problems.

ferritin A protein that functions in cellular iron storage in multiple compartments, including plastids and mitochondria.

The Energetics of Nutrient Assimilation

Nutrient assimilation—particularly of nitrogen and sulfur—requires large amounts of energy to convert stable, low-energy, highly oxidized inorganic compounds into high-energy, highly reduced organic compounds:

- As you have seen, in nitrate (NO_3^-) assimilation, the nitrogen in NO_3^- is converted to a higher-energy (more reduced) form in nitrite (NO_2^-), then to a yet-higher-energy (even more reduced) form in ammonium (NH_4^+), and finally into the amide nitrogen of the amino acid glutamine. This process consumes the equivalent of 12 ATPs per amide nitrogen.

- The process of biological nitrogen fixation, together with the subsequent assimilation of NH_3 into an amino acid, consumes the equivalent of about 16 ATPs per amide nitrogen.

- The assimilation of sulfate (SO_4^{2-}) into the amino acid cysteine consumes about 14 ATPs per S atom assimilated.

For some perspective on the enormous energies involved, consider that if these reactions run rapidly in reverse—say, from NH_4NO_3 (ammonium nitrate) to N_2—they become explosive, liberating vast amounts of energy as motion, heat, and light. Nearly all explosives, including nitroglycerin, TNT (trinitrotoluene), and gunpowder, are based on the rapid oxidation of nitrogen or sulfur compounds.

The consequence of the high cost of nitrate assimilation is that it accounts for about 25% of the total energy expenditure in both roots and shoots. Thus, a plant may use one-fourth of its energy to assimilate nitrogen, a constituent that accounts for less than 2% of the total dry weight of the plant.

Summary

Nutrient assimilation is the often energy-requiring process by which plants incorporate inorganic nutrients into the carbon constituents necessary for growth and development.

Nitrogen in the Environment

- When nitrogen is fixed into ammonia (NH_3) or nitrate (NO_3^-), it passes through several organic or inorganic forms before it eventually returns to molecular nitrogen (N_2) (**Figure 5.1**).

(Continued)

Summary (*continued*)

- At high concentrations, ammonium (NH_4^+) is toxic to living tissues (**Figure 5.2**), whereas nitrate can be safely stored and translocated in plant tissues.

Nitrate Assimilation

- Plant roots actively absorb nitrate, then reduce it to nitrite (NO_2^-) in the cytosol (**Figure 5.3**).

- Nitrate, light, and carbohydrates affect the transcription and translation of nitrate reductase (**Figure 5.4**).

- Darkness and Mg^{2+} can inactivate nitrate reductase. Such inactivation is faster than regulation by decreased synthesis or degradation of the enzyme.

- In chloroplasts and root plastids, the enzyme nitrite reductase reduces nitrite to ammonium (**Figure 5.5**).

- Both roots and shoots assimilate nitrate (**Figure 5.6**).

Ammonium Assimilation

- Plant cells avoid ammonium toxicity by rapidly converting ammonium into amino acids (**Figure 5.7**).

- Nitrogen is incorporated into other amino acids via transamination reactions involving glutamine and glutamate.

- The amino acid asparagine is a key compound for nitrogen transport and storage.

Amino Acid Biosynthesis

- The carbon skeletons for amino acids derive from intermediates of glycolysis and the tricarboxylic acid cycle (**Figure 5.8**).

Biological Nitrogen Fixation

- Biological nitrogen fixation accounts for most of the ammonia formed from atmospheric N_2 (**Figure 5.1; Tables 5.1, 5.2**).

- Several types of nitrogen-fixing bacteria form symbiotic associations with higher plants (**Figures 5.9, 5.10; Table 5.3**).

- Nitrogen fixation requires anaerobic or microaerobic conditions.

- Symbiotic nitrogen-fixing prokaryotes function within specialized structures formed by the plant host (**Figure 5.9**).

- The symbiotic relationship is initiated by the migration of nitrogen-fixing bacteria toward the roots of the host plant, which is mediated by chemical attractants secreted by the roots.

- Attractants activate the rhizobial NodD protein, which then induces the biosynthesis of Nod factors that act as signals for symbiosis (**Figures 5.11, 5.12**).

- Nod factors induce root hair curling, sequestration of rhizobia, cell wall degradation, and bacterial access to the root hair plasma membrane, from which an infection thread forms (**Figure 5.13**).

- Filled with proliferating rhizobia, the infection thread elongates through root tissue in the direction of the developing nodule, which arises from cortical cells (**Figure 5.13**).

- In response to a signal from the plant, the bacteria in the nodule stop dividing and differentiate into nitrogen-fixing bacteroids.

- The reduction of N_2 to NH_3 is catalyzed by the nitrogenase enzyme complex (**Figure 5.14**).

- Fixed nitrogen is transported as amides or ureides (**Figure 5.15**).

Sulfur Assimilation

- Most assimilated sulfur derives from sulfate (SO_4^{2-}) absorbed from the soil solution, but plants can also metabolize gaseous sulfur dioxide (SO_2) entering via stomata.

- Synthesis of sulfur-containing organic compounds begins with the reduction of sulfate to the amino acid cysteine.

- Sulfate is assimilated in leaves and exported as glutathione via the phloem to growing sites.

Phosphate Assimilation

- Roots absorb phosphate (HPO_4^{2-}) from the soil solution, and its assimilation occurs with the formation of ATP.

- From ATP, the phosphate group may be transferred to many different carbon compounds in plant cells.

Iron Assimilation

- Roots use several mechanisms to absorb sufficient amounts of insoluble ferric iron (Fe^{3+}) from the soil solution (**Figure 5.16**).

- Once in the leaves, iron is assimilated by incorporation into heme (**Figure 5.17**).

- To limit the free radical damage that free iron ions can cause, plant cells may store surplus iron as ferritin.

The Energetics of Nutrient Assimilation

- The assimilation of nitrogen and sulfur to highly reduced compounds requires a lot of energy. For instance, nitrate assimilation accounts for about 25% of the total energy expenditure in both roots and shoots.

Suggested Reading

Bloom, A. J., Burger, M., Asensio, J. S. R., and Cousins, A. B. (2010) Carbon dioxide enrichment inhibits nitrate assimilation in wheat and *Arabidopsis*. *Science* 328: 899–903.

Bloom, A. J., Rubio-Asensio, J. S., Randall, L., Rachmilevitch, S., Cousins, A. B., and Carlisle, E. A. (2012) CO_2 enrichment inhibits shoot nitrate assimilation in C_3 but not C_4 plants and slows growth under nitrate in C_3 plants. *Ecology* 93: 355–367.

Geurts, R., Lillo, A., and Bisseling, T. (2012) Exploiting an ancient signalling machinery to enjoy a nitrogen fixing symbiosis. *Curr. Opin. Plant Biol.* 15: 438–443.

Herridge, D. F., Peoples, M. B., and Boddey, R. M. (2008) Global inputs of biological nitrogen fixation in agricultural systems. *Plant Soil* 311: 1–18.

Jeong, J., and Guerinot, M. L. (2009) Homing in on iron homeostasis in plants. *Trends Plant Sci.* 14: 280–285.

Leustek, T., Martin, M. N., Bick, J.-A., and Davies, J. P. (2000) Pathways and regulation of sulfur metabolism revealed through molecular and genetic studies. *Annu. Rev. Plant Physiol. Plant Mol. Biol.* 51: 141–165.

Maillet, F., Poinsot, V., Andre, O., Puech-Pages, V., Haouy, A., Gueunier, M., Cromer, L., Giraudet, D., Formey, D., Niebel, A., et al. (2011) Fungal lipochitooligosaccharide symbiotic signals in arbuscular mycorrhiza. *Nature* 469: 58–63.

Marschner, H., and Marschner, P. (2012) *Marschner's Mineral Nutrition of Higher Plants*, 3rd ed. Elsevier/Academic Press, London.

Mendel, R. R. (2005) Molybdenum: Biological activity and metabolism. *Dalton Trans.* 2005: 3404–3409.

Miller, A. J., Fan, X. R., Orsel, M., Smith, S. J., and Wells, D. M. (2007) Nitrate transport and signalling. *J. Exp. Bot.* 58: 2297–2306.

Oldroyd, G. E., Murray, J. D., Poole, P. S., and Downie, J. A. (2011) The rules of engagement in the legume-rhizobial symbiosis. *Annu. Rev. Gen.* 45: 119–144.

Santi, C., Bogusz, D., and Franche, C. (2013) Biological nitrogen fixation in non-legume plants. *Ann. Bot.* 111: 743–767.

Thomine, S., and Vert, G. (2013) Iron transport in plants: Better be safe than sorry. *Curr. Opin. Plant Biol.* 16: 322–327.

Welch, R. M., and Graham, R. D. (2004) Breeding for micronutrients in staple food crops from a human nutrition perspective. *J. Exp. Bot.* 55: 353–364.

Arco Images GmbH/Alamy Stock Photo

6 Solute Transport

The interior of a plant cell is separated from the plant cell wall and the environment by a plasma membrane that is only two lipid molecules thick. This thin layer separates a relatively constant internal environment from variable external surroundings. In addition to forming a hydrophobic barrier to diffusion, the membrane must facilitate and continuously regulate the inward and outward traffic of selected molecules and ions as the cell takes up nutrients, exports solutes, and regulates its turgor pressure. Similar functions are performed by the internal membranes that separate the various compartments within each cell. The plasma membrane also detects information about the environment, about molecular signals from other cells, and about the presence of invading pathogens. Often these signals are relayed by changes in ion fluxes across the membrane.

Molecular and ionic movement from one location to another is known as **transport**. Local transport of solutes into or within cells is regulated mainly by membrane proteins. Larger-scale transport between plant organs, or between plant and environment, is also controlled by membrane transport at the cellular level. For example, the transport of sucrose from leaf to root through the phloem, referred to as **translocation**, is driven and regulated by membrane transport into the phloem cells of the leaf and from the phloem to the storage cells of the root (see Chapter 10).

In this chapter we consider the physical and chemical principles that govern the movements of molecules in solution. Then we show how these principles apply to membranes and biological systems. We also discuss the molecular mechanisms of transport in living cells and the great variety of membrane transport proteins that are responsible for the particular transport properties of plant cells. Next, we describe a specialized cell type,

transport Molecular or ionic movement from one location to another; may involve crossing a diffusion barrier such as one or more membranes.

translocation The movement of photosynthate from sources to sinks in the phloem.

passive transport   Diffusion across a membrane. The spontaneous movement of a solute across a membrane in the direction of a gradient of (electro) chemical potential (from higher to lower potential). Downhill transport.

active transport   The use of energy to move a solute across a membrane against a concentration gradient, a potential gradient, or both (electrochemical potential gradient). Uphill transport.

chemical potential   The free energy associated with a substance that is available to perform work.

the guard cells regulating stomatal opening, where transmembrane ion transport play a vital role. Finally, we examine the pathways that ions take when they enter the root as well as the mechanism of xylem loading, the process whereby ions are released into the tracheary elements of the stele. Because transported substances, including carbohydrates, amino acids, and metals such as iron and zinc, are vital for human nutrition, understanding and manipulating solute transport in plants can contribute solutions to sustainable food production.

Passive and Active Transport

According to Fick's first law (see Equation 2.1), the movement of molecules by diffusion always proceeds spontaneously, down a gradient of free energy or chemical potential, until equilibrium is reached. The spontaneous "downhill" movement of molecules is termed **passive transport**. At equilibrium, no further net movements of solutes can occur without the application of a driving force.

The movement of substances against a gradient of chemical potential, or "uphill," is termed **active transport**. It is not spontaneous, and it requires that work be done on the system by the application of cellular energy. One common way (but not the only way) of accomplishing this task is to couple transport to the hydrolysis of ATP.

Recall from Chapter 2 that we can calculate the driving force for diffusion or, conversely, the energy input necessary to move substances against a gradient by measuring the potential-energy gradient. For uncharged solutes this gradient is often a simple function of the difference in concentration. Biological transport can be driven by four major forces: concentration, hydrostatic pressure, gravity, and electric fields. (However, recall from Chapter 2 that in small-scale biological systems, gravity seldom contributes substantially to the force that drives transport.)

The **chemical potential** for any solute is defined as the sum of the concentration, electrical, and hydrostatic potentials (and the chemical potential under standard conditions). *The importance of the concept of chemical potential is that it sums all the forces that may act on a molecule to drive net transport.*

$$\tilde{\mu}_j = \mu_j^* + RT \ln C_j + z_j FE + \bar{V}_j P$$

Chemical potential for a given solute, j	Chemical potential of j under standard conditions	Concentration (activity) component	Electric-potential component	Hydrostatic-pressure component	(6.1)

Here $\tilde{\mu}_j$ is the chemical potential of the solute species j in joules per mole (J mol^{-1}), μ_j^* is its chemical potential under standard conditions (a correction factor that will cancel out in future equations and so can be ignored), R is the universal gas constant, T is the absolute temperature, and C_j is the concentration (more accurately the activity) of j.

The electrical term, $z_j FE$, applies only to ions; z is the electrostatic charge of the ion (+1 for monovalent cations, –1 for monovalent anions, +2 for divalent cations, and so on), F is Faraday's constant (96,500 Coulombs, equivalent to the electric charge on 1 mol of H$^+$), and E is the overall electrical potential of the solution (with respect to ground). The final term, $\bar{V}_j P$, expresses the contribution of the partial molal volume of j ($\bar{V}_j$) and pressure (P) to the chemical potential of j. (The partial molal volume of j is the change in volume per mole of substance j added to the system, for an infinitesimal addition.)

This final term, $\bar{V}_j P$, makes a much smaller contribution to $\tilde{\mu}_j$ than do the concentration and electrical terms, except in the very important case of osmotic water movements. As discussed in Chapter 2, when considering water movement at the cellular scale the chemical potential of water (i.e., the water potential) depends on the concentration of dissolved solutes and the hydrostatic pressure on the system.

In general, diffusion (passive transport) always moves molecules energetically downhill from areas of higher chemical potential to areas of lower chemical potential. Movement against a chemical-potential gradient is indicative of active transport (**Figure 6.1**).

If we take the diffusion of sucrose across a plasma membrane as an example, we can accurately approximate the chemical potential of sucrose in any compartment by the concentration term alone (unless a solution is concentrated, causing hydrostatic pressure to build up in the plant cell). From Equation 6.1, the chemical potential of sucrose inside a cell can be described as follows (in the next equations, the subscript s stands for sucrose and the superscripts i and o stand for inside and outside, respectively):

$$\tilde{\mu}_s^i \quad = \quad \mu_s^* \quad + \quad RT \ln C_s^i$$

| Chemical potential of sucrose solution inside the cell | Chemical potential of sucrose solution under standard conditions | Concentration component | (6.2) |

The chemical potential of sucrose outside the cell is calculated as follows:

$$\tilde{\mu}_s^o = \mu_s^* + RT \ln C_s^o \tag{6.3}$$

We can calculate the difference in the chemical potential of sucrose between the solutions inside and outside the cell, $\Delta \tilde{\mu}_s$, regardless of the mechanism of transport. To get the signs right, remember that for inward transport, sucrose

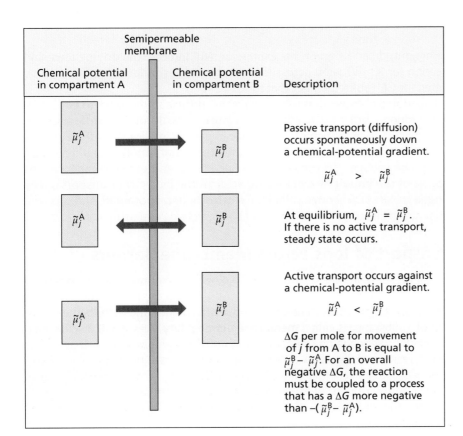

Figure 6.1 Relationship between chemical potential, $\tilde{\mu}$, and the transport of molecules across a permeability barrier. The net movement of molecular species j between compartments A and B depends on the relative magnitude of the chemical potential of j in each compartment, represented here by the size of the boxes. Movement down a chemical gradient occurs spontaneously and is called passive transport; movement against, or up, a gradient requires energy and is called active transport.

electrochemical potential
The chemical potential of an electrically charged solute.

membrane permeability
The extent to which a membrane permits or restricts the movement of a substance.

is being removed (–) from outside the cell and added (+) to the inside, so the change in free energy in joules per mole of sucrose transported will be as follows:

$$\Delta \tilde{\mu}_s = \tilde{\mu}_s^i - \tilde{\mu}_s^o \tag{6.4}$$

Substituting the terms from Equations 6.2 and 6.3 into Equation 6.4, we get the following:

$$\Delta \tilde{\mu}_s = \left(\mu_s^* + RT \ln C_s^i \right) - \left(\mu_s^* + RT \ln C_s^o \right)$$
$$= RT \left(\ln C_s^i - \ln C_s^o \right) = RT \ln \frac{C_s^i}{C_s^o} \tag{6.5}$$

If this difference in chemical potential is negative, sucrose can diffuse inward spontaneously (provided the membrane has a permeability to sucrose; see the next section). In other words, the driving force ($\Delta \tilde{\mu}_s$) for solute diffusion is related to the magnitude of the concentration gradient (C_s^i/C_s^o).

If the solute carries an electric charge (as does, for example, the potassium ion), the electrical component of the chemical potential must also be considered. Suppose the membrane is permeable to K^+ and Cl^- rather than to sucrose. Because the ionic species (K^+ and Cl^-) diffuse independently, each has its own chemical potential. Thus, for inward K^+ diffusion,

$$\Delta \tilde{\mu}_K = \tilde{\mu}_K^i - \tilde{\mu}_K^o \tag{6.6}$$

Substituting the appropriate terms from Equation 6.1 into Equation 6.6, we get

$$\Delta \tilde{\mu}_s = (RT \ln [K^+]^i + zFE^i) - (RT \ln [K^+]^o + zFE^o) \tag{6.7}$$

and because the electrostatic charge of K^+ is +1, $z = +1$, and

$$\Delta \tilde{\mu}_K = RT \ln \frac{[K^+]^i}{[K^+]^o} + F(E^i - E^o) \tag{6.8}$$

The magnitude and sign of this expression will indicate the driving force and direction for K^+ diffusion across the membrane. A similar expression can be written for Cl^- (but remember that for Cl^-, $z = -1$).

Equation 6.8 shows that ions, such as K^+, diffuse in response to both their concentration gradients ($[K^+]^i/[K^+]^o$) and any electrical-potential difference between the two compartments ($E^i - E^o$). One important implication of this equation is that ions can be driven passively against their concentration gradients if an appropriate voltage (electric field) is applied between the two compartments. Because of the importance of electric fields in the biological transport of any charged molecule, $\tilde{\mu}$ is often called the **electrochemical potential**, and $\Delta \tilde{\mu}$ is the difference in electrochemical potential between two compartments.

Transport of Ions across Membrane Barriers

If two ionic solutions are separated by a biological membrane, diffusion is complicated by the fact that the ions must move through the membrane as well as across the open solutions. The extent to which a membrane permits the movement of a substance is called **membrane permeability**. As we will discuss later, permeability depends on the composition of the membrane as well as on the chemical nature of the solute. In a loose sense, permeability can be expressed in terms of a diffusion coefficient for the solute through the membrane. However, permeability is influenced by several additional factors, such as the ability of a substance to enter the membrane, that are difficult to measure.

Despite its theoretical complexity, we can readily measure permeability by determining the rate at which a solute passes through a membrane under a specific set of conditions. Generally, the membrane will hinder diffusion and thus reduce the speed with which equilibrium is reached. For any particular solute, however, the permeability or resistance of the membrane itself cannot alter the final equilibrium conditions. Equilibrium occurs when $\Delta \tilde{\mu}_j = 0$.

In the sections that follow we discuss the factors that influence the distribution of ions across a membrane. These parameters can be used to predict the relationship between the electrical gradient and the concentration gradient of an ion.

Different diffusion rates for cations and anions produce diffusion potentials

When salts diffuse across a membrane, an electrical membrane potential (voltage) can develop. Consider the two KCl solutions separated by a membrane in **Figure 6.2**. The K$^+$ and Cl$^-$ ions will permeate the membrane independently as they diffuse down their respective gradients of electrochemical potential. And unless the membrane is very porous, its permeability to the two ions will differ.

As a consequence of these different permeabilities, K$^+$ and Cl$^-$ will initially diffuse across the membrane at different rates. The result is a slight separation of charge, which instantly creates an electrical potential across the membrane. In biological systems, membranes are usually more permeable to K$^+$ than to Cl$^-$. Therefore, K$^+$ will diffuse out of the cell (see compartment A in Figure 6.2) faster than Cl$^-$, causing the cell to develop a negative electric charge with respect to the extracellular medium. A potential that develops as a result of diffusion is called a **diffusion potential**.

The principle of electrical neutrality must always be kept in mind when the movement of ions across membranes is considered: Bulk solutions always contain equal numbers of anions and cations. The existence of a membrane potential implies that the distribution of charges across the membrane is uneven; however, the actual number of unbalanced ions is negligible in chemical terms. For example, a membrane potential of –100 millivolts (mV), like that found across the plasma membranes of many plant cells, results from the presence of only 1 extra anion out of every 100,000 within the cell—a concentration difference of only 0.001%! As Figure 6.2 shows, all of these extra anions are found immediately adjacent to the surface of the membrane; there is no charge imbalance throughout the bulk of the cell.

In our example of KCl diffusion across a membrane, electrical neutrality is preserved, because as K$^+$ moves ahead of Cl$^-$ in the membrane, the resulting diffusion potential retards the movement of K$^+$ and speeds that of Cl$^-$. Ultimately, both ions diffuse at the same rate, but the diffusion potential persists and can be measured. As the system moves toward equilibrium and the concentration gradient collapses, the diffusion potential also collapses.

Figure 6.2 Development of a diffusion potential and a charge separation between two compartments separated by a membrane that is preferentially permeable to potassium ions. If the concentration of potassium chloride is higher in compartment A ([KCl]$_A$ > [KCl]$_B$), potassium and chloride ions will diffuse into compartment B. If the membrane is more permeable to potassium ions than to chloride, potassium ions will diffuse faster than chloride ions, and a charge separation (+ and –) will develop, resulting in establishment of a diffusion potential.

diffusion potential The potential (voltage) difference that develops across a semipermeable membrane as a result of the differential permeability of solutes with opposite charges (for example K$^+$ and Cl$^-$).

Compartment A   Compartment B

Initial conditions:
[KCl]$_A$ > [KCl]$_B$

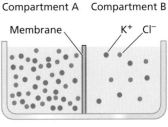

Diffusion potential exists until chemical equilibrium is reached.

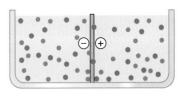

Equilibrium conditions:
[KCl]$_A$ = [KCl]$_B$

At chemical equilibrium, diffusion potential equals zero.

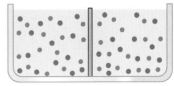

Nernst potential   The electrical potential described by the Nernst equation.

How does membrane potential relate to ion distribution?

Because the membrane in the preceding example is permeable to both K^+ and Cl^- ions, equilibrium will not be reached for either ion until the concentration gradients decrease to zero. However, if the membrane were permeable only to K^+, diffusion of K^+ would carry charges across the membrane until the membrane potential balanced the concentration gradient. Because a change in potential requires very few ions, this balance would be reached instantly. Potassium ions would then be at equilibrium, even though the change in the concentration gradient for K^+ would be negligible.

When the distribution of any solute across a membrane reaches equilibrium, the passive flux, J (i.e., the amount of solute crossing a unit area of membrane per unit of time), is the same in the two directions—outside to inside and inside to outside:

$$J_{o \to i} = J_{i \to o}$$

Fluxes are related to $\Delta \tilde{\mu}$; thus, at equilibrium the electrochemical potentials will be the same:

$$\tilde{\mu}_j^o = \tilde{\mu}_j^i$$

and for any given ion (the ion is symbolized here by the subscript j),

$$\mu_j^* + RT \ln C_j^o + z_j F E^o = \mu_j^* + RT \ln C_j^i + z_j F E^i \tag{6.9}$$

By rearranging Equation 6.9, we obtain the difference in electrical potential between the two compartments at equilibrium ($E^i - E^o$):

$$E^i - E^o = \frac{RT}{z_j F} \left(\ln \frac{C_j^o}{C_j^i} \right)$$

This electrical-potential difference is known as the **Nernst potential** (ΔE_j) for that ion,

$$\Delta E_j = E^i - E^o$$

and

$$\Delta E_j = \frac{RT}{z_j F} \left(\ln \frac{C_j^o}{C_j^i} \right) \tag{6.10}$$

or

$$\Delta E_j = \frac{2.3 RT}{z_j F} \left(\log \frac{C_j^o}{C_j^i} \right)$$

This relationship, known as the *Nernst equation*, states that at equilibrium, the difference in concentration of an ion between two compartments is balanced by the voltage difference between the compartments. The Nernst equation can be further simplified for a univalent cation at 25°C:

$$\Delta E_j = 59 \text{mV} \log \frac{C_j^o}{C_j^i} \tag{6.11}$$

Note that a tenfold difference in concentration corresponds to a Nernst potential of 59 mV ($C_o/C_i = 10/1$; $\log 10 = 1$). That is, a membrane potential of 59 mV would maintain a tenfold concentration gradient of an ion whose movement across the membrane is driven by passive diffusion. Similarly, if a tenfold concentration gradient of an ion existed across the membrane, passive diffusion of that ion down its concentration gradient (if it were allowed to come to equilibrium) would result in a difference of 59 mV across the membrane.

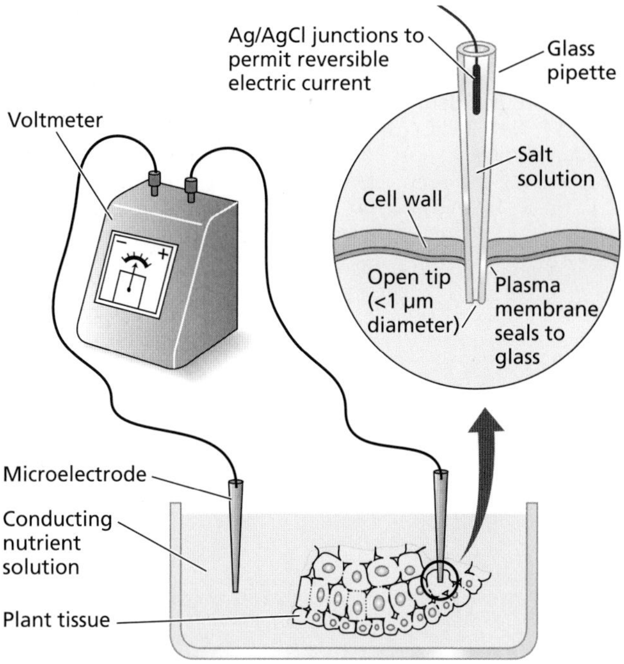

Figure 6.3 Diagram of a pair of microelectrodes used to measure membrane potentials across plasma membranes. One of the glass micropipette electrodes is inserted into the cell compartment under study (usually the vacuole or the cytoplasm), while the other is kept in an electrolytic solution that serves as a reference. The microelectrodes are connected to a voltmeter, which records the electrical-potential difference between the cell compartment and the solution. Typical membrane potentials across plant plasma membranes range from –60 to –240 mV. The insert shows how electrical contact with the interior of the cell is made through the open tip of the glass micropipette, which contains an electrically conducting salt solution.

All living cells exhibit a membrane potential that is due to the asymmetric ion distribution between the inside and outside of the cell. We can determine these membrane potentials by inserting a microelectrode into the cell and measuring the voltage difference between the inside of the cell and the extracellular medium (**Figure 6.3**).

The Nernst equation can be used at any time to determine whether a given ion is at equilibrium across a membrane. However, a distinction must be made between equilibrium and steady state. *Steady state* is the condition in which influx and efflux of a given solute are equal, and therefore the ion concentrations are constant over time. Steady state is not necessarily the same as equilibrium (see Figure 6.1); in steady state, the existence of active transport across the membrane prevents many diffusive fluxes from ever reaching equilibrium.

The Nernst equation distinguishes between active and passive transport

Table 6.1 shows how experimental measurements of ion concentrations at steady state in pea root cells compare with predicted values calculated from the Nernst equation. In this example, the concentration of each ion in the external solution bathing the tissue and the measured membrane potential were substituted into the Nernst equation, and the internal concentration of each ion was predicted.

Prediction using the Nernst equation assumes passive ion distribution, but notice that, of all the ions shown in Table 6.1, only K^+ is at or near equilibrium. The anions NO_3^-, Cl^-, $H_2PO_4^-$, and SO_4^{2-} all have higher internal concentrations than predicted, indicating that their uptake is active. The cations Na^+, Mg^{2+}, and Ca^{2+} have lower internal concentrations than predicted; therefore, these ions enter the cell by diffusion down their electrochemical-potential gradients and are then actively exported.

The example shown in Table 6.1 is an oversimplification; plant cells have several internal compartments, each of which can differ in its ionic composition from the others. The cytosol and the vacuole are the most important intra-

Table 6.1 Comparison of observed and predicted ion concentrations in pea root tissue

Ion	Concentration in external medium (mmol L^{-1})	Internal concentration[a] (mmol L^{-1})	
		Predicted	Observed
K^+	1	74	75
Na^+	1	74	8
Mg^{2+}	0.25	1340	3
Ca^{2+}	1	5360	2
NO_3^-	2	0.0272	28
Cl^-	1	0.0136	7
$H_2PO_4^-$	1	0.0136	21
SO_4^{2-}	0.25	0.00005	19

Source: Data from Higinbotham et al. 1967.

Note: The membrane potential was measured as –110 mV.

[a]Internal concentration values were derived from ion content of hot-water extracts of 1- to 2-cm intact root segments.

Goldman equation   An equation that predicts the diffusion potential across a membrane, as a function of the concentrations and permeabilities of all ions (e.g., K^+, Na^+, and Cl^-) that permeate the membrane.

cellular compartments in determining the ionic relations of plant cells. In most mature plant cells, the central vacuole occupies 90% or more of the cell's volume, and the cytosol is restricted to a thin layer around the periphery of the cell.

Because of its small volume, the cytosol of most angiosperm cells is difficult to assay chemically. For this reason, much of the early work on the ionic relations of plants focused on certain green algae, such as *Chara* and *Nitella*, whose cells are several inches long and may contain an appreciable volume of cytosol. In brief:

- Potassium ions are accumulated passively by both the cytosol and the vacuole. When extracellular K^+ concentrations are very low, K^+ may be taken up actively.

- Sodium ions are pumped actively out of the cytosol into the extracellular space and vacuole.

- Excess protons, generated by intermediary metabolism, are also actively extruded from the cytosol. This process helps maintain the cytosolic pH near neutrality, while the vacuole and the extracellular medium are generally more acidic by one or two pH units.

- Anions are taken up actively into the cytosol.

- Calcium ions are actively transported out of the cytosol at both the plasma membrane and the vacuolar membrane, which is called the tonoplast.

Many different ions permeate the membranes of living cells simultaneously, but K^+ has the highest concentrations in plant cells, and it exhibits high permeabilities. A modified version of the Nernst equation, the **Goldman equation**, includes all permeant ions (all ions for which mechanisms of transmembrane movement exist) and therefore gives a more accurate value for the diffusion potential. When permeabilities and ion gradients are known, it is possible to calculate a diffusion potential across a biological membrane from the Goldman equation. The diffusion potential calculated from the Goldman equation is termed the *Goldman diffusion potential*.

Proton transport is a major determinant of the membrane potential

In most eukaryotic cells, K^+ has both the greatest internal concentration and the highest membrane permeability, so the diffusion potential may approach E_K, the Nernst potential for K^+. In some cells of some organisms—particularly in some mammalian cells such as neurons—the normal resting potential of the cell may also be close to E_K. This is not the case with plants and fungi, however, which often show experimentally measured membrane potentials (often −200 to −100 mV) that are much more negative than those calculated from the Goldman equation, which are usually only −80 to −50 mV. Thus, in addition to the diffusion potential, the membrane potential must have a second component. The excess voltage is provided by the electrogenic plasma membrane H^+-ATPase.

Whenever an ion moves into or out of a cell without being balanced by countermovement of an ion of opposite charge, a voltage is created across the membrane. Any active transport mechanism that results in the movement of a net electric charge will tend to move the membrane potential away from the value predicted by the Goldman equation. Such transport mechanisms are called *electrogenic pumps* and are common in living cells.

The energy required for active transport is often provided by the hydrolysis of ATP. We can study the dependence of the plasma membrane potential on ATP by observing the effect of cyanide (CN^-) on the membrane potential (**Figure 6.4**). Cyanide rapidly poisons the mitochondria (see Chapter 11), and the cell's ATP consequently becomes depleted. As ATP synthesis is inhibited, the membrane potential falls to the level of the Goldman diffusion potential.

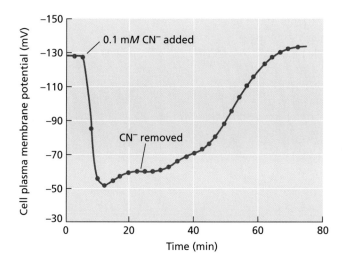

Figure 6.4 The plasma membrane potential of a pea cell collapses when cyanide (CN⁻) is added to the bathing solution. Cyanide blocks ATP production in the cell by poisoning the mitochondria. The collapse of the membrane potential upon addition of cyanide indicates that an ATP supply is necessary for maintenance of the potential. Washing the cyanide out of the tissue results in a slow recovery of ATP production and restoration of the membrane potential. (After Higinbotham et al. 1970.)

Thus, the membrane potentials of plant cells have two components: a diffusion potential and a component resulting from electrogenic ion transport (transport that results in the generation of a membrane potential). When cyanide inhibits electrogenic ion transport, the pH of the external medium increases while the cytosol becomes acidic because protons remain inside the cell. This observation is one piece of evidence that it is the active transport of protons out of the cell that is electrogenic.

A change in membrane potential caused by an electrogenic pump will change the driving forces for diffusion of all ions that cross the membrane. For example, the outward transport of H^+ can create an electrical driving force for the passive diffusion of K^+ into the cell. Protons are transported electrogenically across the plasma membrane not only in plants but also in bacteria, algae, fungi, and some animal cells, such as those of the kidney epithelia.

ATP synthesis in mitochondria and chloroplasts also depends on a H^+-ATPase. In these organelles, this transport protein is usually called an *ATP synthase* because it forms ATP rather than hydrolyzing it (see Chapter 11). We discuss the structure and function of membrane proteins involved in active and passive transport in plant cells in detail later in this chapter.

Membrane Transport Processes

Artificial membranes made of pure phospholipids have been used extensively to study membrane permeability. When the permeability of artificial phospholipid bilayers to ions and molecules is compared with that of biological membranes, important similarities and differences become evident (**Figure 6.5**).

Biological and artificial membranes have similar permeabilities to nonpolar molecules and many small polar molecules. However, biological membranes are much more permeable to ions, to some large polar molecules, such as sugars, and to water than artificial bilayers are. The

Figure 6.5 Typical values for the permeability of a biological membrane to various substances compared with those for an artificial phospholipid bilayer. For nonpolar molecules such as O_2 and CO_2, and for some small uncharged molecules such as glycerol, permeability values are similar in both systems. For ions and selected polar molecules, including water, the permeability of biological membranes is increased by one or more orders of magnitude because of the presence of transport proteins. Note the logarithmic scale.

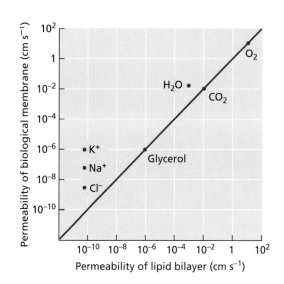

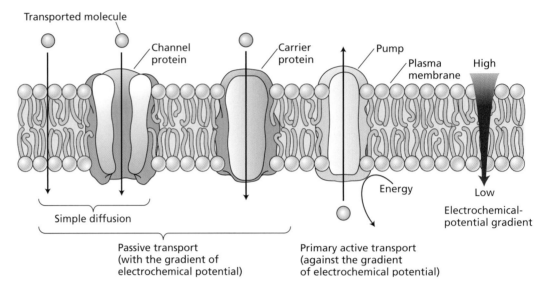

Figure 6.6   Three classes of membrane transport proteins: channels, carriers, and pumps. Channels and carriers can mediate the passive transport of a solute across a membrane (by simple diffusion or facilitated diffusion) down the solute's gradient of electrochemical potential. Channel proteins act as membrane pores, and their specificity is determined primarily by the biophysical properties of the channel. Carrier proteins bind the transported molecule on one side of the membrane and release it on the other side. (The different types of carrier proteins are described in more detail in Figure 6.10.) Primary active transport is carried out by pumps and uses energy directly, usually from ATP hydrolysis, to pump solutes against their gradient of electrochemical potential.

reason is that, unlike artificial bilayers, biological membranes contain **transport proteins** that facilitate the passage of selected ions and other molecules. The general term *transport proteins* encompasses three main categories of proteins: channels, carriers, and pumps (**Figure 6.6**), each of which we describe in more detail later in this section.

Transport proteins exhibit specificity for the solutes they transport, so cells require a great diversity of transport proteins. The simple prokaryote *Haemophilus influenzae*, the first organism for which the complete genome was sequenced, has only 1743 genes, yet more than 200 of those genes (more than 10% of the genome) encode various proteins involved in membrane transport. In Arabidopsis, out of a predicted approximately 27,400 protein-coding genes, as many as 1800 may encode proteins with transport functions.

Although a particular transport protein is usually highly specific for the kinds of substances it will transport, its specificity is often not absolute. In plants, for example, a K^+ transporter in the plasma membrane may transport K^+, Rb^+, and Na^+ with different preferences. In contrast, most K^+ transporters are completely ineffective in transporting anions such as Cl^- or uncharged solutes such as sucrose. Similarly, a protein involved in the transport of neutral amino acids may move glycine, alanine, and valine with equal ease, but may not accept aspartic acid or lysine.

In the next several pages we consider the structures, functions, and physiological roles of the various membrane transporters found in plant cells, especially in the plasma membrane and tonoplast.

Channels enhance diffusion across membranes

Channels are transmembrane proteins that function as selective pores through which ions, and in some cases neutral molecules, can diffuse across the membrane. The size of a pore and the density and nature of the surface charges on its

transport proteins   Transmembrane proteins that are involved in the movement of molecules or ions from one side of a membrane to the other side.

channels   Transmembrane proteins that function as selective pores for passive transport of ions or water across the membrane.

interior lining determine its transport specificity. Transport through channels is always passive, and the specificity of transport depends on pore size and electric charge more than on selective binding (**Figure 6.7**).

As long as the channel pore is open, substances that can penetrate the pore diffuse through it extremely rapidly: about 10^8 ions per second through an ion channel. Channel pores are not open all the time, however. Channel proteins contain particular regions called **gates** that open and close the pore in response to signals. Signals that can regulate channel activity include membrane potential changes, ligands, hormones, light, and posttranslational modifications such as phosphorylation. For example, voltage-gated channels open or close in response to changes in the membrane potential (see Figure 6.7B). Another intriguing regulatory signal is mechanical force, which changes the conformation and thereby controls the gating of mechanosensitive ion channels in plants and other organisms.

Individual ion channels can be studied in detail by an electrophysiological technique called patch clamping (see Figure 6.19A), which can detect the electrical current carried by ions diffusing through a single open channel or a collection of channels. Patch clamp studies reveal that for a given ion, such as K^+, a membrane has a variety of different channels. These channels may open over different voltage ranges, or in response to different signals, which may include K^+ or Ca^{2+} concentrations, pH, reactive oxygen species, and so on. This specificity enables the transport of each ion to be fine-tuned to the prevailing conditions. Thus, the ion permeability of a membrane is a variable that depends on the mix of ion channels that are open at a particular time.

As we saw in the experiment presented in Table 6.1, the distribution of most ions is not close to equilibrium across the membrane. Therefore, we know that channels for most ions are usually closed. Plant cells generally accumulate more anions than could occur via a strictly passive mechanism. Therefore, when anion channels open, anions flow out of the cell, and active mechanisms are required for anion uptake. Calcium ion channels are tightly regulated and essentially open only during signal transduction. Calcium ion channels function only to allow Ca^{2+} flux into the cytosol, and Ca^{2+} must be expelled from the cytosol by active transport. In contrast, K^+ can diffuse either inward or outward through channels, depending on whether the membrane potential is more negative or more positive than E_K, the potassium ion equilibrium potential.

K^+ channels that open only at potentials more negative than the prevailing Nernst potential for K^+ are specialized for inward diffusion of K^+ and are known as **inwardly rectifying**, or simply *inward*, K^+ channels. Conversely, K^+ channels

gate A structural domain of the channel protein that opens or closes the channel in response to external signals such as voltage changes, hormone binding, or light.

inwardly rectifying Refers to ion channels that open only at potentials more negative than the prevailing Nernst potential for a cation, or more positive than the prevailing Nernst potential for an anion, and thus mediate inward current.

(A)

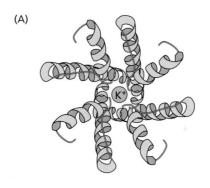

(B)

Figure 6.7 Models of K^+ channels in plants. (A) Top view of a channel, looking through the pore of the protein. Membrane-spanning helices of four subunits come together in an inverted teepee with the pore at the center. The pore-forming regions of the four subunits dip into the membrane, forming a K^+ selectivity finger region at the outer part of the pore. (B) Side view of an inwardly rectifying K^+ channel, showing a polypeptide chain of one subunit, with six membrane-spanning helices (S1–S6). The fourth helix contains positively charged amino acids and acts as a voltage sensor. The pore-forming region (P-domain) is a loop between helices 5 and 6. (A after Leng et al. 2002; B after Buchanan et al. 2000.)

OUTSIDE OF CELL

Plasma membrane

Voltage-sensing region

Pore-forming region (P-domain)

S1 S2 S3 S4 S5 S6

N   C

CYTOSOL

outwardly rectifying   Refers to ion channels that open only at potentials more positive than the prevailing Nernst potential for a cation, or more negative than the prevailing Nernst potential for an anion, and thus mediate outward current.

carriers   Membrane transport proteins that bind to a solute, undergo conformational change, and release the solute on the other side of the membrane.

facilitated diffusion   Passive transport across a membrane using a carrier.

primary active transport   The direct coupling of a metabolic energy source such as ATP hydrolysis, oxidation-reduction reaction, or light absorption to active transport by a carrier protein.

pumps   Membrane proteins that carry out primary active transport across a biological membrane. Most pumps transport ions, such as H^+ or Ca^{2+}.

electrogenic transport   Active ion transport involving the net movement of charge across a membrane.

electroneutral transport   Active ion transport that involves no net movement of charge across a membrane.

that open only at potentials more positive than the Nernst potential for K^+ are **outwardly rectifying**, or *outward*, K^+ channels (**Figure 6.8**). Inward K^+ channels function in the accumulation of K^+ from the apoplast, as occurs, for example, during K^+ uptake by guard cells in the process of stomatal opening (see Figure 6.8). Various outward K^+ channels function in the closing of stomata (see later in this chapter and in the release of K^+ into the xylem or apoplast.

Carriers bind and transport specific substances

Unlike channels, carrier proteins do not have pores that extend completely across the membrane. In transport mediated by a carrier, the substance being transported is initially bound to a specific site on the carrier protein. This requirement for binding allows carriers to be highly selective for a particular substrate to be transported. **Carriers** therefore specialize in the transport of specific inorganic or organic ions as well as other organic metabolites. Binding causes a conformational change in the protein, which exposes the substance to the solution on the other side of the membrane. Transport is complete when the substance dissociates from the carrier's binding site.

Because a conformational change in the protein is required to transport an individual molecule or ion, the rate of transport by a carrier is many orders of magnitude slower than that through a channel. Typically, carriers may transport 100 to 1000 ions or molecules per second, while millions of ions can pass through an open ion channel in the same amount of time. The binding and release of molecules at a specific site on a carrier protein are similar to the binding and release of molecules by an enzyme in an enzyme-catalyzed reaction. As we discuss later in this chapter, enzyme kinetics have been used to characterize transport carrier proteins.

Carrier-mediated transport (unlike transport through channels) can be either passive transport or secondary active transport (secondary active transport is discussed below). Passive transport via a carrier is sometimes called **facilitated diffusion**, although it resembles diffusion only in that it transports substances down their gradient of electrochemical potential, without an additional input of energy. (The term *facilitated diffusion* might seem more appropriately applied to transport through channels, but historically it has not been used in this way.)

Primary active transport requires energy

To carry out active transport, a carrier must couple the energetically uphill transport of a solute with another, energy-releasing event so that the overall free-energy change is negative. **Primary active transport** is coupled directly to a source of energy other than $\Delta\tilde{\mu}_j$, such as ATP hydrolysis, an oxidation–reduction reaction (as in the electron transport chain of mitochondria and chloroplasts), or the absorption of light by the carrier protein (such as bacteriorhodopsin in halobacteria).

Membrane proteins that carry out primary active transport are called **pumps** (see Figure 6.6). Most pumps transport inorganic ions, such as H^+ or Ca^{2+}. However, as we will see later in this chapter, pumps belonging to the ATP-binding cassette (ABC) family of transporters can carry large organic molecules.

Ion pumps can be further characterized as either electrogenic or electroneutral. In general, **electrogenic transport** refers to ion transport involving the net movement of charge across the membrane. In contrast, **electroneutral transport**, as the name implies, involves no net movement of charge. For example, the Na^+/K^+-ATPase of animal cells pumps three Na^+ out for every two K^+ in, resulting in a net outward movement of one positive charge. The Na^+/K^+-ATPase is therefore an electrogenic ion pump. In contrast, the H^+/K^+-ATPase of the animal gastric mucosa pumps one H^+ out of the cell for every one K^+ in, so there is no net movement of charge across the membrane. Therefore, the H^+/K^+-ATPase is an electroneutral pump.

(A)

Equilibrium or Nernst potential for K⁺: by definition, no net flux of K⁺, therefore, no current.

Current carried by the movement of K⁺ out of the cell. By convention, this **outward current** is given a **positive sign**.

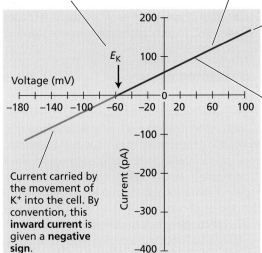

The opening and closing or "gating" of these channels is not regulated by voltage. Therefore, current through the channel is a linear function of voltage.

The slope of the line ($\Delta I/\Delta V$) gives the **conductance** of the channels mediating this K⁺ current.

Current carried by the movement of K⁺ into the cell. By convention, this **inward current** is given a **negative sign**.

$E_K = RT/zF \times \ln \{[K_{out}]/[K_{in}]\}$
$E_K = 0.025 \times \ln \{10/100\}$
$E_K = -59$ mV

(B)

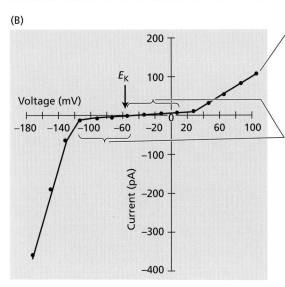

This current–voltage relationship is produced by K⁺ movement through channels that are regulated ("gated") by voltage. Note that the I/V relationship is nonlinear.

There is little or no current over these voltage ranges because the channels are voltage-regulated and the effect of these voltages is to keep the channels in a closed state.

(C)

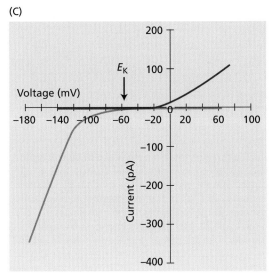

The current response illustrated in (B) is shown here to arise from the activity of two molecularly distinct types of K⁺ channels. The outward K⁺ channels (red) are voltage-gated such that they open only at membrane potentials $>E_K$; thus these channels mediate K⁺ efflux from the cell. The inward K⁺ channels (blue) are voltage-gated such that they open only at membrane potentials $<E_K$; thus these channels mediate K⁺ uptake into the cell.

Figure 6.8 Current–voltage relationships. (A) Diagram showing the current (I) that would result from K⁺ flux through a set of hypothetical plasma membrane K⁺ channels that were not voltage-regulated, given a K⁺ concentration in the cytosol of 100 mM and an extracellular K⁺ concentration of 10 mM. Note that the current would be linear, and that there would be zero current at the equilibrium (Nernst) potential for K⁺ (E_K). (B) Actual K⁺ current data from an Arabidopsis guard cell protoplast, with the same intracellular and extracellular K⁺ concentrations as in (A). These currents result from the activities of voltage-regulated K⁺ channels. Note that, again, there is zero net current at the equilibrium potential for K⁺. However, there is also zero net current over a broader voltage range because the channels are closed over this voltage range in these conditions. When the channels are closed, no K⁺ can flow through them, hence zero current is observed over this voltage range. (C) The current–voltage relationship in (B) actually results from the activities of two sets of channels—the inwardly rectifying K⁺ channels and the outwardly rectifying K⁺ channels—which together produce the current–voltage relationship. (B after L. Perfus-Barbeoch and S. M. Assmann, unpublished data.)

plasma membrane H⁺-ATPase
An ATPase that pumps H⁺ across the plasma membrane energized by ATP hydrolysis.

vacuolar H⁺-ATPase (V-ATPase)
A large, multi-subunit enzyme complex, related to F_oF_1-ATPases, present in endomembranes (tonoplast, Golgi). Acidifies the vacuole and provides the proton motive force for the secondary transport of a variety of solutes into the lumen. V-ATPases also function in the regulation of intracellular protein trafficking.

H⁺-pyrophosphatase
An electrogenic pump that moves protons into the vacuole, energized by the hydrolysis of pyrophosphate.

secondary active transport
Active transport that uses energy stored in the proton motive force or other ion gradient, and operates by symport or antiport.

For the plasma membranes of plants, fungi, and bacteria, as well as for plant tonoplasts and other plant and animal endomembranes, H⁺ is the principal ion that is electrogenically pumped across the membrane. The **plasma membrane H⁺-ATPase** generates the gradient of electrochemical potential of H⁺ across the plasma membrane, while the **vacuolar H⁺-ATPase** (usually called the **V-ATPase**) and the **H⁺-pyrophosphatase** (**H⁺-PPase**) electrogenically pump protons into the lumen of the vacuole and the Golgi cisternae.

Secondary active transport uses stored energy

In plant plasma membranes, the most prominent pumps are those for H⁺ and Ca^{2+}, and the direction of pumping is outward from the cytosol to the extracellular space. Another mechanism is needed to drive the active uptake of mineral nutrients such as NO_3^-, SO_4^{2-}, and $H_2PO_4^-$; the uptake of amino acids, peptides, and sucrose; and the export of Na^+, which at high concentrations is toxic to plant cells. The other important way that solutes are actively transported across a membrane against their gradient of electrochemical potential is by coupling the uphill transport of one solute to the downhill transport of another. This type of carrier-mediated cotransport is termed **secondary active transport** (**Figure 6.9**).

Secondary active transport is driven indirectly by pumps. In plant cells, protons are extruded from the cytosol by electrogenic H⁺-ATPases operating in the plasma membrane and at the vacuolar membrane. Consequently, a mem-

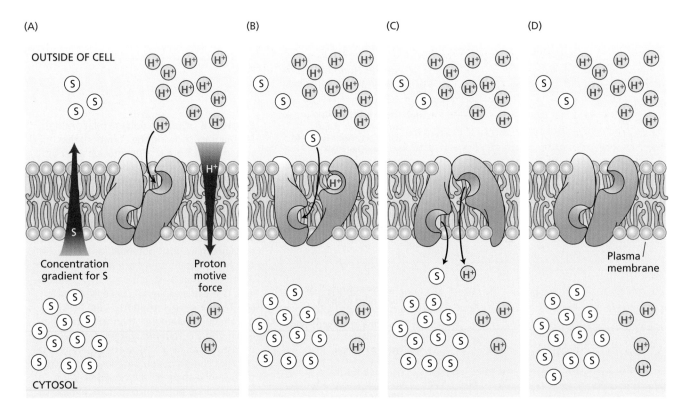

Figure 6.9 Hypothetical model of secondary active transport. In secondary active transport, the energetically uphill transport of one solute is driven by the energetically downhill transport of another solute. In the illustrated example, energy that was stored as proton motive force ($\Delta\tilde{\mu}_{H^+}$, symbolized by the red arrow on the right in [A]) is being used to take up a substrate (S) against its concentration gradient (red arrow on the left). (A) In the initial conformation, the binding sites on the protein are exposed to the outside environment and can bind a proton. (B) This binding results in a conformational change that permits a molecule of S to be bound. (C) The binding of S causes another conformational change that exposes the binding sites and their substrates to the inside of the cell. (D) Release of a proton and a molecule of S to the cell's interior restores the original conformation of the carrier and allows a new pumping cycle to begin.

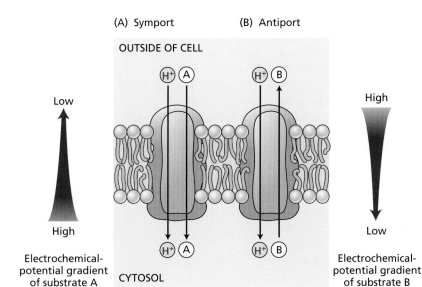

(A) Symport **(B) Antiport**

OUTSIDE OF CELL

Low

High

High

Low

Electrochemical-
potential gradient
of substrate A

Electrochemical-
potential gradient
of substrate B

CYTOSOL

Figure 6.10 Two examples of secondary active transport coupled to a primary proton gradient. (A) In symport, the energy dissipated by a proton moving back into the cell is coupled to the uptake of one molecule of a substrate (e.g., a sugar) into the cell. (B) In antiport, the energy dissipated by a proton moving back into the cell is coupled to the active transport of a substrate (e.g., a sodium ion) out of the cell. In both cases, the substrate under consideration is moving against its gradient of electrochemical potential. Both neutral and charged substrates can be transported by such secondary active transport processes.

brane potential and a pH gradient are created at the expense of ATP hydrolysis. This gradient of electrochemical potential for H^+, referred to as $\Delta \tilde{\mu}_{H^+}$, or (when expressed in other units) the **proton motive force** (**PMF**), represents stored free energy in the form of the H^+ gradient.

The proton motive force generated by electrogenic H^+ transport is used in secondary active transport to drive the transport of many other substances against their gradients of electrochemical potential. Figure 6.9 shows how secondary active transport may involve the binding of a substrate (S) and an ion (usually H^+) to a carrier protein and a conformational change in that protein.

There are two types of secondary active transport: symport and antiport. The example shown in Figure 6.9 is called **symport** (and the proteins involved are called *symporters*) because the two substances move in the same direction through the membrane (see also **Figure 6.10A**). **Antiport** (facilitated by proteins called *antiporters*) refers to coupled transport in which the energetically downhill movement of one solute drives the active (energetically uphill) transport of another solute in the opposite direction (**Figure 6.10B**). Considering the direction of the H^+ gradient, proton-coupled symporters generally function in substrate uptake into the cytosol while proton-coupled antiporters function to export substrates out of the cytosol.

In both types of secondary transport, the ion or solute being transported simultaneously with the protons is moving against its gradient of electrochemical potential, so its transport is active. However, the energy driving this transport is provided by the proton motive force rather than directly by ATP hydrolysis.

Kinetic analyses can elucidate transport mechanisms

Thus far we have described cellular transport in terms of its energetics. However, cellular transport can also be studied by use of enzyme kinetics because it involves the binding and dissociation of molecules at active sites on transport proteins. One advantage of the kinetic approach is that it gives new insights into the regulation of transport.

In kinetic experiments, the effects of external ion (or other solute) concentrations on transport rates are measured. The kinetic characteristics of the transport rates can then be used to distinguish between different transporters. The maximum rate (V_{max}) of carrier-mediated transport, and often of channel transport as well, cannot be exceeded, regardless of the concentration of substrate (**Figure 6.11**).

proton motive force (PMF)
The energetic effect of the electrochemical H^+ gradient across a membrane, expressed in units of electrical potential.

symport A type of secondary active transport in which two substances are moved in the same direction across a membrane.

antiport A type of secondary active transport in which the passive (downhill) movement of protons or other ions drives the active (uphill) transport of a solute in the opposite direction.

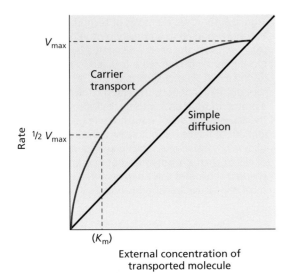

Figure 6.11 Carrier transport often shows enzyme kinetics, including saturation (V_{max}). In contrast, simple diffusion through open channels is ideally directly proportional to the concentration of the transported solute or, for an ion, to the difference in electrochemical potential across the membrane.

V_{max} is approached when the substrate-binding site on the carrier is always occupied or when flux through the channel is at a maximum. The concentration of transporter, not the concentration of solute, becomes rate-limiting. Thus, V_{max} is an indicator of the number of molecules of the specific transport protein that are functioning in the membrane.

The constant K_m (which is numerically equal to the solute concentration that yields half the maximum rate of transport) tends to reflect the properties of the particular binding site. Low K_m values indicate high binding affinity of the transport site for the transported substance. Such values usually imply the operation of a carrier system. Higher values of K_m indicate a lower affinity of the transport site for the solute. The affinity is often so low that in practice V_{max} is never reached. In such cases, kinetics alone cannot distinguish between carriers and channels.

Cells or tissues often display complex kinetics for the transport of a solute. Complex kinetics usually indicate the presence of more than one type of transport mechanism—for example, both high- and low-affinity transporters. **Figure 6.12** shows the rate of sucrose uptake by soybean cotyledon protoplasts as a function of the external sucrose concentration. Uptake increases sharply with concentration and begins to saturate at about 10 mM. At concentrations above 10 mM, uptake becomes linear and nonsaturable within the concentration range tested. Inhibition of ATP synthesis with metabolic poisons blocks the saturable component, but not the linear one.

The interpretation of the pattern shown in Figure 6.12 is that sucrose uptake at low concentrations is an energy-dependent, carrier-mediated process (H^+–sucrose symport). At higher concentrations, sucrose enters the cells by diffusion down its concentration gradient and is therefore insensitive to metabolic poisons. Consistent with these data, both H^+–sucrose symporters and sucrose facilitators (i.e., transport proteins that mediate transmembrane sucrose flux down its free-energy gradient) have been identified at the molecular level.

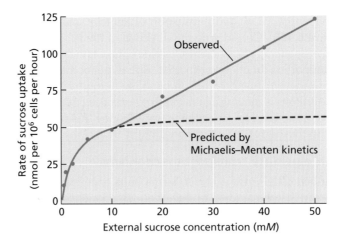

Figure 6.12 The transport properties of a solute can change with solute concentrations. For example, at low concentrations (1–10 mM), the rate of uptake of sucrose by soybean cells shows saturation kinetics typical of carriers. A curve fitted to these data is predicted to approach a maximum rate (V_{max}) of 57 nmol per 10^6 cells per hour. Instead, at higher sucrose concentrations, the uptake rate continues to increase linearly over a broad range of concentrations, consistent with the existence of additional facilitated transport mechanisms for sucrose uptake. (After Lin et al. 1984.)

Membrane Transport Proteins

Numerous representative transport proteins located in the plasma membrane and the tonoplast are illustrated in **Figure 6.13**. Typically, transport across a biological membrane is energized by one primary active transport system coupled to ATP hydrolysis. The transport of one ionic species—for example, H^+—generates an ion gradient and an electrochemical potential. Many other ions or neutral organic substrates can then be transported by a variety of secondary active transport proteins, which energize the transport of their substrates by simultaneously carrying one or two H^+ down their energy gradient. Thus, protons circulate across the membrane, outward through the primary active transport proteins, and back into the cell through the secondary active transport proteins. Most of the ion gradients across membranes of higher plants are generated and maintained by electrochemical-potential gradients of H^+, which are generated by electrogenic proton pumps.

Evidence suggests that in plants, Na^+ is transported out of the cell by a Na^+–H^+ antiporter and that Cl^-, NO_3^-, $H_2PO_4^-$, sucrose, amino acids, and other substances enter the cell via specific H^+ symporters. What about potassium ions? Potassium ions can be taken up from the soil or the apoplast by symport with H^+ (or under some conditions, Na^+). When the free-energy gradient favors passive K^+ uptake, K^+ can enter the cell by flux through specific K^+ channels. However, even influx through channels is driven by the H^+-ATPase, in the sense that K^+ diffusion is driven by the membrane potential, which is maintained at a value more negative than the K^+ equilibrium potential by the action of the electrogenic proton pump. Conversely, K^+ efflux requires the membrane potential to be maintained at a value more positive than E_K, which can be achieved by efflux of Cl^- or other anions through anion channels.

We have seen in preceding sections that some transmembrane proteins operate as channels for the controlled diffusion of ions. Other membrane proteins act as carriers for other substances (uncharged solutes and ions). Active transport uses carrier-type proteins that are energized either directly by ATP hydrolysis or indirectly as in the case of symporters and antiporters. The latter systems use the energy of ion gradients (often a H^+ gradient) to drive the energetically uphill transport of another ion or molecule. In the pages that follow we examine in more detail the molecular properties, cellular locations, and genetic manipulations of some of the transport proteins that mediate the movement of organic and inorganic nutrients, as well as water, across plant plasma membranes.

The genes for many transporters have been identified

Transporter gene identification has revolutionized the study of transporter proteins. One way to identify transporter genes is to screen plant complementary DNA (cDNA) libraries for genes that complement (i.e., compensate for) transport deficiencies in yeast. Many yeast transporter mutants have been used to identify corresponding plant genes by complementation. In the case of genes for ion channels, researchers have also studied the behavior of the channel proteins by expressing the genes in *Xenopus* frog oocytes, which because of their large size are convenient for electrophysiological studies. Genes for both inwardly and outwardly rectifying K^+ channels have been cloned and characterized in this way, and coexpression of ion channels and putative regulatory proteins such as protein kinases in oocytes has provided information on regulatory mechanisms of channel gating. As the number of sequenced genomes has increased, it has become common to identify putative transporter genes by phylogenetic analysis, in which sequence comparison with genes encoding transporters of known function in another organism allows one to predict function in the organism of interest. Based on such analyses, it has become evident that gene families, rather than individual genes, exist in plant genomes for most transport functions.

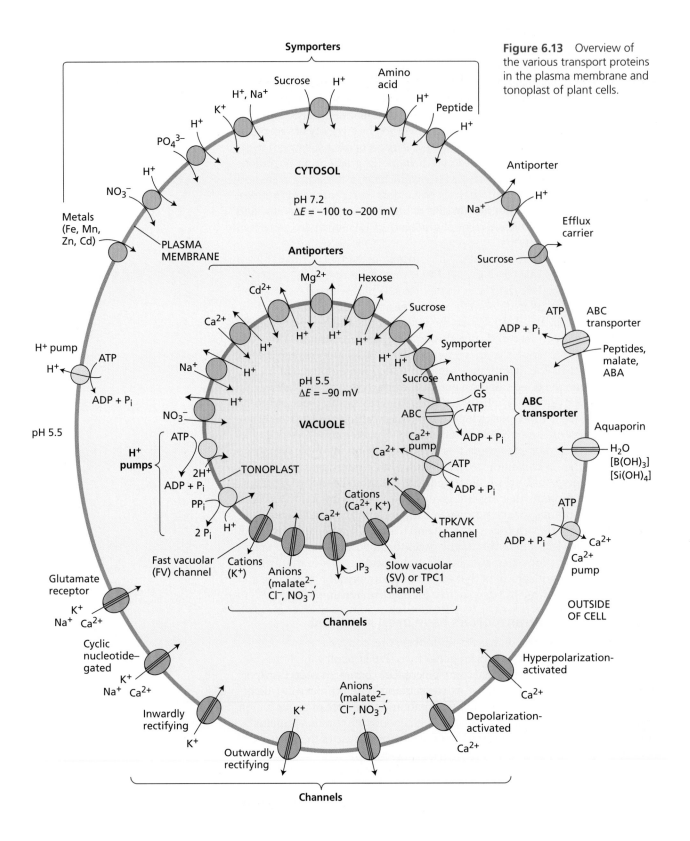

Figure 6.13 Overview of the various transport proteins in the plasma membrane and tonoplast of plant cells.

Within a gene family, variations in transport kinetics, in modes of regulation, and in differential tissue expression give plants a remarkable plasticity to acclimate to and prosper under a broad range of environmental conditions. In the next sections we discuss the functions and diversity of transporters for the major categories of solutes found within the plant body (note that sucrose transport was discussed earlier in this chapter and will also be discussed in Chapter 10).

Transporters exist for diverse nitrogen-containing compounds

Nitrogen, one of the macronutrients, can be present in the soil solution as nitrate (NO_3^-), ammonia (NH_3), or ammonium (NH_4^+). Plant NH_4^+ transporters are facilitators that promote NH_4^+ uptake down its free-energy gradient. Plant NO_3^- transporters are of particular interest because of their complexity. Kinetic analysis shows that NO_3^- transport, like the sucrose transport shown in Figure 6.12, has both high-affinity (low K_m) and low-affinity (high K_m) components. Both of these components are mediated by more than one gene product. In contrast to sucrose, NO_3^- is negatively charged, and such an electric charge imposes an energy requirement for uptake of nitrate. The energy is provided by symport with H^+. Nitrate transport is also strongly regulated according to NO_3^- availability: The enzymes required for NO_3^- transport, as well as for NO_3^- assimilation (see Chapter 5), are induced in the presence of NO_3^- in the environment, and uptake can be repressed if NO_3^- accumulates in the cells.

Mutants with defects in NO_3^- transport or NO_3^- reduction can be selected by growth in the presence of chlorate (ClO_3^-). Chlorate is a NO_3^- analog that is taken up and reduced in wild-type plants to the toxic product chlorite. If plants resistant to ClO_3^- are selected, they are likely to show mutations that block NO_3^- transport or reduction. Several such mutations have been identified in Arabidopsis. The first transporter gene identified in this way, named *CHL1*, encodes an inducible NO_3^-–H^+ symporter that functions as a dual-affinity carrier, with its mode of action (high affinity or low affinity) being switched by its phosphorylation status. Remarkably, this transporter also functions as a NO_3^- sensor that regulates NO_3^--induced gene expression.

Once nitrogen has been incorporated into organic molecules, there are a variety of mechanisms that distribute it throughout the plant. Peptide transporters provide one such mechanism. Peptide transporters are important for mobilizing nitrogen reserves during seed germination and senescence. In the carnivorous pitcher plant *Nepenthes alata*, high levels of expression of a peptide transporter are found in the pitcher, where the transporter presumably mediates uptake of peptides from digested insects into the internal tissues.

Some peptide transporters operate by coupling with the H^+ electrochemical gradient. Other peptide transporters are members of the ABC family of proteins, which directly use energy from ATP hydrolysis for transport; thus, this transport does not depend on a primary electrochemical gradient. The ABC family is an extremely large protein family, and its members transport diverse substrates, ranging from small inorganic ions to macromolecules. For example, large metabolites such as flavonoids, anthocyanins, and secondary products of metabolism are sequestered in the vacuole via the action of specific ABC transporters, while other ABC transporters mediate the transmembrane transport of the hormone abscisic acid.

Amino acids constitute another important category of nitrogen-containing compounds. The plasma membrane amino acid transporters of eukaryotes have been divided into five superfamilies, three of which rely on the proton gradient for coupled amino acid uptake and are present in plants. In general, amino acid transporters can provide high- or low-affinity transport, and they have overlapping substrate specificities. Many amino acid transporters show distinct tissue-specific expression patterns, suggesting specialized functions in different cell types. Amino acids constitute an important form in which nitrogen is distributed long distances in plants, so it is not surprising that the expression patterns of many amino acid transporter genes include expression in living vascular tissue.

Amino acid and peptide transporters have important roles in addition to their function as distributors of nitrogen resources. Because plant hormones are frequently found conjugated with amino acids and peptides, transporters for those molecules may also be involved in the distribution of hormone conjugates throughout the plant body. The hormone auxin is derived from tryptophan, and the genes encoding auxin transporters are related to those for some amino acid transporters. In another example, proline is an amino acid that accumulates under salt stress. This accumulation lowers the water potential of the cell, thereby promoting cellular water retention under stress conditions.

Cation transporters are diverse

Cations are transported by both cation channels and cation carriers. The relative contributions of each type of transport mechanism differ depending on the membrane, cell type, and prevailing conditions.

CATION CHANNELS On the order of 50 genes in the Arabidopsis genome encode channels mediating cation uptake across the plasma membrane or intracellular membranes such as the tonoplast. Some of these channels are highly selective for specific ionic species, such as potassium ions. Others allow passage of a variety of cations, sometimes including Na^+, even though this ion is toxic when overaccumulated. As described in **Figure 6.14**, cation channels are categorized into six types based on their deduced structures and cation selectivity.

Of the six types of plant cation channels, the Shaker channels have been the most thoroughly characterized. These channels are named after a *Drosophila* K^+ channel whose mutation causes the flies to shake or tremble. Plant Shaker chan-

(A) K⁺ channels

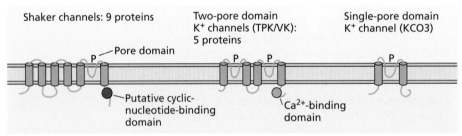

(B) Poorly selective cation channels   (C) Ca²⁺-permeable channels   (D) Cation selective, Ca²⁺ permeable

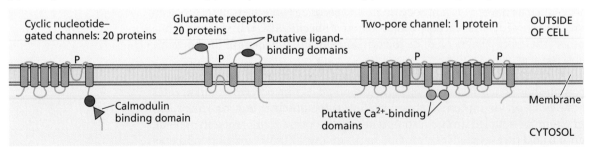

Figure 6.14 Six families of Arabidopsis cation channels. Some channels have been identified from sequence homology with channels of animals, while others have been experimentally verified. (A) K^+-selective channels. (B) Weakly selective cation channels with activity regulated by binding of cyclic nucleotides. (C) Putative glutamate receptors; based on measurements of cytosolic Ca^{2+} changes, these proteins probably function as Ca^{2+}-permeable channels. (D) Two-pore channel: one protein (TPC1) is the sole two-pore channel of this type encoded in the Arabidopsis genome. TPC1 is permeable to mono- and divalent cations, including Ca^{2+}. (A after Lebaudy et al. 2007; B–D after Very and Sentenac 2002.)

nels are highly K$^+$-selective and can be either inwardly or outwardly rectifying or weakly rectifying. Some members of the Shaker family can:

- Mediate K$^+$ uptake or efflux across the guard cell plasma membrane
- Provide a major conduit for K$^+$ uptake from the soil
- Participate in K$^+$ release to dead xylem vessels from the living stelar cells
- Play a role in K$^+$ uptake in pollen, a process that promotes water influx and pollen tube elongation

Some Shaker channels, such as those in roots, can mediate high-affinity K$^+$ uptake, allowing passive K$^+$ uptake at micromolar external K$^+$ concentrations as long as the membrane potential is sufficiently hyperpolarized to drive this uptake.

Not all ion channels are as strongly regulated by membrane potential as the majority of Shaker channels are. Some ion channels, such as the TPK/VK channels (see Figures 6.13 and 6.14A), are not voltage-regulated, and the voltage sensitivity of others, such as the KCO3 channel, has not yet been determined. Cyclic nucleotide–gated cation channels are one example of a ligand-gated channel, with activity promoted by the binding of cyclic nucleotides such as cGMP. These channels exhibit weak selectivity with permeability to K$^+$, Na$^+$, and Ca^{2+}. Cyclic nucleotide–gated cation channels are involved in diverse physiological processes, including disease resistance, senescence, temperature sensing, and pollen tube growth and viability. Another interesting set of ligand-gated channels are the glutamate receptor channels. These channels are homologous to a class of glutamate receptors in the mammalian nervous system that function as glutamate-gated cation channels, and are activated in plants by glutamate and some other amino acids. Plant glutamate receptor channels are permeable to Ca^{2+}, K$^+$, and Na$^+$ to varying extents, but have been particularly implicated in Ca^{2+} uptake and signaling in nutrient acquisition in roots and in guard cell and pollen tube physiology.

Ion fluxes must also occur in and out of the vacuole, and both cation- and anion-permeable channels have been characterized in the vacuolar membrane (see Figure 6.13). Plant vacuolar cation channels include the KCO3 K$^+$ channel (see Figure 6.14A), the Ca^{2+}-activated TPC1/SV cation channel (see Figure 6.13), and most TPK/VK channels (see Figure 6.13), which are highly selective K$^+$ channels that are activated by Ca^{2+}. In addition, Ca^{2+} efflux from internal storage sites such as the vacuole plays an important signaling role. Ca^{2+} release from stores is triggered by several second messenger molecules, including cytosolic Ca^{2+} itself and inositol trisphosphate (InsP$_3$). For a more detailed description of these signal transduction pathways, see Chapter 12.

CATION CARRIERS A variety of carriers also move cations into plant cells. One family of transporters that specializes in K$^+$ transport across plant membranes is the HAK/KT/KUP family (which we will refer to here as the HAK family). The HAK family contains both high-affinity and low-affinity transporters, some of which also mediate Na$^+$ influx at high external Na$^+$ concentrations. High-affinity HAK transporters are thought to take up K$^+$ via H$^+$–K$^+$ symport, and these transporters are particularly important for K$^+$ uptake from the soil when soil K$^+$ concentrations are low. A second family, the cation–H$^+$ antiporters (CPAs), mediates electroneutral exchange of H$^+$ and other cations, including K$^+$ in some cases. A third family consists of the TRK/HKT transporters (which we will refer to here as the HKT transporters), which can operate as K$^+$–H$^+$ or K$^+$–Na$^+$ symporters, or as Na$^+$ channels under high external Na$^+$ concentrations. The importance of the HKT transporters for K$^+$ transport remains incompletely elucidated, but as described below, HKT transporters are central elements in plant tolerance of saline conditions.

Irrigation increases soil salinity, and salinization of croplands is an increasing problem worldwide. Although halophytic plants, such as those found in salt marshes, are adapted to a high-salt environment, such environments are deleterious to other, glycophytic, plant species, including the majority of crop species. Plants have evolved mechanisms to extrude Na^+ across the plasma membrane, to sequester salt in the vacuole, and to redistribute Na^+ within the plant body.

At the plasma membrane, a Na^+–H^+ antiporter was uncovered in a screen to identify Arabidopsis mutants that showed enhanced sensitivity to salt, hence this antiporter was named Salt Overly Sensitive, or SOS1. SOS1-type antiporters in the root extrude Na^+ from the plant, thereby lowering internal concentrations of this toxic ion.

Vacuolar Na^+ sequestration occurs by activity of Na^+–H^+ antiporters—a subset of CPA proteins—which couple the energetically downhill movement of H^+ into the cytosol with Na^+ uptake into the vacuole. When the Arabidopsis *AtNHX1* Na^+–H^+ antiporter gene is overexpressed, it confers greatly increased salt tolerance to both Arabidopsis and crop species such as maize, wheat, and tomato.

Whereas SOS1 and NHX antiporters reduce cytosolic Na^+ concentrations, HKT1 transporters transport Na^+ from the apoplast into the cytosol. However, Na^+ uptake by HKT1 transporters in the plasma membrane of root xylem parenchyma cells is important in retrieving Na^+ from the transpiration stream, thereby reducing Na^+ concentrations and attendant toxicity in photosynthetic tissues. Presumably such Na^+ is then excluded from the root cytosol by the action of SOS1 and NHX transporters. Transgenic expression of a HKT1 transporter in a durum (pasta) variety of wheat greatly increases grain yield in wheat grown on saline soils.

Just as for Na^+, there is a large free-energy gradient for Ca^{2+} that favors its entry into the cytosol from both the apoplast and intracellular stores. This entry is mediated by Ca^{2+}-permeable channels, which we described above. Calcium ion concentrations in the cell wall and in the apoplast are usually in the millimolar range; in contrast, free cytosolic Ca^{2+} concentrations are kept in the hundreds of nanomolar (10^{-9} M) to 1 micromolar (10^{-6} M) range, against the large electrochemical-potential gradient for Ca^{2+} diffusion into the cell. Calcium ion efflux from the cytosol is achieved by Ca^{2+}-ATPases found at the plasma membrane and in some endomembranes such as the tonoplast (see Figure 6.13) and endoplasmic reticulum. Much of the Ca^{2+} inside the cell is stored in the central vacuole, where it is sequestered via Ca^{2+}-ATPases and via Ca^{2+}–H^+ antiporters, which use the electrochemical potential of the proton gradient to energize the vacuolar accumulation of Ca^{2+}.

Because small changes in cytosolic Ca^{2+} concentration drastically alter the activities of many enzymes, cytosolic Ca^{2+} concentration is tightly regulated. The Ca^{2+}-binding protein, calmodulin (CaM), participates in this regulation. Although CaM has no catalytic activity of its own, Ca^{2+}-bound CaM binds to many different classes of target proteins and regulates their activity. Ca^{2+}-permeable cyclic nucleotide–gated channels are CaM-binding proteins, and there is evidence that this CaM binding results in down-regulation of channel activity. One class of Ca^{2+}-ATPases also binds CaM. CaM binding releases these ATPases from autoinhibition, resulting in increased Ca^{2+} extrusion into the apoplast, endoplasmic reticulum, and vacuole. Together, these two regulatory effects of CaM provide a mechanism whereby increases in cytosolic Ca^{2+} concentration initiate a negative feedback loop, via activated CaM, that aids in restoration of resting cytosolic Ca^{2+} levels.

Anion transporters have been identified

Nitrate (NO_3^-), chloride (Cl^-), sulfate (SO_4^{2-}), and phosphate ($H_2PO_4^-$) are the major inorganic ions in plant cells, and malate^{2-} is a major organic anion. The free-energy gradient for all of these anions is in the direction of passive efflux. Several types of plant anion channels have been characterized by electrophysiological techniques, and most anion channels appear to be permeable to a variety of anions. In particular, several anion channels with differential voltage-dependence and anion permeabilities have been shown to be important for anion efflux from guard cells during stomatal closure.

In contrast to the relative lack of specificity of anion channels, anion carriers that mediate the energetically uphill transport of anions into plant cells exhibit selectivity for particular anions. In addition to the transporters for nitrate uptake described above, plants have transporters for various organic anions, such as malate and citrate. As discussed later in this chapter, malate uptake is an important contributor to the increase in intracellular solute concentration that drives the water uptake into guard cells that leads to stomatal opening. One member of the ABC family, AtABCB14, has been assigned this malate import function.

Phosphate availability in the soil solution often limits plant growth (see Chapter 4). In Arabidopsis, a family of about nine plasma membrane phosphate transporters, some high affinity and some low affinity, mediates phosphate uptake in symport with protons. These transporters are expressed primarily in the root epidermis and root hairs, and their expression is induced upon phosphate starvation. Other phosphate–H^+ symporters have also been identified in plants and have been localized to membranes of intracellular organelles such as plastids and mitochondria. Another group of phosphate transporters, the phosphate translocators, are located in the inner plastid membrane, where they mediate exchange of inorganic phosphate with phosphorylated carbon compounds.

Transporters for metal and metalloid ions transport essential micronutrients

Several metals are essential nutrients for plants, although they are required in only trace amounts. One example is iron. Iron deficiency is the most common human nutritional disorder worldwide, so an increased understanding of how plants accumulate iron may also benefit efforts to improve the nutritional value of crops. More than 25 ZIP transporters mediate the uptake of iron, manganese, and zinc ions into plants, and other transporter families that mediate the uptake of copper and molybdenum ions have been identified. Metal ions are usually present at low concentrations in the soil solution, so these transporters are typically high-affinity transporters. Some metal ion transporters mediate the uptake of cadmium or lead ions, which are undesirable in crop species because cadmium and lead ions are toxic to humans. However, this property may prove useful in the detoxification of soils by uptake of contaminants into plants (phytoremediation), which can then be removed and properly discarded.

Once in the plant, metal ions, usually chelated with other molecules, must be transported into the xylem for distribution throughout the plant body via the transpiration stream, and metals must also be sent to their appropriate subcellular destinations. For example, most of the iron in plants is found in chloroplasts, where it is incorporated into chlorophyll and components of the electron transport chain (see Chapter 7). Overaccumulation of ionic forms of metals such as iron and copper can lead to production of toxic reactive oxygen species (ROS). Compounds that chelate metal ions guard against this threat, and transporters that mediate metal uptake into the vacuole are also important in maintaining metal concentrations at nontoxic levels.

Metalloids are elements that have properties of both metals and nonmetals. Boron and silicon are two metalloids that are used by plants. Both play important roles in cell wall structure—boron by cross-linking cell wall polysaccharides, and silicon by increasing structural rigidity. Boron (as boric acid [$B(OH)_3$; also written H_3BO_3]) and silicon (as silicic acid [$Si(OH)_4$; also written H_4SiO_4]) both enter cells via aquaporin-type channels (see below) and are exported via efflux transporters, probably by secondary active transport. Due to similarities in chemical structure, arsenite (a form of arsenic) can also enter plant roots via the silicon channel and be

aquaporins Integral membrane proteins that form channels across a membrane, many of which are selective for water (hence the name). Such channels facilitate water movement across the membrane.

exported to the transpiration stream via the silicon transporter. Rice is particularly efficient at taking up arsenite, and as a result, arsenic poisoning from human consumption of rice is a significant problem in regions of Southeast Asia.

Aquaporins have diverse functions

Aquaporins are a class of transporters that are relatively abundant in plant membranes and are also common in animal membranes (see Chapters 2 and 3). The Arabidopsis genome is predicted to encode approximately 35 aquaporins. As the name implies, many aquaporin proteins mediate the flux of water across membranes, and it has been hypothesized that aquaporins function as sensors of gradients in osmotic potential and turgor pressure. In addition, some aquaporin proteins mediate the influx of mineral nutrients (e.g., boric acid and silicic acid). There is also evidence that aquaporins can act as conduits for carbon dioxide, ammonia (NH_3), and hydrogen peroxide (H_2O_2) movement across plant plasma membranes.

Aquaporin activity is regulated by phosphorylation as well as by pH, Ca^{2+} concentration, heteromerization, and reactive oxygen species. Such regulation may account for the ability of plant cells to quickly alter their water permeability in response to circadian rhythms and to stresses such as salt, chilling, drought, and flooding (anoxia). Regulation also occurs at the level of gene expression. Aquaporins are highly expressed in epidermal and endodermal cells and in the xylem parenchyma, which may be critical points for control of water movement.

Plasma membrane H⁺-ATPases are highly regulated P-type ATPases

As we have seen, the outward active transport of protons across the plasma membrane creates gradients of pH and electrical potential that drive the transport of many other substances (ions and uncharged solutes) through the various secondary active transport proteins. H⁺-ATPase activity is also important for the regulation of cytosolic pH and for the control of cell turgor, which drives organ (leaf and flower) movement, stomatal opening, and cell growth. **Figure 6.15** illustrates how a membrane H⁺-ATPase might work.

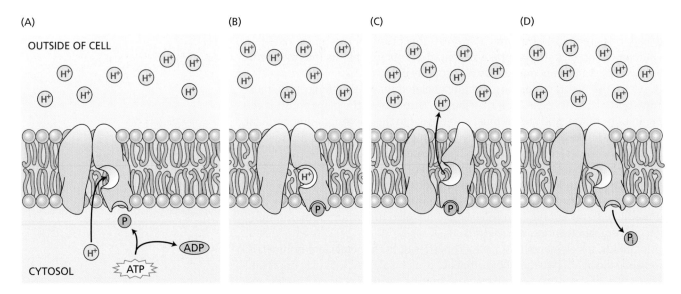

Figure 6.15   Hypothetical steps in the transport of a proton against its chemical gradient by H⁺-ATPase. The pump, embedded in the membrane, (A) binds the proton on the inside of the cell and (B) is phosphorylated by ATP. (C) This phosphorylation leads to a conformational change that exposes the proton to the outside of the cell and makes it possible for the proton to diffuse away. (D) Release of the phosphate ion (P_i) from the pump into the cytosol restores the initial configuration of the H⁺-ATPase and allows a new pumping cycle to begin.

Plant and fungal plasma membrane H⁺-ATPases and Ca²⁺-ATPases are members of a class known as P-type ATPases, which are phosphorylated as part of the catalytic cycle that hydrolyzes ATP. Plant plasma membrane H⁺-ATPases are encoded by a family of about a dozen genes. The roles of each H⁺-ATPase isoform are starting to be understood, based on information from gene expression patterns and functional analysis of Arabidopsis plants harboring null mutations in individual H⁺-ATPase genes. Some H⁺-ATPases exhibit cell-specific patterns of expression. For example, several H⁺-ATPases are expressed in guard cells, where they energize the plasma membrane to drive solute uptake during stomatal opening (see later in this chapter).

In general, H⁺-ATPase expression is high in cells with key functions in nutrient movement, including root endodermal cells and cells involved in nutrient uptake from the apoplast that surrounds the developing seed. In cells in which multiple H⁺-ATPases are coexpressed, they may be differentially regulated or may function redundantly, perhaps providing a "fail-safe" mechanism to this all-important transport function.

Figure 6.16 shows a model of the functional domains of a yeast plasma membrane H⁺-ATPase, which is similar to those of plants. The protein has ten membrane-spanning domains that cause it to loop back and forth across the membrane. Some of the membrane-spanning domains make up the pathway through which protons are pumped. The catalytic domain, which catalyzes ATP hydrolysis, including the aspartic acid residue that becomes phosphorylated during the catalytic cycle, is on the cytosolic face of the membrane.

Like other enzymes, the plasma membrane H⁺-ATPase is regulated by the concentration of substrate (ATP), pH, temperature, and other factors. In addition, H⁺-ATPase molecules can be reversibly activated or deactivated by specific signals, such as light, hormones, or pathogen attack. This type of regulation is mediated by a specialized autoinhibitory domain at the C-terminal end of the polypeptide chain, which acts to regulate the activity of the H⁺-ATPase (see Figure 6.16). If the autoinhibitory domain is removed by a protease, the enzyme becomes irreversibly activated.

The autoinhibitory effect of the C-terminal domain can also be regulated by protein kinases and phosphatases that add phosphate groups to or remove them from serine or threonine residues on this domain. Phosphorylation recruits ubiquitous enzyme-modulating proteins called 14-3-3 proteins, which bind to the phosphorylated region and thus displace the autoinhibitory domain, leading to H⁺-ATPase activation. The fungal toxin **fusicoccin**, which is a strong activator of the H⁺-ATPase, activates this pump by increasing 14-3-3 protein binding affinity even in the absence of phosphorylation. The effect of fusicoccin on the guard cell H⁺-ATPases is so strong that it can lead to irreversible stomatal opening, wilting, and even plant death (see Figure 6.19B).

The tonoplast H⁺-ATPase drives solute accumulation in vacuoles

Plant cells increase their size primarily by taking up water into a large central vacuole. Therefore, the osmotic pressure of the vacuole must be kept

fusicoccin A fungal toxin that induces acidification of plant cell walls by activating H⁺-ATPases in the plasma membrane. Fusicoccin stimulates rapid acid growth in stem and coleoptile sections. It also stimulates stomatal opening by stimulating proton pumping at the guard cell plasma membrane.

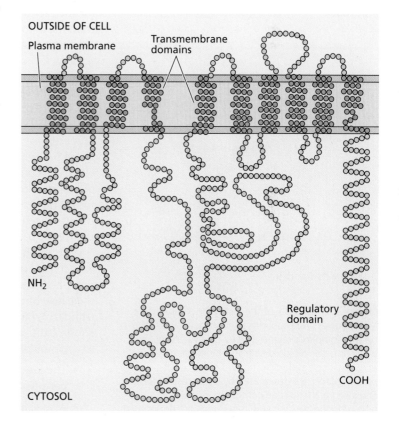

Figure 6.16 Two-dimensional representation of a plasma membrane H⁺-ATPase from yeast. Each small circle represents an amino acid. The H⁺-ATPase protein has ten transmembrane segments. The regulatory domain is an autoinhibitory domain. Posttranslational modifications that lead to displacement of the autoinhibitory domain result in H⁺-ATPase activation. (After Palmgren 2001.)

sufficiently high for water to enter from the cytosol. The tonoplast regulates the traffic of ions and metabolites between the cytosol and the vacuole, just as the plasma membrane regulates their uptake into the cell. Tonoplast transport became a vigorous area of research following the development of methods for the isolation of intact vacuoles and tonoplast vesicles. These studies elucidated a diversity of anion and cation channels in the tonoplast membrane (see Figure 6.13) and led to the discovery of a new type of proton-pumping ATPase, the vacuolar H^+-ATPase, which transports protons into the vacuole (see Figure 6.13).

The vacuolar H^+-ATPase differs both structurally and functionally from the plasma membrane H^+-ATPase. The vacuolar ATPase is more closely related to the F-ATPases of chloroplasts and mitochondria (see Chapters 7 and 11), and the vacuolar ATPase, unlike the plasma membrane ATPases discussed earlier, does not form a phosphorylated intermediate during ATP hydrolysis. Vacuolar ATPases belong to a general class of ATPases that are present on the endomembrane systems of all eukaryotes. They are large enzyme complexes, about 750 kDa, composed of multiple subunits. These subunits are organized into a peripheral complex, V_1, that is responsible for ATP hydrolysis, and an integral membrane channel complex, V_0, that is responsible for H^+ translocation across the membrane (**Figure 6.17**). Because of their similarities to F-ATPases, vacuolar ATPases are assumed to operate like tiny rotary motors (see Chapter 7).

Vacuolar ATPases are electrogenic proton pumps that transport protons from the cytosol to the vacuole and generate a proton motive force across the tonoplast. Electrogenic proton pumping accounts for the fact that the vacuole is typically 20 to 30 mV more positive than the cytosol, although it is still negative relative to the external medium. To allow maintenance of bulk electrical neutrality, anions such as Cl^- or $malate^{2-}$ are transported from the cytosol into the vacuole through channels in the tonoplast. The conservation of bulk electrical neutrality by anion transport makes it possible for the vacuolar H^+-ATPase to generate a large concentration gradient of protons (pH gradient) across the tonoplast. This gradient accounts for the fact that the pH of the vacuolar sap is typically about 5.5, while the cytosolic pH is typically 7.0 to 7.5. Whereas the electrical component of the proton motive force drives the uptake of anions into the vacuole, the electrochemical-potential gradient for H^+ ($\tilde{\mu}_{H^+}$) is harnessed to drive the uptake of cations and sugars into the vacuole via secondary transport (antiporter) systems (see Figure 6.13).

Although the pH of most plant vacuoles is mildly acidic (about 5.5), the pH of the vacuoles of some species is much lower —a phenomenon termed *hyperacidification*. Vacuolar hyperacidification is the cause of the sour taste of certain fruits (lemons) and vegetables (rhubarb). Biochemical studies have suggested that the low pH of lemon fruit vacuoles (specifically, those of the juice sac cells) is due to a combination of factors:

- The low permeability of the vacuolar membrane to protons permits a steeper pH gradient to build up.

- A specialized vacuolar ATPase is able to pump protons more efficiently (with less wasted energy) than normal vacuolar ATPases can.

- Organic acids such as citric, malic, and oxalic acids accumulate in the vacuole and help maintain its low pH by acting as buffers.

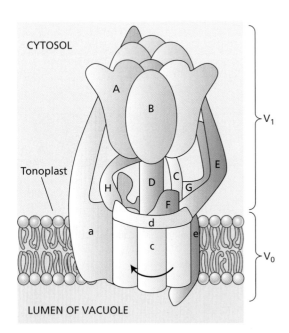

Figure 6.17 Model of the V-ATPase rotary motor. Many polypeptide subunits come together to make up this complex enzyme. The V_1 catalytic complex, which is easily dissociated from the membrane, contains the nucleotide-binding and catalytic sites. Components of V_1 are designated by uppercase letters. The integral membrane complex mediating H^+ transport is designated V_0, and its subunits are designated by lowercase letters. It is proposed that ATPase reactions catalyzed by each of the A subunits, acting in sequence, drive the rotation of the shaft (D) and the six c subunits. The rotation of the c subunits relative to the a subunit is thought to drive the transport of H^+ across the membrane. (After Kluge et al. 2003.)

H⁺-pyrophosphatases also pump protons at the tonoplast

Another type of proton pump, a H⁺-pyrophosphatase (H⁺-PPase), works in parallel with the vacuolar ATPase to create a proton gradient across the tonoplast (see Figure 6.13). This enzyme consists of a single polypeptide that harnesses energy from the hydrolysis of inorganic pyrophosphate (PP_i) to drive H⁺ transport.

The free energy released by PP_i hydrolysis is less than that from ATP hydrolysis. However, the H⁺-PPase transports only one H⁺ ion per PP_i molecule hydrolyzed, whereas the vacuolar ATPase appears to transport two H⁺ ions per ATP hydrolyzed. Thus, the energy available per H⁺ transported appears to be approximately the same, and the two enzymes seem to be able to generate comparable proton gradients. Interestingly, the plant H⁺-PPase is not found in animals or in yeast, although similar enzymes are present in some bacteria and protists.

Both the V-ATPase and the H⁺-PPase are found in other compartments of the endomembrane system in addition to the vacuole. Consistent with this distribution, evidence is emerging that these ATPases regulate not only H⁺ gradients per se, but also vesicle trafficking and secretion. In addition, increased auxin transport and cell division in Arabidopsis plants overexpressing a H⁺-PPase, and the opposite phenotypes in plants with reduced H⁺-PPase activity, indicate connections between H⁺-PPase activity and the synthesis, distribution, and regulation of auxin transporters.

Ion Transport in Stomatal Opening

The opening of stomata illustrates how membrane transport processes control a physiological response. In Chapter 3 we saw that guard cell swelling and stomatal opening were stimulated by light through an effect on guard cell photosynthesis as well as specifically by blue light through an unidentified process. Here we describe the membrane transport processes involved in guard cell opening. In Chapter 13 we will complete the treatment of this important process by describing how the blue light is detected and how the signal is transmitted.

Light stimulates ATPase activity and creates a stronger electrochemical gradient across the guard cell plasma membrane

When guard cell protoplasts are irradiated with blue light under saturating background red-light illumination, the pH of the solution surrounding the cells becomes more acidic and the protoplasts swell (**Figure 6.18A**). Such blue light–induced swelling is blocked by inhibitors of the plasma membrane proton-pumping H⁺-ATPase, such as orthovanadate (**Figure 6.18B**).

(A)

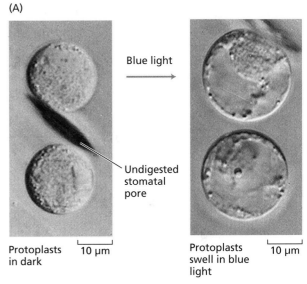

(B)

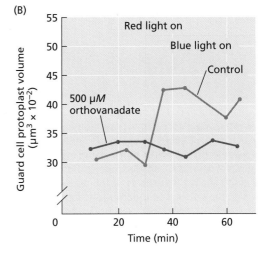

Figure 6.18 Blue light–stimulated swelling of guard cell protoplasts. (A) In the absence of a rigid cell wall, guard cell protoplasts of onion (*Allium cepa*) swell. (B) Blue light stimulates the swelling of guard cell protoplasts of broad bean (*Vicia faba*), and orthovanadate, an inhibitor of the H⁺-ATPase, inhibits this swelling. Blue light stimulates ion and water uptake in the guard cell protoplasts, which in the intact guard cells provides a mechanical force working against the rigid cell wall that distorts the guard cells and drives increases in stomatal apertures. (A, left courtesy of E. Zeiger; right from Zeiger and Hepler 1977; B after Amodeo et al. 1992.)

In the intact leaf, the blue-light stimulation of a proton-pumping ATPase in the guard cell plasma membrane lowers the pH of the apoplastic space surrounding the guard cells by as much as 0.5–1.0 pH unit as well as hyperpolarizes the plasma membrane (making it as much as 64 mV more negative), thereby generating the driving force needed for ion uptake and stomatal opening.

The activity of electrogenic pumps such as the H^+-ATPase can be measured in whole-cell patch clamp experiments, in which a hollow electrode (a pipette) is attached to the protoplast in such a way that the inside of the electrode is continuous with the cytoplasm of the cell (**Figure 6.19A**). Activation of electrogenic pumps in this experimental setup results in measurable electric currents across the plasma membrane. **Figure 6.19B** shows a patch clamp recording of a guard cell protoplast treated in the dark with the fungal toxin fusicoccin to activate the plasma membrane H^+-ATPase. Exposure to fusicoccin stimulates an outward electric current, which generates a proton gradient. This proton gradient is abolished by carbonyl cyanide m-chlorophenylhydrazone (CCCP), a proton ionophore (uncoupler) that makes the plasma membrane highly permeable to protons, thus collapsing the proton gradient across the membrane.

The relationship between proton pumping at the guard cell plasma membrane and stomatal opening is evident from the observations that (1) fusicoccin stimulates both proton extrusion from guard cell protoplasts and stomatal opening, and (2) CCCP inhibits the fusicoccin-stimulated opening. A pulse of blue light given under a saturating red-light background can also stimulate an outward

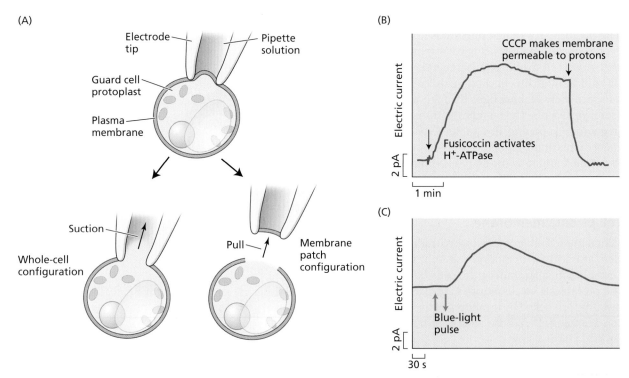

Figure 6.19 Activation of the H^+-ATPase at the plasma membrane of guard cell protoplasts by fusicoccin and blue light can be measured as electric current in patch clamp experiments. (A) For patch clamp experiments, an electrode consisting of a glass pipette is pressed against the outside of the protoplast to make a tight seal with the membrane. Applying suction creates a hole in the membrane that connects the electrode solution to the cell's cytoplasm (left). If instead the pipette is pulled away from the protoplast, this detaches a patch of membrane that remains tightly sealed to the end of the pipette (right). The experiments shown in (B) and (C) were done in the whole-cell configuration. (B) Outward electric current (measured in picoamperes, pA) at the plasma membrane of a guard cell protoplast stimulated by the fungal toxin fusicoccin, an activator of the H^+-ATPase. The current is abolished by the proton ionophore carbonyl cyanide m-chlorophenylhydrazone (CCCP). (C) Outward electric current at the plasma membrane of a guard cell protoplast stimulated by a blue-light pulse. These results indicate that blue light stimulates the H^+-ATPase. (B after Serrano et al. 1988; C after Assmann et al. 1985.)

electric current from guard cell protoplasts (**Figure 6.19C**), again showing that blue light stimulates an electrogenic plasma membrane ATPase.

Hyperpolarization of the guard cell plasma membrane leads to uptake of ions and water

Blue light modulates guard cell osmoregulation by means of its activation of proton pumping, solute uptake, and stimulation of the synthesis of organic solutes (**Figure 6.20**). The K^+ concentration in guard cells increases from 100 mM in the closed state to 400 to 800 mM in the open state. The electrical component of the proton gradient provides the driving force for the passive uptake of K^+ via voltage-regulated K^+ channels discussed earlier in this chapter. These large concentration changes in K^+ are electrically balanced by varying amounts of the anions Cl^- and malate^{2-}. Chloride anions are taken up into the guard cells from the apoplast during stomatal opening and extruded during stomatal closing. Malate anions, by contrast, are synthesized in the guard cell cytosol, in a metabolic pathway that uses carbon skeletons generated by starch hydrolysis. The

Figure 6.20 Light-stimulated plasma membrane ATPase activity provides the driving force for guard cell swelling and stomatal opening. (After Inoue and Kinoshita 2017.)

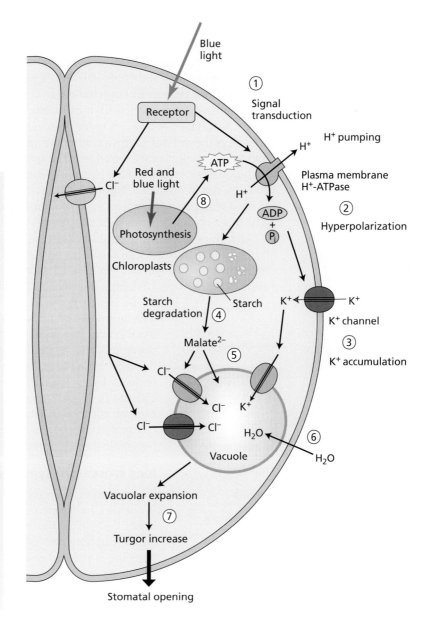

1. Blue light activates the plasma membrane H$^+$-ATPase via a signal transduction pathway (described in Chapter 13).

2. Activation of the H$^+$-ATPase hyperpolarizes the plasma membrane.

3. K$^+$ movement into the guard cell increases.

4. Malate accumulates in the cytosol as a result of chloroplast starch degradation.

5. Vacuolar K$^+$ uptake is accompanied by uptake of Cl$^-$ and malate^{2-}.

6. These ion movements lower the water potential of the vacuole (see Chapter 2), driving water uptake.

7. The vacuoles expand, and increased turgor causes stomatal opening.

8. Photosynthetically active radiation increases cellular ATP generation by chloroplasts, which supports increased rates of cellular ATPase activity.

extracellular space In plants, the space continuum outside the plasma membrane made up of interconnecting cell walls through which water and mineral nutrients readily diffuse.

apoplast The mostly continuous system of cell walls, intercellular air spaces, and xylem vessels in a plant.

symplast The continuous system of cell protoplasts interconnected by plasmodesmata.

decreasing water potential caused by the influx of ions causes water to move into the guard cells, thus increasing the osmotic pressure (see Figure 6.20) and causing the guard cells to swell and the stomata to open (see Figure 3.15).

Ion Transport in Roots

Mineral nutrients absorbed by the root are carried to the shoot by the transpiration stream moving through the xylem (see Chapter 3). Both the initial uptake of nutrients and water and the subsequent movement of these substances from the root surface across the cortex and into the xylem are highly specific, well-regulated processes.

Ion transport across the root obeys the same biophysical laws that govern cellular transport. However, as you have seen in the case of water movement (see Chapter 3), the anatomy of roots imposes some special constraints on the pathway of ion movement. In this section we discuss the pathways and mechanisms involved in the radial movement of ions from the root surface to the tracheary elements of the xylem.

Solutes move through both apoplast and symplast

Thus far, our discussion of cellular ion transport has not included the cell wall. In terms of the transport of small molecules, the cell wall is a fluid-filled lattice of polysaccharides through which mineral nutrients diffuse readily. Because all plant cells are separated by cell walls, ions can diffuse across a tissue (or be carried passively by water flow) entirely through the cell wall space without ever entering a living cell. This continuum of cell walls is called the **extracellular space**, or **apoplast** (see Figure 3.4). Typically, 5 to 20% of the plant tissue volume is occupied by cell walls.

Just as the cell walls form a continuous phase, so do the cytoplasms of neighboring cells, collectively referred to as the **symplast**. Plant cells are interconnected by cytoplasmic bridges called plasmodesmata (see Chapter 1), cylindrical pores 20 to 60 nm in diameter (**Figure 6.21**; also see Figure 1.8). Each plasmodesma is lined with plasma membrane and contains a narrow tubule, the *desmotubule*, which is a continuation of the endoplasmic reticulum.

In tissues where significant amounts of intercellular transport occur, neighboring cells contain large numbers of plasmodesmata, up to 15 per square micrometer of cell surface. Specialized secretory cells, such as floral nectaries and leaf salt glands, have high densities of plasmodesmata.

By injecting dyes or by making electrical-resistance measurements on cells containing large numbers of plasmodesmata, investigators have shown that inorganic ions, water, and small organic molecules can move from cell to cell through these pores. Because each plasmodesma is partly occluded by the desmotubule and its associated proteins (see Chapter 1), the movement of large molecules such as proteins through plasmodesmata requires special mechanisms. Ions, by contrast, appear to move symplastically through the plant by simple diffusion through plasmodesmata.

Ions cross both symplast and apoplast

Ion absorption by the root (see Chapter 4) is more pronounced in the root hair zone than in the meristem and elongation zones. Cells in the root hair zone have completed their elongation, but have not yet begun secondary growth. The root hairs are simply extensions of specific epidermal cells that greatly increase the surface area available for ion absorption.

An ion that enters a root may immediately enter the symplast by crossing the plasma membrane of an epider-

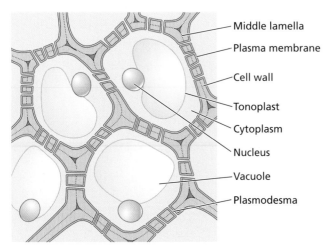

Middle lamella

Plasma membrane

Cell wall

Tonoplast

Cytoplasm

Nucleus

Vacuole

Plasmodesma

Figure 6.21 Plasmodesmata (singular "plasmodesma") connect the cytoplasms of neighboring cells, thereby facilitating cell-to-cell communication.

mal cell, or it may enter the apoplast and diffuse between the epidermal cells through the cell walls. From the apoplast of the cortex, an ion (or other solute) may either be transported across the plasma membrane of a cortical cell, thus entering the symplast, or diffuse radially all the way to the endodermis via the apoplast. The apoplast forms a continuous phase from the root surface through the cortex. However, in all cases, ions must enter the symplast before they can enter the stele, because of the presence of the Casparian strip. As discussed in Chapters 3 and 4, the Casparian strip is a lignified or suberized layer that forms as rings around the specialized cells of the endodermis (**Figure 6.22**) and effectively blocks the entry of water and solutes into the stele via the apoplast.

The stele consists of dead tracheary elements surrounded by living pericycle and xylem parenchyma cells. Once an ion has entered the stele through the symplastic connections across the endodermis, it continues to diffuse through the living cells. Finally, the ion is released into the apoplast and diffuses into the conducting cells of the xylem—because these cells are dead, their interiors are continuous with the apoplast. The Casparian strip allows nutrient uptake to be selective; it also prevents ions from diffusing back out of the root through the apoplast. Thus, the presence of the Casparian strip allows the plant to maintain a higher ion concentration in the xylem than exists in the soil water surrounding the roots.

(A)

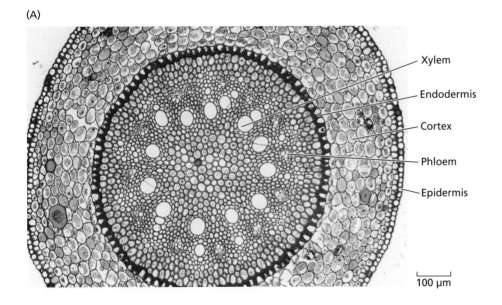

(B)

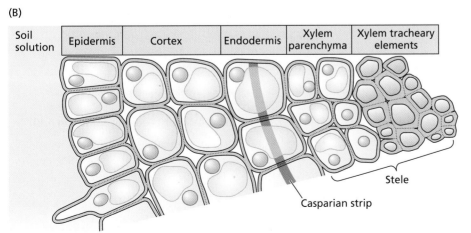

Figure 6.22 Tissue organization in roots. (A) Cross section through a root of carrion flower (genus *Smilax*), a monocot, showing the epidermis, cortex parenchyma, endodermis, xylem, and phloem. (B) Schematic diagram of a root cross section, illustrating the cell layers through which solutes pass from the soil solution to the xylem tracheary elements. (A © Biodisc/Visuals Unlimited.)

xylem loading   The process whereby ions exit the symplast and enter the conducting cells of the xylem.

Xylem parenchyma cells participate in xylem loading

The process whereby ions exit the symplast of a xylem parenchyma cell and enter the conducting cells of the xylem for translocation to the shoot is called **xylem loading**. Xylem loading is a highly regulated process. Xylem parenchyma cells, like other living plant cells, maintain plasma membrane H$^+$-ATPase activity and a negative membrane potential. Transporters that specifically function in the transfer of solutes to the tracheary elements have been identified by electrophysiological and genetic approaches. The plasma membranes of xylem parenchyma cells contain proton pumps, aquaporins, and a variety of ion channels and carriers specialized for influx or efflux.

In Arabidopsis xylem parenchyma, the stelar outwardly rectifying K$^+$ channel (SKOR) is expressed in cells of the pericycle and xylem parenchyma, where it functions as an efflux channel, transporting K$^+$ from the living cells out to the tracheary elements. In mutant Arabidopsis plants lacking the SKOR channel protein, or in plants in which SKOR has been pharmacologically inactivated, K$^+$ transport from the root to the shoot is severely reduced, confirming the function of this channel protein.

Several types of anion-selective channels have also been identified that participate in efflux of Cl$^-$ and NO$_3^-$ from the xylem parenchyma. Drought, abscisic acid (ABA) treatment, and elevation of cytosolic Ca^{2+} concentrations (which often occurs as a response to ABA) all reduce the activity of SKOR and anion channels of the root xylem parenchyma, a response that could help maintain cellular hydration in the root under desiccating conditions.

Other, less selective ion channels found in the plasma membrane of xylem parenchyma cells are permeable to K$^+$, Na$^+$, and anions. Other transport molecules have also been identified that mediate loading of boron (as boric acid [B(OH)$_3$] or borate [B(OH)$_4^-$]), Mg^{2+}, and H$_2$PO$_4^{2-}$. Thus, the flux of ions from the xylem parenchyma cells into the xylem tracheary elements is under tight metabolic control through the regulation of plasma membrane H$^+$-ATPases, ion efflux channels, and carriers.

Summary

The biologically regulated movement of molecules and ions from one location to another is known as transport. Plants exchange solutes within their cells, with their local environment, and among their tissues and organs. Both local and long-distance transport processes in plants are controlled largely by cellular membranes. Ion transport in plants is vital to their mineral nutrition and stress tolerance, and modulation of plant transport components and properties has potential to improve the nutritive value, stress tolerance, and yield of crops.

Passive and Active Transport

- Movement of solutes across membranes down their free-energy gradient is facilitated by passive transport mechanisms, whereas movement of solutes against their free-energy gradient is known as active transport and requires energy input (**Figure 6.1**).

- Concentration gradients and electrical-potential gradients, the main forces that drive transport across biological membranes, are integrated by a term called the electrochemical potential (**Equation 6.8**).

Transport of Ions across Membrane Barriers

- The extent to which a membrane permits the movement of a substance is a property known as membrane permeability (**Figure 6.5**).

- Permeability depends on the lipid composition of the membrane, the chemical properties of the solutes, and particularly on the membrane proteins that facilitate the transport of specific substances.

- For each permeant ion, the distribution of that particular ionic species across a membrane that would occur at equilibrium is described by the Nernst equation (**Equation 6.10**).

- Transport of H$^+$ across the plant plasma membrane by H$^+$-ATPases is a major determinant of the membrane potential (**Figures 6.15, 6.16**).

(Continued)

Summary (*continued*)

Membrane Transport Processes

- Biological membranes contain specialized proteins—channels, carriers, and pumps—that facilitate solute transport (**Figure 6.6**).

- The net result of membrane transport processes is that most ions are maintained in disequilibrum with their surroundings.

- Channels are regulated protein pores that, when open, greatly enhance fluxes of ions and, in some cases, neutral molecules across membranes (**Figures 6.6, 6.7**).

- Organisms have a great diversity of ion channel types. Depending on the channel type, channels can be nonselective or highly selective for just one ionic species. Channels can be regulated by many parameters, including voltage, intracellular signaling molecules, ligands, hormones, and light (**Figures 6.8, 6.13, 6.14**).

- Carriers bind specific substances and transport them at a rate several orders of magnitude lower than that of channels (**Figures 6.6, 6.11**).

- Pumps require energy for transport. Active transport of H^+ and Ca^{2+} across plant plasma membranes is mediated by pumps (**Figure 6.6**).

- Secondary active transporters in plants harness energy from energetically downhill movement of protons to mediate energetically uphill transport of another solute (**Figure 6.9**).

- In symport, both transported solutes move in the same direction across the membrane, whereas in antiport, the two solutes move in opposite directions (**Figures 6.9, 6.10**).

Membrane Transport Proteins

- Many channels, carriers, and pumps of the plant plasma membrane and tonoplast have been identified at the molecular level (**Figure 6.13**) and characterized using electrophysiological (**Figure 6.8**) and biochemical techniques.

- Transporters exist for diverse nitrogenous compounds, including NO_3^-, amino acids, and peptides.

- Plants have a great variety of cation channels that can be classified according to their ionic selectivity and regulatory mechanisms (**Figure 6.14**).

- Several different classes of cation carriers mediate K^+ uptake into the cytosol (**Figure 6.13**).

- Na^+–H^+ antiporters on the tonoplast and plasma membrane extrude Na^+ into the vacuole and apoplast, respectively, thereby opposing accumulation of toxic levels of Na^+ in the cytosol (**Figure 6.13**).

- Ca^{2+} is an important second messenger in signal transduction cascades, and its cytosolic concentration is tightly regulated. Ca^{2+} enters the cytosol passively, via Ca^{2+}-permeable channels, and is actively removed from the cytosol by Ca^+ pumps and Ca^{2+}–H^+ antiporters (**Figure 6.13**).

- Selective carriers that mediate NO_3^-, Cl^-, SO_4^-, and $H_2PO_4^-$ uptake into the cytosol, and anion channels that nonselectively mediate anion efflux from the cytosol, regulate cellular concentrations of these macronutrients (**Figure 6.13**).

- Both essential and toxic metal ions are transported by high-affinity transport proteins in the plasma membrane (**Figure 6.13**).

- Aquaporins facilitate flux of water and other specific molecules, including boric acid and silicic acid across plant plasma membranes, and their regulation allows for rapid changes in water permeability in response to environmental stimuli (**Figure 6.13**).

- Plasma membrane H^+-ATPases are encoded by a multigene family, and their activity is reversibly controlled by an autoinhibitory domain (**Figure 6.16**).

- Like the plasma membrane, the tonoplast also contains both cation and anion channels, as well as a diversity of other transporters (**Figure 6.13**).

- Two types of proton pumps found in the vacuolar membrane, V-ATPases and H^+-pyrophosphatases, regulate the proton motive force across the tonoplast, which in turn drives the movement of other solutes across this membrane via antiport mechanisms (**Figures 6.13, 6.17**).

Ion Transport in Stomatal Opening

- Light-stimulated stomatal movements are driven by changes in the osmoregulation of guard cells. Blue light stimulates a H^+-ATPase at the guard cell plasma membrane, generating an electrochemical-potential gradient that drives ion uptake (**Figures 6.18–6.20**).

- Blue light also stimulates starch degradation and malate biosynthesis. Accumulation of malate and K^+ and its counterions within guard cells leads to stomatal opening (**Figure 6.20**).

Ion Transport in Roots

- Solutes such as mineral nutrients move between cells either through the extracellular space (the apoplast) or from cytoplasm to cytoplasm (via the symplast). The cytoplasm of neighboring cells is connected by plasmodesmata, which facilitate symplastic transport (**Figure 6.21**).

(Continued)

Summary (continued)

- When a solute enters the root, it may be taken up into the cytosol of an epidermal cell, or it may diffuse through the apoplast into the root cortex and then enter the symplast through a cortical or endodermal cell.

- The presence of the Casparian strip prevents apoplastic diffusion of solutes into the stele. Solutes enter the stele via diffusion from endodermal cells to pericycle and xylem parenchyma cells.

- During xylem loading, solutes are released from xylem parenchyma cells to the conducting cells of the xylem, and then move to the shoot in the transpiration stream (**Figure 6.22**).

Suggested Reading

Barbier-Brygoo, H., Vinauger, M., Colcombet, J., Ephritikhine, G., Frachisse, J., and Maurel, C. (2000) Anion channels in higher plants: Functional characterization, molecular structure and physiological role. *Biochim. Biophys. Acta* 1465: 199–218.

Buchanan, B. B., Gruissem, W., and Jones, R. L., eds. (2000) *Biochemistry and Molecular Biology of Plants.* American Society of Plant Physiologists, Rockville, MD.

Burch-Smith, T. M., and Zambryski, P. C. (2012) Plasmodesmata paradigm shift: Regulation from without versus within. *Annu. Rev. Plant. Biol.* 63: 239–260.

Harold, F. M. (1986) *The Vital Force: A Study of Bioenergetics.* W. H. Freeman, New York.

Inoue, S.-I., and Kinoshita, T. (2017) Blue light regulation of stomatal opening and the plasma membrane H+-ATPase. *Plant Physiol.* 174: 531–538.

Jammes, F., Hu, H. C., Villiers, F., Bouten, R., and Kwak, J. M. (2011) Calcium-permeable channels in plant cells. *FEBS J.* 278: 4262–4276.

Li, G., Santoni, V., and Maurel, C. (2013) Plant aquaporins: Roles in plant physiology. *Biochim. Biophys. Acta* 1840: 1574–1582.

Marschner, H. (1995) *Mineral Nutrition of Higher Plants.* Academic Press, London.

Martinoia, E., Meyer, S., De Angeli, A., and Nagy, R. (2012) Vacuolar transporters in their physiological context. *Annu. Rev. Plant Biol.* 63: 183–213.

Munns, R., James, R. A., Xu, B., Athman, A., Conn, S. J., Jordans, C., Byrt, C. S., Hare, R. A., Tyerman, S. D., Tester, M., et al. (2012) Wheat grain yield on saline soils is improved by an ancestral Na+ transporter gene. *Nat. Biotechnol.* 30: 360–364.

Nicholls, D. G., and Ferguson, S. J. (2013) *Bioenergetics,* 4th ed. Academic Press, Amsterdam.

Nobel, P. (1991) *Physicochemical and Environmental Plant Physiology.* Academic Press, San Diego, CA.

Palmgren, M. G. (2001) Plant plasma membrane H+-ATPases: Powerhouses for nutrient uptake. *Annu. Rev. Plant Physiol. Plant Mol. Biol.* 52: 817–845.

Roelfsema, M. R., and Hedrich, R. (2005) In the light of stomatal opening: New insights into "the Watergate." *New Phytol.* 167: 665–691.

Schroeder, J. I., Delhaize, E., Frommer, W. B., Guerinot, M. L., Harrison, M. J., Herrera-Estrella, L., Horie, T., Kochian, L. V., Munns, R., Nishizawa, N. K., et al. (2013) Using membrane transporters to improve crops for sustainable food production. *Nature* 497: 60–66.

Ward, J. M., Mäser, P., and Schroeder, J. I. (2009). Plant ion channels: Gene families, physiology, and functional genomics analyses. *Annu. Rev. Plant Biol.* 71: 59–82.

Yamaguchi, T., Hamamoto, S., and Uozumi, N. (2013) Sodium transport system in plant cells. *Front. Plant Sci.* 4: 410.

Yazaki, K., Shitan, N., Sugiyama, A., and Takanashi, K. (2009) Cell and molecular biology of ATP-binding cassette proteins in plants. *Int. Rev. Cell Mol. Biol.* 276: 263–299.

7 Photosynthesis: The Light Reactions

Life on Earth ultimately depends on energy derived from the sun. Photosynthesis is the only process of biological importance that can harvest this energy. A large fraction of the planet's energy resources results from photosynthetic activity in either recent or ancient times (fossil fuels). This chapter introduces the basic physical principles that underlie photosynthetic energy storage and the current understanding of the structure and function of the photosynthetic apparatus.

The term *photosynthesis* means literally "synthesis using light." As we will see in this chapter, photosynthetic organisms use solar energy to synthesize complex carbon compounds. More specifically, light energy drives the synthesis of carbohydrates and generation of oxygen from carbon dioxide and water:

$$6 \, CO_2 + 6 \, H_2O \xrightarrow{\text{Light}} C_6H_{12}O_6 + 6 \, O_2 \qquad (7.1)$$

$$\text{Carbon dioxide} \qquad \text{Water} \qquad \text{Carbohydrate} \qquad \text{Oxygen}$$

Where $C_6H_{12}O_6$ represents a simple sugar such as glucose. As we will discuss in Chapter 8, glucose is not the actual product of the carbon fixation reactions, so this part of the equation should not be taken literally. However, the energetics for the actual reaction are approximately the same as represented here. Energy stored in these carbohydrate molecules can be used later to power cellular processes in the plant and can serve as the energy source for all forms of life.

This chapter deals with the role of light in photosynthesis, the structure of the photosynthetic apparatus, and the processes that begin with the excitation of chlorophyll by light and culminate in the synthesis of ATP and NADPH.

Photosynthesis in Higher Plants

The most active photosynthetic tissue in higher plants is the mesophyll of leaves. Mesophyll cells have many chloroplasts, which contain the specialized light-absorbing green pigments, the **chlorophylls**. In photosynthesis, the plant uses solar energy to oxidize water, thereby releasing oxygen, and to reduce carbon dioxide, thereby forming large carbon compounds, primarily sugars. The complex series of reactions that culminate in the reduction of CO_2 include the thylakoid reactions and the carbon fixation reactions.

The **thylakoid reactions** of photosynthesis take place in the specialized internal membranes of the chloroplast called thylakoids (see Chapter 1). The end products of these thylakoid reactions are the high-energy compounds ATP and NADPH, which are used for the synthesis of sugars in the **carbon fixation reactions**. These synthetic processes take place in the stroma of the chloroplast, the aqueous region that surrounds the thylakoids. The thylakoid reactions, also called the "light reactions" of photosynthesis, are the subject of this chapter; the carbon fixation reactions will be discussed in Chapter 8.

In the chloroplast, light energy is converted into chemical energy by two different functional units called *photosystems*. The absorbed light energy is used to power the transfer of electrons through a series of compounds that act as electron donors and electron acceptors. The majority of electrons are extracted from H_2O, which is oxidized to O_2, and ultimately reduce $NADP^+$ to NADPH. Light energy is also used to generate a proton motive force (see Chapter 6) across the thylakoid membrane; this proton motive force is used to synthesize ATP.

General Concepts

In this section we explore the essential concepts that provide a foundation for an understanding of photosynthesis. These concepts include the nature of light, the properties of pigments, and the various roles of pigments.

Light has characteristics of both a particle and a wave

A triumph of physics in the early twentieth century was the realization that light has properties of both particles and waves. A wave (**Figure 7.1**) is characterized by a **wavelength**, denoted by the Greek letter lambda (λ), which is the distance between successive wave crests. The **frequency**, represented by the Greek letter nu (ν), is the number of wave crests that pass an observer in a given time. A simple equation relates the wavelength, the frequency, and the speed of any wave:

$$c = \lambda\nu \qquad (7.2)$$

where c is the speed of the wave—in the present case, the speed of light (3.0×10^8 m s^{-1}). The light wave is a transverse (side-to-side) electromagnetic wave, in which both electric and magnetic fields oscillate perpendicularly to the direction of propagation of the wave and at 90° with respect to each other.

Light is also a particle, which we call a **photon**. Each photon contains an amount of energy that is called a **quantum** (plural *quanta*). The energy content of light is not continuous but rather is delivered in discrete packets, the quanta. The energy (E) of a photon depends on the frequency of the light according to a relation known as Planck's law:

$$E = h\nu \qquad (7.3)$$

where h is Planck's constant (6.626×10^{-34} J s).

chlorophylls   A group of light-absorbing green pigments active in photosynthesis.

thylakoid reactions   The chemical reactions of photosynthesis that occur in the specialized internal membranes of the chloroplast (called thylakoids). Include photosynthetic electron transport and ATP synthesis.

carbon fixation reactions   The synthetic reactions occurring in the stroma of the chloroplast that use the high-energy compounds ATP and NADPH for the incorporation of CO_2 into carbon compounds.

wavelength (λ)   A unit of measurement for characterizing light energy. The distance between successive wave crests. In the visible spectrum, it corresponds to a color.

frequency (ν)   A unit of measurement that characterizes waves, in particular light energy. The number of wave crests that pass an observer in a given time.

photon   A discrete physical unit of radiant energy.

quantum   (plural *quanta*) A discrete packet of energy contained in a photon.

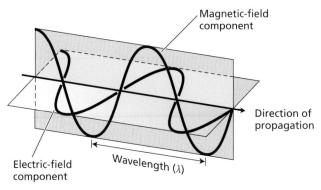

Figure 7.1   Light is a transverse electromagnetic wave, consisting of oscillating electric and magnetic fields that are perpendicular to each other and to the direction of propagation of the light. Light moves at a speed of 3.0×10^8 m s^{-1}. The wavelength (λ) is the distance between successive crests of the wave.

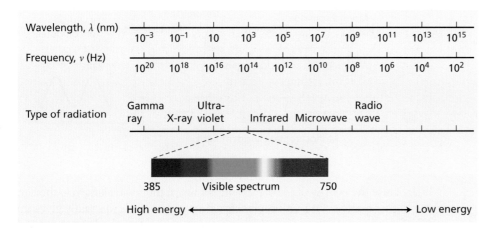

absorption spectrum
A graphic representation of the amount of light energy absorbed by a substance plotted against the wavelength of the light.

light energy   The energy associated with photons.

Figure 7.2   Electromagnetic spectrum. Wavelength (λ) and frequency (v) are inversely related. Our eyes are sensitive to only a narrow range of wavelengths of radiation, the visible region, which extends from about 385 nm (violet) to about 750 nm (red). Short-wavelength (high-frequency) light has a high energy content; long-wavelength (low-frequency) light has a low energy content.

Sunlight is like a rain of photons of different frequencies. Our eyes are sensitive to only a small range of frequencies—the visible-light region of the electromagnetic spectrum (**Figure 7.2**). Light of slightly higher frequencies (or shorter wavelengths) is in the ultraviolet region of the spectrum, and light of slightly lower frequencies (or longer wavelengths) is in the infrared region. The output of the sun is shown in **Figure 7.3**, along with the energy density that strikes the surface of Earth. The **absorption spectrum** (plural *spectra*) of chlorophyll *a* (green curve in Figure 7.3) indicates the approximate portion of the solar output that is used by plants.

An absorption spectrum provides information about the amount of **light energy** taken up or absorbed by a molecule or substance as a function of the wavelength of the light. The absorption spectrum for a particular substance in a nonabsorbing solvent can be determined by a spectrophotometer, as illustrated in **Figure 7.4**. Spectrophotometry is the technique used to measure the absorption of light by a sample.

When molecules absorb or emit light, they change their electronic state

Chlorophyll appears green to our eyes because it absorbs light mainly in the red and blue parts of the spectrum, so only some of the light enriched in green wavelengths (about 550 nm) is reflected into our eyes (see Figure 7.3).

The absorption of light is represented by Equation 7.4, in which chlorophyll (Chl) in its lowest energy, or ground, state absorbs a photon (represented by hv) and makes a transition to a higher energy, or excited, state (Chl*):

$$\text{Chl} + hv \rightarrow \text{Chl*} \tag{7.4}$$

The distribution of electrons in the excited molecule is somewhat different from the distribution in the ground-state molecule (**Figure 7.5**). Absorption of blue light excites the chlorophyll to a higher energy state than absorption of red

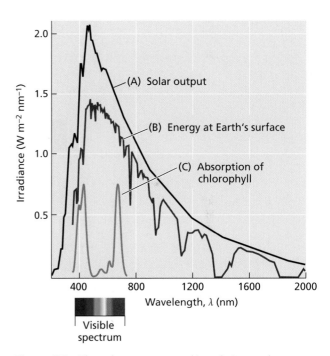

Figure 7.3   The solar spectrum and its relation to the absorption spectrum of chlorophyll. Curve A is the energy output of the sun as a function of wavelength. Curve B is the energy that strikes Earth's surface. The sharp valleys in the infrared region beyond 700 nm represent the absorption of solar energy by molecules in the atmosphere, chiefly water vapor. Curve C is the absorption spectrum of chlorophyll, which absorbs strongly in the blue (about 430 nm) and red (about 660 nm) portions of the spectrum. Because the green light in the middle of the visible spectrum is not efficiently absorbed, some of it is reflected into our eyes and gives plants their characteristic green color.

Figure 7.4 Schematic diagram of a spectrophotometer. The instrument consists of a light source, a monochromator that contains a wavelength selection device such as a prism, a sample holder, a photodetector, and a recorder or computer. The output wavelength of the monochromator can be changed by rotating the prism; the graph of absorbance (A) versus wavelength (λ) is called a spectrum.

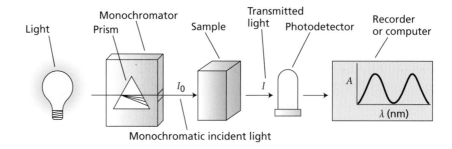

light, because the energy of photons is higher when their wavelength is shorter. In the higher excited state, chlorophyll is extremely unstable; it rapidly gives up some of its energy to the surroundings as heat, and enters the **lowest excited state**, where it can be stable for a maximum of several nanoseconds (10^{-9} s). Because of the inherent instability of the excited state, any process that captures its energy must be extremely rapid.

In the lowest excited state, the excited chlorophyll has four alternative pathways for disposing of its available energy:

1. Excited chlorophyll can re-emit a photon and thereby return to its ground state—a process known as **fluorescence**. When it does so, the wavelength of fluorescence is slightly longer (and of lower energy) than the wavelength of absorption, because a portion of the excitation energy is converted into heat before the fluorescent photon is emitted. Chlorophylls fluoresce in the red region of the spectrum.

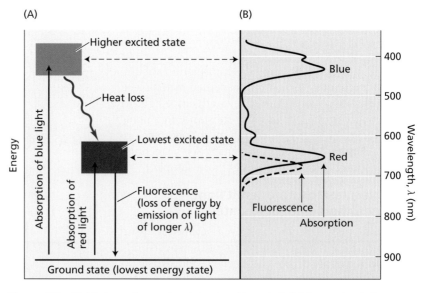

Figure 7.5 Light absorption and emission by chlorophyll. (A) Energy-level diagram. Absorption or emission of light is indicated by vertical arrows that connect the ground state with excited electron states. The blue and red absorption bands of chlorophyll (which absorb blue and red photons, respectively) correspond to the upward vertical arrows, signifying that energy absorbed from light causes the molecule to change from the ground state to an excited state. The downward-pointing arrow indicates fluorescence, in which the molecule goes from the lowest excited state to the ground state while re-emitting energy as a photon. (B) Spectra of absorption and fluorescence. The long-wavelength (red) absorption band of chlorophyll corresponds to light that has the energy required to cause the transition from the ground state to the first excited state. The short-wavelength (blue) absorption band corresponds to a transition to a higher excited state.

lowest excited state The excited state with the lowest energy attained when a chlorophyll molecule in a higher energy state gives up some of its energy to the surroundings as heat.

fluorescence Following light absorption, the emission of light at a slightly longer wavelength (lower energy) than the wavelength of the absorbed light.

2. The excited chlorophyll can return to its ground state by directly converting its excitation energy into heat, with no emission of a photon.

3. Chlorophyll may participate in **energy transfer**, during which an excited chlorophyll transfers its energy to another molecule.

4. A fourth process is **photochemistry**, in which the energy of the excited state causes chemical reactions to occur. The photochemical reactions of photosynthesis are among the fastest known chemical reactions. This extreme speed is necessary for photochemistry to compete with the three other possible reactions of the excited state just described.

Photosynthetic pigments absorb the light that powers photosynthesis

The energy of sunlight is first absorbed by the pigments of the plant. All pigments active in photosynthesis are found in the chloroplast. Structures and absorption spectra of several photosynthetic pigments are shown in **Figure 7.6** and **Figure 7.7**, respectively. The chlorophylls and **bacteriochlorophylls** (pigments found in certain bacteria) are the typical pigments of photosynthetic organisms.

Chlorophylls a and b are abundant in green plants, and c, d, and f are found in some protists and cyanobacteria. Several different types of bacteriochlorophyll have been found; type a is the most widely distributed.

All chlorophylls have a complex ring structure that is chemically related to the porphyrin-like groups found in hemoglobin and cytochromes (see Figure 7.6A).

energy transfer In the light reactions of photosynthesis, the direct transfer of energy from an excited molecule, such as β-carotene, to another molecule, such as chlorophyll. Energy transfer can also take place between chemically identical molecules, as in chlorophyll-to-chlorophyll transfer.

photochemistry The very rapid chemical reactions in which light energy absorbed by a molecule causes a chemical reaction to occur.

bacteriochlorophylls Light-absorbing pigments active in photosynthesis in anoxygenic photosynthetic organisms.

(A) Chlorophylls

Chlorophyll a

Chlorophyll b

Bacteriochlorophyll a

(B) Carotenoids

β-Carotene

(C) Bilin pigments

Phycoerythrobilin

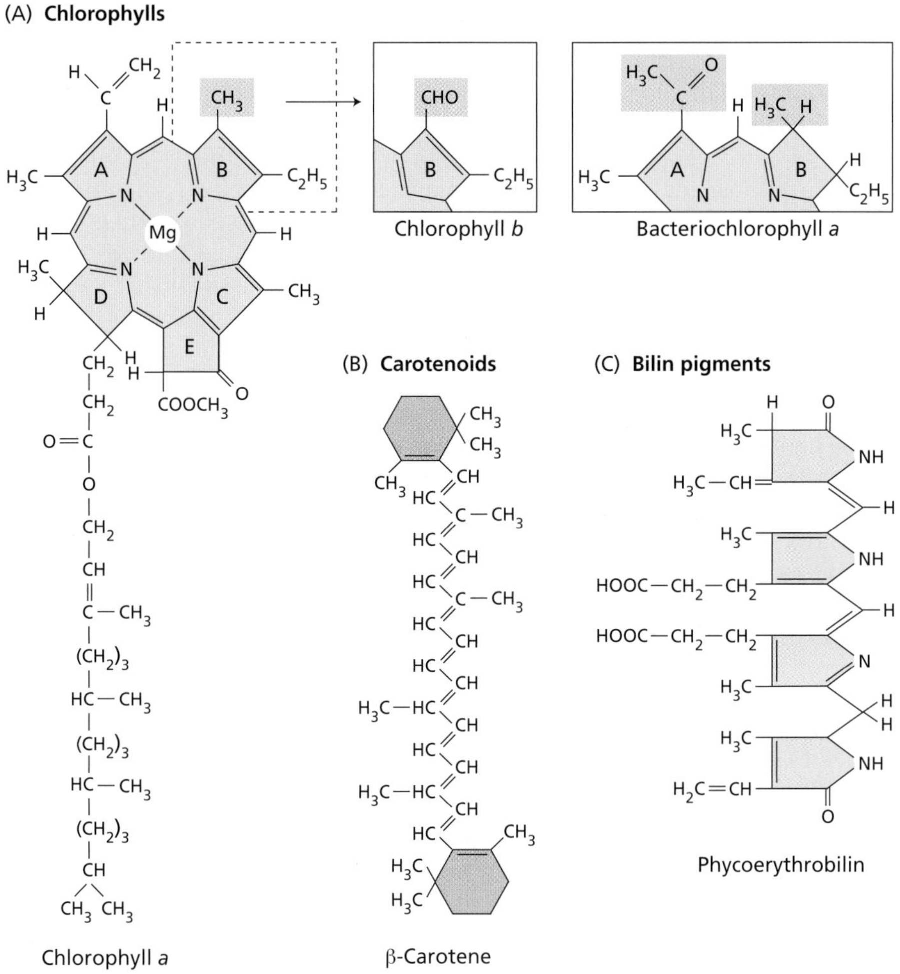

Figure 7.6 Molecular structure of some photosynthetic pigments. (A) The chlorophylls have a porphyrin-like ring structure with a magnesium ion (Mg) coordinated in the center and a long hydrophobic hydrocarbon tail that anchors them in the photosynthetic membrane. The porphyrin-like ring is the site of the electron rearrangements that occur when the chlorophyll is excited, and of the unpaired electrons when the chlorophyll is either oxidized or reduced. Various chlorophylls differ chiefly in the substituents around the rings and the pattern of double bonds. (B) Carotenoids are linear polyenes that serve as both antenna pigments and photoprotective agents. (C) Bilin pigments are open-chain tetrapyrroles found in antenna structures known as phycobilisomes that occur in cyanobacteria and red algae.

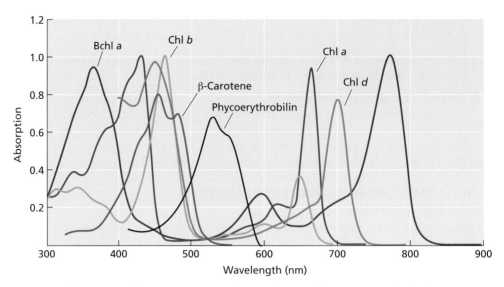

Figure 7.7   Absorption spectra of some photosynthetic pigments, including β-carotene, chlorophyll *a* (Chl *a*), chlorophyll *b* (Chl *b*), bacteriochlorophyll *a* (Bchl *a*), chlorophyll *d* (Chl *d*), and phycoerythrobilin. The absorption spectra shown are for pure
pigments dissolved in nonpolar solvents, except phycoerythrin, a protein from cyanobacteria that contains a phycoerythrobilin chromophore covalently attached to the peptide chain. In many cases the spectra of photosynthetic pigments in vivo are substantially affected by the environment of the pigments in the photosynthetic membrane.

A long hydrocarbon tail is almost always attached to the ring structure. The tail
anchors the chlorophyll to the hydrophobic portion of its environment. The ring
structure contains some loosely bound electrons and is the part of the molecule
involved in electronic transitions and redox (reduction–oxidation) reactions.

The different types of **carotenoids** found in photosynthetic organisms are
all linear molecules with multiple conjugated double bonds (see Figure 7.6B).
Absorption bands in the 400- to 500-nm region give carotenoids their characteristic orange color. The color of carrots, for example, is due to the carotenoid
β-carotene, whose structure and absorption spectrum are shown in Figures 7.6
and 7.7, respectively.

Carotenoids are found in all natural photosynthetic organisms. Carotenoids
are integral constituents of the thylakoid membrane and are usually associated
intimately with many of the proteins that make up the photosynthetic apparatus.
The light energy absorbed by the carotenoids is transferred to chlorophyll for
photosynthesis; because of this role they are called **accessory pigments**. Carotenoids
also help protect the organism from damage caused by light (see Chapter 9).

Key Experiments in Understanding Photosynthesis

Here we describe the relationship between photosynthetic activity and the spectrum of absorbed light. We also discuss some of the critical experiments that have
contributed to our present understanding of photosynthesis, and we consider
equations for the essential chemical reactions of photosynthesis.

Action spectra relate light absorption to photosynthetic activity

The use of action spectra has been central to the development of our current understanding of photosynthesis. An **action spectrum** depicts the magnitude of a
response of a biological system to light as a function of wavelength. For example,
an action spectrum for photosynthesis can be constructed from measurements of
oxygen evolution (the photosynthetic production of O_2) at different wavelengths

carotenoids   Linear polyenes arranged
as a planar zigzag chain, with conjugated double bonds. These orange pigments serve both as antenna pigments
and photoprotective agents.

accessory pigments   Light-absorbing
molecules in photosynthetic organisms
that work with chlorophyll *a* in the
absorption of light used for photosynthesis. They include carotenoids, other
chlorophylls, and phycobiliproteins.

action spectrum   A graphic representation of the magnitude of a biological response to light as a function of
wavelength.

antenna complex   A group of
pigment molecules that cooperate to
absorb light energy and transfer it to a
reaction center complex.

reaction center complex   A group
of electron transfer proteins that receive
energy from the antenna complex and
convert it into chemical energy using
oxidation–reduction reactions.

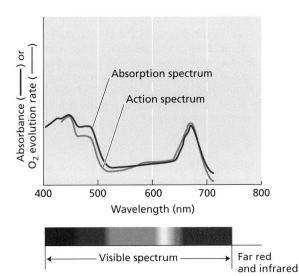

Figure 7.8   Action spectrum compared with an absorption spectrum. The absorption spectrum is measured as shown in Figure 7.4. An action spectrum is measured by plotting a response to light, such as oxygen evolution, as a function of wavelength. If the pigment used to obtain the absorption spectrum is the same as that which causes the response, the absorption and action spectra will match. In the example shown here, the action spectrum for oxygen evolution matches the absorption spectrum of intact chloroplasts quite well, indicating that light absorption by the chlorophylls mediates oxygen evolution. Discrepancies are found in the region of carotenoid absorption, from 450 to 550 nm, indicating that energy transfer from carotenoids to chlorophylls is not as effective as energy transfer between chlorophylls.

(**Figure 7.8**). Often an action spectrum can identify the *chromophore* (pigment) responsible for a particular light-induced phenomenon.

Some of the first action spectra were measured by T. W. Engelmann in the late 1800s (**Figure 7.9**). Engelmann used a prism to disperse sunlight into a rainbow that was allowed to fall on an aquatic algal filament. A population of O_2-seeking bacteria was introduced into the system. The bacteria congregated in the regions of the filaments that evolved the most O_2. These were the regions illuminated by blue light and red light, which are strongly absorbed by chlorophyll. Today, action spectra can be measured in room-sized spectrographs in which a huge monochromator bathes the experimental samples in monochromatic light. The technology is more sophisticated, but the principle is the same as that of Engelmann's experiments.

Action spectra were very important for the discovery of two distinct photosystems operating in O_2-evolving photosynthetic organisms. Before we introduce the two photosystems, however, we need to describe the light-gathering antennas and the energy needs of photosynthesis.

Photosynthesis takes place in complexes containing light-harvesting antennas and photochemical reaction centers

A portion of the light energy absorbed by chlorophylls and carotenoids is eventually stored as chemical energy via the formation of chemical bonds. This conversion of energy from one form to another is a complex process that depends on cooperation between many pigment molecules and a group of electron transfer proteins.

The majority of the pigments serve as an **antenna complex**, collecting light and transferring the energy to the **reaction center complex**, where the chemical oxidation and reduction reactions leading to long-term

(A)

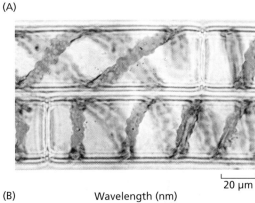

(B)

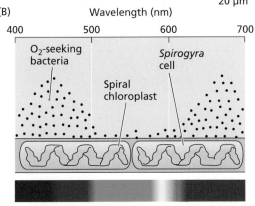

Figure 7.9   The action spectrum measurements by T. W. Engelmann. (A) Four cells of the filamentous green alga *Spirogyra* with its spiral chloroplast. (B) Schematic diagram of how Engelmann projected a spectrum of light onto *Spirogyra* and observed that O_2-seeking bacteria, introduced into the system, collected in the region of the spectrum where chlorophyll pigments absorb. This action spectrum gave the first indication of the effectiveness of light absorbed by pigments in driving photosynthesis. (A © Biophoto Associates/ Science Source.)

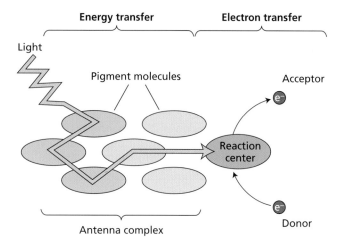

Energy transfer   **Electron transfer**

Light

Pigment molecules

Acceptor

Reaction center

Antenna complex

Donor

Figure 7.10 Basic concept of energy transfer during photosynthesis. Many pigments together serve as an antenna, collecting light and transferring its energy to the reaction center, where chemical reactions store some of the energy by transferring electrons from a chlorophyll pigment to an electron acceptor molecule. An electron donor then reduces the chlorophyll again. The transfer of energy in the antenna is a purely physical phenomenon and involves no chemical changes.

energy storage take place (**Figure 7.10**). Molecular structures of some of the antenna and reaction center complexes are discussed later in this chapter.

How does the plant benefit from this division of labor between antenna and reaction center pigments? Even in bright sunlight, a single chlorophyll molecule absorbs only a few photons each second. If there were a reaction center associated with each chlorophyll molecule, the reaction center enzymes would be idle most of the time, only occasionally being activated by photon absorption. However, if a reaction center receives energy from many pigments at once, the system is kept active a large fraction of the time.

In 1932, Robert Emerson and William Arnold performed a key experiment that provided the first evidence for the cooperation of many chlorophyll molecules in energy conversion during photosynthesis. They delivered very brief (10^{-5} s) flashes of light to a suspension of the green alga *Chlorella pyrenoidosa* and measured the amount of oxygen produced. The flashes were spaced about 0.1 s apart, a time that Emerson and Arnold had determined in earlier work was long enough for the enzymatic steps of the process to be completed before the arrival of the next flash. The investigators varied the energy of the flashes and found that at high energies the oxygen production did not increase when a more intense flash was given: The photosynthetic system was saturated with light (**Figure 7.11**).

In their measurement of the relationship of oxygen production to flash energy, Emerson and Arnold were surprised to find that under saturating conditions, only 1 molecule of oxygen was produced for each 2500 chlorophyll molecules in the sample. We know now that several hundred pigments are associated with each reaction center and that each reaction center must operate four times to produce 1 molecule of oxygen—hence the value of 2500 chlorophylls per O_2.

The reaction centers and most of the antenna complexes are integral components of the photosynthetic membrane. In eukaryotic photosynthetic organisms, these membranes are found within the chloroplast; in photosynthetic prokaryotes, the site of photosynthesis is the plasma membrane or membranes derived from it.

The graph shown in Figure 7.11 permits us to calculate another important parameter of the light reactions of photosynthesis, the quantum yield. The **quantum yield** of photochemistry (Φ) is defined as follows:

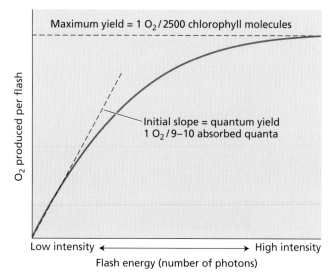

Maximum yield = 1 O_2/2500 chlorophyll molecules

Initial slope = quantum yield 1 O_2/9–10 absorbed quanta

O_2 produced per flash

Low intensity ⟷ High intensity

Flash energy (number of photons)

Figure 7.11 Relationship of oxygen production to flash energy, the first evidence for the interaction between the antenna pigments and the reaction center. At saturating energies, the maximum amount of O_2 produced is 1 molecule per 2500 chlorophyll molecules.

$$\Phi = \frac{\text{Number of photochemical products}}{\text{Total number of quanta absorbed}} \qquad (7.5)$$

In the linear portion (low light intensity) of the curve, an increase in the number of photons stimulates a proportional increase in oxygen evolution. Thus, the slope of the curve measures the quantum yield for oxygen production. The quantum yield for a particular process can range from 0 (if that process does not respond to light) to 1.0 (if every photon absorbed contributes to the process by forming a product).

In functional chloroplasts kept in dim light, the quantum yield of photochemistry is approximately 0.95, the quantum yield of fluorescence is 0.05 or lower, and the quantum yields of other processes are negligible. Thus, the most common result of chlorophyll excitation is photochemistry. Products of photosynthesis such as O_2 require more than a single photochemical event to be formed, and therefore have a lower quantum yield of formation than the photochemical quantum yield. It takes about ten photons to produce one molecule of O_2, so the quantum yield of O_2 production is about 0.1, even though the photochemical quantum yield for each step in the process is nearly 1.0.

The chemical reaction of photosynthesis is driven by light

It is important to realize that equilibrium for the chemical reaction shown in Equation 7.1 lies very far in the direction of the reactants. The equilibrium constant for Equation 7.1, calculated from tabulated free energies of formation for each of the compounds involved, is about 10^{-500}. This number is so close to zero that one can be quite confident that in the entire history of the universe no molecule of glucose has formed spontaneously from H_2O and CO_2 without external energy being provided. The energy needed to drive the photosynthetic reaction comes from light. Here's a simpler form of Equation 7.1:

$$CO_2 + H_2O \xrightarrow{\text{Light, plant}} (CH_2O) + O_2 \qquad (7.6)$$

where (CH_2O) is one-sixth of a glucose molecule. About nine or ten photons of light are required to drive the reaction of Equation 7.6.

Although the photochemical quantum yield under optimum conditions is nearly 100%, the *efficiency* of the conversion of light into chemical energy is much less. If red light of wavelength 680 nm is absorbed, the total energy input (see Equation 7.3) is 1760 kJ per mole of oxygen formed. This amount of energy is more than enough to drive the reaction in Equation 7.6, which has a standard-state free-energy change of +467 kJ mol^{-1}. The efficiency of conversion of light energy at the optimal wavelength into chemical energy is therefore about 27%. Most of this stored energy is used for cellular maintenance processes; the amount diverted to the formation of biomass is much less (see Chapter 9).

There is no conflict in the fact that the photochemical quantum efficiency (quantum yield) is nearly 1.0 (100%), the energy conversion efficiency is only 27%, and the overall efficiency of conversion of solar energy is only a few percent. The *quantum efficiency* is a measure of the fraction of absorbed photons that engage in photochemistry; the *energy efficiency* is a measure of how much energy in the absorbed photons is stored as chemical products; and the *solar energy storage efficiency* is a measure of how much of the energy in the entire solar spectrum is converted to usable form. The numbers indicate that almost all the absorbed photons engage in photochemistry, but only about one-fourth of the energy in each photon is stored, the remainder being converted to heat, and only about half of the solar spectrum is absorbed by the plant. The overall energy conversion efficiency into biomass, including all loss processes and considering the entire solar spectrum as energy source, is significantly lower still—approximately 4.3% for C_3 plants and 6% for C_4 plants.

Light drives the reduction of NADP$^+$ and the formation of ATP

The overall process of photosynthesis is a redox chemical reaction, in which electrons are removed from one chemical species, thereby oxidizing it, and added to another species, thereby reducing it. In 1937, Robert Hill found that in

quantum yield (Φ) The ratio of the yield of a particular product of a photochemical process to the total number of quanta absorbed.

the light, isolated chloroplast thylakoids reduce a variety of compounds, such as iron salts. These compounds serve as oxidants in place of CO_2, as the following equation shows:

$$4 Fe^{3+} + 2 H_2O \rightarrow 4 Fe^{2+} + O_2 + 4 H^+ \qquad (7.7)$$

Many compounds have since been shown to act as artificial electron acceptors in what has come to be known as the Hill reaction. The use of artificial electron acceptors has been invaluable in elucidating the reactions that precede carbon reduction. The demonstration of oxygen evolution linked to the reduction of artificial electron acceptors provided the first evidence that oxygen evolution could occur in the absence of carbon dioxide and led to the now accepted and proven idea that the oxygen in photosynthesis originates from water, not from carbon dioxide.

We now know that during the normal functioning of the photosynthetic system, light reduces nicotinamide adenine dinucleotide phosphate ($NADP^+$), which in turn serves as the reducing agent for carbon fixation in the Calvin–Benson cycle (see Chapter 8). ATP is also formed during the electron flow from water to $NADP^+$, and it too is used in carbon reduction.

The chemical reactions in which water is oxidized to oxygen, $NADP^+$ is reduced to NADPH, and ATP is formed are known as the *thylakoid reactions* because almost all the reactions up to $NADP^+$ reduction take place in the thylakoids. The carbon fixation and reduction reactions are called the *stroma reactions* because the carbon reduction reactions take place in the aqueous region of the chloroplast, the stroma. Although this division is somewhat arbitrary, it is conceptually useful.

Oxygen-evolving organisms have two photosystems that operate in series

By the late 1950s, several experiments were puzzling the scientists who studied photosynthesis. One of these experiments, carried out by Emerson, measured the quantum yield of photosynthesis as a function of wavelength and revealed an effect known as the red drop (**Figure 7.12**).

If the quantum yield is measured for the wavelengths at which chlorophyll absorbs light, the values found throughout most of the range are fairly constant, indicating that any photon absorbed by chlorophyll or other pigments is as

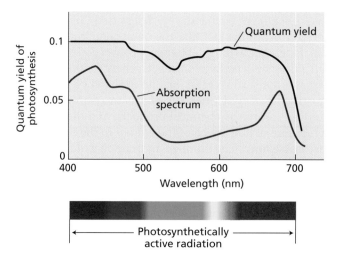

Figure 7.12 Red drop effect. The quantum yield of oxygen evolution (upper, black curve) falls off drastically for far-red light of wavelengths greater than 680 nm, indicating that far-red light alone is inefficient in driving photosynthesis. The slight dip near 500 nm reflects the somewhat lower efficiency of photosynthesis using light absorbed by accessory pigments, carotenoids.

effective as any other photon in driving photosynthesis. However, the yield drops dramatically in the far-red region of chlorophyll absorption (greater than 680 nm).

This drop cannot be caused by a decrease in chlorophyll absorption, because the quantum yield measures only light that has actually been absorbed. Thus, light with a wavelength greater than 680 nm is much less efficient than light of shorter wavelengths.

Another puzzling experimental result was the **enhancement effect**, also discovered by Emerson. He measured the rate of photosynthesis separately with light of two different wavelengths and then used the two beams simultaneously. When red and far-red light were given together, the rate of photosynthesis was greater than the sum of the individual rates, a startling and surprising observation. These and others observations were eventually explained by experiments performed in the 1960s that led to the discovery that two photochemical complexes, now known as **photosystems I** and **II** (**PSI** and **PSII**), operate in series to carry out the early energy-storage reactions of photosynthesis.

PSI preferentially absorbs far-red light of wavelengths greater than 680 nm; PSII preferentially absorbs red light of 680 nm and is driven very poorly by far-red light. This wavelength dependence explains the enhancement effect and the red drop effect. Another difference between the photosystems is that:

- PSI produces a strong reductant, capable of reducing NADP⁺, and a weak oxidant.

- PSII produces a very strong oxidant, capable of oxidizing water, and a weaker reductant than the one produced by PSI.

The reductant produced by PSII re-reduces the oxidant produced by PSI. These properties of the two photosystems are shown schematically in **Figure 7.13**.

The scheme of photosynthesis depicted in Figure 7.13, called the Z (for *zigzag*) *scheme,* has become the basis for understanding O₂-evolving (oxygenic) photosynthetic organisms. It accounts for the operation of two physically and chemically distinct photosystems (I and II), each with its own antenna

enhancement effect The synergistic (higher) effect of red and far-red light on the rate of photosynthesis, as compared with the sum of the rates when the two different wavelengths are delivered separately.

photosystem I (PSI) A system of photoreactions that absorbs maximally far-red light (700 nm), oxidizes plastocyanin, and reduces ferredoxin.

photosystem II (PSII) A system of photoreactions that absorbs maximally red light (680 nm), oxidizes water, and reduces plastoquinone. Operates very poorly under far-red light.

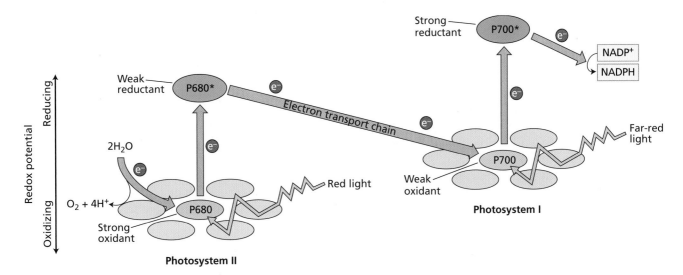

Figure 7.13 Z scheme of photosynthesis. Red light absorbed by photosystem II (PSII) produces a strong oxidant and a weak reductant. Far-red light absorbed by photosystem I (PSI) produces a weak oxidant and a strong reductant. The strong oxidant generated by PSII oxidizes water, while the strong reductant produced by PSI reduces NADP⁺. This scheme is basic to an understanding of photosynthetic electron transport. P680 and P700 refer to the wavelengths of maximum absorption of the reaction center chlorophylls in PSII and PSI, respectively.

thylakoids   The specialized, internal chlorophyll-containing membranes of the chloroplast where light absorption and the chemical reactions of photosynthesis take place.

stroma   The fluid component surrounding the thylakoid membranes of a chloroplast.

grana lamellae   Stacked thylakoid membranes within the chloroplast. Each stack is called a granum.

stroma lamellae   Unstacked thylakoid membranes within the chloroplast.

chloroplast envelope   The double-membrane system surrounding the chloroplast.

pigments and photochemical reaction center. The two photosystems are linked by an electron transport chain.

Organization of the Photosynthetic Apparatus

The previous section explained some of the physical principles underlying photosynthesis, some aspects of the functional roles of various pigments, and some of the chemical reactions carried out by photosynthetic organisms. We now turn to the architecture of the photosynthetic apparatus and the structure of its components, and discuss how the molecular structure of the system leads to its functional characteristics.

The chloroplast is the site of photosynthesis

In photosynthetic eukaryotes, photosynthesis takes place in the subcellular organelle known as the chloroplast. **Figure 7.14** shows a transmission electron micrograph of a thin section from a pea chloroplast. The most striking aspect of the structure of the chloroplast is the extensive system of internal membranes known as **thylakoids**. All the chlorophyll is contained within this membrane system, which is the site of the light reactions of photosynthesis.

The carbon reduction reactions, which are catalyzed by water-soluble enzymes, take place in the **stroma**, the region of the chloroplast outside the thylakoids. Most of the thylakoids appear to be very closely associated with each other. These stacked membranes are known as **grana lamellae** (singular *lamella*; each stack is called a *granum*), and the exposed membranes in which stacking is absent are known as **stroma lamellae**.

Two separate membranes, each composed of a lipid bilayer and together known as the **chloroplast envelope**, surround most types of chloroplasts (**Figure 7.15**). This double-membrane system contains a variety of metabolite transport systems. The chloroplast also contains its own DNA, RNA, and ribosomes. Some of the chloroplast proteins are products of transcription and translation within the chloroplast itself, whereas most of the others are encoded by nuclear DNA, synthesized on cytoplasmic ribosomes, and then imported into the chloroplast. This remarkable division of labor, extending in many cases to different subunits of the same enzyme complex, is discussed in more detail later in this chapter.

Thylakoids contain integral membrane proteins

A wide variety of proteins essential to photosynthesis are embedded in the thylakoid membranes. In many cases, portions of these proteins extend into the aqueous regions on both sides of the thylakoids. These integral membrane proteins contain a large proportion of hydrophobic amino acids and are therefore much more stable in a nonaqueous medium such as the hydrocarbon portion of the membrane (see Figure 1.12A).

The reaction centers, the antenna pigment–protein complexes, and most of the electron carrier proteins are all integral membrane proteins. In all known cases, integral

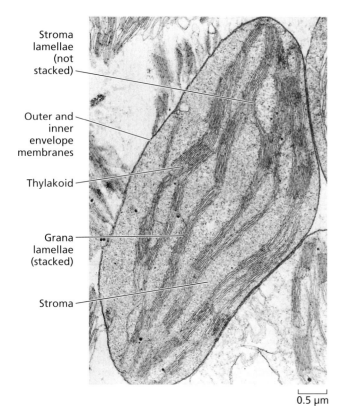

Stroma lamellae (not stacked)

Outer and inner envelope membranes

Thylakoid

Grana lamellae (stacked)

Stroma

0.5 μm

Figure 7.14   Transmission electron micrograph of a chloroplast from pea (*Pisum sativum*) fixed in glutaraldehyde and OsO_4, embedded in plastic resin, and thin-sectioned with an ultramicrotome. (Courtesy of J. Swafford.)

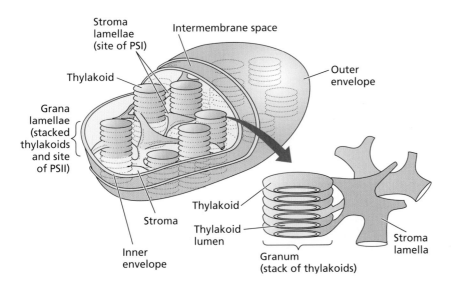

Figure 7.15 Schematic picture of the overall organization of the membranes in the chloroplast. The chloroplast of higher plants is surrounded by the inner and outer membranes (envelope). The region of the chloroplast that is inside the inner membrane and surrounds the thylakoid membranes is known as the stroma. It contains the enzymes that catalyze carbon fixation and other biosynthetic pathways. The thylakoid membranes are highly folded and appear in many pictures to be stacked like coins (the granum), although in reality they form one or a few large interconnected membrane systems, with a well-defined interior and exterior with respect to the stroma. (After Becker 1986.)

membrane proteins of the chloroplast have a unique orientation within the membrane. Thylakoid membrane proteins have one region pointing toward the stromal side of the membrane and the other oriented toward the interior space of the thylakoid, known as the *lumen* (see Figure 7.15).

The chlorophylls and accessory light-gathering pigments in the thylakoid membrane are always associated in a noncovalent, but highly specific, way with proteins, thereby forming pigment–protein complexes. Both antenna and reaction center chlorophylls are associated with proteins that are organized within the membrane so as to optimize energy transfer in antenna complexes and electron transfer in reaction centers, while at the same time minimizing wasteful processes.

Photosystems I and II are spatially separated in the thylakoid membrane

The PSII reaction center, along with its antenna chlorophylls and associated electron transport proteins, is located predominantly in the grana lamellae (**Figure 7.16A**). The PSI reaction center and its associated antenna pigments and electron transfer proteins, as well as the ATP synthase enzyme that catalyzes the formation of ATP, are found almost exclusively in the stroma lamellae and at the edges of the grana lamellae. The cytochrome $b_6 f$ complex of the electron transport chain that connects the two photosystems is evenly distributed between stroma and grana lamellae. The structures of all these complexes are shown in **Figure 7.16B**.

Thus, the two photochemical events that take place in O_2-evolving photosynthesis are spatially separated. This separation implies that one or more of the electron carriers that function between the photosystems diffuses from the grana region of the membrane to the stroma region, where electrons are delivered to PSI. These diffusible carriers are the blue-colored copper protein plastocyanin (PC) and the organic redox cofactor plastoquinone (PQ). These carriers are discussed in more detail later in this chapter.

In PSII, the oxidation of two water molecules produces four electrons, four protons, and a single O_2 (see the section *Mechanisms of Electron Transport* for details). The protons produced by this oxidation of water must also be able to diffuse to the stroma region, where ATP is synthesized. The functional role of this large separation (many tens of nanometers) between photosystems I and II is not entirely clear but is thought to improve the efficiency of energy distribution between the two photosystems.

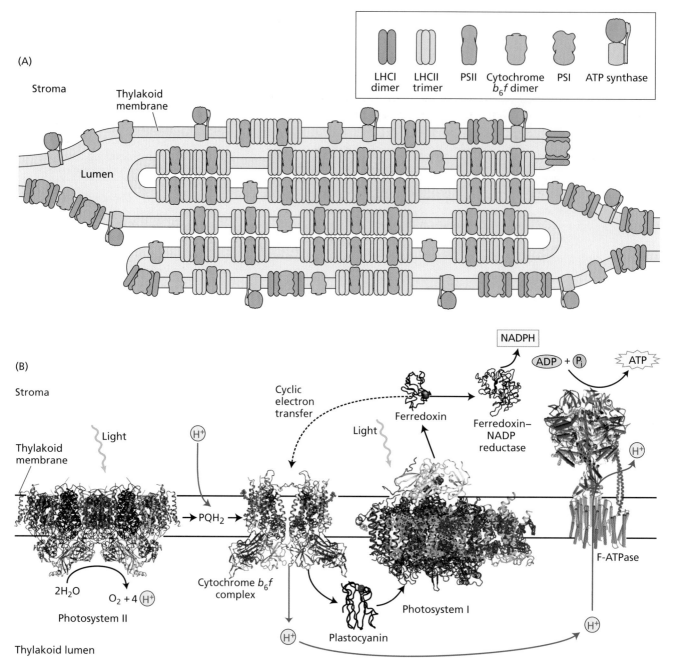

Figure 7.16 Organization and structure of the four major protein complexes of the thylakoid membrane. (A) PSII is located predominantly in the stacked regions of the thylakoid membrane; PSI and ATP synthase are found in the unstacked regions protruding into the stroma. Cytochrome b_6f complexes are evenly distributed. This lateral separation of the two photosystems requires that electrons and protons produced by PSII be transported a considerable distance before they can be acted on by PSI and the ATP-coupling enzyme. (B) Structures of the four main protein complexes of the thylakoid membrane. Shown also are the two diffusible electron carriers—plastocyanin, which is located in the thylakoid lumen, and plastohydroquinone (PQH₂) in the membrane. The lumen has a positive electric charge with respect to the stroma. (A after Allen and Forsberg 2001; B after Nelson and Ben-Shem 2004.)

The spatial separation between photosystems I and II indicates that a strict one-to-one stoichiometry between the two photosystems is not required. Instead, PSII reaction centers feed reducing equivalents into a common intermediate pool of lipid-soluble electron carriers (plastoquinone). The PSI reaction centers remove the reducing equivalents from the common pool, rather than from any specific PSII reaction center complex.

Most measurements of the relative quantities of photosystems I and II have shown that there is an excess of PSII in chloroplasts. Most commonly, the ratio of PSII to PSI is about 1.5:1, but it can change when plants are grown under different light conditions. In contrast to the situation in chloroplasts of eukaryotic photosynthetic organisms, cyanobacteria usually have an excess of PSI over PSII.

Organization of Light-Absorbing Antenna Systems

The antenna systems of different classes of photosynthetic organisms are remarkably varied, in contrast to the reaction centers, which appear to be similar in even distantly related organisms. The variety of antenna complexes reflects evolutionary adaptation to the diverse environments in which different organisms live, as well as the need in some organisms to balance energy input to the two photosystems. In this section we learn how energy transfer processes absorb light and deliver energy to the reaction center.

Antenna systems contain chlorophyll and are membrane-associated

Antenna systems function to deliver energy efficiently to the reaction centers with which they are associated. The size of the antenna system varies considerably in different organisms, ranging from a low of 20 to 30 bacteriochlorophylls per reaction center in some photosynthetic bacteria, to generally 200 to 300 chlorophylls per reaction center in higher plants, to a few thousand pigments per reaction center in some types of algae and bacteria. The molecular structures of antenna pigments are also quite diverse, although all of them are associated in some way with the photosynthetic membrane. In almost all cases, the antenna pigments are associated with proteins to form pigment–protein complexes.

The physical mechanism by which excitation energy is conveyed from the chlorophyll that absorbs the light to the reaction center is thought to be **fluorescence resonance energy transfer**, often abbreviated as FRET. By this mechanism the excitation energy is transferred from one molecule to another by a non-radiative process.

A useful analogy for resonance transfer is the transfer of energy between two tuning forks. If one tuning fork is struck and properly placed near another, the second tuning fork receives some energy from the first and begins to vibrate. The efficiency of energy transfer between the two tuning forks depends on their distance from each other and their relative orientation, as well as on their vibrational frequencies, or pitches. Similar parameters affect the efficiency of energy transfer in antenna complexes, with energy substituted for pitch.

Energy transfer in antenna complexes is usually very efficient: Approximately 95 to 99% of the photons absorbed by the antenna pigments have their energy transferred to the reaction center, where it can be used for photochemistry. There is an important difference between energy transfer among pigments in the antenna and the electron transfer that occurs in the reaction center: Whereas energy transfer is a purely physical phenomenon, electron transfer involves chemical (redox) reactions.

The antenna funnels energy to the reaction center

The sequence of pigments within the antenna that funnel absorbed energy toward the reaction center has absorption maxima that are progressively shifted toward longer red wavelengths (**Figure 7.17**). This red shift in absorption maximum means that the energy of the excited state is somewhat lower nearer the reaction center than in the more peripheral portions of the antenna system.

fluorescence resonance energy transfer (FRET) The physical mechanism by which excitation energy is conveyed from a molecule that absorbs light to an adjacent molecule.

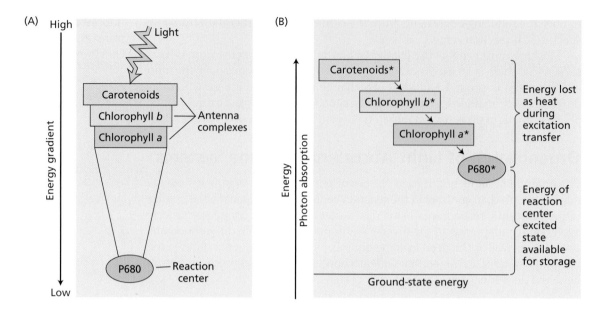

Figure 7.17 Funneling of excitation energy from the antenna system toward the reaction center. (A) The excited-state energy of pigments increases with distance from the reaction center; that is, pigments closer to the reaction center are lower in energy than those farther from the reaction center. This energy gradient ensures that excitation transfer toward the reaction center is energetically favorable and that excitation transfer back out to the peripheral portions of the antenna is energetically unfavorable. (B) Some energy is lost as heat to the environment by this process, but under optimal conditions almost all the excitation energy absorbed in the antenna complexes can be delivered to the reaction center. The asterisks denote excited states.

As a result of this arrangement, when excitation is transferred, for example, from a chlorophyll *b* molecule absorbing maximally at 650 nm to a chlorophyll *a* molecule absorbing maximally at 670 nm, the difference in energy between these two excited chlorophylls is lost to the environment as heat.

For the excitation to be transferred back to the chlorophyll *b*, the energy lost as heat would have to be resupplied. The probability of reverse transfer is therefore smaller simply because thermal energy is not sufficient to make up the deficit between the lower-energy and higher-energy pigments. This effect gives the energy-trapping process a degree of directionality or irreversibility and makes the delivery of excitation to the reaction center more efficient. In essence, the system sacrifices some energy from each quantum so that nearly all of the quanta can be trapped by the reaction center.

Many antenna pigment–protein complexes have a common structural motif

In all eukaryotic photosynthetic organisms that contain both chlorophyll *a* and chlorophyll *b*, the most abundant antenna proteins are members of a large family of structurally related proteins. Some of these proteins are associated primarily with PSII and are called **light-harvesting complex II** (**LHCII**) proteins; others are associated with PSI and are called LHCI proteins. These antenna complexes are also known as **chlorophyll *a/b* antenna proteins**.

The structure of one of the LHCII proteins has been determined (**Figure 7.18**). The protein contains three α-helical regions and binds 14 chlorophyll *a* and *b* molecules, as well as four carotenoids. The structure of the LHCI proteins is generally similar to that of the LHCII proteins. All of these proteins have significant sequence similarity and are almost certainly descendants of a common ancestral protein.

light-harvesting complex II (LHCII)
The most abundant antenna protein complex, associated primarily with photosystem II.

chlorophyll *a/b* antenna proteins
Chlorophyll-containing proteins associated with one or the other of the two photosystems in eukaryotic organisms. Also known as light-harvesting complex proteins (LHC proteins).

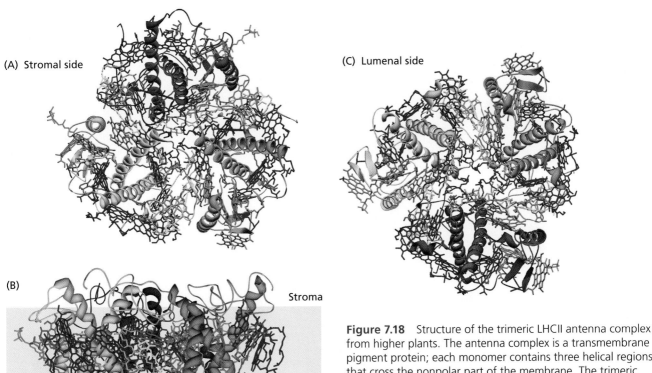

(A) Stromal side

(C) Lumenal side

(B)

Stroma

Lumen

Figure 7.18 Structure of the trimeric LHCII antenna complex from higher plants. The antenna complex is a transmembrane pigment protein; each monomer contains three helical regions that cross the nonpolar part of the membrane. The trimeric complex is shown (A) from the stromal side, (B) from within the membrane, and (C) from the lumenal side. Gray, polypeptide; dark blue, Chl *a*; green, Chl *b*; dark orange, lutein; light orange, neoxanthin; yellow, violaxanthin; pink, lipids. (After Barros and Kühlbrandt 2009.)

Light absorbed by carotenoids or chlorophyll *b* in the LHC proteins is rapidly transferred to chlorophyll *a* and then to other antenna pigments that are intimately associated with the reaction center. The LHCII complex is also involved in regulatory processes, which we discuss later in the chapter.

Mechanisms of Electron Transport

Some of the evidence that led to the idea of two photochemical reactions operating in series was discussed earlier in this chapter. In this section we consider in detail the chemical reactions involved in electron transfer during photosynthesis. We discuss the excitation of chlorophyll by light and the reduction of the first electron acceptor, the flow of electrons through photosystems II and I, the oxidation of water as the primary source of electrons, and the reduction of the final electron acceptor ($NADP^+$). The chemiosmotic mechanism that mediates ATP synthesis is discussed in detail later in the chapter (see the section *Proton Transport and ATP Synthesis in the Chloroplast*).

Electrons from chlorophyll travel through the carriers organized in the Z scheme

Figure 7.19 shows a current version of the Z scheme, in which all the electron carriers known to function in electron flow from H_2O to $NADP^+$ are arranged vertically at their midpoint redox potentials. Components known to react with each other are connected by arrows, so the Z scheme is really a synthesis of both kinetic and thermodynamic information. The large vertical arrows represent the input of light energy into the system.

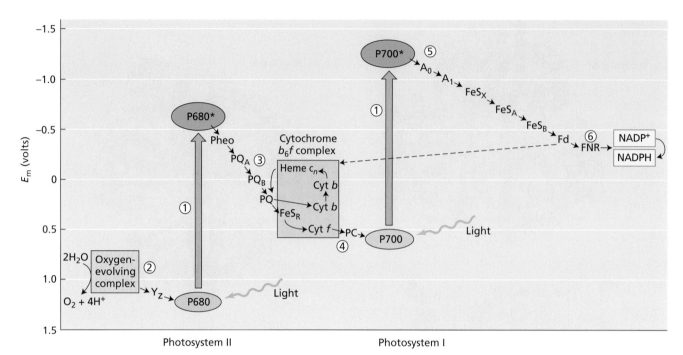

Figure 7.19 Detailed Z scheme for O_2-evolving photosynthetic organisms. The redox carriers are placed at their midpoint redox potentials (at pH 7). (1) The vertical arrows represent photon absorption by the reaction center chlorophylls: P680 for photosystem II (PSII) and P700 for photosystem I (PSI). The excited PSII reaction center chlorophyll, P680*, transfers an electron to pheophytin (Pheo). (2) On the oxidizing side of PSII (to the left of the arrow joining P680 with P680*), P680 oxidized by pheophytin after light excitation is re-reduced by Y_z, which has received electrons from oxidation of water. (3) On the reducing side of PSII (to the right of the arrow joining P680 with P680*), pheophytin transfers electrons to the acceptors PQ_A and PQ_B, which are plastoquinones. (4) The cytochrome b_6f complex transfers electrons to plastocyanin (PC), a soluble protein, which in turn reduces P700$^+$ (oxidized P700). (5) The acceptor of electrons from P700* (A_0) is thought to be a chlorophyll, and the next acceptor (A_1) is a quinone. A series of membrane-bound iron–sulfur proteins (FeS_X, FeS_A, and FeS_B) transfers electrons to soluble ferredoxin (Fd). (6) The soluble flavoprotein ferredoxin–NADP$^+$ reductase (FNR) reduces NADP$^+$ to NADPH, which is used in the Calvin–Benson cycle to reduce CO_2 (see Chapter 8). The dashed line indicates cyclic electron flow around PSI. (After Blankenship and Prince 1985.)

Photons excite the specialized chlorophyll of the reaction centers (P680 for PSII; P700 for PSI), and an electron is ejected. The electron then passes through a series of electron carriers and eventually reduces P700 (for electrons from PSII) or NADP$^+$ (for electrons from PSI). Much of the following discussion describes the journeys of these electrons and the nature of their carriers.

Almost all the chemical processes that make up the light reactions of photosynthesis are carried out by four major protein complexes: PSII, the cytochrome b_6f complex, PSI, and ATP synthase. These four integral membrane complexes are vectorially oriented in the thylakoid membrane to function as follows (**Figure 7.20**; also see Figure 7.16):

- PSII oxidizes water to O_2 in the thylakoid lumen and in the process releases protons into the lumen. The reduced product of photosystem II is plastohydroquinone (PQH_2).

- Cytochrome b_6f oxidizes PQH_2 molecules that were reduced by PSII and delivers electrons to PSI via the soluble copper protein plastocyanin. The oxidation of PQH_2 is coupled to proton transfer into the lumen from the stroma, generating a proton motive force.

- PSI reduces NADP$^+$ to NADPH in the stroma by the action of ferredoxin (Fd) and the flavoprotein ferredoxin–NADP$^+$ reductase (FNR).

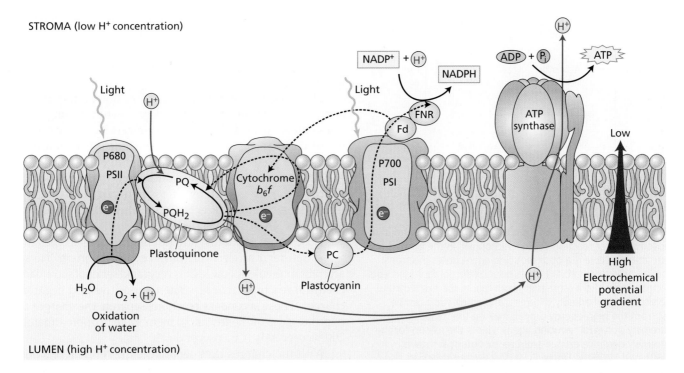

Figure 7.20   Transfer of electrons and protons in the thylakoid membrane is carried out vectorially by four protein complexes (see Figure 7.16B for structures). Water is oxidized and protons are released in the lumen by PSII. PSI reduces $NADP^+$ to NADPH in the stroma, via the action of ferredoxin (Fd) and the flavoprotein ferredoxin–$NADP^+$ reductase (FNR). Protons are also transported into the lumen by the action of the cytochrome b_6f complex and contribute to the electrochemical proton gradient. These protons must then diffuse to the ATP synthase enzyme, where their diffusion down the electrochemical potential gradient is used to synthesize ATP in the stroma. Reduced plastoquinone (PQH_2) and plastocyanin transfer electrons to cytochrome b_6f and to PSI, respectively. The dashed lines represent electron transfer; solid lines represent proton movement.

- ATP synthase produces ATP as protons diffuse back through it from the lumen into the stroma.

Energy is captured when an excited chlorophyll reduces an electron acceptor molecule

As discussed earlier, the function of light is to excite a specialized chlorophyll in the reaction center, either by direct absorption or, more frequently, via energy transfer from an antenna pigment. This excitation process can be envisioned as the promotion of an electron from the highest-energy filled orbital of the chlorophyll to the lowest-energy unfilled orbital (**Figure 7.21**). The electron in the upper orbital is only loosely bound to the chlorophyll and is easily lost if a molecule that can accept the electron is nearby.

The first reaction that converts electron energy into chemical energy—that is, the primary photochemical event—is the transfer of an electron from the excited state of a chlorophyll in the reaction center to an acceptor molecule. An equivalent way to view this process is that the absorbed photon causes an electron rearrangement in the reaction center chlorophyll, followed by an electron transfer process in which part of the energy in the photon is captured in the form of redox energy.

Immediately after the photochemical event, the reaction center chlorophyll is in an oxidized state (electron deficient, or positively charged), and the nearby electron acceptor molecule is reduced (electron rich, or negatively charged). The system is now at a critical juncture. The lower-energy orbital of the positively charged oxidized reaction center chlorophyll shown in Figure 7.21 has a vacancy

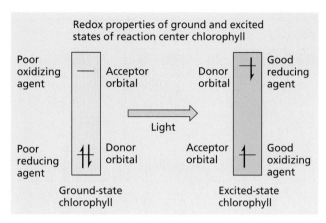

Redox properties of ground and excited states of reaction center chlorophyll

Ground-state chlorophyll

Excited-state chlorophyll

Figure 7.21   Orbital occupation diagram for the ground and excited states of reaction center chlorophyll. In the ground state the molecule is a poor reducing agent (loses electrons from a low-energy orbital) and a poor oxidizing agent (accepts electrons only into a high-energy orbital). In the excited state the situation is markedly different, and an electron can be lost from the high-energy orbital, making the molecule an extremely powerful reducing agent. This is the reason for the extremely negative excited-state redox potential shown by P680* and P700* in Figure 7.19. The excited state can also act as a strong oxidant by accepting an electron into the lower-energy orbital, although this pathway is not significant in reaction centers. (After Blankenship and Prince 1985.)

bleaching   The loss of chlorophyll's characteristic absorbance due to its conversion into another structural state, often by oxidation.

P700   The chlorophyll of the photosystem I reaction center that absorbs maximally at 700 nm in its neutral state. The P stands for pigment.

P680   The chlorophyll of the photosystem II reaction center that absorbs maximally at 680 nm in its neutral state.

and can accept an electron. If the acceptor molecule donates its electron back to the reaction center chlorophyll, the system will be returned to the state that existed before the light excitation, and all the absorbed energy will be converted into heat.

This wasteful *recombination* process, however, does not appear to occur to any substantial degree in functioning reaction centers. Instead, the acceptor transfers its extra electron to a secondary acceptor and so on down the electron transport chain. The oxidized reaction center of the chlorophyll that had donated an electron is re-reduced by a secondary donor, which in turn is reduced by a tertiary donor. In plants, the ultimate electron donor is H_2O, and the ultimate electron acceptor is $NADP^+$ (see Figure 7.19).

The essence of photosynthetic energy storage is thus the initial transfer of an electron from an excited chlorophyll to an acceptor molecule, followed by a very rapid series of secondary chemical reactions that separate the positive and negative charges. These secondary reactions separate the charges to opposite sides of the thylakoid membrane in approximately 200 picoseconds (1 picosecond = 10^{-12} s).

With the charges thus separated, the reversal reaction is many orders of magnitude slower, and the energy has been captured. Each of the secondary electron transfers is accompanied by a loss of some energy, thus making the process effectively irreversible. The quantum yield for the production of stable products in purified reaction centers from photosynthetic bacteria has been measured as 1.0; that is, every photon produces stable products, and no reversal reactions occur.

Measured quantum requirements for O_2 production in higher plants under optimal conditions (low-intensity light) indicate that the values for the primary photochemical events are also very close to 1.0. The structure of the reaction center appears to be extremely fine-tuned for maximum rates of productive reactions and minimum rates of energy-wasting reactions.

The reaction center chlorophylls of the two photosystems absorb at different wavelengths

As discussed earlier in the chapter, PSI and PSII have distinct absorption characteristics. Precise measurements of absorption maxima are made possible by optical changes in the reaction center chlorophylls in the reduced and oxidized states. The reaction center chlorophyll is transiently in an oxidized state after losing an electron and before being re-reduced by its electron donor.

In the oxidized state, chlorophylls lose their characteristic strong light absorbance in the red region of the spectrum; they become **bleached**. It is therefore possible to monitor the redox state of these chlorophylls by time-resolved optical absorbance measurements in which this bleaching is monitored directly.

Using such techniques, it was found that the reaction center chlorophyll of PSI absorbs maximally at 700 nm in its reduced state. Accordingly, this chlorophyll is named **P700** (the P stands for *pigment*). The analogous optical transient of PSII is at 680 nm, so its reaction center chlorophyll is known as **P680**. The primary donor of PSI, P700, is also a dimer of chlorophyll *a* molecules. PSII also contains a dimer of chlorophylls, although the primary electron transfer event may not originate from these pigments. In the oxidized state, reaction center chlorophylls contain an unpaired electron.

Molecules with unpaired electrons can often be detected by a magnetic-resonance technique known as **electron spin resonance (ESR) spectroscopy**. ESR studies, along with the spectroscopic measurements already described, have led to the discovery of many intermediate electron carriers in the photosynthetic electron transport system.

The PSII reaction center is a multi-subunit pigment–protein complex

PSII is contained in a multi-subunit protein supercomplex. In higher plants, the supercomplex has two complete reaction centers and some antenna complexes. The core of the reaction center consists of two membrane proteins known as D1 and D2, as well as other proteins, as shown in **Figure 7.22**.

The primary donor chlorophyll, additional chlorophylls, carotenoids, pheophytins, and plastoquinones (two electron acceptors described below) are

electron spin resonance (ESR) spectroscopy   A magnetic-resonance technique that detects unpaired electrons in molecules. Instrumental measurements that identify intermediate electron carriers in the photosynthetic or respiratory electron transport system.

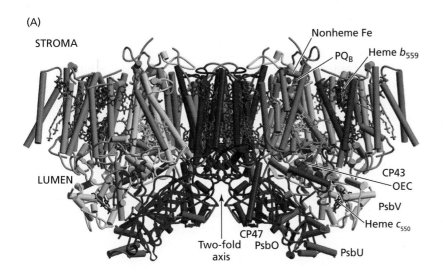

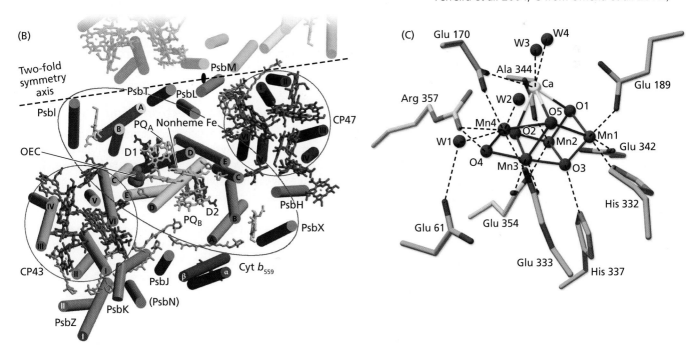

Figure 7.22   Structure of the PSII reaction center from the cyanobacterium *Thermosynechococcus elongatus*, resolved at 0.35 nm. The structure includes the D$_1$ (yellow) and D$_2$ (orange) core reaction center proteins, the CP43 (green) and CP47 (red) antenna proteins, cytochromes b_{559} and c_{550}, the extrinsic 33-kDa protein PsbO (dark blue), and the pigments and other cofactors. (A) Side view parallel to the membrane plane. (B) View from the lumenal surface, perpendicular to the plane of the membrane. (C) Detail of the Mn-containing oxygen-evolving complex (OEC). (A and B from Ferreira et al. 2004; C from Umena et al. 2011.)

pheophytin   A chlorophyll in which the central magnesium atom has been replaced by two hydrogen atoms.

plastohydroquinone (PQH₂)
The fully reduced form of plastoquinone.

cytochrome *b₆f* complex   A large multi-subunit protein complex containing two *b*-type hemes, one *c*-type heme (cytochrome *f*), and a Rieske iron–sulfur protein. A relatively immobile complex distributed equally between the grana and the stroma regions of the thylakoid membranes.

cytochrome *f*   A subunit in the cytochrome *b₆f* complex that plays a role in electron transport between photosystems I and II.

Rieske iron–sulfur protein
A protein subunit in the cytochrome *b₆f* complex, in which two iron atoms are bridged by two sulfur atoms, with two histidine and two cysteine ligands.

bound to the membrane proteins D1 and D2. Other proteins serve as antenna complexes or are involved in oxygen evolution. Some, such as cytochrome b_{559}, have no known function but may be involved in a protective cycle around PSII.

Water is oxidized to oxygen by PSII

Water is oxidized according to the following chemical reaction:

$$2\,H_2O \rightarrow O_2 + 4\,H^+ + 4\,e^- \tag{7.8}$$

This equation indicates that four electrons are removed from two water molecules, generating an oxygen molecule and four hydrogen ions.

Water is a very stable molecule. Oxidation of water to form molecular oxygen is very difficult: The photosynthetic oxygen-evolving complex (OEC) is the only known biochemical system that carries out this reaction, and is the source of almost all the oxygen in Earth's atmosphere.

The protons produced by water oxidation are released into the lumen of the thylakoid, not directly into the stromal compartment (see Figure 7.20). They are released into the lumen because of the vectorial nature of the membrane and the fact that the oxygen-evolving complex is localized near the interior surface of the thylakoid membrane (see Figure 7.22A). These protons are eventually transferred from the lumen to the stroma by translocation through the ATP synthase. In this way, the protons released during water oxidation contribute to the electrochemical potential driving ATP formation (see Figure 7.20).

It has been known for many years that manganese (Mn) is an essential cofactor in the water-oxidizing process (see Chapter 4), and a classic hypothesis in photosynthesis research postulates that Mn ions undergo a series of oxidations—known as *S states* and labeled S_0, S_1, S_2, S_3, and S_4—that are perhaps linked to H_2O oxidation and the generation of O_2. This hypothesis has received strong support from a variety of experiments, most notably X-ray absorption and ESR studies, both of which detect the manganese ions directly. Analytical experiments indicate that four Mn ions are associated with each oxygen-evolving complex. Other experiments have shown that Cl^- and Ca^{2+} ions are essential for O_2 evolution. The detailed chemical mechanism of the oxidation of water to O_2 is not yet well understood, but with structural information now available, rapid progress is being made in this area.

One electron carrier, generally identified as Y_z, functions between the oxygen-evolving complex and P680 (see Figure 7.19). To function in this region, Y_z needs to have a very strong tendency to retain its electrons. This species has been identified as a radical formed from a tyrosine residue in the D1 protein of the PSII reaction center.

Pheophytin and two quinones accept electrons from PSII

Spectral and ESR studies have revealed the structural arrangement of the carriers in the electron acceptor complex. **Pheophytin**, a chlorophyll in which the central magnesium ion has been replaced by two hydrogen ions, acts as an early acceptor in PSII. The structural change gives pheophytin chemical and spectral properties that are slightly different from those of Mg-based chlorophylls. Pheophytin passes electrons to a complex of two plastoquinones in close proximity to an iron ion.

The two plastoquinones, PQ_A and PQ_B, are bound to the reaction center and receive electrons from pheophytin in a sequential fashion. Transfer of the two electrons to PQ_B reduces it to PQ_B^{2-}, and the reduced PQ_B^{2-} takes two protons from the stroma side of the medium, yielding a fully reduced **plastohydroquinone (PQH₂)** (**Figure 7.23**). The PQH_2 then dissociates from the reaction center complex and enters the hydrocarbon portion of the membrane, where it in turn transfers its electrons to the cytochrome b_6f complex. Unlike the large protein complexes of the thylakoid membrane, PQH_2 is a small, nonpolar molecule that diffuses readily in the nonpolar core of the membrane bilayer.

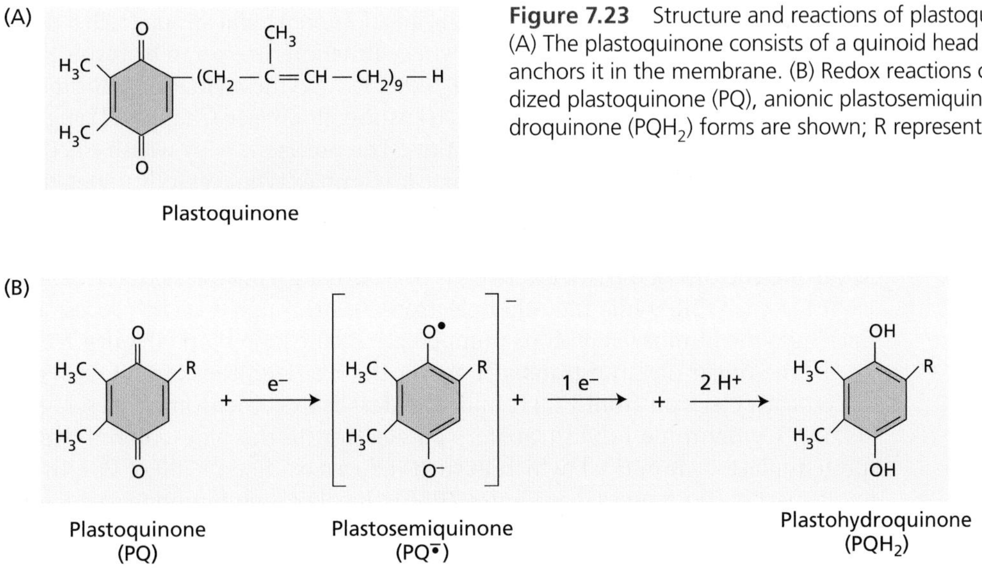

(A)

Plastoquinone

Figure 7.23 Structure and reactions of plastoquinones that operate in PSII. (A) The plastoquinone consists of a quinoid head and a long nonpolar tail that anchors it in the membrane. (B) Redox reactions of plastoquinone. The fully oxidized plastoquinone (PQ), anionic plastosemiquinone (PQ•), and reduced plastohydroquinone (PQH$_2$) forms are shown; R represents the side chain.

(B)

Plastoquinone (PQ)	Plastosemiquinone (PQ•⁻)	Plastohydroquinone (PQH$_2$)

Electron flow through the cytochrome b_6f complex also transports protons

The **cytochrome b_6f complex** is a large multi-subunit protein with several prosthetic groups (**Figure 7.24**). It contains two b-type hemes and one c-type heme (**cytochrome f**). In c-type cytochromes the heme is covalently attached to the peptide; in b-type cytochromes the chemically similar protoheme group is not covalently attached. In addition, the complex contains a **Rieske iron–sulfur protein** (named for the scientist who discovered it), in which two iron ions are bridged by two sulfide ions. The functional roles of all these cofactors are reasonably well understood, as described below. However, the cytochrome b_6f complex also

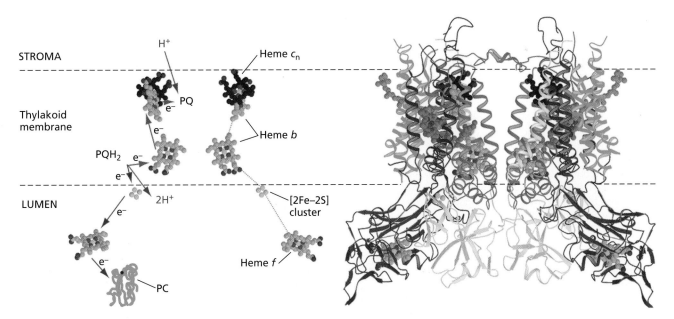

Figure 7.24 Structure of the cytochrome b_6f complex from cyanobacteria. The diagram on the right shows the arrangement of the proteins and cofactors in the complex. Cytochrome b_6 protein is shown in blue, cytochrome f protein in red, Rieske iron–sulfur protein in yellow, and other smaller subunits in green and purple. On the left, the proteins have been omitted to more clearly show the positions of the cofactors. [2 Fe–2S] cluster, part of the Rieske iron–sulfur protein; PC, plastocyanin; PQ, plastoquinone; PQH$_2$, plastohydroquinone. (After Kurisu et al. 2003.)

Q cycle A mechanism for oxidation of plastohydroquinone (reduced plastoquinone, also called plastoquinol) in chloroplasts and of ubihydroquinone (reduced ubiquinone, also called ubiquinol) in mitochondria.

FeS$_R$ An iron- and sulfur-containing subunit of the cytochrome b_6f complex, involved in electron and proton transfer.

contains additional cofactors—including an additional heme group (called heme c_n), a chlorophyll, and a carotenoid—whose functions are yet to be resolved.

The structures of the cytochrome b_6f complex and the related cytochrome bc_1 complex in the mitochondrial electron transport chain (see Chapter 11) suggest a mechanism for electron and proton flow. The precise way by which electrons and protons flow through the cytochrome b_6f complex is not yet fully understood, but a mechanism known as the **Q cycle** accounts for most of the observations. In this mechanism, plastohydroquinone (also called plastoquinol) (PQH$_2$) is oxidized, and one of the two electrons is passed along a linear electron transport chain toward PSI, while the other electron goes through a cyclic process that increases the number of protons pumped across the membrane (**Figure 7.25**).

In the linear electron transport chain, the oxidized Rieske protein (**FeS$_R$**) accepts an electron from PQH$_2$ and transfers it to cytochrome f (see Figure 7.25A). Cytochrome f then transfers an electron to the blue-colored copper protein plastocyanin (PC), which in turn reduces oxidized P700 of PSI. In the cyclic part of the process (see Figure 7.25B), the plastosemiquinone (see Figure 7.23) transfers its other electron to one of the b-type hemes, releasing both of its protons to the lumenal side of the membrane.

The first b-type heme transfers its electron through the second b-type heme to an oxidized plastoquinone molecule, reducing it to the semiquinone form near the stromal surface of the complex. Another similar sequence of electron flow (see Figure 7.25B) fully reduces the plastoquinone, which picks up protons from the stromal side of the membrane and is released from the b_6f complex as plastohydroquinone.

Figure 7.25 Mechanism of electron and proton transfer in the cytochrome b_6f complex. This complex contains two b-type cytochromes (Cyt b), a c-type cytochrome (Cyt c, historically called cytochrome f), a Rieske Fe–S protein (FeS$_R$), and two quinone oxidation–reduction sites. (A) The noncyclic or linear processes: A plastohydroquinone (PQH$_2$) molecule produced by the action of PSII (see Figures 7.20, 7.23) is oxidized near the lumenal side of the complex, transferring its two electrons to the Rieske Fe–S protein and one of the b-type cytochromes and simultaneously expelling two protons to the lumen. The electron transferred to FeS$_R$ is passed to cytochrome f (Cyt f) and then to plastocyanin (PC), which reduces P700 of PSI. The reduced b-type cytochrome transfers an electron to the other b-type cytochrome, which reduces a plastoquinone (PQ) to the plastosemiquinone (PQ•) state (see Figure 7.23). (B) The cyclic processes: A second PQH$_2$ is oxidized, with one electron going from FeS$_R$ to PC and finally to P700. The second electron goes through the two b-type cytochromes and reduces the plastosemiquinone to the plastohydroquinone, at the same time picking up two protons from the stroma. Overall, four protons are transported across the membrane for every two electrons delivered to P700.

(A) First QH$_2$ oxidized

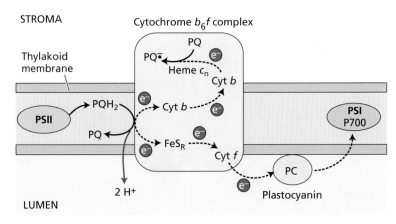

(B) Second QH$_2$ oxidized

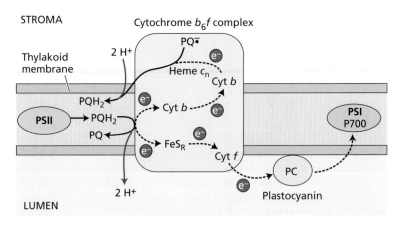

The overall result of two turnovers of the complex is that two electrons are transferred to P700, two plastohydroquinones are oxidized to the plastoquinone form, and one oxidized plastoquinone is reduced to the plastohydroquinone form. In the process of oxidizing the plastohydroquinones, four protons are transferred from the stromal to the lumenal side of the membrane.

By this mechanism, electron flow connecting the acceptor side of the PSII reaction center to the donor side of the PSI reaction center also gives rise to an electrochemical potential across the membrane, due in part to H^+ concentration differences on the two sides of the membrane. This electrochemical potential is used to power the synthesis of ATP. The cyclic electron flow through the cytochrome b and plastoquinone increases the number of protons pumped per electron beyond what could be achieved in a strictly linear sequence.

Plastoquinone and plastocyanin carry electrons between photosystem II and photosystem I

The location of the two photosystems at different sites on the thylakoid membranes (see Figure 7.16) requires that at least one component is capable of moving along or within the membrane in order to deliver electrons produced by PSII to PSI. The cytochrome b_6f complex is distributed equally between the grana and the stroma regions of the membranes, but its large size makes it unlikely that it is the mobile carrier. Instead, plastoquinone or plastocyanin or possibly both are thought to serve as mobile carriers to connect the two photosystems.

Plastocyanin (**PC**) is a small (10.5 kDa), water-soluble, copper-containing protein that transfers electrons between the cytochrome b_6f complex and P700. This protein is found in the lumenal space (see Figure 7.25).

The PSI reaction center reduces NADP⁺

The PSI reaction center complex is a large multi-subunit complex (**Figure 7.26**). Unlike in PSII, in which the antenna chlorophylls are associated with the reaction center but are present on separate pigment–proteins, a core antenna consisting of about 100 chlorophylls is an integral part of the PSI reaction center. The core antenna and P700 are bound to two proteins, PsaA and PsaB, with molecular masses in the range of 66 to 70 kDa. The PSI reaction center complex from pea contains four LHCI complexes in addition to the core structure similar to that found in cyanobacteria (see Figure 7.26). The total number of chlorophyll molecules in this complex is nearly 200.

The core antenna pigments form a bowl surrounding the electron transfer cofactors, which are in the center of the complex. In their reduced form, the electron carriers that function in the acceptor region of PSI are all extremely strong reducing agents. These reduced species are very unstable and thus difficult to identify. Evidence indicates that one of these early acceptors is a chlorophyll molecule, and another is a quinone species, phylloquinone, also known as vitamin K_1.

Additional electron acceptors include a series of three membrane-associated iron–sulfur proteins, also known as **Fe–S centers**: **FeS$_X$**, **FeS$_A$**, and **FeS$_B$** (see Figure 7.26). FeS$_X$ is part of the P700-binding protein; FeS$_A$ and FeS$_B$ reside on an 8-kDa protein that is part of the PSI reaction center complex. Electrons are transferred through FeS$_A$ and FeS$_B$ to **ferredoxin** (**Fd**), a small, water-soluble iron–sulfur protein (see Figures 7.19 and 7.26). The membrane-associated flavoprotein **ferredoxin–NADP⁺ reductase** (**FNR**) reduces NADP⁺ to NADPH, thus completing the sequence of noncyclic electron transport that begins with the oxidation of water.

In addition to the reduction of NADP⁺, reduced ferredoxin produced by PSI has several other functions in the chloroplast, such as supplying reductant for nitrate reduction and regulating some of the carbon fixation enzymes (see Chapter 8).

plastocyanin (PC) A small (10.5 kDa), water-soluble, copper-containing protein that transfers electrons between the cytochrome b_6f complex and P700. This protein is found in the lumenal space.

Fe–S centers Prosthetic groups consisting of inorganic iron and sulfur that are abundant in proteins in respiratory and photosynthetic electron transport.

FeS$_X$, FeS$_A$, FeS$_B$ Membrane-bound iron–sulfur proteins that transfer electrons between photosystem I and ferredoxin.

ferredoxin (Fd) A small, water-soluble iron–sulfur protein involved in electron transport in photosystem I.

ferredoxin–NADP⁺ reductase (FNR) A membrane-associated flavoprotein that receives electrons from photosystem I and reduces NADP⁺ to NADPH.

(A)

(B)

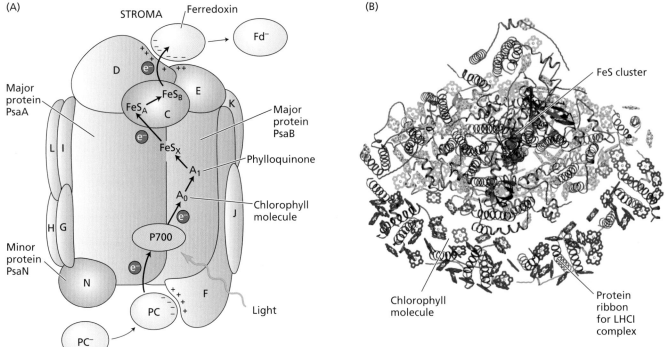

Figure 7.26 Structure of PSI. (A) Structural model of the PSI reaction center from higher plants. Components of the PSI reaction center are organized around two major core proteins, PsaA and PsaB. Minor proteins PsaC to PsaN are labeled C to N. Electrons are transferred from plastocyanin (PC) to P700 (see Figures 7.19 and 7.20) and then to a chlorophyll molecule (A_0), to phylloquinone (A_1), to the Fe–S centers FeS_X, FeS_A, and FeS_B, and finally to the soluble iron–sulfur protein ferredoxin (Fd). (B) Structure of the PSI reaction center complex from pea at 0.44 nm resolution, including the LHCI antenna complexes. This is viewed from the stromal side of the membrane. (A after Buchanan et al. 2000; B after Nelson and Ben-Shem 2004.)

Cyclic electron flow generates ATP but no NADPH

Some of the cytochrome $b_6 f$ complexes are found in the stroma region of the membrane, where PSI is located. Under certain conditions, **cyclic electron flow** is known to occur from the reducing side of PSI via plastohydroquinone and the $b_6 f$ complex and back to P700. This cyclic electron flow is coupled to proton pumping into the lumen, which can be used for ATP synthesis but does not oxidize water or reduce $NADP^+$ (see Figure 7.16B). Cyclic electron flow is especially important as an ATP source in the bundle sheath chloroplasts of some plants that carry out C_4 carbon fixation (see Chapter 8). The molecular mechanism of cyclic electron flow is not well understood.

Some herbicides block photosynthetic electron flow

The use of herbicides to kill unwanted plants is widespread in modern agriculture. Many different classes of herbicides have been developed. Some act by blocking amino acid, carotenoid, or lipid biosynthesis or by disrupting cell division. Other herbicides, such as dichlorophenyldimethylurea (DCMU, also known as diuron) and paraquat, block photosynthetic electron flow (**Figure 7.27**).

DCMU blocks electron flow at the quinone acceptors of PSII, by competing for the binding site of plastoquinone that is normally occupied by PQ_B. Paraquat accepts electrons from the early acceptors of PSI and then reacts with oxygen to form superoxide, $O_2^{\bullet}$, a reactive oxygen species that is very damaging to chloroplast components.

cyclic electron flow In photosystem I, the flow of electrons from the electron acceptors through the cytochrome $b_6 f$ complex and back to P700, coupled to proton pumping into the lumen. This electron flow energizes ATP synthesis but does not oxidize water or reduce $NADP^+$.

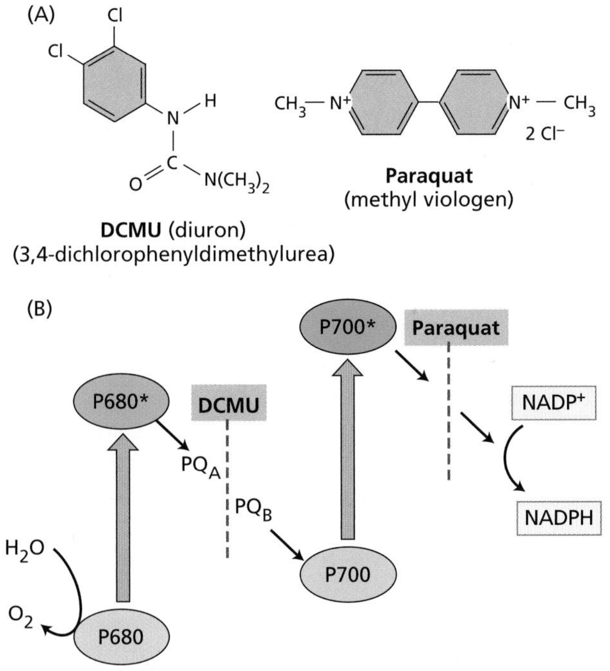

Figure 7.27 Chemical structure and mechanism of action of two important herbicides. (A) Chemical structure of dichlorophenyldimethylurea (DCMU) and methyl viologen (paraquat, a cloride salt), two herbicides that block photosynthetic electron flow. DCMU is also known as diuron. (B) Sites of action of the two herbicides. DCMU blocks electron flow at the plastoquinone acceptors of PSII by competing for the binding site of plastoquinone. Paraquat acts by accepting electrons from the early acceptors of PSI.

Proton Transport and ATP Synthesis in the Chloroplast

In the preceding sections we learned how captured light energy is used to reduce $NADP^+$ to NADPH. Another fraction of the captured light energy is used for light-dependent ATP synthesis, which is known as **photophosphorylation**. This process was discovered by Daniel Arnon and his coworkers in the 1950s. Under normal cellular conditions, photophosphorylation requires electron flow, although under some conditions electron flow and photophosphorylation can take place independently of each other. Electron flow without accompanying phosphorylation is said to be **uncoupled**.

It is now widely accepted that photophosphorylation works via the chemiosmotic mechanism. This mechanism was first proposed in the 1960s by Peter Mitchell. The same general mechanism drives phosphorylation during aerobic respiration in bacteria and mitochondria (see Chapter 11), as well as the transfer of many ions and metabolites across membranes (see Chapter 6). Chemiosmosis appears to be a unifying aspect of membrane processes in all forms of life.

In Chapter 6 we discussed the role of ATPases in chemiosmosis and ion transport at the cell's plasma membrane. The ATP used by the plasma membrane ATPase is synthesized by photophosphorylation in the chloroplast and oxidative phosphorylation in the mitochondrion. Here we are concerned with chemiosmosis and transmembrane proton concentration differences used to make ATP in the chloroplast.

The basic principle of chemiosmosis is that ion concentration differences and electrical potential differences across membranes are sources of free energy

photophosphorylation The formation of ATP from ADP and inorganic phosphate (P_i), catalyzed by the CF_0F_1-ATP synthase and using light energy stored in the proton gradient across the thylakoid membrane.

uncoupling A process by which coupled reactions are separated in such a way that the free energy released by one reaction is not available to drive the other reaction.

that can be used by the cell. As described by the second law of thermodynamics, any nonuniform distribution of matter or energy represents a source of energy. Differences in **chemical potential** of any molecular species whose concentrations are not the same on opposite sides of a membrane provide such a source of energy.

The asymmetric nature of the photosynthetic membrane and the fact that proton flow from one side of the membrane to the other accompanies electron flow were discussed earlier. The direction of proton translocation is such that the stroma becomes more alkaline (fewer H+ ions) and the lumen becomes more acidic (more H+ ions) as a result of electron transport (see Figures 7.20 and 7.25).

Some of the early evidence supporting a chemiosmotic mechanism of photosynthetic ATP formation was provided by an elegant experiment carried out by André Jagendorf and coworkers (**Figure 7.28**). They suspended chloroplast thylakoids in a pH 4 buffer, and the buffer diffused across the membrane, causing the interior, as well as the exterior, of the thylakoid to equilibrate at this acidic pH. They then rapidly transferred the thylakoids to a pH 8 buffer, thereby creating a pH difference of four units across the thylakoid membrane, with the inside acidic relative to the outside.

They found that large amounts of ATP were formed from ADP and P_i by this process, with no light input or electron transport. This result supports the predictions of the chemiosmotic hypothesis, described in the paragraphs that follow.

Mitchell proposed that the total energy available for ATP synthesis, which he called the **proton motive force** (Δp), is the sum of a proton chemical potential and a transmembrane electrical potential. These two components of the proton motive force from the outside of the membrane to the inside are given by the following equation:

$$\Delta p = \Delta E - 59(pH_i - pH_o) \tag{7.9}$$

where ΔE is the transmembrane electrical potential, and $pH_i - pH_o$ (or ΔpH) is the pH difference across the membrane. The constant of proportionality (at

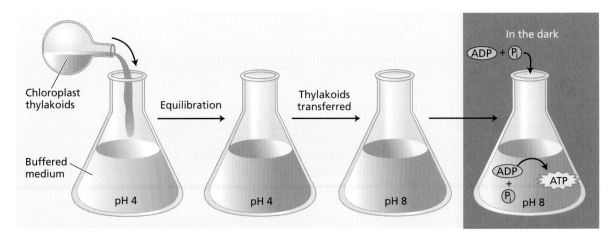

Figure 7.28   Summary of the experiment carried out by Jagendorf and coworkers. Isolated chloroplast thylakoids kept previously at pH 8 were equilibrated in an acid medium at pH 4. The thylakoids were then transferred to a buffer at pH 8 that contained ADP and P_i. The proton gradient generated by this manipulation provided a driving force for ATP synthesis in the absence of light. This experiment verified a prediction of the chemiosmotic theory stating that a chemical potential across a membrane can provide energy for ATP synthesis. (After Jagendorf 1967.)

25°C) is 59 mV per pH unit, so a transmembrane pH difference of one pH unit is equivalent to a membrane potential of 59 mV. Most evidence suggests that the steady-state electrical potential is relatively small in chloroplasts, so that most of the proton motive force is derived from the pH gradient.

In addition to the need for mobile electron carriers discussed earlier, the uneven distribution of photosystems II and I, and of ATP synthase in the thylakoid membrane (see Figure 7.16A), poses some challenges for the formation of ATP. ATP synthase is found only in the stroma lamellae and at the edges of the grana stacks. Protons pumped across the membrane by the cytochrome b_6f complex or protons produced by water oxidation in the middle of the grana must move laterally up to several tens of nanometers to reach an ATP synthase.

The ATP is synthesized by an enzyme complex (mass ~400 kDa) known by several names: **ATP synthase**, ATPase (after the reverse reaction of ATP hydrolysis), and CF_0–CF_1. This enzyme consists of two parts: a hydrophobic membrane-bound portion called CF_0 and a portion that sticks out into the stroma called CF_1 (**Figure 7.29**). CF_0 appears to form a channel across the membrane through which protons can pass. CF_1 is made up of several peptides, including three copies of each of the α and β peptides arranged alternately much like the sections of an orange. Whereas the catalytic sites are located largely on the β polypeptide, many of the other peptides are thought to have primarily regulatory functions. CF_1 is the portion of the complex that synthesizes ATP.

The molecular structure of the mitochondrial ATP synthase has been determined by X-ray crystallography. Although there are significant differ-

ATP synthase The multi-subunit protein complex that synthesizes ATP from ADP and phosphate (P_i). F_0F_1 and CF_0-CF_1 types are present in mitochondria and chloroplasts, respectively. Also called ATPase.

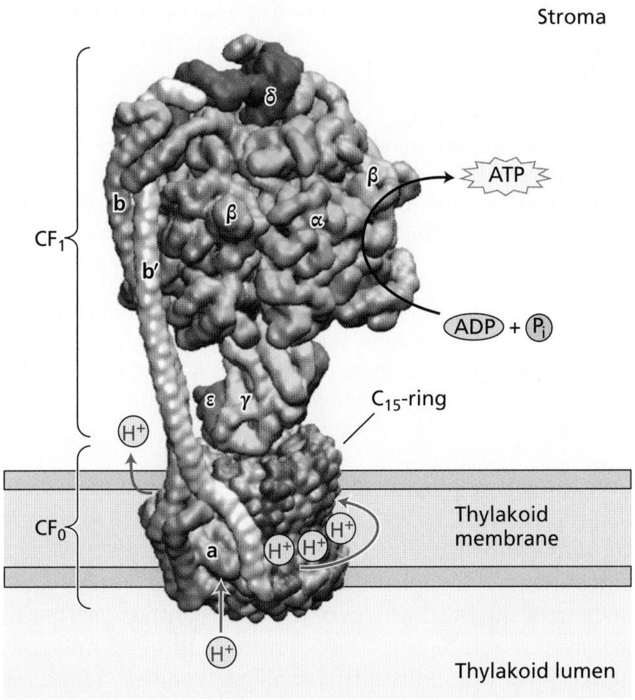

Figure 7.29 Structure of chloroplast F_0F_1-ATP synthase. This enzyme consists of a large multi-subunit complex, CF_1, attached on the stromal side of the membrane to an integral membrane portion, known as CF_0. CF_1 consists of five different polypeptides, with a stoichiometry of $\alpha_3 \beta_3 \gamma \delta \epsilon$. CF_0 contains four different polypeptides, with a stoichiometry of a b b′ c_{14}. Protons from the lumen are transported by the rotating c polypeptide and ejected on the stroma side. The structure closely resembles that of the mitochondrial F_0F_1-ATP synthase (see Chapter 11) and the vacuolar V-ATPase (see Chapter 6). (Figure courtesy of W. Frasch.)

(A) Chloroplasts

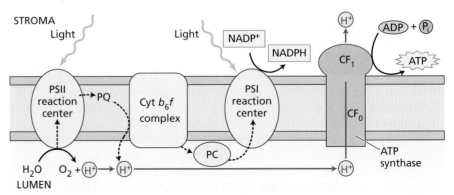

(B) Mitochondria

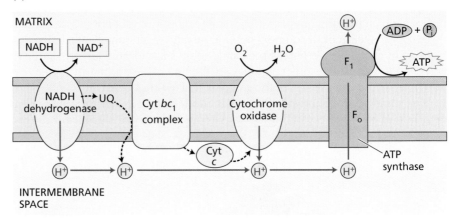

Figure 7.30   Similarities of photosynthetic and respiratory electron flow in chloroplasts and mitochondria. In both, electron flow is coupled to proton translocation, creating a transmembrane proton motive force (Δp). The energy in the proton motive force is then used for the synthesis of ATP by ATP synthase. (A) Chloroplasts carry out noncyclic electron flow, oxidizing water and reducing $NADP^+$. Protons are produced by the oxidation of water and by the oxidation of PQH_2 (labeled "PQ" in the illustration) by the cytochrome b_6f complex. (B) Mitochondria oxidize NADH to NAD^+ and reduce oxygen to water. Protons are pumped by the enzyme NADH dehydrogenase, the cytochrome bc_1 complex, and cytochrome oxidase. The ATP synthases in the two systems are very similar in structure. UQ, ubiquinone.

ences between the chloroplast and mitochondrial enzymes, they have the same overall architecture and probably nearly identical catalytic sites. In fact, there are remarkable similarities in the way electron flow is coupled to proton translocation in chloroplasts and mitochondria (**Figure 7.30**). Another remarkable aspect of the mechanism of the ATP synthase is that the internal stalk and probably much of the CF_0 portion of the enzyme rotate during catalysis. The enzyme is actually a tiny molecular motor. Three molecules of ATP are synthesized for each rotation of the enzyme.

Direct microscopic imaging of the CF_0 part of the chloroplast ATP synthase indicates that it contains 14 copies of the integral membrane subunit. Each subunit can translocate one proton across the membrane each time the complex rotates. This suggests that the stoichiometry of protons translocated to ATP formed is 14/3, or 4.67. Measured values of this parameter are usually somewhat lower than this value, and the reasons for this discrepancy are not yet understood.

Summary

Photosynthesis in plants uses light energy for the synthesis of carbohydrates and generation of oxygen from carbon dioxide and water. Energy stored in carbohydrates is used to power cellular processes in the plant and serves as energy resources for all forms of life.

Photosynthesis in Higher Plants

- Within chloroplasts, chlorophylls absorb light energy for the oxidation of water, releasing oxygen and generating NADPH and ATP (thylakoid reactions).
- NADPH and ATP are used to reduce carbon dioxide to form sugars (carbon fixation reactions).

General Concepts

- Light behaves as both a particle and a wave, delivering energy as photons, some of which are absorbed and used by plants (**Figures 7.1–7.3**).
- Light-energized chlorophyll may fluoresce, transfer energy to another molecule, or use its energy to drive chemical reactions (**Figures 7.5, 7.10**).
- All photosynthetic organisms contain a mixture of pigments with distinct structures and light-absorbing properties (**Figures 7.6, 7.7**).

Key Experiments in Understanding Photosynthesis

- An action spectrum for photosynthesis shows algal oxygen evolution at certain wavelengths (**Figures 7.8, 7.9**).
- Antenna pigment–protein complexes collect light energy and transfer it to the reaction center complexes (**Figure 7.10**).
- Light drives the reduction of NADP$^+$ and the formation of ATP. Oxygen-evolving organisms have two photosystems (PSI and PSII) that operate in series (**Figures 7.12, 7.13**).

Organization of the Photosynthetic Apparatus

- Within the chloroplast, thylakoid membranes contain the reaction centers, the light-harvesting antenna complexes, and most of the electron carrier proteins. PSI and PSII are spatially separated in thylakoids (**Figure 7.16**).

Organization of Light-Absorbing Antenna Systems

- The antenna system funnels energy to the reaction center (**Figure 7.17**).

- Light-harvesting proteins of both photosystems are structurally similar (**Figure 7.18**).

Mechanisms of Electron Transport

- The Z scheme identifies the flow of electrons through carriers in PSII and PSI from H_2O to NADP$^+$ (**Figures 7.13, 7.19**).
- Four large protein complexes transfer electrons: PSII, the cytochrome b_6f complex, PSI, and ATP synthase (**Figures 7.16, 7.20**).
- PSI reaction center chlorophyll has maximum absorption at 700 nm; PSII reaction center chlorophyll absorbs maximally at 680 nm.
- The PSII reaction center is a multi-subunit protein–pigment complex (**Figure 7.22**).
- Manganese ions are required to oxidize water.
- Two hydrophobic plastoquinones accept electrons from PSII (**Figures 7.20, 7.23**).
- Protons are transported into the thylakoid lumen when electrons pass through the cytochrome b_6f complex (**Figures 7.20, 7.24**).
- Plastoquinone and plastocyanin carry electrons between PSII and PSI (**Figure 7.25**).
- NADP$^+$ is reduced by the PSI reaction center, using three Fe–S centers and ferredoxin as electron carriers (**Figures 7.20, 7.26**).
- Cyclic electron flow generates ATP, but no NADPH, by proton pumping.
- Herbicides may block photosynthetic electron flow (**Figure 7.27**).

Proton Transport and ATP Synthesis in the Chloroplast

- In vitro transfer of pH 4–equilibrated chloroplast thylakoids to a pH 8 buffer resulted in the formation of ATP from ADP and P$_i$, supporting the chemiosmotic hypothesis (**Figure 7.28**).
- Protons move down an electrochemical gradient (proton motive force), passing through an ATP synthase and forming ATP (**Figures 7.20, 7.29**).
- During catalysis, the CF$_0$ portion of the ATP synthase rotates like a miniature motor.
- Proton translocation in chloroplasts and mitochondria shows significant similarities (**Figure 7.30**).

Suggested Reading

Blankenship, R. E. (2014) *Molecular Mechanisms of Photosynthesis*, 2nd ed. Wiley-Blackwell, Oxford, UK.

Hohmann-Marriott, M. F., and Blankenship, R. E. (2011) Evolution of photosynthesis. *Annu. Rev. Plant Biol.* 62: 515–548.

Ke, B. (2001) *Photosynthesis Photobiochemistry and Photobiophysics* (*Advances in Photosynthesis*, vol. 10). Kluwer, Dordrecht, Netherlands.

Nicholls, D. G., and Ferguson, S. J. (2013) *Bioenergetics*, 4th ed. Academic Press, Amsterdam.

Ort, D. R., and Yocum, C. F., eds. (1996) *Oxygenic Photosynthesis: The Light Reactions* (*Advances in Photosynthesis*, vol. 4). Kluwer, Dordrecht, Netherlands.

Zhu, X.-G., Long, S. P., and Ort, D. R. (2010) Improving photosynthetic efficiency for greater yield. *Ann. Rev. Plant Biol.* 61: 235–261.

8 Photosynthesis
The Carbon Reactions

Chapter 4 examined the requirements of plants for mineral nutrients and light in order to grow and complete their life cycle. Because the amount of matter in our planet remains constant, the transformation and circulation of molecules through the biosphere demand a continuous flux of energy. Otherwise, entropy would increase and the flow of matter would ultimately stop. The ultimate source of energy for sustaining life in the biosphere is the solar radiant energy that strikes Earth's surface. Photosynthetic organisms capture approximately 3×10^{21} Joules per year of sunlight energy and use it for the fixation of approximately 2×10^{11} tonnes of carbon per year.

More than 1 billion years ago, heterotrophic cells dependent on abiotically produced organic molecules acquired the ability to convert sunlight into chemical energy through primary endosymbiosis with an ancient cyanobacterium. This endosymbiotic event entailed the gain of new metabolic pathways: The ancestral endosymbiont conveyed the ability not only to carry out oxygenic photosynthesis, but also to synthesize novel compounds, such as *starch*.

In Chapter 7 we saw how the energy associated with the photochemical oxidation of water to molecular oxygen at thylakoid membranes generates ATP, reduced ferredoxin, and NADPH. Subsequently, the products of the light reactions, ATP and NADPH, flow from thylakoid membranes to the surrounding fluid phase (stroma) and drive the enzyme-catalyzed reduction of atmospheric CO_2 to carbohydrates and other cell components

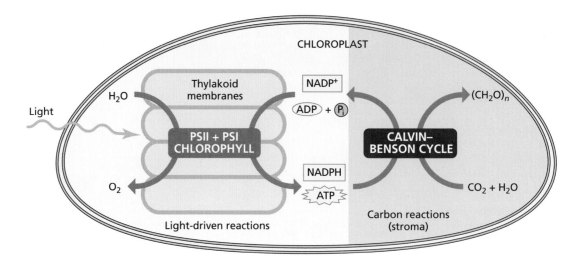

Figure 8.1 Light and carbon reactions of photosynthesis in chloroplasts of land plants. In thylakoid membranes, the excitation of chlorophyll in the photosynthetic electron transport chain (photosystem II [PSII)] + photosystem I [PSI]) by light elicits the formation of ATP and NADPH (see Chapter 7). In the stroma, both ATP and NADPH are consumed by the Calvin–Benson cycle in a series of enzyme-driven reactions that reduce the atmospheric CO_2 to carbohydrates (triose phosphates).

Calvin–Benson cycle The biochemical pathway for the reduction of CO_2 to carbohydrate. The cycle involves three phases: the carboxylation of ribulose 1,5-bisphosphate with atmospheric CO_2, catalyzed by Rubisco; the reduction of the formed 3-phosphoglycerate to triose phosphates; and the regeneration of ribulose 1,5-bisphosphate through the concerted action of ten enzymatic reactions.

starch A polyglucan consisting of long chains of 1,4-linked glucose molecules and branch points where 1,6-linkages are used. Starch is the carbohydrate storage form in most plants.

sucrose A disaccharide consisting of a glucose and a fructose molecule linked via an ether bond between C-1 on the glucosyl subunit and C-2 on the fructosyl unit. Sucrose is the carbohydrate transport form (e.g., in the phloem between source and sink).

(**Figure 8.1**). The latter reactions in the stroma of chloroplasts were long thought to be independent of light and, as a consequence, were for many years referred to as the dark reactions. However, these stroma-localized reactions are more properly referred to as the carbon reactions of photosynthesis because products of the photochemical processes not only provide substrates for enzymes, but also control their catalytic rates.

We begin this chapter by analyzing the metabolic cycle that incorporates atmospheric CO_2 into organic compounds appropriate for life: the **Calvin–Benson cycle**. Next we consider how the unavoidable phenomenon of photorespiration—a consequence of a side reaction with molecular oxygen—releases part of the assimilated CO_2. Because photorespiration decreases the efficiency of photosynthetic CO_2 assimilation, we also examine biochemical mechanisms for mitigating the loss of CO_2: CO_2 pumps, C_4 metabolism, and crassulacean acid metabolism. Finally, we briefly consider the formation of the two major products of the photosynthetic CO_2 fixation: **starch**, the reserve polysaccharide that accumulates transiently in chloroplasts; and **sucrose**, the disaccharide that is exported from leaves to developing and storage organs of the plant.

The Calvin–Benson Cycle

A requisite for maintaining life in the biosphere is the fixation of atmospheric CO_2 into skeletons of organic compounds that are compatible with the needs of the cell. These endergonic (energy-consuming) transformations are driven by the energy coming from physical and chemical sources. The predominant pathway of autotrophic CO_2 fixation is the Calvin–Benson cycle, which is found in many prokaryotes and in all photosynthetic eukaryotes, from the most primitive algae to the most advanced angiosperms. This pathway decreases the oxidation state of carbon from the highest value, found in CO_2 (+4), to levels found in sugars (e.g., +2 in keto groups [—CO—]; 0 in secondary alcohols [—CHOH—]). In view of its notable capacity for lowering the carbon oxidation state, the Calvin–Benson

Figure 8.2 The Calvin–Benson cycle proceeds in three phases: (1) *carboxylation*, which covalently links atmospheric carbon (CO_2) to a carbon skeleton; (2) *reduction*, which forms a carbohydrate (triose phosphate) at the expense of photochemically generated ATP and reducing equivalents in the form of NADPH; and (3) *regeneration*, which restores the CO_2 acceptor ribulose 1,5-bisphosphate. At steady state, the input of CO_2 equals the output of triose phosphates. The latter either serve as precursors of starch biosynthesis in the chloroplast or flow to the cytosol for sucrose biosynthesis and other metabolic reactions. Sucrose is loaded into the phloem sap (see Chapter 10) and used for growth or polysaccharide biosynthesis in other parts of the plant.

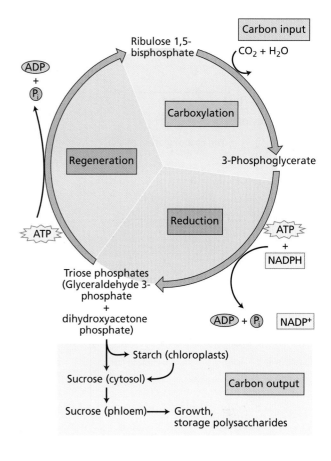

cycle is also aptly named the *reductive pentose phosphate cycle* and *photosynthetic carbon reduction cycle*.

The Calvin–Benson cycle has three phases: carboxylation, reduction, and regeneration

In the 1950s, a series of ingenious experiments carried out by M. Calvin, A. Benson, J. A. Bassham, and their colleagues provided convincing evidence for the Calvin–Benson cycle. The cycle proceeds in three phases that are highly coordinated in the chloroplast (**Figure 8.2**):

1. *Carboxylation* of the CO_2 acceptor molecule. The first committed enzymatic step in the cycle is the reaction of CO_2 and water with a five-carbon acceptor molecule (ribulose 1,5-bisphosphate) to generate two molecules of a three-carbon intermediate (3-phosphoglycerate).

2. *Reduction* of 3-phosphoglycerate. The 3-phosphoglycerate is converted to three-carbon carbohydrates (triose phosphates) by enzymatic reactions driven by photochemically generated ATP and NADPH.

3. *Regeneration* of the CO_2 acceptor ribulose 1,5-bisphosphate. The cycle is completed by regeneration of ribulose 1,5-bisphosphate through a series of ten enzyme-catalyzed reactions, one requiring ATP.

The carbon output as triose phosphates balances the carbon input provided by atmospheric CO_2. Triose phosphates generated by the Calvin–Benson cycle are converted to starch in the chloroplast or exported to the cytosol for the formation of sucrose. Sucrose is transported in the phloem to heterotrophic plant organs for sustaining growth and the synthesis of storage products.

The fixation of CO_2 via carboxylation of ribulose 1,5-bisphosphate and the reduction of the product 3-phosphoglycerate yield triose phosphates

In the carboxylation step of the Calvin–Benson cycle, one molecule of CO_2 and one molecule of H_2O react with one molecule of ribulose 1,5-bisphosphate to yield two molecules of 3-phosphoglycerate (**Figure 8.3** and **Table 8.1**, reaction 1). This reaction is catalyzed by the chloroplast enzyme ribulose 1,5-bisphosphate carboxylase/oxygenase, referred to as **Rubisco**. In the first partial reaction, an H^+ is extracted from carbon 3 of ribulose 1,5-bisphosphate (**Figure 8.4**). The addition of CO_2 to the unstable Rubisco-bound enediol intermediate drives the second partial reaction to the irreversible formation of 2-carboxy-3-ketoarabinitol 1,5-bisphosphate (see Figure 8.4, upper branch of pathway). The

Rubisco The acronym for the chloroplast enzyme *ribulose 1,5-bis*phosphate carboxylase/oxygenase. In a carboxylase reaction, Rubisco uses atmospheric CO_2 and ribulose 1,5-bisphosphate to form two molecules of 3-phosphoglycerate. It also functions as an oxygenase that adds O_2 to ribulose 1,5-bisphosphate to yield 3-phosphoglycerate and 2-phosphoglycolate. The competition between CO_2 and O_2 for ribulose 1,5-bisphosphate limits net CO_2 fixation.

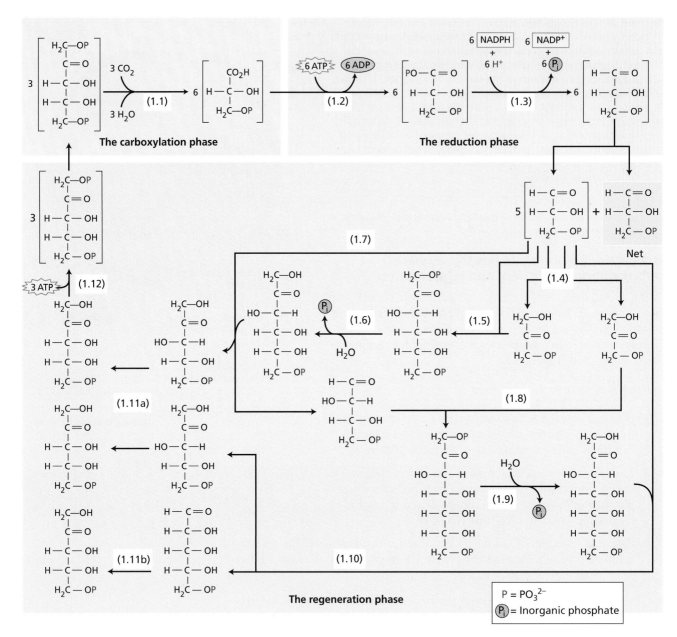

Figure 8.3 Calvin–Benson cycle. The carboxylation of three molecules of ribulose 1,5-bisphosphate yields six molecules of 3-phosphoglycerate (carboxylation phase). After phosphorylation of the carboxylic group, 1,3-bisphosphoglycerate is reduced to six molecules of glyceraldehyde 3-phosphate with the concurrent release of six molecules of inorganic phosphate (reduction phase). From the total six molecules of glyceraldehyde 3-phosphate, one (shaded) represents the net assimilation of the three molecules of CO_2 while the other five undergo a series of reactions that finally regenerate the starting three molecules of ribulose 1,5-bisphosphate (regeneration phase). See Table 8.1 for a description of each numbered reaction.

hydration of 2-carboxy-3-ketoarabinitol 1,5-bisphosphate yields two molecules of 3-phosphoglycerate.

In the reduction phase of the Calvin–Benson cycle, two successive reactions reduce the carbon of the 3-phosphoglycerate produced by the carboxylation phase (see Figure 8.3 and Table 8.1, reactions 2 and 3):

1. First, ATP formed by the light reactions phosphorylates 3-phosphoglycerate at the carboxyl group, yielding a mixed anhydride, 1,3-bisphosphoglycerate, in a reaction catalyzed by 3-phosphoglycerate kinase.

Table 8.1   Reactions of the Calvin–Benson cycle

Enzyme	Reaction
1. Ribulose 1,5-bisphosphate carboxylase/oxygenase (Rubisco)	Ribulose 1,5-bisphosphate + CO_2 + H_2O → 2 3-phosphoglycerate
2. 3-Phosphoglycerate kinase	3-Phosphoglycerate + ATP → 1,3-bisphosphoglycerate + ADP
3. NADP–glyceraldehyde-3-phosphate dehydrogenase	1,3-Bisphosphoglycerate + NADPH + H^+ → glyceraldehyde 3-phosphate + $NADP^+$ + P_i
4. Triose phosphate isomerase	Glyceraldehyde 3-phosphate → dihydroxyacetone phosphate
5. Aldolase	Glyceraldehyde 3-phosphate + dihydroxyacetone phosphate → fructose 1,6-bisphosphate
6. Fructose 1,6-bisphosphatase	Fructose 1,6-bisphosphate + H_2O → fructose 6-phosphate + P_i
7. Transketolase	Fructose 6-phosphate + glyceraldehyde 3-phosphate → erythrose 4-phosphate + xylulose 5-phosphate
8. Aldolase	Erythrose 4-phosphate + dihydroxyacetone phosphate → sedoheptulose 1,7-bisphosphate
9. Sedoheptulose 1,7-bisphosphatase	Sedoheptulose 1,7-bisphosphate + H_2O → sedoheptulose 7-phosphate + P_i
10. Transketolase	Sedoheptulose 7-phosphate + glyceraldehyde 3-phosphate → ribose 5-phosphate + xylulose 5-phosphate
11a. Ribulose 5-phosphate epimerase	Xylulose 5-phosphate → ribulose 5-phosphate
11b. Ribose 5-phosphate isomerase	Ribose 5-phosphate → ribulose 5-phosphate
12. Phosphoribulokinase (ribulose 5-phosphate kinase)	Ribulose 5-phosphate + ATP → ribulose 1,5-bisphosphate + ADP + H^+

Note: P_i stands for inorganic phosphate.

2. Next, NADPH, also generated by the light reactions, reduces 1,3-bisphosphoglycerate to glyceraldehyde 3-phosphate, in a reaction catalyzed by the chloroplast enzyme NADP–glyceraldehyde-3-phosphate dehydrogenase.

The operation of three carboxylation and reduction phases yields six molecules of glyceraldehyde 3-phosphate (6 molecules × 3 carbons/molecule = 18 carbons total) when three molecules of ribulose 1,5-bisphosphate (3 molecules × 5 carbons/molecule = 15 carbons total) react with three molecules of CO_2 (3 carbons total) and the six molecules of 3-phosphoglycerate are reduced (see Figure 8.3).

The regeneration of ribulose 1,5-bisphosphate ensures the continuous assimilation of CO_2

In the regeneration phase, the Calvin–Benson cycle facilitates the continuous uptake of atmospheric CO_2 by restoring the CO_2 acceptor ribulose 1,5-bisphosphate. To this end, three molecules of ribulose 1,5-bisphosphate (3 molecules × 5 carbons/molecule = 15 carbons total) are formed by reactions that reshuffle the carbons from five molecules of glyceraldehyde 3-phosphate (5 molecules × 3 carbons/molecule = 15 carbons) (see Figure 8.3). The sixth molecule of glyceraldehyde 3-phosphate (1 molecule × 3 carbons/molecule = 3 carbons total) represents the net assimilation of three molecules of CO_2 and becomes available for the carbon metabolism of the plant. The reshuffling of the other five molecules of glyceraldehyde 3-phosphate to yield three molecules of ribulose 1,5-bisphosphate proceeds through reactions 4 to 12 in Table 8.1 and Figure 8.3:

- Two molecules of glyceraldehyde 3-phosphate are converted to dihydroxyacetone phosphate in the reaction catalyzed by triose phosphate isomerase (see Table 8.1, reaction 4). Glyceraldehyde 3-phosphate and dihydroxyacetone phosphate are collectively designated *triose phosphates*.

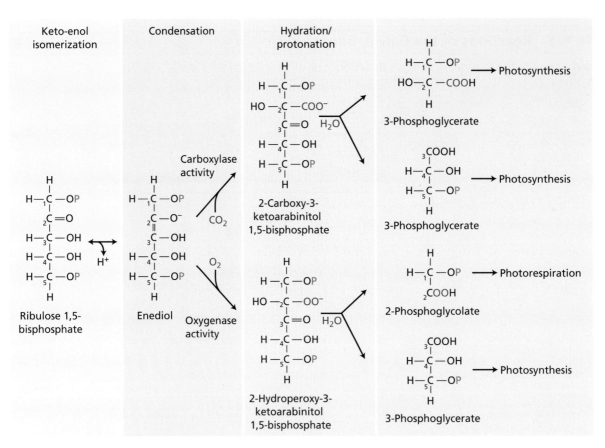

Figure 8.4 Carboxylation and oxygenation of ribulose 1,5-bis-phosphate catalyzed by Rubisco. The binding of ribulose 1,5-bis-phosphate to Rubisco facilitates the formation of an enzyme-bound enediol intermediate that can be attacked by CO_2 or O_2 at carbon 2. With CO_2 (upper branch of pathway), the product is a six-carbon intermediate (2-carboxy-3-ketoarabinitol 1,5-bis-phosphate); with O_2 (lower branch of pathway), the product is a five-carbon reactive intermediate (2-hydroperoxy-3-ketoarabinitol 1,5-bisphosphate). The hydration of these intermediates at carbon 3 triggers the cleavage of the carbon–carbon bond between carbons 2 and 3, yielding two molecules of 3-phosphoglycerate (carboxylase activity) or one molecule of 2-phosphoglycolate and one molecule of 3-phosphoglyerate (oxygenase activity). The important physiological effect of the oxygenase activity is described in the section *Photorespiration: The C_2 Oxidative Photosynthetic Carbon Cycle.*

- One molecule of dihydroxyacetone phosphate undergoes aldol condensation with a third molecule of glyceraldehyde 3-phosphate, a reaction catalyzed by aldolase, to give fructose 1,6-bisphosphate (see Table 8.1, reaction 5).

- Fructose 1,6-bisphosphate is hydrolyzed to fructose 6-phosphate in a reaction catalyzed by a specific chloroplast fructose 1,6-bisphosphatase (see Table 8.1, reaction 6).

- A two-carbon unit of the fructose 6-phosphate molecule (carbons 1 and 2) is transferred via the enzyme transketolase to a fourth molecule of glyceraldehyde 3-phosphate to form xylulose 5-phosphate. The other four carbons of the fructose 6-phosphate molecule (carbons 3, 4, 5, and 6) form erythrose 4-phosphate (see Table 8.1, reaction 7).

- The erythrose 4-phosphate then combines, via aldolase, with the remaining molecule of dihydroxyacetone phosphate to yield the seven-carbon sugar sedoheptulose 1,7-bisphosphate (see Table 8.1, reaction 8).

- Sedoheptulose 1,7-bisphosphate is then hydrolyzed to sedoheptulose 7-phosphate by a specific chloroplast sedoheptulose 1,7-bisphosphatase (see Table 8.1, reaction 9).

- Sedoheptulose 7-phosphate donates a two-carbon unit (carbons 1 and 2) to the fifth (and last) molecule of glyceraldehyde 3-phosphate, via transketolase, producing xylulose 5-phosphate. The remaining five carbons (carbons 3–7) of the sedoheptulose 7-phosphate molecule become ribose 5-phosphate (see Table 8.1, reaction 10).

- Two molecules of xylulose 5-phosphate are converted to two molecules of ribulose 5-phosphate by a ribulose 5-phosphate epimerase (see Table 8.1, reaction 11a), while a third molecule of ribulose 5-phosphate originates from ribose 5-phosphate by the action of ribose 5-phosphate isomerase (see Table 8.1, reaction 11b).

- Finally, phosphoribulokinase (also called ribulose 5-phosphate kinase) catalyzes the phosphorylation of three molecules of ribulose 5-phosphate with ATP, thus regenerating the three molecules of ribulose 1,5-bisphosphate needed for restarting the cycle (see Table 8.1, reaction 12).

In summary, triose phosphates are formed in the carboxylation and reduction phases of the Calvin–Benson cycle using energy (ATP) and reducing equivalents (NADPH) generated by the illuminated photosystems of chloroplast thylakoid membranes:

$$3\ CO_2 + 3\ \text{ribulose 1,5-bisphosphate} + 3\ H_2O + 6\ NADPH + 6\ H^+ + 6\ ATP$$
$$\downarrow$$
$$6\ \text{Triose phosphates} + 6\ NADP^+ + 6\ ADP + 6\ P_i$$

From these six triose phosphates, five are used in the regeneration phase that restores the CO_2 acceptor (ribulose 1,5-bisphosphate) for the continuous functioning of the Calvin–Benson cycle:

$$5\ \text{Triose phosphates} + 3\ ATP + 2\ H_2O$$
$$\downarrow$$
$$3\ \text{Ribulose 1,5-bisphosphate} + 3\ ADP + 2\ P_i$$

The sixth triose phosphate represents the net synthesis of an organic compound from CO_2 that is used as a building block for stored carbon or for other metabolic processes. Hence, the fixation of three CO_2 into one triose phosphate uses 6 NADPH and 9 ATP:

$$3\ CO_2 + 5\ H_2O + 6\ NADPH + 9\ ATP$$
$$\downarrow$$
$$\text{Glyceraldehyde 3-phosphate} + 6\ NADP^+ + 9\ ADP + 8\ P_i$$

The Calvin–Benson cycle uses two molecules of NADPH and three molecules of ATP to assimilate a single molecule of CO_2.

An induction period precedes the steady state of photosynthetic CO_2 assimilation

In the dark, both the activity of photosynthetic enzymes and the concentration of intermediates of the Calvin–Benson cycle are low. Therefore, enzymes of the Calvin–Benson cycle and most of the triose phosphates are committed to restore adequate concentrations of metabolic intermediates when leaves receive light. The rate of CO_2 fixation increases with time in the first few minutes after the onset of illumination—a time lag called the **induction period**. The photosynthesis rate accelerates both because several of the Calvin–Benson cycle enzymes are activated by light (discussed later in this chapter) and because the concentrations of intermediates of the cycle increase. In short, the six triose phosphates formed in the carboxylation and reduction phases of the Calvin–Benson cycle during the induction period are used mainly for the regeneration of the CO_2 acceptor ribulose 1,5-bisphosphate.

induction period   The period of time (time lag) between the perception of a signal and the activation of the response. In the Calvin–Benson cycle, the time elapsed between the onset of illumination and the full activation of the cycle.

When photosynthesis reaches a steady state, five of the six triose phosphates formed contribute to the regeneration of the CO_2 acceptor ribulose 1,5-bisphosphate, while a sixth triose phosphate is used in the chloroplast for starch formation and in the cytosol for sucrose synthesis and other metabolic processes (see Figure 8.2).

Many mechanisms regulate the Calvin–Benson cycle

The efficient use of energy in the Calvin–Benson cycle requires the existence of specific regulatory mechanisms which ensure not only that all intermediates in the cycle are present at adequate concentrations in the light, but also that the cycle is turned off in the dark. Chloroplasts adjust the rates of Calvin–Benson cycle reactions through the modification of enzyme levels (moles of enzyme/chloroplast) and catalytic activities (moles of substrate converted/[minute × mole of enzyme]).

Gene expression and protein biosynthesis determine the concentrations of enzymes in cell compartments. Because the small subunit of Rubisco is encoded in the nuclear genome and the large subunit in the plastid genome, Rubisco biosynthesis requires coordinated expression of two sets of genes. Nucleus-encoded proteins are translated on 80S ribosomes in the cytosol and subsequently transported into the plastid. Plastid-encoded proteins are translated in the stroma on prokaryotic-like 70S ribosomes.

Light modulates the expression of stromal enzymes encoded by the nuclear genome via specific photoreceptors (e.g., phytochrome and blue-light receptors) (see Chapter 13). However, nuclear gene expression needs to be synchronized with the expression of other components of the photosynthetic apparatus in the organelle. Most of the regulatory signaling between nucleus and plastids is anterograde—that is, the products of nuclear genes control the transcription and translation of plastid genes. Such is the case, for example, in the assembly of stromal Rubisco from eight nucleus-encoded small subunits (S) and eight plastid-encoded large subunits (L). However, in some cases (e.g., the synthesis of proteins associated with chlorophyll), regulation can be retrograde—that is, the regulatory signal flows from the plastid to the nucleus.

In contrast to slow changes in catalytic rates caused by variations in the concentration of enzymes, posttranslational modifications rapidly change the specific activity of chloroplast enzymes (μmoles of substrate converted/[minute × μmole of enzyme]). Two general mechanisms accomplish the light-mediated modification of the kinetic properties of stromal enzymes:

1. Change in covalent bonds that results in a chemically modified enzyme, such as the reduction of disulfide bonds on the enzyme (Enz) with electrons donated by another protein (Prot):

$$Enz-(S)_2 + Prot-(SH)_2 \leftrightarrow Enz-(SH)_2 + Prot-(S)_2$$

2. Modification of noncovalent interactions caused by changes in (a) ionic composition of the cellular milieu (e.g., pH, Mg^{2+}), (b) binding of enzyme effectors, (c) close association with regulatory proteins in supramolecular complexes, or (d) interaction with thylakoid membranes. Only ionic interactions are described in the following sections.

In our further discussion of regulation, we will examine light-dependent mechanisms that regulate the specific activity of five pivotal enzymes within minutes of the light–dark transition:

1. Rubisco

2. Fructose 1,6-bisphosphatase

3. Sedoheptulose 1,7-bisphosphatase

4. Phosphoribulokinase

5. NADP–glyceraldehyde-3-phosphate dehydrogenase

Rubisco activase regulates the catalytic activity of Rubisco

Because Rubisco has such a central role in carbon fixation, its activity is very tightly regulated. Rubisco that has been inactivated (e.g., at night) is reactivated when conditions become favorable for photosynthesis. First, the enzyme Rubisco activase removes bound sugar phosphates from Rubisco, and then Rubisco is activated by binding of a CO_2 molecule (activator CO_2). After this activation, the enzyme can start its catalytic cycle, whereby CO_2 reacts with ribulose 1,5-bisphosphate. The activity of Rubisco activase is, in turn, regulated by the ferredoxin–thioredoxin system.

Light regulates the Calvin–Benson cycle via the ferredoxin–thioredoxin system

Light regulates the catalytic activity of four enzymes of the Calvin–Benson cycle directly via the **ferredoxin–thioredoxin system**. This mechanism uses ferredoxin reduced by the photosynthetic electron transport chain together with two chloroplast proteins (ferredoxin–thioredoxin reductase and thioredoxin) to regulate fructose 1,6-bisphosphatase, sedoheptulose 1,7-bisphosphatase, phosphoribulokinase, and NADP–glyceraldehyde-3-phosphate dehydrogenase (**Figure 8.5**).

Light transfers electrons from water to ferredoxin via the photosynthetic electron transport chain (see Chapter 7). Reduced ferredoxin converts the disulfide bond of the regulatory protein thioredoxin (—S—S—) to the reduced state (—SH HS—) with the iron–sulfur enzyme ferredoxin–thioredoxin reductase.

ferredoxin–thioredoxin system
Three chloroplast proteins (ferredoxin, ferredoxin–thioredoxin reductase, thioredoxin). The three proteins use reducing power from the photosynthetic electron transport chain to reduce protein disulfide bonds in a cascade of thiol–disulfide exchanges. As a result, light controls the activity of several enzymes of the Calvin–Benson cycle.

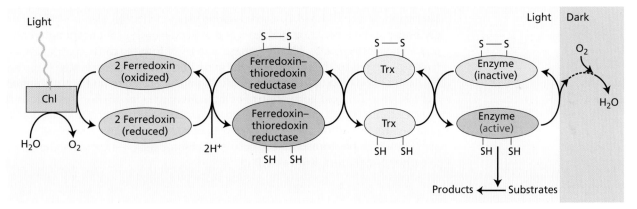

Figure 8.5 Ferredoxin–thioredoxin system. The ferredoxin–thioredoxin system links the light signal sensed by thylakoid membranes to the activity of enzymes in the chloroplast stroma. The activation of enzymes of the Calvin–Benson cycle starts in the light with the reduction of ferredoxin by the photosynthetic electron transport chain (Chl) (see Chapter 7). Reduced ferredoxin, together with two protons, is used to reduce a catalytically active disulfide bond (—S—S—) of the iron–sulfur enzyme ferredoxin–thioredoxin reductase, which in turn reduces the unique disulfide bond (—S—S—) of the regulatory protein thioredoxin (Trx). The reduced form of thioredoxin (—SH HS—) then reduces regulatory disulfide bonds of the target enzyme, triggering its conversion to the catalytically active state that catalyzes the transformation of substrates into products. Darkness halts the electron flow from ferredoxin to the enzyme, and thioredoxin becomes oxidized. Although the mechanism for the deactivation of thioredoxin-activated enzymes in the dark is not fully clear, it appears that O_2-actuated oxidations cause the formation of oxidized thioredoxin. Next, the unique disulfide bond (—S—S—) of thioredoxin brings the reduced form (—SH HS—) of the enzyme back to the oxidized form (—S—S—) with the concurrent loss of the catalytic capacity. In contrast to thioredoxin-activated enzymes, a chloroplast enzyme of the oxidative pentose phosphate cycle, glucose 6-phosphate dehydrogenase, does not operate in the light but is functional in the dark, because thioredoxin reduces the disulfide critical for the activity of the enzyme. The ability of thioredoxin to regulate functional enzymes in different pathways minimizes futile cycling.

photorespiration   Uptake of atmospheric O_2 with a concomitant release of CO_2 by illuminated leaves. Molecular oxygen serves as substrate for Rubisco, producing 2-phosphoglycolate that enters the photorespiratory cycle (the oxidative photosynthetic carbon cycle). The activity of the cycle recovers some of the carbon found in 2-phosphoglycolate, but some is lost to the atmosphere.

Subsequently, reduced thioredoxin cleaves a specific disulfide bridge (oxidized cysteines) of the target enzyme, forming free (reduced) cysteines. The cleavage of the enzyme disulfide bonds causes a conformational change that increases catalytic activity (see Figure 8.5). Deactivation of thioredoxin-activated enzymes takes place when darkness relieves the "electron pressure" from photosynthetic electron transport. However, details of the deactivation process are unknown.

Proteomic studies have shown that the ferredoxin–thioredoxin system regulates enzymes functional in numerous chloroplast processes other than carbon fixation. Thioredoxin also protects proteins against damage caused by reactive oxygen species, such as hydrogen peroxide (H_2O_2), the superoxide anion ($O_2^{\bullet}$), and the hydroxyl radical ($HO^{\bullet}$).

Light-dependent ion movements modulate enzymes of the Calvin–Benson cycle

Upon illumination, the flow of protons from the stroma into the thylakoid lumen (see Figure 7.20) is coupled to the release of Mg^{2+} from the intrathylakoid space to the stroma. These light-driven ion fluxes decrease the stromal concentration of protons (the pH increases from 7 to 8) and increase that of Mg^{2+} by 2 to 5 mM. The light-mediated increase of pH and the concentration of Mg^{2+} activate enzymes of the Calvin–Benson cycle that require Mg^{2+} for catalysis and are more active at pH 8 than at pH 7: Rubisco, fructose 1,6-bisphosphatase, sedoheptulose 1,7-bisphosphatase, and phosphoribulokinase. The changes in ionic composition of the chloroplast stroma are reversed rapidly upon darkening, decreasing the activity of these four enzymes.

Photorespiration: The C_2 Oxidative Photosynthetic Carbon Cycle

Rubisco catalyzes both the carboxylation and the oxygenation of ribulose 1,5-bisphosphate (see Figure 8.4). Carboxylation yields two molecules of 3-phosphoglycerate, and oxygenation produces one molecule each of 3-phosphoglycerate and 2-phosphoglycolate. The oxygenase activity of Rubisco causes partial loss of the carbon fixed by the Calvin–Benson cycle and yields 2-phosphoglycolate, an inhibitor of two chloroplast enzymes: triose phosphate isomerase and phosphofructokinase. To avoid both drain of carbon out of the Calvin–Benson cycle and enzyme inhibition, 2-phosphoglycolate is metabolized through the C_2 oxidative photosynthetic carbon cycle. This network of coordinated enzymatic reactions, also known as **photorespiration**, occurs in chloroplasts, leaf peroxisomes, and mitochondria (**Figure 8.6, Table 8.2**).

Figure 8.6   Operation of the C_2 oxidative photosynthetic carbon cycle. Enzymatic reactions are distributed in three organelles: chloroplasts, peroxisomes, and mitochondria. *In chloroplasts*, the oxygenase activity of Rubisco yields two molecules of 2-phosphoglycolate, which, upon the action of phosphoglycolate phosphatase, form two molecules of glycolate and two molecules of inorganic phosphate. Two molecules of glycolate (four carbons) flow concurrently with one molecule of glutamate from chloroplasts to peroxisomes. *In peroxisomes*, the glycolate is oxidized by O_2 to glyoxylate in a reaction catalyzed by the enzyme glycolate oxidase. The glutamate:glyoxylate aminotransferase catalyzes the conversion of glyoxylate and glutamate to glycine and 2-oxoglutarate. The amino acid glycine diffuses from peroxisomes to mitochondria. *In mitochondria*, two molecules of glycine (four carbons) yield a molecule of serine (three carbons) with the concurrent release of CO_2 (one carbon) and NH_4^+ by the successive action of the glycine decarboxylase complex and serine hydroxymethyltransferase. The amino acid serine is then transported back to the peroxisome and transformed to glycerate (three carbons) by the successive action of serine:2-oxoglutarate aminotransferase and hydroxypyruvate reductase. Glycerate and 2-oxoglutarate from peroxisomes, and NH_4^+ from mitochondria, return to chloroplasts in a process that recovers part of the carbon (three carbons) and all the nitrogen lost in photorespiration. Glycerate is phosphorylated to 3-phosphoglycerate and incorporated back into the Calvin–Benson cycle. The NH_4^+ is converted back into glutamate in the chloroplast stroma by the successive action of glutamine synthetase and ferredoxin-dependent glutamate synthase (GOGAT). See Table 8.2 for a description of each numbered reaction. Reactions in black are part of the carbon cycle, those in red are part of the nitrogen cycle, and those in blue are part of the oxygen cycle.

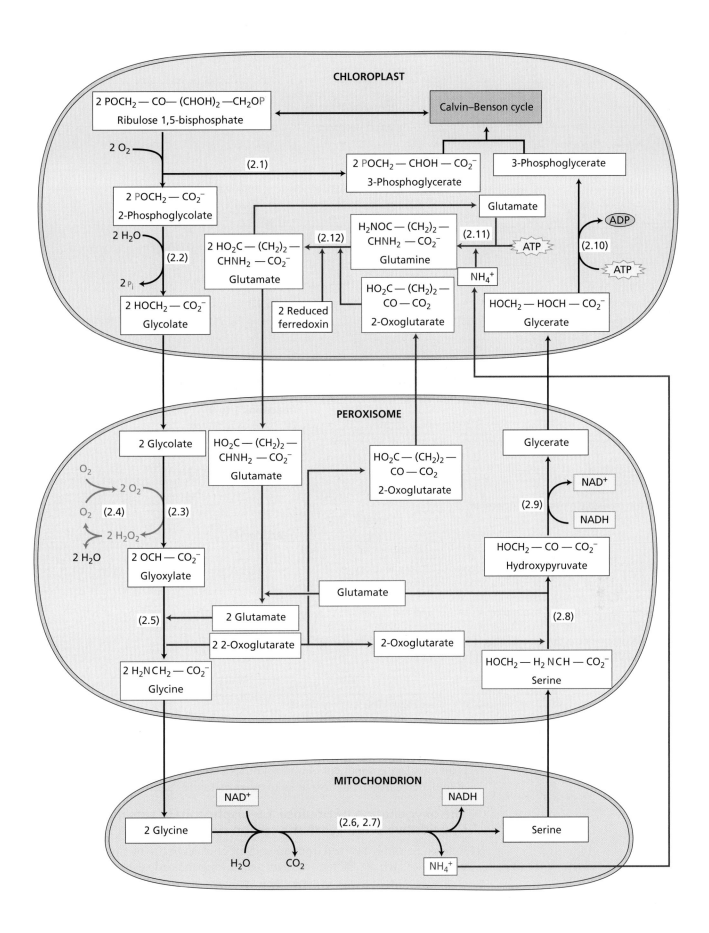

Table 8.2   Reactions of the C$_2$ oxidative photosynthetic carbon cycle

Enzyme[a]	Reaction
1. Rubisco	2 Ribulose 1,5-bisphosphate + 2 O$_2$ → 2 2-phosphoglycolate + 2 3-phosphoglycerate
2. Phosphoglycolate phosphatase	2 2-Phosphoglycolate + 2 H$_2$O → 2 glycolate + 2 P$_i$
3. Glycolate oxidase	2 Glycolate + 2 O$_2$ → 2 glyoxylate + 2 H$_2$O$_2$
4. Catalase	2 H$_2$O$_2$ → 2 H$_2$O + O$_2$
5. Glutamate:glyoxylate aminotransferase	2 Glyoxylate + 2 glutamate → 2 glycine + 2 2-oxoglutarate
6. Glycine decarboxylase complex (GDC)	Glycine + NAD$^+$ + [GDC] → CO$_2$ + NH$_4^+$ + NADH + [GDC-THF-CH$_2$]
7. Serine hydroxymethyltransferase	[GDC-THF-CH$_2$] + glycine + H$_2$O → serine + [GDC]
8. Serine:2-oxoglutarate aminotransferase	Serine + 2-oxoglutarate → hydroxypyruvate + glutamate
9. Hydroxypyruvate reductase	Hydroxypyruvate + NADH + H$^+$ → glycerate + NAD$^+$
10. Glycerate kinase	Glycerate + ATP → 3-phosphoglycerate + ADP
11. Glutamine synthetase	Glutamate + NH$_4^+$ + ATP → glutamine + ADP + P$_i$
12. Ferredoxin-dependent glutamate synthase (GOGAT)	2-Oxoglutarate + glutamine + 2 Fd$_{red}$ + 2 H$^+$ → 2 glutamate + 2 Fd$_{oxid}$

Net reaction of the C$_2$ oxidative photosynthetic carbon cycle

2 Ribulose 1,5-bisphosphate + 3 O$_2$ + H$_2$O + glutamate

↓   (**reactions 1 to 9**)

Glycerate + 2 3-phosphoglycerate + NH$_4^+$ + CO$_2$ + 2 P$_i$ + 2-oxoglutarate

Two reactions in the chloroplasts restore the molecule of glutamate:

2-Oxoglutarate + NH$_4^+$ + [(2 Fd$_{red}$ + 2 H$^+$), ATP]

↓   (**reactions 11 and 12**)

Glutamate + H$_2$O + [(2 Fd$_{oxid}$), ADP + P$_i$]

and the molecule of 3-phosphoglycerate:

Glycerate + ATP

↓   (**reaction 10**)

3-Phosphoglycerate + ADP

Hence, the consumption of three molecules of atmospheric oxygen in the C$_2$ oxidative photosynthetic carbon cycle (two in the oxygenase activity of Rubisco and one in peroxisomal oxidations) elicits
- the release of one molecule of CO$_2$ and
- the consumption of two molecules of ATP and two molecules of reducing equivalents (2 Fd$_{red}$ + 2 H$^+$)

for
- incorporating a three-carbon skeleton back into the Calvin–Benson Cycle, and
- restoring glutamate from NH$_4^+$ and 2-oxoglutarate.

---

[a]Locations: Chloroplasts; peroxisomes; mitochondria. Fd: ferredoxin; THF, tetrahydrofolate.

Recent studies have shown that the C$_2$ oxidative photosynthetic carbon cycle is an ancillary component of photosynthesis that not only salvages part of the assimilated carbon, but also links to other metabolic pathways of land plants.

The oxygenation of ribulose 1,5-bisphosphate sets in motion the C$_2$ oxidative photosynthetic carbon cycle

Rubisco probably evolved from an ancient enolase in the methionine-salvage pathway of archaea. Billions of years ago, during the early evolution of Rubisco and before the advent of oygenic photosynthesis in cyanobacteria, the enzyme's ability to oxygenate ribulose 1,5-bisphosphate was insignificant because the lack of O_2 and the high concentrations of CO_2 in the ancient atmosphere prevented the oxygenation reaction. Atmospheric O_2 concentrations are much higher now, and CO_2 concentrations much lower despite recent increases, boosting the oxygenase activity of Rubisco and making the formation of the toxic 2-phosphoglycolate inevitable. All Rubiscos catalyze the incorporation of O_2 into ribulose 1,5-bisphosphate. Even homologs from anaerobic autotrophic bacteria exhibit the oxygenase activity, demonstrating that the oxygenase reaction is intrinsically linked to the active site of Rubisco and not an adaptive response to the appearance of O_2 in the biosphere.

The oxygenation of the 2,3-enediol isomer of ribulose 1,5-bisphosphate with one molecule of O_2 yields an unstable intermediate that rapidly splits into one molecule each of 3-phosphoglycerate and 2-phosphoglycolate (see Figures 8.4 and 8.6, and Table 8.2, reaction 1). In chloroplasts of land plants, 2-phosphoglycolate phosphatase catalyzes the rapid hydrolysis of 2-phosphoglycolate to glycolate (see Figure 8.6 and Table 8.2, reaction 2). The subsequent transformations of glycolate take place in the peroxisomes and mitochondria (see Chapter 1). Glycolate leaves the chloroplasts through a specific transporter in the envelope inner membrane and diffuses to the peroxisomes (see Figure 8.6). In peroxisomes, glycolate oxidase catalyzes the oxidation of glycolate by O_2, producing H_2O_2 and glyoxylate (see Table 8.2, reaction 3). The peroxisomal catalase breaks down the H_2O_2, releasing O_2 and H_2O (see Figure 8.6 and Table 8.2, reaction 4). The glutamate:glyoxylate aminotransferase catalyzes the transamination of glyoxylate with glutamate, yielding the amino acid glycine (see Figure 8.6 and Table 8.2, reaction 5).

Glycine exits the peroxisomes and enters the mitochondria, where a multienzyme complex of glycine decarboxylase (GDC) and serine hydroxymethyltransferase catalyzes the conversion of two molecules of glycine and one molecule of NAD^+ into one molecule each of serine, NADH, NH_4^+, and CO_2 (see Figure 8.6 and Table 8.2, reactions 6 and 7). First, GDC uses one molecule of NAD^+ for the oxidative decarboxylation of one molecule of glycine, yielding one molecule each of NADH, NH_4^+, and CO_2 and the activated one-carbon unit methylene tetrahydrofolate (THF) bound to GDC (GDC-THF-CH_2):

$$Glycine + NAD^+ + GDC\text{-}THF \rightarrow NADH + NH_4^+ + CO_2 + GDC\text{-}THF\text{-}CH_2$$

Next, serine hydroxymethyltransferase catalyzes the addition of the methylene unit to a second molecule of glycine, forming serine and regenerating THF to ensure high levels of glycine decarboxylase activity:

$$Glycine + GDC\text{-}THF\text{-}CH_2 \rightarrow Serine + GDC\text{-}THF$$

The oxidation of carbon atoms (two molecules of glycine [oxidation states C1: +3; C2: –1] → serine [oxidation states C1: +3; C2: 0; C3: –1] and CO_2 [oxidation state C: +4]) drives the reduction of oxidized pyridine nucleotide:

$$NAD^+ + H^+ + 2\,e^- \rightarrow NADH$$

The reaction products of the enzyme glycine decarboxylase are metabolized at different locations in leaf cells. NADH is oxidized to NAD^+ in the mitochondria. NH_4^+ and CO_2 are exported to chloroplasts, where they are assimilated to form glutamate (see below) and 3-phosphoglycerate, respectively.

The newly formed serine diffuses from the mitochondria back to the peroxisomes for donation of its amino group to 2-oxoglutarate via transamination catalyzed by serine:2-oxoglutarate aminotransferase, forming glutamate and hydroxypyruvate (see Figure 8.6 and Table 8.2, reaction 8). Next, an NADH-dependent reductase catalyzes the transformation of hydroxypyruvate into glycerate (see Figure 8.6 and Table 8.2, reaction 9). Finally, glycerate reenters the chloroplast, where it is phosphorylated by ATP to yield 3-phosphoglycerate and ADP (see Figure 8.6 and Table 8.2, reaction 10). Hence, the formation of 2-phosphoglycolate (via Rubisco) and the phosphorylation of glycerate (via glycerate kinase) metabolically link the Calvin–Benson cycle to the C_2 oxidative photosynthetic carbon cycle.

The NH_4^+ released in the oxidation of glycine diffuses rapidly from the matrix of the mitochondria to the chloroplasts (see Figure 8.6). In the chloroplast stroma, glutamine synthetase catalyzes the ATP-dependent incorporation of NH_4^+ into glutamate, yielding glutamine, ADP, and inorganic phosphate (see Figure 8.6 and Table 8.2, reaction 11). Subsequently, glutamine and 2-oxoglutarate are substrates of ferredoxin-dependent glutamate synthase (GOGAT) for the production of two molecules of glutamate (see Table 8.2, reaction 12). The reassimilation of NH_4^+ into the photorespiratory cycle restores glutamate for the action of the peroxisomal glutamate:glyoxylate aminotransferase in the conversion of glyoxylate to glycine (see Table 8.2, reaction 5).

Carbon, nitrogen, and oxygen atoms circulate through photorespiration (**Figure 8.7**).

- In the *carbon* cycle, the chloroplasts transfer two molecules of glycolate (four carbon atoms) to peroxisomes and recover one molecule of glycerate (three carbon atoms). The mitochondria release one molecule of CO_2 (one carbon atom).

- In the *nitrogen* cycle, the chloroplasts transfer one molecule of glutamate (one nitrogen atom) and recover one molecule of NH_4^+ (one nitrogen atom).

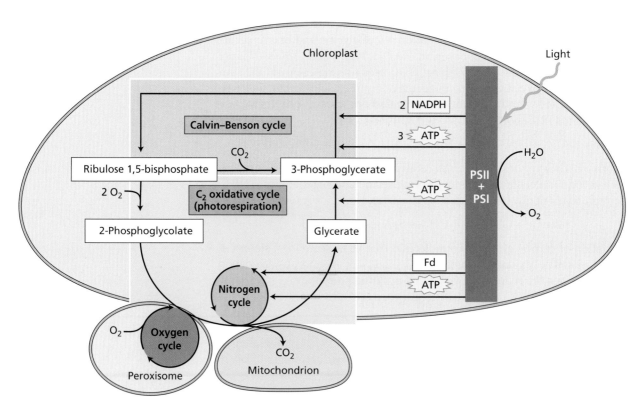

Figure 8.7 Dependence of the C_2 oxidative photosynthetic carbon cycle on chloroplast metabolism. The supply of ATP and reducing equivalents from light reactions in thylakoid membranes is needed for the functioning of the C_2 oxidative photosynthetic cycle in three compartments: chloroplasts, peroxisomes, and mitochondria. The *carbon cycle* uses (1) NADPH and ATP to maintain the adequate level of ribulose 1,5-bisphosphate in the Calvin–Benson cycle and (2) ATP to convert glycerate to 3-phosphoglycerate in the C_2 oxidative photosynthetic carbon cycle. The *nitrogen cycle* employs ATP and reducing equivalents to restore glutamate from NH_4^+ and 2-oxoglutarate coming from the photorespiratory cycle. In the peroxisome, the *oxygen cycle* contributes to the removal of H_2O_2 formed in the oxidation of glycolate by O_2.

- In the *oxygen* cycle, Rubisco and glycolate oxidase catalyze the incorporation of two molecules of O_2 each (eight oxygen atoms) when two molecules of ribulose 1,5-bisphosphate enter the C_2 oxidative photosynthetic carbon cycle (see Table 8.2, reactions 1 and 3). However, catalase releases one molecule of O_2 from two molecules of H_2O_2 (two oxygen atoms) (see Table 8.2, reaction 4). Hence, three molecules of O_2 (six oxygen atoms) are reduced in the photorespiratory cycle.

In vivo, three factors regulate the distribution of metabolites between the Calvin–Benson cycle and the C_2 oxidative photosynthetic carbon cycle: one inherent to the plant (the kinetic properties of Rubisco) and two linked to the environment (the ratio of the atmospheric concentrations of CO_2 and O_2, and temperature).

The specificity factor (Ω) estimates the preference of Rubisco for CO_2 versus O_2:

$$\Omega = (V_C/K_C)/(V_O/K_O)$$

where V_C and V_O are the maximum velocities for carboxylation and oxygenation, respectively, and K_C and K_O are the Michaelis–Menten constants for CO_2 and O_2, respectively. Ω settles the ratio of the velocity of carboxylation (v_C) to that of oxygenation (v_O) at environmental concentrations of CO_2 and O_2:

$$\Omega = (v_C/v_O) \times ([O_2]/[CO_2])$$

The specificity factor estimates the relative capacity of Rubisco for carboxylation and oxygenation (v_C/v_O) when the concentration of CO_2 around the active site is equal to that of O_2 ($[O_2]/[CO_2] = 1$). Ω is a constant for every Rubisco that denotes the relative efficiency with which O_2 competes with CO_2 at a given temperature. Rubiscos from different organisms exhibit variations in the value of Ω: the Ω of Rubisco from cyanobacteria (Ω ~40) is lower than that from C_3 plants (Ω ~82–90) and from C_4 species (Ω ~70–82).

The ambient temperature exerts an important influence over Ω and the concentrations of CO_2 and O_2 around the active site of Rubisco. Warmer environments have the effect of:

- Increasing the oxygenase activity of Rubisco more than the carboxylase activity. The greater increase of K_C for CO_2 than of K_O for O_2 diminishes the Ω of Rubisco.

- Lowering the solubility of CO_2 to a greater extent than O_2. The increase of $[O_2]/[CO_2]$ decreases the v_C/v_O ratio; that is, the oxygenase activity of Rubisco prevails over the carboxylase activity.

- Reducing the stomatal aperture to conserve water. Stomatal closure lowers the uptake of atmospheric CO_2, thereby decreasing CO_2 at the active site of Rubisco.

Overall, warmer environments significantly limit the efficiency of photosynthetic carbon assimilation because a progressive increase in temperature tilts the balance away from photosynthesis (carboxylation) and toward photorespiration (oxygenation) (see Chapter 9).

Photorespiration is linked to the photosynthetic electron transport chain

Photosynthetic carbon metabolism in intact leaves reflects the competition for ribulose 1,5-bisphosphate between two mutually opposing cycles, the Calvin–Benson cycle and the C_2 oxidative photosynthetic carbon cycle. These cycles are interlocked with the photosynthetic electron transport chain for the supply of ATP and reducing equivalents (reduced ferredoxin and NADPH) (see Figure

8.7). For salvaging two molecules of 2-phosphoglycolate by conversion to one molecule of 3-phosphoglycerate, photophosphorylation provides one molecule of ATP (necessary for the transformation of glycerate to 3-phosphoglycerate; see Table 8.2, reaction 10), while the consumption of NADH by hydroxypyruvate reductase (see Table 8.2, reaction 9) is counterbalanced by its production by glycine decarboxylase (see Table 8.2, reaction 6).

Nitrogen enters the photorespiratory cycle in the peroxisome through the transamination step catalyzed by the glutamate:glyoxylate aminotransferase (two nitrogen atoms) (see Table 8.2, reaction 5). Nitrogen leaves the photorespiratory cycle in two steps: (1) in the mitochondria as NH_4^+ (one nitrogen atom), in the reaction catalyzed by the glycine decarboxylase–serine hydroxymethyltransferase complex (see Table 8.2, reactions 6 and 7) and (2) in the peroxisomes in the transamination step catalyzed by serine:2-oxoglutarate aminotransferase (one nitrogen atom) (see Table 8.2, reaction 8).

The photosynthetic electron transport chain supplies one molecule of ATP and two molecules of reduced ferredoxin needed for salvaging one molecule of NH_4^+ through its incorporation into glutamate via glutamine synthetase (see Table 8.2, reaction 11) and ferredoxin-dependent glutamate synthase (GOGAT) (see Table 8.2, reaction 12).

In summary,

$$2 \text{ Ribulose 1,5-bisphosphate} + 3\,O_2 + H_2O + ATP + [2\text{ ferredoxin}_{red} + 2\,H^+ + ATP]$$
$$\downarrow$$
$$3\text{ 3-Phosphoglycerate} + CO_2 + 2\,P_i + ADP + [2\text{ ferredoxin}_{oxid} + ADP + P_i]$$

Because of the additional provision of ATP and reducing power for the operation of the photorespiratory cycle, the quantum requirement for CO_2 fixation under photorespiratory conditions (high $[O_2]$ and low $[CO_2]$) is higher than under nonphotorespiratory conditions (low $[O_2]$ and high $[CO_2]$).

Inorganic Carbon–Concentrating Mechanisms

Except for some photosynthetic bacteria, photoautotrophic organisms in the biosphere use the Calvin–Benson cycle to assimilate atmospheric CO_2. The pronounced reduction in CO_2 concentration and rise in O_2 concentration that commenced about 350 million years ago triggered a series of adaptations to handle an environment that promoted photorespiration in photosynthetic organisms. These adaptations include various strategies for active uptake of CO_2 and HCO_3^- from the surrounding environment and accumulation of inorganic carbon near Rubisco. The immediate consequence of higher concentrations of CO_2 around Rubisco is a decrease in the oxygenation reaction. CO_2 and HCO_3^- pumps at the plasma membrane have been studied extensively in prokaryotic cyanobacteria, eukaryotic algae, and aquatic plants.

In land plants, the diffusion of CO_2 from the atmosphere to the chloroplast plays a crucial role in net photosynthesis. To be incorporated into sugar compounds, inorganic carbon has to cross four barriers: the cell wall, plasma membrane, cytoplasm, and chloroplast envelope. Recent evidence has revealed that pore-forming membrane proteins (aquaporins) function as diffusion facilitators for various small molecules, decreasing the resistance of the mesophyll to the diffusion of CO_2.

Land plants evolved two carbon-concentrating mechanisms for increasing the concentration of CO_2 at the Rubisco carboxylation site:

1. C_4 photosynthetic carbon fixation (C_4)

2. Crassulacean acid metabolism (CAM)

The uptake of atmospheric CO_2 by these carbon-concentrating mechanisms precedes CO_2 assimilation through the Calvin–Benson cycle.

Inorganic Carbon–Concentrating Mechanisms: The C₄ Carbon Cycle

C₄ photosynthesis has evolved as one of the major carbon-concentrating mechanisms used by land plants to compensate for limitations associated with low concentrations of atmospheric CO_2. Some of the most productive crops on the planet (e.g., maize [corn; *Zea mays*], sugarcane, and sorghum) use this mechanism to enhance the catalytic capacity of Rubisco. In this section we examine:

- The biochemical and anatomical attributes of C₄ photosynthesis that minimize the oxygenase activity of Rubisco and the concurrent loss of carbon through the photorespiratory cycle

- The concerted action of different types of cells for the incorporation of inorganic carbon into carbon skeletons

- The light-mediated regulation of enzyme activities, and

- The importance of C₄ photosynthesis for sustaining plant growth in many tropical areas

Malate and aspartate are the primary carboxylation products of the C₄ cycle

In the late 1950s, H. P. Kortschack and Y. Karpilov observed that ^{14}C label appeared initially in the four-carbon acids malate and aspartate when $^{14}CO_2$ was provided to leaves of sugarcane and maize in the light. This finding was unexpected because a three-carbon acid, 3-phosphoglycerate, is the first labeled product in the Calvin–Benson cycle. M. D. Hatch and C. R. Slack explained that particular distribution of radioactive carbon by proposing an alternative mechanism to the Calvin–Benson cycle. This pathway is named the *C₄ photosynthetic carbon cycle* (also known as the Hatch–Slack cycle or the C₄ cycle).

Hatch and Slack found that (1) malate and aspartate are the first stable intermediates of photosynthesis and (2) that the carbon 4 of these four-carbon acids subsequently becomes carbon 1 of 3-phosphoglycerate. These transformations take place in two morphologically distinct cell types—the mesophyll and bundle sheath cells—that are separated by their respective walls and membranes.

In the C₄ cycle, the enzyme phospho*enol*pyruvate carboxylase (PEPCase), rather than Rubisco, catalyzes the initial carboxylation in mesophyll cells close to the external atmosphere (**Table 8.3**, reaction 1). Unlike Rubisco, O_2 does not compete with HCO_3^- in the carboxylation catalyzed by PEPCase. The four-carbon acids formed in mesophyll cells move to the bundle sheath cells, where they are decarboxylated, releasing CO_2 that is refixed by Rubisco via the Calvin–Benson cycle. Although all C₄ plants share primary carboxylation via PEPCase, the other enzymes used to concentrate CO_2 in the vicinity of Rubisco vary among different C₄ species. There are three types of C₄ cycles, named after their main decarboxylating enzyme(s): NADP–ME (using NADP–malic enzyme; illustrated in **Figure 8.8**), NAD–ME (using NAD–malic enzyme), and PEP carboxykinase (PEPCK; using both NAD–malic enzyme and phospho*enol*pyruvate carboxykinase). The reactions catalyzed by these enzymes can be found in Table 8.3.

The C₄ cycle assimilates CO₂ by the concerted action of two different types of cells

The key features of the C₄ cycle were initially described in leaves of plants such as maize whose vascular tissues are surrounded by two distinctive photosynthetic

C₄ photosynthesis Photosynthetic carbon metabolism in which the initial fixation of CO_2 is catalyzed by phospho*enol*pyruvate carboxylase (not by Rubisco as in C₃ photosynthesis), producing a four-carbon compound (oxaloacetate). The fixed carbon is subsequently released and refixed by the Calvin–Benson cycle.

cell types. In this anatomical context, the transport of CO_2 from the external atmosphere to the bundle sheath cells proceeds through five successive stages. Here we describe the reactions associated with each step in the NADP–ME C_4 cycle (see Figure 8.8 and Table 8.3):

1. *Fixation* of the HCO_3^- in phospho*enol*pyruvate by PEPCase in the mesophyll cell (see Table 8.3, reaction 1). The reaction product, oxaloacetate, is subsequently reduced to malate by NADP–malate dehydrogenase in the mesophyll chloroplasts (see Table 8.3, reaction 2).

2. *Transport* of the four-carbon acid (in this case malate) to the bundle sheath cells.

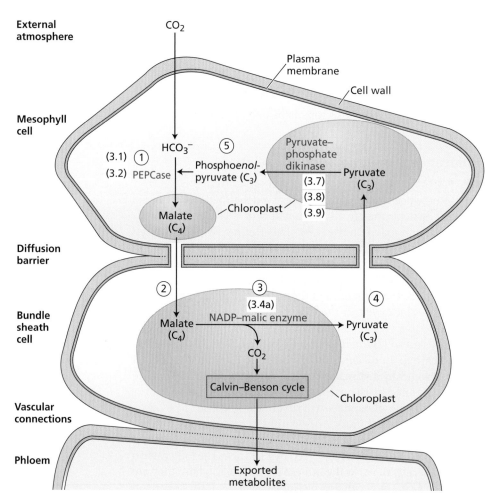

Figure 8.8   The C_4 photosynthetic carbon cycle involves five successive stages in two distinct cell types. ① In the mesophyll cells, the enzyme phospho*enol*pyruvate carboxylase (PEPCase) catalyzes the reaction of HCO_3^-, provided by the uptake of atmospheric CO_2, with phospho*enol*pyruvate, a three-carbon compound. The reaction product, oxaloacetate, a four-carbon compound, is further transformed to malate by the action of NADP–malate dehydrogenase (see Table 8.3, reaction 2). ② The four-carbon acid moves to the bundle sheath cell, close to vascular connections. ③ The decarboxylating enzyme (here, NADP–malic enzyme in the chloroplast; see Table 8.3, reaction 4a) releases the CO_2 from the four-carbon acid, yielding a three-carbon acid (e.g., pyruvate). The released CO_2 in the bundle sheath chloroplast builds up a large excess of CO_2 over O_2 around Rubisco, thereby facilitating the assimilation of CO_2 through the Calvin–Benson cycle. ④ The residual three-carbon acid (pyruvate) flows back to the mesophyll cell. ⑤ To complete the C_4 cycle, the enzyme pyruvate–phosphate dikinase catalyzes the regeneration of phospho*enol*pyruvate, the acceptor of the HCO_3^-, for another turn of the cycle. The consumption of two molecules of ATP per molecule of fixed CO_2 (see Table 8.3, reactions 7, 8, and 9) drives the C_4 cycle in the direction of the arrows, thus pumping CO_2 from the atmosphere to the Calvin–Benson cycle. The assimilated carbon leaves the chloroplast and, after transformation to sucrose in the cytoplasm, enters the phloem for translocation to other parts of the plant.

Table 8.3 Reactions of C_4 and CAM photosynthesis

Enzyme	Compartment	Reaction
1. PEPCase	Cytosol	Phospho*enol*pyruvate + HCO_3^- → oxaloacetate + P_i
2. NADP–malate dehydrogenase	Chloroplast	Oxaloacetate + NADPH + H^+ → malate + $NADP^+$
3. Aspartate aminotransferase	Cytosol/mitochondrion	Oxaloacetate + glutamate → aspartate + 2-oxoglutarate
Decarboxylating enzymes		
4a. NADP–malic enzyme	Chloroplast	Malate + $NADP^+$ → pyruvate + CO_2 + NADPH + H^+
4b. NAD–malic enzyme	Mitochondrion	Malate + NAD^+ → pyruvate + CO_2 + NADH + H^+
5. Phospho*enol*pyruvate carboxykinase	Cytosol	Oxaloacetate + ATP → phospho*enol*pyruvate + CO_2 + ADP
6. Alanine aminotransferase	Cytosol	Pyruvate + glutamate → alanine + 2-oxoglutarate
7. Pyruvate–phosphate dikinase	Chloroplast	Pyruvate + P_i + ATP → phospho*enol*pyruvate + AMP + PP_i
8. Adenylate kinase	Chloroplast	AMP + ATP → 2 ADP
9. Pyrophosphatase	Chloroplast	PP_i + H_2O → 2 P_i

Note: P_i and PP_i stand for inorganic phosphate and pyrophosphate, respectively.

3. *Decarboxylation* of the four-carbon acid in the bundle sheath cell (here by chloroplast NADP–malic enzyme; see Table 8.3, reaction 4a) and generation of CO_2, which is then reduced to carbohydrate via the Calvin–Benson cycle.

4. *Transport* of the three-carbon backbone (in this case pyruvate) formed by the decarboxylation step back to the mesophyll cell.

5. *Regeneration* of phospho*enol*pyruvate, the HCO_3^- acceptor. ATP and inorganic phosphate convert pyruvate to phospho*enol*pyruvate, releasing AMP and pyrophosphate (see Table 8.3, reaction 7). Two molecules of ATP are consumed in the conversion of pyruvate to phospho*enol*pyruvate: one in the reaction catalyzed by pyruvate–phosphate dikinase (see Table 8.3, reaction 7) and another in the transformation of AMP to ADP catalyzed by adenylate kinase (see Table 8.3, reaction 8).

In all three types of the C_4 cycle, the compartmentation of enzymes ensures that inorganic carbon from the surrounding atmosphere can be taken up initially by mesophyll cells, fixed subsequently by the Calvin–Benson cycle of bundle sheath cells, and finally exported to the phloem.

Bundle sheath cells and mesophyll cells exhibit anatomical and biochemical differences

Originally described for tropical grasses and *Atriplex*, the C_4 cycle is now known to occur in at least 62 independent lineages of angiosperms distributed across 19 different families. C_4 plants evolved from C_3 ancestors about 30 million years ago in response to multiple environmental stimuli such as atmospheric changes (decline of CO_2, increase of O_2), modification of global weather, periods of drought, and intense solar radiation. The transition from C_3 to C_4 plants requires the coordinated modification of genes that affect leaf anatomy, cell ultrastructure, metabolite transport, and regulation of metabolic enzymes. The analyses of (i) specific genes and elements that control their expression; (ii) mRNAs and the deduced amino acid sequences; and (iii) C_3 and C_4 genomes and transcriptomes indicate that convergent evolution underlies the multiple origins of C_4 plants.

Since the seminal studies of the 1950s and 1960s, the C_4 cycle has been associated with a particular leaf structure, called **Kranz anatomy** (*Kranz*, German

Kranz anatomy (German *kranz*, "wreath.") The wreathlike arrangement of mesophyll cells around a layer of bundle sheath cells. The two concentric layers of photosynthetic tissue surround the vascular bundle. This anatomical feature is typical of leaves of many C_4 plants.

for "wreath"). Typical Kranz anatomy exhibits an inner ring of bundle sheath cells around vascular tissues and an outer layer of mesophyll cells (**Figure 8.9A**). This particular leaf anatomy generates a diffusion barrier that (1) separates the uptake of atmospheric carbon in mesophyll cells from CO_2 assimilation by Rubisco in bundle sheath cells and (2) limits the leakage of CO_2 from bundle sheath to mesophyll cells (**Figure 8.9B**, left panel). Thus, diffusion gradients guide the shuttling of metabolites between the two cell types that operate the C_4 cycle. However, there are now clear examples of single-cell C_4 photosynthesis in several green algae, diatoms, and aquatic and land plants. We will turn later to the mechanisms used to set up CO_2 diffusion gradients within single cells carrying out C_4 photosynthesis.

Except in three terrestrial plants (see below), the distinctive Kranz anatomy increases the concentration of CO_2 in bundle sheath cells to almost tenfold higher than in the external atmosphere. The efficient accumulation of CO_2 in the vicinity of chloroplast Rubisco reduces the rate of photorespiration to 2 to 3% of photosynthesis. Mesophyll and bundle sheath cells show large biochemical differences. PEPCase and Rubisco are located in mesophyll and bundle sheath cells, respectively, while the decarboxylases are found in different intracellular compartments of bundle sheath cells: NADP–ME in chloroplasts, NAD–ME in mitochondria, and PEPCK in the cytosol. In addition, mesophyll cells contain randomly arranged chloroplasts with stacked thylakoid membranes, while chloroplasts in bundle sheath cells are concentrically arranged and exhibit unstacked thylakoids. These chloroplasts correlate with the energy requirements of the type of C_4 photosynthesis. For example, C_4 species of the NADP–ME type, in which malate is shuttled from mesophyll chloroplasts to bundle sheath cells (see Figure 8.8), exhibit functional photosystems II and I in mesophyll chloroplasts, whereas bundle sheath chloroplasts are deficient in photosystem II. Since the water-splitting enzyme that produces oxygen is associated with photosystem II (see Figure 7.20), there is no oxygen production in these bundle sheath chloroplasts, which also contributes to a greatly improved $[CO_2]:[O_2]$ ratio.

The C_4 cycle also concentrates CO_2 in single cells

The finding of C_4 photosynthesis in organisms devoid of Kranz anatomy disclosed a much greater diversity in modes of C_4 carbon fixation than had previously been thought to exist. Three plants that grow in Asia, *Suaeda aralocaspica* (formerly *Borszczowia aralocaspica*) and two *Bienertia* species, perform complete C_4 photosynthesis within single chlorenchyma cells (see Figure 8.9B, right panel, and **Figure 8.9C**). The external region, proximal to the external atmosphere, carries out the initial carboxylation and regeneration of phospho*enol*pyruvate, whereas the internal region functions in the decarboxylation of four-carbon acids and the refixation of the liberated CO_2 via Rubisco. The cytosol of these Chenopodiaceae species houses dimorphic chloroplasts with different subsets of enzymes.

Diatoms—photosynthetic eukaryotic algae found in marine and freshwater systems—also accomplish C_4 photosynthesis within a single cell. The importance of the C_4 pathway in carbon fixation was confirmed by using inhibitors specific for PEPCase and by identifying nucleotide sequences encoding enzymes essential for C_4 metabolism (PEPCase, PEPCK, and pyruvate–phosphate dikinase) in the genomes of two diatoms, *Thalassiosira pseudonana* and *Phaeodactylum tricornutum*. Although the discovery of these genes suggests that carbon is assimilated through the C_4 pathway, diatoms also possess bicarbonate transporters and carbonic anhydrases that may function to elevate the concentration of CO_2 at the active site of Rubisco. Biochemical analyses of C_4-essential enzymes and HCO_3^- transporters will be required to assess the functional importance of the different concentrating mechanisms in diatoms.

(A) Kranz anatomy

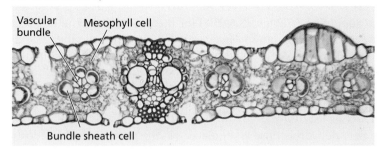

(B)

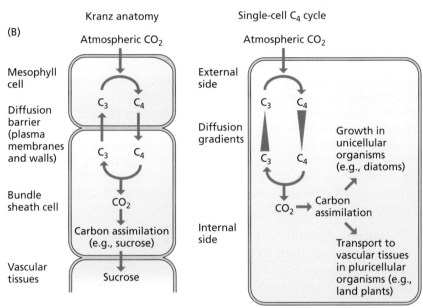

(C) Single-cell C$_4$ cycle

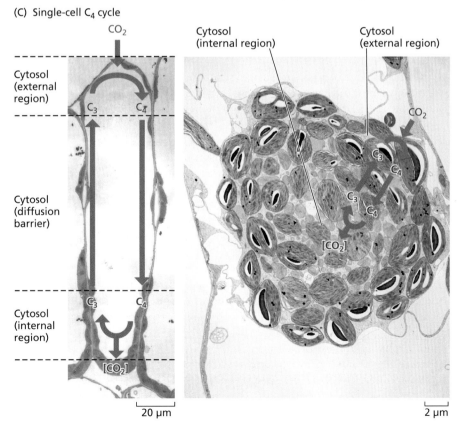

20 μm   2 μm

Figure 8.9 C$_4$ photosynthetic pathway in leaves of different plants. (A) Kranz anatomy. Light micrograph of transverse section of leaf blade of maize (NAD–ME type C$_4$ photosynthesis). This anatomical feature compartmentalizes photosynthetic reactions in two distinct types of cells that are arranged concentrically around the veins: mesophyll and bundle sheath cells. Bundle sheath cells surround the vascular tissue, while an outer ring of mesophyll cells is peripheral to the bundle sheath and adjacent to intercellular spaces. (B) In almost all known C$_4$ plants, photosynthetic CO$_2$ assimilation requires the development of Kranz anatomy (left panel). Membranes that separate cells assigned to CO$_2$ fixation from cells destined to reduce carbon form a diffusion barrier that is essential for the efficient function of C$_4$ photosynthesis in land plants. Some unicellular organisms (e.g., diatoms) and a few land plants (typified by *Suaeda aralocaspica* [formerly *Borszczowia aralocaspica*] and two *Bienertia* species) contain the equivalents of the C$_4$ compartmentation in a single cell (right panel). Studies of the key photosynthetic enzymes of these plants also indicate two dimorphic chloroplasts located in different cytoplasmic compartments having functions analogous to mesophyll and bundle sheath cells in Kranz anatomy. Products of CO$_2$ assimilation sustain growth in unicellular organisms and leave the cytosol for vascular tissues in multicellular organisms. (C) Single-cell C$_4$ photosynthesis. Diagrams of the C$_4$ cycle are superimposed on electron micrographs of *Suaeda aralocaspica* (left) and *Bienertia cycloptera* (right). (A © Dr. John Cunningham/Visuals Unlimited, Inc.; C from Edwards et al. 2004.)

Light regulates the activity of key C_4 enzymes

In addition to supplying ATP and NADPH for the operation of the C_4 cycle, light is essential for the regulation of several participating enzymes. Variations in photon flux density elicit changes in the activities of NADP–malate dehydrogenase, PEPCase, and pyruvate–phosphate dikinase by two different mechanisms: thiol–disulfide exchange [Enz–(Cys-S)$_2$ ↔ Enz–(Cys-SH)$_2$] and phosphorylation–dephosphorylation of specific amino acid residues (e.g., serine, Enz–Ser-OH ↔ Enz–Ser-OP).

NADP–malate dehydrogenase is regulated via the ferredoxin–thioredoxin system as in C_3 plants (see Figure 8.5). The enzyme is reduced (activated) by thioredoxin when leaves are illuminated, and is oxidized (deactivated) in the dark. The diurnal phosphorylation of PEPCase by a specific kinase, named PEPCase kinase, increases the uptake of ambient CO_2, and the nocturnal dephosphorylation by protein phosphatase 2A brings PEPCase back to low activity. A highly unusual enzyme regulates the dark–light activity of pyruvate–phosphate dikinase. Pyruvate–phosphate dikinase is modified by a bifunctional threonine kinase–phosphatase that catalyzes both ADP-dependent phosphorylation and P_i-dependent dephosphorylation of pyruvate–phosphate dikinase. Darkness promotes the phosphorylation of pyruvate–phosphate dikinase (PPDK) by the regulatory kinase–phosphatase [(PPDK)$_{active}$ + ADP → (PPDK-P)$_{inactive}$ + AMP], causing the loss of enzyme activity. The phosphorolytic cleavage of the phosphoryl group in the light by the same enzyme restores the catalytic capacity of PPDK [(PPDK-P)$_{inactive}$ + P_i → (PPDK)$_{active}$ + PP_i].

Photosynthetic assimilation of CO_2 in C_4 plants demands more transport processes than in C_3 plants

The chloroplasts export part of the fixed carbon to the cytosol during active photosynthesis while importing the phosphate released from biosynthetic processes to replenish ATP and other phosphorylated metabolites in the stroma. In C_3 plants, the major factors that modulate the partitioning of assimilated carbon between the chloroplast and cytosol are the relative concentrations of triose phosphates and inorganic phosphate. Triose phosphate isomerases rapidly interconvert dihydroxyacetone phosphate and glyceraldehyde 3-phosphate in the plastid and cytosol. The triose phosphate translocator—a protein complex in the inner membrane of the chloroplast envelope—exchanges chloroplast triose phosphates for cytosol phosphate. Thus, C_3 plants require one transport process across the chloroplast envelope to export triose phosphates (three molecules of CO_2 assimilated) from the chloroplasts to the cytosol.

In C_4 plants, the distribution of photosynthetic CO_2 assimilation over two different cells entails a massive flux of metabolites between mesophyll cells and bundle sheath cells. Moreover, three different pathways accomplish the assimilation of inorganic carbon in C_4 photosynthesis. In this context, different metabolites flow from the cytosol of leaf cells to chloroplasts, mitochondria, and conducting tissues. Therefore, the composition and the function of translocators in organelles and plasma membrane of C_4 plants depend on the pathway used for CO_2 assimilation. For example, mesophyll cells of NADP–ME type C_4 photosynthesis use four transport steps across the chloroplast envelope to fix one molecule of atmospheric CO_2: (1) import of cytosolic pyruvate (unknown transporter); (2) export of stromal phospho*enol*pyruvate (phospho*enol*pyruvate phosphate translocator); (3) import of cytosolic oxaloacetate (dicarboxylate transporter); and (4) export of stromal malate (dicarboxylate transporter).

In hot, dry climates, the C_4 cycle reduces photorespiration

As noted earlier in this chapter, elevated temperatures limit the rate of photosynthetic CO_2 assimilation in C_3 plants by decreasing the solubility of CO_2, and the ratio of the carboxylation to oxygenation reactions of Rubisco. Because of the decrease in the photosynthetic activity of Rubisco, the energy demands associated with photorespiration increase in warmer areas of the world. In C_4 plants, two features contribute to overcome the deleterious effects of high temperature:

1. First, atmospheric CO_2 enters the cytoplasm of mesophyll cells where carbonic anhydrase rapidly and reversibly converts CO_2 into bicarbonate ($CO_2 + H_2O \rightarrow HCO_3^- + H^+$) ($K_{eq} = 1.7 \times 10^{-4}$). Warm climates decrease the concentration of CO_2, but the low concentrations of cytosolic HCO_3^- saturate PEPCase because the affinity of the enzyme for its substrate is sufficiently high. Thus, the high activity of PEPCase enables C_4 plants to reduce their stomatal aperture at high temperatures and thereby conserve water while fixing CO_2 at rates equal to or greater than those of C_3 plants.

2. Second, the high concentration of CO_2 in bundle sheath chloroplasts minimizes the oxygenase activity of Rubisco and thus photorespiration.

The response of net CO_2 assimilation to temperature controls the distribution of C_3 and C_4 species on Earth. The optimal photosynthetic efficiency of C_3 species generally occurs at temperatures lower than for C_4 species: approximately 20–25°C and 25–35°C, respectively. Because C_4 photosynthesis enables more efficient assimilation of CO_2 at higher temperatures, C_4 species are more abundant in the tropics and subtropics and less abundant farther from the equator. Although C_4 photosynthesis is commonly dominant in warm environments, a group of perennial grasses (*Miscanthus*, *Spartina*) are chilling-tolerant C_4 crops that thrive in areas where the weather is moderately cold.

Inorganic Carbon–Concentrating Mechanisms: Crassulacean Acid Metabolism (CAM)

Another mechanism for concentrating CO_2 around Rubisco is present in many plants that inhabit arid environments with seasonal water availability, including commercially important plants such as pineapple (*Ananas comosus*), agave (*Agave* spp.), cacti (Cactaceae), and orchids (Orchidaceae). This important variant of photosynthetic carbon fixation was historically named **crassulacean acid metabolism (CAM)**, to recognize its initial observation in *Bryophyllum calycinum*, a succulent member of the Crassulaceae. Like the C_4 mechanism, CAM appears to have originated during the last 35 million years to conserve water in habitats where rainfall is insufficient for plant growth. The leaves of CAM plants have traits that minimize water loss, such as thick cuticles, large vacuoles, and stomata with small apertures.

Tight packing of the mesophyll cells enhances CAM performance by restricting CO_2 loss during the day. In all CAM plants, the initial capture of CO_2 into four-carbon acids takes place at night, and the subsequent incorporation of CO_2 into carbon skeletons occurs during the day (**Figure 8.10**). At night, cytosolic PEPCase fixes atmospheric and respiratory CO_2 into oxaloacetate using phospho*enol*pyruvate formed via the glycolytic breakdown of stored carbohydrates (see Table 8.3, reaction 1). A cytosolic NADP–malate dehydrogenase converts the oxaloacetate to malate, which is stored in the acidic solution of vacuoles for the remainder of the night (see Table 8.3, reaction 2). During the day, the stored malate exits the vacuole for decarboxylation by mechanisms similar to those in C_4 plants—that is, by a cytosolic NADP–ME or mitochondrial NAD–ME (see Table 8.3, reactions 4a and 4b). The latter is shown operating in Figure 8.10. The released CO_2 is made available to chloroplasts for fixation via Rubisco, while the coproduced three-carbon acid is converted to triose phosphate and subsequently to starch or sucrose via gluconeogenesis (see Figure 8.10).

crassulacean acid metabolism (CAM)
A biochemical process for concentrating CO_2 at the carboxylation site of Rubisco. Found in the family Crassulaceae (*Crassula*, *Kalanchoe*, *Sedum*) and numerous other families of angiosperms. In CAM, CO_2 uptake and initial fixation take place at night, and decarboxylation and reduction of the internally released CO_2 occur during the day.

Dark: Stomata opened

Light: Stomata closed

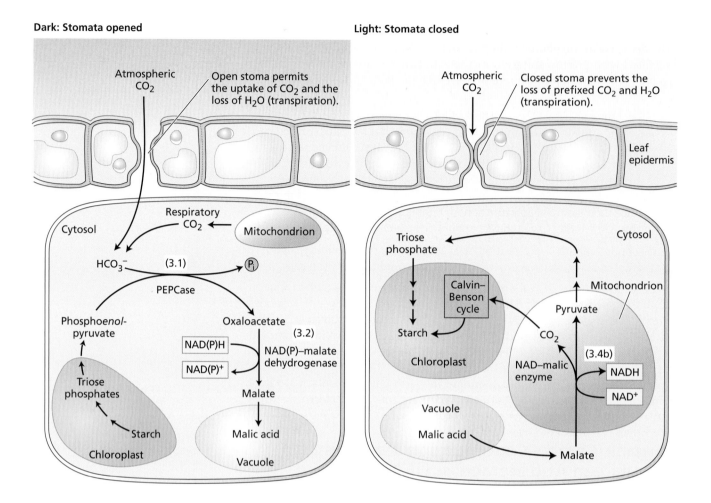

Figure 8.10 Crassulacean acid metabolism (CAM) in *Kalanchoe*. In CAM metabolism, CO_2 uptake is separated temporally from refixation via the Calvin–Benson cycle. The uptake of atmospheric CO_2 takes place at night when stomata are open. At this stage, gaseous CO_2 in the cytosol, coming from the external atmosphere and mitochondrial respiration, increases the levels of HCO_3^- ($CO_2 + H_2O \leftrightarrow HCO_3^- + H^+$). Then cytosolic PEPCase catalyzes a reaction between HCO_3^- and phospho*enol*pyruvate provided by the nocturnal breakdown of chloroplast starch. The four-carbon acid formed, oxaloacetate, is reduced to malate, which in turn diffuses into the acidic vacuole and becomes protonated to malic acid. During the day, the malic acid stored in the vacuole flows back to the cytosol and dissociates back to malate. Mitochondrial NAD–malic enzyme decarboxylates the malate, releasing CO_2 that is refixed into carbon skeletons by the Calvin–Benson cycle. In essence, the diurnal accumulation of starch in the chloroplast constitutes the net gain of the nocturnal uptake of inorganic carbon. The adaptive advantage of the stomatal closure during the day is that it prevents not only water loss by transpiration, but also the diffusion of internal CO_2 to the external atmosphere. See Table 8.3 for a description of numbered reactions.

Changes in the rate of carbon uptake and in enzyme regulation throughout the day create a 24-h CAM cycle. Four distinct phases encompass the temporal control of C_4 and C_3 carboxylations within the same cellular environment: phase I (night), phase II (early morning), phase III (daytime), and phase IV (late afternoon). During the nocturnal phase I, when stomata are open, CO_2 is captured and stored as malic acid in the vacuole. CO_2 uptake by PEPCase dominates phase I. In the diurnal phase III, when stomata are closed and leaves are photosynthesizing, the stored malic acid is decarboxylated. This results in high concentrations of CO_2 around the active site of Rubisco, thereby alleviating the adverse effects of photorespiration. The transient phases II and IV shift the metabolism in preparation for phases III and I, respectively. In phase II, Rubisco activity increases, whereas it decreases in phase IV. In contrast, the activity of PEPCase increases in phase IV but declines in phase II. The contribution of each phase to the overall carbon balance varies considerably among different CAM plants and is sensitive to environmental conditions. Constitutive CAM plants use the nocturnal uptake of CO_2 at all times, while their facultative counterparts resort to the CAM pathway only when induced by water or salt stress.

Whether the triose phosphates produced by the Calvin–Benson cycle are stored as starch in chloroplasts or used for the synthesis of sucrose depends on the plant species. However, these carbohydrates ultimately ensure not only plant growth, but also the supply of substrates for the next nocturnal carboxylation phase. To sum up, the temporal separation of nocturnal initial carboxylation from diurnal decarboxylation increases the concentration of CO_2 near Rubisco and reduces the rate of the oxygenase activity, thereby increasing the efficiency of photosynthesis.

Different mechanisms regulate C_4 PEPCase and CAM PEPCase

Comparative analysis of photosynthetic PEPCases provides a remarkable example of the adaptation of enzyme regulation to particular metabolisms. As we mentioned earlier in connection with C_4 photosynthesis, phosphorylation of plant PEPCases by PEPCase kinase converts the inactive nonphosphorylated form into the active phosphorylated counterpart:

$$PEPCase_{inactive} + ATP \xrightarrow{\text{PEPCase kinase}} PEPCase\text{--}P_{active} + ADP$$

Dephosphorylation of PEPCase by protein phosphatase 2A returns the enzyme to the inactive form. C_4 PEPCase is functional during the day and inactive at night, and CAM PEPCase operates at night and has reduced activity in the daytime. Thus, diurnal C_4 PEPCase and nocturnal CAM PEPCase are phosphorylated. The contrasting responses of photosynthetic PEPCases to light are conferred by regulatory elements that control the synthesis and degradation of PEPCase kinases. The synthesis of PEPCase kinase is mediated by light-sensing mechanisms in C_4 leaves and by endogenous circadian rhythms in CAM leaves.

CAM is a versatile mechanism sensitive to environmental stimuli

The high efficiency of water use in CAM plants likely accounts for their extensive diversification and speciation in water-limited environments. CAM plants that grow in deserts, such as cacti, open their stomata during the cool nights and close them during the hot, dry days. The potential advantage of terrestrial CAM plants in arid environments is well illustrated by the unintentional introduction of African prickly pear (*Opuntia stricta*) into the Australian ecosystem. From a few plants in 1840, the population of *O. stricta* progressively expanded to occupy 25 million ha in less than a century.

Closing the stomata during the day minimizes the loss of water in CAM plants, but because H_2O and CO_2 share the same diffusion pathway, CO_2 must then be taken up by the open stomata at night (see Figure 8.10). The availability of light mobilizes the reserves of vacuolar malic acid for the action of specific decarboxylating enzymes—NAD–ME, NADP–ME, and PEPCK—and the assimilation of the resulting CO_2 via the Calvin–Benson cycle. CO_2 released by decarboxylation does not escape from the leaf because stomata are closed during the day. As a consequence, the internally generated CO_2 is fixed by Rubisco and converted to carbohydrates by the Calvin–Benson cycle. Thus, stomatal closure not only helps conserve water, but also assists in the buildup of the elevated internal concentration of CO_2 that enhances the photosynthetic carboxylation of ribulose 1,5-bisphosphate.

Genotypic attributes and environmental factors modulate the extent to which the biochemical and physiological capacity of CAM plants is expressed. Although many species of succulent ornamental houseplants in the family Crassulaceae (e.g., *Kalanchoe*) are obligate CAM plants that exhibit circadian rhythmicity, others (e.g., *Clusia*) show C_3 photosynthesis and CAM simultaneously in distinct leaves. The proportion of CO_2 taken up by PEPCase at night or by Rubisco during the day (net CO_2 assimilation) is adjusted by (1) stomatal behavior, (2) fluctuations in organic acid and storage carbohydrate accumulation, (3) the activity of primary (PEPCase) and secondary (Rubisco) CO_2-fixing enzymes, (4) the activity of decarboxylating enzymes, and (5) synthesis and breakdown of three-carbon skeletons.

Many CAM representatives are able to adjust their pattern of CO_2 uptake in response to longer-term variations of environmental conditions. The ice plant (*Mesembryanthemum crystallinum* L.), agave, and *Clusia* are among the plants that use CAM when water is scarce but undergo a gradual transition to C_3 when water becomes abundant. Other environmental conditions, such as salinity, temperature, and light, also contribute to the extent of CAM induction in these species. This form of regulation requires the expression of numerous CAM genes in response to stress signals.

The water-conserving closure of stomata in arid lands may not be the unique basis of CAM evolution, because paradoxically, CAM species are also found among aquatic plants. Perhaps this mechanism also enhances the acquisition of inorganic carbon (as HCO_3^-) in aquatic habitats, where high resistance to gas diffusion restricts the availability of CO_2.

Accumulation and Partitioning of Photosynthates— Starch and Sucrose

Metabolites accumulated in the light—photosynthates—become the ultimate source of energy for plant development. The photosynthetic assimilation of CO_2 by most leaves yields sucrose in the cytosol and starch in the chloroplasts. During the day, sucrose flows continuously from the leaf cytosol to heterotrophic sink tissues, while starch accumulates as dense granules in chloroplasts (**Figure 8.11**). The onset of darkness not only stops the assimilation of CO_2, but also starts the degradation of chloroplast starch. The content of starch in the chloroplasts falls through the night because breakdown products flow to the cytosol to sustain the export of sucrose to other organs. The large fluctuation of stromal starch in the light versus the dark is why the polysaccharide stored in chloroplasts is called *transitory starch*. Transitory starch functions as (1) an overflow mechanism that stores photosynthate when the synthesis and transport of sucrose are limited during the day and (2) an energy reserve to provide an adequate supply of carbohydrate at night when sugars are not formed by photosynthesis. Plants vary widely in the extent to which they accumulate starch and sucrose in leaves (see Figure 8.11). In some species (e.g., soybean, sugar beet, Arabidopsis), the proportion of starch to sucrose in the leaf is almost constant throughout the day. In others (e.g., spinach, French

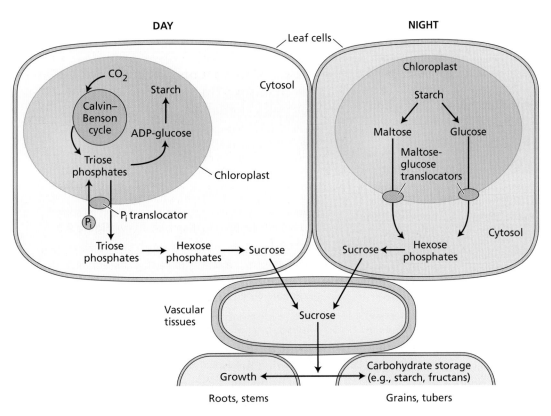

Figure 8.11 Carbon mobilization in land plants. During the day, carbon assimilated photosynthetically is used either for the formation of starch in the chloroplast or is exported to the cytosol for the synthesis of sucrose. External and internal stimuli control the partitioning between starch and sucrose. Triose phosphates from the Calvin–Benson cycle may be used for (1) the synthesis of chloroplast ADP-glucose (the glucosyl donor for starch synthesis) or (2) translocation to the cytosol for the synthesis of sucrose. At night, the breakdown of starch releases both maltose and glucose, which flow across the chloroplast envelope to supplement the hexose phosphate pool and contribute to sucrose synthesis. Transport across the chloroplast envelope, carried out by translocators for phosphate and maltose and glucose, conveys information between the two compartments. As a consequence of the diurnal synthesis and nocturnal breakdown, the levels of chloroplast starch are maximal during the day and minimal at night. This transitory starch serves as the nocturnal energy reserve that provides an adequate supply of carbohydrates to land plants, and also as a diurnal overflow that accepts the carbon excess when photosynthetic CO_2 assimilation proceeds faster than the synthesis of sucrose. Daily, sucrose links the assimilation of inorganic carbon (CO_2) in leaves to the utilization of organic carbon for growth and storage in nonphotosynthetic parts of the plant.

beans), starch accumulates when sucrose exceeds the storage capacity of the leaf or the demand of sink tissues.

The carbon metabolism of leaves also responds to the requirements of sink tissues for energy and growth. Regulatory mechanisms ensure that physiological processes in the chloroplast are synchronized not only with the cytoplasm of the leaf cell but also with other parts of the plant during the day–night cycle. An abundance of sugars in leaves promotes plant growth and carbohydrate storage in reserve organs, while low levels of sugars in sink tissues stimulate the rate of photosynthesis. Sucrose transport links the availability of carbohydrates in source leaves to the use of energy and the formation of reserve polysaccharides in sink tissues (see Chapter 10).

Summary

Sunlight ultimately provides energy for the assimilation of inorganic carbon into organic material (autotrophy). The Calvin–Benson cycle is the predominant pathway for this conversion in many prokaryotes and in all plants.

The Calvin–Benson Cycle

- NADPH and ATP generated by light in chloroplast thylakoids drive the endergonic fixation of atmospheric CO_2 through the Calvin–Benson cycle in the chloroplast stroma (**Figure 8.1**).

- The Calvin–Benson cycle has three phases:
(1) carboxylation of ribulose 1,5-bisphosphate with CO_2 catalyzed by Rubisco, yielding 3-phosphoglycerate;
(2) reduction of the 3-phosphoglycerate to triose phosphates using ATP and NADPH; and (3) regeneration of the CO_2 acceptor molecule ribulose 1,5-bisphosphate (**Figures 8.2, 8.3**).

- CO_2 and O_2 compete in the carboxylation and oxygenation reactions catalyzed by Rubisco (**Figure 8.4**).

- Rubisco activase controls the activity of Rubisco wherein CO_2 functions as both activator and substrate.

- Light regulates the activity of Rubisco activase and four enzymes of the Calvin–Benson cycle via the ferredoxin–thioredoxin system and changes in Mg^{2+} concentration and pH (**Figure 8.5**).

Photorespiration: The C_2 Oxidative Photosynthetic Carbon Cycle

- The C_2 oxidative photosynthetic carbon cycle (photorespiration) minimizes the loss of fixed CO_2 by the oxygenase activity of Rubisco (**Figure 8.6, Table 8.2**).

- Chloroplasts, peroxisomes, and mitochondria participate in the movement of carbon, nitrogen, and oxygen atoms through photorespiration (**Figures 8.6, 8.7**).

- Kinetic properties of Rubisco, temperature, and concentrations of atmospheric CO_2 and O_2 control the balance between the Calvin–Benson and the C_2 oxidative photosynthetic carbon cycles.

Inorganic Carbon–Concentrating Mechanisms

- Land plants have two carbon-concentrating mechanisms that precede CO_2 assimilation through the Calvin–Benson cycle: C_4 photosynthetic carbon fixation (C_4) and crassulacean acid metabolism (CAM).

Inorganic Carbon–Concentrating Mechanisms: The C_4 Carbon Cycle

- The C_4 photosynthetic carbon cycle fixes atmospheric CO_2 via PEPCase into carbon skeletons in one compartment. The four-carbon acid products flow to another compartment where CO_2 is released and refixed via Rubisco (**Figure 8.8, Table 8.3**).

- The C_4 cycle may be driven by diffusion gradients within a single cell as well as by gradients between mesophyll and bundle sheath cells (Kranz anatomy) (**Figure 8.9, Table 8.3**).

- Light regulates the activity of key C_4 cycle enzymes: NADP–malate dehydrogenase, PEPCase, and pyruvate–phosphate dikinase.

- The C_4 cycle reduces photorespiration and water loss in hot, dry climates.

Inorganic Carbon–Concentrating Mechanisms: Crassulacean Acid Metabolism (CAM)

- CAM photosynthesis captures atmospheric CO_2 and scavenges respiratory CO_2 in arid environments.

- CAM is generally associated with anatomical features that minimize water loss.

- In CAM plants, the initial capture of CO_2 (occurring during the night) and its final incorporation into carbon skeletons (in the daytime) are temporally separated (**Figure 8.10**).

- Genetics and environmental factors determine CAM expression.

Accumulation and Partitioning of Photosynthates—Starch and Sucrose

- In most leaves, sucrose in the cytosol and starch in chloroplasts are the end products of photosynthetic CO_2 assimilation (**Figure 8.11**).

- During the day, sucrose flows from the leaf cytosol to sink tissues, while starch accumulates as granules in chloroplasts. At night, the starch content of chloroplasts falls to provide carbon skeletons for sucrose synthesis in the cytosol to nourish heterotrophic tissues.

Suggested Reading

Balsera, M., Uberegui, E., Schürmann, P., and Buchanan, B. B. (2014) Evolutionary development of redox regulation in chloroplasts. *Antioxid. Redox Signal.* 21: 1327–1355.

Bordych, C., Eisenhut, M., Pick, T. R., Kuelahoglu, C., and Weber, A. P. M. (2013) Co-expression analysis as tool for the discovery of transport proteins in photorespiration. *Plant Biol.* 15: 686–693.

Christin, P. A., Arakaki, M., Osborne, C. P., Bräutigam, A., Sage, R. F., Hibberd, J. M., Kelly, S., Covshoff, S., Wong, G. S., Hancock, L., et al. (2014) Shared origins of a key enzyme during the evolution of C4 and CAM metabolism. *J. Exp. Bot.* 65: 3609–3621.

Denton, A. K., Simon, R., and Weber, A. P. M. (2013) C4 photosynthesis: From evolutionary analyses to strategies for synthetic reconstruction of the trait. *Curr. Opin. Plant Biol.* 16: 315–321.

Dever, L. V., Boxall, S. F., Knerová, J., and Hartwell, J. (2015) Transgenic perturbation of the decarboxylation phase of Crassulacean acid metabolism alters physiology and metabolism but has only a small effect on growth. *Plant Physiol.* 167: 44–59.

Ducat, D. C., and Silver, P. A. (2012) Improving carbon fixation pathways. *Curr. Opin. Chem. Biol.* 16: 337–344.

Florian, A., Araújo, W. L., and Fernie, A. R. (2013) New insights into photorespiration obtained from metabolomics. *Plant Biol.* 15: 656–666.

Hagemann, M., Fernie, A. R., Espie, G. S., Kern, R., Eisenhut, M., Reumann, S., Bauwe, H., and Weber, A. P. M. (2013) Evolution of the biochemistry of the photorespiratory C2 cycle. *Plant Biol.* 15: 639–647.

Hibberd, J. M., and Covshoff, S. (2010) The regulation of gene expression required for C4 photosynthesis. *Annu. Rev. Plant Biol.* 61: 181–207.

Sage, R. F., Christin, P. A., and Edwards, E. J. (2011) The C4 plant lineages of planet Earth. *J. Exp. Bot.* 62: 3155–3169.

Sage, R. F., Khoshravesh, R., and Sage, T. L. (2014) From proto-Kranz to C4 Kranz: Building the bridge to C4 photosynthesis. *J. Exp. Bot.* 65: 3341–3356.

Timm, S., and Bauwe, H. (2013) The variety of photorespiratory phenotypes—Employing the current status for future research directions on photorespiration. *Plant Biol.* 15: 737–747.

9 Photosynthesis: Physiological and Ecological Considerations

The conversion of solar energy to the chemical energy of organic compounds is a complex process that includes electron transport and photosynthetic carbon metabolism (see Chapters 7 and 8). This chapter addresses some of the photosynthetic responses of the intact leaf to its environment. Additional photosynthetic responses to different types of stress will be covered in Chapter 19. When discussing photosynthesis in this chapter, we are referring to the rate of net photosynthesis, the difference between photosynthetic carbon assimilation and loss of CO_2 via mitochondrial respiration.

The impact of the environment on photosynthesis is of broad interest, especially to physiologists, ecologists, evolutionary biologists, climate change scientists, and agronomists. From a physiological standpoint, we wish to understand the direct responses of photosynthesis to environmental factors such as light, ambient CO_2 concentrations, and temperature, as well as the indirect responses (mediated through the effects of stomatal control) to environmental factors such as humidity and soil moisture. The dependence of photosynthetic processes on environmental conditions is also important to agronomists because plant productivity, and hence crop yield, depends strongly on prevailing photosynthetic rates in a dynamic environment. To the ecologist, photosynthetic variation among different environments is of great interest in terms of adaptation and evolution.

In studying the environmental dependence of photosynthesis, a central question arises: How many environmental factors can limit photosynthesis at one time? The British plant physiologist F. F. Blackman hypothesized in 1905 that, under any particular conditions, the rate of photosynthesis is limited by the slowest step in the process, the so-called *limiting factor*. The implication of this hypothesis is that at any given time, photosynthesis can be limited either by light or by CO_2 concentration, for instance, but not by both factors. This hypothesis has had a marked influence on the approach used by plant physiologists to study photosynthesis—that is, varying one factor and keeping all other environmental conditions constant. In the intact leaf, three major metabolic processes have been identified as important for photosynthetic performance:

- Rubisco capacity
- Regeneration of ribulose 1,5-bisphosphate (RuBP)
- Metabolism of the triose phosphates

Graham Farquhar and Tom Sharkey added a fundamentally new perspective to our understanding of photosynthesis by pointing out that we should think of the controls on the overall net photosynthetic rates of leaves in economic terms, considering "supply" and "demand" functions for carbon dioxide. The metabolic processes referred to above take place in the palisade cells and spongy mesophyll of the leaf (**Figure 9.1**). These biochemical activities describe the "demand" for CO_2 by photosynthetic metabolism in the cells. However, the rate of CO_2 "supply" to these cells is largely determined by diffusion limitations resulting from stomatal regulation and subsequent resistance in the mesophyll. The coordinated actions of "demand" by photosynthetic cells and "supply" by guard cells affect the leaf photosynthetic rate as measured by net CO_2 uptake.

In the following sections we focus on how naturally occurring variation in light and temperature influences photosynthesis in leaves and how leaves in turn adjust or acclimate to such variation. We also explore how atmospheric carbon dioxide influences photosynthesis, an especially important consideration in a world where CO_2 concentrations are rapidly increasing as humans continue to burn fossil fuels for energy production.

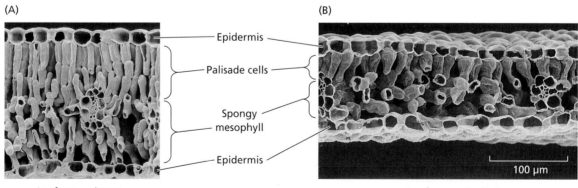

(A)

(B)

Epidermis

Palisade cells

Spongy mesophyll

Epidermis

100 μm

Leaf grown in sun

Leaf grown in shade

Figure 9.1 Scanning electron micrographs of the leaf anatomy of a legume (*Thermopsis montana*) grown in different light environments. Note that the sun leaf (A) is much thicker than the shade leaf (B) and that the palisade (columnlike) cells are much longer in the leaves grown in sunlight. Layers of spongy mesophyll cells can be seen below the palisade cells. The scale is the same for both micrographs. (Courtesy of T. Vogelmann.)

The Effect of Leaf Properties on Photosynthesis

Scaling up from the chloroplast (the focus of Chapters 7 and 8) to the leaf adds new levels of complexity to photosynthesis. At the same time, the structural and functional properties of the leaf make possible other levels of regulation.

We start by examining the capture of light, and how leaf anatomy and leaf orientation maximize light absorption for photosynthesis. Then we describe how leaves acclimate to their light environment. We will see that the photosynthetic response of leaves grown under different light conditions also reflects the ability of a plant to grow under different light environments. However, there are limits to the extent to which photosynthesis in a species can acclimate to very different light environments. For example, in some situations photosynthesis is limited by an inadequate supply of light. In other situations, absorption of too much light would cause severe problems if special mechanisms did not protect the photosynthetic system from excessive light. While plants have multiple levels of control over photosynthesis that allow them to grow successfully in constantly changing environments, there are ultimately limits to what is possible.

Consider the many ways in which leaves are exposed to different spectra (qualities) and quantities of light that result in photosynthesis. Plants grown outdoors are exposed to sunlight, and the spectrum of that sunlight depends on whether it is measured in full sunlight or under the shade of a canopy. Plants grown indoors may receive either incandescent or fluorescent lighting, each of which is spectrally different from sunlight. To account for these differences in spectral quality and quantity, we need uniformity in how we measure and express the light that affects photosynthesis.

The light reaching the plant is a flux, and that flux can be measured in either energy or photon flux units. **Irradiance** is the amount of energy that falls on a flat sensor of known area per unit of time, expressed in watts per square meter (W m^{-2}). Recall that time (seconds) is contained within the term watt: 1 W = 1 joule (J) per second (s^{-1}). Quantum flux, or **photon flux density (PFD)**, is the number of incident **quanta** (singular *quantum*) striking the leaf, expressed in moles per square meter per second (mol m^{-2} s^{-1}), where *moles* refers to the number of photons (1 mol of light = 6.02 × 10^{23} photons, Avogadro's number). Quanta and energy units for sunlight can be interconverted relatively easily, provided that the wavelength of the light, λ, is known. The energy of a photon is related to its wavelength as follows:

$$E = \frac{hc}{\lambda}$$

where c is the speed of light (3 × 10^8 m s^{-1}), h is Planck's constant (6.63 × 10^{-34} J s), and λ is the wavelength of light, usually expressed in nanometers (1 nm = 10^{-9} m). From this equation it can be shown that a photon at 400 nm has twice the energy of a photon at 800 nm.

When considering photosynthesis and light, it is appropriate to express light as photosynthetic photon flux density (PPFD)—the flux of light (usually expressed as micromoles per square meter per second [µmol m^{-2} s^{-1}]) within the photosynthetically active range (400–700 nm). How much light is there on a sunny day? Under direct sunlight on a clear day, PPFD is about 2000 µmol m^{-2} s^{-1} at the top of a dense forest canopy, but may be only 10 µmol m^{-2} s^{-1} at the bottom of the canopy because of light absorption by the leaves overhead.

Leaf anatomy and canopy structure maximize light absorption

On average, about 340 W of radiant energy from the sun reach each square meter of Earth's surface. When this sunlight strikes the vegetation, only 5% of the energy is ultimately converted into carbohydrates by photosynthesis

irradiance The amount of energy that falls on a flat surface of known area per unit of time. Expressed as watts per square meter (W m^{-2}). Time (seconds) is contained within the term watt: 1 W = 1 joule (J) s^{-1}, or as micromoles of quanta per square meter per second (µmol m^{-2}s^{-1}), also referred to as fluence rate.

photon flux density (PFD)
The amount of energy striking a leaf per unit of time, expressed as moles of quanta per square meter per second (mol m^{-2} s^{-1}). Also referred to as fluence rate.

quantum (plural *quanta*) A discrete packet of energy contained in a photon.

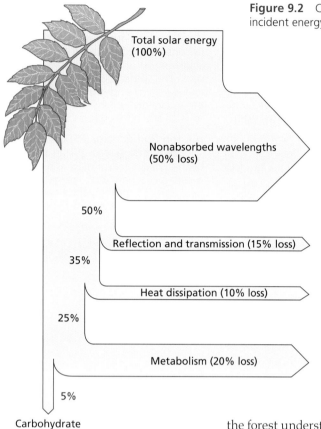

Figure 9.2 Conversion of solar energy into carbohydrates by a leaf. Of the total incident energy, only 5% is converted into carbohydrates.

Total solar energy (100%)

Nonabsorbed wavelengths (50% loss)

50%

Reflection and transmission (15% loss)

35%

Heat dissipation (10% loss)

25%

Metabolism (20% loss)

5%

Carbohydrate

(**Figure 9.2**). The reason this percentage is so low is that a major percentage of the light is of a wavelength either too short or too long to be absorbed by the photosynthetic pigments (**Figure 9.3**). Furthermore, of the photosynthetically active radiation (400–700 nm) that is incident on a leaf, a small percentage is transmitted through the leaf and some is also reflected from its surface. Because chlorophyll absorbs strongly in the blue and red regions of the spectrum (see Figure 7.3), green wavelengths are the ones most dominant in the transmitted and reflected light (see Figure 9.3)—hence the green color of vegetation. Lastly, a percentage of the photosynthetically active radiation that is initially absorbed by the leaf is lost through metabolism and a smaller amount is lost as heat (see Chapter 7).

The anatomy of the leaf is highly specialized for light absorption. The outermost cell layer, the epidermis, is typically transparent to visible light, and the individual cells are often convex. Convex epidermal cells can act as lenses and focus light so that the amount reaching some of the chloroplasts can be many times greater than the amount of ambient light. Epidermal focusing is common among herbaceous plants and is especially prominent among tropical plants that grow in the forest understory, where light levels are very low.

Below the epidermis, the top layers of photosynthetic cells are called **palisade cells**; they are shaped like pillars that stand in parallel columns one to three layers deep (see Figure 9.1). Some leaves have several layers of columnar palisade cells, and we may wonder how efficient it is for a plant to invest energy in developing multiple cell layers when the high chlorophyll content of the first layer would appear to allow little transmission of incident light to the leaf interior. In fact, more light than might be expected penetrates the first layer of palisade cells because of the *sieve effect* and *light channeling*.

The **sieve effect** occurs because chlorophyll is not uniformly distributed throughout cells but instead is confined to the chloroplasts. This packaging of chlorophyll results in shading between the chlorophyll molecules and creates gaps between the chloroplasts where light is not absorbed—hence the reference to a sieve. Because of the sieve effect, the total absorption of light by a given amount of chlorophyll in chloroplasts of a palisade cell is less than the light that would be absorbed by the same amount of chlorophyll were it uniformly distributed in solution.

Light channeling occurs when some of the incident light is propagated through the central vacuoles of the palisade cells and through the air spaces between the cells, an arrangement that results in the transmission of light into the leaf interior.

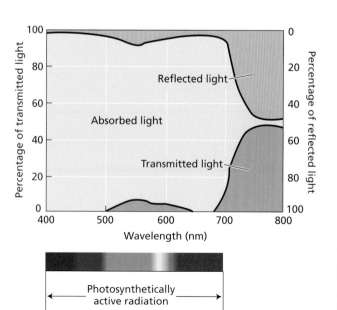

Figure 9.3 Optical properties of a bean leaf. Shown here are the percentages of light absorbed, reflected, and transmitted, as a function of wavelength. The transmitted and reflected green light in the wave band at 500–600 nm gives leaves their green color. Note that most of the light above 700 nm is not absorbed by the leaf. (After Smith 1986.)

In the interior, below the palisade layers, is the **spongy mesophyll**, where the cells are very irregular in shape and are surrounded by large air spaces (see Figure 9.1). The large air spaces generate many interfaces between air and water that reflect and refract the light, thereby randomizing its direction of travel. This phenomenon is called **interface light scattering**.

Light scattering is especially important in leaves because the multiple refractions between cell–air interfaces greatly increase the length of the path over which photons travel, thereby increasing the probability of absorption. In fact, photon path lengths within leaves are commonly four times longer than the thickness of the leaf. Thus, the palisade cell properties that allow light to pass through, and the spongy mesophyll cell properties that are conducive to light scattering, result in more uniform light absorption throughout the leaf.

In some environments, such as deserts, there is so much light that it is potentially harmful to the photosynthetic machinery of leaves. In these environments leaves often have special anatomic features, such as hairs, salt glands, and epicuticular wax, that increase the reflection of light from the leaf surface, thereby reducing light absorption. Such adaptations can decrease light absorption by as much as 60%, thereby reducing overheating and other problems associated with the absorption of too much solar energy.

At the whole-plant level, leaves at the top of a canopy absorb most of the sunlight, and reduce the amount of radiation that reaches leaves lower down in the canopy. Leaves that are shaded by other leaves experience lower light levels and different light quality than the leaves above them and have much lower photosynthetic rates. However, like the layers of an individual leaf, the structure of most plants, and especially of trees, represents an outstanding adaptation for light interception. The elaborate branching structure of trees vastly increases the interception of sunlight. In addition, leaves at different levels of the canopy have varied morphology and physiology that help improve light capture. The result is that very little PPFD penetrates all the way to the bottom of the forest canopy; almost all of the PPFD is absorbed by leaves before reaching the forest floor (**Figure 9.4**).

The deep shade of a forest floor thus makes for a challenging growth environment for plants. However, in many shady habitats **sunflecks** are a common

palisade cells   Below the leaf upper epidermis, the top one to three layers of pillar-shaped photosynthetic cells.

sieve effect   The penetration of photosynthetically active light through several layers of cells due to the gaps between chloroplasts permitting the passage of light.

light channeling   In photosynthetic cells, the propagation of some of the incident light through the central vacuole of the palisade cells and through the air spaces between the cells.

spongy mesophyll   Mesophyll cells of very irregular shape located below the palisade cells and surrounded by large air spaces. Functions in photosynthesis and gas exchange.

interface light scattering   The randomization of the direction of photon movement within plant tissues due to the reflecting and refracting of light from the many air–water interfaces. Greatly increases the probability of photon absorption within a leaf.

sunflecks   Patches of sunlight that pass through openings in a forest canopy to the forest floor. A major source of incident radiation for plants growing under the forest canopy.

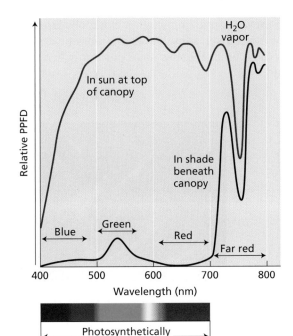

Figure 9.4   Relative spectral distributions of sunlight at the top of a canopy and in the shade under the canopy. Most photosynthetically active radiation is absorbed by leaves in the canopy. (After Smith 1994.)

solar tracking The movement of leaf blades throughout the day so that the planar surface of the blade remains perpendicular to the sun's rays.

environmental feature that brings high light levels deep into the canopy. These are patches of sunlight that pass through small gaps in the leaf canopy; as the sun moves, the sunflecks move across the normally shaded leaves. In spite of the short, ephemeral nature of sunflecks, the photons in them account for nearly 50% of the total light energy available during the day. In a dense forest, sunflecks can change the sunlight impinging on a shade leaf by more than tenfold within seconds. This critical energy is available for only a few minutes, and in a very high dose. Many deep-shade species that experience sunflecks have physiological mechanisms for taking advantage of this burst of light when it occurs. Sunflecks also play a role in the carbon metabolism of densely planted crops, where the lower leaves are shaded by leaves higher up on the plant.

Leaf angle and leaf movement can control light absorption

The angle of the leaf relative to the sun determines the amount of sunlight incident on it. Incoming sunlight can strike a flat leaf surface at a variety of angles depending on the time of day and the orientation of the leaf. Maximum incident radiation occurs when sunlight strikes a leaf perpendicular to its surface. When the rays of light deviate from perpendicular, however, the incident sunlight on a leaf is proportional to the angle at which the rays hit the surface.

Under natural conditions, leaves exposed to full sunlight at the top of the canopy tend to have steep leaf angles so that less than the maximum amount of sunlight is incident on the leaf blade; this allows more sunlight to penetrate into the canopy. For this reason, it is common to see the angle of leaves within a canopy decrease (become more horizontal) with increasing depth in the canopy.

Some leaves maximize light absorption by **solar tracking**; that is, they continuously adjust the orientation of their laminae (blades) such that they remain perpendicular to the sun's rays (**Figure 9.5**). Many species, including alfalfa, cotton, soybean, bean, and lupine, have leaves capable of solar tracking.

Solar-tracking leaves present a nearly vertical position at sunrise, facing the eastern horizon. Individual leaf blades then begin to track the rising sun, following its movement across the sky with an accuracy of ±15° until sunset, when the laminae are nearly vertical, facing the west. During the night the leaves take a horizontal position and reorient just before dawn so that they face the eastern horizon, ready for another sunrise. Leaves track the sun only on clear days, and they stop moving when a cloud obscures the sun. In the case of intermittent cloud cover, some leaves can reorient as rapidly as 90° per hour and thus can catch up to the new solar position when the sun emerges from behind a cloud.

(A)

(B)

Figure 9.5 Leaf movement in a sun-tracking plant. (A) Initial leaf orientation in the lupine *Lupinus succulentus*, with no direct sunlight. (B) Leaf orientation 4 h after exposure to oblique light. Arrows indicate the direction of the light beam. Movement is generated by asymmetric swelling of a pulvinus, found at the junction between the lamina and the petiole. In natural conditions, the leaves track the sun's trajectory in the sky. (From Vogelmann and Björn 1983, courtesy of T. Vogelmann.)

Solar tracking is a blue-light response (see Chapter 13), and the sensing of blue light in solar-tracking leaves occurs in specialized regions of the leaf or stem. In species of *Lavatera* (Malvaceae), the photosensitive region is located in or near the major leaf veins, but in many species, notably legumes, leaf orientation is controlled by a specialized organ called the **pulvinus** (plural *pulvini*), found at the junction between the blade and the petiole. In lupines (*Lupinus*, Fabaceae), for example, leaves consist of five or more leaflets, and the photosensitive region is in a pulvinus located at the basal part of each leaflet lamina (see Figure 9.5). The pulvinus contains motor cells that change their osmotic potential and generate mechanical forces that determine laminar orientation. In other plants, leaf orientation is controlled by small mechanical changes along the length of the petiole and by movements of the younger parts of the stem.

Heliotropism is another term used to describe leaf orientation by solar tracking. Leaves that maximize light interception by solar tracking are referred to as *diaheliotropic*. Some solar-tracking plants can also move their leaves so that they *avoid* full exposure to sunlight, thus minimizing heating and water loss. These sun-avoiding leaves are called *paraheliotropic*. Some plant species, such as soybean, have leaves that can display diaheliotropic movements when they are well watered and paraheliotropic movements when they experience water stress.

Leaves acclimate to sun and shade environments

Acclimation is a developmental process in which leaves express a set of biochemical and morphological adjustments that are suited to the particular environment in which the leaves are exposed. Acclimation can occur in mature leaves and in newly developing leaves. **Plasticity** is the term we use to define how much adjustment can take place. Many plants have developmental plasticity to respond to a range of light regimes, growing as sun plants in sunny areas and as shade plants in shady habitats. The ability to acclimate is important, given that shady habitats can receive less than 20% of the PPFD available in an exposed habitat, and deep-shade habitats receive less than 1% of the PPFD incident at the top of the canopy.

In some plant species, individual leaves that develop under very sunny or deep shady environments are often unable to persist when transferred to the other type of habitat. In such cases, the mature leaf will abscise and a new leaf will develop that is better suited for the new environment. You may notice this if you take a plant that developed indoors and transfer it outdoors; after some time, if it is the right type of plant, a new set of leaves will develop that are better suited to bright sunlight. However, some plant species are not able to acclimate when transferred from a sunny to a shady environment, or vice versa. The lack of acclimation indicates that these plants are specialized for either a sunny or a shady environment. When plants adapted to deep-shade conditions are transferred into full sunlight, the leaves experience chronic photoinhibition and leaf bleaching, and they eventually die. We discuss photoinhibition later in this chapter.

Sun and shade leaves have contrasting biochemical and morphological characteristics:

- Shade leaves increase light capture by having more total chlorophyll per reaction center, a higher ratio of chlorophyll *b* to chlorophyll *a*, and usually thinner laminae than sun leaves.

- Sun leaves increase CO_2 assimilation by having more Rubisco and can dissipate excess light energy by having a large pool of xanthophyll-cycle components (see Figures 9.11 and 9.12). Morphologically they have thicker leaves and a larger palisade layer than shade leaves (see Figure 9.1).

These morphological and biochemical modifications are associated with specific acclimation responses to the *amount* of sunlight in a plant's habitat, but light *quality* can also influence such responses. For example, far-red light, which is

pulvinus   (plural *pulvini*) A turgor-driven organ found at the junction between the blade and the petiole of the leaf that provides the mechanical force for leaf movements.

heliotropism   The movements of leaves toward or away from the sun.

acclimation   The increase in plant stress tolerance due to exposure to prior stress. May involve changes in gene expression.

plasticity   The ability to adjust morphologically, physiologically, and biochemically in response to changes in the environment.

light compensation point
The amount of light reaching a photosynthesizing leaf at which photosynthetic CO_2 uptake exactly balances respiratory CO_2 release.

absorbed primarily by photosystem I (PSI) (see Chapter 7) but also by phytochrome (see Chapter 13), is proportionally more abundant in shady habitats than in sunny ones. To better balance the flow of energy through PSII and PSI, the adaptive response of some shade plants is to produce a higher ratio of PSII to PSI reaction centers, compared with that found in sun plants. Other shade plants, rather than altering the ratio of PSII to PSI reaction centers, add more antenna chlorophyll to PSII to increase absorption by this photosystem. These changes appear to enhance light absorption and energy transfer in shady environments.

Effects of Light on Photosynthesis in the Intact Leaf

Light is a critical resource that limits plant growth, but at times leaves can be exposed to too much rather than too little light. In this section we describe typical photosynthetic responses to light as measured by light-response curves. We also consider how features of a light-response curve can help explain contrasting physiological properties between sun and shade plants, and between C_3 and C_4 species. The section concludes with descriptions of how leaves respond to excess light.

Light-response curves reveal photosynthetic properties

Measuring net CO_2 fixation in intact leaves across varying PPFD levels generates light-response curves (**Figure 9.6**). In near darkness there is little photosynthetic carbon assimilation, but because mitochondrial respiration continues, CO_2 is given off by the plant (see Chapter 11). CO_2 uptake is negative in this part of the light-response curve. At higher PPFD levels, photosynthetic CO_2 assimilation eventually reaches a point at which CO_2 uptake exactly balances CO_2 evolution. This is called the **light compensation point**. The PPFD at which different leaves reach the light compensation point can vary among species and developmental conditions. One of the more interesting differences is found between plants that normally grow in full sunlight and those that grow in the shade (**Figure 9.7**). Light compensation points of sun plants range from 10 to 20 μmol m^{-2} s^{-1}, whereas corresponding values for shade plants are 1 to 5 μmol m^{-2} s^{-1}.

Why are light compensation points lower for shade plants? For the most part, this is because respiration rates in shade plants are very low; therefore only a little photosynthesis is necessary to bring the net rates of CO_2 exchange to zero. Low respiratory rates allow shade plants to survive in light-limited environments through their ability to achieve positive CO_2 uptake rates at lower PPFD values than sun plants.

A linear relationship between PPFD and photosynthetic rate persists at light levels above the light compensation point (see Figure 9.6). Throughout this linear portion of the light-response curve, photosynthesis is light-limited; more

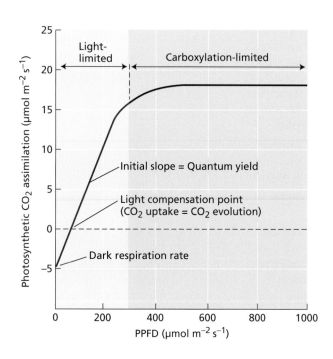

Figure 9.6 Response of photosynthesis to light in a C_3 plant. In darkness, respiration causes a net efflux of CO_2 from the plant. The light compensation point is reached when photosynthetic CO_2 assimilation equals the amount of CO_2 evolved by respiration. Increasing light above the light compensation point proportionally increases photosynthesis, indicating that photosynthesis is limited by the rate of electron transport, which in turn is limited by the amount of available light. This portion of the curve is referred to as light-limited. Further increases in photosynthesis are eventually limited by the carboxylation capacity of Rubisco or the metabolism of triose phosphates. This part of the curve is referred to as carboxylation-limited.

Figure 9.7 Light-response curves of photosynthetic carbon fixation in sun and shade plants. Triangle orache (*Atriplex triangularis*) is a sun plant, and wild ginger (*Asarum caudatum*) is a shade plant. Typically, shade plants have a low light compensation point and have lower maximum photosynthetic rates than sun plants. The dashed line has been extrapolated from the measured part of the curve. (After Harvey 1979.)

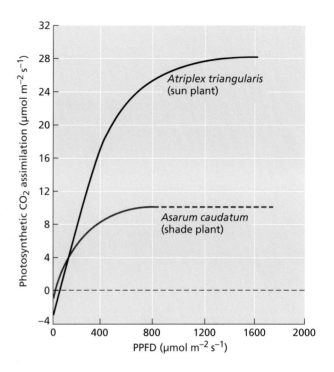

light stimulates proportionately more photosynthesis. When corrected for light absorption, the slope of this linear portion of the curve provides the **maximum quantum yield** of photosynthesis for the leaf. Leaves of sun and shade plants show very similar quantum yields despite their different growth habitats. This is because the basic biochemical processes that determine quantum yield are the same for these two types of plants. But quantum yield can vary among plants with different photosynthetic pathways.

Quantum yield is the ratio of a given light-dependent product to the number of absorbed photons (see Equation 7.5). Photosynthetic quantum yield can be expressed on either a CO_2 or an O_2 basis, and as explained in Chapter 7, the quantum yield of photochemistry is about 0.95. However, the maximum photosynthetic quantum yield of an integrated process such as photosynthesis is lower than the theoretical yield when measured in chloroplasts (organelles) or whole leaves. Based on the biochemistry discussed in Chapters 7 and 8, we expect the theoretical maximum quantum yield for photosynthesis to be 0.125 for C_3 plants (one CO_2 molecule fixed per eight photons absorbed). But under today's atmospheric conditions (400 ppm CO_2, 21% O_2), the quantum yields for CO_2 of C_3 and C_4 leaves vary between 0.04 and 0.07 mole of CO_2 per mole of photons.

In C_3 plants the reduction from the theoretical maximum is caused primarily by energy loss through photorespiration. In C_4 plants the reduction is caused by the additional energy requirements of the CO_2-concentrating mechanism and potential cost of refixing CO_2 that has diffused out from within the bundle sheath cells. If C_3 leaves are exposed to low O_2 concentrations, photorespiration is minimized and the maximum quantum yield increases to about 0.09 mole of CO_2 per mole of photons. In contrast, if C_4 leaves are exposed to low O_2 concentrations, the quantum yields for CO_2 fixation remain constant at about 0.05 to 0.06 mole of CO_2 per mole of photons. This is because the carbon-concentrating mechanism in C_4 photosynthesis eliminates nearly all CO_2 evolution via photorespiration.

At higher PPFD along the light-response curve, the photosynthetic response to light starts to level off (see Figures 9.6 and 9.7) and eventually approaches *saturation*. Beyond the light saturation point, net photosynthesis no longer increases, indicating that factors other than incident light, such as electron transport rate, Rubisco activity, or the metabolism of triose phosphates, have become limiting to photosynthesis. Light saturation levels for shade plants are substantially lower than those for sun plants (see Figure 9.7). This is also true for leaves of the same plant species when grown in sun versus shade (**Figure 9.8**). These levels usually reflect the maximum PPFD to which a leaf was exposed during growth.

The light-response curve of most leaves saturates between 500 and 1000 μmol m^{-2} s^{-1}, well below full sunlight (which is about 2000 μmol m^{-2} s^{-1}). An exception to this is well-fertilized crop leaves, which often saturate above 1000 μmol m^{-2} s^{-1}. Although individual leaves are rarely able to use full sunlight, whole plants

maximum quantum yield The ratio between photosynthetic product and the number of photons absorbed by a photosynthetic tissue. In a graphic plot of photon flux and photosynthetic rate, the maximum quantum yield is given by the slope of the linear portion of the curve.

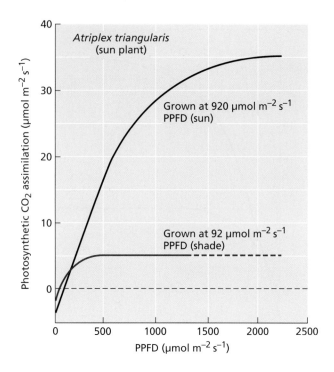

Figure 9.8 Light-response curve of photosynthesis of a sun plant grown under sun versus shade conditions. The upper curve represents an *Atriplex triangularis* leaf grown at a PPFD level ten times higher than that of the lower curve. In the plant grown at the lower light levels, photosynthesis saturates at a substantially lower PPFD, indicating that the photosynthetic properties of a leaf depend on its growing conditions. The dashed red line has been extrapolated from the measured part of the curve. (After Björkman 1981.)

usually consist of many leaves that shade each other. Thus, at any given time of the day only a small proportion of the leaves are exposed to full sun, especially in plants with dense canopies. The rest of the leaves receive subsaturating photon fluxes that come from sunflecks that pass through gaps in the leaf canopy, diffuse light, and light transmitted through other leaves.

Because the photosynthetic response of the intact plant is the sum of the photosynthetic activity of all the leaves, only rarely is photosynthesis light-saturated at the level of the whole plant (**Figure 9.9**). It is for this reason that crop productivity is usually related to the total amount of light received during the growing season, rather than to single-leaf photosynthetic capacity. Given enough water and nutrients, the more light a crop receives, the higher the biomass produced.

Leaves must dissipate excess light energy

When exposed to excess light, leaves must dissipate the surplus absorbed light energy to prevent damage to the photosynthetic apparatus (**Figure 9.10**). There are several routes for energy dissipation that involve *nonphotochemical quenching*, the quenching of chlorophyll fluorescence by mechanisms other than photochemistry (described in Chapter 7). The most important example involves the transfer of absorbed light energy away from electron transport toward heat production. Although the molecular mechanisms are not yet fully understood, the xanthophyll cycle is an important avenue for dissipation of excess light energy.

THE XANTHOPHYLL CYCLE The xanthophyll cycle, which comprises the three carotenoids violaxanthin, antheraxanthin, and zeaxanthin, establishes an ability to dissipate excess light energy in the leaf (**Figure 9.11**). Under high light, violaxanthin is converted to antheraxanthin and then to zeaxanthin. This interconversion consumes reducing equivalents in the form of ascorbate, and the reverse reactions consume reducing equivalents in the form of NADPH. Thus, the xanthophyll cycle prevents over-reduction of the chloroplasts, which would otherwise occur when photochemical processes—leading to the formation of NADPH and ascorbate—outrun the carbon fixation reactions

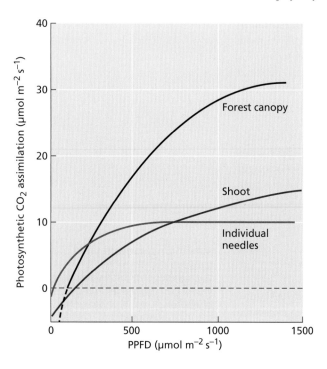

Figure 9.9 Changes in photosynthesis (expressed on a per-square-meter basis) in individual needles, a complex shoot, and a forest canopy of Sitka spruce (*Picea sitchensis*) as a function of PPFD. Complex shoots consist of groupings of needles that often shade each other, similar to the situation in a canopy where branches often shade other branches. As a result of shading, much higher PPFD levels are needed to saturate photosynthesis. The dashed portion of the forest canopy trace has been extrapolated from the measured part of the curve. (After Jarvis and Leverenz 1983.)

Figure 9.10 Excess light energy in relation to a light-response curve of photosynthetic oxygen evolution in a shade leaf. The broken line shows theoretical oxygen evolution in the absence of any rate limitation to photosynthesis. At PPFD levels up to 150 μmol m^{-2} s^{-1}, a shade plant is able to use the absorbed light. Above 150 μmol m^{-2} s^{-1}, however, photosynthesis saturates, and an increasingly larger amount of the absorbed light energy must be dissipated. At higher PPFD levels there is a large difference between the fraction of light used by photosynthesis versus that which must be dissipated (excess light energy). The differences are much greater in a shade plant than in a sun plant. (After Osmond 1994.)

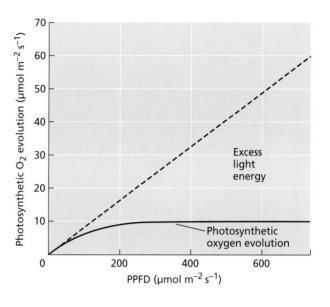

(see Chapters 7 and 8). Zeaxanthin is the most effective of the three xanthophylls in energy dissipation because its formation from violaxanthin uses 2 NADPH and the reverse process another 2 NADPH. Antheraxanthin is only half as effective (see Figure 9.11). Whereas the level of antheraxanthin remains relatively constant throughout the day, the zeaxanthin content increases at high PPFD and decreases at low PPFD.

In leaves growing under full sunlight, zeaxanthin and antheraxanthin can make up 40% of the total xanthophyll-cycle pool at maximum PPFD levels attained at midday (**Figure 9.12**). In these conditions a substantial amount of excess light energy absorbed by the thylakoid membranes can be dissipated as heat (due to the reoxidation of NADPH), thus preventing damage to the photosynthetic machinery of the chloroplast. Leaves that grow in full sunlight contain a substantially larger xanthophyll pool than do shade leaves, so they can dissipate higher amounts of excess light energy. Nevertheless, the xanthophyll cycle also operates in plants that grow in the low light of the forest understory, where they are occasionally exposed to sunflecks. Exposure to just one sunfleck results in the conversion of much of the violaxanthin in the leaf to zeaxanthin.

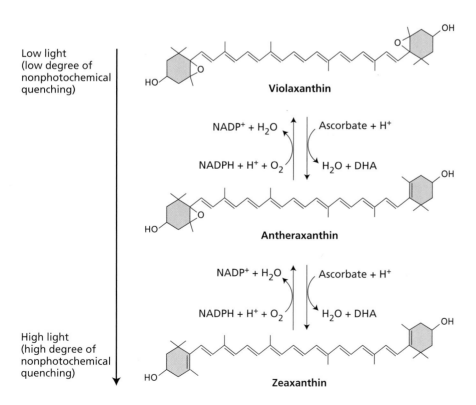

Figure 9.11 Chemical structures of violaxanthin, antheraxanthin, and zeaxanthin. The highly quenched state of PSII is associated with zeaxanthin, the unquenched state with violaxanthin. Enzymes interconvert these two carotenoids, with antheraxanthin as the intermediate, in response to changing conditions, especially changes in light intensity. Zeaxanthin formation uses ascorbate as a cofactor, and violaxanthin formation requires NADPH. DHA, dehydroascorbate.

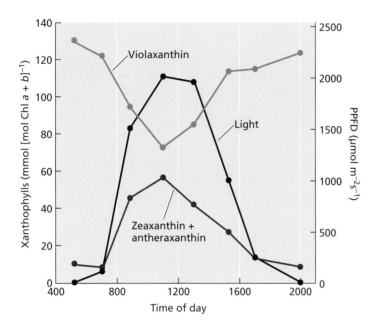

 (chart)

Figure 9.12 Diurnal changes in xanthophyll content as a function of PPFD in sunflower (*Helianthus annuus*). As the amount of light incident to a leaf increases, a greater proportion of violaxanthin is converted to antheraxanthin and zeaxanthin, thereby dissipating excess excitation energy and protecting the photosynthetic apparatus. (After Adams and Demmig-Adams 1992.)

The xanthophyll cycle is also important in species that remain green during winter, when photosynthetic rates are very low yet light absorption remains high. Unlike in the diurnal cycling of the xanthophyll pool observed in the summer, zeaxanthin levels remain high all day during the winter. This mechanism maximizes dissipation of light energy, thereby protecting the leaves against photooxidation when winter cold prevents carbon assimilation.

CHLOROPLAST MOVEMENTS An alternative means of reducing excess light energy is to move the chloroplasts so that they are no longer exposed to high light. Chloroplast movement is widespread among algae, mosses, and leaves of higher plants. If chloroplast orientation and location are controlled, leaves can regulate how much incident light is absorbed. In darkness or weak light (**Figure 9.13A and B**), chloroplasts gather at the cell surfaces parallel to the plane of the leaf so that they are aligned perpendicularly to the incident light—a position that maximizes absorption of light.

Under high light (**Figure 9.13C**), the chloroplasts move to the cell surfaces that are parallel to the incident light, thus avoiding excess absorption of light. Such chloroplast rearrangement can decrease the amount of light absorbed by the leaf

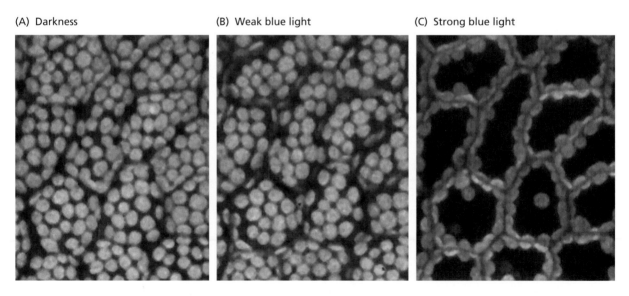

(A) Darkness   **(B) Weak blue light**   **(C) Strong blue light**

Figure 9.13 Chloroplast distribution in photosynthesizing cells of the duckweed *Lemna*. These surface views show the same cells under three conditions: (A) darkness, (B) weak blue light, and (C) strong blue light. In (A) and (B), chloroplasts are positioned near the upper surface of the cells, where they can absorb maximum amounts of light. When the cells are irradiated with strong blue light (C), the chloroplasts move to the side walls, where they shade each other, thus minimizing the absorption of excess light. (Data from Tlalka and Fricker 1999. Images courtesy of M. Tlalka and M. D. Fricker.)

by about 15%. Chloroplast movement in leaves is a typical blue-light response (see Chapter 13). Blue light also controls chloroplast orientation in many of the lower plants, but in some algae, chloroplast movement is controlled by phytochrome (see Chapter 13). In leaves, chloroplasts move along actin microfilaments in the cytoplasm, and calcium ions regulate their movement.

LEAF MOVEMENTS   Plants have also evolved responses that reduce the excess radiation load on whole leaves during high sunlight periods, especially when transpiration and its cooling effects are reduced because of water stress. These responses often involve changes in the leaf orientation relative to the incoming sunlight. For example, heliotropic leaves of both alfalfa and lupine track the sun, but at the same time can reduce incident light levels by folding leaflets together so that the leaf laminae become nearly parallel to the sun's rays (paraheliotropic). These movements are accomplished by changes in the turgor pressure of pulvinus cells at the tip of the petiole. Another common response is mild wilting, as seen in many sunflowers, whereby a leaf droops to a vertical orientation, again effectively reducing the incident heat load and reducing transpiration and incident light levels. Many grasses are able to effectively "twist" through loss of turgor in bulliform cells, resulting in reduced incident PPFD.

Absorption of too much light can lead to photoinhibition

When leaves are exposed to more light than they can use (see Figure 9.10), the reaction center of PSII is inactivated and often damaged in a phenomenon called **photoinhibition**. The characteristics of photoinhibition in the intact leaf depend on the amount of light to which the plant is exposed. The two types of photoinhibition are dynamic photoinhibition and chronic photoinhibition.

Under moderate excess light, **dynamic photoinhibition** is observed. Quantum yield decreases, but the maximum photosynthetic rate remains unchanged. Dynamic photoinhibition is caused by the diversion of absorbed light energy toward photoprotective heat dissipation (e.g., the xanthophyll cycle)—hence the decrease in quantum yield. This decrease is often temporary, and quantum yield can return to its initial higher value when PPFD decreases below saturation levels. **Figure 9.14** shows how photons from sunlight are allocated to photosynthetic reactions versus being thermally dissipated as excess energy over the course of a day under favorable and stressed environmental conditions.

Chronic photoinhibition results from exposure to high levels of excess light that damage the photosynthetic system and decrease both instantaneous quantum yield and maximum photosynthetic rate. This would happen if the stress condition in Figure 9.14B persisted because photoprotection was not possible. Chronic photoinhibition is associated with damage to the D1 protein from the reaction center of PSII (see Chapter 7). In contrast to the transient effects of dynamic photoinhibition, the effects of chronic photoinhibition are relatively long-lasting, persisting for weeks or months.

How significant is photoinhibition in nature? Dynamic photoinhibition appears to occur daily, when leaves are exposed to maximum amounts of light and there is a corresponding reduction

photoinhibition   The inhibition of photosynthesis by excess light.

dynamic photoinhibition
Photoinhibition of photosynthesis in which quantum efficiency decreases but the maximum photosynthetic rate remains unchanged. Occurs under moderate, not high, excess light.

chronic photoinhibition
Photoinhibition of photosynthetic activity in which both quantum efficiency and the maximum rate of photosynthesis are decreased. Occurs under high levels of excess light.

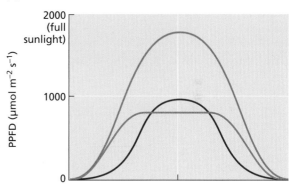

(A) Favorable environmental conditions

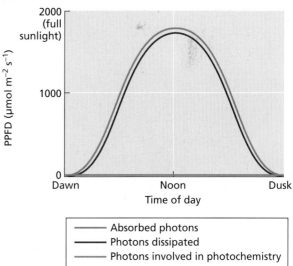

(B) Environmental stress conditions

Figure 9.14   Changes over the course of a day in the allocation of photons absorbed from sunlight. Shown here are contrasts in how the photons striking a leaf are either involved in photochemistry or thermally dissipated as excess energy under (A) favorable conditions and (B) stress conditions. (After Demmig-Adams and Adams 2000.)

Bowen ratio   The ratio of sensible heat loss to evaporative heat loss, the two most important processes in the regulation of leaf temperature.

in carbon fixation. Photoinhibition is more pronounced at low temperatures, and becomes chronic under more extreme climatic conditions.

Effects of Temperature on Photosynthesis in the Intact Leaf

Photosynthesis (CO_2 uptake) and transpiration (H_2O loss) share a common pathway. That is, CO_2 diffuses into the leaf, and H_2O diffuses out, through the stomatal opening regulated by the guard cells (see Chapter 3). While these are independent processes, vast quantities of water are lost during photosynthetic periods, with the molar ratio of H_2O loss to CO_2 uptake often exceeding 250. This high water-loss rate also removes heat from leaves through evaporative cooling, keeping them relatively cool even under full sunlight conditions. Transpirational cooling is important, since photosynthesis is a temperature-dependent process, but the concurrent water loss means that cooling comes at a cost, especially in arid and semiarid ecosystems.

Leaves must dissipate vast quantities of heat

The heat load on a leaf exposed to full sunlight is very high. In fact, under normal sunny conditions with moderate air temperatures, a leaf would warm up to a dangerously high temperature if all incident solar energy were absorbed and none of the heat was dissipated. However, this does not occur because leaves absorb only about 50% of the total solar energy (300–3000 nm), with most of the absorption occurring in the visible portion of the spectrum (see Figures 9.2 and 9.3). This amount is still large. The typical heat load of a leaf is dissipated through three processes (**Figure 9.15**):

- Radiative heat loss: All objects emit long-wave radiation (at about 10,000 nm) in proportion to their temperature to the fourth power (Stephan Boltzman equation). However, the maximum emitted wavelength is inversely proportional to the leaf temperature, and leaf temperatures are low enough that the wavelengths emitted are not visible to the human eye.

- Sensible heat loss: If the temperature of the leaf is higher than that of the air circulating around the leaf, the heat is convected (transferred) away from the leaf to the air. The size and shape of a leaf influence the amount of sensible heat loss.

- Latent heat loss: Because the evaporation of water requires energy, when water evaporates from a leaf (transpiration), it removes large amounts of heat from the leaf and thus cools it. The human body is cooled by the same principle, through perspiration.

Sensible heat loss and evaporative heat loss are the most important processes in the regulation of leaf temperature, and the ratio of the two fluxes is called the **Bowen ratio**:

$$\text{Bowen ratio} = \frac{\text{Sensible heat loss}}{\text{Evaporative heat loss}}$$

In well-watered crops, transpiration (see Chapter 3), and hence water evaporation from the leaf, is high, so the Bowen ratio is low. Conversely, when evaporative cooling is

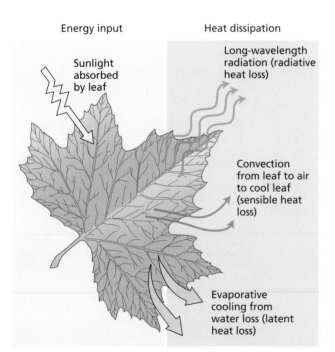

Energy input   Heat dissipation

Sunlight absorbed by leaf

Long-wavelength radiation (radiative heat loss)

Convection from leaf to air to cool leaf (sensible heat loss)

Evaporative cooling from water loss (latent heat loss)

Figure 9.15   Absorption and dissipation of energy from sunlight by the leaf. The imposed heat load must be dissipated in order to avoid damage to the leaf. The heat load is dissipated by emission of long-wavelength radiation, by sensible heat loss to the air surrounding the leaf, and by the evaporative cooling caused by transpiration.

limited, the Bowen ratio is high. For example, in a water-stressed crop, partial stomatal closure reduces evaporative cooling and the Bowen ratio is increased. The amount of evaporative heat loss (and thus the Bowen ratio) is influenced by the degree to which stomata remain open.

Plants with very high Bowen ratios conserve water, but consequently may also experience high leaf temperatures. However, the temperature difference between the leaf and the air does increase the amount of sensible heat loss. Reduced growth is usually correlated with high Bowen ratios, because a high Bowen ratio is indicative of at least partial stomatal closure.

There is an optimal temperature for photosynthesis

Maintaining favorable leaf temperatures is crucial to plant growth because maximum photosynthesis occurs within a relatively narrow temperature range. The peak photosynthetic rate across a range of temperatures is the *photosynthetic thermal optimum*. When the optimal temperature for a given plant is exceeded, photosynthetic rates decrease. The photosynthetic thermal optimum reflects biochemical, genetic (adaptation), and environmental (acclimation) components.

Species adapted to different thermal regimes usually have an optimal temperature range for photosynthesis that reflects the temperatures of the environment in which they evolved. A contrast is especially clear between the C_3 plant *Atriplex glabriuscula*, which commonly grows in cool coastal environments, and the C_4 plant *Tidestromia oblongifolia*, from a hot desert environment (**Figure 9.16**). The ability to acclimate or biochemically adjust to temperature can also be found within species. When plants of the same species are grown at different temperatures and then tested for their photosynthetic response, they show photosynthetic thermal optima that correlate with the temperature at which they were grown. That is, plants of the same species grown at low temperatures have higher photosynthetic rates at low temperatures, whereas those same plants grown at high temperatures have higher photosynthetic rates at high temperatures. This phenomenon is another example of plasticity. Plants with a high thermal plasticity are capable of growing over a wide range of temperatures.

Changes in photosynthetic rates in response to temperature play an important role in plant adaptations to different environments and contribute to plants being productive even in some of the most extreme thermal habitats. In the lower temperature range, plants growing in alpine areas of Colorado and arctic regions in Alaska are capable of net CO_2 uptake at temperatures close to 0°C. At the other extreme, plants living in Death Valley, California, one of the hottest places on Earth, can achieve positive photosynthetic rates at temperatures approaching 50°C.

Photosynthesis is sensitive to both high and low temperatures

When photosynthetic rates are plotted as a function of temperature, the temperature-response curve has an asymmetric bell-type shape (see Figure 9.16). In spite of some differences in shape, the temperature-response curve of photosynthesis among and within species has many common features. The ascending portion of the curve represents a temperature-dependent stimulation of enzymatic activities; the flat top is the temperature range that is optimal for photosynthesis; and the descending portion of the curve is associated with

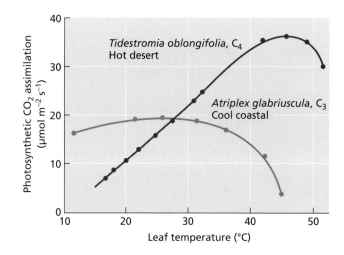

Figure 9.16 Photosynthesis as a function of leaf temperature at normal atmospheric CO_2 concentrations for a C_3 plant grown in its natural cool habitat and a C_4 plant growing in its natural hot habitat. (After Björkman et al. 1975.)

temperature-sensitive deleterious effects, some of which are reversible while others are not.

What factors are associated with the decline in photosynthesis above the photosynthetic temperature optimum? Temperature affects all biochemical reactions of photosynthesis as well as membrane integrity in chloroplasts, so it is not surprising that the responses to temperature are complex. Cellular respiration rates increase as a function of temperature, but they are not the primary reason for the sharp decrease in net photosynthesis at high temperatures. A major impact of high temperature is on membrane-bound electron transport processes, which become uncoupled or unstable at high temperatures. This cuts off the supply of reducing power needed to fuel net photosynthesis and leads to a sharp overall decrease in photosynthesis.

Under ambient CO_2 concentrations and with favorable light and soil moisture conditions, the photosynthetic thermal optimum is often limited by the activity of Rubisco. In leaves of C_3 plants, the response to increasing temperature reflects conflicting processes: an increase in carboxylation rate and a decrease in the affinity of Rubisco for CO_2 with a corresponding increase in photorespiration (see Chapter 8). (There is also evidence that Rubisco activity decreases because of negative heat effects on an activator of Rubisco, Rubisco activase (see Chapter 8), at higher [>35°C] temperatures.) The reduction in the affinity for CO_2 and the increase in photorespiration attenuate the potential temperature response of photosynthesis under ambient CO_2 concentrations. By contrast, in plants with C_4 photosynthesis, the leaf interior is CO_2-saturated, or nearly so (as we discussed in Chapter 8), and the negative effect of high temperature on Rubisco affinity for CO_2 is not realized. This is one reason that leaves of C_4 plants tend to have a higher photosynthetic temperature optimum than do leaves of C_3 plants (see Figure 9.16).

At low temperatures, C_3 photosynthesis can also be limited by factors such as phosphate availability in the chloroplast. When triose phosphates are exported from the chloroplast to the cytosol, an equimolar amount of inorganic phosphate is taken up via translocators in the chloroplast membrane. If the rate of triose phosphate use in the cytosol decreases, phosphate uptake into the chloroplast is inhibited and photosynthesis becomes phosphate-limited. Starch synthesis and sucrose synthesis decrease rapidly with decreasing temperature, reducing the demand for triose phosphates and causing the phosphate limitation observed at low temperatures.

Photosynthetic efficiency is temperature-sensitive

Photorespiration (see Chapter 8) and the quantum yield (light-use efficiency) differ between C_3 and C_4 photosynthesis, with changes particularly noticeable as temperatures vary. **Figure 9.17** illustrates quantum yield for photosynthesis as a function of leaf temperature in C_3 plants and C_4 plants in

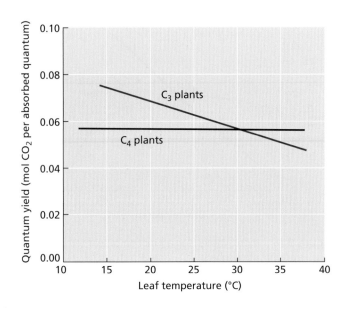

Figure 9.17   Quantum yield of photosynthetic carbon fixation in C_3 and C_4 plants as a function of leaf temperature at today's atmospheric CO_2 concentration of 400 ppm. Photorespiration increases with temperature in C_3 plants, and the energy cost of net CO_2 fixation increases accordingly. This higher energy cost is reflected in lower quantum yields at higher temperatures. In contrast, photorespiration is very low in C_4 plants and the quantum yield does not show a temperature dependence. Note that, at lower temperatures, the quantum yield of C_3 plants is higher than that of C_4 plants, indicating that C_3 photosynthesis is more efficient at lower temperatures. (After Ehleringer and Björkman 1977.)

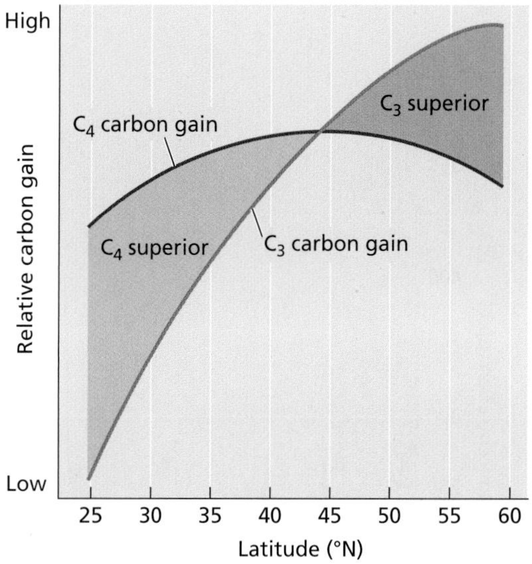

Figure 9.18 Relative rates of photosynthetic carbon gain predicted for identical C_3 and C_4 grass canopies as a function of latitude across the Great Plains of North America. (After Ehleringer 1978.)

today's atmosphere of 400 ppm CO_2. In the C_4 plants the quantum yield remains constant with temperature, reflecting low rates of photorespiration. In the C_3 plants the quantum yield is higher than in C_4 plants at low temperatures, reflecting the lower intrinsic costs of the C_3 pathway. However, the C_3 quantum yield decreases with temperature, reflecting a stimulation of photorespiration by temperature and an ensuing higher energy cost for net CO_2 fixation.

The combination of reduced quantum yield and increased photorespiration leads to expected differences in the photosynthetic capacities of C_3 and C_4 plants in habitats with different temperatures. The predicted relative rates of primary productivity of C_3 and C_4 grasses along a latitudinal transect in the Great Plains of North America from southern Texas in the United States to Manitoba in Canada are shown in **Figure 9.18**. This decline in C_4 relative to C_3 productivity moving northward closely parallels the shift in abundance of plants with these pathways in the Great Plains: C_4 species are more common below 40°N, and C_3 species dominate above 45°N.

Effects of Carbon Dioxide on Photosynthesis in the Intact Leaf

We have discussed how light and temperature influence leaf physiology and anatomy. Now we turn our attention to how CO_2 concentration affects photosynthesis. CO_2 diffuses from the atmosphere into leaves—first through stomata, then through the intercellular air spaces, and ultimately into cells and chloroplasts. In the presence of adequate amounts of light, higher CO_2 concentrations support higher photosynthetic rates. The reverse is also true: Low CO_2 concentrations can limit the amount of photosynthesis in C_3 plants.

In this section we discuss the concentration of atmospheric CO_2 in recent history, and its availability for carbon-fixing processes. Then we consider the limitations that CO_2 places on photosynthesis and the impact of the CO_2-concentrating mechanisms of C_4 plants.

Atmospheric CO_2 concentration keeps rising

Carbon dioxide presently accounts for about 0.040%, or 400 ppm, of air. The partial pressure of ambient CO_2 (c_a) varies with atmospheric pressure and is approximately 40 pascals (Pa) at sea level. Water vapor usually accounts for up to 2% of the atmosphere and O_2 for about 21%. The largest constituent in the atmosphere is diatomic nitrogen, at about 77%.

Today the atmospheric concentration of CO_2 is almost twice the concentration that prevailed over the last 400,000 years, as measured from air bubbles trapped in glacial ice in Antarctica (**Figure 9.19A and B**), and it is higher than any experienced on Earth in the last 2 million years. Most extant plant taxa are therefore thought to have evolved in a low-CO_2 world (~180–280 ppm CO_2). Only when one looks back about 35 million years does one find CO_2 concentrations of much higher levels (>1000 ppm). Thus, the geologic trend over these many millions of years was one of decreasing atmospheric CO_2 concentrations.

Currently, the CO_2 concentration of the atmosphere is increasing by about 1 to 3 ppm each year, primarily because of the burning of fossil fuels (e.g., coal, oil, and natural gas) and deforestation (**Figure 9.19C**). Since 1958, when C. David Keeling began systematic measurements of CO_2 in the clean air at Mauna Loa,

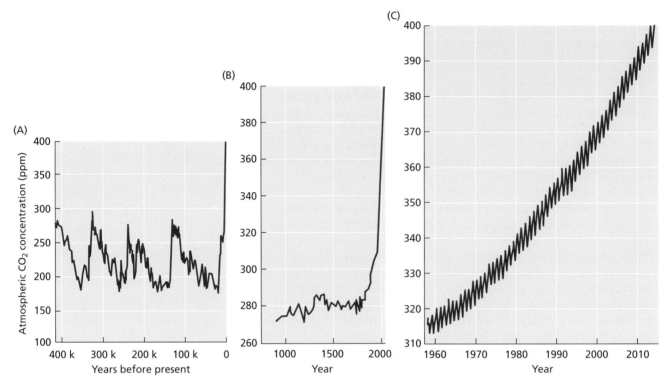

Figure 9.19 Concentration of atmospheric CO_2 from 420,000 years ago to the present. (A) Past atmospheric CO_2 concentrations, determined from bubbles trapped in glacial ice in Antarctica, were much lower than current levels. (B) In the last 1000 years, the rise in atmospheric CO_2 concentration coincides with the Industrial Revolution and the increased burning of fossil fuels. (C) Current atmospheric concentrations of CO_2 measured at Mauna Loa, Hawaii, continue to rise. The wavy nature of the trace is caused by changes in atmospheric CO_2 concentrations associated with seasonal changes in relative balance between photosynthesis and respiration rates. Each year the highest CO_2 concentration is observed in May, just before the Northern Hemisphere growing season, and the lowest concentration is observed in October. (A after Barnola et al. 2003; B after Etheridge et al. 1998; C after Keeling and Whorf 1994, updated using data from www.esrl.noaa.gov/gmd/ccgg/trends/ and scrippsco2.ucsd.edu/.)

Hawaii, atmospheric CO_2 concentrations have increased by more than 25%. By 2100, the atmospheric CO_2 concentration could reach 600 to 750 ppm unless fossil fuel emissions and deforestation are diminished.

CO_2 diffusion to the chloroplast is essential to photosynthesis

For photosynthesis to occur, CO_2 must diffuse from the atmosphere into the leaf and to the carboxylation site of Rubisco. The diffusion rate depends on the CO_2 concentration gradient in the leaf (see Chapters 3 and 6) and resistances along the diffusion pathway. The cuticle that covers the leaf is nearly impermeable to CO_2, so the main port of entry of CO_2 into the leaf is the stomatal pore. (The same path is traveled in the reverse direction by H_2O.) CO_2 diffuses through the pore into the substomatal cavity and into the intercellular air spaces between the mesophyll cells. This portion of the diffusion path of CO_2 into the chloroplast is a gaseous phase. The remainder of the diffusion path to the chloroplast is a liquid phase, which begins at the water layer that wets the walls of the mesophyll cells and continues through the plasma membrane, the cytosol, and the chloroplast.

The sharing of the stomatal entry pathway by CO_2 and H_2O presents the plant with a functional dilemma. In air of high relative humidity, the diffusion gradient that drives water loss is about 50 times larger than the gradient that drives CO_2 uptake. In drier air, this difference can be much larger. Therefore, a decrease in stomatal resistance through the opening of stomata facilitates higher CO_2 uptake, but is unavoidably accompanied by substantial water loss.

Not surprisingly, many adaptive features help counteract this water loss in plants in arid and semiarid regions of the world.

Each portion of the CO_2 diffusion pathway imposes a resistance to CO_2 diffusion, so the supply of CO_2 for photosynthesis meets a series of different points of resistance. The gaseous phase of CO_2 diffusion into the leaf can be divided into three components—the boundary layer, the stomata, and the intercellular spaces of the leaf—each of which imposes a resistance to CO_2 diffusion (**Figure 9.20**). An evaluation of the magnitude of each point of resistance is helpful for understanding CO_2 limitations to photosynthesis.

The boundary layer consists of relatively unstirred air near the leaf surface, and its resistance to diffusion is called the **boundary layer resistance**. The boundary layer resistance affects all diffusive processes, including water and CO_2 diffusion as well as sensible heat loss, discussed earlier. The boundary layer resistance decreases with smaller leaf size and greater wind speed. Smaller leaves thus have a lower resistance to CO_2 and water diffusion, and to sensible heat loss. Leaves of desert plants are usually small, facilitating sensible heat loss. In contrast, large leaves are often found in the humid tropics, especially in the shade. These leaves have large boundary layer resistances, but they can dissipate the radiation heat load by evaporative cooling made possible by the abundant water supply in these habitats.

After diffusing through the boundary layer, CO_2 enters the leaf through the stomatal pores, which impose the next type of resistance in the diffusion pathway, **stomatal resistance**. Under most conditions in nature, in which the air around a leaf is seldom completely still, the boundary layer resistance is much smaller than the stomatal resistance, and the main limitation to CO_2 diffusion into the leaf is imposed by the stomatal resistance.

There are two additional resistances within the leaf. The first is resistance to CO_2 diffusion in the air spaces that separate the substomatal cavity from the walls of the mesophyll cells. This is called the **intercellular air space resistance**. The second is the **mesophyll resistance**, which is resistance to CO_2 diffusion in the liquid phase in C_3 leaves. Localization of chloroplasts near the cell periphery minimizes the distance that CO_2 must diffuse through liquid to reach carboxylation sites within the chloroplast. The mesophyll resistance to CO_2 diffusion is thought to be approximately 1.4 times the combined boundary layer resistance and stomatal resistance when the stomata are fully open. Because the stomatal guard cells can impose a variable and potentially large resistance to CO_2 influx and water loss in the diffusion pathway, regulating the stomatal aperture provides the plant with an effective way to control gas exchange between the leaf and the atmosphere (see Chapter 3).

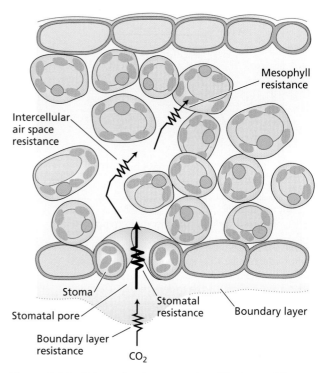

Figure 9.20 Points of resistance to the diffusion and fixation of CO_2 from outside the leaf to the chloroplasts. The stomatal aperture is the major point of resistance to CO_2 diffusion into the leaf.

CO_2 imposes limitations on photosynthesis

For C_3 plants growing with adequate light, water, and nutrients, CO_2 enrichment above natural atmospheric concentrations results in increased photosynthesis and enhanced productivity. Expressing the photosynthetic rate as a function of the partial pressure of CO_2 in the intercellular air space (c_i) within the leaf makes it possible to evaluate limitations to photosynthesis imposed by CO_2 supply. At low c_i concentrations, photosynthesis is strongly limited by the low CO_2. In the absence of atmospheric CO_2, leaves release CO_2 because of mitochondrial respiration (see Chapter 11).

boundary layer resistance
The resistance to the diffusion of water vapor, CO_2, and heat due to the layer of unstirred air next to the leaf surface. A component of diffusional resistance.

stomatal resistance A measurement of the limitation to the free diffusion of gases from and into the leaf posed by the stomatal pores. The inverse of stomatal conductance.

intercellular air space resistance
The resistance or hindrance that slows down the diffusion of CO_2 inside a leaf, from the substomatal cavity to the walls of the mesophyll cells.

mesophyll resistance The resistance to CO_2 diffusion imposed by the liquid phase inside leaves. The liquid phase includes diffusion from the intercellular leaf spaces to the carboxylation sites in the chloroplast.

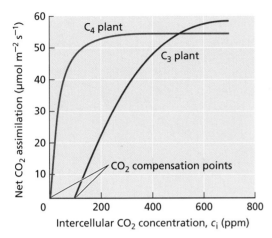

Figure 9.21 Changes in photosynthesis as a function of intercellular CO_2 concentrations in Arizona honeysweet (*Tidestromia oblongifolia*), a C_4 plant, and creosote bush (*Larrea tridentata*), a C_3 plant. Photosynthetic rate is plotted against calculated intercellular CO_2 concentration inside the leaf. The intercellular CO_2 concentration at which net CO_2 assimilation is zero defines the CO_2 compensation point. (After Berry and Downton 1982.)

Increasing c_i to the concentration at which photosynthesis and respiration balance each other defines the **CO_2 compensation point**. This is the point at which net assimilation of CO_2 by the leaf is zero (**Figure 9.21**). This concept is analogous to that of the light compensation point discussed earlier in this chapter (see Figures 9.6–9.8). The CO_2 compensation point reflects the balance between photosynthesis and respiration as a function of CO_2 concentration, whereas the light compensation point reflects that balance as a function of PPFD under constant CO_2 concentration.

C_3 VERSUS C_4 PLANTS   In C_3 plants, increasing c_i above the compensation point increases photosynthesis over a wide concentration range (see Figure 9.21). At low to intermediate, but still subambient, CO_2 concentrations, photosynthesis is limited by the carboxylation capacity of Rubisco. At higher c_i concentrations, photosynthesis begins to saturate as the net photosynthetic rate becomes limited by another factor (remember Blackman's concept of limiting factors). At these higher c_i levels, net photosynthesis becomes limited by the capacity of the light reactions to generate sufficient NADPH and ATP to regenerate the acceptor molecule ribulose 1,5-bisphosphate. Most leaves appear to regulate their c_i values by controlling stomatal opening, so that c_i remains at an intermediate, but still subambient, concentration between the limits imposed by carboxylation capacity and the capacity to regenerate ribulose 1,5-bisphosphate. In this way, both light capture and carbon fixation reactions of photosynthesis are co-limiting. A plot of net CO_2 assimilation as a function of c_i tells us how photosynthesis is regulated by CO_2, independent of the functioning of stomata (see Figure 9.21).

Comparing such a plot for C_3 and C_4 plants reveals interesting differences between the two pathways of carbon metabolism:

- In C_4 plants, photosynthetic rates saturate at c_i values of about 100–200 ppm, reflecting the effective CO_2-concentrating mechanisms operating in these plants (see Chapter 8).

- In C_3 plants, increasing c_i levels continue to stimulate photosynthesis over a much broader CO_2 range than for C_4 plants.

- In C_4 plants, the CO_2 compensation point is zero or nearly zero, reflecting their very low levels of photorespiration (see Chapter 8).

- In C_3 plants, the CO_2 compensation point is about 50–100 ppm at 25°C, reflecting CO_2 production because of photorespiration (see Chapter 8).

These responses reveal that C_3 plants will be more likely than C_4 plants to benefit from ongoing increases in today's atmospheric CO_2 concentrations (see Figure 9.21). Because photosynthesis in C_4 plants is saturated at low CO_2 concentrations, C_4 plants do not benefit much from increases in atmospheric CO_2 concentrations.

From an evolutionary perspective, the ancestral photosynthetic pathway is C_3 photosynthesis, and C_4 photosynthesis is a derived pathway. During earlier geologic time periods when atmospheric CO_2 concentrations were much higher than they are today, CO_2 diffusion through stomata into leaves would have resulted in higher c_i values and therefore higher photosynthetic rates in C_3 plants, but not in C_4 plants. The evolution of C_4 photosynthesis is one biochemical adaptation to a CO_2-limited atmosphere. Our current understanding is that C_4 photosynthesis evolved recently in geological terms, more than 20 million years ago.

If the ancient Earth of more than 50 million years ago had atmospheric CO_2 concentrations that were well above those of today, under what atmospheric conditions might we expect C_4 photosynthesis to become a major photosyn-

CO_2 compensation point   The CO_2 concentration at which the rate of respiration balances the photosynthetic rate.

Figure 9.22 Predicted crossover temperatures of the quantum yield for CO_2 uptake as a function of atmospheric CO_2 concentrations. At any given CO_2 concentration, the crossover temperature is defined as the temperature at which the quantum yields for CO_2 uptake for C_3 and the C_4 plants are equal. At temperatures above the crossover temperature, C_4 plants will have a higher quantum yield than C_3 plants and thus be favored; the opposite is the case at temperatures below the crossover temperature. At any point in time, Earth is at a single atmospheric CO_2 concentration, and the curve shows that C_4 plants would be more efficient, and therefore most common, in habitats with the warmest growing seasons. (After Ehleringer et al. 1997.)

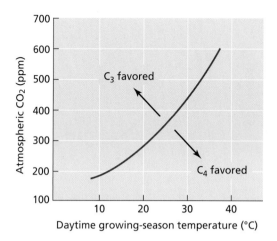

thetic pathway found in Earth's ecosystems? Jim Ehleringer's group suggests that C_4 photosynthesis first became a prominent component of terrestrial ecosystems in the warmest growing regions of Earth when global CO_2 concentrations decreased below some critical and, as yet, unknown threshold (**Figure 9.22**). Concurrently, the negative impacts of high photorespiration and CO_2 limitation on C_3 photosynthesis would have been greatest under these warm to hot growing conditions, and low atmospheric CO_2. C_4 plants would have been most favored during periods of Earth's history when CO_2 levels were lowest. There are now data to indicate that C_4 photosynthesis was more prominent during the glacial periods when atmospheric CO_2 levels were below 200 ppm (see Figure 9.19). Other factors may have contributed to the spread of C_4 plants, but certainly low atmospheric CO_2 was one important factor favoring their evolution and ultimately geographic expansion.

Because of the CO_2-concentrating mechanisms in C_4 plants, CO_2 concentration at the carboxylation sites within C_4 chloroplasts is typically close to saturating for Rubisco activity. As a result, plants with C_4 metabolism need less Rubisco than C_3 plants to achieve a given rate of photosynthesis, and thus require less nitrogen to grow. In addition, the CO_2-concentrating mechanism allows the leaf to maintain high photosynthetic rates at lower c_i values. This permits stomates to remain relatively closed, resulting in less water loss for a given rate of photosynthesis. Thus, the CO_2-concentrating mechanism helps C_4 plants use water and nitrogen more efficiently than C_3 plants. However, the additional energy cost required by the CO_2-concentrating mechanism (see Chapter 8) reduces the light-use efficiency of C_4 photosynthesis. This is probably one reason that relatively few shade-adapted plants in temperate regions are C_4 plants.

CAM PLANTS Plants with crassulacean acid metabolism (CAM), including many cacti, orchids, bromeliads, and other succulents, have stomatal activity patterns that contrast with those found in C_3 and C_4 plants. CAM plants open their stomata at night and close them during the day, exactly the opposite of the pattern observed in leaves of C_3 and C_4 plants (**Figure 9.23**). At night, atmospheric CO_2 diffuses into CAM plants where it is combined with phospho*enol*pyruvate and fixed into oxaloacetate, which is reduced to malate (see Chapter 8). Because stomata are open primarily at night, when lower temperatures and higher humidity reduce transpiration demand, the ratio of water loss to CO_2 uptake is much lower in CAM plants than it is in either C_3 or C_4 plants.

The main photosynthetic constraint on CAM metabolism is that the capacity to store malic acid is limited, and this limitation restricts the total amount of CO_2 uptake. However, the daily cycle of CAM photosynthesis can be very flexible. Some CAM plants are able to enhance total photosynthesis during wet conditions by fixing CO_2 via the Calvin–Benson cycle at the end of the day, when temperature gradients are less extreme. Other plants may use CAM

(A)

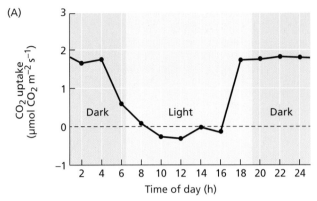

(B)

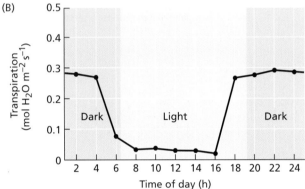

(C)

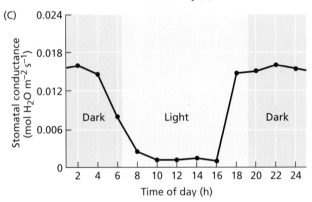

Figure 9.23 Photosynthetic net CO_2 assimilation, H_2O evaporation, and stomatal conductance of a CAM plant, the orchid *Doritaenopsis*, during a 24-h period. Whole plants were kept in a gas-exchange chamber in the laboratory. The dark period is indicated by dark brown areas. Three parameters were measured over the study period: (A) photosynthetic rate, (B) water loss, and (C) stomatal conductance. In contrast to plants with C_3 or C_4 metabolism, CAM plants open their stomata and fix CO_2 at night. (After Jeon et al. 2006.)

as a survival mechanism during severe water limitations. For example, cladodes (flattened stems) of cacti can survive after detachment from the plant for several months without water. Their stomata are closed all the time, and the CO_2 released by respiration is refixed into malate. This process, which has been called *CAM idling*, also allows the intact plant to survive for prolonged drought periods while losing remarkably little water.

How will photosynthesis and respiration change in the future under elevated CO_2 conditions?

The consequences of increasing global atmospheric CO_2 are under intense scrutiny by scientists and government agencies, particularly because of predictions that the greenhouse effect is altering the world's climate. The **greenhouse effect** refers to the warming of Earth's climate that is caused by the trapping of long-wavelength radiation by the atmosphere.

A greenhouse roof transmits visible light, which is absorbed by plants and other surfaces inside the greenhouse. Some of the absorbed light energy is converted to heat, and some of it is re-emitted as long-wavelength radiation. Because glass transmits long-wavelength radiation very poorly, this radiation cannot leave the greenhouse through the glass roof, and the greenhouse heats up. Certain gases in the atmosphere, particularly CO_2 and methane, play a role similar to that of the glass roof in a greenhouse. The increased CO_2 concentrations, and elevated temperatures associated with the greenhouse effect, have multiple influences on photosynthesis and plant growth. At today's atmospheric CO_2 concentrations, photosynthesis in C_3 plants is CO_2-limited, but this situation will change as atmospheric CO_2 concentrations continue to rise.

A central question in plant physiology today is: How will photosynthesis and respiration differ by the year 2100 when global CO_2 levels have reached 500 ppm, 600 ppm, or even higher? This question is particularly relevant as humans continue to add CO_2 derived from fossil fuel combustion to Earth's atmosphere. Under well-watered and highly fertilized laboratory conditions, most C_3 plants grow about 30% faster when the CO_2 concentration reaches 600 to 750 ppm than they do today; above that atmospheric CO_2 concentration, the growth rate becomes more limited by the nutrients available to the plant. To study this question in the field, scientists need to be able to create realistic simulations of future environments. A promising approach to the study of plant physiology and ecology in environments with elevated CO_2 levels has been the use of *Free Air CO_2 Enrichment* (FACE) experiments.

For FACE experiments, entire fields of plants or natural ecosystems are encircled by emitters that add CO_2 to the air to create the high-CO_2 environment

greenhouse effect The warming of Earth's climate, caused by the trapping of long-wavelength radiation by CO_2 and other gases in the atmosphere. The term is derived from the heating of a greenhouse that results from the penetration of long-wavelength radiation through the glass roof, the conversion of the long-wave radiation to heat, and the blocking of the heat escape by the glass roof.

(A)

Figure 9.24 Free Air CO_2 Enrichment (FACE) experiments are used to study how plants and ecosystems will respond to future CO_2 levels. Shown here are FACE experiments in (A) a deciduous forest and (B) a crop canopy. (C) Under elevated CO_2 levels, leaf stomata are more closed, resulting in higher leaf temperatures, as shown by the infrared image of a crop canopy. (A courtesy of D. Karnosky; B courtesy of the USDA/ARS; C from Long et al. 2006.)

(B)

(C)

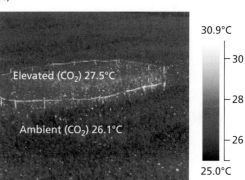

we might expect to have 25 to 50 years from now. **Figure 9.24** shows FACE experiments in two different vegetation types.

FACE experiments have provided key new insights into how plants and ecosystems will respond to the CO_2 levels expected in the future. One key observation is that plants with the C_3 photosynthetic pathway are much more responsive than C_4 plants under well-watered conditions, with net photosynthetic rates increasing 20% or more in C_3 plants and little to not at all in C_4 plants. Photosynthesis increases in C_3 plants because c_i levels increase (see Figure 9.21). At the same time, there is a down-regulation of photosynthetic capacity manifested in reduced activity of the enzymes associated with the carbon reactions of photosynthesis.

Elevated CO_2 levels will affect many plant processes. For instance, leaves tend to keep their stomata more closed under elevated CO_2 levels. As a direct consequence of reduced transpiration, leaf temperatures are higher (see Figure 9.24C), which may feed back on basic mitochondrial respiration. This is indeed an exciting and promising area of current research. From FACE studies, it has become increasingly clear that an acclimation process occurs under higher CO_2 levels in which respiration rates are different than they would be under today's atmospheric conditions, but not as high as would have been predicted without the down-regulation acclimation response.

While CO_2 is indeed important for photosynthesis and respiration, other factors are important for growth under elevated CO_2. For example, a common FACE observation is that plant growth under elevated CO_2 levels quickly becomes constrained by nutrient availability (remember Blackman's rule of limiting factors). A second, surprising, observation is that the presence of pollutant trace gases, such as ozone, can reduce the net photosynthetic response below the maximum values predicted from initial FACE and greenhouse studies of a decade ago.

As a result of increased atmospheric CO_2, warmer and drier conditions are also predicted to occur in the near future, and nutrient limitations are predicted to increase. Important progress is being made by studying how growth of irrigated and fertilized crops compares with that of plants in natural ecosystems in a world with elevated CO_2. Understanding these responses is crucial as society looks for increased agricultural outputs to support rising human populations and to provide raw materials for biofuels.

Summary

In considering photosynthetic performance, both the limiting factor hypothesis and an "economic perspective" emphasizing CO_2 "supply" and "demand" have guided research.

The Effect of Leaf Properties on Photosynthesis

- Leaf anatomy is highly specialized for light absorption (**Figure 9.1**).

- About 5% of the solar energy reaching Earth is converted into carbohydrates by photosynthesis. Much absorbed light is lost in reflection and transmission, in metabolism, and as heat (**Figures 9.2, 9.3**).

- In dense forests, almost all photosynthetically active radiation is absorbed by leaves in the canopy and little reaches the forest floor (**Figure 9.4**).

- Leaves of some plants maximize light absorption by solar tracking (**Figure 9.5**).

- Some plant species respond to a range of light regimes. Sun and shade leaves have contrasting morphological and biochemical characteristics.

- To increase light absorption, some shade plants produce a higher ratio of PSII to PSI reaction centers, while others add antenna chlorophyll to PSII.

Effects of Light on Photosynthesis in the Intact Leaf

- Light-response curves show the PPFD where photosynthesis is limited by light or by carboxylation capacity. The slope of the linear portion of the light-response curve measures the maximum quantum yield (**Figure 9.6**).

- Light compensation points for shade plants are lower than for sun plants because respiration rates in shade plants are very low (**Figures 9.7, 9.8**).

- Beyond the saturation point, factors other than incident light, such as electron transport, Rubisco activity, or triose phosphate metabolism, limit photosynthesis. Rarely is an entire plant light-saturated (**Figure 9.9**).

- The xanthophyll cycle dissipates excess absorbed light energy to avoid damaging the photosynthetic apparatus (**Figures 9.10–9.12**). Chloroplast movements also limit excess light absorption (**Figure 9.13**).

- Dynamic photoinhibition temporarily diverts excess light absorption to heat but maintains maximum photosynthetic rate (**Figure 9.14**). Chronic photoinhibition is irreversible.

Effects of Temperature on Photosynthesis in the Intact Leaf

- Plants are remarkably plastic in their adaptations to temperature. Optimal photosynthetic temperatures have strong biochemical, genetic (adaptation), and environmental (acclimation) components.

- Leaf absorption of light energy generates a heat load that must be dissipated (**Figure 9.15**).

- Temperature-sensitivity curves identify (a) a temperature range where enzymatic events are stimulated, (b) a range for optimal photosynthesis, and (c) a range where deleterious events occur (**Figure 9.16**).

- Below 30°C, the quantum yield of C_3 plants is higher than that of C_4 plants; above 30°C, the situation is reversed (**Figure 9.17**). Because of photorespiration, the quantum

(Continued)

Summary *(continued)*

yield is strongly dependent on temperature in C_3 plants, but is nearly independent of temperature in C_4 plants.

- Reduced quantum yield and increased photorespiration due to temperature effects lead to differences in the photosynthetic capacities of C_3 and C_4 plants and result in a shift of species dominance across different latitudes (**Figure 9.18**).

Effects of Carbon Dioxide on Photosynthesis in the Intact Leaf

- Atmospheric CO_2 levels have been increasing since the Industrial Revolution because of human use of fossil fuels and deforestation (**Figure 9.19**).

- Concentration gradients drive the diffusion of CO_2 from the atmosphere to the carboxylation site in the leaf, using both gaseous and liquid routes. There are multiple resis-

tances along the CO_2 diffusion pathway, but under most conditions stomatal resistance has the greatest effect on CO_2 diffusion into a leaf (**Figure 9.20**).

- Enrichment of CO_2 above natural atmospheric levels results in increased photosynthesis and productivity (**Figure 9.21**).

- C_4 photosynthesis may have become prominent in Earth's warmest regions when global atmospheric CO_2 concentrations fell below a threshold value (**Figure 9.22**).

- Stomata in CAM plants open at night and close during the day, which is the opposite pattern to that found in C_3 and C_4 plants (**Figure 9.23**).

- Free Air CO_2 Enrichment (FACE) experiments show that C_3 plants are more responsive to elevated CO_2 than are C_4 plants (**Figure 9.24**).

Suggested Reading

Adams, W. W., Zarter, C. R., Ebbert, V., and Demmig-Adams, B. (2004) Photoprotective strategies of overwintering evergreens. *Bioscience* 54: 41–49.

Koller, D. (2000) Plants in search of sunlight. *Adv. Bot. Res.* 33: 35–131.

Long, S. P., Ainsworth, E. A., Leakey, A. D., Nosberger, J., and Ort, D. R. (2006) Food for thought: Lower-than-expected crop stimulation with rising CO_2 concentrations. *Science* 312: 1918–1921.

Long, S. P., Ainsworth, E. A., Rogers, A., and Ort, D. R. (2004) Rising atmospheric carbon dioxide: Plants FACE the future. *Annu. Rev. Plant Biol.* 55: 591–628.

Sharkey, T. D. (1996) Emission of low molecular mass hydrocarbons from plants. *Trends Plant Sci.* 1: 78–82.

Terashima, I., and Hikosaka, K. (1995) Comparative ecophysiology of leaf and canopy photosynthesis. *Plant Cell Environ.* 18: 1111–1128.

Vogelmann, T. C. (1993) Plant tissue optics. *Annu. Rev. Plant Physiol. Plant Mol. Biol.* 44: 231–251.

von Caemmerer, S. (2000) *Biochemical models of leaf photosynthesis.* CSIRO, Melbourne, Australia.

Zhu, X. G., Long, S. P., and Ort, D. R. (2010) Improving photosynthetic efficiency for greater yield. *Annu. Rev. Plant Biol.* 61: 235–261.

10 Translocation in the Phloem

S urvival on land poses some serious challenges to terrestrial plants; foremost among these challenges is the need to acquire and retain water. In response to such environmental pressures, plants evolved roots and leaves. Roots anchor the plant and absorb water and nutrients; leaves absorb light and exchange gases. As plants increased in size, the roots and leaves became increasingly separated from each other in space. Thus, systems evolved for long-distance transport that allowed the shoot and the root to efficiently exchange products of absorption and assimilation.

You will recall from Chapters 3 and 6 that the xylem is the tissue that transports water and minerals from the root system to the aerial portions of the plant. The **phloem** is the tissue that transports (*translocates*) the products of photosynthesis—particularly sugars—from mature leaves to areas of growth and storage, including the roots.

Along with sugars, the phloem also transmits signals in the form of regulatory molecules, and redistributes water and various compounds throughout the plant body. All of these molecules appear to move with the transported sugars. The compounds to be redistributed, some of which initially arrive in the mature leaves via the xylem, can be either transferred out of the leaves without modification or metabolized before redistribution. The fluid that flows through the phloem—the water plus all its solutes—is called *phloem sap*. (*Sap* is a general term used to refer to the fluid contents of plant cells.)

phloem The tissue that transports the products of photosynthesis from mature leaves (or storage organs) to areas of growth and storage, including the roots.

source Any organ that is capable of exporting photosynthetic products in excess of its own needs, such as a mature leaf or a storage organ.

sink Any organ that imports photosynthate, including nonphotosynthetic organs and organs that do not produce enough photosynthetic products to support their own growth or storage needs, such as roots, tubers, developing fruits, and immature leaves.

collection phloem Sieve elements of sources.

release phloem Sieve elements of sinks where sugars and other photosynthetic products are unloaded into sink tissues.

photosynthate A carbon-containing product of photosynthesis.

The discussion that follows describes translocation in the phloem of angiosperms, because most of the research has been conducted on that group of plants. The translocation mechanism may be different in gymnosperms, but it is less well understood and will not be covered here.

Patterns of Translocation: Source to Sink

The two long-distance transport pathways—the phloem and the xylem—extend throughout the plant body. Unlike the xylem, the phloem does not translocate materials exclusively in either an upward or a downward direction, and translocation in the phloem is not defined with respect to gravity. Rather, sap is translocated from areas of supply, called **sources**, to areas of metabolism or storage, called **sinks** (**Figure 10.1**). Because of their roles in sugar transport, the sieve elements (the conducting cells of the phloem; see next section) of sources are often referred to as **collection phloem**, the sieve elements of the connecting pathway as *transport phloem*, and the sieve elements of sinks as **release phloem**.

Sources include exporting organs, typically mature leaves that are capable of producing photosynthate in excess of their own needs. The term **photosynthate** refers to products of photosynthesis. Another type of source is a storage organ during the exporting phase of its development. For example, the storage root of the biennial wild beet (*Beta maritima*) is a sink during the growing season of the first year, when it accumulates sugars received from the source leaves. During the second growing season, the same root becomes a source; the sugars are remobilized and used to produce a new shoot, which ultimately becomes reproductive.

Sinks include all nonphotosynthetic organs of the plant and organs that do not produce enough photosynthetic products to support their own growth or storage needs. Roots, tubers, developing fruits, and immature leaves, which must import carbohydrate for normal development, are all examples of sink tissues. Both girdling and labeling studies support the source-to-sink pattern of translocation in the phloem (**Figure 10.2A**).

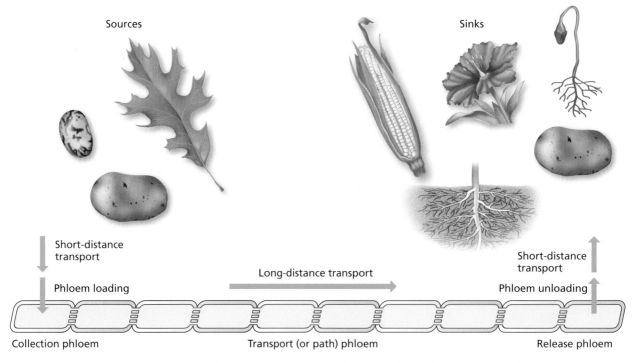

Figure 10.1 Phloem translocates materials from sources to sinks.

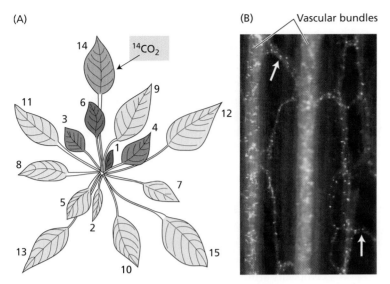

Figure 10.2 Source-to-sink patterns of phloem translocation. (A) Distribution of radio-activity from a single labeled source leaf in an intact plant. The distribution of radioactivity in leaves of a sugar beet plant (*Beta vulgaris*) was determined 1 week after $^{14}CO_2$ was supplied for 4 h to a single source leaf (leaf 14, arrow). The degree of radioactive labeling is indicated by the intensity of shading of the leaves. Leaves are numbered according to their age; the youngest, newly emerged leaf is designated 1. The ^{14}C label was trans-located mainly to the sink leaves directly above the source leaf (that is, sink leaves with the most direct vascular connections to the source; for example, leaves 1 and 6 are sink leaves directly above source leaf 14). (B) Longitudinal view of a typical three-dimensional structure of the phloem in a thick section (from an internode of dahlia [*Dahlia pinnata*]), viewed here after clearing, staining with aniline blue, and observing under an epifluo-rescence microscope. The sieve plates of the phloem are seen as numerous small yellow dots because of the yellow staining of callose in the sieve elements (see Figure 10.7). Two large longitudinal vascular bundles are prominent. This staining reveals the delicate sieve tubes forming the phloem network; two phloem anastomoses (vascular interconnections) are marked by arrows. (A after Joy 1964; B courtesy of R. Aloni.)

Photosynthate enters the phloem by **phloem loading** (see Figure 10.1), a general term for a variety of different mechanisms for sugar uptake into the phloem. Phloem loading is preceded by **short-distance transport** from the photosynthate-exporting cells within the source tissue to the sites of seive cell loading. Translocation through the vascular system from a source to a sink occurs via **long-distance transport**. Once photosynthate arrives at the sink, it exits the phloem, a process called **phloem unloading**. Short-distance transport carries photosynthate from the seive cells to the sink cells outside the phloem.

Although the overall pattern of transport in the phloem can be stated simply as source-to-sink movement, the specific pathways involved are often more complex, depending on proximity, development, vascular connections (**Figure 10.2B**), and modification of translocation pathways. Not all sources supply all sinks on a plant; rather, certain sources preferentially supply specific sinks.

Pathways of Translocation

The phloem is generally found on the outer side of both primary and secondary vascular tissues (**Figures 10.3 and 10.4**). In plants with secondary growth, the phloem constitutes the inner bark. Although phloem is commonly found in a position external to the xylem, it is *also* found on the inner side in many eudicot

phloem loading The movement of photosynthetic products into the sieve elements of mature leaves.

short-distance transport Transport over a distance of only two or three cell diameters. Precedes phloem loading, when sugars move from the mesophyll to the vicinity of the smallest veins of the source leaf, and follows phloem unloading, when sugars move from the veins to the sink cells.

long-distance transport Transloca-tion through the phloem to the sink.

phloem unloading The movement of photosynthates from the sieve ele-ments to neighboring cells that store or metabolize them or pass them on to other sink cells via short-distance transport.

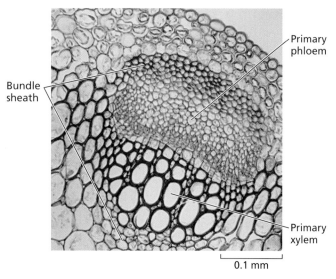

Figure 10.3 Transverse section of a vascular bundle of trefoil, a clover (*Trifolium*). The primary phloem is toward the outside of the stem. Both the primary phloem and the primary xylem are surrounded by a bundle sheath of thick-walled sclerenchyma cells, which isolate the vascular tissue from the ground tissue. Fibers and xylem vessels are stained red. (© J.N.A. Lott/Biological Photo Service.)

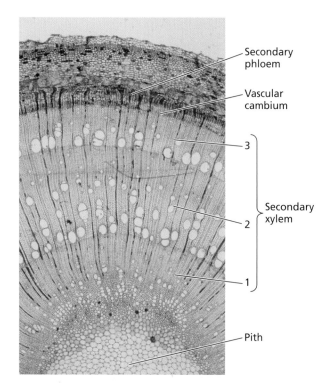

Figure 10.4 Transverse section of a 3-year-old stem of an ash (*Fraxinus excelsior*) tree. The numbers 1, 2, and 3 indicate growth rings in the secondary xylem. The old (outer) secondary phloem has been crushed by expansion of the xylem. Only the most recent (innermost) layer of secondary phloem is functional. (© P. Gates/ Biological Photo Service.)

families. In these families the phloem in the two positions is called external and internal phloem, respectively.

The cells of the phloem that conduct sugars and other organic materials throughout the plant are called **sieve elements**. In angiosperms, the highly differentiated *sieve elements* are often called **sieve tube elements**. In addition to sieve elements, the phloem tissue contains companion cells (discussed below) and parenchyma cells (which store and release food molecules). In some cases the phloem tissue also includes fibers and sclereids (for protection and strengthening of the tissue) and laticifers (latex-containing cells). However, only the sieve elements are directly involved in translocation.

The small veins of leaves and the primary vascular bundles of stems are often surrounded by a **bundle sheath** (see Figure 10.3), which consists of one or more layers of compactly arranged cells. (You will recall the bundle sheath cells involved in C$_4$ metabolism discussed in Chapter 8.) In the vascular tissue of leaves, the bundle sheath surrounds the small veins all the way to their ends, isolating the veins from the intercellular spaces of the leaf.

We begin our discussion of translocation pathways with the experimental evidence demonstrating that the sieve elements are the conducting cells in the phloem. Then we examine the structure and physiology of these unusual plant cells.

Sugar is translocated in phloem sieve elements

Early experiments on phloem transport date back to the nineteenth century, indicating the importance of long-distance transport in plants. These classic experiments demonstrated that removal of a ring of bark around the trunk of a tree, which removes the phloem, effectively stops sugar transport from the leaves to the roots without altering water transport through the xylem. When radioactive compounds became available, $^{14}CO_2$ was used to show that sugars made in the photosynthetic process are translocated through the phloem sieve elements (see Figure 10.2A).

Mature sieve elements are living cells specialized for translocation

Detailed knowledge of the ultrastructure of sieve elements is critical to any discussion of the mechanism of translocation in the phloem. Mature sieve elements are unique among living plant cells (**Figures 10.5 and 10.6**). They lack many structures normally found in living cells, even in the undifferentiated cells from which they are formed. For example, sieve elements lose their nucleus and tonoplast (vacuolar membrane) during development. Microfilaments, microtubules, Golgi bodies, and ribosomes are also generally absent from the mature cells. In addition to the plasma membrane, organelles that are retained include somewhat modified mitochondria, plastids, and smooth endoplasmic reticulum. The walls are nonlignified, though they are secondarily thickened in some cases.

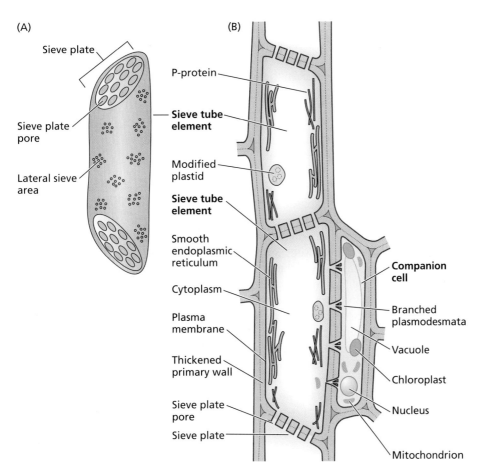

Figure 10.5 Schematic drawings of mature sieve elements (sieve tube elements) joined together to form a sieve tube. (A) External view of a single sieve element, showing sieve plates and lateral sieve areas. (B) Longitudinal section, showing two sieve tube elements joined together to form a sieve tube. The pores in the sieve plates between the sieve tube elements are open channels for transport through the sieve tube. The plasma membrane of a sieve tube element is continuous with that of its neighboring sieve tube element. Each sieve tube element is associated with one or more companion cells, which take over some of the essential metabolic functions that are reduced or lost during differentiation of the sieve tube elements. Note that the companion cell has many cytoplasmic organelles, whereas the sieve tube element has relatively few organelles. An ordinary companion cell of a leaf is depicted here.

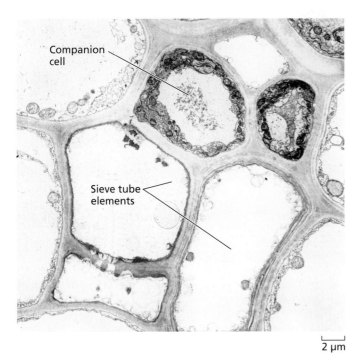

2 μm

Figure 10.6 Electron micrograph of a transverse section of ordinary companion cells and mature sieve tube elements. The cellular components are distributed along the walls of the sieve tube elements, where they offer less resistance to mass flow. (From Warmbrodt 1985.)

sieve elements Cells of the phloem that conduct sugars and other organic materials throughout the plant. Refers to both sieve tube elements (angiosperms) and sieve cells (gymnosperms).

sieve tube elements The highly differentiated sieve elements typical of the angiosperms.

bundle sheath One or more layers of closely packed cells surrounding the small veins of leaves and the primary vascular bundles of stems.

sieve plates Sieve areas found in angiosperm sieve tube elements; they have larger pores (sieve plate pores) than other sieve areas and are generally found in end walls of sieve tube elements.

sieve tube Tube formed by the joining together of individual sieve tube elements at their end walls.

Thus, the cellular structure of sieve elements is different from that of tracheary elements of the xylem, which lack a plasma membrane, have lignified secondary walls, and are dead at maturity. As we will see, living cells are critical to the mechanism of translocation in the phloem.

Large pores in cell walls are the prominent feature of sieve elements

Sieve elements (sieve cells and sieve tube elements) have characteristic sieve areas in their cell walls, where pores interconnect the conducting cells (**Figure 10.7**). The sieve area pores range in diameter from less than 1 μm to approximately 15 μm. The sieve areas of angiosperms can differentiate into **sieve plates** (see Figures 10.5 and 10.7 and Table 10.1). Sieve plates have larger pores than the other sieve areas in the cell and are generally found on the end walls of sieve tube elements, where the individual cells are joined together to form a longitudinal series called a **sieve tube** (see Figure 10.5B).

The distribution of sieve tube contents, especially within the sieve plate pores, has been debated for many years and is a critical question when considering the mechanism of phloem transport. Early micrographs showed blocked or occluded pores, thought to be artifacts due to damage as the tissues were prepared for observation. (See next section, *Damaged sieve elements are sealed off.*) Later, less invasive techniques often showed the sieve plate pores of sieve tube elements to be open channels that allow unfettered transport between cells (see Figure 10.7A–C). A later section (*Sieve plate pores appear to be open channels*) further considers the distribution of sieve element contents within the cells and within the sieve plate pores.

Table 10.1 lists characteristics of sieve tube elements.

(A)

(B)

(C)

(D)

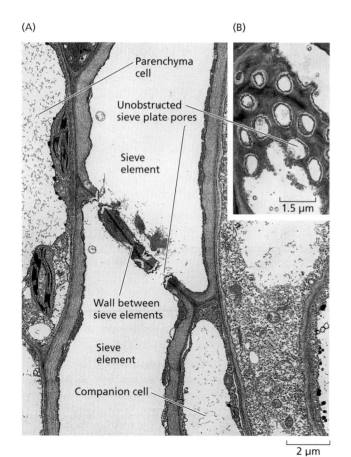

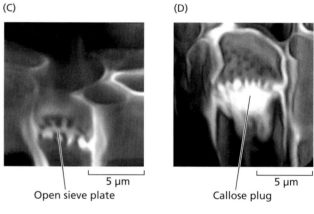

Figure 10.7 Sieve elements and sieve plate pores. In images A, B, and C, the sieve plate pores are open—that is, unobstructed by P-protein or callose. Open pores provide a low-resistance pathway for transport between sieve elements. (A) Electron micrograph of a longitudinal section of two mature sieve elements (sieve tube elements), showing the wall between the sieve elements (called a sieve plate) in the hypocotyl of winter squash (*Cucurbita maxima*). (B) The inset shows sieve plate pores in face view. (C and D) Three-dimensional reconstructions of Arabidopsis sieve plates using a staining technique that can be used to image entire plant organs with confocal laser scanning microscopy. Open sieve pores are visible in (C), while a callose plug, such as that formed in response to sieve tube damage, is visible in (D). (A and B from Evert 1982; C and D from Truernit et al. 2008.)

Table 10.1 Characteristics of sieve elements in angiosperms

Sieve tube elements found in angiosperms

1. Some sieve areas are differentiated into sieve plates; individual sieve tube elements are joined together into a sieve tube.
2. Sieve plate pores are open channels.
3. P-protein is present in all eudicots and many monocots.
4. Companion cells are sources of ATP and perhaps other compounds.

Damaged sieve elements are sealed off

Sieve element sap is rich in sugars and other organic molecules. These molecules represent an energy investment for the plant, and their loss must be prevented when sieve elements are damaged. Short-term sealing mechanisms involve sap proteins, while the principal long-term mechanism for preventing sap loss entails closing sieve plate pores with callose, a glucose polymer.

The main phloem proteins involved in sealing damaged sieve elements are structural proteins called **P-proteins** (see Figure 10.5B). (In earlier scientific literature, P-protein was called *slime*.) The sieve tube elements of most angiosperms, including those of all eudicots and many monocots, are rich in P-protein, which occurs in several different forms (tubular, fibrillar, granular, and crystalline), depending on the species and maturity of the cell.

P-protein appears to function in sealing off damaged sieve elements by plugging sieve plate pores. Sieve tubes are under very high internal turgor pressure, and the sieve elements in a sieve tube are connected through sieve plate pores that appear to be open. When a sieve tube is cut or punctured, the release of pressure causes the contents of the sieve elements to surge toward the cut end, from which the plant could lose much sugar-rich phloem sap if there were no sealing mechanism. When surging occurs, however, P-protein is trapped on the sieve plate pores, helping to seal the sieve element and prevent further loss of sap. Direct support for the sealing function of P-protein has been found in both tobacco and Arabidopsis, in which mutants lacking P-protein lose significantly more transport sugar by sap exudation after wounding than do wild-type plants. (See *Phloem sap can be collected and analyzed* below for more information about exudation.) No visible phenotypic differences were observed between the mutant and wild-type plants.

Protein crystals released from ruptured plastids may play a similar sealing role in some monocots as P-protein does in eudicots. The sieve element organelles (mitochondria, plastids, and ER), however, appear to be anchored to each other or to the sieve element plasma membrane by minute protein "clamps." Which organelles are anchored depends on the species.

Another mechanism for blocking wounded sieve tubes with proteins occurs in plants in the legume family (Fabaceae). These plants contain large crystalloid P-proteins that do not disperse during development. However, following damage or osmotic shock, the P-proteins rapidly disperse and block the sieve tube. The process is reversible and controlled by calcium ions. These P-proteins occur only in certain legumes.

A longer-term solution to sieve tube damage is the production of the glucose polymer **callose** in the sieve pores (see Figure 10.7D). Callose, a β-1,3-glucan, is synthesized by an enzyme in the plasma membrane (callose synthase) and is deposited between the plasma membrane and the cell wall. Callose is synthesized in functioning sieve elements in response to damage

P-proteins Phloem proteins that act to seal damaged sieve elements by plugging the sieve element pores. Abundant in the sieve elements of most angiosperms, but absent from gymnosperms.

callose A β-1,3-glucan synthesized at the plasma membrane and deposited between the plasma membrane and the cell wall. Synthesized by sieve elements in response to damage, stress, or as part of a normal developmental process.

wound callose Callose deposited in the sieve pores of damaged sieve elements that seals them off from surrounding intact tissue. As sieve elements recover, the callose disappears from pores.

companion cells In angiosperms, metabolically active cells that are connected to their sieve element by large, branched plasmodesmata and that take over many of the metabolic activities of the sieve element. In source leaves, they function in the transport of photosynthate into the sieve elements.

and other stresses, such as mechanical stimulation and high temperatures, or in preparation for normal developmental events, such as dormancy. The deposition of **wound callose** in the sieve pores efficiently seals off damaged sieve elements from surrounding intact tissue, with complete occlusion occurring about 20 min after injury. In all cases, as sieve elements recover from damage or break dormancy, the callose disappears from the sieve pores; its dissolution is mediated by a callose-hydrolyzing enzyme. As mentioned earlier, Arabidopsis and tobacco mutants lacking P-protein show no visible phenotypic changes, but Arabidopsis mutants lacking one callose synthase show reduced inflorescence growth, apparently due to reduced assimilate transport to the inflorescence.

Callose deposition is induced and callose synthase genes are up-regulated in rice (*Oryza sativa*) plants attacked by a phloem-feeding insect (brown planthopper); this occurs both in plants resistant to the insect and in susceptible plants. In the susceptible plants, however, feeding by the insects also activates genes for a callose-hydrolyzing enzyme. This unplugs the pores, allows continued feeding, and results in decreased sucrose and starch levels in the leaf sheath being attacked. Sealing off sieve elements that have been penetrated by insect mouthparts can thus play a key role in herbivore resistance.

Companion cells aid the highly specialized sieve elements

Each sieve tube element is usually associated with one or more **companion cells** (see Figures 10.5B, 10.6, and 10.7A). The division of a single mother cell forms the sieve tube element and the companion cell. Numerous plasmodesmata (see Chapter 1) penetrate the walls between sieve tube elements and their companion cells; the plasmodesmata are often complex and branched on the companion cell side. The presence of abundant plasmodesmata suggests a close functional relationship between a sieve element and its companion cell, an association that is demonstrated by the rapid exchange of solutes, such as fluorescent dyes, between the two cells.

Companion cells play a role in the transport of photosynthetic products from producing cells in mature leaves to the sieve elements in the minor (small) veins of the leaf. They also take over some of the critical metabolic functions, such as protein synthesis, that are reduced or lost during differentiation of the sieve elements. In addition, the numerous mitochondria in companion cells may supply energy as ATP to the sieve elements.

Materials Translocated in the Phloem

Water is the most abundant substance in the phloem. Dissolved in the water are the translocated solutes, including carbohydrates, amino acids, hormones, some inorganic ions, RNAs and proteins, and some secondary compounds involved in defense and protection. Carbohydrates are the most significant and concentrated solutes in phloem sap (**Table 10.2**), with sucrose being the sugar most commonly transported in sieve elements. There is always some sucrose in sieve element sap, and it can reach concentrations of 0.3 to 0.9 M. Sugars, potassium ions, and amino acids and their amides are the principal molecules contributing to the osmotic potential of the phloem.

Complete identification of solutes that are mobile in the phloem and that have a significant function has been difficult; no one method of sampling phloem sap is free of artifacts or provides a complete picture of mobile solutes. We begin this discussion with a brief examination of the available sampling methods, and then continue with a description of the solutes that are currently accepted as significant mobile substances in the phloem.

Table 10.2 **The composition of phloem sap from castor bean (*Ricinus communis*), collected as an exudate from cuts in the phloem**

Component	Concentration (mg mL^{-1})
Sugars	80.0–106.0
Amino acids	5.2
Organic acids	2.0–3.2
Protein	1.45–2.20
Potassium ions	2.3–4.4
Chloride	0.355–0.675
Phosphate	0.350–0.550
Magnesium ions	0.109–0.122

Source: Hall and Baker 1972.

Phloem sap can be collected and analyzed

The collection of phloem sap is experimentally challenging because of the high turgor pressure in the sieve elements and the wound reactions described previously. Because of processes that plug sieve plate pores, only a few species exude phloem sap from wounds that sever sieve elements. Considerable challenges and problems present themselves when exuded sap is collected from cuts or wounds:

- The initial samples may be contaminated by the contents of surrounding damaged cells.

- In addition to plugging the sieve plate pores, sudden pressure release in sieve elements can disrupt cellular organelles and proteins and even pull substances from surrounding cells, especially the companion cells. Some materials, such as the Rubisco small subunit, are expected to be present only in tissues surrounding the phloem; failure to detect these materials in collected sap provides evidence that no significant contamination from surrounding tissues has occurred.

- The exudate is substantially diluted by the influx of water from the xylem and surrounding cells when the pressure/tension in the vascular tissue is released.

- The sap of cucurbit species such as cucumber (*Cucumis sativus*) and pumpkin (*Cucurbita maxima*) has been used in many studies of translocated materials. These species have a complex phloem, including both internal and external sieve tubes (see the section *Pathways of Translocation* above), as well as sieve tubes outside the vascular bundles. In addition to the concerns listed above, the source of exudate in these species could be any of the sieve tubes present and could differ among species.

Exudation of sap from cut petioles or stems, enhanced by the inclusion of EDTA in the collection fluid, has also been used in several studies. Chelating agents such as EDTA bind calcium ions, thus inhibiting callose synthesis (which requires calcium ions) and allowing exudation to occur for extended periods. However, exudation into EDTA is subject to several additional technical problems, such as the leakage of solutes, including hexoses, from the affected tissues and is not a reliable method of obtaining phloem sap for analysis.

A preferable approach for collecting exuded sap is to use an aphid stylet as a "natural syringe." Aphids are small insects that feed by inserting their mouthparts, consisting of four tubular stylets, into a single sieve element of a leaf or stem. Sap can be collected from aphid stylets cut from the body of the insect, usually with a laser, after the aphid has been anesthetized with CO_2. The high turgor pressure in the sieve element forces the cell contents through the stylet to the cut end, where they can be collected. However, quantities of collected sap are small, and the method is technically difficult. Furthermore, exudation from severed stylets can continue for hours, suggesting that substances in aphid saliva prevent the plant's normal sealing mechanisms from operating and potentially alter the sap contents. Nonetheless, this method is thought to yield relatively pure sap from the sieve elements and companion cells and to provide a fairly accurate picture of the composition of phloem sap.

Sugars are translocated in a nonreducing form

Results from many analyses of collected sap indicate that the translocated carbohydrates are nonreducing sugars. Reducing sugars, such as the hexoses glucose and fructose, contain an exposed aldehyde or ketone group (**Figure 10.8A**). In a nonreducing sugar, such as sucrose, the ketone or aldehyde group is reduced to an alcohol or combined with a similar group on another sugar, eliminating both groups (**Figure 10.8B**).

Most researchers believe that the nonreducing sugars are the major compounds translocated in the phloem because they are less reactive than their reducing counterparts. In fact, reducing sugars such as hexoses are quite reactive and may be as much of a threat as reactive oxygen and nitrogen species. Animals can tolerate transporting glucose because it is present in fairly low concentrations in the blood, but hexoses cannot be tolerated in the phloem, in which very high sugar levels are maintained. Mechanistically, hexoses are sequestered in the vacuoles of plant cells and so have no direct access to the phloem.

Sucrose is the most commonly translocated sugar; many of the other mobile carbohydrates contain sucrose bound to varying numbers of galactose molecules. Raffinose consists of sucrose and one galactose molecule, stachyose consists of sucrose and two galactose molecules, and verbascose consists of sucrose and three galactose molecules (see Figure 10.8B). Translocated sugar alcohols include mannitol and sorbitol.

Other solutes are translocated in the phloem

Nitrogen is found in the phloem largely in amino acids—especially glutamate and aspartate and their respective amides, glutamine and asparagine (also amino acids). Reported levels of amino acids and organic acids vary widely, even for the same species, but they are usually low compared with carbohydrates (see Table 10.2). A variety of proteins and RNAs occur in phloem sap in relatively low concentrations. RNAs found in the phloem include mRNAs, pathogenic RNAs, and small regulatory RNA molecules.

Almost all the endogenous plant hormones, including auxin, gibberellins, cytokinins, and abscisic acid, have been found in sieve elements. The long-distance transport of hormones, especially auxin, is thought to occur at least partly in the sieve elements. Nucleotide phosphates have also been found in phloem sap.

Some inorganic ions move in the phloem, including potassium, magnesium, phosphate, and chloride (see Table 10.2). In contrast, nitrate, calcium, sulfur, and iron are relatively immobile in the phloem.

Proteins found in the phloem include structural P-proteins involved in the sealing of wounded sieve elements, as well as several water-soluble proteins. The function of many of the proteins commonly found in phloem sap is related to

(A) Reducing sugars, which are not generally translocated in the phloem

The reducing groups are aldehyde (glucose and mannose) and ketone (fructose) groups.

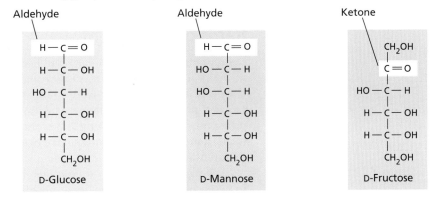

(B) Compounds commonly translocated in the phloem

Sucrose is a disaccharide made up of one glucose and one fructose molecule. Raffinose, stachyose, and verbascose contain sucrose bound to one, two, or three galactose molecules, respectively.

Mannitol is a sugar alcohol formed by the reduction of the aldehyde group of mannose.

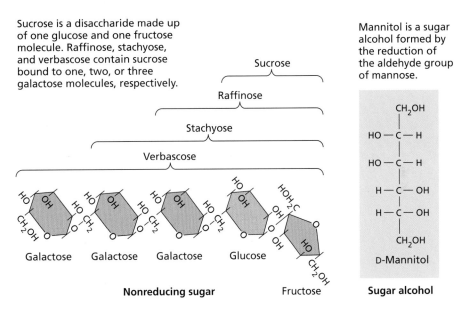

Glutamic acid, an amino acid, and glutamine, its amide, are important nitrogenous compounds in the phloem, in addition to aspartate and asparagine.

Species with nitrogen-fixing nodules also utilize ureides as transport forms of nitrogen.

Figure 10.8 Structures of (A) compounds not normally translocated in the phloem and (B) compounds commonly translocated in the phloem.

mass transfer rate   The quantity of material passing through a given cross section of phloem or sieve elements per unit of time.

stress and defense reactions. The possible roles of RNAs and proteins as signal molecules is further discussed at the end of the chapter.

Rates of Movement

It should be noted that in early publications reporting on rates of transport in the phloem, the units of velocity were centimeters per hour (cm h^{-1}), and the units of mass transfer were grams per hour per square centimeter (g h^{-1} cm^{-2}) of phloem or sieve elements. However, the currently preferred units (SI units) are meters (m) for length, seconds (s) for time, and kilograms (kg) for mass. Rates from earlier works have been converted to SI units and are found in parentheses below.

The rate of movement of materials in the sieve elements can be expressed in two ways: as velocity, the linear distance traveled per unit of time, or as **mass transfer rate**, the quantity of material passing through a given cross section of phloem or sieve elements per unit of time. Mass transfer rates based on the cross-sectional area of the sieve elements are preferred because the sieve elements are the conducting cells of the phloem. Values for mass transfer rate range from 1 to 15 g h^{-1} cm^{-2} of sieve elements (in SI units, 2.8–41.7 µg s^{-1} mm^{-2}).

Both velocities and mass transfer rates can be measured with radioactive tracers. In the simplest type of experiment for measuring velocity, ^{11}C- or ^{14}C-labeled CO_2 is applied for a brief period of time to a source leaf (pulse labeling), and the arrival of label at a sink tissue or at a particular point along the pathway is monitored with an appropriate detector.

In general, velocities measured by a variety of conventional techniques average about 1.0 m h^{-1} (0.28 mm s^{-1}) and range from 0.3 to 1.5 m h^{-1} (in SI units, 0.08–0.42 mm s^{-1}). More recent measurements of velocity using NMR spectrometry and magnetic resonance imaging yielded an average velocity for castor bean of 0.25 mm s^{-1}, which is remarkably close to the average obtained using older methods. Transport velocities in the phloem are clearly quite high and exceed the rate of diffusion by many orders of magnitude over distances exceeding a few millimeters (see Chapter 2). Any proposed mechanism of phloem translocation must account for these high velocities.

The Pressure-Flow Model, a Passive Mechanism for Phloem Transport

The most widely accepted mechanism of phloem translocation in angiosperms is the pressure-flow model. The pressure-flow model explains phloem translocation as a flow of solution (mass flow or bulk flow) driven by an osmotically generated pressure gradient between source and sink. This section describes the pressure-flow model, predictions arising from mass flow, and data, both supporting and challenging.

In early research on phloem translocation, both active and passive mechanisms were considered. All theories, both active and passive, assume an energy requirement in both sources and sinks. In sources, energy is necessary to synthesize the materials for transport and, in some cases, to move photosynthates into the sieve elements by active membrane transport. We discuss phloem loading in detail later in the chapter. In sinks, energy is essential for some aspects of movement (phloem unloading) from sieve elements to sink cells, which store or metabolize the sugar. Phloem unloading is also discussed in detail later.

The passive mechanisms of phloem transport further assume that energy is required in the sieve elements of the path between sources and sinks simply to maintain structures such as the cell plasma membrane and to recover sugars lost from the phloem by leakage. The pressure-flow model is an example of a passive mechanism. The active theories, by contrast, postulate an additional expenditure of energy by sieve elements of the transport phloem in order to drive translocation itself. While the active theories have largely been discounted, interest in certain aspects of these models may be revived, based on observations of pressures present in large plants, such as trees. (See the discussion below, *Pressure gradients in the sieve elements may be modest.*)

An osmotically generated pressure gradient drives translocation in the pressure-flow model

Diffusion is far too slow to account for the velocities of solute movement observed in the phloem. Translocation velocities average 1 m h^{-1}; the rate of diffusion would be 1 m per 32 years! (See Chapter 2 for a discussion of diffusion velocities and the distances over which diffusion is an effective transport mechanism.)

The **pressure-flow model**, first proposed by Ernst Münch in 1930, states that a flow of solution in the sieve elements is driven by an osmotically generated *pressure gradient* between source and sink (Ψ_p). Phloem loading at the source and phloem unloading at the sink establish the pressure gradient.

As we will see later (see the section *Phloem loading can occur via the apoplast or symplast*), three different mechanisms exist to generate high concentrations of sugars in the sieve elements of the source: photosynthetic metabolism in the mesophyll, conversion of photoassimilate to transport sugars in intermediary cells (polymer trapping), and active membrane transport. Recall from Chapter 2 (Equation 2.5) that $\Psi = \Psi_s + \Psi_p$; that is, $\Psi_p = \Psi - \Psi_s$. In source tissues, an accumulation of sugars in the sieve elements generates a low (negative) solute potential (Ψ_s) and causes a steep drop in the water potential (Ψ). In response to the water potential gradient, water enters the sieve elements and causes the turgor pressure (Ψ_p) to increase.

At the receiving end of the translocation pathway, phloem unloading leads to a lower sugar concentration in the sieve elements, generating a higher (less negative) solute potential in the sieve elements of sink tissues. As the water potential of the phloem rises above that of the xylem, water tends to leave the phloem in response to the water potential gradient, causing a decrease in turgor pressure in the sieve elements of the sink. **Figure 10.9** illustrates the pressure-flow hypothesis; the figure specifically shows the case in which active membrane transport from the apoplast generates a high sugar concentration in the source sieve elements.

The phloem sap moves by mass flow rather than by osmosis. That is, no membranes are crossed during transport from one sieve tube to another, and solutes move at the same rate as the water molecules. Since this is the case, mass flow can occur from a source organ with a lower water potential to a sink organ with a higher water potential, or vice versa, depending on the identities of the source and sink organs. In fact, Figure 10.9 illustrates an example in which the flow is against the water potential gradient. Such water movement does not violate the laws of thermodynamics, because it is an example of mass flow, which is driven by a pressure gradient, as opposed to osmosis, which is driven by a water potential gradient.

According to the pressure-flow model, movement in the translocation pathway is driven by transport of solutes and water into source sieve elements and out of sink sieve elements. The passive, pressure-driven, long-distance translocation in the sieve tubes ultimately depends on the mechanisms involved in phloem loading and unloading. These mechanisms are responsible for setting up the pressure gradient.

pressure-flow model A widely accepted model of phloem translocation in angiosperms. It states that transport in the sieve elements is driven by a pressure gradient between source and sink. The pressure gradient is osmotically generated and results from the loading at the source and unloading at the sink.

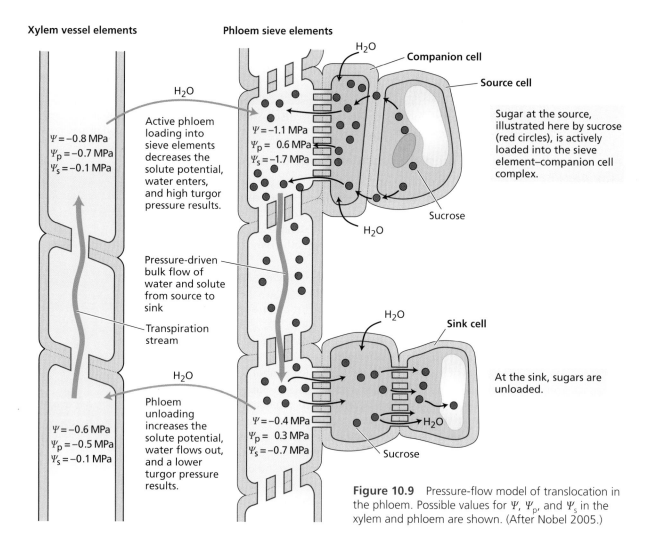

Xylem vessel elements

Phloem sieve elements

$\Psi = -0.8$ MPa
$\Psi_p = -0.7$ MPa
$\Psi_s = -0.1$ MPa

H_2O

Active phloem loading into sieve elements decreases the solute potential, water enters, and high turgor pressure results.

$\Psi = -1.1$ MPa
$\Psi_p = 0.6$ MPa
$\Psi_s = -1.7$ MPa

H_2O

Companion cell

Source cell

Sugar at the source, illustrated here by sucrose (red circles), is actively loaded into the sieve element–companion cell complex.

Sucrose

H_2O

Pressure-driven bulk flow of water and solute from source to sink

Transpiration stream

H_2O

Sink cell

Phloem unloading increases the solute potential, water flows out, and a lower turgor pressure results.

$\Psi = -0.6$ MPa
$\Psi_p = -0.5$ MPa
$\Psi_s = -0.1$ MPa

$\Psi = -0.4$ MPa
$\Psi_p = 0.3$ MPa
$\Psi_s = -0.7$ MPa

H_2O

At the sink, sugars are unloaded.

Sucrose

Figure 10.9 Pressure-flow model of translocation in the phloem. Possible values for Ψ, Ψ_p, and Ψ_s in the xylem and phloem are shown. (After Nobel 2005.)

Some predictions of pressure flow have been confirmed, while others require further experimentation

Some important predictions emerge from the model of phloem translocation as mass flow described above:

- No true bidirectional transport (i.e., simultaneous transport in both directions) in a single sieve element can occur. A mass flow of solution precludes such bidirectional movement because a solution can flow in only one direction in a pipe at any one time. Solutes within the phloem can move bidirectionally, but in different sieve elements or at different times. Furthermore, water and solutes must move at the same velocity in a flowing solution.

- No great expenditures of energy are required in order to drive long-distance translocation. Therefore, treatments that restrict the supply of ATP along the path, such as low temperature, anoxia, and metabolic inhibitors, should not stop translocation. However, energy *is* required to maintain the structure of the sieve elements, to reload and retrieve any sugars lost to the apoplast by leakage, and perhaps to reload sugars at the termination of sieve tubes.

- The sieve tube lumen and the sieve plate pores must be largely unobstructed. If P-protein or other materials block the pores, the resistance to flow of the sieve element sap might be too great.

- The pressure-flow hypothesis predicts the presence of a positive pressure gradient, with turgor pressure higher in sieve elements of sources than in those of sinks. According to the traditional picture of mass flow, the pressure difference must be large enough to overcome the resistance of the pathway and to maintain flow at the observed velocities. Thus, pressure gradients should be larger in long transport pathways, for example, in trees, than in short transport pathways, as in herbaceous plants.

The available evidence testing these predictions is presented below.

There is no bidirectional transport in single sieve elements, and solutes and water move at the same velocity

Researchers have investigated bidirectional transport by applying two different radiotracers to two source leaves, one above the other. Each leaf receives one of the tracers, and a point between the two sources is monitored for the presence of both tracers.

Transport in two directions has often been detected in sieve elements of different vascular bundles in stems. Transport in two directions has also been seen in adjacent sieve elements of the same bundle in petioles. Bidirectional transport in adjacent sieve elements can occur in the petiole of a leaf that is undergoing the transition from sink to source and simultaneously importing and exporting photosynthates through its petiole. However, simultaneous bidirectional transport in a single sieve element has never been demonstrated.

Measured velocities for transport in the phloem are remarkably similar, whether measured using carbon-labeled solutes or using NMR techniques, which detect water flow. Solutes and water move at the same velocity.

Both of these observations—the lack of bidirectional transport in a single sieve element and similar velocities for solutes and water—support the existence of mass flow in the sieve elements of the phloem.

The energy requirement for transport through the phloem pathway is small in herbaceous plants

In herbaceous plants that can survive periods of low temperature, such as sugar beet (*Beta vulgaris*), rapidly chilling a short segment of the petiole of a source leaf to approximately 1°C does not cause sustained inhibition of mass transport out of the leaf (**Figure 10.10**). Rather, there is a brief period of inhibition (minutes to a few hours), after which transport slowly returns to the control rate. Chilling reduces respiration rate and both the synthesis and the consumption of ATP in the petiole by about 90%, at a time when translocation has recovered and is proceeding normally. These experiments show that the energy requirement for long-distance transport through the pathway of these herbaceous plants is small, consistent with mass flow.

Chilling experiments in large plants, such as trees, generally extend over longer time periods (days to a few weeks). Chilling of the stem in such experiments often inhibits phloem transport over the treatment period. However, the methods used to evaluate transport, such as radial growth rates below the treatment zone or soil CO_2 efflux, do not permit short-term, transient changes in transport to be observed.

It should be noted that extreme treatments that inhibit all energy metabolism do inhibit translocation even in herbaceous plants. For example, in bean, treating the petiole of

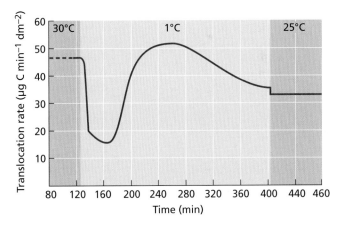

Figure 10.10 The energy requirement for translocation in the path is small in herbaceous plants. Loss of metabolic energy resulting from the chilling of a source leaf petiole partially reduces the rate of translocation in sugar beet. However, translocation rates recover with time despite the fact that ATP production and use are still largely inhibited by chilling. $^{14}CO_2$ was supplied to a source leaf, and a 2-cm portion of its petiole was chilled to 1°C. Translocation was monitored by the arrival of ^{14}C at a sink leaf. (1 dm [decimeter] = 0.1 m) (After Geiger and Sovonick 1975.)

a source leaf with a metabolic inhibitor (cyanide) inhibited translocation out of the leaf. However, examination of the treated tissue by electron microscopy revealed blockage of the sieve plate pores by cellular debris. Clearly, these results do not bear on the question of whether energy is required for translocation along the pathway.

Sieve plate pores appear to be open channels

Ultrastructural studies of sieve elements are challenging because of the high internal pressure in these cells. When the phloem is excised or killed slowly with chemical fixatives, the turgor pressure in the sieve elements is released. The contents of the cells, particularly P-protein, surge toward the point of pressure release and, in the case of sieve tube elements, accumulate on the sieve plates. This accumulation is probably the reason that many earlier electron micrographs show sieve plates that are obstructed.

Newer, rapid freezing and fixation techniques provide reliable pictures of undisturbed sieve elements. The use of confocal laser scanning microscopy, which allows for the direct observation of translocation through living sieve elements, addresses the additional question of whether the sieve plate pores and sieve element lumen are open in intact, translocating tissues.

When young Arabidopsis plants are rapidly frozen by plunging them in liquid nitrogen, then freeze-substituted and fixed, sieve plate pores are often unobstructed in the tissue. The sieve plate pores of living, translocating sieve elements of broad bean were also observed to be mostly open. The open condition of the pores seen in many species, such as cucurbits, sugar beet, bean (*Phaseolus vulgaris*), and Arabidopsis (see Figure 10.7), is consistent with mass flow.

What about the distribution of P-protein in the sieve tube lumen? Electron micrographs of sieve tube members prepared by rapid freezing and fixation have often shown P-protein along the periphery of the sieve tube members or evenly distributed throughout the lumen of the cell. Furthermore, the sieve plate pores often contain P-protein in similar positions, lining the pore or in a loose network.

Pressure gradients in the sieve elements may be modest; pressures in herbaceous plants and trees appear to be similar

Mass flow or bulk flow is the combined movement of all the molecules in a solution, driven by a pressure gradient. What are the pressure values in sieve elements, and how can they be determined? Does a pressure gradient exist between sources and sinks, and if so, is the gradient modest or substantial? Do large plants, such as trees, have proportionally higher pressures in the phloem than small, herbaceous species?

Turgor pressure in sieve elements can either be calculated from the water potential and solute potential ($\Psi_p = \Psi - \Psi_s$) or measured directly. The most effective technique uses micromanometers or pressure transducers sealed over exuding aphid stylets. The data obtained are accurate because aphids pierce only a single sieve element, and the plasma membrane apparently seals well around the aphid stylet. Pressures measured using the aphid stylet technique range from approximately 0.7 to 1.5 MPa in both herbaceous plants and small trees.

Studies using calculated turgor pressures have detected pressure gradients sufficient to drive mass flow in a few herbaceous plants such as soybean. However, no systematic studies have been made of turgor gradients measured using aphid stylets in any plant. The data are critical to any evaluation of the pressure-flow hypothesis. Ideally, techniques that can measure turgor differences along the same continuous sieve tube, both in herbaceous plants and in large plants such as trees, must be developed. This will be an enormous technical challenge.

One observation is fairly certain, however, and that is that turgor pressures in trees are not proportionally higher than those in herbaceous plants. One study compared calculated turgor pressures (the technique often used in trees) and pressures measured using aphid stylets (a technique used in herbaceous plants) in small willow saplings. The two techniques yielded comparable values, averaging 0.6 MPa for the calculated pressures and 0.8 MPa for the measured pressures. Calculated pressures were as high as 2.0 MPa in large white ash trees. These values are not substantially different from those measured in herbaceous plants, as noted above. (Herbaceous plants and trees do often differ in their phloem loading strategies, in a way that is consistent with the relatively low pressures in trees; see the section *Phloem loading is passive in several tree species.*)

Although the relatively low turgor pressures measured in the phloem of trees are inconsistent with one of the predictions of the pressure-flow theory, the theory is still the best available to describe the experimental observations at least for angiosperms. Several modifications to the theory have been proposed to explain this deviation, but they are not treated here.

Phloem Loading

Several transport steps are involved in the movement of photosynthates from the mesophyll chloroplasts to the sieve elements of mature leaves:

1. Triose phosphate formed by photosynthesis during the day (see Chapter 8) is transported from the chloroplast to the cytosol, where it is converted to sucrose. During the night, carbon from stored starch exits the chloroplast primarily in the form of maltose and is converted to sucrose. (Other transport sugars are later synthesized from sucrose in some species, while sugar alcohols are synthesized using hexose phosphate and in some cases hexose as the starting molecules.)

2. Sucrose moves from producing cells in the mesophyll to cells in the vicinity of the sieve elements in the smallest veins of the leaf. This short-distance transport pathway usually covers a distance of only a few cell diameters.

3. In phloem loading, sugars are transported into the sieve elements and companion cells. Note that with respect to loading, the sieve elements and companion cells are often considered a functional unit, called the *sieve element–companion cell complex*. Once inside the sieve elements, sucrose and other solutes are translocated away from the source, a process known as **export**.

As discussed earlier, the processes of loading at the source and perhaps unloading at the sink provide the driving force for long-distance transport and are thus of considerable basic, as well as agricultural, importance. A thorough understanding of these mechanisms should provide the basis for technology aimed at enhancing crop productivity by increasing the accumulation of photosynthates in edible sink tissues, such as cereal grains.

Phloem loading can occur via the apoplast or symplast

We have seen that solutes (mainly sugars) in source leaves must move from the photosynthesizing cells in the mesophyll to the sieve elements. The initial short-distance pathway is probably symplastic (**Figure 10.11**). However, sugars might move entirely through the symplast (cytoplasm) to the sieve elements via the plasmodesmata (see Figure 10.11A), or they might enter the apoplast prior to phloem loading (see Figure 10.11B). (See Figure 3.4 for a general description of the symplast and apoplast.) One of the two routes, apoplastic or symplastic, is dominant in some species; many species, however, show evidence of being able to use more than one loading mechanism. For simplicity's sake, we will initially consider the pathways separately, then return to the subject of loading diversity.

export The movement of photosynthates in sieve elements away from the source tissue.

Figure 10.11 Schematic diagram of pathways of phloem loading in source leaves. (A) In the totally symplastic pathway, sugars move from one cell to another through the plasmodesmata, all the way from the mesophyll to the sieve elements. (B) In the partly apoplastic pathway, sugars initially move through the symplast, but enter the apoplast just prior to loading into the companion cells and sieve elements. Sugars loaded into the companion cells are thought to move through plasmodesmata into the sieve elements.

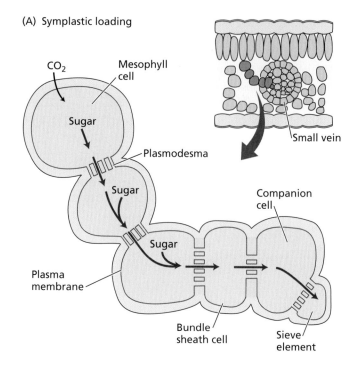

(A) Symplastic loading

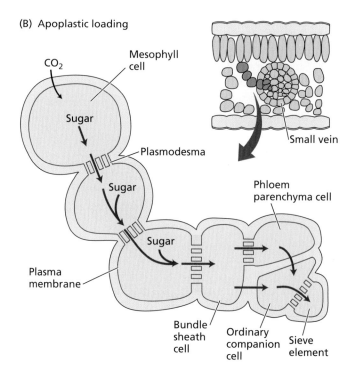

(B) Apoplastic loading

Several mechanisms for phloem loading are now recognized: apoplastic loading, symplastic loading with polymer trapping, and passive symplastic loading. Early research on phloem loading focused on the apoplastic pathway, probably because it is very common in herbaceous plants and therefore crops. (In fact, much of our knowledge of plant physiology is probably slanted by its primary focus on herbaceous crops. As it turns out, the apoplastic pathway apparently is the most common mechanism.) In this section, we discuss apoplastic loading first, and then introduce the two types of symplastic loading (polymer trapping and passive symplastic loading) in the order in which their importance was recognized.

Abundant data support the existence of apoplastic loading in some species

In the case of apoplastic loading, the sugars enter the apoplast quite near the sieve element–companion cell complex. Sugars are then actively transported from the apoplast into the sieve elements and companion cells by an energy-driven, selective transporter located in the plasma membranes of these cells. Efflux into the apoplast is highly localized, probably into the walls of phloem parenchyma cells. The sucrose transporters that mediate the efflux of sucrose, most likely from the phloem parenchyma to the apoplast near the sieve element–companion cell complexes, have recently been identified in Arabidopsis and rice as a subfamily of the SWEET transporters.

Apoplastic phloem loading leads to three basic predictions:

1. Transported sugars should be found in the apoplast.

2. In experiments in which sugars are supplied to the apoplast, the exogenously supplied sugars should accumulate in sieve elements and companion cells.

3. Inhibition of sugar efflux from the phloem parenchyma or of uptake from the apoplast should result in inhibition of export from the leaf.

Many studies devoted to testing these predictions have provided solid evidence for apoplastic loading in several species.

Sucrose uptake in the apoplastic pathway requires metabolic energy

In many of the species initially studied, sugars become more concentrated in the sieve elements and companion cells than in the mesophyll. This difference in solute concentration can be demonstrated through measurement of the osmotic potential (Ψ_s) of the various cell types in the leaf (see Chapter 2).

In sugar beet, the osmotic potential of the mesophyll is approximately –1.3 MPa, and the osmotic potential of the sieve elements and companion cells is about –3.0 MPa. Most of this difference in osmotic potential is thought to result from accumulated sugar, specifically sucrose, because sucrose is the major transport sugar in this species. Experimental studies have also demonstrated that both externally supplied sucrose and sucrose made from photosynthetic products accumulate in the sieve elements and companion cells of the minor veins of sugar beet source leaves (**Figure 10.12**).

The fact that sucrose is at a higher concentration in the sieve element–companion cell complex than in surrounding cells indicates that sucrose is actively transported against its chemical-potential gradient. The dependence of sucrose accumulation on active transport is supported by the fact that treating source tissue with respiratory inhibitors both decreases ATP concentration and inhibits loading of exogenous sugar.

Plants that load sugars apoplastically into the phloem may also load amino acids and sugar alcohols (sorbitol and mannitol) actively. In contrast, other metabolites, such as organic acids and hormones, may enter sieve elements passively.

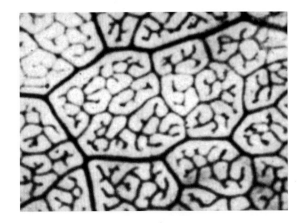

Figure 10.12 This autoradiograph shows that labeled sugar moves against its concentration gradient from the apoplast into sieve elements and companion cells of a sugar beet source leaf. A solution of ^{14}C-labeled sucrose was applied for 30 min to the upper surface of a sugar beet leaf that had previously been kept in darkness for 3 h. The leaf cuticle was removed to allow penetration of the solution to the interior of the leaf. The sieve elements and companion cells of the small veins in the source leaf contain high concentrations of labeled sugar, shown by the black accumulations, indicating that sucrose is actively transported against its concentration gradient. (From Fondy 1975, courtesy of D. Geiger.)

intermediary cell   A type of companion cell with numerous plasmodesmatal connections to surrounding cells, particularly to the bundle sheath cells.

Phloem loading in the apoplastic pathway involves a sucrose–H⁺ symporter

A sucrose–H⁺ symporter is thought to mediate the transport of sucrose from the apoplast into the sieve element–companion cell complex. Recall from Chapter 6 that symport is a secondary transport process that uses the energy generated by the proton pump (see Figure 6.10A). The energy dissipated by protons moving back into the cell is coupled to the uptake of a substrate, in this case sucrose (**Figure 10.13**).

Several sucrose–H⁺ symporters have been cloned and localized in the phloem. SUT1 and SUC2 appear to be the major sucrose transporters in phloem loading into either companion cells or sieve elements. Data from several other studies also support the operation of a sucrose–H⁺ symporter in phloem loading.

Phloem loading is symplastic in some species

Many results point to apoplastic phloem loading in species that transport only sucrose and that have few plasmodesmata leading into the minor vein phloem. However, many other species have numerous plasmodesmata at the interface between the sieve element–companion cell complex and the surrounding cells, which seems inconsistent with apoplastic loading. The operation of a symplastic pathway requiring the presence of open plasmodesmata between the different cells in the pathway has been implicated in these species.

The polymer-trapping model explains symplastic loading in plants with intermediary-type companion cells

A symplastic pathway has become evident in species that transport raffinose and stachyose, in addition to sucrose, in the phloem. They have a special kind of companion cells, **intermediary cells**, with many plasmodesmata leading into the minor veins. Some examples of such species are common coleus (*Coleus blumei*), pumpkin and squash (*Cucurbita pepo*), and melon (*Cucumis melo*). Remember that intermediary cells are specialized companion cells; see *Companion cells aid the highly specialized sieve elements.*

Two majors questions emerge concerning symplastic loading:

1. In many species the composition of sieve element sap is different from the solute composition in tissues surrounding the phloem. This difference indicates that certain sugars are specifically selected for transport in the source leaf. The involvement of symporters in apoplastic phloem loading provides a clear mechanism for selectivity, because symporters are specific for certain sugar molecules. Symplastic loading, in contrast, depends on the diffusion of sugars from the mesophyll to the sieve elements via the plasmodesmata. How can diffusion through plasmodesmata during symplastic loading be selective for certain sugars?

2. Data from several species showing symplastic loading indicate that sieve elements and companion cells have a higher osmolyte content (more negative osmotic potential) than the mesophyll. How can diffusion-dependent symplastic loading account for the observed selectivity for transported molecules and the accumulation of sugars against a concentration gradient?

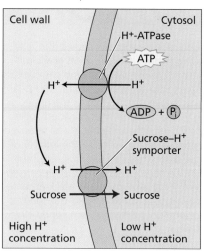

Sieve element–companion cell complex

Figure 10.13   ATP-dependent sucrose transport in apoplastic sieve element loading. In the cotransport model of sucrose loading into the symplast of the sieve element–companion cell complex, the plasma membrane ATPase pumps protons out of the cell into the apoplast, establishing a higher proton concentration in the apoplast and a membrane potential of approximately –120 mV. The energy in the proton gradient is then used to drive the transport of sucrose into the symplast of the sieve element–companion cell complex through a sucrose–H⁺ symporter.

The **polymer-trapping model** (**Figure 10.14**) has been developed to address these questions in species such as coleus and cucurbits. This model states that the sucrose synthesized in the mesophyll diffuses from the bundle sheath cells into the intermediary cells through the abundant plasmodesmata that connect the two cell types. In the intermediary cells, raffinose and stachyose (polymers made of three and four hexose sugars, respectively; see Figure 10.8B) are synthesized from the transported sucrose and from galactinol (a metabolite of galactose). Because of the anatomy of the tissue and the relatively large size of raffinose and stachyose, the polymers cannot diffuse back into the bundle sheath cells, but they can diffuse into the sieve element. Sugar concentrations in the sieve elements of these plants can reach levels equivalent to those in plants that load apoplastically. Sucrose can continue to diffuse into the intermediary cells, because its synthesis in the mesophyll and its utilization in the intermediary cells maintain the concentration gradient (see Figure 10.14).

The polymer-trapping model makes three predictions:

1. Sucrose should be more concentrated in the mesophyll than in the intermediary cells.

2. The enzymes for raffinose and stachyose synthesis should be preferentially located in the intermediary cells.

3. The plasmodesmata linking the bundle sheath cells and the intermediary cells should exclude molecules larger than sucrose. Plasmodesmata between the intermediary cells and sieve elements must be wider to allow passage of raffinose and stachyose.

Several studies support the polymer-trapping model of symplastic loading in some species. However, recent modeling results suggest that additional, unknown factors must be present to enable plasmodesmata to block diffusion

polymer-trapping model   A model that explains the specific accumulation of sugars in the sieve elements of symplastically loading species.

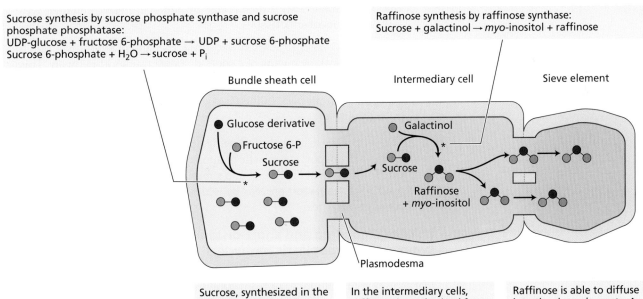

Sucrose synthesis by sucrose phosphate synthase and sucrose phosphate phosphatase:
UDP-glucose + fructose 6-phosphate → UDP + sucrose 6-phosphate
Sucrose 6-phosphate + H_2O → sucrose + P_i

Raffinose synthesis by raffinose synthase:
Sucrose + galactinol → *myo*-inositol + raffinose

Bundle sheath cell   Intermediary cell   Sieve element

Glucose derivative
Fructose 6-P
Sucrose

Galactinol
Sucrose
Raffinose + *myo*-inositol

Plasmodesma

Sucrose, synthesized in the mesophyll, diffuses from the bundle sheath cells into the intermediary cells through the abundant plasmodesmata.

In the intermediary cells, raffinose is synthesized from sucrose and galactinol, thus maintaining the diffusion gradient for sucrose. Because of its larger size, raffinose is not able to diffuse back into the mesophyll.

Raffinose is able to diffuse into the sieve elements. As a result, the concentration of transport sugar rises in the intermediary cells and the sieve elements. Note that stachyose is not shown here for clarity.

Figure 10.14   Polymer-trapping model of phloem loading. For simplicity, the trisaccharide stachyose is omitted. (After van Bel 1992.)

of oligosaccharides, such as raffinose and stachyose, back into the mesophyll, while permitting sufficient sucrose flux into the intermediary cells to maintain observed transport rates.

Phloem loading is passive in several tree species

Passive symplastic phloem loading has recently been recognized as a mechanism that is widespread among plant species. While the data supporting this mechanism are relatively recent, passive symplastic loading was actually a part of Münch's original conception of pressure flow.

It has become apparent that several tree species possess abundant plasmodesmata between the sieve element–companion cell complex and surrounding cells, but do not have intermediary-type companion cells and do not transport raffinose and stachyose. Willow (*Salix babylonica*) and apple (*Malus domestica*) trees are among the species that fall into this category. These plants have no concentrating step in the pathway from the mesophyll into the sieve element–companion cell complex. Since a concentration gradient from the mesophyll into the phloem drives diffusion along this short-distance pathway, the absolute levels of sugars in the source leaves of these species must be high in order to maintain the required high solute concentrations and the resulting high turgor pressures in the sieve elements. Although there is wide variation (over 50-fold) and considerable overlap among groups of plants with different loading mechanisms, source leaf sugar concentrations are generally higher in the tree species that load passively.

Phloem Unloading and Sink-to-Source Transition

Now that we have learned about the events leading up to the export of sugars from sources, let's take a look at **import** into sinks such as developing roots, tubers, and reproductive structures. In many ways the events in sink tissues are simply the reverse of the events in sources. The following steps are involved in the import of sugars into sink cells.

1. *Phloem unloading.* This is the process by which imported sugars leave the sieve elements of sink tissues.

2. *Short-distance transport.* After unloading, the sugars are transported to cells in the sink by means of a short-distance transport pathway. This pathway has also been called post–sieve element transport.

3. *Storage and metabolism.* In the final step, sugars are stored or metabolized in sink cells.

In this section we discuss the following questions: Are phloem unloading and short-distance transport symplastic or apoplastic? Is sucrose hydrolyzed during the process? Do phloem unloading and subsequent steps require energy? Finally, we examine the transition process by which a young, importing leaf becomes an exporting source leaf.

Phloem unloading and short-distance transport can occur via symplastic or apoplastic pathways

In sink organs, sugars move from the sieve elements to the cells that store or metabolize them. Sinks vary from growing vegetative organs (root tips and young leaves) to storage tissues (roots and stems) to organs of reproduction and dispersal (fruits and seeds). Because sinks vary so greatly in structure and function, there is no single mechanism of phloem unloading and short-distance transport. Differences in import pathways due to differences in sink types are emphasized in this section; however, the pathway often depends on the stage of sink development as well.

import   The movement of photosynthate in sieve elements into sink organs.

As in sources, the sugars may move entirely through the symplast via the plasmodesmata in sinks, or they may enter the apoplast at some point. Both unloading and the short-distance pathway appear to be completely symplastic in some young eudicot leaves, such as sugar beet and tobacco. Meristematic and elongating regions of primary root tips also appear to unload symplastically.

While symplastic import predominates in most sink tissues, part of the short-distance pathway is apoplastic in some sink organs at some stages of development—for example, in fruits, seeds, and other storage organs that accumulate high concentrations of sugars. The pathway can switch between symplastic and apoplastic in these sinks, with an apoplastic step being required when sink sugar concentrations are high. The apoplastic step could be located at the site of unloading itself or farther removed from the sieve elements. The latter arrangement, typical of developing seeds, appears to be the most common in apoplastic pathways.

An apoplastic step is required in developing seeds because there are no symplastic connections between the maternal tissues and the tissues of the embryo. Sugars exit the sieve elements (phloem unloading) via a symplastic pathway and are transferred from the symplast to the apoplast at some point removed from the sieve element–companion cell complex. The apoplastic step permits membrane control over the substances that enter the embryo, because two membranes must be crossed in the process.

When an apoplastic step occurs in the import pathway, the transport sugar can be partly metabolized in the apoplast, or it can cross the apoplast unchanged. For example, sucrose can be hydrolyzed into glucose and fructose in the apoplast by invertase, a sucrose-splitting enzyme, and glucose and/or fructose would then enter the sink cells. Such sucrose-cleaving enzymes play a role in the control of phloem transport by sink tissues.

Transport into sink tissues requires metabolic energy

Inhibitor studies have shown that import into sink tissues is energy dependent. Growing leaves, roots, and storage sinks in which carbon is stored as starch or in protein appear to use symplastic phloem unloading and short-distance transport. Transport sugars are used as substrates for respiration and are metabolized into storage polymers and into compounds needed for growth. Sucrose metabolism thus results in a low sucrose concentration in the sink cells, maintaining a concentration gradient for sugar uptake. In this pathway, no membranes are crossed during sugar uptake into the sink cells, and transport is passive: Transport sugars move from a high concentration in the sieve elements to a low concentration in the sink cells. Metabolic energy is thus required in these sink organs mainly for respiration and for biosynthetic reactions.

In apoplastic import, sugars must cross at least two membranes: the plasma membrane of the cell that is releasing the sugar and the plasma membrane of the sink cell. When sugars are transported into the vacuole of the sink cell, they must also traverse the tonoplast. As discussed earlier, transport across membranes in an apoplastic pathway may be energy dependent. While some evidence indicates that both efflux and uptake of sucrose can be active, the transporters have yet to be completely characterized.

Since these transporters have been shown to be bidirectional in some studies, some of the same sucrose transporters described above for sucrose loading could also be involved in sucrose unloading; the direction of transport would depend on the sucrose gradient, the pH gradient, and the membrane potential. Furthermore, symporters important in phloem loading have been found in some sink tissues—for example, SUT1 in potato tubers. The symporter may function in sucrose retrieval from the apoplast, in import into sink cells, or in both. Monosaccharide transporters must be involved in uptake into sink cells when sucrose is hydrolyzed in the apoplast.

The transition of a leaf from sink to source is gradual

Leaves of eudicots such as tomato or bean begin their development as sink organs. A transition from sink to source status occurs later in development, when the leaf is approximately 25% expanded, and it is usually complete when the leaf is 40 to 50% expanded. Export from the leaf begins at the tip or apex of the blade and progresses toward the base until the whole leaf becomes a sugar exporter. During the transition period, the tip exports sugar, while the base imports it from the other source leaves (**Figure 10.15**).

The maturation of leaves is accompanied by a large number of functional and anatomic changes, resulting in a reversal of transport direction from importing to exporting. In general, the cessation of import and the initiation of export are independent events. In albino leaves of tobacco, which have no chlorophyll and therefore are incapable of photosynthesis, import stops at the same developmental stage as in green leaves, even though export is not possible. Therefore, some change besides the initiation of export must occur in developing leaves of tobacco that causes them to cease importing sugars.

Sugars are unloaded and loaded almost entirely via different veins in tobacco leaves (**Figure 10.16**), contributing to the conclusion that import cessation and export initiation are two separate events. The minor veins that are eventually responsible for most of the loading in tobacco and other *Nicotiana* species do not mature until about the time import ceases and cannot play a role in unloading.

The change that stops import must thus involve blockage of unloading from the large veins at some point in the development of mature leaves. Factors that could account for the cessation of unloading include plasmodesmatal closure and a decrease in plasmodesmatal frequency. Experimental data have shown that both plasmodesmatal closure and elimination of plasmodesmata can occur.

Export of sugars begins when events have occurred that close the importing pathway and activate apoplastic loading and when loading has accumulated sufficient photosynthates in the sieve elements to drive translocation out of the leaf. The following conditions are necessary for export to begin:

- The leaf is synthesizing photosynthates in sufficient quantity that some is available for export. The sucrose-synthesizing genes are being expressed.

- Minor veins responsible for loading have matured. A regulatory element (enhancer) has been identified in the DNA of Arabidopsis that acts as part of a cascade of events leading to minor vein maturation. The enhancer can activate a reporter gene fused to a companion cell–specific promoter and does so in the same tip-to-base pattern as in the sink-to-source transition.

- The sucrose–H$^+$ symporter is expressed and in place in the plasma membrane of the sieve element–companion cell complex. Regulation

Figure 10.15 Autoradiographs of a leaf of summer squash (*Cucurbita pepo*), showing the transition of the leaf from sink to source status. In each case, the leaf imported ^{14}C from the source leaf on the plant for 2 h. Radioactive label is visible as black accumulations. (A) The entire leaf is a sink, importing sugar from the source leaf. (B–D) The base is still a sink. As the tip of the leaf loses the ability to unload and stops importing sugar (as shown by the loss of black accumulations), it gains the ability to load and to export sugar. (From Turgeon and Webb 1973.)

(A)

(B)

(C)

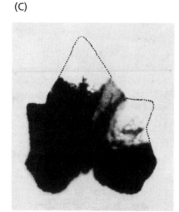

(D)

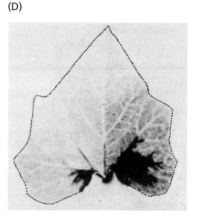

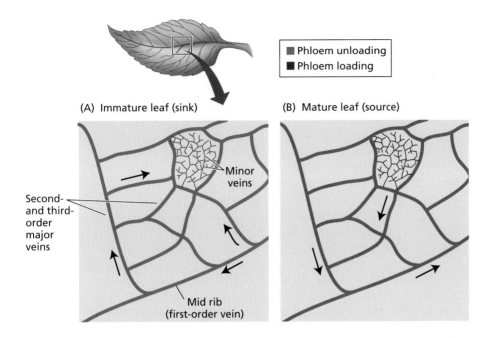

(A) Immature leaf (sink)   (B) Mature leaf (source)

■ Phloem unloading
■ Phloem loading

Minor veins

Second- and third-order major veins

Mid rib (first-order vein)

Figure 10.16 Division of labor in the veins of a tobacco leaf. (A) When the leaf is immature and still in its sink phase, photosynthate is imported from mature leaves and distributed (arrows) throughout the blade (or lamina) via the larger, major veins (thicker lines). The imported photosynthate unloads from the major veins into the mesophyll. The smallest, minor veins are shown as thinner lines within the areas enclosed by the third-order veins. The minor veins do not function in import and unloading because they are immature. (B) In a mature source leaf, import has ceased and export has begun. Photosynthate loads into the minor veins, while the larger veins serve only in export (arrows); they can no longer unload. (A) is drawn to scale from an autoradiograph but (B) is for illustration purposes only; in reality, the portion of leaf shown in (A) would have grown considerably as the leaf matured. (From Turgeon 2006.)

of these events is being investigated. For example, in Arabidopsis the promoter of the *SUC2* gene, which encodes a sucrose–H$^+$ symporter, becomes active in companion cells in a pattern that corresponds to that in the sink-to-source transition (**Figure 10.17**). Binding sites for transcription factors have been identified within the *SUC2* promoter that mediate this source-specific and companion cell–specific gene expression.

In leaves of plants such as sugar beet and tobacco, the ability to accumulate exogenous sucrose in the sieve element–companion cell complex is acquired as the leaves undergo the sink-to-source transition, suggesting that the symporter required for loading has become functional. In developing leaves of Arabidopsis, expression of the symporter that is thought to transport sugars during loading begins in the tip and proceeds to the base during a sink-to-source transition. This is the same basipetal pattern that is seen in the development of export capacity.

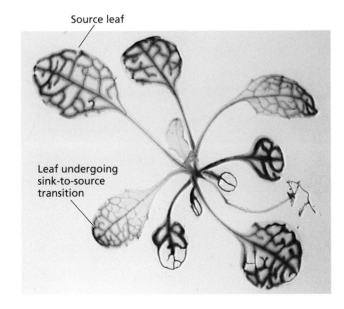

Source leaf

Leaf undergoing sink-to-source transition

Figure 10.17 Export from source tissue depends on the placement and activity of active sucrose transporters. The figure shows an Arabidopsis rosette transformed with a construct consisting of a reporter gene under control of the *AtSUC2* promoter. SUC2, a sucrose–H$^+$ symporter, is one of the major sucrose transporters functioning in phloem loading. The reporter system used (GUS) forms a visible product (blue) where the promoter is active. Staining is visible only in the vascular tissue of source leaves and in the tips of leaves undergoing the sink-to-source transition. (From Schneidereit et al. 2008.)

allocation   The regulated diversion of photosynthates into storage, utilization, and/or transport.

partitioning   The differential distribution of photosynthates to multiple sinks within the plant.

Photosynthate Distribution: Allocation and Partitioning

The photosynthetic rate determines the total amount of fixed carbon available to the leaf. However, the amount of fixed carbon available for translocation depends on subsequent metabolic events. The regulation of the distribution of fixed carbon into various metabolic pathways is termed **allocation** in this chapter (**Figure 10.18**).

The vascular bundles in a plant form a system of "pipes" that can direct the flow of photosynthates to various sinks: young leaves, stems, roots, fruits, or seeds. However, the vascular system is highly interconnected, forming an open network that allows source leaves to communicate with multiple sinks. Under these conditions, what determines the volume of flow to any given sink? The differential distribution of photosynthates within the plant is termed **partitioning** in this chapter. (The terms *allocation* and *partitioning* are sometimes used interchangeably in current publications.)

Throughout our overview of allocation and partitioning, keep in mind that a limited number of species have been studied, mainly those that load sucrose actively from the apoplast. It is likely that the mechanism of phloem loading affects the regulation of allocation, and so studies of allocation will have to be extended to a wider range of species. We will conclude by discussing how sinks compete, how sink demand might regulate photosynthetic rate in the source leaf, and how sources and sinks communicate with each other.

Allocation includes storage, utilization, and transport

The carbon fixed in a source cell can be used for storage, metabolism, and transport:

- *Synthesis of storage compounds.* Starch is synthesized and stored within chloroplasts and, in most species, is the primary storage form that is mobilized for translocation during the night. Plants that store carbon primarily as starch are called *starch storers.*

- *Metabolic utilization.* Fixed carbon can be utilized within various compartments of the photosynthesizing cell to meet the energy needs of the

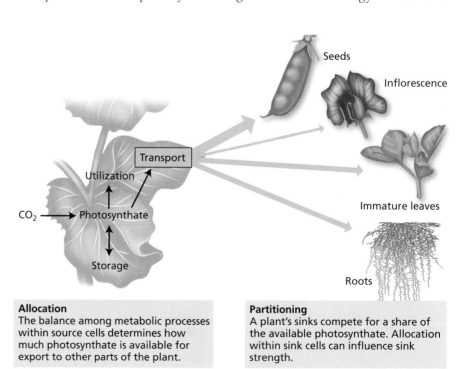

Figure 10.18   Allocation and partitioning determine the distribution of photosynthate within a plant.

Allocation
The balance among metabolic processes within source cells determines how much photosynthate is available for export to other parts of the plant.

Partitioning
A plant's sinks compete for a share of the available photosynthate. Allocation within sink cells can influence sink strength.

cell or to provide carbon skeletons for the synthesis of other compounds required by the cell.

- *Synthesis of transport compounds.* Fixed carbon can be incorporated into transport sugars for export to various sink tissues. A portion of the transport sugar can also be stored temporarily in the vacuole.

Allocation is also a key process in sink tissues. Once the transport sugars have been unloaded and enter the sink cells, they can remain as such or can be transformed into various other compounds. In storage sinks, fixed carbon can be accumulated as sucrose or hexose in vacuoles or as starch in amyloplasts. In growing sinks, sugars can be used for respiration and for the synthesis of other molecules required for growth.

Various sinks partition transport sugars

Sinks compete for the photosynthate being exported by the sources (see Figure 10.18). Such competition determines the partitioning of transport sugars among the various sink tissues of the plant, at least in the short term. The allocation of sugar within a sink (storage or metabolism) affects its ability to compete for available sugars. In this way, the processes of partitioning and allocation interact.

Of course, events in sources and sinks must be synchronized. Partitioning determines the patterns of growth, and growth must be balanced between shoot growth (photosynthetic productivity) and root growth (water and mineral uptake) in such a way that the plant can respond to the challenges of a variable environment. The goal is *not* a constant root-to-shoot ratio, but one that secures a supply of carbon and mineral nutrients appropriate to the needs of the plant.

So an additional level of control lies in the interaction between areas of supply and demand. Turgor pressure in the sieve elements could be an important means of communication between sources and sinks, acting to coordinate rates of loading and unloading. Chemical messengers are also important in communicating the status of one organ to the other organs in the plant. Such chemical messengers include plant hormones and nutrients, such as potassium and phosphate ions, and even the transport sugars themselves. Recent findings suggest that macromolecules (RNA and protein) may also play a role in photosynthate partitioning, perhaps by influencing transport through plasmodesmata.

Attainment of higher yields of crop plants is one goal of research on photosynthate allocation and partitioning. Whereas grains and fruits are examples of edible yields, total yield includes inedible portions of the shoot. Harvest index, the ratio of economical yield (edible grain) to total aboveground biomass, has increased over the years largely due to the efforts of plant breeders. One goal of modern plant physiology is to further increase yield based on a fundamental understanding of metabolism, development, and in the present context, partitioning.

However, allocation and partitioning in the whole plant must be coordinated such that increased transport to edible tissues does not occur at the expense of other essential processes and structures. Crop yield may also be improved if photosynthates that are normally "lost" by the plant are retained. For example, losses due to nonessential respiration or exudation from roots could be reduced. In the latter case, care must be taken not to disrupt essential processes outside the plant, such as growth of beneficial microbial species in the vicinity of the root that obtain nutrients from the root exudate.

Source leaves regulate allocation

Increases in the rate of photosynthesis in a source leaf generally result in an increase in the rate of translocation from the source. Control points for the allocation

Figure 10.19 A simplified scheme for starch and sucrose synthesis during the day. Triose phosphate, formed in the Calvin–Benson cycle, can either be used in starch formation in the chloroplast or transported into the cytosol in exchange for inorganic phosphate (P_i) via the phosphate translocator in the inner chloroplast envelope. The outer chloroplast envelope (omitted here for clarity) is permeable to small molecules. In the cytosol, triose phosphate can be converted to sucrose for storage in the vacuole or for transport or can be degraded via glycolysis. Key enzymes involved are starch synthetase (1), fructose 1,6-bisphosphatase (2), and sucrose phosphate synthase (3). The second and third enzymes, along with ADP-glucose pyrophosphorylase, which forms adenosine diphosphate glucose (ADPG), are regulated enzymes in sucrose and starch synthesis. UDPG, uridine diphosphate glucose. (After Preiss 1982.)

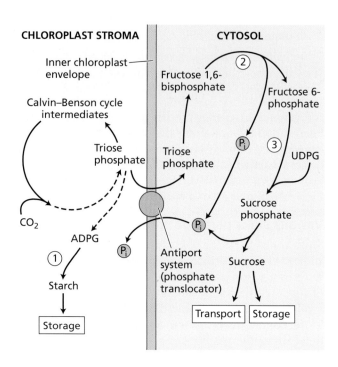

of photosynthate (**Figure 10.19**) include the distribution of triose phosphates to the following processes:

- Regeneration of intermediates in the C_3 photosynthetic carbon reduction cycle (the Calvin–Benson cycle; see Chapter 8)
- Starch synthesis
- Sucrose synthesis, as well as distribution of sucrose between transport and temporary storage pools

Various enzymes operate in the pathways that process the photosynthate, and the control of these steps is complex. The research described below focuses on species that load sucrose actively from the apoplast, specifically during the daylight hours. Further studies will be needed to extend our knowledge to plants using other loading strategies and to the regulation of allocation in those species.

During the day the rate of starch synthesis in the chloroplast must be coordinated with sucrose synthesis in the cytosol. Triose phosphates (glyceraldehyde 3-phosphate and dihydroxyacetone phosphate) produced in the chloroplast by the Calvin–Benson cycle (see Chapter 8) can be used for either starch or sucrose synthesis or in respiration. Sucrose synthesis in the cytosol diverts triose phosphate away from starch synthesis and storage. For example, it has been shown that when the demand for sucrose by other parts of a soybean plant is high, less carbon is stored as starch by the source leaves. The key enzymes involved in the regulation of sucrose synthesis in the cytosol and of starch synthesis in the chloroplast are sucrose phosphate synthase and fructose 1,6-bisphosphatase in the cytosol and ADP-glucose pyrophosphorylase in the chloroplast (see Figure 10.19).

However, there is a limit to the amount of carbon that normally can be diverted from starch synthesis in species that store carbon primarily as starch. Studies of allocation between starch and sucrose under different conditions suggest that a fairly steady rate of translocation throughout the 24-h period is a priority for most plants.

Sink tissues compete for available translocated photosynthate

As discussed earlier, translocation to sink tissues depends on the position of the sink in relation to the source and on the vascular connections between source and sink. Another factor determining the pattern of transport is competition between sinks, for example, between terminal sinks or between terminal sinks and axial sinks along the transport pathway. For example, young leaves might compete with roots for photosynthates in the translocation stream. Competition has been shown by numerous experiments in which removal of a sink tissue from a plant generally results in increased translocation to alternative, and hence competing, sinks. Conversely, increased sink size, for example, increased fruit load, decreases translocation to other sinks, especially the roots.

In the reverse type of experiment, the source supply can be altered while the sink tissues are left intact. When the supply of photosynthates from sources to competing sinks is suddenly and drastically reduced by shading of all the source leaves but one, the sink tissues become dependent on a single source. In sugar beet and bean plants, the rates of photosynthesis and export from the single remaining source leaf usually do not change over the short term (approximately 8 h). However, the roots receive less sugar from the single source, while the young leaves receive relatively more. Shading in general decreases partitioning to roots. Presumably, the young leaves can deplete the sugar content of the sieve elements more readily and thus increase the pressure gradient and the rate of translocation toward themselves.

Treatments such as making the sink water potential more negative increase the pressure gradient and enhance transport to the sink. Treatment of the root tips of pea (*Pisum sativum*) seedlings with mannitol solutions increased the import of sucrose over the short term by more than 300%, possibly because of a turgor decrease in the sink cells. Longer-term experiments show the same trend. Moderate water stress induced with polyethylene glycol treatment of the roots increased the proportion of assimilates transported to the roots of apple plants over a period of 15 days, but decreased the proportion transported to the shoot apex. This contrasts with shading treatments (above) in which source limitation diverts more sugar to the young leaves.

Sink strength depends on sink size and activity

The ability of a sink to mobilize photosynthate toward itself is often described as **sink strength**. Sink strength depends on two factors—sink size and sink activity—as follows:

$$\text{Sink strength} = \text{sink size} \times \text{sink activity}$$

Sink size is the total biomass of the sink tissue, and **sink activity** is the rate of uptake of photosynthates per unit of biomass of sink tissue. Altering either the size or the activity of the sink results in changes in translocation patterns. For example, the ability of a pea pod to import carbon depends on the dry weight of that pod as a proportion of the total number of pods.

Changes in sink activity can be complex, because various activities in sink tissues can potentially limit the rate of uptake by the sink. These activities include unloading from the sieve elements, metabolism in the cell wall, uptake from the apoplast, and metabolic processes that use the photosynthate for either growth or storage.

Experimental treatments to manipulate sink strength have often been unspecific. For example, cooling a sink tissue, which would be expected to inhibit all activities that require metabolic energy, often results in a decrease in the velocity of transport toward the sink. More recent experiments take advantage of our ability to specifically over- or underexpress enzymes related to sink activity—for example, those involved in sucrose metabolism in the sink. The two major enzymes that split sucrose are acid invertase and sucrose synthase, both of which can catalyze the first step in sucrose utilization.

sink strength The ability of a sink organ to mobilize assimilates toward itself. Depends on two factors: sink size and sink activity.

sink size The total weight of the sink.

sink activity The rate of uptake of photosynthate per unit of weight of sink tissue.

The source adjusts over the long term to changes in the source-to-sink ratio

If all but one of the source leaves of a soybean plant are shaded for an extended period (e.g., 8 days), many changes occur in the single remaining source leaf. These changes include a decrease in starch concentration and increases in photosynthetic rate, Rubisco activity, sucrose concentration, transport from the source, and orthophosphate concentration. Thus, in addition to the observed short-term changes in the distribution of photosynthate among different sinks, there are adjustments in the source leaf's metabolism in response to altered conditions over a longer term.

Photosynthetic rate (the net amount of carbon fixed per unit of leaf area per unit of time) often increases over several days when sink demand increases, and it decreases when sink demand decreases. An accumulation of photosynthate (sucrose or hexoses) in the source leaf can account for the linkage between sink demand and photosynthetic rate in starch-storing plants. Sugars act as signaling molecules that regulate many metabolic and developmental processes in plants. In general, carbohydrate depletion enhances the expression of genes for photosynthesis, reserve mobilization, and export processes, while abundant carbon resources favor expression of genes for storage and utilization.

Sucrose or hexoses that would accumulate as a result of decreased sink demand are well known to repress photosynthetic genes. Interestingly, the genes for invertase and sucrose synthase, both of which can catalyze the first step in sucrose utilization, and genes for sucrose–H^+ symporters, which play a key role in apoplastic loading, are also among those regulated by carbohydrate supply.

Such regulation of photosynthesis by sink demand suggests that sustained increases in photosynthesis in response to elevated CO_2 in the atmosphere may depend on increasing sink strength (increasing the strength of existing sinks or developing new sinks). See Chapter 9 for a discussion of the results of increased CO_2 levels in the atmosphere on photosynthesis and growth of plants.

Transport of Signaling Molecules

Besides its major function in the long-distance transport of photosynthate, the phloem is also a conduit for the transport of signaling molecules from one part of the plant to another. Such long-distance signals coordinate the activities of sources and sinks and regulate plant growth and development. As indicated earlier, the signals between sources and sinks might be physical or chemical. Physical signals such as turgor change are transmitted rapidly via the interconnecting system of sieve elements. Molecules traditionally considered to be chemical signals, such as proteins and plant hormones, are found in the phloem sap, as are mRNAs and small RNAs, which have more recently been added to the list of signal molecules. The translocated carbohydrates themselves may also act as signals.

Turgor pressure and chemical signals coordinate source and sink activities

Turgor pressure may play a role in coordinating the activities of sources and sinks. For example, if phloem unloading were rapid under conditions of rapid sugar utilization at the sink tissue, turgor pressures in the sieve elements of sinks would be reduced, and this reduction would be transmitted to the sources. If loading were controlled in part by sieve element turgor, loading would increase in response to this signal from the sinks. The opposite response would be seen when unloading was slow in the sinks. Some data suggest that cell turgor can modify the activity of the proton-pumping ATPase at the plasma membrane and therefore alter membrane transport rates.

Shoots produce growth regulators such as auxin, which can be rapidly transported to the roots via the phloem, and roots produce cytokinins, which move to the shoots through the xylem. Gibberellins (GAs) and abscisic acid (ABA) are also transported throughout the plant in the vascular system (see Chapter 12). Plant hormones play a role in regulating source–sink relationships. They affect photosynthate partitioning in part by controlling sink growth, leaf senescence, and other developmental processes. Plant defense responses against herbivores and pathogens can also change allocation and partitioning of photoassimilates, with plant defense hormones such as jasmonic acid mediating the responses.

Loading of sucrose has been shown to be stimulated by exogenous auxin, but inhibited by ABA in some source tissues, while exogenous ABA enhances, and auxin inhibits, sucrose uptake by some sink tissues. Hormones might regulate apoplastic loading and unloading by influencing the levels of active transporters in plasma membranes. Other potential sites of hormone regulation of unloading include tonoplast transporters, enzymes for metabolism of incoming sucrose, wall extensibility, and plasmodesmatal permeability in the case of symplastic unloading (see the next section).

Proteins and RNAs function as signal molecules in the phloem to regulate growth and development

It has long been known that viruses can move in the phloem, traveling as complexes of proteins and nucleic acids or as intact virus particles. More recently, endogenous RNA molecules and proteins have been found in phloem sap, and at least some of these can function as signal molecules or generate phloem-mobile signals.

At least some proteins synthesized in companion cells can clearly enter the sieve elements through the plasmodesmata that connect the two cell types and can move with the translocation stream to sink tissues. In Chapter 17 we will discuss the role of phloem in translocating a protein signaling molecule from leaves to the shoot apex to induce flowering. RNAs transported in the phloem consist of endogenous mRNAs, pathogenic RNAs, and small RNAs associated with gene silencing. Most of these RNAs appear to travel in the phloem as complexes of RNA and protein (ribonucleoproteins [RNPs]).

Plasmodesmata function in phloem signaling

Plasmodesmata (see Figure 1.8) have been implicated in nearly every aspect of phloem translocation, from loading to long-distance transport (pores in sieve areas and sieve plates are modified plasmodesmata) to allocation and partitioning. What role might plasmodesmata play in macromolecular signaling in the phloem?

The mechanism of plasmodesmatal transport (called trafficking) can be either passive (nontargeted) or selective and regulated. When a molecule moves passively, its size must be smaller than the size exclusion limit (SEL) of the plasmodesmata (see Chapter 1). When a molecule moves in a selective fashion, it must possess a trafficking signal or be targeted in some other way to the plasmodesmata. The transport of some developmental transcription factors and of viral movement proteins appears to occur by means of a selective mechanism.

Viral **movement proteins** interact directly with plasmodesmata to increase the SEL and allow the viral genome to move between cells, which spreads the infection through the plant. Movement proteins increase the SEL by one of two mechanisms. Movement proteins from some viruses coat the surface of the viral genome (typically RNA), forming ribonucleoprotein complexes. Movement proteins enable tobacco mosaic virus to move between cells in leaves that are susceptible to the virus, where it recruits other proteins in the cell that increase the size of the plasmodesmatal pore. Other viruses, such as cowpea mosaic virus and tomato spotted wilt virus, encode movement proteins that form a transport tubule within the plasmodesmatal channel that enhances the passage of mature virus particles through plasmodesmata.

movement proteins Nonstructural proteins encoded by a virus's genome that facilitate movement of that virus through the symplast.

It is fitting to end this chapter with research topics that will continue to engage plant physiologists of the future: regulation of growth and development via the transport of endogenous RNA and protein signals, the nature of the proteins that facilitate the transport of signals through plasmodesmata, and the possibility of targeting signals to specific sinks in contrast to the relative nonspecificity of mass flow. Many other potential areas of inquiry have been indicated in this chapter as well, such as the mechanism of phloem transport in gymnosperms, the nature and role of proteins in the lumen of the sieve elements, and the magnitude of pressure gradients in the sieve elements, especially in trees. As always in science, an answer to one question generates more questions!

Summary

Phloem translocation moves the products of photosynthesis from mature leaves to areas of growth and storage. It also transmits chemical signals and redistributes ions and other substances throughout the plant body.

Patterns of Translocation: Source to Sink

- Phloem translocation is not defined by gravity. Sap is translocated from sources to sinks, and the pathways involved are often complex (**Figures 10.1, 10.2**).

Pathways of Translocation

- Sieve elements of the phloem conduct sugars and other organic materials throughout the plant (**Figures 10.3–10.5**).

- During development, sieve elements lose many organelles, retaining only the plasma membrane and modified mitochondria, plastids, and smooth endoplasmic reticulum (**Figures 10.5, 10.6**).

- Sieve elements are interconnected through pores in their cell walls (**Figure 10.7**).

- P-proteins and callose seal off damaged sieve elements to limit loss of sap.

- Companion cells aid transport of photosynthetic products to the sieve elements. They also supply proteins and ATP to the sieve elements (**Figures 10.5–10.7**).

Materials Translocated in the Phloem

- The composition of sap has been determined; nonreducing sugars are the main transported molecules (**Table 10.2; Figure 10.8**).

- Sap includes proteins, many of which may have functions related to stress and defense reactions.

Rates of Movement

- Transport velocities in the phloem are high and exceed the rate of diffusion over long distances by many orders of magnitude.

The Pressure-Flow Model, a Passive Mechanism for Phloem Transport

- The pressure-flow model explains phloem translocation as a bulk flow of solution driven by an osmotically generated pressure gradient between source and sink.

- Phloem loading at the source and phloem unloading at the sink establish the pressure gradient for passive, long-distance bulk flow (**Figure 10.9**).

- Pressure gradients in the phloem sieve elements may be modest; pressures in herbaceous plants and trees appear to be similar.

Phloem Loading

- The export of sugars from sources involves allocation of photosynthate to transport, short-distance transport, and phloem loading.

- Phloem loading can occur by way of the symplast or apoplast (**Figure 10.11**).

- Sucrose is actively transported into the sieve element–companion cell complex in the apoplastic pathway (**Figures 10.12, 10.13**).

- The polymer-trapping model holds that polymers are synthesized from sucrose in the intermediary cells; the larger oligosaccharides can only diffuse into the sieve elements (**Figure 10.14**).

Phloem Unloading and Sink-to-Source Transition

- The import of sugars into sink cells involves phloem unloading, short-distance transport, and storage or metabolism.

- Phloem unloading and short-distance transport may operate by symplastic or apoplastic pathways in different sinks.

- Transport into sink tissues is energy dependent.

(Continued)

Summary *(continued)*

- Import cessation and export initiation are separate events, and there is a gradual transition from sink to source (**Figures 10.15, 10.16**).

- The transition from sink to source requires several conditions, including the expression and localization of the sucrose–H⁺ symporter (**Figure 10.17**).

Photosynthate Distribution: Allocation and Partitioning

- Allocation in source leaves includes synthesis of storage compounds, metabolic utilization, and synthesis of transport compounds (**Figure 10.18**).

- The regulation of allocation controls the distribution of fixed carbon to the Calvin–Benson cycle, starch synthesis, sucrose synthesis, and respiration (**Figure 10.19**).

- A variety of chemical and physical signals are involved in partitioning resources among the various sinks.

- In competing for photosynthate, sink strength depends on sink size and sink activity.

- In response to altered conditions, short-term changes alter the distribution of photosynthate among different sinks, while long-term changes take place in source metabolism and alter the amount of photosynthate available for transport.

Transport of Signaling Molecules

- Turgor pressure, sugars, cytokinins, gibberellins, and abscisic acid have signaling roles in coordinating source and sink activities.

- Some proteins can move from companion cells into sieve elements of source leaves, and through the phloem to sink leaves.

- Proteins and RNAs transported in the phloem can alter cellular functions.

- Changes in the size exclusion limit (SEL) may control what passes through plasmodesmata.

Suggested Reading

Holbrook, N. M., and Zwieniecki, M. A., eds. (2005) *Vascular Transport in Plants*. Elsevier Academic Press, Boston, MA.

Jekat, S. B., Ernst, A. M., von Bohl, A., Zielonka, S., Twyman, R. M., Noll, G. A., and Prufer, D. (2013) P-proteins in *Arabidopsis* are heteromeric structures involved in rapid sieve tube sealing. *Front. Plant Sci.* 4: 225. DOI: 10.3389/fpls.2013.00225

Jensen, K. H., Berg-Sørensen, K., Bruus, H., Holbrook, N. M., Liesche, J., Schulz, A., Zwieniecki, M. A., and Bohr, T. (2016) Sap flow and sugar transport in plants. *Reviews of Modern Physics* 88(3): [035007]. DOI: 10.1103/RevModPhys.88.035007

Knoblauch, M., and Oparka, K. (2012) The structure of the phloem—still more questions than answers. *Plant J.* 70: 147–156.

Liesche, J., and Schulz, A. (2013) Modeling the parameters for plasmodesmatal sugar filtering in active symplasmic phloem loaders. *Front. Plant Sci.* 4: 207. DOI: 10.3389/fpls.2013.00207

Mullendore, D. L., Windt, C. W., Van As, H., and Knoblauch, M. (2010) Sieve tube geometry in relation to phloem flow. *Plant Cell* 22: 579–593.

Patrick, J. W. (2013) Does Don Fisher's high-pressure manifold model account for phloem transport and resource partitioning? *Front. Plant Sci.* 4: 184.

Slewinski, T. L., Zhang, C., and Turgeon, R. (2013) Structural and functional heterogeneity in phloem loading and transport. *Front. Plant Sci.* 4: 244. DOI: 10.3389/fpls.2013.00244

Thompson, G. A., and van Bel, A. J. E., eds. (2013) *Phloem: Molecular Cell Biology, Systemic Communication, Biotic Interactions*. Wiley-Blackwell, Ames, IA.

Turgeon, R. (2010) The puzzle of phloem pressure. *Plant Physiol.* 154: 578–581.

Yoo, S.-C., Chen, C., Rojas, M., Daimon, Y., Ham, B.-K., Araki, T., and Lucas, W. J. (2013) Phloem long-distance delivery of FLOWERING LOCUS T (FT) to the apex. *Plant J.* 75: 456–468.

Zhang, C., Yu, X., Ayre, B. G., and Turgeon, R. (2012) The origin and composition of cucurbit "phloem" exudate. *Plant Physiol.* 158: 1873–1882.

Photo by M. H. Siddall

11 Respiration and Lipid Metabolism

Photosynthesis provides the organic building blocks that plants (and nearly all other organisms) depend on. Respiration, with its associated carbon metabolism, releases the energy stored in carbon compounds in a controlled manner for cellular use. At the same time, it generates many carbon precursors for biosynthesis.

We begin this chapter by reviewing respiration in its metabolic context, emphasizing the interconnections among the processes involved and the special features that are peculiar to plants. We also relate respiration to recent developments in our understanding of the biochemistry and molecular biology of plant mitochondria and respiratory fluxes in intact plant tissues. Then we describe the pathways of lipid biosynthesis that lead to the accumulation of fats and oils, which many plant species use for energy and carbon storage. Finally, we discuss the catabolic pathways involved in the breakdown of lipids and the conversion of their degradation products into sugars that occurs during the germination of fat-storing seeds.

Overview of Plant Respiration

Aerobic (oxygen-requiring) respiration is common to nearly all eukaryotic organisms, and in its broad outlines the respiratory process in plants is similar to that found in animals and other aerobic eukaryotes. However, some specific aspects of plant respiration distinguish it from its animal counterpart. **Aerobic respiration** is the biological process by which reduced organic compounds are oxidized in a controlled manner. During respiration, energy is released and transiently stored in a compound, **adenosine triphosphate** (**ATP**), which is used by the cellular reactions for maintenance and development.

aerobic respiration The complete oxidation of carbon compounds to CO_2 and H_2O, using oxygen as the final electron acceptor. Energy is released and conserved as ATP.

adenosine triphosphate (ATP) The major carrier of chemical energy in the cell, which by hydrolysis is converted to adenosine diphosphate (ADP) or adenosine monophosphate (AMP) with release of energy.

Gibbs free energy The energy that is available to do work; in biological systems, the work of synthesis, transport, and movement.

Glucose is usually cited as the substrate for respiration. In most plant cell types, however, reduced carbon is derived from sources such as the disaccharide sucrose, other sugars, organic acids, triose phosphates from photosynthesis, and metabolites from lipid and protein degradation (**Figure 11.1**).

From a chemical standpoint, plant respiration can be expressed as the oxidation of the 12-carbon molecule sucrose and the reduction of 12 molecules of O_2:

$$C_{12}H_{22}O_{11} + 13\ H_2O \rightarrow 12\ CO_2 + 48\ H^+ + 48\ e^- \tag{11.1}$$

$$12\ O_2 + 48\ H^+ + 48\ e^- \rightarrow 24\ H_2O \tag{11.2}$$

giving the following net reaction:

$$C_{12}H_{22}O_{11} + 12\ O_2 \rightarrow 12\ CO_2 + 11\ H_2O \tag{11.3}$$

This reaction is the reversal of the photosynthetic process; it represents a coupled redox reaction in which sucrose is completely oxidized to CO_2 while oxygen serves as the ultimate electron acceptor and is reduced to water in the process. The change in standard **Gibbs free energy** ($\Delta G^{0'}$) for the net reaction is −5760 kJ per mole (342 g) of sucrose oxidized. This large negative value means that the equilibrium point is strongly shifted to the right, and much energy is therefore released by sucrose degradation. The controlled release of this free energy, along with its coupling to the synthesis of ATP, is the primary, although by no means only, role of respiratory metabolism.

To prevent damage by heating of cellular structures, the cell oxidizes sucrose in a series of step-by-step reactions. The reactions can be grouped into four major

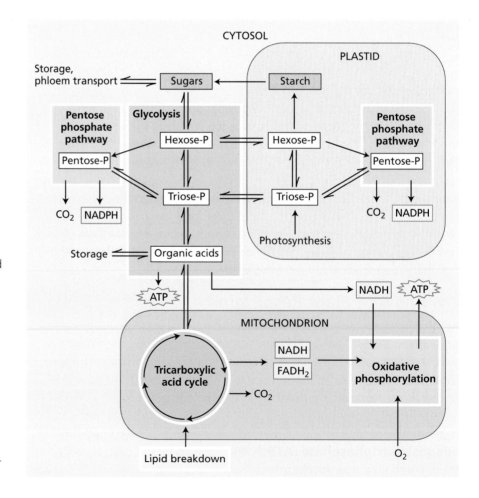

Figure 11.1 Overview of respiration. Substrates for respiration are generated by other cellular processes and enter the respiratory pathways. Glycolysis and the oxidative pentose phosphate pathways in the cytosol and plastids convert sugars into organic acids such as pyruvate, via hexose phosphates and triose phosphates, generating NADH or NADPH, and ATP. The organic acids are oxidized in the mitochondrial TCA cycle, and the NADH and $FADH_2$ produced provide the energy for ATP synthesis by the electron transport chain and ATP synthase in oxidative phosphorylation. In gluconeogenesis, carbon from lipid breakdown is broken down in the glyoxysomes, metabolized in the TCA cycle, and then used to synthesize sugars in the cytosol by reverse glycolysis.

processes: glycolysis, the oxidative pentose phosphate pathway, the TCA cycle, and oxidative phosphorylation. These pathways do not function in isolation, but exchange metabolites at several levels. The substrates of respiration enter the respiratory process at different points in the pathways, as summarized in Figure 11.1:

- **Glycolysis** involves a series of reactions catalyzed by enzymes located in both the cytosol and the plastids. A sugar—for example, sucrose—is partly oxidized via six-carbon sugar phosphates (hexose phosphates) and three-carbon sugar phosphates (triose phosphates) to produce an organic acid—mainly pyruvate. The process yields a small amount of energy as ATP and reducing power in the form of a reduced nicotinamide nucleotide, NADH.

- In the **oxidative pentose phosphate pathway**, also located in both the cytosol and the plastids, the six-carbon glucose 6-phosphate is initially oxidized to the five-carbon ribulose 5-phosphate. Carbon is lost as CO_2, and reducing power is conserved in the form of another reduced nicotinamide nucleotide, NADPH. In subsequent near-equilibrium reactions of the pentose phosphate pathway, ribulose 5-phosphate is converted into sugar phosphates containing three to seven carbon atoms. These intermediates can be used in biosynthetic pathways or reenter glycolysis.

- In the **tricarboxylic acid** (**TCA**) **cycle**, pyruvate is oxidized completely to CO_2, via stepwise oxidations of organic acids in the innermost compartment of the mitochondrion—the matrix. This process mobilizes the major amount of reducing power (16 NADH + 4 $FADH_2$ per sucrose) and a small amount of energy (ATP) from the breakdown of sucrose.

- In **oxidative phosphorylation**, electrons are transferred along an electron transport chain consisting of a series of protein complexes embedded in the inner of the two mitochondrial membranes. This system transfers electrons from NADH (and related species)—produced by glycolysis, the oxidative pentose phosphate pathway, and the TCA cycle—to oxygen. This electron transfer releases a large amount of free energy, much of which is conserved through the synthesis of ATP from ADP and P_i (inorganic phosphate), catalyzed by the enzyme ATP synthase. Collectively, the redox reactions of the electron transport chain and the synthesis of ATP are called oxidative phosphorylation.

glycolysis A series of reactions in which a sugar is oxidized to produce two molecules of pyruvate. A small amount of ATP and NADH is produced.

oxidative pentose phosphate pathway A cytosolic and plastidic pathway that oxidizes glucose and produces NADPH and several sugar phosphates.

tricarboxylic acid (TCA) cycle A cycle of reactions catalyzed by enzymes localized in the mitochondrial matrix that leads to the oxidation of pyruvate to CO_2. ATP and NADH are generated in the process. Also called the citric acid cycle.

oxidative phosphorylation The transfer of electrons to oxygen in the mitochondrial electron transport chain that is coupled to ATP synthesis from ADP and phosphate by ATP synthase.

Nicotinamide adenine dinucleotide (NAD) is an organic cofactor (coenzyme) associated with many enzymes that catalyze cellular redox reactions. NAD^+ is the oxidized form that undergoes a reversible two-electron reduction to yield NADH (**Figure 11.2**). The standard reduction potential for the NAD^+/NADH redox couple is about −320 mV. This tells us that NADH is a relatively strong reductant (i.e., electron donor), which can conserve the free energy carried by the electrons released during the stepwise oxidations of glycolysis and the TCA cycle. A related compound, nicotinamide adenine dinucleotide phosphate (NADP/$NADP^+$/NADPH), has a similar function in photosynthesis (see Chapters 7 and 8) and the oxidative pentose phosphate pathway, and also takes part in mitochondrial metabolism. These roles are discussed later in the chapter.

The oxidation of NADH by oxygen via the electron transport chain releases free energy (220 kJ mol^{-1}) that drives the synthesis of approximately 60 ATP per sucrose molecule oxidized (as we will see later). We can now formulate a more complete picture of respiration as related to its role in cellular energy metabolism by coupling the following two reactions:

$$C_{12}H_{22}O_{11} + 12\ O_2 \rightarrow 12\ CO_2 + 11\ H_2O \tag{11.3}$$

$$60\ ADP + 60\ P_i \rightarrow 60\ ATP + 60\ H_2O \tag{11.4}$$

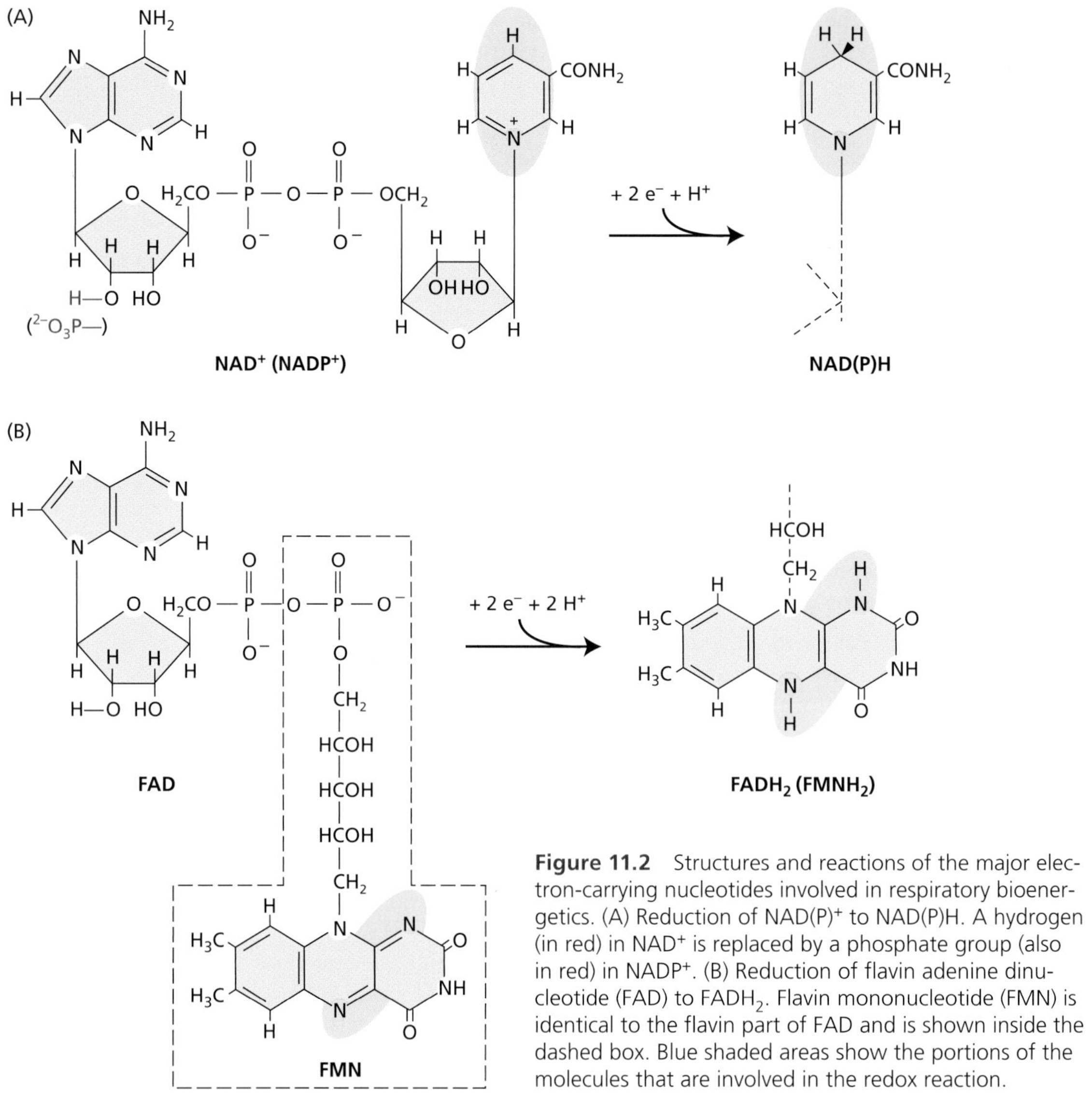

Figure 11.2 Structures and reactions of the major electron-carrying nucleotides involved in respiratory bioenergetics. (A) Reduction of NAD(P)$^+$ to NAD(P)H. A hydrogen (in red) in NAD$^+$ is replaced by a phosphate group (also in red) in NADP$^+$. (B) Reduction of flavin adenine dinucleotide (FAD) to FADH$_2$. Flavin mononucleotide (FMN) is identical to the flavin part of FAD and is shown inside the dashed box. Blue shaded areas show the portions of the molecules that are involved in the redox reaction.

Keep in mind that not all the carbon that enters the respiratory pathway ends up as CO$_2$. Many respiratory carbon intermediates are the starting points for pathways that synthesize amino acids, nucleotides, lipids, and many other compounds.

Glycolysis

In the early steps of glycolysis (from the Greek words *glykos*, "sugar," and *lysis*, "splitting"), carbohydrates are converted into hexose phosphates, each of which is then split into two triose phosphates. In a subsequent energy-conserving phase, each triose phosphate is oxidized and rearranged to yield one molecule of pyruvate, an organic acid. Besides preparing the substrate for oxidation in the TCA cycle, glycolysis yields a small amount of chemical energy in the form of ATP and NADH.

When molecular oxygen is unavailable—for example, in plant roots in flooded soils—glycolysis can be the main source of energy for cells. For this to work, the *fermentative pathways*, which are carried out in the cytosol, must reduce pyruvate to recycle the NADH produced by glycolysis. In this section we describe the basic glycolytic and fermentative pathways, emphasizing features that are specific to plant cells. In the next section we discuss the pentose phosphate pathway, another pathway for sugar oxidation in plants.

Glycolysis metabolizes carbohydrates from several sources

Glycolysis occurs in all living organisms (prokaryotes and eukaryotes). The principal reactions associated with the classic glycolytic pathway in plants are almost identical to those in animal cells (**Figure 11.3**). However, plant glycolysis has unique regulatory features, alternative enzymatic routes for several steps, and a parallel partial glycolytic pathway in plastids.

In animals, the substrate of glycolysis is glucose, and the end product is pyruvate. Because sucrose is the major translocated sugar in most plants, and is therefore the form of carbon that most nonphotosynthetic tissues import, sucrose (not glucose) can be argued to be the true sugar substrate for plant glycolysis. The end products of plant glycolysis include another organic acid, malate.

In the early steps of glycolysis, sucrose is split into its two monosaccharide units—glucose and fructose—which can readily enter the glycolytic pathway. Two pathways for the splitting of sucrose are known in plants, both of which take part in the use of sucrose from phloem unloading (see Chapter 10): the invertase pathway and the sucrose synthase pathway.

Invertases hydrolyze sucrose in the cell wall, vacuole, or cytosol into its two component hexoses (glucose and fructose). The hexoses are then phosphorylated in the cytosol by a hexokinase that uses ATP to form **hexose phosphates**. Alternatively, *sucrose synthase* combines sucrose with UDP to produce fructose and UDP-glucose in the cytosol. UDP-glucose pyrophosphorylase can then convert UDP-glucose and pyrophosphate (PP_i) into UTP and glucose 6-phosphate (see Figure 11.3). While the sucrose synthase reaction is close to equilibrium, the invertase reaction is essentially irreversible, driving the flux in the forward direction.

Through studies of transgenic plants lacking specific invertases or sucrose synthase, each enzyme has been found to be essential for specific life processes, but differences are observed among plant tissues and species. For example, sucrose synthase and cell wall invertase are needed for normal fruit development in several crop species, whereas the cytosolic invertase is necessary for optimum root cell wall integrity and leaf respiration in *Arabidopsis thaliana*. Both sucrose synthase and invertases can degrade sucrose for glycolysis, and if one of the enzymes is absent, for example in a mutant, the other enzyme(s) can still maintain respiration. The existence of different pathways that serve a similar function and can replace each other without a clear loss in function is called **metabolic redundancy**; it is a common feature in plant metabolism. In plastids, a partial glycolysis occurs that produces metabolites for plastidic biosynthetic reactions, for example synthesis of fatty acids, tetrapyrroles, and aromatic amino acids. Starch is both synthesized and catabolized only in plastids, and carbon obtained from starch degradation (for example, in a chloroplast at night) enters the glycolytic pathway in the cytosol primarily as glucose (see Chapter 8). In the light, photosynthetic products can enter the glycolytic pathway directly as triose phosphate. In overview, glycolysis works like a funnel with an initial phase collecting carbon from different carbohydrate sources, depending on the physiological situation.

In the initial phase of glycolysis, each hexose unit is phosphorylated twice and then split, producing two molecules of **triose phosphate**. This series of reactions consumes two to four molecules of ATP per sucrose unit, depending on whether the sucrose is split by sucrose synthase or invertase. The initial phase also includes two of the three essentially irreversible reactions of the glycolytic pathway, which are catalyzed by hexokinase and phosphofructokinase (see Figure 11.3). As we will see later, the phosphofructokinase reaction is one of the control points of glycolysis in both plants and animals.

hexose phosphates   Six-carbon sugars with phosphate groups attached.

metabolic redundancy   A common feature of plant metabolism in which different pathways serve a similar function. They can therefore replace each other without apparent loss in function.

triose phosphate   A three-carbon sugar phosphate.

(A)

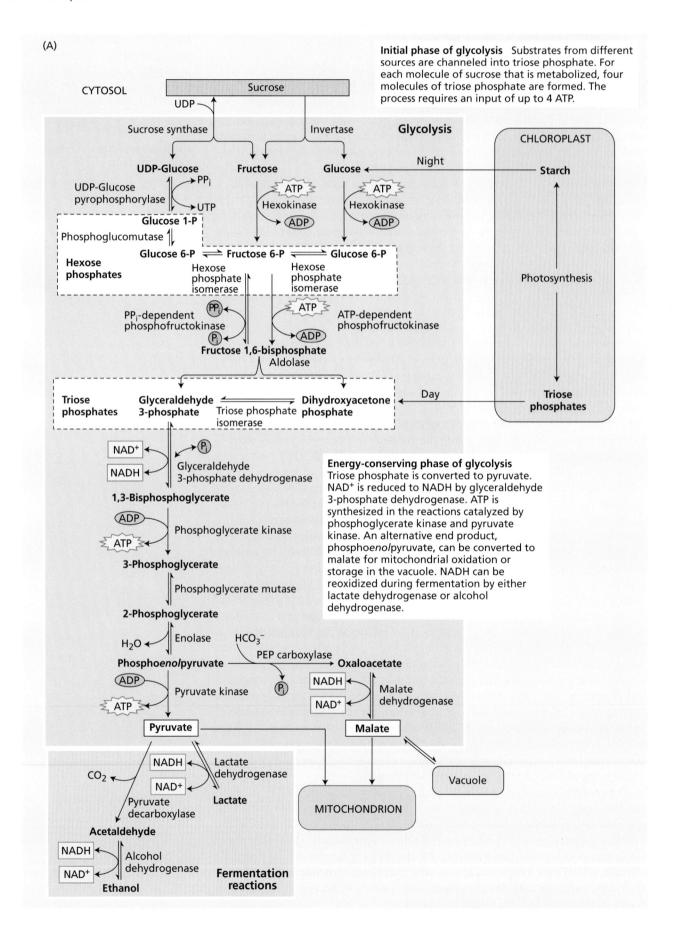

Initial phase of glycolysis Substrates from different sources are channeled into triose phosphate. For each molecule of sucrose that is metabolized, four molecules of triose phosphate are formed. The process requires an input of up to 4 ATP.

Energy-conserving phase of glycolysis
Triose phosphate is converted to pyruvate. NAD^+ is reduced to NADH by glyceraldehyde 3-phosphate dehydrogenase. ATP is synthesized in the reactions catalyzed by phosphoglycerate kinase and pyruvate kinase. An alternative end product, phospho*enol*pyruvate, can be converted to malate for mitochondrial oxidation or storage in the vacuole. NADH can be reoxidized during fermentation by either lactate dehydrogenase or alcohol dehydrogenase.

(B)

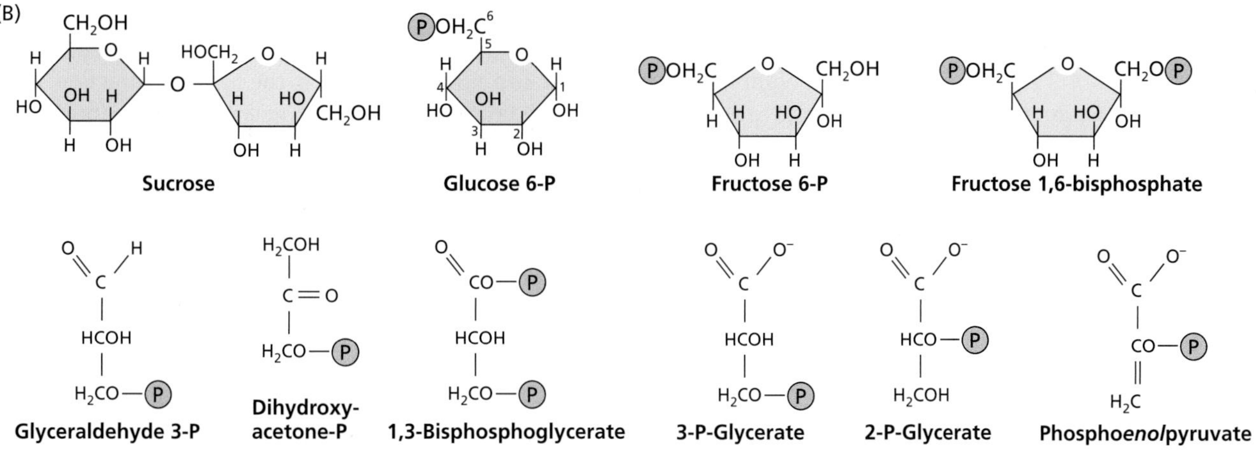

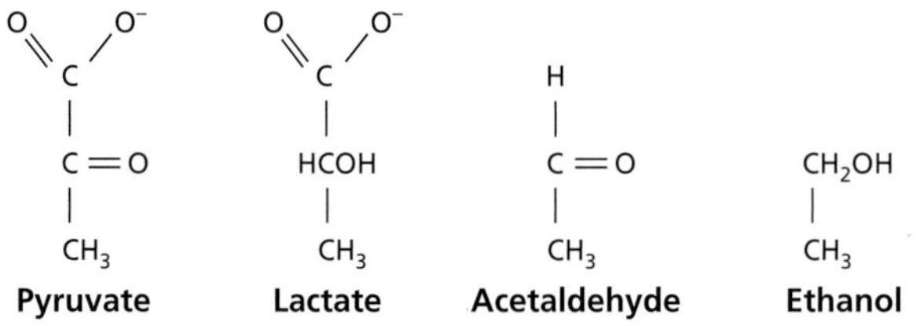

Figure 11.3 Reactions of plant glycolysis and fermentation. (A) In the main glycolytic pathway, sucrose is oxidized via hexose phosphates and triose phosphates to the organic acid pyruvate, but plants also carry out alternative reactions. The double arrows denote reversible reactions; the single arrows, essentially irreversible reactions. (B) The structures of the carbon intermediates. P, phosphate group.

The energy-conserving phase of glycolysis extracts usable energy

The reactions discussed thus far convert carbon from the various substrate pools to triose phosphates. Once *glyceraldehyde 3-phosphate* is formed, the glycolytic pathway can begin to extract usable energy in the energy-conserving phase. The enzyme *glyceraldehyde 3-phosphate dehydrogenase* catalyzes the oxidation of the aldehyde to a carboxylic acid, reducing NAD^+ to NADH. This reaction releases sufficient free energy to allow the phosphorylation (using inorganic phosphate) of glyceraldehyde 3-phosphate to produce 1,3-bisphosphoglycerate. The phosphorylated carboxylic acid on carbon 1 of 1,3-bisphosphoglycerate (see Figure 11.3B) has a large standard free-energy change ($\Delta G^{0'}$) of hydrolysis (-49.3 kJ mol^{-1}). Thus 1,3-bisphosphoglycerate is a strong donor of phosphate groups.

In the next step of glycolysis, catalyzed by *phosphoglycerate kinase,* the phosphate on carbon 1 is transferred to a molecule of ADP, yielding ATP and 3-phosphoglycerate. For each sucrose entering the pathway, four ATPs are generated by this reaction—one for each molecule of 1,3-bisphosphoglycerate.

This type of ATP synthesis, traditionally referred to as **substrate-level phosphorylation**, involves the direct transfer of a phosphate group from a substrate molecule to ADP to form ATP. ATP synthesis by substrate-level phosphorylation is mechanistically distinct from ATP synthesis by the ATP synthases involved in oxidative phosphorylation in mitochondria (which we describe later in this chapter) or in photophosphorylation in chloroplasts (see Chapter 7).

In the subsequent two reactions, the phosphate on 3-phosphoglycerate is transferred to carbon 2, and then a molecule of water is removed, yielding the compound *phospho*enol*pyruvate* (PEP). The phosphate group on PEP has a high $\Delta G^{0'}$ of hydrolysis (-61.9 kJ mol^{-1}), which makes PEP an extremely good phosphate donor for ATP formation. Using PEP as substrate, the enzyme *pyruvate kinase* catalyzes a second substrate-level phosphorylation to yield ATP and pyruvate. This final step, which is the third essentially irreversible step in glycolysis, yields four additional molecules of ATP for each sucrose molecule that enters the pathway.

substrate-level phosphorylation
A process that involves the direct transfer of a phosphate group from a substrate molecule to ADP to form ATP.

gluconeogenesis   The synthesis of carbohydrates through the reversal of glycolysis.

PEP carboxylase   A cytosolic enzyme that forms oxaloacetate by the carboxylation of phospho*enol*pyruvate.

fermentation   The metabolism of pyruvate in the absence of oxygen, leading to the oxidation of the NADH generated in glycolysis to NAD⁺. Allows glycolytic ATP production to function in the absence of oxygen.

Plants have alternative glycolytic reactions

The glycolytic degradation of sugars to pyruvate occurs in most organisms, but many organisms can also operate a similar pathway in the opposite direction. This process, to synthesize sugars from organic acids, is known as **gluconeogenesis**.

Gluconeogenesis is particularly important in plants (such as castor bean [*Ricinus communis*] and sunflower) that store carbon in the form of oils (triacylglycerols) in the seeds. Plants cannot transport lipids, so when such a seed germinates, the oil is converted by gluconeogenesis into sucrose, which is transported to the growing cells in the seedling. In the initial phase of glycolysis, gluconeogenesis overlaps with the pathway for synthesis of sucrose from photosynthetic triose phosphate, which is typical of leaf cells.

Because the glycolytic reaction catalyzed by *ATP-dependent phosphofructokinase* is essentially irreversible (see Figure 11.3), an additional enzyme, *fructose 1,6-bisphosphate phosphatase*, converts fructose 1,6-bisphosphate irreversibly into fructose 6-phosphate and P_i during gluconeogenesis. ATP-dependent phosphofructokinase and fructose 1,6-bisphosphate phosphatase represent a major control point of carbon flux through the glycolytic/gluconeogenic pathways of both plants and animals as well as in sucrose synthesis in plants (see Chapter 8).

In plants, the interconversion of fructose 6-phosphate and fructose 1,6-bisphosphate is made more complex by the presence of an additional (cytosolic) enzyme, *PP_i-dependent phosphofructokinase* (pyrophosphate:fructose 6-phosphate 1-phosphotransferase), which catalyzes the following reversible reaction (see Figure 11.3):

$$\text{Fructose 6-P} + PP_i \rightarrow \text{fructose 1,6-bisphosphate} + P_i \qquad (11.5)$$

where -P represents bound phosphate. PP_i-dependent phosphofructokinase is found in the cytosol of most plant tissues at levels that are considerably higher than those of ATP-dependent phosphofructokinase. The reaction catalyzed by PP_i-dependent phosphofructokinase is readily reversible, but it is unlikely to operate in sucrose synthesis. Suppression of PP_i-dependent phosphofructokinase in transgenic plants has shown that it contributes to glycolytic conversion of hexose phosphates to triose phosphates, but that it is not essential for plant survival, indicating that the ATP-dependent phosphofructokinase can take over its function. The three enzymes that interconvert fructose 6-phosphate and fructose 1,6-bisphosphate are all regulated to match the plant demands for both respiration and synthesis of sucrose and polysaccharides. As a consequence, the operation of the glycolytic pathway in plants has several unique characteristics.

At the end of the glycolytic process, plants have alternative pathways for metabolizing PEP. In one pathway PEP is carboxylated by the ubiquitous cytosolic enzyme **PEP carboxylase** to form the organic acid oxaloacetate. The oxaloacetate is then reduced to malate by the action of *malate dehydrogenase,* which uses NADH as a source of electrons (see Figure 11.3). The resulting malate can be stored by export to the vacuole or transported to the mitochondrion, where it can be used in the TCA cycle (discussed later). Thus, the action of pyruvate kinase and PEP carboxylase can produce pyruvate or malate for mitochondrial respiration, although pyruvate dominates in most tissues.

In the absence of oxygen, fermentation regenerates the NAD⁺ needed for glycolytic ATP production

Oxidative phosphorylation does not function in the absence of oxygen. Glycolysis then cannot continue because the cell's supply of NAD⁺ is limited and once the NAD⁺ becomes tied up in the reduced state (NADH), the glyceraldehyde 3-phosphate dehydrogenase comes to a halt. To overcome this limitation, plants and other organisms can further metabolize pyruvate by carrying out one or more forms of **fermentation** (see Figure 11.3).

Alcoholic fermentation is common in plants, although more widely known from brewer's yeast. Two enzymes, pyruvate decarboxylase and alcohol dehydrogenase, act on pyruvate, ultimately producing ethanol and CO_2 and oxidizing NADH in the process. In lactic acid fermentation (common in mammalian muscle, but also found in plants), the enzyme lactate dehydrogenase uses NADH to reduce pyruvate to lactate, thus regenerating NAD^+.

Plant tissues may be subjected to low (hypoxic) or zero (anoxic) concentrations of ambient oxygen. The best-studied example involves flooded or waterlogged soils in which the diffusion of oxygen is sufficiently reduced for root tissues to become hypoxic (see also Chapter 19). Such conditions force the tissues to carry out fermentative metabolism. In maize (corn; *Zea mays*), the initial metabolic response to low oxygen concentrations is lactic acid fermentation, but the subsequent response is alcoholic fermentation. Ethanol is thought to be a less toxic end product of fermentation because it can diffuse out of the cell, whereas lactate accumulates and promotes acidification of the cytosol. In numerous other cases, plants or plant parts function under near-anoxic conditions by carrying out some form of fermentation.

It is important to consider the efficiency of fermentation. *Efficiency* is defined here as the energy conserved as ATP relative to the energy potentially available in a molecule of sucrose. The standard free-energy change ($\Delta G^{0'}$) for the complete oxidation of sucrose to CO_2 is -5760 kJ mol^{-1}. The $\Delta G^{0'}$ for the synthesis of ATP is 32 kJ mol^{-1}. However, under the nonstandard conditions that normally exist in both mammalian and plant cells, the synthesis of ATP requires an input of free energy of approximately 50 kJ mol^{-1}. Normal glycolysis—with ethanol or lactate as the final product—leads to a net synthesis of four ATP molecules (cost ~200 kJ mol^{-1}) for each sucrose molecule converted into pyruvate. The efficiency of fermentation is therefore only about 4% (200 kJ mol^{-1}/5760 kJ mol^{-1}). Most of the energy available in sucrose remains in the ethanol or lactate.

Changes in the glycolytic pathway under oxygen deficiency can increase the ATP yield. This is the case when sucrose is degraded via sucrose synthase instead of invertase, avoiding ATP consumption by the hexokinase in the initial phase of glycolysis. Such modifications emphasize the importance of energetic efficiency for plant survival in the absence of oxygen.

Because of the low energy recovery of fermentation, an increased rate of carbohydrate breakdown is needed to sustain the ATP production necessary for cell survival. The increase in glycolytic rate is called the *Pasteur effect* after the French microbiologist Louis Pasteur, who first noted it when yeast switched between aerobic respiration and fermentation. Glycolysis is up-regulated by changes in metabolite levels and by the induction of genes encoding the enzymes of glycolysis and fermentation.

In contrast to the products of fermentation, the pyruvate produced by glycolysis during aerobic respiration is further oxidized by mitochondria, resulting in a much more efficient use of the free energy available in sucrose.

The Oxidative Pentose Phosphate Pathway

The glycolytic pathway is not the only route available for the oxidation of sugars in plant cells. The oxidative pentose phosphate pathway can also accomplish this task (**Figure 11.4**). The reactions are carried out by soluble enzymes present in the cytosol and in plastids. Under most conditions, the pathway in plastids predominates over that in the cytosol.

The first two reactions of this pathway involve the oxidative events that convert the six-carbon molecule glucose 6-phosphate into the five-carbon unit **ribulose 5-phosphate**, with loss of a CO_2 molecule and generation of two molecules of NADPH (not NADH). The remaining reactions of the pathway convert ribulose 5-phosphate into the glycolytic intermediates glyceraldehyde 3-phosphate and

ribulose 5-phosphate In the pentose phosphate pathway, the initial five-carbon product of the oxidation of glucose 6-phosphate; in subsequent reactions, it is converted into sugars containing three to seven carbon atoms.

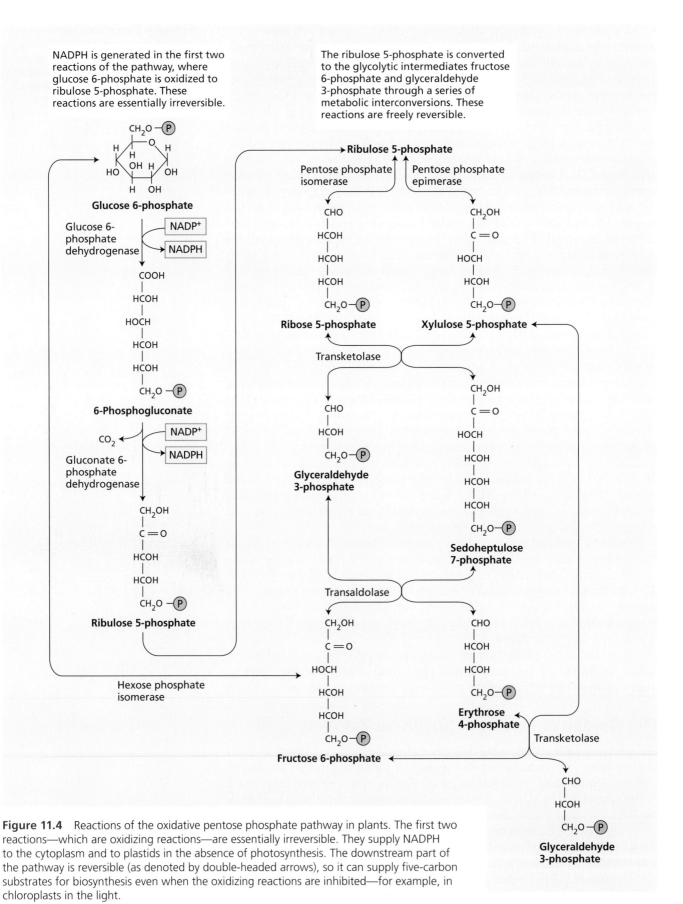

NADPH is generated in the first two reactions of the pathway, where glucose 6-phosphate is oxidized to ribulose 5-phosphate. These reactions are essentially irreversible.

The ribulose 5-phosphate is converted to the glycolytic intermediates fructose 6-phosphate and glyceraldehyde 3-phosphate through a series of metabolic interconversions. These reactions are freely reversible.

Figure 11.4 Reactions of the oxidative pentose phosphate pathway in plants. The first two reactions—which are oxidizing reactions—are essentially irreversible. They supply NADPH to the cytoplasm and to plastids in the absence of photosynthesis. The downstream part of the pathway is reversible (as denoted by double-headed arrows), so it can supply five-carbon substrates for biosynthesis even when the oxidizing reactions are inhibited—for example, in chloroplasts in the light.

fructose 6-phosphate by using three of the Calvin–Benson cycle enzymes in reverse (see Chapter 8). These products can be further metabolized by glycolysis to yield pyruvate. Alternatively, glucose 6-phosphate can be regenerated from glyceraldehyde 3-phosphate and fructose 6-phosphate by glycolytic enzymes. For six turns of this cycle, we can write the reaction as follows:

$$6 \text{ Glucose 6-P} + 12 \text{ NADP}^+ + 7 \text{ H}_2\text{O} \rightarrow$$
$$5 \text{ glucose 6-P} + 6 \text{ CO}_2 + \text{P}_i + 12 \text{ NADPH} + 12 \text{ H}^+ \tag{11.6}$$

The net result is the complete oxidation of one glucose 6-phosphate molecule to CO_2 (five molecules are regenerated) with the concomitant synthesis of 12 NADPH molecules.

Studies of the release of CO_2 from isotopically labeled glucose indicate that the pentose phosphate pathway accounts for 10 to 25% of the glucose breakdown, with the rest occurring mainly via glycolysis. As we will see, the contribution of the pentose phosphate pathway changes during development and with changes in growth conditions as the plant's requirements for specific products vary.

The oxidative pentose phosphate pathway produces NADPH and biosynthetic intermediates

The oxidative pentose phosphate pathway plays several roles in plant metabolism:

- *NADPH supply in the cytosol.* The product of the two oxidative steps is NADPH. This NADPH drives reductive steps associated with biosynthetic and stress defense reactions and is a substrate for reactions that remove reactive oxygen species (ROS). Because plant mitochondria possess an NADPH dehydrogenase located on the external surface of the inner membrane, the reducing power generated by the pentose phosphate pathway can be balanced by mitochondrial NADPH oxidation. The pentose phosphate pathway may therefore also contribute to cellular energy metabolism; that is, electrons from NADPH may end up reducing O_2 and generating ATP through oxidative phosphorylation.

- *NADPH supply in plastids.* In nongreen plastids, such as amyloplasts in the root, and in chloroplasts functioning in the dark, the pentose phosphate pathway is the major supplier of NADPH. The NADPH is used for biosynthetic reactions such as lipid synthesis and nitrogen assimilation. The formation of NADPH by glucose 6-phosphate oxidation in amyloplasts may also signal sugar status to the thioredoxin system (see Chapter 8) for control of starch synthesis.

- *Supply of substrates for biosynthetic processes.* In most organisms, the pentose phosphate pathway produces ribose 5-phosphate, which is a precursor of the ribose and deoxyribose needed in the synthesis of nucleic acids. In plants, however, ribose appears to be synthesized by another, as yet unknown, pathway. Another intermediate in the pentose phosphate pathway, the four-carbon erythrose 4-phosphate, combines with PEP in the initial reaction of the shikimic acid pathway that produces plant phenolic compounds, including aromatic amino acids and the precursors of lignin and flavonoids. This role of the pentose phosphate pathway is supported by the observation that its enzymes are induced by stress conditions such as wounding, under which biosynthesis of aromatic compounds is needed for reinforcing and protecting the tissue.

The oxidative pentose phosphate pathway is redox-regulated

Each enzymatic step in the oxidative pentose phosphate pathway is catalyzed by a group of isoenzymes that vary in their abundance and regulatory properties among plant cell types. The initial reaction of the pathway, catalyzed by

glucose 6-phosphate dehydrogenase
A cytosolic and plastidic enzyme that catalyzes the initial reaction of the oxidative pentose phosphate pathway.

mitochondrion   (plural *mitochondria*)
The organelle that is the site for most reactions in the respiratory process in eukaryotes.

outer mitochondrial membrane
The outer of the two mitochondrial membranes, which appears to be freely permeable to all small molecules.

inner mitochondrial membrane
The inner of the two mitochondrial membranes, containing the electron transport chain, the F_oF_1-ATP synthase, and numerous transporters.

cristae   Folds in the inner mitochondrial membrane that project into the mitochondrial matrix.

intermembrane space   The fluid-filled space between the two mitochondrial membranes or between the two chloroplast envelope membranes.

glucose 6-phosphate dehydrogenase, is in many cases inhibited by a high ratio of NADPH to $NADP^+$.

In the light, little operation of the pentose phosphate pathway occurs in chloroplasts. Glucose 6-phosphate dehydrogenase is inhibited by a reductive inactivation involving the *ferredoxin–thioredoxin system* (see Chapter 8) and by the NADPH to $NADP^+$ ratio. Moreover, the end products of the pathway, fructose 6-phosphate and glyceraldehyde 3-phosphate, are being synthesized by the Calvin–Benson cycle. Thus, mass action will drive the non-oxidative reactions of the pathway in the reverse direction. In this way, synthesis of erythrose 4-phosphate can be maintained in the light. In nongreen plastids, the glucose 6-phosphate dehydrogenase is less sensitive to inactivation by reduced thioredoxin and NADPH, and can therefore reduce $NADP^+$ to maintain a high reduction of plastid components in the absence of photosynthesis.

The Tricarboxylic Acid Cycle

During the nineteenth century, biologists discovered that in the absence of air, cells produce ethanol or lactic acid, whereas in the presence of air, cells consume O_2 and produce CO_2 and H_2O. In 1937 the German-born British biochemist Hans A. Krebs reported the discovery of the citric acid cycle—also called the *Krebs cycle* or perhaps more often the tricarboxylic acid (TCA) cycle. The elucidation of the TCA cycle not only explained how pyruvate is broken down into CO_2 and H_2O, but also highlighted the key concept of cycles in metabolic pathways. For his discovery, Hans Krebs was awarded the Nobel Prize in physiology or medicine in 1953.

Because the TCA cycle occurs in the mitochondrial matrix, we begin with a general description of mitochondrial structure and function, the knowledge of which was obtained mainly through experiments on isolated mitochondria. We then review the steps of the TCA cycle, emphasizing the features that are specific to plants and how they affect respiratory function.

Mitochondria are semiautonomous organelles

The breakdown of sucrose into pyruvate releases less than 25% of the total energy in sucrose; the remaining energy is stored in the four molecules of pyruvate. The next two stages of respiration (the TCA cycle and oxidative phosphorylation) take place within an organelle enclosed by a double membrane, the **mitochondrion** (plural *mitochondria*).

Plant mitochondria are usually spherical or rodlike and range from 0.5 to 1.0 μm in diameter and up to 3 μm in length (**Figure 11.5**), but they can also be branched or reticulated. The number and sizes of mitochondria in a cell can vary dynamically due to mitochondrial fission, fusion, and degradation (see Figure 11.5C) while keeping up with cell division. Like chloroplasts, mitochondria are semiautonomous organelles, because they contain ribosomes, RNA, and DNA, which encodes a limited number of mitochondrial proteins. Plant mitochondria are thus able to carry out the various steps of protein synthesis and to transmit their genetic information. Metabolically active tissues usually contain more mitochondria than less active cells, reflecting the mitochondrial role in energy metabolism. Guard cells, for example, are unusually rich in mitochondria.

The ultrastructural features of plant mitochondria are similar to those of mitochondria in other organisms (see Figure 11.5A and B). Plant mitochondria have two membranes: a smooth **outer mitochondrial membrane** completely surrounds a highly invaginated **inner mitochondrial membrane**. The invaginations of the inner membrane are known as **cristae** (singular *crista*). As a consequence of its greatly enlarged surface area, the inner membrane can contain more than 50% of the total mitochondrial protein. The region between the two mitochondrial membranes is known as the **intermembrane space**. The compartment enclosed

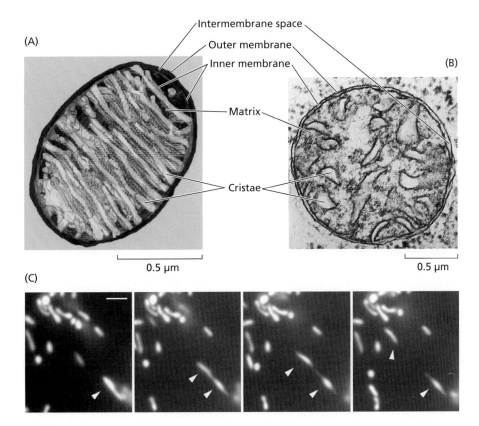

(A)

Intermembrane space

Outer membrane

Inner membrane

Matrix

Cristae

0.5 μm

(B)

0.5 μm

(C)

Figure 11.5 Structure of mitochondria from animals and plants. (A) Three-dimensional tomography picture of a chicken brain mitochondrion, showing the invaginations of the inner membrane, called cristae, as well as the locations of the matrix and intermembrane space. (B) Electron micrograph of a mitochondrion in a mesophyll cell of broad bean (*Vicia faba*). Typically, individual mitochondria are 1 to 3 μm long in plant cells, which means they are substantially smaller than nuclei and plastids. (C) Time-lapse pictures showing a dividing mitochondrion in an Arabidopsis epidermal cell (arrowheads). All the visible organelles are mitochondria labeled with green fluorescent protein. The pictures shown were taken 2 s apart. Scale bar = 1 μm. (A from Perkins et al. 1997; B from Gunning and Steer 1996; C time-lapse photos courtesy of David C. Logan.)

by the inner membrane is referred to as the mitochondrial **matrix**. It has a very high content of macromolecules, approximately 50% by weight. Because there is little water in the matrix, mobility is restricted, and it is likely that matrix proteins are organized into multienzyme complexes (so-called metabolons) to facilitate substrate channeling.

Intact mitochondria are osmotically active; that is, they take up water and swell when placed in a hypoosmotic medium. Ions and polar molecules are generally unable to diffuse freely through the inner membrane, which functions as the osmotic barrier. The outer membrane is permeable to solutes that have a molecular mass of less than approximately 10,000 Da—that is, most cellular metabolites and ions, but not proteins. The lipid fraction of both membranes is made up primarily of phospholipids, 80% of which are either phosphatidylcholine or phosphatidylethanolamine. About 15% of the inner mitochondrial membrane lipids is diphosphatidylglycerol (also called cardiolipin), which occurs in that membrane.

Pyruvate enters the mitochondrion and is oxidized via the TCA cycle

The TCA cycle got its name because of the importance of the tricarboxylic acids, citric acid (citrate) and isocitric acid (isocitrate) as early intermediates (**Figure 11.6**). This cycle constitutes the second stage in respiration and takes place in the mitochondrial matrix. Its operation requires that the pyruvate generated in the cytosol during glycolysis be transported through the inner mitochondrial membrane barrier via a specific transport protein (as we will describe shortly).

Once inside the mitochondrial matrix, pyruvate is decarboxylated in an oxidation reaction catalyzed by **pyruvate dehydrogenase**, a large complex consisting of several enzymes. The products are NADH, CO_2, and acetyl-CoA, in which the acetyl group derived from pyruvate is linked by a thioester bond to a cofactor, coenzyme A (CoA) (see Figure 11.6).

matrix The aqueous, gel-like phase of a mitochondrion that occupies the internal space into which the cristae extend. Contains the DNA, ribosomes, and soluble enzymes required for the tricarboxylic acid cycle, oxidative phosphorylation, and other metabolic reactions.

pyruvate dehydrogenase
An enzyme in the mitochondrial matrix that decarboxylates pyruvate, producing NADH (from NAD+), CO_2, and acetic acid in the form of acetyl-CoA (acetic acid bound to coenzyme A).

Malic enzyme can decarboxylate malate to pyruvate. In combination with malate dehydrogenase, this enables plant mitochondria to oxidize malate to CO_2.

One molecule of ATP is synthesized by a substrate-level phosphorylation during the reaction catalyzed by succinyl-CoA synthetase.

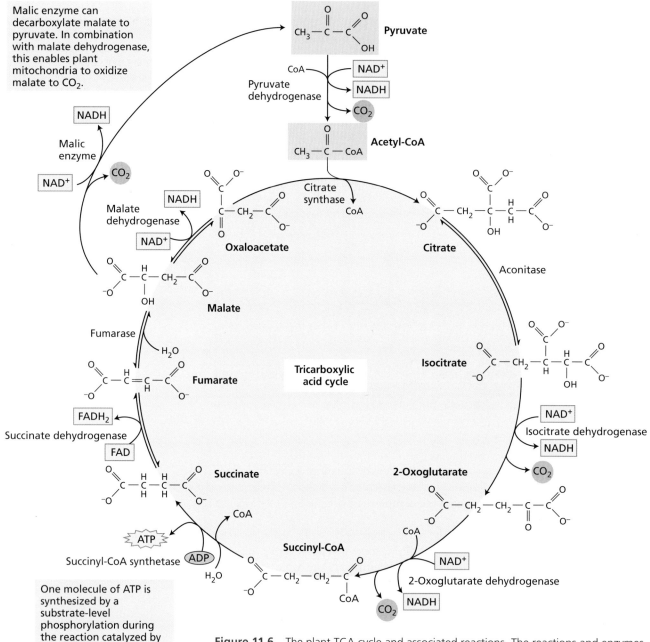

Figure 11.6 The plant TCA cycle and associated reactions. The reactions and enzymes of the TCA cycle are displayed, along with the associated reactions of pyruvate dehydrogenase and malic enzyme. Pyruvate is completely oxidized to three molecules of CO_2, and in combination, malate dehydrogenase and malic enzyme enable plant mitochondria to completely oxidize malate. The electrons released during these oxidations are used to reduce four molecules of NAD^+ to NADH and one molecule of FAD to $FADH_2$.

In the next reaction, the enzyme citrate synthase, formally the first enzyme in the TCA cycle, combines the acetyl group of acetyl-CoA with a four-carbon dicarboxylic acid (*oxaloacetate*) to give a six-carbon tricarboxylic acid (citrate). Citrate is then isomerized to isocitrate by the enzyme aconitase.

The following two reactions are successive oxidative decarboxylations, each of which produces one NADH and releases one molecule of CO_2, yielding a four-carbon product bound to CoA, succinyl-CoA. At this point, three molecules of CO_2 have been produced for each three carbons that entered the mitochondrion as pyruvate, or 12 CO_2 for each molecule of sucrose oxidized.

In the remainder of the TCA cycle, succinyl-CoA is oxidized to oxaloacetate, allowing the continued operation of the cycle. Initially the large amount of free energy available in the thioester bond of succinyl-CoA is conserved through the synthesis of ATP from ADP and P_i via a substrate-level phosphorylation catalyzed by *succinyl-CoA synthetase*. (Recall that the free energy available in the thioester bond of acetyl-CoA was used to form a carbon–carbon bond in the step catalyzed by citrate synthase.) The resulting succinate is oxidized to fumarate by *succinate dehydrogenase*, which is the only membrane-associated enzyme of the TCA cycle and also part of the electron transport chain.

The electrons and protons removed from succinate end up not on NAD^+, but on another cofactor involved in redox reactions: **flavin adenine dinucleotide** (**FAD**). FAD is covalently bound to the active site of succinate dehydrogenase and undergoes a reversible two-electron reduction to produce $FADH_2$ (see Figure 11.2B).

In the final two reactions of the TCA cycle, fumarate is hydrated to produce malate, which is subsequently oxidized by *malate dehydrogenase* to regenerate oxaloacetate and produce another molecule of NADH. The oxaloacetate produced is now able to react with another acetyl-CoA and continue the cycling.

The stepwise oxidation of one molecule of pyruvate in the mitochondrion gives rise to three molecules of CO_2, and much of the free energy released during these oxidations is conserved in the form of four NADH and one $FADH_2$. In addition, one molecule of ATP is produced by a substrate-level phosphorylation.

The TCA cycle of plants has unique features

The TCA cycle reactions outlined in Figure 11.6 are not all identical to those carried out by animal mitochondria. For example, the step catalyzed by succinyl-CoA synthetase produces ATP in plants and GTP in animals. These nucleotides are energetically equivalent.

A feature of the plant TCA cycle that is absent in many other organisms is the presence of **malic enzyme** in the mitochondrial matrix of plants. This enzyme catalyzes the oxidative decarboxylation of malate:

$$\text{Malate} + NAD^+ \rightarrow \text{pyruvate} + CO_2 + NADH \qquad (11.7)$$

The activity of malic enzyme enables plant mitochondria to operate alternative pathways for the metabolism of PEP derived from glycolysis. As already described, malate can be synthesized from PEP in the cytosol via the enzymes PEP carboxylase and malate dehydrogenase (see Figure 11.3). For degradation, malate is transported into the mitochondrial matrix, where malic enzyme can oxidize it to pyruvate. This reaction makes possible the oxidation of TCA cycle intermediates such as malate (**Figure 11.7A**) or citrate (**Figure 11.7B**) to CO_2. Many plant tissues, not only those that carry out crassulacean acid metabolism (see Chapter 8), store significant amounts of malate or other organic acids in their vacuoles. Degradation of malate via mitochondrial malic enzyme is important for regulating levels of organic acids in cells—for example, during fruit ripening.

Instead of being degraded, the malate produced via PEP carboxylase can replace TCA cycle intermediates used in biosynthesis. Reactions that replenish intermediates in a metabolic cycle are known as *anaplerotic*. For example, export of 2-oxoglutarate for nitrogen assimilation in the chloroplast causes a shortage of malate for the citrate synthase reaction. This malate can be replaced through the PEP carboxylase pathway (**Figure 11.7C**).

Gamma-aminobutyric acid (GABA) is an amino acid that accumulates under several stress conditions in plants and that may have a role as a signal. GABA is synthesized from 2-oxoglutarate and degraded into succinate by the so-called **GABA shunt**, which bypasses two TCA cycle enzymes. The functional relationship between GABA accumulation and stress remains poorly understood.

flavin adenine dinucleotide (FAD)
A riboflavin-containing cofactor that undergoes a reversible two-electron reduction to produce $FADH_2$.

malic enzyme An enzyme that catalyzes the oxidation of malate to pyruvate, permitting plant mitochondria to oxidize malate or citrate to CO_2 without involving pyruvate generated by glycolysis.

GABA shunt A pathway supplementing the tricarboxylic acid cycle with the ability to form and degrade GABA.

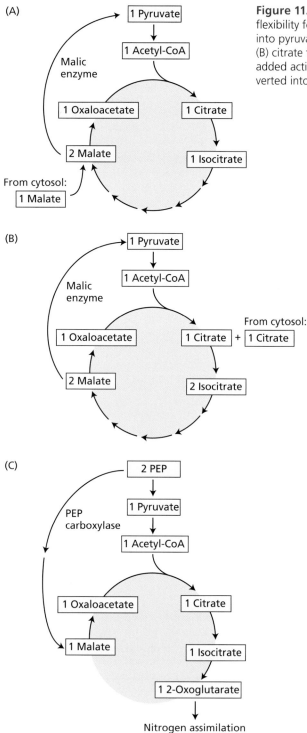

Figure 11.7 Malic enzyme and PEP carboxylase provide plants with metabolic flexibility for the metabolism of PEP and pyruvate. Malic enzyme converts malate into pyruvate and thus allows plant mitochondria to oxidize both (A) malate and (B) citrate to CO_2 without involving pyruvate delivered by glycolysis. (C) With the added action of PEP carboxylase to the standard pathway, glycolytic PEP is converted into 2-oxoglutarate, which is used for nitrogen assimilation.

Mitochondrial Electron Transport and ATP Synthesis

ATP is the energy carrier used by cells to drive life processes, so chemical energy conserved during the TCA cycle in the form of NADH and $FADH_2$ must be converted into ATP to perform useful work in the cell. This O_2-dependent process, called oxidative phosphorylation, occurs in the inner mitochondrial membrane.

In this section we describe the process by which the energy level of the electrons from NADH and $FADH_2$ is lowered in a stepwise fashion and conserved in the form of an electrochemical proton gradient across the inner mitochondrial membrane. Although fundamentally similar in all aerobic cells, the electron transport chain of plants (and many fungi and protists) contains multiple NAD(P)H dehydrogenases and an alternative oxidase, none of which are found in mammalian mitochondria.

We also examine the enzyme that uses the energy of the proton gradient to synthesize ATP, the F_oF_1-ATP synthase. After examining the various stages in the production of ATP, we summarize the energy conservation steps at each stage, as well as the regulatory mechanisms that coordinate the different pathways.

The electron transport chain catalyzes a flow of electrons from NADH to O_2

For each molecule of sucrose oxidized through glycolysis and the TCA cycle, 4 molecules of NADH are generated in the cytosol, and 16 molecules of NADH plus 4 molecules of $FADH_2$ (associated with succinate dehydrogenase) are generated in the mitochondrial matrix. These reduced compounds must be reoxidized, or the entire respiratory process will come to a halt.

The electron transport chain catalyzes a transfer of two electrons from NADH (or $FADH_2$) to oxygen, the final electron acceptor of the respiratory process. For the oxidation of NADH, the reaction can be written as

$$NADH + H^+ + \tfrac{1}{2} O_2 \rightarrow NAD^+ + H_2O \qquad (11.8)$$

From the reduction potentials for the NADH–NAD^+ pair (–320 mV) and the H_2O–$\tfrac{1}{2} O_2$ pair (+810 mV), it can be calculated that the standard free energy released during this overall reaction ($-nF\Delta E^{0\prime}$) is about 220 kJ per mole of NADH. Because the succinate–fumarate reduction potential is higher (+30 mV), only 152 kJ per mole of succinate is released. The role of the electron transport chain is to bring about the oxidation of NADH (and $FADH_2$) and, in the process, use some of the free energy released to generate an electrochemical proton gradient, $\Delta\tilde{\mu}_{H^+}$, across the inner mitochondrial membrane.

The electron transport chain of plants contains the same set of electron carriers found in the mitochondria of other organisms (**Figure 11.8**). The individual

INTERMEMBRANE SPACE

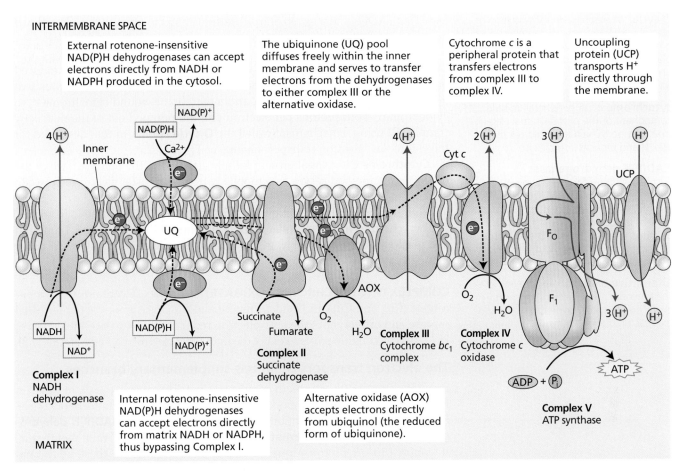

External rotenone-insensitive NAD(P)H dehydrogenases can accept electrons directly from NADH or NADPH produced in the cytosol.

The ubiquinone (UQ) pool diffuses freely within the inner membrane and serves to transfer electrons from the dehydrogenases to either complex III or the alternative oxidase.

Cytochrome c is a peripheral protein that transfers electrons from complex III to complex IV.

Uncoupling protein (UCP) transports H⁺ directly through the membrane.

Internal rotenone-insensitive NAD(P)H dehydrogenases can accept electrons directly from matrix NADH or NADPH, thus bypassing Complex I.

Alternative oxidase (AOX) accepts electrons directly from ubiquinol (the reduced form of ubiquinone).

Figure 11.8 Organization of the electron transport chain and ATP synthesis in the inner membrane of the plant mitochondrion. Mitochondria from nearly all eukaryotes contain the four standard protein complexes: I, II, III, and IV. The structures of all complexes have been determined, but they are shown here as simplified shapes. The electron transport chain of the plant mitochondrion contains additional enzymes (both external and internal NAD[P] dehydrogenases and the alternative oxidase; depicted in green) that do not pump protons. Additionally, uncoupling proteins directly bypass the ATP synthase by allowing passive proton influx. This multiplicity of bypasses in plants, whereas mammals have only the uncoupling protein, gives a greater flexibility to plant energy coupling.

electron transport proteins are organized into four transmembrane multiprotein complexes (identified by Roman numerals I through IV), all of which are localized in the inner mitochondrial membrane. Three of these complexes (I, III, and IV) are engaged in proton pumping.

COMPLEX I (NADH DEHYDROGENASE) Electrons from NADH generated by the TCA cycle in the mitochondrial matrix are oxidized by **complex I** (an **NADH dehydrogenase**). The electron carriers in complex I include a tightly bound cofactor (**flavin mononucleotide**, or **FMN**, which is chemically similar to FAD; see Figure 11.2B) and several iron–sulfur centers. Complex I then transfers these electrons to ubiquinone. Four protons are pumped from the matrix into the intermembrane space for every electron pair passing through the complex.

Ubiquinone, a small lipid-soluble electron and proton carrier, is localized within the inner membrane. It is not tightly associated with any protein, and it can diffuse within the hydrophobic core of the membrane bilayer.

COMPLEX II (SUCCINATE DEHYDROGENASE) Oxidation of succinate in the TCA cycle is catalyzed by this complex, and the reducing equivalents are trans-

NADH dehydrogenase (complex I)
A multi-subunit protein complex in the mitochondrial electron transport chain that catalyzes oxidation of NADH and reduction of ubiquinone linked to the pumping of protons from the matrix to the intermembrane space.

flavin mononucleotide (FMN)
A riboflavin-containing cofactor that undergoes a reversible one- or two-electron reduction to produce FMNH or FMNH$_2$.

ubiquinone A mobile electron carrier of the mitochondrial electron transport chain. Chemically and functionally similar to plastoquinone in the photosynthetic electron transport chain.

Q cycle A mechanism for oxidation of plastohydroquinone (reduced plastoquinone, also called plastoquinol) in chloroplasts and of ubihydroquinone (reduced ubiquinone, also called ubiquinol) in mitochondria.

cytochrome c A peripheral, mobile component of the mitochondrial electron transport chain that oxidizes complex III and reduces complex IV.

NAD(P)H dehydrogenases A collective term for membrane-bound enzymes that oxidize NADH or NADPH, or both, and reduce quinone. Several are present in the electron transport chain of mitochondria; for example, the proton-pumping complex I, but also simpler non-proton-pumping enzymes.

alternative oxidase An enzyme in the mitochondrial electron transport chain that reduces oxygen to water and oxidizes ubiquinol (ubihydroquinone).

ferred via $FADH_2$ and a group of iron–sulfur centers to ubiquinone. Complex II does not pump protons.

COMPLEX III (CYTOCHROME bc_1 COMPLEX) Complex III oxidizes reduced ubiquinone (ubiquinol) and transfers the electrons via an iron–sulfur center, two b-type cytochromes (b_{565} and b_{560}), and a membrane-bound cytochrome c_1 to cytochrome c. Four protons per electron pair are pumped out of the matrix by complex III using a mechanism called the **Q cycle** similar to that described for the chloroplast electron transport chain (see Figure 7.25).

Cytochrome c is a small protein loosely attached to the outer surface of the inner membrane and serves as a mobile carrier to transfer electrons between complexes III and IV.

Both structurally and functionally, ubiquinone, the cytochrome bc_1 complex, and cytochrome c are very similar to plastoquinone, the cytochrome b_6f complex, and plastocyanin, respectively, in the photosynthetic electron transport chain (see Chapter 7).

COMPLEX IV (CYTOCHROME c OXIDASE) Complex IV contains two copper centers (Cu_A and Cu_B) and cytochromes a and a_3. This complex is the terminal oxidase and brings about the four-electron reduction of O_2 to two molecules of H_2O. Two protons are pumped out of the matrix per electron pair (see Figure 11.8).

The electron transport chain has supplementary branches

In addition to the set of protein complexes described above, the plant electron transport chain contains components not found in mammalian mitochondria (see Figure 11.8). Especially, additional non-energy-conserving **NAD(P)H dehydrogenases** and a so-called **alternative oxidase** are bound to the inner membrane. In contrast to the proton-pumping complexes I, III, and IV, these additional enzymes do not pump protons. The consequence is that when they are used, a smaller part of the energy released from oxidation of NADH (or succinate) is conserved as ATP.

- Plant mitochondria have two pathways for oxidizing matrix NADH. Electron flow through complex I, described above, is sensitive to inhibition by several compounds, including rotenone and piericidin. In addition, plant mitochondria have a rotenone-insensitive NADH dehydrogenase, $ND_{in}(NADH)$, on the matrix surface of the inner mitochondrial membrane. This enzyme oxidizes NADH derived from the TCA cycle, and may also be a bypass engaged when complex I is overloaded, as we will see shortly. An NADPH dehydrogenase, $ND_{in}(NADPH)$, is also present on the matrix surface, but very little is known about this enzyme.

- Rotenone-insensitive NAD(P)H dehydrogenases, mostly Ca^{2+}-dependent, are also attached to the outer surface of the inner membrane facing the intermembrane space. They oxidize either NADH or NADPH from the cytosol. Electrons from these external NAD(P)H dehydrogenases—$ND_{ex}(NADH)$ and $ND_{ex}(NADPH)$—enter the main electron transport chain at the level of the ubiquinone pool.

- Most, if not all, plants have an additional respiratory pathway for the oxidation of ubiquinol and reduction of oxygen. This pathway involves the alternative oxidase, which, unlike cytochrome c oxidase, is insensitive to inhibition by cyanide, carbon monoxide, and the signal molecule nitric oxide.

The physiological significance of these supplementary electron transport enzymes is considered more fully later in the chapter.

ATP synthesis in the mitochondrion is coupled to electron transport

In oxidative phosphorylation, the transfer of electrons to oxygen via complexes I, III, and IV is coupled to the synthesis of ATP from ADP and P_i via the F_oF_1-ATP synthase (complex V). The number of ATPs synthesized depends on the nature of the electron donor.

In experiments conducted on isolated mitochondria, electrons donated to complex I (e.g., generated by malate oxidation) give ADP:O ratios (the number of ATPs synthesized per two electrons transferred to oxygen) of 2.4 to 2.7 (**Table 11.1**). Electrons donated to complex II (from succinate) and to the external NADH dehydrogenase give values in the range of 1.6 to 1.8, while electrons donated directly to cytochrome c oxidase (complex IV) via artificial electron carriers give values of 0.8 to 0.9. Results such as these have led to the general concept that there are three sites of energy conservation along the electron transport chain, at complexes I, III, and IV.

The experimental ADP:O ratios agree quite well with the values calculated on the basis of the number of H^+ pumped by complexes I, III, and IV and the cost of 4 H^+ for producing 1 ATP (see next section and Table 11.1). For instance, electrons from external NADH pass only complexes III and IV, so a total of 6 H^+ are pumped, giving 1.5 ATP (when the alternative oxidase pathway is not used).

The mechanism of mitochondrial ATP synthesis is based on the **chemiosmotic hypothesis**, described in Chapter 7, which was first proposed in 1961 by Nobel laureate Peter Mitchell as a general mechanism of energy conservation across biological membranes. According to the chemiosmotic hypothesis, the orientation of electron carriers within the inner mitochondrial membrane allows for the transfer of protons across the inner membrane during electron flow (see Figure 11.8).

Because the inner mitochondrial membrane is highly impermeable to protons, an **electrochemical proton gradient** can build up. The free energy associated with an electrochemical proton gradient, $\Delta\tilde{\mu}_{H^+}$ (expressed in kJ mol^{-1}), is also referred to as the *proton motive force*, Δp, when expressed in units of volts. The transfer of an H^+ from the intermembrane space to the matrix releases energy due to the Δp, which is the sum of an electrical transmembrane potential component (ΔE) and a chemical-potential component (ΔpH) according to the following approximate equation:

$$\Delta p = \Delta E - 59\Delta pH \text{ (at 25°C)} \tag{11.9}$$

where

$$\Delta E = E_{inside} - E_{outside} \tag{11.10}$$

and

$$\Delta pH = pH_{inside} - pH_{outside} \tag{11.11}$$

ΔE results from the asymmetric distribution of a charged species (H^+ and other ions) across the membrane, and ΔpH is due to the H^+ concentration difference across the membrane. Because protons are translocated from the mitochondrial matrix to the intermembrane space, the resulting ΔE across the inner mitochondrial membrane has a negative value. Under normal conditions, the ΔpH is approximately 0.5 and the ΔE approximately 0.2 V. Because the membrane is only 7 to 8 nm thick, this ΔE corresponds to an electric field of at least 25 million V/m (or ten times the field generating a lightning flash in a thunderstorm), emphasizing the enormous forces involved in electron transport.

Table 11.1 Theoretical and experimental ADP:O ratios in isolated plant mitochondria

Electrons feeding into	ADP:O ratio	
	Theoretical[a]	Experimental
Complex I	2.5	2.4–2.7
Complex II	1.5	1.6–1.8
External NADH dehydrogenase	1.5	1.6–1.8
Complex IV	1.0[b]	0.8–0.9

[a]It is assumed that complexes I, III, and IV pump 4, 4, and 2 H^+ per 2 electrons, respectively; that the cost of synthesizing 1 ATP and exporting it to the cytosol is 4 H^+; and that the nonphosphorylating pathways are not active.

[b]Cytochrome c oxidase (complex IV) pumps only 2 protons. However, 2 electrons move from the outer surface of the inner membrane (where the electrons are donated) across the inner membrane to the inner, matrix side. As a result, 2 H^+ are consumed on the matrix side. This means that the net movement of H^+ and charges is equivalent to the movement of a total of 4 H^+, giving a theoretical ADP:O ratio of 1.0.

chemiosmotic hypothesis
The mechanism whereby the electrochemical gradient of protons established across a membrane by an electron transport process is used to drive energy-requiring ATP synthesis. It operates in mitochondria and chloroplasts.

electrochemical proton gradient
The sum of the electrical charge gradient and the pH gradient across the membrane, resulting from a concentration gradient of protons.

F$_o$F$_1$-ATP synthase   A multi-subunit protein complex associated with the inner mitochondrial membrane that couples the passage of protons across the membrane to the synthesis of ATP from ADP and phosphate. The subscript "o" in F$_o$ refers to the binding of the inhibitor oligomycin. Similar to CF$_0$CF$_1$-ATP synthase in photophosphorylation, to which oligomycin does not bind and inhibit (hence the subscript is "0").

F$_o$ The integral membrane part of the F$_o$F$_1$-ATP synthase.

F$_1$ The ATP-binding, matrix-facing part of the F$_o$F$_1$-ATP synthase.

uncoupler   A chemical compound that increases the proton permeability of membranes and thus uncouples the formation of the proton gradient from ATP synthesis.

As this equation shows, both ΔE and ΔpH contribute to the proton motive force in plant mitochondria, although ΔpH constitutes the smaller part, probably because of the large buffering capacity of both cytosol and matrix, which prevents large pH changes. This situation contrasts with that in the chloroplast, where almost all of the proton motive force across the thylakoid membrane is due to ΔpH (see Chapter 7).

The free-energy input required to generate $\Delta\tilde{\mu}_{H^+}$ comes from the free energy released during electron transport. How electron transport is coupled to proton translocation is not completely understood in all cases. Because of the low permeability (conductance) of the inner membrane to protons, the proton electrochemical gradient can be consumed to carry out chemical work (ATP synthesis). The $\Delta\tilde{\mu}_{H^+}$ is coupled to the synthesis of ATP by an additional protein complex associated with the inner membrane, the F$_o$F$_1$-ATP synthase.

The **F$_o$F$_1$-ATP synthase** (also called *complex V*) consists of two major components, F$_o$ and F$_1$ (see Figure 11.8). **F$_o$** (subscript "o" for oligomycin-sensitive) is an integral membrane protein complex of at least three different polypeptides. They form the channel through which protons cross the inner membrane. The other component, **F$_1$**, is a peripheral membrane protein complex that is composed of at least five different subunits and contains catalytic sites for converting ADP and P$_i$ into ATP. This complex is attached to the matrix side of F$_o$.

The passage of protons through the channel is coupled to the catalytic cycle of the F$_1$ component of the ATP synthase, allowing the ongoing synthesis of ATP and the simultaneous use of the $\Delta\tilde{\mu}_{H^+}$. For each ATP synthesized, 3 H$^+$ pass through the F$_o$ component from the intermembrane space to the matrix, down the electrochemical proton gradient.

A high-resolution structure for the F$_1$ component of the mammalian ATP synthase provided evidence for a model in which a part of F$_o$ rotates relative to F$_1$ to couple H$^+$ transport to ATP synthesis. The structure and function of the mitochondrial ATP synthase are similar to those of the CF$_0$CF$_1$-ATP synthase in chloroplasts (see Figure 7.29).

The operation of a chemiosmotic mechanism of ATP synthesis has several implications. First, the true site of ATP formation on the inner mitochondrial membrane is the ATP synthase, not complex I, III, or IV. These complexes serve as sites of energy conservation whereby electron transport is coupled to the generation of a $\Delta\tilde{\mu}_{H^+}$. The synthesis of ATP decreases the $\Delta\tilde{\mu}_{H^+}$ and, as a consequence, its restriction on the electron transport complexes. Electron transport is therefore stimulated by a large supply of ADP.

The chemiosmotic hypothesis also explains the action mechanism of **uncouplers**. These are a wide range of chemically unrelated, artificial compounds (including 2,4-dinitrophenol and *p*-trifluoromethoxycarbonylcyanide phenylhydrazone [FCCP]) that decrease mitochondrial ATP synthesis but stimulate the rate of electron transport. All uncouplers make the inner membrane leaky to protons and prevent the buildup of a sufficiently large $\Delta\tilde{\mu}_{H^+}$ to drive ATP synthesis or restrict electron transport.

Transporters exchange substrates and products

The electrochemical proton gradient also plays a role in the movement of the organic acids of the TCA cycle, and of the substrates and products of ATP synthesis, into and out of mitochondria (**Figure 11.9**). Although ATP is synthesized in the mitochondrial matrix, most of it is used outside the mitochondrion, so an efficient mechanism is needed for moving ADP into and ATP out of the organelle.

The ADP/ATP (adenine nucleotide) transporter performs the active exchange of ADP and ATP across the inner membrane. The movement of the more negatively charged ATP^{4-} out of the mitochondrion in exchange for ADP^{3-}—that is, one net negative charge out—is driven by the electrical-potential gradient (ΔE, positive outside) generated by proton pumping.

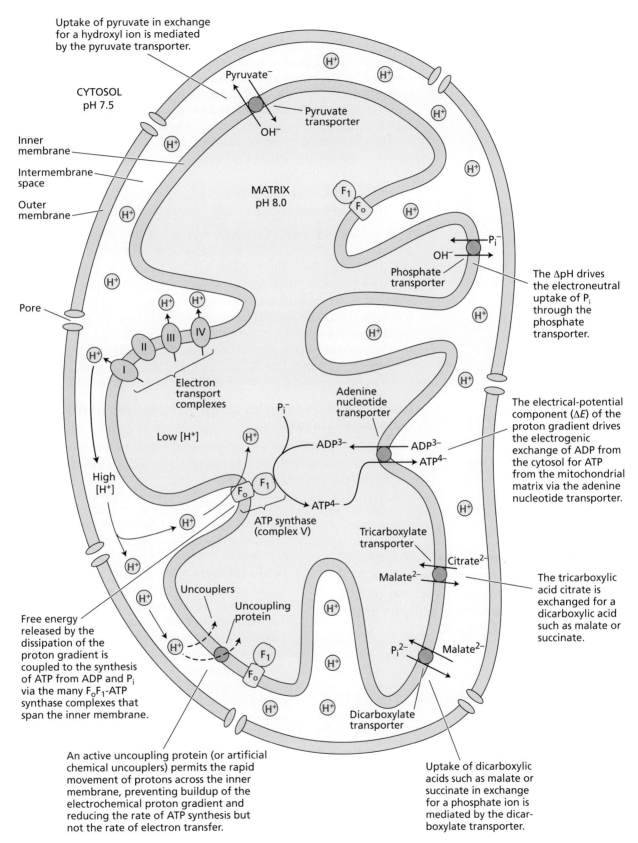

Uptake of pyruvate in exchange for a hydroxyl ion is mediated by the pyruvate transporter.

CYTOSOL
pH 7.5

Inner membrane

Intermembrane space

Outer membrane

MATRIX
pH 8.0

Pore

Pyruvate⁻

Pyruvate transporter

OH⁻

F_1

F_o

Electron transport complexes

II   III   IV

I

Low [H⁺]

High [H⁺]

F_o F_1

ATP synthase (complex V)

Free energy released by the dissipation of the proton gradient is coupled to the synthesis of ATP from ADP and P_i via the many F_oF_1-ATP synthase complexes that span the inner membrane.

Uncouplers

Uncoupling protein

F_1

F_o

An active uncoupling protein (or artificial chemical uncouplers) permits the rapid movement of protons across the inner membrane, preventing buildup of the electrochemical proton gradient and reducing the rate of ATP synthesis but not the rate of electron transfer.

P_i^-

OH⁻

Phosphate transporter

The ΔpH drives the electroneutral uptake of P_i through the phosphate transporter.

Adenine nucleotide transporter

P_i^-

ADP³⁻

ADP³⁻

ATP⁴⁻

ATP⁴⁻

The electrical-potential component (ΔE) of the proton gradient drives the electrogenic exchange of ADP from the cytosol for ATP from the mitochondrial matrix via the adenine nucleotide transporter.

Tricarboxylate transporter

Citrate²⁻

Malate²⁻

The tricarboxylic acid citrate is exchanged for a dicarboxylic acid such as malate or succinate.

P_i^{2-} Malate²⁻

Dicarboxylate transporter

Uptake of dicarboxylic acids such as malate or succinate in exchange for a phosphate ion is mediated by the dicarboxylate transporter.

Figure 11.9 Transmembrane transport in plant mitochondria. An electrochemical proton gradient, $\Delta\tilde{\mu}_{H^+}$, consisting of an electrical-potential component (ΔE, –200 mV, negative inside) and a chemical-potential component (ΔpH, alkaline inside), is established across the inner mitochondrial membrane during electron transport. The $\Delta\tilde{\mu}_{H^+}$ is used by specific transporters that move metabolites across the inner membrane.

The uptake of inorganic phosphate (P_i) involves an active phosphate transporter protein that uses the chemical-potential component (ΔpH) of the proton motive force to drive the electroneutral exchange of P_i^- (in) for OH^- (out). As long as a ΔpH is maintained across the inner membrane, the P_i content within the matrix remains high. Similar reasoning applies to the uptake of pyruvate, which is driven by the electroneutral exchange of pyruvate for OH^-, leading to continued uptake of pyruvate from the cytosol (see Figure 11.9).

The total energetic cost of taking up one phosphate and one ADP into the matrix and exporting one ATP is the movement of one H^+ from the intermembrane space into the matrix:

- Moving one OH^- out in exchange for P_i^- is equivalent to one H^+ in, so this electroneutral exchange consumes the ΔpH, but not the ΔE.

- Moving one negative charge out (ADP^{3-} entering the matrix in exchange for ATP^{4-} leaving) is the same as moving one positive charge in, so this transport lowers only the ΔE.

This proton, which drives the exchange of ATP for ADP and P_i, should also be included in our calculation of the cost of synthesizing one ATP. Thus, the total cost is 3 H^+ used by the ATP synthase plus 1 H^+ for the exchange across the membrane, or a total of 4 H^+.

The inner membrane also contains transporters for dicarboxylic acids (malate or succinate) exchanged for P_i^{2-} and for the tricarboxylic acids (citrate, aconitate, or isocitrate) exchanged for dicarboxylic acids (see Figure 11.9).

Aerobic respiration yields about 60 molecules of ATP per molecule of sucrose

The complete oxidation of a sucrose molecule leads to the net formation of:

- Eight molecules of ATP by substrate-level phosphorylation (four from glycolysis and four from the TCA cycle)

- Four molecules of NADH in the cytosol

- Sixteen molecules of NADH plus four molecules of $FADH_2$ (via succinate dehydrogenase) in the mitochondrial matrix

On the basis of theoretical ADP:O values (see Table 11.1), we can estimate that 52 ATP molecules will be generated per molecule of sucrose by oxidative phosphorylation. The complete aerobic oxidation of sucrose (including substrate-level phosphorylation) results in a total of about 60 ATPs synthesized per sucrose molecule (**Table 11.2**).

Using 50 kJ mol^{-1} as the actual free energy of formation of ATP in vivo, we find that about 3010 kJ mol^{-1} of free energy is conserved in the form of ATP per mole of sucrose oxidized during aerobic respiration. This amount represents about 52% of the standard free energy available from the complete oxidation of sucrose; the rest is lost as heat. It also represents a vast improvement over fermentative metabolism, in which only 4% of the energy available in sucrose is converted into ATP.

Plants have several mechanisms that lower the ATP yield

As we have seen, a complex machinery is required for conserving energy in oxidative phosphorylation. So it is perhaps surprising that plant mitochondria have several

Table 11.2 Maximum yield of cytosolic ATP from the complete oxidation of sucrose to CO_2 via aerobic glycolysis and the TCA cycle

Part reaction	ATP per sucrose[a]
Glycolysis	
4 substrate-level phosphorylations	4
4 NADH	$4 \times 1.5 = 6$
TCA cycle	
4 substrate-level phosphorylations	4
4 $FADH_2$	$4 \times 1.5 = 6$
16 NADH	$16 \times 2.5 = 40$
Total	60

Source: Adapted from Brand 1994.

Note: Cytosolic NADH is assumed to be oxidized by the external NADH dehydrogenase. The other nonphorphorylating pathways (e.g., the alternative oxidase) are assumed not to be engaged.

[a]Calculated using the theoretical ADP:O values from Table 11.1.

functional proteins that reduce this efficiency. Plants are probably less limited by energy supply (sunlight) than by other factors in the environment (e.g., access to water and nutrients). As a consequence, metabolic flexibility may be more important to them than energetic efficiency.

In the following subsections we discuss the role of three non-energy-conserving pathways and their possible usefulness in the life of the plant: the alternative oxidase, the uncoupling protein, and the rotenone-insensitive NAD(P)H dehydrogenases.

THE ALTERNATIVE OXIDASE   Most plants display a capacity for *cyanide-resistant respiration* that is comparable to the capacity of the cyanide-sensitive cytochrome *c* oxidase pathway. The cyanide-resistant oxygen uptake is catalyzed by the alternative oxidase.

Electrons feed off the main electron transport chain into this alternative pathway at the level of the ubiquinone pool (see Figure 11.8). The alternative oxidase, the only component of the alternative pathway, catalyzes a four-electron reduction of oxygen to water and is specifically inhibited by several compounds, most notably salicylhydroxamic acid. When electrons pass to the alternative pathway from the ubiquinone pool, two sites of proton pumping (at complexes III and IV) are bypassed. Because there is no energy conservation site in the alternative pathway between ubiquinone and oxygen, the free energy that would normally be conserved as ATP is lost as heat when electrons are shunted through this pathway.

How can a process as seemingly energetically wasteful as the alternative pathway contribute to plant metabolism? One example of the functional usefulness of the alternative oxidase is its activity in so-called thermogenic flowers of several plant families—for example, the voodoo lily (*Sauromatum guttatum*). Just before pollination, parts of the inflorescence exhibit a dramatic increase in the rate of respiration caused by a greatly increased expression of alternative oxidase or uncoupling protein (depending on the species). As a result, the temperature of the upper appendix increases by as much as 25°C over the ambient temperature. During this extraordinary burst of heat production the plant can actually melt surrounding snow. Simultaneously, certain amines, indoles, and terpenes are volatilized, and the plant therefore gives off a putrid odor that attracts insect pollinators. Salicylic acid was identified as the signal initiating this thermogenic event in the voodoo lily and was later found also to be involved in plant pathogen defense (see Chapter 18).

In most plants, the respiratory rates are too low to generate sufficient heat to raise the temperature significantly. What other role(s) does the alternative pathway play? To answer that question, we need to consider the regulation of the alternative oxidase: Its transcription is often specifically induced, for example, by various types of abiotic and biotic stress. The activity of the alternative oxidase, which functions as a dimer, is regulated by reversible oxidation–reduction of an intermolecular sulfhydryl bridge, by the reduction level of the ubiquinone pool, and by pyruvate. The first two factors ensure that the enzyme is most active under reducing conditions, while the last factor ensures that the enzyme has high activity when there is plenty of substrate for the TCA cycle.

If the respiration rate exceeds the cell's demand for ATP (i.e., if ADP levels are very low), the reduction level in the mitochondrion will be high, and the alternative oxidase will be activated. Thus, the alternative oxidase makes it possible for the mitochondrion to adjust the relative rates of ATP production and synthesis of carbon skeletons for use in biosynthetic reactions.

Another possible function of the alternative pathway is in the response of plants to a variety of stresses (phosphate deficiency, chilling, drought, osmotic stress, and so on), many of which can inhibit mitochondrial respiration (see Chapter 19). In response to stress, the electron transport chain leads to increased formation of reactive oxygen species (ROS), initially superoxide but also hydrogen peroxide and the hydroxyl radical, which act as a signal for the activation of alternative oxidase expression. By draining off electrons from the ubiquinone pool (see Figure 11.8), the alternative pathway prevents overreduction, thus limiting the production of ROS and minimizing the detrimental effects of stress on respiration.

THE UNCOUPLING PROTEIN A protein found in the inner membrane of mammalian mitochondria, the **uncoupling protein**, can dramatically increase the proton permeability of the membrane and thus act as an uncoupler. As a result, less ATP and more heat are generated. Heat production appears to be one of the uncoupling protein's main functions in mammalian cells.

It had long been thought that the alternative oxidase in plants and the uncoupling protein in mammals were simply two different means of achieving the same end. It was therefore surprising when a protein similar to the uncoupling protein was discovered in plant mitochondria. This protein is induced by stress and stimulated by ROS. In knockout mutants, photosynthetic carbon assimilation and growth were decreased, consistent with the interpretation that the uncoupling protein, like the alternative oxidase, functions to prevent overreduction of the electron transport chain and formation of ROS.

ROTENONE-INSENSITIVE NAD(P)H DEHYDROGENASES Multiple rotenone-insensitive dehydrogenases oxidizing NADH or NADPH are found in plant mitochondria (see Figure 11.8). The internal, rotenone-insensitive NADH dehydrogenase (ND_{in}[NADH]) may work as a non-proton-pumping bypass when complex I is overloaded. Complex I has a higher affinity (ten times lower K_m) for NADH than ND_{in}(NADH). At lower NADH levels in the matrix, typically when ADP is available, complex I dominates, whereas when ADP is rate-limiting, the NADH concentration increases and ND_{in}(NADH) becomes more active. ND_{in}(NADH) and the alternative oxidase probably recycle the NADH into NAD^+ to maintain pathway activity. Since reducing power can be shuttled from the matrix to the cytosol by the exchange of different organic acids, external NADH dehydrogenases can have bypass functions similar to those of ND_{in}(NADH). Taken together, these NADH dehydrogenases and the NADPH dehydrogenases are likely to make plant respiration more flexible and allow control of specific redox homeostasis of NADH and NADPH in mitochondria and cytosol.

Short-term control of mitochondrial respiration occurs at different levels

The respiratory pathways are controlled by changes in gene expression. Several genes encoding enzymes of the TCA cycle and electron transport chain, especially the non-energy-conserving enzymes, are up-regulated by light or sugars, indicating an increase in respiratory capacity in response to carbon status. Plant respiratory rates are also under important *allosteric control* from the "bottom up" (**Figure 11.10**). The substrates of ATP synthesis—ADP and P_i—appear to be key short-term regulators of the rates of glycolysis in the cytosol and of the TCA cycle and oxidative phosphorylation in the mitochondria. Control points exist at all three stages of respiration. Here we give a brief overview of some major features of respiratory control, starting with the TCA cycle.

The best-characterized site of posttranslational regulation of mitochondrial respiratory metabolism is the pyruvate dehydrogenase complex, which is phosphorylated by a *regulatory protein kinase* and dephosphorylated by a *protein phosphatase*. Pyruvate dehydrogenase is inactive in the phosphorylated state, and the regulatory protein kinase is inhibited by pyruvate, allowing the enzyme to be active when substrate is available (**Figure 11.11**). Pyruvate dehydrogenase forms the entry point to the TCA cycle, so this regulation adjusts the activity of the cycle to the cellular demand.

Figure 11.10 Model of bottom-up regulation of plant respiration. Several substrates for respiration (e.g., ADP) stimulate enzymes in early steps of the pathways (green arrows). In contrast, accumulation of products (e.g., ATP) inhibits upstream reactions (red lines) in a stepwise fashion. For instance, ATP inhibits the electron transport chain, leading to an accumulation of NADH. NADH inhibits TCA cycle enzymes such as isocitrate dehydrogenase and 2-oxoglutarate dehydrogenase. TCA cycle intermediates such as citrate inhibit the PEP-metabolizing enzymes in the cytosol. Finally, PEP inhibits the conversion of fructose 6-phosphate into fructose 1,6-bisphosphate and restricts carbon flow into glycolysis. In this way, respiration can be up- or down-regulated in response to changing demands for either of its products: ATP and organic acids.

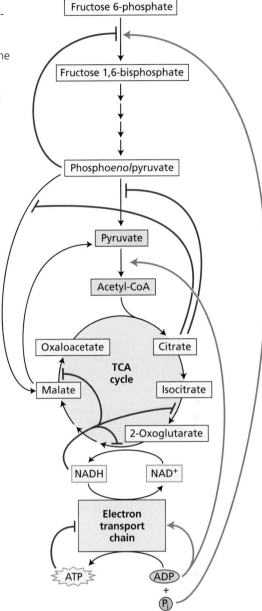

The TCA cycle oxidations, and subsequently respiration, are dynamically controlled by the cellular level of adenine nucleotides (see Figure 11.10). As the cell's demand for ATP in the cytosol decreases relative to the rate of synthesis of ATP in the mitochondria, less ADP is available, and the electron transport chain operates at a reduced rate. This slowdown leads to an increase in matrix NADH, which inhibits the activity of several TCA cycle dehydrogenases, and in matrix NADPH, which is required for thioredoxin turnover. In photosynthesis, thioredoxins control many enzymes by reversible redox dimerization of cysteine residues (see Chapter 8). Carbon flow through the TCA cycle appears to be regulated by a deactivation of succinate dehydrogenase and fumarase. Although the detailed mechanisms are still to be elucidated, it is clear that mitochondrial redox status exerts further controls on respiratory processes. The buildup of TCA cycle intermediates (such as citrate) and their derivatives (such as glutamate) inhibits the action of cytosolic pyruvate kinase and PEP carboxylase, increasing the cytosolic PEP concentration, which in turn reduces the rate of conversion of fructose 6-phosphate into fructose 1,6-bisphosphate, thus inhibiting glycolysis. This bottom-up inhibitory effect of PEP on phosphofructokinase is strongly decreased by inorganic phosphate, making the cytosolic ratio of PEP to P_i a critical factor in the control of plant glycolytic activity. In contrast, regulation in animals operates from the "top down," with a primary activation occurring at the phosphofructokinase and secondary activation at the pyruvate kinase.

One possible benefit of bottom-up control of glycolysis is that it permits plants to regulate net glycolytic flux to pyruvate independently of related metabolic processes such as the Calvin–Benson cycle and

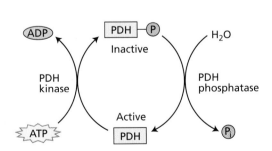

Pyruvate + CoA + NAD^+ ⟶ Acetyl-CoA + CO_2 + NADH + H^+

Figure 11.11 Metabolic regulation of pyruvate dehydrogenase (PDH) activity, directly and by reversible phosphorylation. Upstream and downstream metabolites regulate PDH activity by direct actions on the enzyme itself or by regulating its protein kinase or protein phosphatase.

Effect on PDH activity	Mechanism
Activating	
Pyruvate	Inhibits kinase
ADP	Inhibits kinase
Mg^{2+} (or Mn^{2+})	Stimulates phosphatase
Inactivating	
NADH	Inhibits PDH Stimulates kinase
Acetyl-CoA	Inhibits PDH Stimulates kinase
NH_4^+	Inhibits PDH Stimulates kinase

sucrose–triose phosphate–starch interconversion. Another benefit of this control mechanism is that glycolysis can adjust to the demand for removal of biosynthetic precursors from the TCA cycle.

A consequence of bottom-up control of glycolysis is that its rate can influence cellular concentrations of sugars, in combination with sugar-supplying processes such as phloem transport (see Chapter 10). Glucose and sucrose are potent signaling molecules that make the plant adjust its growth and development to its carbohydrate status.

Respiration is tightly coupled to other pathways

In addition to providing 2-oxoglutarate for nitrogen assimilation as mentioned earlier (see Figure 11.7C), glycolysis, the oxidative pentose phosphate pathway, and the TCA cycle produce building blocks for the synthesis of many central plant metabolites, including amino acids, lipids, and nucleotides and their related compounds (**Figure 11.12**). Indeed, much of the reduced carbon that is metabolized by glycolysis and the TCA cycle is diverted to biosynthetic purposes and not oxidized to CO_2.

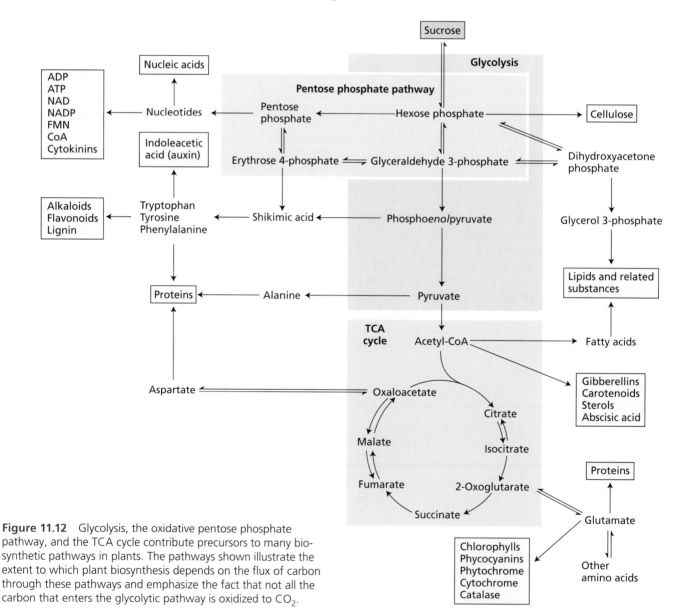

Figure 11.12 Glycolysis, the oxidative pentose phosphate pathway, and the TCA cycle contribute precursors to many biosynthetic pathways in plants. The pathways shown illustrate the extent to which plant biosynthesis depends on the flux of carbon through these pathways and emphasize the fact that not all the carbon that enters the glycolytic pathway is oxidized to CO_2.

Mitochondria are also integrated into the cellular redox network. Variations in consumption or production of redox and energy-carrying compounds such as NAD(P)H and organic acids are likely to affect metabolic pathways in the cytosol and in plastids. Of special importance is the synthesis of ascorbic acid, a central redox and stress defense molecule in plants, by the electron transport chain. Mitochondria also carry out steps in the biosynthesis of coenzymes necessary for many metabolic enzymes in other cell compartments.

Respiration in Intact Plants and Tissues

Many rewarding studies of plant respiration and its regulation have been carried out on isolated organelles and on cell-free extracts of plant tissues. But how does this knowledge relate to the function of the whole plant in a natural or agricultural setting?

In this section we examine respiration and mitochondrial function in the context of the whole plant under a variety of conditions. First we explore what happens when green organs are exposed to light: Respiration and photosynthesis operate simultaneously and are functionally integrated in the cell. Next we discuss rates of respiration in different tissues, which may be under developmental control. Finally, we look at the influence of various environmental factors on respiration rates.

Plants respire roughly half of the daily photosynthetic yield

Many factors can affect the respiration rate of an intact plant or of its individual organs. Relevant factors include the species and growth habit of the plant, the type and age of the specific organ, and environmental variables such as light, external O_2 and CO_2 concentrations, temperature, and nutrient and water supply (see Chapter 19). By measuring different oxygen isotopes, it is possible to measure the in vivo activities of the alternative oxidase and cytochrome c oxidase simultaneously. Therefore, we know that a significant part of respiration in most tissues takes place via the "energy-wasting" alternative pathway.

Whole-plant respiration rates, particularly when considered on a fresh-weight basis, are generally lower than respiration rates reported for animal tissues. This difference is mainly due to the presence in plant cells of a large vacuole and a cell wall, neither of which contains mitochondria. Nonetheless, respiration rates in some plant tissues are as high as those observed in actively respiring animal tissues, so the respiratory process in plants is not inherently slower than in animals. In fact, isolated plant mitochondria respire as fast as or faster than mammalian mitochondria.

The contribution of respiration to the overall carbon economy of the plant can be substantial. Whereas only green tissues photosynthesize, all tissues respire, and they do so 24 h a day. Even in photosynthetically active tissues, respiration, if integrated over the entire day, uses a substantial fraction of gross photosynthesis. A survey of several herbaceous species indicated that 30 to 60% of the daily gain in photosynthetic carbon is lost to respiration, although these values tend to decrease in older plants. Trees respire a similar fraction of their photosynthetic production, but their respiratory loss increases with age as the ratio of photosynthetic to nonphotosynthetic tissue decreases. In general, unfavorable growth conditions will increase respiration relative to photosynthesis, and thus lower the overall carbon yield of the plant.

Respiratory processes operate during photosynthesis

Mitochondria are involved in the metabolism of photosynthesizing leaves in several ways. The glycine generated by photorespiration is oxidized to serine in the mitochondrion in a reaction involving mitochondrial oxygen consumption (see Chapter 8). At the same time, mitochondria in photosynthesizing tissue car-

maintenance respiration   The respiration needed to support the function and turnover of existing tissue.

growth respiration   The respiration that provides the energy needed for converting sugars into the building blocks that make up new tissue.

respiratory quotient (RQ)   The ratio of CO_2 evolution to O_2 consumption.

ry out normal mitochondrial respiration (i.e., via the TCA cycle). Relative to the maximum rate of photosynthesis, rates of mitochondrial respiration measured in green tissues in the light are far slower, generally by a factor of 6- to 20-fold. Given that rates of photorespiration can often reach 20 to 40% of the gross photosynthetic rate, daytime photorespiration is a larger provider of NADH for the mitochondrial electron transport chain than the normal respiratory pathways.

The activity of pyruvate dehydrogenase, one of the ports of entry into the TCA cycle, decreases in the light to 25% of its activity in darkness. Consistently, the overall rate of mitochondrial respiration decreases in the light, but the extent of the decrease remains uncertain at present. It is clear, however, that the mitochondrion is a major supplier of ATP to the cytosol (e.g., for driving biosynthetic pathways), especially under conditions where photosynthetically produced ATP is used in the chloroplasts for carbon fixation. Nonetheless, a basic flux through the respiratory pathways is needed during photosynthesis to supply precursors for biosynthetic reactions, such as the 2-oxoglutarate needed for nitrogen assimilation (see Figures 11.7C and 11.12). The formation of 2-oxoglutarate also produces NADH in the matrix, linking the process to oxidative phosphorylation or to non-energy-conserving electron transport chain activities.

Additional evidence for the involvement of mitochondrial respiration in photosynthesis has been obtained in studies with mitochondrial mutants defective in respiratory complexes. Compared with the wild type, these plants have slower leaf development and photosynthesis because changes in levels of redox-active metabolites are communicated between mitochondria and chloroplasts, negatively affecting photosynthetic function.

Different tissues and organs respire at different rates

Respiration is often considered to have two components of comparable magnitude. **Maintenance respiration** is needed to support the function and turnover of the tissues already present. **Growth respiration** provides the energy needed for converting sugars into the building blocks that make up new tissues. A useful rule of thumb is that the greater the overall metabolic activity of a given tissue, the higher its respiration rate. Developing buds usually show very high rates of respiration, and respiration rates of vegetative organs usually decrease from the point of growth (e.g., the leaf tip in eudicots and the leaf base in monocots) to more differentiated regions. A well-studied example is the growing barley leaf.

In mature vegetative organs, stems generally have the lowest respiration rates, whereas leaf and root respiration varies with the plant species and the conditions under which the plants are growing. Low availability of soil nutrients, for example, increases the demand for respiratory ATP production in the root. This increase reflects increased energy costs for active ion uptake and root growth in search of nutrients.

When a plant organ has reached maturity, its respiration rate either remains roughly constant or decreases slowly as the tissue ages and ultimately senesces. An exception to this pattern is the marked rise in respiration, known as the *climacteric*, that accompanies the onset of ripening in many fruits (e.g., avocado, apple, and banana) and senescence in detached leaves and flowers. During fruit ripening, massive conversion of, for example, starch (banana) or organic acids (tomato and apple) into sugars occurs, accompanied by a rise in the hormone ethylene and the activity of the cyanide-resistant alternative pathway.

Different tissues can use different substrates for respiration. Sugars dominate overall, but in specific organs other compounds, such as organic acids in maturing apples or lemons and lipids in germinating sunflower or canola seedlings, may provide the carbon for respiration. These compounds are built with different ratios of carbon to oxygen atoms. Therefore, the ratio of CO_2 release to O_2 consumption, which is called the **respiratory quotient**, or **RQ**, varies with the substrate oxidized. Lipids, sugars, and organic acids represent a series of increasing RQs because lipids contain little oxygen per carbon, and organic acids much. Alcoholic fermentation releases CO_2 without consuming O_2, so a high RQ is also a marker for fermentation. Since RQ can be determined in the field, it is an important parameter in analyses of carbon metabolism on a larger scale.

Environmental factors alter respiration rates

Many environmental factors can alter the operation of metabolic pathways and change respiratory rates. Disadvantageous growth conditions will generally increase respiration to supply energy for protective reactions. However, environmental oxygen (O_2), temperature, and carbon dioxide (CO_2) can either stimulate or suppress respiration.

OXYGEN   Plant respiration can be limited by supply of oxygen, its terminal substrate. At 25°C, the equilibrium concentration of O_2 in an air-saturated (21% O_2) aqueous solution is about 250 μM. The K_m value for O_2 in the reaction catalyzed by cytochrome c oxidase is well below 1 μM, so there should be no apparent dependence of the respiration rate on external O_2 concentrations. However, respiration rates decrease if the atmospheric O_2 concentration is below 5% for whole organs or below 2 to 3% for tissue slices. These findings show that oxygen supply can impose a limitation on plant respiration.

Oxygen diffuses slowly in aqueous solutions. Compact organs such as seeds and potato tubers have a noticeable O_2 concentration gradient from the surface to the center, which restricts the ATP/ADP ratio. Diffusion limitation is even more significant in seeds with a thick seed coat or in plant organs submerged in water. When plants are grown hydroponically, the solutions must be aerated to keep oxygen levels high in the vicinity of the roots (see Chapter 4). The problem of oxygen supply is particularly important in plants growing in very wet or flooded soils (see also Chapter 19).

Some plants, particularly trees, have a restricted geographic distribution because of the need to maintain a supply of oxygen to their roots. For instance, the dogwood *Cornus florida* and tulip tree (*Liriodendron tulipifera*) can survive only in well-drained, aerated soils. However, many plant species are adapted to grow in flooded soils. For example, rice and sunflower rely on a network of intercellular air spaces (called **aerenchyma**) running from the leaves to the roots to provide a continuous gaseous pathway for the movement of oxygen to the flooded roots. If this gaseous diffusion pathway throughout the plant did not exist, the respiration rates of many plants would be limited by an insufficient oxygen supply.

Limitation in oxygen supply can be more severe for trees with very deep roots that grow in wet soils. Such roots must survive on anaerobic (fermentative) metabolism (see Figure 11.3) or develop structures that facilitate the movement of oxygen to the roots. Examples of such structures are outgrowths of the roots, called *pneumatophores*, that protrude out of the water and provide a gaseous pathway for oxygen diffusion into the roots. Pneumatophores are found in *Avicennia* and *Rhizophora*, both trees that grow in mangrove swamps under continuously flooded conditions.

TEMPERATURE   Respiration operates over a wide temperature range. Over short time, it typically increases with temperatures between 0 and 30°C and reaches a plateau at 40 to 50°C. At higher temperatures, it again decreases because of inactivation of the respiratory machinery. The increase in respiration rate for every 10°C increase in temperature is commonly called the **temperature coefficient, Q_{10}.** This coefficient describes how respiration responds to short-term temperature changes, and it varies with plant development and external factors. On a longer time scale, plants acclimate to low temperatures by increasing their respiratory capacity so that ATP production can be maintained.

aerenchyma   An anatomical feature of roots found in hypoxic conditions, showing large, gas-filled intercellular spaces in the root cortex.

temperature coefficient (Q_{10})
The increase in the rate of a process (e.g., respiration) for every 10°C increase in temperature.

Low temperatures are used to retard postharvest respiration during the storage of fruits and vegetables, but those temperatures must be adjusted with care. For instance, when potato tubers are stored at temperatures above 10°C, respiration and ancillary metabolic activities are sufficient to allow sprouting. Below 5°C, respiration rates and sprouting are reduced, but the breakdown of stored starch and its conversion into sucrose impart an unwanted sweetness to the tubers. Therefore, potatoes are best stored at 7 to 9°C, which prevents the breakdown of starch while minimizing respiration and germination.

CARBON DIOXIDE   It is common practice in commercial storage of fruits to take advantage of the effects of oxygen concentration and temperature on respiration by storing fruits at low temperatures under 2 to 3% O_2 and 3 to 5% CO_2 concentrations. The reduced temperature lowers the respiration rate, as does the reduced O_2 level. Low levels of oxygen, instead of anoxic conditions, are used to avoid lowering tissue oxygen tensions to the point at which fermentative metabolism sets in. Carbon dioxide has a limited direct inhibitory effect on respiration at the artificially high concentration of 3 to 5%.

The atmospheric CO_2 concentration is currently (2018) around 400 ppm, but it is increasing as a result of human activities, and it is projected to increase to 700 ppm before the end of the twenty-first century (see Chapter 9). The flux of CO_2 between plants and the atmosphere by photosynthesis and respiration is much larger than the flux of CO_2 to the atmosphere caused by the burning of fossil fuels. Therefore, an effect of elevated CO_2 concentrations on plant respiration will strongly influence future global atmospheric changes. Within the range relevant for atmospheric changes, there is no direct effect of CO_2 concentration on plant respiration. Over longer time scales, at higher atmospheric CO_2 concentrations dark respiration per biomass unit may decrease indirectly, due to changes in plant structure and metabolism, whereas the effect on respiration in the light is uncertain. It is presently not possible to predict fully the global effect of anthropogenic CO_2 on plant respiration.

Lipid Metabolism

Whereas animals use fats for energy storage, plants use them for both energy and carbon storage. Fats and oils are important storage forms of reduced carbon in many seeds, including those of agriculturally important species such as soybean, sunflower, canola, peanut, and cotton. Oils serve a major storage function in many nondomesticated plants that produce small seeds. Some fruits, such as olives and avocados, also store fats and oils.

In this final part of the chapter we describe the biosynthesis of two types of glycerolipids: the *triacylglycerols* (the fats and oils stored in seeds) and the *polar glycerolipids* (which form the lipid bilayers of cellular membranes) (**Figure 11.13**). We will see that the biosynthesis of triacylglycerols and polar glycerolipids requires the cooperation of two organelles: the plastid and the endoplasmic reticulum. We will also examine the complex process by which germinating seeds obtain carbon skeletons and metabolic energy from the oxidation of fats and oils.

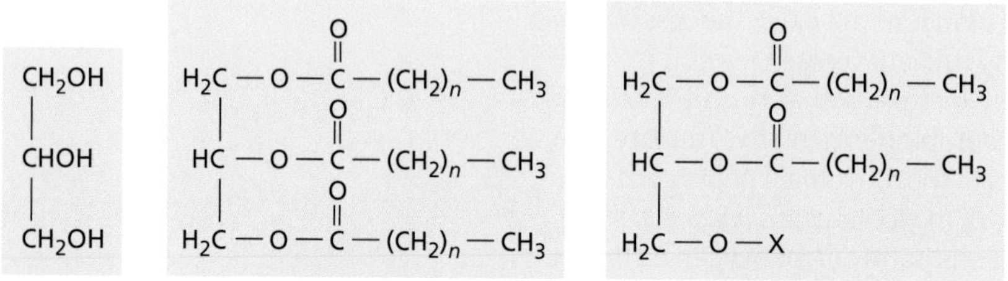

X = H	Diacylglycerol (DAG)
X = HPO_3^-	Phosphatidic acid
X = PO_3^- — CH_2 — CH_2 — $\overset{+}{N}(CH_3)_3$	Phosphatidylcholine
X = PO_3^- — CH_2 — CH_2 — NH_2	Phosphatidylethanolamine
X = galactose	Galactolipids

Figure 11.13   Structural features of triacylglycerols and polar glycerolipids in higher plants. The carbon chain lengths of the fatty acids, which always have an even number of carbons, are typically 16 or 18. Thus, the value of *n* is usually 14 or 16.

Fats and oils store large amounts of energy

Fats and oils belong to the general class termed *lipids,* a structurally diverse group of hydrophobic compounds that are soluble in organic solvents and highly insoluble in water. Lipids represent a more reduced form of carbon than carbohydrates, so the complete oxidation of 1 g of fat or oil (which contains about 40 kJ of energy) can produce considerably more ATP than the oxidation of 1 g of starch (about 15.9 kJ). Conversely, the biosynthesis of lipids requires a correspondingly large investment of metabolic energy.

Other lipids are important for plant structure and function but are not used for energy storage. These lipids include the phospholipids and galactolipids that are components of plant membranes, as well as sphingolipids, which are also important membrane components; waxes, which make up the protective cuticle that reduces water loss from exposed plant tissues; and terpenoids (also known as isoprenoids), which include carotenoids involved in photosynthesis and sterols present in many plant membranes.

Triacylglycerols are stored in oil bodies

Fats and oils exist mainly in the form of **triacylglycerols** (*acyl* refers to the fatty acid portion), in which fatty acid molecules are linked by ester bonds to the three hydroxyl groups of glycerol (see Figure 11.13).

The fatty acids in plants are usually straight-chain carboxylic acids having an even number of carbon atoms. They are synthesized in the plastids and in the endoplasmic reticulum. The carbon chains can be as short as 12 units and as long as 30 or more, but the most commonly found are 16 or 18 carbons long. *Oils* are liquid at room temperature, primarily because of the presence of carbon–carbon double bonds (unsaturation) in their component fatty acids; *fats,* which have a higher proportion of saturated fatty acids, are solid at room temperature. The major fatty acids in plant lipids are shown in **Table 11.3**.

The proportions of fatty acids in plant lipids vary with the plant species. For example, peanut oil is about 9% palmitic acid, 59% oleic acid, and 21% linoleic acid, and soybean oil is 13% palmitic acid, 7% oleic acid, 51% linoleic acid, and 23% linolenic acid. In most seeds, triacylglycerols are stored in the cytoplasm of either cotyledon or endosperm cells in organelles known as **oil bodies** (also called *spherosomes, oleosomes,* or *lipid droplets*) (see Chapter 1). The oil-body membrane is a single layer of phospholipids (i.e., a half-bilayer) with the hydrophilic ends of the phospholipids exposed to the cytosol and the hydrophobic acyl hydrocarbon chains facing the triacylglycerol interior (see Chapter 1). The oil body is stabilized by the

triacylglycerols Three fatty acyl groups in ester linkage to the three hydroxyl groups of glycerol. Fats and oils.

oil bodies Organelles that accumulate and store triacylglycerols. They are bounded by a single phospholipid leaflet ("half–unit membrane" or "phospholipid monolayer") derived from the endoplasmic reticulum. Also known as oleosomes or spherosomes.

Table 11.3 Common fatty acids in higher plant tissues

Name[a]	Structure
Saturated fatty acids	
Lauric acid (12:0)	$CH_3(CH_2)_{10}CO_2H$
Myristic acid (14:0)	$CH_3(CH_2)_{12}CO_2H$
Palmitic acid (16:0)	$CH_3(CH_2)_{14}CO_2H$
Stearic acid (18:0)	$CH_3(CH_2)_{16}CO_2H$
Unsaturated fatty acids	
Oleic acid (18:1)	$CH_3(CH_2)_7CH{=}CH(CH_2)_7CO_2H$
Linoleic acid (18:2)	$CH_3(CH_2)_4CH{=}CH{-}CH_2{-}CH{=}CH(CH_2)_7CO_2H$
Linolenic acid (18:3)	$CH_3CH_2CH{=}CH{-}CH_2{-}CH{=}CH{-}CH_2{-}CH{=}CH{-}(CH_2)_7CO_2H$

[a]Each fatty acid has a numerical abbreviation. The number before the colon represents the total number of carbons; the number after the colon is the number of double bonds.

polar glycerolipids The main structural lipids in membranes, in which the hydrophobic portion consists of two 16-carbon or 18-carbon fatty acid chains esterified to positions 1 and 2 of a glycerol.

glyceroglycolipids Glycerolipids in which sugars form the polar head group. Glyceroglycolipids are the most abundant glycerolipids in chloroplast membranes.

glycerophospholipids Polar glycerolipids in which the hydrophobic portion consists of two 16-carbon or 18-carbon fatty acid chains esterified to positions 1 and 2 of a glycerol backbone. The phosphate-containing polar head group is attached to position 3 of the glycerol.

phosphatidylinositol bisphosphate (PIP$_2$) A phosphorylated derivative of phosphatidylinositol.

presence of specific proteins, called oleosins, that coat its outer surface and prevent the phospholipids of adjacent oil bodies from coming in contact and fusing with it.

The unique membrane structure of oil bodies results from the pattern of triacylglycerol biosynthesis. Triacylglycerol synthesis is completed by enzymes located in the membranes of the endoplasmic reticulum (ER), and the resulting fats accumulate between the two monolayers of the ER membrane bilayer. The bilayer swells apart as more fats are added to the growing structure, and ultimately a mature oil body buds off from the ER.

Polar glycerolipids are the main structural lipids in membranes

As outlined in Chapter 1, each membrane in the cell is a bilayer of *amphipathic* (i.e., having both hydrophilic and hydrophobic regions) lipid molecules in which a polar head group interacts with the aqueous environment while hydrophobic fatty acid chains form the core of the membrane. This hydrophobic core prevents unregulated diffusion of solutes between cell compartments and thereby allows the biochemistry of the cell to be organized.

The main structural lipids in membranes are the **polar glycerolipids** (see Figure 11.13), in which the hydrophobic portion consists of two 16-carbon or 18-carbon fatty acid chains esterified to positions 1 and 2 of a glycerol backbone. The polar head group is attached to position 3 of the glycerol. There are two categories of polar glycerolipids:

1. **Glyceroglycolipids**, in which sugars form the head group (**Figure 11.14A**)

2. **Glycerophospholipids**, in which the head group contains phosphate (**Figure 11.14B**)

Plant membranes have additional structural lipids, including sphingolipids and sterols, but these are minor components. Other lipids perform specific roles in photosynthesis and other processes. Included among these lipids are chlorophylls, plastoquinone, carotenoids, and tocopherols, which together account for about one-third of the lipids in plant leaves.

Figure 11.14 shows the nine major polar glycerolipid classes in plants, each of which can be associated with many different fatty acid combinations. The structures shown in Figure 11.14 illustrate some of the more common molecular species.

Chloroplast membranes, which account for 70% of the membrane lipids in photosynthetic tissues, are dominated by glyceroglycolipids; other membranes of the cell contain glycerophospholipids (**Table 11.4**). In nonphotosynthetic tissues, glycerophospholipids are the major membrane glycerolipids.

Membrane lipids are precursors of important signaling compounds

Plants, animals, and microbes all use membrane lipids as precursors for compounds that are used for intracellular or long-range signaling. For example, jasmonate hormone—derived from linolenic acid (see Table 11.3)—activates plant defenses against insects and many fungal pathogens (see Chapter 18). In addition, jasmonate regulates other aspects of plant growth, including the development of anthers and pollen.

Phosphatidylinositol 4,5-bisphosphate (PIP$_2$) is the most important of several phosphorylated derivatives of phosphatidylinositol known as *phosphoinositides*. In animals, receptor-mediated activation of phospholipase C leads to the hydrolysis

Figure 11.14 Major polar glycerolipid classes found in plant membranes: (A) glyceroglycolipids and a sphingolipid and (B) glycerophospholipids. Two of at least six different fatty acids may be attached to the glycerol backbone. One of the more common molecular species is shown for each glycerolipid class. The numbers given below each name refer to the number of carbons (number before the colon) and the number of double bonds (number after the colon).

(A) Glyceroglycolipids

(B) Glycerophospholipids

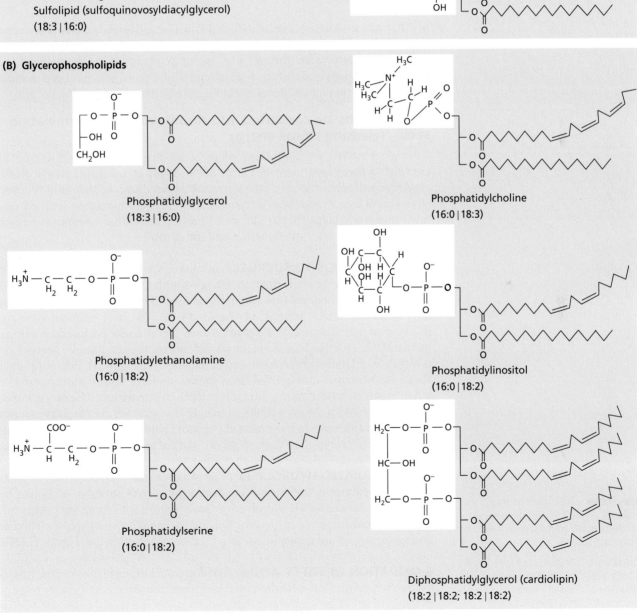

Monogalactosyldiacylglycerol
(18:3 | 16:3)

Glucosylceramide

Sulfolipid (sulfoquinovosyldiacylglycerol)
(18:3 | 16:0)

Digalactosyldiacylglycerol
(16:0 | 18:3)

Phosphatidylglycerol
(18:3 | 16:0)

Phosphatidylcholine
(16:0 | 18:3)

Phosphatidylethanolamine
(16:0 | 18:2)

Phosphatidylinositol
(16:0 | 18:2)

Phosphatidylserine
(16:0 | 18:2)

Diphosphatidylglycerol (cardiolipin)
(18:2 | 18:2; 18:2 | 18:2)

Table 11.4   Glycerolipid components of cellular membranes

Lipid	Lipid composition (percentage of total)		
	Chloroplast	Endoplasmic reticulum	Mitochondrion
Phosphatidylcholine	4	47	43
Phosphatidylethanolamine	—	34	35
Phosphatidylinositol	1	17	6
Phosphatidylglycerol	7	2	3
Diphosphatidylglycerol	—	—	13
Monogalactosyldiacylglycerol	55	—	—
Digalactosyldiacylglycerol	24	—	—
Sulfolipid	8	—	—

of PIP_2 into inositol trisphosphate ($InsP_3$) and diacylglycerol, both of which act as intracellular second messengers (see Chapter 12). The action of $InsP_3$ in releasing Ca^{2+} into the cytoplasm (through Ca^{2+}-sensitive channels in the tonoplast and other membranes), and thereby regulating cellular processes, has been demonstrated in several plant systems, including the stomatal guard cells.

Storage lipids are converted into carbohydrates in germinating seeds, releasing stored energy

After germinating, oil-containing seeds metabolize stored triacylglycerols by converting them into sucrose. Plants are not able to transport fats from the cotyledons to other tissues of the germinating seedling, so they must convert stored lipids into a more mobile form of carbon, generally sucrose. This process involves several steps that are located in different cellular compartments: oil bodies, glyoxysomes, mitochondria, and the cytosol.

OVERVIEW: LIPIDS TO SUCROSE   In oilseeds, the conversion of lipids into sucrose is triggered by germination. It begins with the hydrolysis of triacylglycerols stored in oil bodies into free fatty acids, followed by oxidation of those fatty acids to produce acetyl-CoA (**Figure 11.15**). The fatty acids are oxidized in a type of peroxisome called a **glyoxysome**, an organelle enclosed by a single membrane bilayer that is found in the oil-rich storage tissues of seeds. Acetyl-CoA is metabolized in the glyoxysome and cytoplasm (see Figure 11.15A) to produce succinate, which is transported from the glyoxysome to the mitochondrion, where it is converted first into fumarate and then into malate. The process ends in the cytosol with the conversion of malate into glucose via gluconeogenesis, and then into sucrose. In most oilseeds, approximately 30% of the acetyl-CoA is used for energy production via respiration, and the rest is converted into sucrose.

LIPASE-MEDIATED HYDROLYSIS   The initial step in the conversion of lipids into carbohydrates is the breakdown of triacylglycerols stored in oil bodies by the enzyme lipase, which hydrolyzes triacylglycerols into three fatty acid molecules and one molecule of glycerol. During the breakdown of lipids, oil bodies and glyoxysomes are generally in close physical association (see Figure 11.15B).

β-OXIDATION OF FATTY ACIDS   The fatty acid molecules enter the glyoxysome, where they are activated by conversion into fatty-acyl-CoA by the enzyme *fatty-acyl-CoA synthetase*. Fatty-acyl-CoA is the initial substrate for the

glyoxysome   An organelle found in the oil-rich storage tissues of seeds in which fatty acids are oxidized. A type of microbody.

β-oxidation   Oxidation of fatty acids into fatty acyl-CoA, and the sequential breakdown of the fatty acids into acetyl-CoA units. NADH is also produced.

(A)

Fatty acids are metabolized by β-oxidation to acetyl-CoA in the glyoxysome.

Lipase — OIL BODY — Triacylglycerols are hydrolyzed to yield fatty acids.

OIL BODY
Triacylglycerols

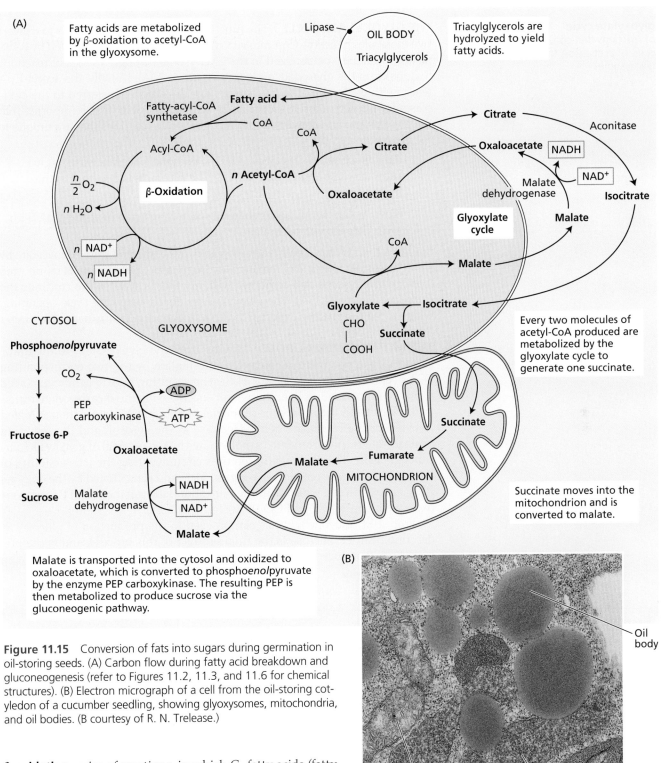

Figure 11.15   Conversion of fats into sugars during germination in oil-storing seeds. (A) Carbon flow during fatty acid breakdown and gluconeogenesis (refer to Figures 11.2, 11.3, and 11.6 for chemical structures). (B) Electron micrograph of a cell from the oil-storing cotyledon of a cucumber seedling, showing glyoxysomes, mitochondria, and oil bodies. (B courtesy of R. N. Trelease.)

β-oxidation series of reactions, in which C_n fatty acids (fatty acids composed of n carbons) are sequentially broken down into $n/2$ molecules of acetyl-CoA (see Figure 11.15A). This reaction sequence involves the reduction of $\frac{1}{2} O_2$ to H_2O and the formation of one NADH for each acetyl-CoA produced.

In mammalian tissues, the four enzymes associated with β-oxidation are present in the mitochondrion. In plant seed storage tissues, they are located exclusively in the glyoxysome or the equivalent organelle in vegetative tissues, the peroxisome (see Chapter 1).

glyoxylate cycle The sequence of reactions that convert two molecules of acetyl-CoA to succinate in the glyoxysome.

THE GLYOXYLATE CYCLE   The function of the **glyoxylate cycle** is to convert two molecules of acetyl-CoA into succinate. The acetyl-CoA produced by β-oxidation is further metabolized in the glyoxysome through a series of reactions that make up the glyoxylate cycle (see Figure 11.15A). Initially, the acetyl-CoA reacts with oxaloacetate to give citrate, which is then transferred to the cytoplasm for isomerization to isocitrate by aconitase. Isocitrate is reimported into the glyoxysome and converted into malate by two reactions that are unique to the glyoxylate cycle:

1. First, isocitrate (C_6) is cleaved by the enzyme isocitrate lyase to give succinate (C_4) and glyoxylate (C_2). The succinate is exported to the mitochondria.

2. Next, malate synthase combines a second molecule of acetyl-CoA with glyoxylate to produce malate.

Malate is then transferred to the cytoplasm and converted into oxaloacetate by the cytoplasmic isozyme of malate dehydrogenase. Oxaloacetate is reimported into the glyoxysome and combines with another acetyl-CoA to continue the cycle (see Figure 11.15A). The glyoxylate produced keeps the cycle operating, whereas the succinate is exported to the mitochondria for further processing.

THE MITOCHONDRIAL ROLE   Moving from the glyoxysomes to the mitochondria, the succinate is converted into malate by the two corresponding TCA cycle reactions (see Figure 11.6). The resulting malate can be exported from the mitochondria in exchange for succinate via the dicarboxylate transporter located in the inner mitochondrial membrane. Malate is then oxidized to oxaloacetate by malate dehydrogenase in the cytosol, and the resulting oxaloacetate is converted into carbohydrates by the reversal of glycolysis (gluconeogenesis). This conversion requires circumventing the irreversibility of the pyruvate kinase reaction (see Figure 11.3) and is facilitated by the enzyme PEP carboxykinase, which uses the phosphorylating ability of ATP to convert oxaloacetate into PEP and CO_2 (see Figure 11.15A).

From PEP, gluconeogenesis can proceed to the production of glucose, as described earlier. Sucrose is the final product of this process, and is the primary form of reduced carbon translocated from the cotyledons to the growing seedling tissues.

Summary

Using the building blocks provided by photosynthesis, respiration releases the energy stored in carbon compounds in a controlled manner for cellular use. At the same time, it generates many carbon precursors for biosynthesis.

Overview of Plant Respiration

- In plant respiration, reduced cellular carbon generated by photosynthesis is oxidized to CO_2 and water, and this oxidation is coupled to the synthesis of ATP.

- Respiration takes place by four main processes: glycolysis, the oxidative pentose phosphate pathway, the TCA cycle, and oxidative phosphorylation (the electron transport chain and ATP synthesis) (**Figure 11.1**).

Glycolysis

- In glycolysis, carbohydrates are converted into pyruvate in the cytosol, and a small amount of ATP is synthesized via substrate-level phosphorylation. NADH is also produced (**Figure 11.3**).

- Plant glycolysis has alternative enzymes for several steps. These allow differences in substrates used, products made, and the direction of the pathway.

- When insufficient O_2 is available, fermentation regenerates NAD^+ for glycolytic ATP production Only a minor fraction of the energy available in sugars is conserved by fermentation (**Figure 11.3**).

(Continued)

Summary *(continued)*

The Oxidative Pentose Phosphate Pathway

- Carbohydrates can be oxidized via the oxidative pentose phosphate pathway, which provides building blocks for biosynthesis and reducing power as NADPH (**Figure 11.4**).

The Tricarboxylic Acid Cycle

- Pyruvate is oxidized to CO_2 within the mitochondrial matrix through the TCA cycle, generating a large number of reducing equivalents in the form of NADH and $FADH_2$ (**Figures 11.5, 11.6**).

- In plants, the TCA cycle is involved in alternative pathways that allow oxidation of malate or citrate and export of intermediates for biosynthesis (**Figures 11.6, 11.7**).

Mitochondrial Electron Transport and ATP Synthesis

- Electron transport from NADH and $FADH_2$ to oxygen is coupled by enzyme complexes to proton transport across the inner mitochondrial membrane. This generates an electrochemical proton gradient used for powering synthesis and export of ATP (**Figures 11.8, 11.9**).

- During aerobic respiration, up to 60 molecules of ATP are produced per molecule of sucrose (**Table 11.2**).

- Typical for plant respiration is the presence of several proteins (alternative oxidase, NAD[P]H dehydrogenases, and uncoupling protein) that lower the energy recovery (**Figure 11.8**).

- The main products of the respiratory process are ATP and metabolic intermediates used in biosynthesis. The cellular demand for these compounds regulates respiration via control points in the electron transport chain, the TCA cycle, and glycolysis (**Figures 11.10–11.12**).

Respiration in Intact Plants and Tissues

- More than 50% of the daily photosynthetic yield may be respired by a plant.

- Many factors can affect the respiration rate observed at the whole-plant level. These factors include the nature and age of the plant tissue and environmental factors such as light, temperature, nutrient and water supply, and O_2 and CO_2 concentrations.

Lipid Metabolism

- Triacylglycerols (fats and oils) are efficient forms for storage of reduced carbon, particularly in seeds. Polar glycerolipids are the primary structural components of membranes (**Figures 11.13, 11.14; Tables 11.3, 11.4**).

- Triacylglycerols are synthesized in the endoplasmic reticulum and accumulate within the phospholipid bilayer, forming oil bodies.

- Some lipid derivatives, such as jasmonate, are important plant hormones.

- During germination in oil-storing seeds, the stored lipids are metabolized to carbohydrates in a series of reactions that include the glyoxylate cycle. The glyoxylate cycle takes place in glyoxysomes, and subsequent steps occur in the mitochondria (**Figure 11.15**).

- The reduced carbon generated during lipid breakdown in the glyoxysomes is ultimately converted into carbohydrates in the cytosol by gluconeogenesis (**Figure 11.15**).

Suggested Reading

Atkin, O. K., Meir, P., and Turnbull, M. H. (2014) Improving representation of leaf respiration in large-scale predictive climate-vegetation models. *New Phytol.* 202: 743–748.

Bates, P. D., Stymne, S., and Ohlrogge, J. (2013) Biochemical pathways in seed oil synthesis. *Curr. Opin. Plant Biol.* 16: 358–364.

Brennicke, A., and Leaver, C. J. (Jan 2007) Mitochondrial Genome Organization and Expression in Plants. In: eLS. John Wiley & Sons Ltd, Chichester. http://www.els. net [doi: 10.1002/9780470015902.a0003825].

Markham, J. E., Lynch, D. V., Napier, J. A., Dunn, T. M., and Cahoon, E. B. (2013) Plant sphingolipids: function follows form. *Curr. Opin. Plant Biol.* 16: 350–357.

Millar, A. H., Siedow, J. N., and Day, D. A. (2015) Respiration and photorespiration. In *Biochemistry and Molecular Biology of Plants*, 2nd ed., B. B. Buchanan, W. Gruissem, and R. L. Jones, eds., Wiley, Somerset, NJ, pp. 610–655.

Millar, A. H., Whelan, J., Soole, K. L., and Day, D. A. (2011) Organization and regulation of mitochondrial respiration in plants. *Annu. Rev. Plant Biol.* 62: 79–104.

Møller, I. M. (2001) Plant mitochondria and oxidative stress. Electron transport, NADPH turnover and metabolism of reactive oxygen species. *Annu. Rev. Plant Physiol. Plant Mol. Biol.* 52: 561–591.

Nicholls, D. G., and Ferguson, S. J. (2013) *Bioenergetics*, 4th ed. Academic Press, San Diego, CA.

O'Leary, B. M. and Plaxton, W. C. (2016) Plant Respiration. eLS. 1–11. DOI: 10.1002/9780470015902.a0001301.pub3

Rasmusson, A. G., Geisler, D. A., and Møller, I. M. (2008) The multiplicity of dehydrogenases in the electron transport chain of plant mitochondria. *Mitochondrion* 8: 47–60.

Sweetlove, L. J., Beard, K. F. M., Nunes-Nesi, A., Fernie, A. R., and Ratcliffe, R. G. (2010) Not just a circle: Flux modes in the plant TCA cycle. *Trends Plant Sci.* 15: 462–470.

van Dongen, J. T., and Licausi, F. (2015) Oxygen sensing and signaling. *Annu. Rev. Plant Biol.* 66: 345–367.

Vanlerberghe, G. C. (2013) Alternative oxidase: A mitochondrial respiratory pathway to maintain metabolic and signaling homeostasis during abiotic and biotic stress in plants. *Int. J. Mol. Sci.* 14: 6805–6847.

Wallis, J. G., and Browse, J. (2010) Lipid biochemists salute the genome. *Plant J.* 61: 1092–1106.

12 Signals and Signal Transduction

As sessile organisms, plants constantly make adjustments in response to their environment, either to take advantage of favorable conditions or to survive unfavorable ones. To facilitate such adjustments, plants have evolved sophisticated sensory systems to optimize water and nutrient usage; to monitor light quantity, quality, and directionality; and to defend themselves from biotic and abiotic threats. Charles and Francis Darwin performed pioneering studies on signal transduction during the light-induced bending growth of grass coleoptiles, tubular, pointed sheaths that protect emerging leaves during germination under the soil surface. They observed that a unidirectional light source was perceived at the coleoptile tip, yet the bending response took place farther back along the shoot tissue. This led the Darwins to conclude that there must be a mobile signal that transferred information from one region of the coleoptile tissue to another and elicited the bending response. The mobile signal was later identified as auxin, indole-3-acetic acid, the first plant hormone to be discovered.

In general, an environmental input that initiates one or more plant responses is referred to as a signal, and the physical component that biochemically responds to that signal is designated a receptor. Receptors are either proteins or, in the case of light receptors, pigments associated with proteins. Once receptors sense their specific signal, they must *transduce* the signal (i.e., convert it from one form to another) in order to amplify the signal and trigger the cellular response. Receptors often do this by modifying

second messenger A small intracellular molecule (e.g., cyclic AMP, cyclic GMP, calcium, IP_3, or diacylglycerol) whose concentration increases or decreases in response to the activation of a receptor by an external signal, such as hormones or light. It then diffuses intracellularly to the target enzymes or intracellular receptor to produce and amplify the physiological response.

signal transduction pathway A sequence of biochemical processes by which an extracellular signal (typically light or a hormone) interacts with a receptor, causing a change in the level of a second messenger and ultimately a change in cell function.

cell autonomous response A response to an environmental stimulus or genetic mutation that is localized to a particular cell.

non–cell autonomous response A cellular response to an environmental stimulus or genetic mutation that is induced by other cells.

the activity of other proteins or by employing intracellular signaling molecules called **second messengers**—small molecules and ions that are rapidly produced or mobilized at relatively high levels after signal perception, and which can modify the activity of target signaling proteins. Second messengers then alter cellular processes such as gene transcription. Hence, all signal transduction pathways typically involve the following chain of events:

$$\text{Signal} \rightarrow \text{receptor} \rightarrow \text{signal transduction} \rightarrow \text{response}$$

In many cases the initial response is the production of secondary signals, such as hormones, which are then transported to the site of action to evoke the main physiological response. Many of the specific events and intermediate steps involved in plant signal transduction have now been identified, and these intermediates constitute the **signal transduction pathways**.

We begin this chapter by providing a brief overview of the types of external cues that direct plant growth. Next we discuss how plants employ signal transduction pathways to regulate physiological responses. Signal amplification via second messengers is required, as well as mechanisms for signal transmission to coordinate responses throughout the plant. Finally, we examine how individual stimulus-response cascades are often integrated with other signaling pathways, termed *cross-regulation*, to shape plant responses to their environment in time and space.

Temporal and Spatial Aspects of Signaling

Plant signal transduction mechanisms may be relatively rapid or extremely slow (**Figure 12.1**). When some carnivorous plants, most notably Venus flytrap (*Dionaea muscipula*), catch insects, they use modified leaf traps that close within milliseconds after touch stimulation. Similarly, the sensitive plant (*Mimosa pudica*) folds its leaflets rapidly upon being touched. Young seedlings reorient themselves with respect to gravity minutes after being placed horizontally. In general, such rapid response mechanisms involve electrochemical responses to transduce signals, since gene transcription and protein translation mechanisms are too slow. In contrast, plants attacked by insect herbivores may emit volatiles to attract insect predators within a few hours. Processes occurring on this timescale often involve changes in gene expression (see Chapter 18).

Longer-term environmental responses modify developmental programs to shape plant architecture over the entire life of the plant. Examples of long-term responses include modulation of root branching in response to nutrient availability, growth of sun or shade leaves to adjust for light conditions (see Chapter 9), and activation of lateral bud outgrowth when the shoot apex is damaged by grazing herbivores. Long-term plant responses can operate over timescales of months or years. For example, a long period of low temperature, termed *vernalization*, is required by many plant species for flowering to occur (see Chapter 17).

Plant responses to environmental signals also differ spatially. In a **cell autonomous response** to an environmental signal, both signal reception and response occur in the same cell. In contrast, a **non–cell autonomous response** is one in which signal reception occurs in one cell and the response occurs in distal cells, tissues, or organs. An example of cell autonomous signaling is the opening of guard cells, where blue light activates membrane ion transporters to swell guard cells via the phototropin blue-light photoreceptors (see Chapter 13). An example of non–cell autonomous signaling in the same organs would be the formation of additional stomata when mature leaves are exposed to high light intensity, in a process that requires transmission of information from one organ to another (see Chapter 16).

(A)

(B)

(C)

(D)

(E)

Figure 12.1 Timing of plant responses to the environment ranges from very rapid to extremely slow. (A) Insect movements on modified leaves of a Venus flytrap (*Dionaea muscipula*) activate trigger hairs, inducing rapid closure of the leaf lobes. (B) The leaves of the sundew plant *Drosera anglica* capture insects in a sticky fluid produced by stalked glands called tentacles, then roll up to secure the prey and begin digestion. (C) A hawthorn tree (*Crataegus* sp.) subjected to prevailing onshore winds responds slowly by growing away from the wind. (D) Tree trunks and branches can respond slowly to mechanical stress by producing reaction wood. In this case the tree is an angiosperm, which produces *tension wood* on the upper surface. Gymnosperms produce *compression wood* on the lower surface. (E) Cross section through a gymnosperm tree branch with compression wood (arrow), creating an asymmetric ring structure. (A Nigel Cattlin/Alamy; B blickwinkel/Alamy; C Steven Sheppardson/Alamy; D and E by David McIntyre.)

Signal Perception and Amplification

Although highly varied in nature and makeup, all signal transduction pathways share common features: An initial stimulus is perceived by a receptor and transmitted via intermediate processes to sites where physiological responses are initiated (**Figure 12.2**). The stimulus may derive from developmental programming or from the external environment. When the response mechanism reaches an optimal point, feedback mechanisms attenuate the processes and reset the sensor mechanism. Receptors can be located at the plasma membrane, cytosol, endomembrane system, or nucleus (**Figure 12.3**). Some receptors are found in more than one cellular location, as is seen with mechanosensitive stretch receptors that help cells and chloroplasts adjust to osmotically induced swelling. In some cases, receptors move from one compartment to another.

Figure 12.2 General scheme for signal transduction. Environmental or developmental signals are perceived by specialized receptors. A signaling cascade is then activated that involves second messengers and leads to a response by the plant cell. When an optimal response has been achieved, feedback mechanisms attenuate the signal.

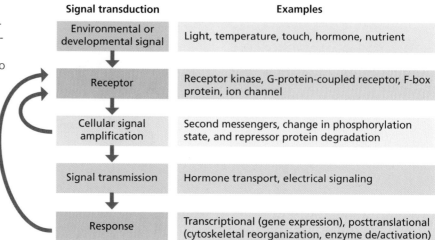

Signal transduction	Examples
Environmental or developmental signal	Light, temperature, touch, hormone, nutrient
Receptor	Receptor kinase, G-protein-coupled receptor, F-box protein, ion channel
Cellular signal amplification	Second messengers, change in phosphorylation state, and repressor protein degradation
Signal transmission	Hormone transport, electrical signaling
Response	Transcriptional (gene expression), posttranslational (cytoskeletal reorganization, enzyme de/activation)

Signal attenuation, e.g., receptor degradation, dephosphorylation, ion homeostasis, repressor synthesis

Signals must be amplified intracellularly to regulate their target molecules

If a receptor is considered the gateway through which a signal enters a signaling network, the receptor location to some extent prescribes the length of the subsequent signaling pathway; such pathways can consist either of a few signaling steps or an elaborate cascade of signaling events. Perception of signals at the plasma membrane often activates transduction pathways with many intermediates. The most common of these modifications is the transfer of a phosphate from ATP to a protein. This transfer, called phosphorylation, is catalyzed by enzymes classified as **kinases**. Some kinases are components of receptor complexes. Other kinases function in amplification cascades, either to elevate weak initial signaling events above the threshold of detection or to propagate them throughout the cytoplasm. Some signal amplification cascades, such as the **MAP (mitogen-activated protein) kinase cascade**, which is present in plants and all other eukaryotes, can involve multiple levels of amplification.

Ca^{2+} is the most ubiquitous second messenger in plants and other eukaryotes

Second messengers represent another strategy to enhance or propagate signals. Probably the most ubiquitous second messenger in all eukaryotes is the divalent cation Ca^{2+}, which in plants is involved in a vast number of different signaling pathways, including symbiotic interactions (see Figure 5.12), plant defense responses, and responses to various hormones and abiotic stresses. Cytosolic Ca^{2+} levels can increase rapidly when Ca^{2+} is taken up into the cytosol, either from external stores, such as the cell wall, or from internal stores, such as vacuoles. In both cases Ca^{2+} influx into the cytosol is mediated primarily by Ca^{2+}-permeable ion channels (**Figure 12.4**). Channel activity must be tightly regulated to maintain precise control over the timing and duration of cytosolic Ca^{2+} elevation. Generally, ion channels are gated, meaning the channel pores are opened or closed by changes in transmembrane electrical potential, membrane tension, posttranslational modification, or binding of a ligand (see also Chapter 6). Several families of Ca^{2+}-permeable channels have been identified in plants; these include plasma membrane–localized glutamate-like receptors and cyclic nucleotide-gated channels. Electrophysiological and other evidence supports the presence of Ca^{2+}-permeable channels at the tonoplast and the endoplasmic reticulum.

Once receptor-mediated signaling activates Ca^{2+}-permeable channels, Ca^{2+} sensor proteins play a pivotal role as signaling intermediaries, linking Ca^{2+}

kinases Enzymes that have the capacity to transfer phosphate groups from ATP to other molecules.

MAP (mitogen-activated protein) kinase cascade A series of protein kinases that transmit an activation signal from a cell surface receptor to DNA in the nucleus.

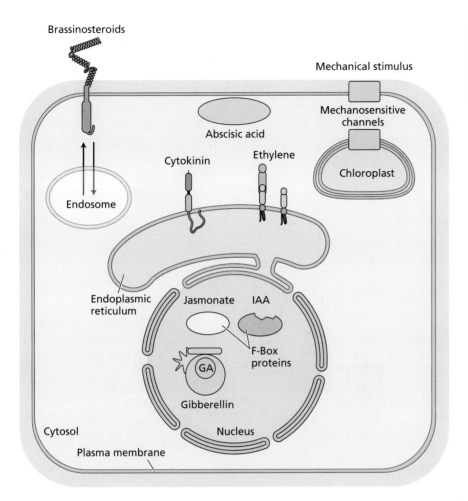

Figure 12.3 Primary locations of plant hormone receptors and mechanosensitive receptors in the cell. The individual receptors are discussed later in the chapter. IAA, indole-3-acetic acid (auxin); GA, gibberellin. (After Santer and Estelle 2009.)

signals to changes in cellular activities. Most plant genomes contain four major multigene families of Ca^{2+} sensors: the calmodulin (CaM) and calmodulin-like proteins, the Ca^{2+}-dependent protein kinases, Ca^{2+}/calmodulin-dependent protein kinases, and calcineurin-B like (CBL) proteins, which function in concert with CBL-interacting protein kinases. Members of these sensor families modulate the activity of target proteins either by binding to CaM or by phosphorylating the target protein in a Ca^{2+}-dependent manner (see Figure 12.4). Target proteins include transcription factors, various protein kinases, Ca^{2+}-ATPases, ion channels, and other enzymes. Finally, Ca^{2+} pumps and Ca^{2+} exchangers in organelle and plasma membranes actively remove Ca^{2+} from the cytosol to terminate Ca^{2+} signaling (see Figure 12.4). Signaling molecules derived from cellular lipids (diacylglycerol, inositol 1,4,5-trisphosphate, and phosphatidic acid; **Figure 12.5**) are enzymatically generated from membrane lipids and are thought to also regulate Ca^{2+} fluxes in response to environmental cues.

Changes in the cytosolic or cell wall pH can serve as second messengers for hormonal and stress responses

Plant cells use the **proton motive force (PMF)** (i.e., the electrochemical proton gradient) across cellular membranes to drive ATP synthesis (see Chapters 7 and 11) and to energize secondary active transport (see Figure 12.4; also see Chapter 6). In addition to having such "housekeeping" activity, protons also appear to have signaling activity and function as second messengers. In a resting cell, cytosolic pH is typically kept constant at approximately pH 7.5, whereas the cell wall is acidified to pH 5.5 or lower. Extracellular pH can

proton motive force (PMF)
The energetic effect of the electrochemical H^+ gradient across a membrane, expressed in units of electrical potential.

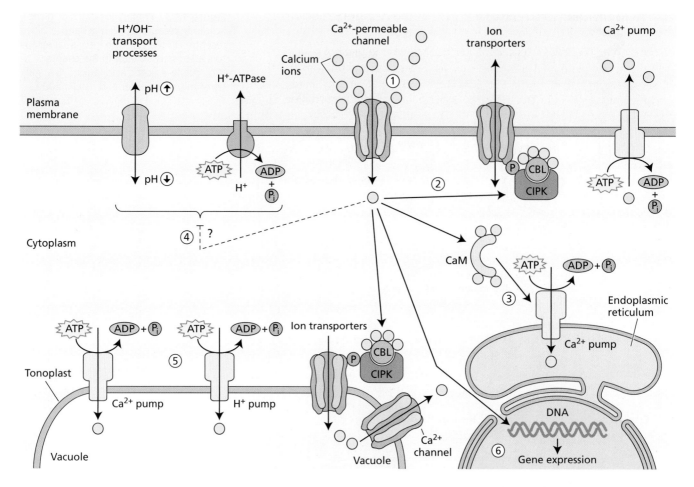

1. Signal-induced activation of Ca²⁺-permeable ion channels leads to an increase in the concentration of free cytosolic Ca²⁺.

2. Activated CBL/CIPK Ca²⁺ sensor proteins interact with ion transporters on the plasma membrane and the vacuolar membrane.

3. Activated calmodulin (CaM) stimulates Ca²⁺ pumps on the endoplasmic reticulum.

4. Ca²⁺ may affect membrane transporters, triggering changes in the intracellular and extracellular pH.

5. Pumps on vacuolar membrane create H⁺ and Ca²⁺ gradients between cytoplasm and vacuole.

6. Increased Ca²⁺ in nucleus activates Ca²⁺-dependent transcription.

Figure 12.4   Calcium ions and pH function as second messengers that amplify signals and regulate the activity of target signaling proteins to trigger physiological responses. An increase in the $[Ca^{2+}]_{cyt}$ activates calcium sensor proteins (calmodulins [CaMs] and calcineurin-B like (CBL) proteins/CBL-interacting protein kinases [CBL/CIPKs]), which are located at different subcellular sites. Six types of activation are shown in the figure.

change rapidly in response to a variety of different endogenous and environmental signals, while intracellular pH changes occur more slowly because of the very high cellular buffering capacity. In growing hypocotyls, for example, the plant hormone auxin triggers activation of the plasma membrane H⁺-ATPase (see Figure 12.4). As will be discussed in Chapter 15, activation of the plasma membrane H⁺-ATPase, causes the cell wall to become more acidic, which promotes cell expansion. In roots, however, where auxin inhibits cell expansion, auxin triggers the rapid alkalinization of the cell wall, a process that has been shown to be Ca²⁺-dependent (see Figure 12.4, number 4). Similar Ca²⁺-dependent pH changes are observed in many environmental stress responses of plants (see Chapter 19).

Reactive oxygen species act as second messengers mediating both environmental and developmental signals

In recent years, **reactive oxygen species** (**ROS**) have emerged not just as cytotoxic by-products of metabolic processes such as respiration and photosynthesis, but as signaling molecules regulating plant responses to various environmental and endogenous signals. ROS result from the partial reduction of oxygen and are formed predominantly in mitochondria, chloroplasts, peroxisomes, and the cell wall. In the context of cell signaling, plasma membrane–localized NADPH oxidases make up the best-understood family of ROS-producing enzymes. NADPH oxidases (or respiratory burst oxidase homologs, RBOHs) transfer electrons from the cytosolic electron donor NADPH across the membrane to reduce extracellular molecular oxygen. The resulting ROS, superoxide, can dismutate to hydrogen peroxide, a more membrane-permeable ROS that can also enter cells through specific aquaporin channels.

Targets of ROS signaling are just beginning to be identified. For example, the thiol side chain of cysteine amino acid residues can be oxidized by ROS to form intramolecular (within the polypeptide/protein) or intermolecular (between different polypeptides/proteins) disulfide bonds. Direct redox regulation of this type has been shown to alter the DNA-binding activity or cellular localization of several transcription factors and transcriptional activators. In the cell wall, tyrosine residues of structural proteins, feruloyl (ferulic acid) conjugates of poly-saccharides, and monolignols can be oxidatively cross-linked by ROS, thereby modifying the strength or the barrier properties of the cell wall.

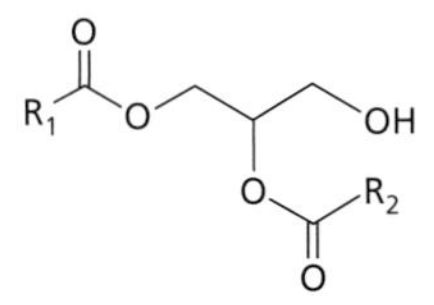

Hormones and Plant Development

The form and function of multicellular organisms would not be possible without efficient communication among cells, tissues, and organs. In higher plants, regulation and coordination of metabolism, growth, and morphogenesis often depend on chemical signals from one part of the plant to another. In the latter part of the nineteenth century, the great German plant physiologist Julius von Sachs proposed that "organ-forming substances" were synthesized by the plant and transported to different parts of the plant, where they regulated growth and development. He also suggested that environmental factors such as gravity could affect the distribution of these substances within a plant. It has since become apparent that the majority of signaling networks that translate environmental cues into growth and developmental responses regulate the metabolism or re-distribution of these endogenous chemical messengers. Although Sachs did not know the identity of these messengers, his ideas led to their eventual discovery.

Hormones are chemical messengers that are produced in one cell and mod-ulate cellular processes in another cell by interacting with specific proteins that function as receptors linked to cellular signal transduction pathways. As is the case with animal hormones, most plant hormones (phytohormones) are capable of activating responses in target cells at nanomolar concentrations. Although the details of hormonal control of development are quite diverse, all of the basic hormonal pathways share common features (**Figure 12.6**). For example, signal perception (environmental input) or developmental programming often results in increases in hormone biosynthesis. The hormone is then transported to a site of action. Perception of the hormone by a receptor results in transcriptional or posttranscriptional (e.g., phosphorylation, protein turnover, ion extrusion) events that ultimately induce a physiological or developmental response. In addition, the response can be attenuated by negative feedback mechanisms that repress hormone synthesis and by catabolism or sequestration, which combine to cause the return of the active hormone concentration to pre-signal levels. In this way the plant reacquires the ability to respond to the next signal input.

1,2-diacylglycerol

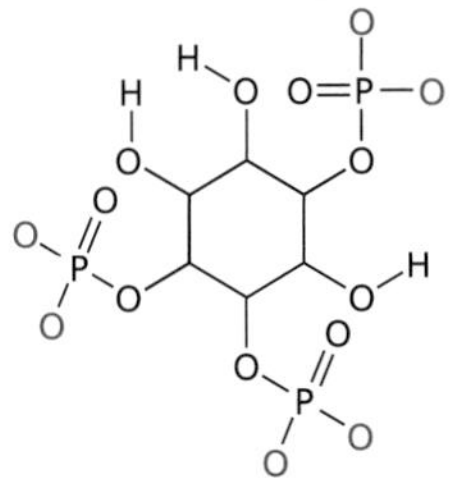

Inositol 1,4,5-triphosphate

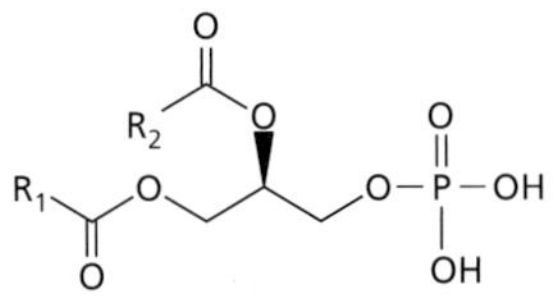

Phosphatidic acid

Figure 12.5 Signaling molecules derived from membrane lipids. R_1 and R_2 represent acyl side chains.

reactive oxygen species (ROS)
These include the superoxide anion ($O_2^{\cdot-}$), hydrogen peroxide (H_2O_2), the hydroxyl radical ($HO\cdot$), and singlet oxygen. They are generated in several cell compart-ments and can act as signals or cause damage to cellular components.

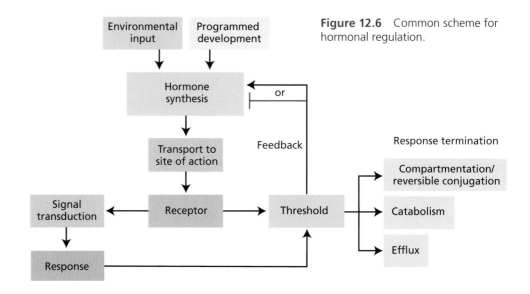

Figure 12.6 Common scheme for hormonal regulation.

Plant development is regulated by nine major hormones or hormone families: auxins, gibberellins, cytokinins, ethylene, abscisic acid, brassinosteroids, jasmonates, salicylic acid, and strigolactones (**Figure 12.7**). In addition, several small peptides function in cell-to-cell communication in plant development and in signaling in response to mineral nutrient deficiencies.

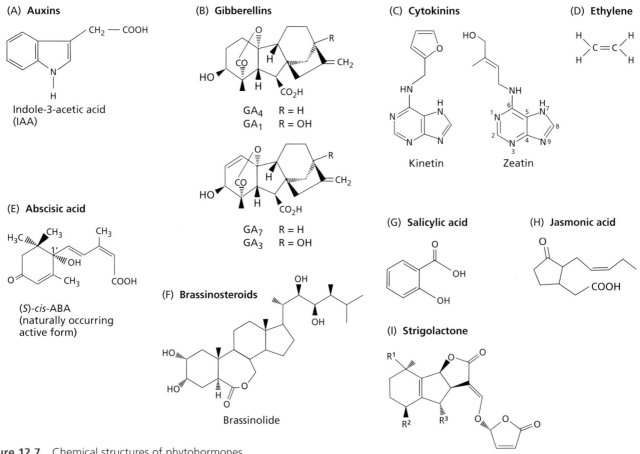

Figure 12.7   Chemical structures of phytohormones.

Auxin was discovered in early studies of coleoptile bending during phototropism

Auxin is essential to plant growth, and auxin signaling functions in virtually every aspect of plant development. Auxin was the first growth hormone to be studied in plants, and was discovered after the prediction of its existence by Charles and Francis Darwin in their book, *The Power of Movement in Plants*, which was published in 1880, the same year that Sachs proposed his theory of "organ-forming substances." The Darwins studied the bending of coleoptiles of canary grass (*Phalaris canariensis*) and seedling hypocotyls of other species in response to unidirectional light, and concluded that a signal produced at the apex travels downward and causes lower cells on the shaded side to grow faster than on the illuminated side. Subsequently it was shown that the signal was a chemical that could diffuse through gelatin blocks (**Figure 12.8**). Plant physiologists named the chemical signal auxin from the Greek *auxein*, meaning "to increase" or "to grow," and identified indole-3-acetic acid (IAA) (see Figure 12.7A) as the primary plant auxin. We will have more to say about the role of auxin in cell elongation growth in Chapter 15.

In some species, 4-chloro-IAA and phenylacetic acid (PAA) function as natural auxins, but IAA is by far the most abundant and physiologically important form. Because the structure of IAA is relatively simple, researchers were quickly able to synthesize a wide array of molecules with auxin activity. Some of these compounds, such as 1-naphthalene acetic acid (NAA), 2,4-dichlorophenoxyacetic acid (2,4-D), and 2-methoxy-3,6-dichlorobenzoic acid (dicamba), are now used widely as growth regulators and herbicides in horticulture and agriculture.

Gibberellins promote stem growth and were discovered in relation to the "foolish seedling disease" of rice

A second group of plant hormones are the **gibberellins** (abbreviated GA and numbered in the chronological sequence of their discovery). This group comprises a large number of compounds, all of which are tetracyclic (four-ringed) diterpenoid acids, but only a few of which, primarily GA_1, GA_3, GA_4, and GA_7, have intrinsic biological activity (see Figure 12.7B). One of the most striking effects of biologically active gibberellins, achieved through their role in promot-

auxins   Major plant hormones involved in numerous developmental processes, including cell elongation, organogenesis, apical–basal polarity, differentiation, apical dominance, and tropic responses. The most abundant chemical form is indole-3-acetic acid.

gibberellins (GAs)   A large group of chemically related plant hormones synthesized by a branch of the terpenoid pathway and associated with the promotion of stem growth (especially in dwarf and rosette plants), seed germination, and many other functions.

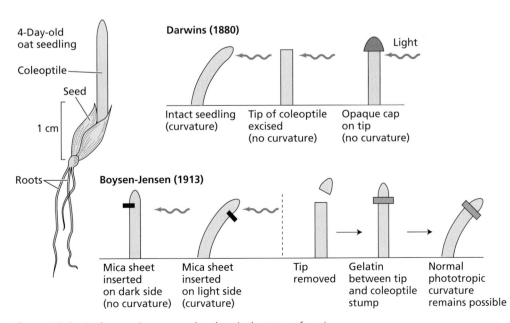

From experiments on coleoptile phototropism, the Darwins concluded in 1880 that a growth stimulus is produced in the coleoptile tip and is transmitted to the growth zone.

In 1913, P. Boysen-Jensen discovered that the growth stimulus passes through gelatin but not through water-impermeable barriers such as mica.

Figure 12.8   Early experiments on the chemical nature of auxin.

ing cell elongation, is the induction of internode elongation in dwarf seedlings. Gibberellins have other diverse roles during the plant life cycle; for example, they can promote seed germination (see Chapter 15), the transition to flowering, and fruit development (see Chapter 17).

Gibberellins were first isolated in Japan by Teijiro Yabuta and Yusuke Sumuki in the 1930s as natural products in the fungus *Gibberella fujikuroi* (renamed *Fusarium fujikuroi*), from which the hormones derive their unusual name. Rice plants infected with *F. fujikuroi* grow abnormally tall, which leads to lodging (falling over) and reduced yield; hence the name *bakanae*, or "foolish seedling disease." Such overgrowth can be reproduced by applying gibberellins to uninfected rice seedlings. *Fusarium fujikuroi* produces several different gibberellins, the most abundant of which is GA_3, also called gibberellic acid, which can be obtained commercially for horticultural and agronomic use. For example, GA_3 is sprayed onto grape vines to produce the large, seedless grapes we now routinely purchase in the grocery store (**Figure 12.9A**). Spectacular responses were obtained in the stem elongation of dwarf and rosette plants, particularly in genetically dwarf peas (*Pisum sativum*), dwarf maize (corn; *Zea mays*) (**Figure 12.9B**), and many rosette plants (**Figure 12.9C**).

Shortly after the first characterization of gibberellins from *F. fujikuroi*, it was discovered that plants also contain gibberellin-like substances but at much lower concentrations than in the fungus. The first plant gibberellin to be identified was GA_1, which was discovered in extracts of runner bean seeds in 1958. We now know that gibberellins are ubiquitous in plants and are also present in several fungi in addition to *F. fujikuroi*. The majority of plants studied to date contain GA_1 and/or GA_4, so these are the gibberellins to which we assign "hormonal" function. In addition to GA_1 and GA_4, plants contain many inactive gibberellins that are precursors or deactivation products of the bioactive gibberellins.

(A)

(B)

(C)

Figure 12.9   (A) Gibberellin induces growth in Thompson Seedless grapes. Untreated grapes normally remain small because of natural seed abortion. The bunch on the left is untreated. The bunch on the right was sprayed with GA_3 during fruit development, leading to larger fruits and elongation of the pedicels (fruit stalks). (B) The effect of exogenous GA_1 on wild-type (labeled as "normal" in the photograph) and dwarf mutant (*d1*) maize. Gibberellin stimulates dramatic stem elongation in the dwarf mutant, but has little or no effect on the tall, wild-type plant. (C) Under short-day conditions, cabbage remains a low-growing, vegetative rosette, but it can be induced to grow long internodes and flower by applications of GA. The effect of day length on flowering will be discussed in Chapter 17. (A, C © Sylvan Wittwer/Visuals Unlimited; B courtesy of B. Phinney.)

(A)

Auxin

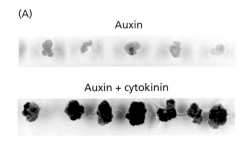

Auxin + cytokinin

(B)

Figure 12.10 Cytokinin enhances cell division and greening. (A) Wild-type Arabidopsis leaf explants were induced to form callus (undifferentiated cells) by culturing in the presence of auxin alone (top) or auxin plus cytokinin (bottom). Cytokinin was required for callus growth and greening in the presence of light. (B) Tumor that formed on a tomato stem infected with the crown gall bacterium, *Agrobacterium tumefaciens*. Two months before this photo was taken, the stem was wounded and inoculated with a virulent strain of the crown gall bacterium. (A from Riou-Khamlichi et al. 1999; B from Aloni et al. 1998, courtesy of R. Aloni.)

Cytokinins were discovered as cell division–promoting factors in tissue culture experiments

Cytokinins were discovered in a search for factors that stimulated plant cells to divide (i.e., undergo cytokinesis) in conjunction with the phytohormone auxin. A small molecule was identified that could, in the presence of auxin, stimulate tobacco pith parenchyma tissue to proliferate into an amorphous structure, known as a *callus*, in culture (**Figure 12.10A**). The cytokinesis-inducing molecule was named *kinetin*. While kinetin is a synthetic cytokinin, its structure is similar to that of naturally occurring cytokinins, such as zeatin (see Figure 12.7C).

After their discovery, cytokinins were rapidly shown to be produced in callus structures formed in plants that are infected with pathogenic bacteria, nematodes, and viruses (**Figure 12.10B**). As we will see in later chapters, cytokinins have been shown to have effects on many physiological and developmental processes, including leaf senescence, shoot branching, breaking of bud dormancy, and the formation of apical meristems (see Chapter 16). In addition, cytokinins are produced during mycorrhizal formation and when nitrogen-fixing bacteria stimulate root nodulation (see Chapters 5 and 18), as well as in plant responses to salinity, drought, and macronutrient (nitrate, phosphorus, iron, and sulfate) deficiencies (see Chapter 19).

Ethylene is a gaseous hormone that promotes fruit ripening and other developmental processes

Ethylene is a gas with a simple chemical structure (see Figure 12.7D) and was first identified as a plant growth regulator in 1901 by Dimitry Neljubov when he demonstrated its ability to alter the growth of etiolated pea seedlings in the laboratory, causing the **triple response** (**Figure 12.11A**). Subsequently, ethylene was identified as a natural product synthesized by plant tissues.

Ethylene regulates a wide range of responses in plants, including seed germination and seedling growth, cell expansion and differentiation, leaf and flower senescence and abscission (see Chapters 15 and 16), fruit ripening (see Chapter 17), and responses to biotic and abiotic stresses (see Chapters 18 and 19), including the downward bending of leaves known as *epinasty* (**Figure 12.11B**).

Abscisic acid regulates seed maturation and stomatal closure in response to water stress

Abscisic acid (**ABA**) is a ubiquitous hormone in vascular plants and has also been found in mosses, some phytopathogenic fungi, and a wide range of animals. ABA

cytokinins   A class of plant hormones that function in cell division and differentiation, as well as axillary bud growth and leaf senescence. Cytokinins are adenine derivatives, the most common form of which is zeatin.

ethylene   A gaseous plant hormone involved in fruit ripening, abscission, and the growth of etiolated seedlings. The chemical formula of ethylene is C_2H_4.

triple response   A response to ethylene of etiolated eudicot seedlings consisting of a short, thick hypocotyl or epicotyl, a pronounced apical hook, and horizontal growth.

abscisic acid   A plant hormone that functions in regulation of seed dormancy as well as in stress responses such as stomatal closure during water deficit and responses to cold and heat. ABA is derived from carotenoid precursors.

(A)

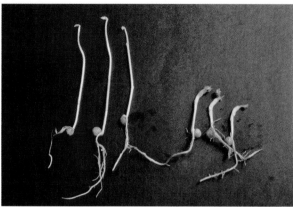

(B)

Figure 12.11 Ethylene responses. (A) Triple response of etiolated pea seedlings. Six-day-old pea seedlings were grown in the dark in the presence of 10 ppm (parts per million) ethylene (right) or left untreated (left). The treated seedlings show radial swelling, inhibition of elongation of the epicotyl, and horizontal growth of the epicotyl (diagravitropism). (B) Leaf epinasty in tomato. Epinasty, or downward bending of the tomato leaves (right), is caused by ethylene treatment. An untreated tomato is shown on the left. Epinasty results when the cells on the upper side of the petiole grow faster than those on the bottom. (Courtesy of S. Gepstein.)

is a 15-carbon terpenoid (see Figure 12.7E) that was identified in the 1960s as a growth-inhibiting compound associated with the onset of bud dormancy and promotion of cotton fruit abscission. However, later work showed that ABA promotes senescence, the process preceding abscission, rather than abscission itself. Since then, ABA has also been shown to be a hormone that regulates salinity, dehydration, and temperature stress responses, including stomatal closure (**Figure 12.12**) (see Chapter 19). ABA also promotes seed maturation and dormancy (see Chapter 15) and regulates the growth of roots and shoots, the production of different leaf types on an individual plant, flowering, and some responses to pathogens (see Chapter 18).

Brassinosteroids regulate floral sex determination, photomorphogenesis, and germination

Brassinosteroids (**BRs**) were first discovered as growth-promoting substances present in the pollen of *Brassica napus* (rape plant, some varieties of which are called canola in North America). Subsequent X-ray analysis showed that the most bioactive brassinosteroid in eudicots, which was named **brassinolide**, is a polyhydroxylated steroid similar to animal steroid hormones (see Figure 12.7F).

Many BRs have been identified, primarily intermediates of the brassinolide biosynthetic or catabolic pathways. Of these, the two known active brassinosteroid forms are brassinolide and its immediate precursor castasterone, which is more active in monocots such as maize than it is in eudicots. BRs are ubiquitous plant hormones that, like auxins and gibberellins, appear to predate the evolution of land plants.

brassinosteroids (BRs) A group of plant steroid hormones that play important roles in many developmental processes, including cell division and cell elongation in stems and roots, photomorphogenesis, reproductive development, leaf senescence, and stress responses.

brassinolide A plant steroid hormone with growth-promoting activity, first isolated from *Brassica napus* pollen. One of a group of plant hormones with similar structures and activities, called brassinosteroids.

(A)

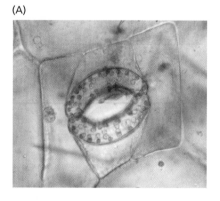

(B)

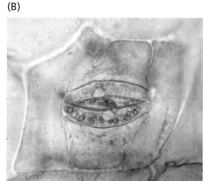

Figure 12.12 Stomatal closure in response to ABA. Stomata are open in the light for gas exchange with the environment (left). ABA treatment closes stomata in the light (right). This reduces water loss during the day under drought stress conditions. (Photos © Ray Simons/Science Source.)

In the angiosperms, BRs are found at low levels in various organs (e.g., flowers, leaves, roots) and at higher relative levels in pollen, immature seeds, and fruits.

Brassinosteroids play pivotal roles in a wide range of developmental phenomena in plants, including cell division, cell elongation, cell differentiation, photomorphogenesis, reproductive development, germination, leaf senescence, and stress responses. Mutants deficient in BR synthesis show growth and developmental abnormalities, including dwarfism, reduced apical dominance, and photomorphogenesis in the dark. BRs play a major role in sex determination of flowers. This role is demonstrated by BR-deficient maize mutants, in which male flowers (tassels) exhibit feminine traits such as carpels and silk production (**Figure 12.13**).

Salicylic acid and jasmonates function in defense responses

Salicylic acid (see Figure 12.7G) and conjugated compound derived from **jasmonic acid** (see Figure 12.7H) are the primary signaling compounds that function in plant defense responses to herbivory and pathogen infection. The role of these two hormones in defense responses is discussed in Chapter 18. Jasmonic acid and its conjugated forms also function in abiotic stress responses (to salt) and interact with auxin and possibly other hormone signaling pathways in environmental regulation of plant development.

Strigolactones suppress branching and promote rhizosphere interactions

Strigolactones, which occur in about 80% of plant species, are a group of terpenoid lactones (see Figure 12.7I) that were originally discovered as host-derived germination stimulants for root parasitic plants, such as the witchweeds (*Striga* spp.) and broomrapes (*Orobanche* and *Phelipanche* spp.). They also promote symbiotic interactions with arbuscular mycorrhizal fungi, which facilitate phosphate uptake from the soil. In addition, strigolactones suppress shoot branching and stimulate activity of the cambium and secondary growth (see Chapter 16). Strigolactones have analogous functions in roots, where they reduce adventitious and lateral root formation and promote root hair growth.

Phytohormone Metabolism and Homeostasis

To be effective signals, the concentrations of plant hormones must be tightly regulated. In the simplest terms, a hormone's concentration in any given tissue or cell is determined by the balance between the rate of increase in its concentration (e.g., by local synthesis/activation or by import from elsewhere in the plant) and the rate of decrease in its concentration (e.g., by inactivation, degradation, sequestration, or efflux) (**Figure 12.14**). However, the regulation of hormone levels is complicated by many factors. First, primary hormone biosynthetic pathways may be augmented by secondary or intersecting biosynthetic mechanisms.

Figure 12.13 Loss of active brassinosteroid in the *nana1* mutant of maize results in altered floral sex determination. The staminate wild-type tassel is shown on the left, and the feminized *nana1* mutant tassel is shown on the right. (From Hartwig et al. 2011.)

salicylic acid A benzoic acid derivative that serves as an endogenous signal for systemic acquired resistance.

jasmonic acid A plant hormone that functions in plant defenses to biotic and abiotic stresses, as well as some other aspects of development. Jasmonic acid is derived from the octadecanoid pathway. Can function as a volatile signal when methylated (methyl jasmonate) and is also active in some plants when conjugated to amino acids.

strigolactones Carotenoid-derived plant hormones that inhibit shoot branching. They also play roles in the soil by stimulating the growth of arbuscular mycorrhizae and the germination of parasitic plant seeds, such as those of *Striga* (the source of their name).

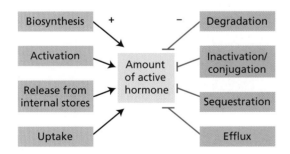

Figure 12.14 Homeostatic regulatory mechanisms that influence hormone concentration. Both positive and negative factors work in concert to maintain hormone homeostasis.

Second, there may be multiple structural variants of a hormone, which vary widely in biological activity. Finally, as we will see later, there may be multiple mechanisms for removing active hormone from a system.

In this section we discuss mechanisms for modulating hormone concentrations locally (within a cell or tissue).

Indole-3-pyruvate is the primary intermediate in auxin biosynthesis

IAA is structurally related to the amino acid tryptophan, and is primarily synthesized in a two-step process using indole-3-pyruvate (IPyA) as the intermediate (**Figure 12.15A**). The second step in the pathway is carried out by the YUCCA family of flavin monooxygenases.

IAA biosynthesis is associated with rapidly dividing and growing tissues, especially in shoots. Although virtually all plant tissues appear to be capable of producing low levels of IAA, shoot apical meristems, young leaves, and young fruits are the primary sites of auxin synthesis. In plants that produce indole glucosinolate defense compounds (see Chapter 18), IAA can also be synthesized from tryptophan via a pathway with indole acetonitrile as an intermediate. In maize kernels, IAA also appears to be synthesized by a tryptophan-independent pathway.

Auxin is toxic at high cellular concentrations, and without homeostatic controls the hormone could easily build up to toxic levels. Auxin catabolism by conjugation to hexose sugars and oxidative degradation ensures the permanent removal of active hormone when the concentration exceeds the optimal level or when the response to the hormone is complete. Covalent conjugation of amino acids to IAA can also result in permanent inactivation. However, most amino acyl conjugates serve as storage forms from which IAA can be rapidly released by enzymatic processes. Indole-3-butyric acid (IBA) is a compound that is routinely used in horticulture to promote rooting of cuttings and is rapidly converted by β-oxidation in the peroxisome to IAA. Both free and conjugated IBA are thought to occur naturally in plants and to serve as auxin sources for specific developmental processes. Auxin has also been shown to be conjugated to peptides, complex glycans (multiple sugar units), or glycoproteins in some plant species, but the precise physiological role of these conjugates is still unknown.

Sequestration of auxin in endomembrane compartments, primarily the endoplasmic reticulum, also appears to regulate the amount of auxin available for signaling. Proteins that mediate IAA movement across the endoplasmic reticulum membrane have been identified, and a large store of AUXIN BINDING PROTEIN1 (ABP1) is found in the endoplasmic reticulum lumen, where it may help sequester auxin.

The well-documented toxicity of exogenously applied auxin, especially on eudicot species, forms the basis for the use of a family of synthetic auxins, such as 2,4-dichlorophenoxyacetic acid (2,4-D), as herbicides. Mutations causing auxin overexpression would tend to be lethal if it were not for the homeostatic control of auxin levels. The reason that synthetic auxins are more effective as herbicides than natural auxins is that the synthetic auxins are much less subject to homeostatic control—degradation, conjugation, transport, and sequestration—than natural auxins are.

Gibberellins are synthesized by oxidation of the diterpene *ent*-kaurene

Gibberellins are synthesized in several parts of a plant, including developing seeds, germinating seeds, developing leaves, and elongating internodes. The biosynthetic pathway, which starts in plastids, leads to the production of a linear (straight chain) precursor molecule containing 20 carbon atoms, geranylgeranyl diphosphate (GGPP). GGPP gets converted into *ent*-kaurene in the first committed step of GA biosynthesis (**Figure 12.15B**). This compound is oxidized sequentially by enzymes associated with the endoplasmic reticulum to active forms of gibberellin. Dioxygenase enzymes in the cytosol are able to oxidize GA_{12} to all other gibberellins found in plants.

The pathways involved in gibberellin biosynthesis and catabolism are under tight genetic control. Several regulatory mechanisms have been described to date. These include gibberellin inactivation via a family of enzymes termed GA 2-oxidases, methylation via a methyl transferase, and conjugation to sugars. Gibberellin biosynthesis is also regulated by feedback inhibition when cellular gibberellin exceeds threshold levels.

Cytokinins are adenine derivatives with isoprene side chains

Cytokinins are adenine derivatives, and the most common class of cytokinins has isoprenoid side chains, including isopentenyl adenine (iP), dihydrozeatin (DHZ), and the most abundant cytokinin in higher plants, zeatin. Cytokinins are made from ADP/ATP and dimethylallyl diphosphate (DMAPP), primarily in plastids (**Figure 12.15C**).

In addition to the free bases, which are the only active forms, cytokinins are also present in the plant as ribosides (in which a ribose sugar is attached to the 9 nitrogen of the purine ring), ribotides (in which the ribose sugar moiety contains a phosphate group), or glycosides (in which a sugar molecule is attached to the 3, 7, or 9 nitrogen of the purine ring, or to the oxygen of the zeatin or dihydrozeatin side chain). In addition to this glycosylation-mediated inactivation, active cytokinin levels are also decreased catabolically through irreversible cleavage by cytokinin oxidases.

Ethylene is synthesized from methionine via the intermediate ACC

Ethylene can be produced by almost all parts of higher plants, although the rate of production depends on the type of tissue, the stage of development, and environmental inputs. For example, certain mature fruits undergo a respiratory burst in response to ethylene, and ethylene levels increase in these fruits at the time of ripening (see Chapter 17). Ethylene is derived from the amino acid methionine and the intermediate *S*-adenosylmethionine (**Figure 12.15D**). The first committed and generally rate-limiting step in the biosynthesis is the conversion of *S*-adenosylmethionine to 1-aminocyclopropane-1-carboxylic acid (ACC) by the enzyme **ACC synthase**. ACC is then converted to ethylene by enzymes called **ACC oxidases**. As ethylene is a gaseous hormone, there is no evidence of ethylene catabolism in plants. However, if its biosynthesis is blocked by an exogenous inhibitor, such as aminoethoxy-vinylglycine (AVG), which inhibits ACC synthase, the rapid diffusion of ethylene out of the plant depletes cellular ethylene levels.

Abscisic acid is synthesized from a carotenoid intermediate

ABA is synthesized in almost all cells that contain chloroplasts or amyloplasts and has been detected in every major organ and tissue. ABA is a 15-carbon terpenoid, or sesquiterpenoid, which is synthesized in plants by an indirect pathway via 40-carbon carotenoid intermediates (**Figure 12.15E**). Early steps of this pathway occur in plastids. Cleavage of the resulting carotenoid by 9-*cis*-epoxycarotenoid dioxygenase results in production of xanthoxin, which then undergoes a series of oxidative reactions to form ABA. Further oxidation by ABA-8'-hydroxylases leads to ABA inactivation. ABA can also be inactivated by conjugation, but this is reversible. Both types of inactivation are also tightly regulated.

ABA concentrations can fluctuate dramatically in specific tissues during development or in response to changing environmental conditions. In developing

ACC synthase The enzyme that catalyzes the synthesis of 1-aminocyclopropane-1-carboxylic acid (ACC) from *S*-adenosylmethionine.

ACC oxidase The enzyme that catalyzes the conversion of 1-aminocyclopropane-1-carboxylic acid (ACC) to ethylene, the last step in ethylene biosynthesis.

cytochrome P450 monooxygenase (CYP)   A generic term for a large number of related, but distinct, mixed-function oxidative enzymes localized on the endoplasmic reticulum. CYPs participate in a variety of oxidative processes, including steps in the biosynthesis of gibberellins and brassinosteroids.

seeds, for example, ABA levels can increase 100-fold within a few days, reaching average concentrations in the micromolar range, and then decline to very low levels as maturation proceeds (see Chapter 15). Under conditions of water stress (i.e., dehydration stress), ABA in the leaves can increase 50-fold within 4 to 8 h (see Chapter 19).

Brassinosteroids are derived from the sterol campesterol

Brassinosteroids are synthesized from the plant sterol campesterol, which is similar in structure to cholesterol. Members of the **cytochrome P450 monooxygenase (CYP)** enzyme family that are associated with the endoplasmic reticulum catalyze most of the reactions in the brassinosteroid biosynthesis pathway (**Figure 12.15F**). Bioactive brassinosteroid levels are also modulated by a variety of inactivation or catabolic reactions, including epimerization, oxidation, hydroxylation, sulfonation, and conjugation to glucose or lipids.

The levels of active brassinosteroids are also regulated by brassinosteroid-dependent negative feedback mechanisms in which hormone concentrations above a certain threshold cause a decrease in brassinosteroid biosynthesis. This attenuation is brought about by the down-regulation of brassinosteroid biosynthetic genes and the up-regulation of genes involved in brassinosteroid catabolism.

(A) **Auxin (indole-3-acetic acid) biosynthesis**

(B) **Gibberellic acid (GA) biosynthesis**

(C) **Cytokinin biosynthesis**

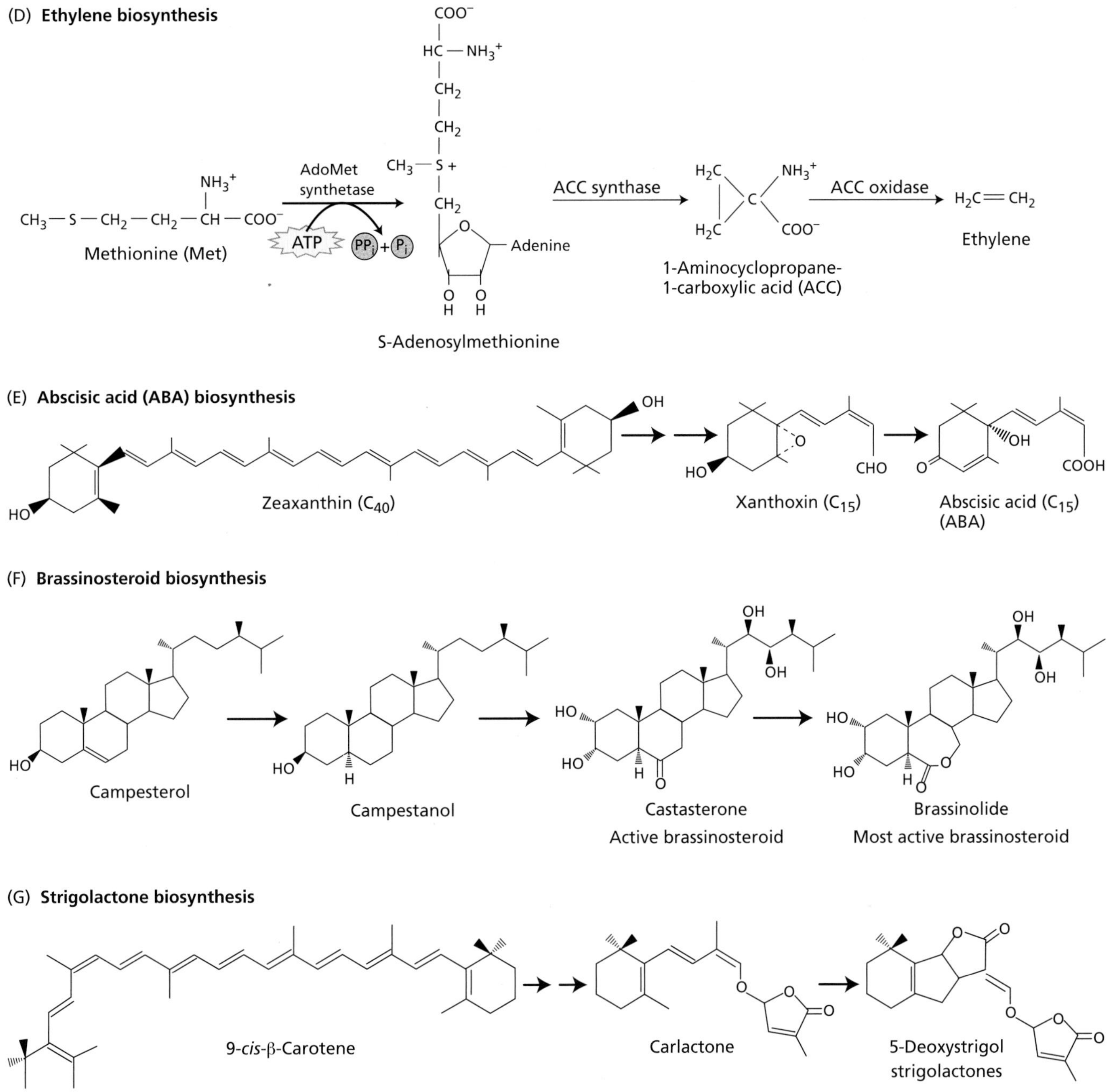

(D) Ethylene biosynthesis

Methionine (Met)

AdoMet synthetase

ATP → PP$_i$ + P$_i$

S-Adenosylmethionine

Adenine

ACC synthase →

1-Aminocyclopropane-
1-carboxylic acid (ACC)

ACC oxidase → H$_2$C=CH$_2$

Ethylene

(E) Abscisic acid (ABA) biosynthesis

Zeaxanthin (C$_{40}$)

Xanthoxin (C$_{15}$)

Abscisic acid (C$_{15}$)
(ABA)

(F) Brassinosteroid biosynthesis

Campesterol

Campestanol

Castasterone
Active brassinosteroid

Brassinolide
Most active brassinosteroid

(G) Strigolactone biosynthesis

9-*cis*-β-Carotene

Carlactone

5-Deoxystrigol
strigolactones

Figure 12.15 Abbreviated biosynthetic pathways of phytohormones. (A) Auxin biosynthesis from tryptophan (Trp). In the first step, Trp is converted to indole-3-pyruvate (IPyA) by the TAA family of tryptophan amino transferases. Subsequently, IAA is produced from IPyA by the YUCCA (YUC) family of flavin monooxygenases. (B) Gibberellin (GA) biosynthesis. In the plastid, geranylgeranyl diphosphate (GGPP) is converted to *ent*-kaurene. After this committed step, *ent*-kaurene is converted in multiple steps to active forms of GA. (C) Cytokinin biosynthesis. An isopentenyl side chain from dimethylallyl diphosphate (DMAPP) is transferred by isopentenyl transferase to an adenosine moiety (ATP or ADP) and eventually is converted to zeatin. (D) Ethylene biosynthesis. The amino acid methionine is converted to *S*-adenosylmethionine, which is converted to 1-aminocyclopropane-1-carboxylic acid (ACC) by the enzyme ACC synthase. Oxidation of ACC via ACC oxidase produces ethylene. (E) ABA biosynthesis via the terpenoid pathway. The initial stages occur in plastids, where isopentenyl diphosphate (IPP) is converted to zeaxanthin, which is modified and cleaved to form xanthoxin. Xanthoxin then is converted to ABA in the cytosol. (F) Brassinosteroid synthesis. The primary precursor for brassinosteroid biosynthesis is campesterol. In different branches of the pathway, cholesterol and sitosterol can also serve as precursors. Active brassinosteroids, castasterone and brassinolide, are derived from the immediate precursor campestanol after multiple C-6 oxidation steps. (G) Strigolactone biosynthesis. Cleavage of 9-*cis*-β-carotene produces the intermediate carlactone. The final conversion to strigolactones occurs in the cytosol.

action potential A transient event in which the membrane potential difference rapidly decreases (depolarizes) and abruptly increases (hyperpolarizes). Action potentials, which are triggered by the opening of ion channels, can be self-propagating along linear files of cells, especially in the vascular systems of plants.

Thus, mutants impaired in their ability to respond to brassinolide accumulate high levels of the active brassinosteroids compared with wild-type plants.

Strigolactones are synthesized from β-carotene

Like ABA, strigolactones are derived from carotenoid precursors in plastids in a pathway that is conserved up to synthesis of the intermediate carlactone, beyond which strigolactone biosynthesis diverges in a species-specific manner (**Figure 12.15G**). This divergence is attributed to the functional diversity of cytochrome P450 isoforms, which act on carlactone. The role of strigolactones in regulation of shoot branching will be discussed in Chapter 16.

The biosynthesis of jasmonic acid and salicylic acid are discussed in the context of biotic interactions in Chapter 18.

Signal Transmission and Cell–Cell Communication

Hormonal signaling typically involves the transmission of the hormone from its site of synthesis to its site of action. In general, hormones that are transported to sites of action in tissues distant from their site of synthesis are referred to as *endocrine* hormones, while those that act on cells adjacent to the source of synthesis are referred to as *paracrine* hormones. Hormones can also function in the same cells in which they are synthesized, in which case they are referred to as *autocrine effectors*. Most plant hormones have paracrine activities, since plants lack the fast-moving circulatory systems found in animals associated with classic endocrine hormones. However, slower, long-distance hormone transport via the vascular system is a common feature in plants, despite the absence of hormone-secreting glands like those in animal endocrine systems.

For example, the polar transport of auxin via highly regulated cellular uptake and efflux mechanisms is essential to auxin's role in establishing and maintaining polar plant growth and organogenesis. The cellular mechanisms that control polar auxin transport will be described in Chapters 14 and 15. Lipophilic hormones such as ABA and strigolactones can diffuse across membranes, but are actively transported across membranes by ATP-binding cassette subfamily G (ABCG) transporters in some tissues. Polarized transport of strigolactone out of the root apex by an ABCG protein has recently been demonstrated. Cytokinins can move over long distances in the transpiration streams of the xylem and have recently been shown to be actively transported into the vascular system in the root. Auxins and cytokinins can also move with source–sink fluxes in the phloem. Recent research suggests that gibberellin levels in root tissues are controlled via an active transport mechanism, resulting in accumulation of this growth hormone in expanding endodermal cells that control root elongation. As a gaseous compound, ethylene is more soluble in lipid bilayers than in the aqueous phase and can freely pass through the plasma membrane. In contrast, its precursor, ACC, is water soluble and is thought to be transported via the xylem to shoot tissues. Brassinosteroids are synthesized throughout the plant and are thought to function within the cells where they are synthesized. Brassinosteroids do not seem to undergo root-to-shoot and shoot-to-root translocation, as shown by experiments in pea, tomato, and Arabidopsis, in which reciprocal stock/scion grafting of wild-type plants to brassinosteroid-deficient mutants does not rescue the phenotype of the latter.

Although plants lack nervous systems, like animals they employ long-distance electrical signaling to communicate between distant parts of the plant body. The most common type of electrical signaling in plants is the **action potential**, the transient depolarization of the plasma membrane of a cell generated by voltage-gated ion channels (see Chapter 6). Action potentials have been shown to mediate touch-induced leaflet closure in sensitive plant (*Mimosa pudica*), as well as the rapid closure (~0.1 s) of leaves of Venus flytrap, which occurs when

(A)

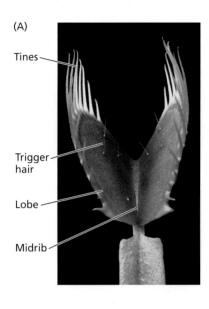

Tines

Trigger hair

Lobe

Midrib

(B)

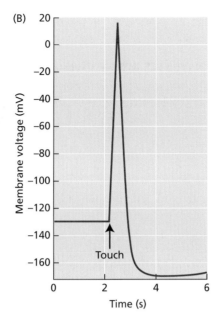

Touch

Figure 12.16 Electrical signaling in Venus flytrap (*Dionaea muscipula*). (A) Illustration of the snap-trap leaves with needle-like tines and touch-sensitive trigger hairs. (B) Action potential in response to the displacement of a single trigger hair. Mechanical stimulation of the trigger hairs by a prey activates mechanosensitive ion channels, which leads to the induction of action potentials that cause the lobes of the leaf to close and secrete digestive enzymes. (A © blickwinkel/Alamy; B after Escalante-Pérez et al. 2011.)

an insect touches the sensitive hairs on the upper sides of the traplike leaf lobes (**Figure 12.16A**). For the response to be activated, either two hairs must be touched within 20 s of each other, or one hair must be touched twice in rapid succession. Since each displacement elicits an action potential (**Figure 12.16B**), the leaf must have a mechanism for counting action potentials. Propagation of electrical signals within the vascular system of plants has been implicated in responses to herbivores as well (see Chapter 18). However, unlike in animal nervous systems, plants lack synapses that transmit electrical signals from neuron to neuron via the secretion of neurotransmitters. The mechanism of the much slower electrical signal transmission along the vascular systems of plants is still poorly understood.

Hormonal Signaling Pathways

The sites of action of hormones are cells that possess specific receptors that can bind the hormone and initiate a signal transduction cascade. Plants employ large numbers of receptor kinases and signal transduction kinases to bring about the physiological responses of hormone target cells. In this section, we examine the types of receptors and signal transduction pathways that are associated with each of the main plant hormones.

The cytokinin and ethylene signal transduction pathways are derived from the bacterial two-component regulatory system

The cellular receptors for ethylene and cytokinins resemble the **two-component regulatory systems** that function in bacterial responses to environmental stimuli (**Figure 12.17**). The two components of this signaling system consist of a membrane-bound histidine kinase **sensor protein** and a soluble **response regulator** protein. Sensor proteins receive the input signal, undergo autophosphorylation on a histidine residue, and pass the signal on to response regulators by transferring the phosphoryl group to a conserved aspartate residue on the response regulator. The phosphorylation-activated response regulators, many of which function as transcription factors, then bring about the cellular response. Sensor proteins have two domains, an *input domain*, which receives the environmental signal, and a *transmitter domain*, which transmits the signal to the response regulator. The response regulator proteins also have two domains, a *receiver domain*,

two-component regulatory systems Signaling pathways common in prokaryotes. They typically involve a membrane-bound histidine kinase sensor protein that senses environmental signals and a response regulator protein that mediates the response. Although rare in eukaryotes, systems resembling bacterial two-component systems are involved in both ethylene and cytokinin signaling.

sensor proteins Bacterial receptor proteins that perceive external or internal signals as part of two-component regulatory systems. They consist of two domains, an input domain, which receives the environmental signal, and a transmitter domain, which transmits the signal to the response regulator.

response regulator One component of the two-component regulatory systems that are composed of a histidine kinase sensor protein and a response regulator protein. Response regulators have a receiver domain, which is phosphorylated by the sensor protein, and an output domain, which carries out the response.

(A) Prokaryotic two-component system

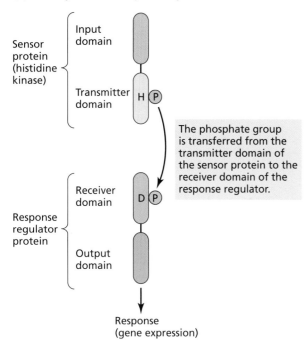

(B) Multistep version of the prokaryotic two-component system

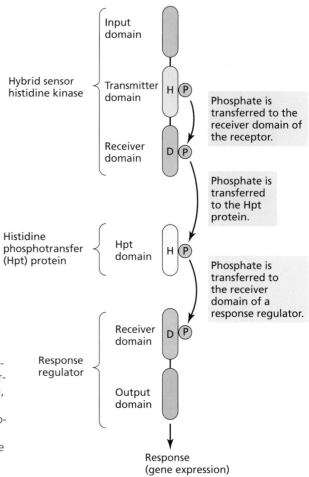

Figure 12.17 Two-component signaling systems of bacteria and plants. (A) The bacterial two-component system, consisting of a sensor protein and a response regulator protein, is found only in prokaryotes. (B) A derived multistep version of the two-component system, involving a phospho-relay protein intermediate, is found in both prokaryotes and eukaryotes. The plant two-component receptor protein includes a receiver domain fused to the transmitter domain. A separate histidine phosphotransfer (Hpt) protein transfers phosphate from the receiver domain of the receptor to the receiver domain of the response regulator. H, histidine residue; D, aspartate residue.

which receives the signal from the transmitter domain of the sensor protein, and an *output domain,* which mediates the response.

Cytokinin signaling is mediated by a phosphorylation relay system that consists of a transmembrane cytokinin receptor, a phosphotransfer protein, and a nuclear response regulator (**Figure 12.18A**). The endoplasmic reticulum–localized cytokinin receptors transmit cytokinin signals to two types of nuclear-localized transcriptional response regulators. Type A response regulators negatively regulate cytokinin signaling, while Type B response regulators positively regulate cytokinin signaling. Both Type A and Type B response regulators are themselves negatively regulated by ubiquitin-mediated protein degradation.

Ethylene receptors are also evolutionarily related to bacterial two-component histidine kinases, but the kinase activity does not appear to function in ethylene signaling. Like the cytokinin receptors, ethylene receptors are also localized to the endoplasmic reticulum membrane and interact with two downstream signaling proteins, which in turn regulate transcriptional regulators. Ethylene receptors function as *negative regulators* that actively repress the hormone response in the absence of the hormone.

Receptor-like kinases mediate brassinosteroid signaling

The largest class of plant receptor kinases consists of the receptor-like serine/threonine kinases (RLKs). Many RLKs localize to the plasma membrane as transmembrane proteins possessing extracellular ligand-binding domains and

(A) Cytokinin signaling

1. Cytokinin binds to the input domain of a receptor dimer in the endoplasmic reticulum membrane.

2. Cytokinin binding activates a phosphorylation relay involving a his kinase transmitter domain and a receiver domain.

3. The receiver domain phosphorylates a histidine phosphotransfer (Hpt) protein, triggering downstream signaling cascades.

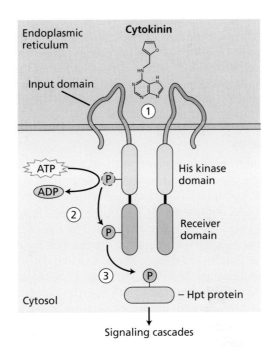

Figure 12.18 Cytokinins, brassinosteroids, and ABA trigger kinase/phosphatase signaling cascades.

(B) Brassinosteroid signaling

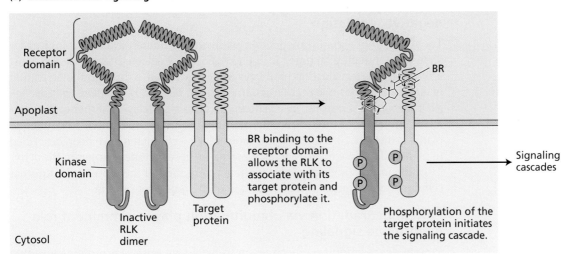

BR binding to the receptor domain allows the RLK to associate with its target protein and phosphorylate it.

Phosphorylation of the target protein initiates the signaling cascade.

(C) ABA signaling

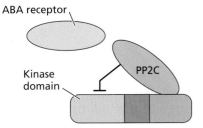

Dephosphorylated SnRK2 (inactive)

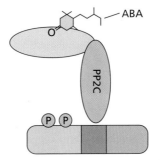

Phosphorylated SnRK2 (active)

In the absence of ABA, the phosphatase PP2C keeps an associated kinase (SnRk2) dephosphorylated and thereby inactivated.

When ABA is present, its receptor prevents dephosphorylation of the SnRk2 kinase by PP2C. The kinase is phosphorylated and thereby activated, initiating the signaling cascades.

ubiquitin–proteasome pathway
A mechanism for the specific degradation of cellular proteins involving two discrete steps: the polyubiquitination of proteins via the E3 ubiquitin ligase and the degradation of the tagged protein by the 26S proteasome.

cytoplasmic kinase domains, which relay information to the cell interior via the phosphorylation of serine or threonine residues of target proteins. Some RLKs have also been shown to phosphorylate tyrosine residues. The RLK-mediated brassinosteroid (BR) signaling pathway combines signal amplification and repressor inactivation mechanisms to transduce an extracellular brassinosteroid hormone signal into a transcriptional response (**Figure 12.18B**).

The core ABA signaling components include phosphatases and kinases

In addition to protein kinases, protein phosphatases (enzymes that remove phosphate groups from proteins) play important roles in signal transduction pathways. A well-described example is the signal transduction pathway of the hormone ABA (**Figure 12.18C**). ABA first binds to soluble receptor proteins in the cytosol near the plasma membrane. The ABA-receptor complex then binds to a membrane-associated regulatory protein on the cytosolic face of the plasma membrane. Finally, the membrane-associated ABA-protein complex regulates protein phosphatase 2C (PP2C) activity, either to activate or repress multiple transcriptional and non-transcriptional ABA response mechanisms. The same PP2C also interacts with other proteins implicated in cellular ABA responses, including other protein kinases, Ca^{2+} sensor proteins, transcription factors, and ion channels, presumably regulating their activity by dephosphorylating specific serine or threonine residues.

Plant hormone signaling pathways generally employ negative regulation

The majority of signal transduction pathways ultimately elicit a biological response by inducing changes in the expression of selected target genes. Most animal signal transduction pathways induce a response through the activation of a cascade of positive regulators. In contrast, *the majority of plant transduction pathways induce a response by inactivating repressor proteins.* Whether or not negative regulation of their signaling pathways confers an adaptive advantage to plants is unclear. Several different molecular mechanisms have been described in plant cells to inactivate repressor proteins; these include changes in the activity of the repressor via phosphorylation, retargeting of the repressor to another cellular compartment, and degradation of the repressor protein.

Protein degradation via ubiquitination plays a prominent role in hormone signaling

Protein degradation as a mechanism to inactivate repressor proteins was first described as part of the auxin signaling pathway. Since then, the **ubiquitin–proteasome pathway** has been shown to be central to most, if not all, hormone signaling pathways (**Figure 12.19**). In this process, one or more small proteins called ubiquitin are ligated (covalently linked) to a repressor protein to mark the protein for degradation. Such polyubiquitination targets the protein for degradation by the 26S proteasome complex (see Figure 1.16).

In hormone signaling, proteins that bind and repress transcriptional activators are recruited to ubiquitination complexes by specific F-box proteins when auxin, jasmonate, or GA are present. The hormones bring together F-box protein–repressor protein complexes to recruit the repressor for ubiquitination and proteolytic degradation. The degradation of the repressor protein releases the transcriptional activators associated with each hormone to activate gene transcription. In auxin signaling, the transcriptional activators are known as auxin response factors (ARFs; see Figure 12.19A). JRF transcription factors activate jasmonate-responsive gene transcription, and phytochrome interacting factors (PIFs) activate GA responses (see Figure 12.19B and C; see also Chapter 13).

As the above discussion indicates, auxin, jasmonate, and gibberellins signal by *directly* targeting the stability of nuclear-localized repressor proteins and thereby inducing a transcriptional response. Such a short signal transduction pathway provides the means for a very rapid change in nuclear gene expression. However, there is no opportunity for signal amplification as in the case of a signaling pathway involving a kinase cascade or second messengers. Instead, any resulting transcriptional response is directly related to the abundance of the signal molecule, since this determines the number of repressor molecules that are degraded. This important feature in the organization of signal transduction pathways may help explain why comparatively high concentrations of signals such as auxin and gibberellin are required to elicit a biological response.

Plants have mechanisms for switching off or attenuating signaling responses

Arguably, the ability to switch off a response to a signal is just as important as the ability to initiate it. Plants terminate signaling through a variety of mechanisms.

As previously discussed, chemical signals such as plant hormones can be degraded or rendered inactive by oxidation or conjugation to sugars or amino acids. They may also be sequestered into other cellular compartments to spatially separate them from receptors. In general, oxidative processes slowly inactivate hormones to prevent their binding to their target repressor proteins. This terminates recruitment to receptor complexes and restores binding and repression of transcriptional activators. Receptors and signaling intermediates can also be inactivated by dephosphorylation. Similarly, ion transporters and cellular scavengers can quickly lower elevated concentrations of second messengers to switch off signal amplification.

Hormone signaling pathways are often subject to several loops of negative feedback regulation. Combinations of positive and negative feedback loops can help ensure that the appropriate hormone levels and responses are maintained during plant development.

The cellular response output to a signal is often tissue-specific

Many environmental and endogenous signals can trigger diverse plant responses. Typically, particular tissue or cell types do not display the entire range of potential responses when exposed to a signal, but exhibit distinct response specificity. The plant hormone auxin, for example, promotes cell expansion in growing aerial tissues while inhibiting cell expansion in roots. It elicits lateral root initiation in a subset of cells of the root pericycle while inducing leaf primordia at the shoot apical meristem and controls vascular differentiation in developing plant organs. Although the central signal transduction mechanisms are generally present in all tissues and cell types of the plant, different components are present in specific tissues and developmental stages. This allows for developmentally programmed responses and differential responses to environmental inputs.

Cross-regulation allows signal transduction pathways to be integrated

Within plant cells, signal transduction pathways never function in isolation, but operate as part of a complex web of signaling interactions. These interactions account for the fact that plant hormones often exhibit *agonistic* (additive or positive) or *antagonistic* (inhibitory or negative) interactions with other signals. Classic examples include the antagonistic interaction between gibberellin and ABA in the control of seed germination (see Chapter 15).

(A) Auxin response

(B) Jasmonate response

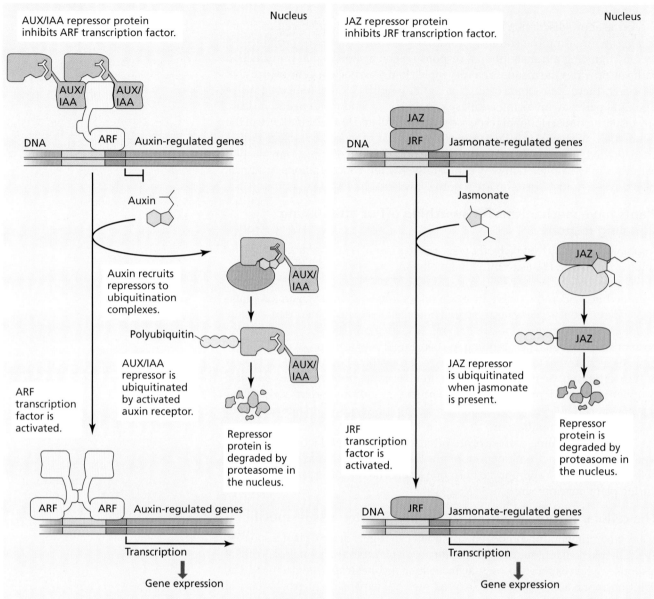

The interaction between signaling pathways has been termed **cross-regulation**, and three categories have been proposed (**Figure 12.20**):

- **Primary cross-regulation** involves distinct signaling pathways regulating a shared transduction component in a positive or a negative manner. An example of primary cross-regulation is the interaction of GA and light in the regulation of hypocotyl elongation in Arabidopsis. The signaling pathways of GA and light share a common signaling component, PIF3/4.

- **Secondary cross-regulation** involves the output of one signaling pathway regulating the abundance or perception of a second signal. An example of secondary cross-regulation is the interaction between auxin and ethylene in regulating root growth. In Arabidopsis ethylene inhibits root cell elongation by promoting auxin biosynthesis.

cross-regulation The interaction of two or more signaling pathways.

primary cross-regulation Involves distinct signaling pathways regulating a shared transduction component in a positive or a negative manner.

secondary cross-regulation Regulation by the output of one signal pathway of the abundance or perception of a second signal.

(C) Gibberellin response

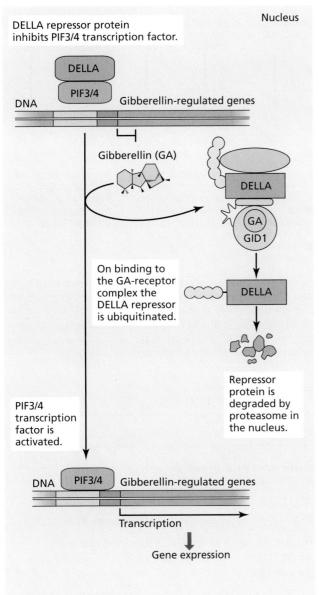

DELLA repressor protein inhibits PIF3/4 transcription factor.

Nucleus

DELLA

PIF3/4

DNA

Gibberellin-regulated genes

Gibberellin (GA)

DELLA

GA
GID1

On binding to the GA-receptor complex the DELLA repressor is ubiquitinated.

DELLA

Repressor protein is degraded by proteasome in the nucleus.

PIF3/4 transcription factor is activated.

DNA

PIF3/4

Gibberellin-regulated genes

Transcription

Gene expression

Figure 12.19 Signal transduction pathways in plants often function by inactivating repressor proteins. Several plant hormone receptors are part of ubiquitination complexes. (A) Auxin, (B) jasmonate (JA), and (C) gibberellin (GA) signal by promoting interaction between components of the ubiquitination machinery and repressor proteins (AUX/IAA, JAZ, and DELLA, respectively) operating in each hormone's signal transduction pathway. (A) Auxin and (B) JA directly promote interaction between the ubiquitin ligase complexes and the AUX/IAA and JAZ transcriptional repressors, respectively. The structural characteristics of the auxin response factors (ARFs) and AUX/IAA proteins that function in auxin signaling have been determined by X-ray crystallography and are reflected in the figure. The structural characteristics of the JAZ repressor protein have not yet been determined. (C) In contrast, gibberellin additionally requires a receptor protein, GID1 (from the name of the Arabidopsis loss-of-function mutant *gibberellin insensitive dwarf1* that lacks this receptor), to form the receptor complex. Responses to all three hormones involve the addition of multiple ubiquitins to repressor proteins, marking them for degradation. This activates auxin response factor (ARF), jasmonate response factor (JRF), and phytochrome interacting factor (PIF) transcription factors, resulting in auxin-, jasmonate-, and gibberellin-induced changes in gene expression.

- **Tertiary cross-regulation** involves the outputs of two distinct pathways exerting influences on one another. For example, cross-talk between auxin and cytokinin plays an important role in organogenesis. Each hormone influences the signaling pathway of the other in complex ways, with higher auxin concentrations promoting the differentiation of roots and higher cytokinin concentrations promoting the differentiation of shoots.

Thus, plant signaling is not based on a simple linear sequence of transduction events, but involves cross-regulation among many pathways. Understanding how such complex signaling pathways operate will demand a new scientific approach. This approach is often referred to as **systems biology** and employs mathematical and computational models to simulate these nonlinear biological networks and better predict their outputs.

tertiary cross-regulation Regulation that involves the outputs of two distinct signaling pathways exerting influences on one another.

systems biology An approach to examining complex living processes that employs mathematical and computational models to simulate nonlinear biological networks and to better predict their operation.

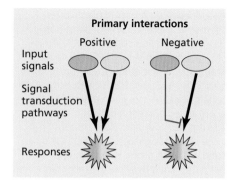

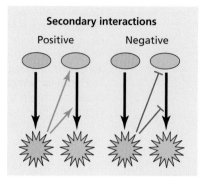

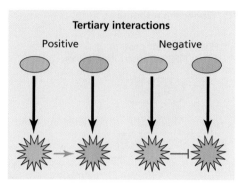

Two input pathways regulate a single shared protein or multiple shared proteins controlling a response. Both pathways have the same effect on the response.

Two input pathways converge on shared protein(s), but one of the pathways inhibits the effect of the other.

Two input pathways regulate separate responses. In addition, one pathway enhances the input levels or perception of the other pathway.

As in the positive interaction, except that one pathway represses the input levels or perception of the other pathway.

The response of one of the signaling pathways promotes the response of the other pathway.

The response of one of the signaling pathways inhibits the response of the other pathway.

Figure 12.20 Signal transduction pathways operate as part of a complex web of signaling interactions. Three types of cross-regulation have been proposed: primary, secondary, and tertiary. Input signals are shown here as ovals, signal transduction pathways are indicated by heavy arrows, and responses (pathway outputs) are shown as stars. Green (positive) or red (negative) colored lines indicate where one pathway influences the other pathway. The three types of cross-regulation can be either positive or negative.

Summary

Both short- and long-term physiological responses to external and internal signals arise from the transformation (transduction) of signals into mechanistic pathways. In order to activate areas that may be distal from the initial signaling location, signaling intermediates are amplified before dissemination (transmission). Once in play, signaling pathways often overlap into complex signaling networks, a phenomenon termed cross-regulation, to coordinate integrated physiological responses.

Temporal and Spatial Aspects of Signaling

• Plants use signal transduction to coordinate both rapid and slow responses to stimuli (**Figures 12.1, 12.2**).

Signal Perception and Amplification

• Receptors are present throughout cells (**Figure 12.3**).

• Signals can be amplified by phosphorylation and by second messengers such as Ca^{2+}, H^+, reactive oxygen species (ROS), and other small molecules (**Figures 12.4, 12.5**).

Hormones and Plant Development

• Hormones are conserved chemical messengers that can, at very low concentrations, transmit signals between cells and initiate physiological responses (**Figures 12.6, 12.7**).

• The first growth hormone to be identified was auxin, during studies of coleoptile bending due to phototropism (**Figure 12.8**).

• Studies on the "foolish seedling disease" of rice led to the discovery of the gibberellin group of growth hormones.

• Gibberellins can be used as growth-promoting regulators (**Figure 12.9**).

• Tissue-culture experiments revealed the role of cytokinins as cell division–promoting factors (**Figure 12.10**).

• Ethylene is a gaseous hormone that promotes fruit ripening and other developmental processes (**Figure 12.11**).

• Abscisic acid regulates seed maturation and stomatal closure in response to water stress (**Figure 12.12**).

• Brassinosteroids are lipid-soluble hormones that regulate many processes, including sex determination of flowers, elongation growth, and photomorphogenesis (**Figure 12.13**).

(Continued)

Summary (*continued*)

- Strigolactones are terpenoid lactone hormones that suppress branching and promote rhizosphere interactions.

- Salicylic acid and jasmonate are plant signaling molecules that are involved in plant defense responses.

Phytohormone Metabolism and Homeostasis

- The concentration of hormones is tightly regulated so that signals produce timely responses without compromising sensitivity to the same signal in the future (**Figure 12.14**).

- Most hormones are derived from multiple precursors and biosynthetic pathways (**Figure 12.15**).

Signal Transmission and Cell–Cell Communication

- Hormones can signal cells within, nearby, or far away from their site of synthesis.

- Plants can also employ fast-acting, long-distance electrical signaling using action potentials, though the transmission of such signals is poorly understood (**Figure 12.16**).

Hormonal Signaling Pathways

- Cytokinin and ethylene signaling pathways utilize two-component regulatory systems derived from bacterial signaling systems (**Figure 12.17**).

- Cytokinins, brassinosteroids, and ABA act through kinases and phosphatases to initiate signaling cascades (**Figure 12.18**).

- In contrast to animal hormone pathways, plant hormone pathways generally employ negative regulators (inactivating repressors), allowing faster activation of downstream response genes (**Figure 12.19**).

- Switching off signaling pathways is accomplished by degradation or sequestration of chemical signals via feedback mechanisms.

- Although hormones can effect a wide variety of responses, tissues exhibit response specificity.

- Integration of signal transduction pathways is accomplished through cross-regulation (**Figure 12.20**).

Suggested Reading

Davière, J.-M., and Achard, P. (2013) Gibberellin signaling in plants. *Development* 140: 1147–1151.

Hwang, I., Sheen, J., and Müller, B. (2012) Cytokinin signaling networks. *Annu. Rev. Plant Biol.* 63: 353–380.

Jiang, J., Zhang, C., and Wang, X. (2013) Ligand perception, activation, and early signaling of plant steroid receptor brassinosteroid insensitive 1. *J. Integr. Plant Biol.* 55: 1198–1211.

Ju, C., and Chang, C. (2012) Advances in ethylene signalling: Protein complexes at the endoplasmic reticulum membrane. *AoB Plants 2012: pls031.* DOI: 10.1093/aobpla/pls031

Santner, A., and Estelle, M. (2009) Recent advances and emerging trends in plant hormone signaling. *Nature* (Lond.) 459: 1071–1078.

Suarez-Rodriguez, M. C., Petersen, M., and Mundy, J. (2010) Mitogen-activated protein kinase signaling in plants. *Annu. Rev. Plant Biol.* 61: 621–649.

Wang, X. M. (2004) Lipid signaling. *Curr. Opin. Plant Biol.* 7: 329–336.

13 Signals from Sunlight

Sunlight serves not only as an energy source for photosynthesis, but also as a signal that regulates various developmental processes, from seed germination to fruit development and senescence (**Figure 13.1**). In addition, sunlight provides directional cues for plant growth as well as nondirectional cues for plant movements. We have already touched on several light-sensing mechanisms in the preceding chapters. In Chapter 9 we saw that chloroplasts move within leaf palisade cells to optimize light absorption (see Figure 9.13), a blue-light response. The leaves of many species are able to bend toward the sun during its progress across the sky, a phenomenon known as **solar tracking** (see Figure 9.5). As discussed in Chapters 3 and 6, stomata use blue light as a signal for opening, a sensory response that enables CO_2 to enter the leaf and initiates transpiration.

In later chapters we will encounter examples of light-regulated plant development. For example, many seeds require light to germinate. Sunlight inhibits stem growth and stimulates leaf expansion in growing seedlings, two of several light-induced phenotypic changes collectively referred to as **photomorphogenesis (Figure 13.2)**. Most of us are familiar with the observation that the branches of houseplants placed near a window grow toward the incoming light. This phenomenon, called **phototropism**, is an example of how plants alter their growth patterns in response to the direction of incident radiation. This response is stimulated primarily by blue light and is particularly important for seedling establishment (**Figure 13.3**; also see Chapter 15). In some species the leaves fold up at night (**nyctinasty**) and open at dawn (**photonasty**). Photonastic movements

solar tracking The movement of leaf blades throughout the day so that the planar surface of the blade remains perpendicular to the sun's rays.

photomorphogenesis The influence and specific roles of light on plant development. In the seedling, light-induced changes in gene expression that support aboveground growth in the light rather than belowground growth in the dark.

phototropism The alteration of plant growth patterns in response to the direction of incident radiation, especially blue light.

nyctinasty Sleep movements of leaves. Leaves extend horizontally to face the light during the day and fold together vertically at night.

photonasty Plant movements in response to nondirectional light.

Figure 13.1 Sunlight exerts multiple influences on plants. Plants expose their leaves to sunlight to transform solar energy into chemical energy, and they also use sunlight for a wide range of developmental signals that optimize photosynthesis and detect seasonal changes. (© Shosei/Corbis.)

are plant movements in response to nondirectional light. As we will discuss in Chapter 17, many plants flower at specific times of the year in response to changing day length, a phenomenon called **photoperiodism**.

In addition to visible light (**Figure 13.4**), sunlight also contains infrared and ultraviolet (UV) radiation, which can damage plant tissues. Many plants can sense the presence of UV radiation and protect themselves against cellular damage by synthesizing simple phenolics, flavonoids, and carotenoid compounds that act as sunscreens and remove damaging oxidants and free radicals generated by high-energy UV photons.

All of the photoresponses to visible light noted above, as well as the responses to UV radiation, involve receptors that detect specific wavelengths of light and induce developmental or physiological changes. As we saw in Chapter 12, hormone signal transduction involves a chain of reactions beginning with a hormone receptor and ending with a physiological response. The receptor molecules that plants use to detect sunlight are termed

Figure 13.2 Comparison of seedlings grown in the light versus the dark. (Left) Cress seedlings grown in the light. (Right) Cress seedlings grown in the dark. The light-grown seedlings exhibit photomorphogenesis. The dark-grown seedlings undergo etiolation, characterized by elongated hypocotyls and a lack of chlorophyll. (© Nigel Cattlin/Alamy.)

photoperiodism A biological response to the length and timing of day and night, making it possible for an event to occur at a particular time of year.

photoreceptors Proteins that sense the presence of light and initiate a response via a signaling pathway.

Figure 13.3 Time-lapse photograph of a maize (corn; *Zea mays*) coleoptile growing toward unilateral blue light given from the right. In the first image on the left, the coleoptile is about 3 cm long. The consecutive exposures were made 30 min apart. Note the increasing angle of curvature as the coleoptile bends. (Courtesy of M. A. Quiñones.)

photoreceptors. Like hormone receptors, photoreceptors undergo a conformational change when irradiated by a particular wavelength of light (perceived by the eye as color) to initiate signaling reactions that typically involve second messengers and phosphorylation cascades collectively referred to as photoresponses.

In this chapter we discuss the signaling mechanisms involved in light-regulated growth and development, focusing primarily on the receptors for red light (620–700 nm), far-red light (710–850 nm), blue light (350–500 nm), and UV-B radiation (290–320 nm).

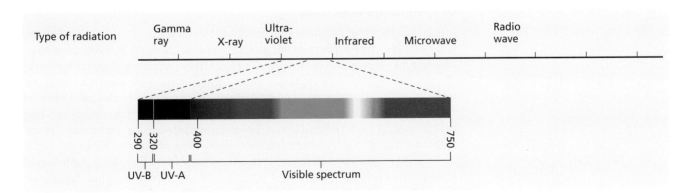

Figure 13.4 Plants can use visible light and UV-A and UV-B radiation as developmental signals (all wavelengths in nm).

chromophore   A light-absorbing pigment molecule that is usually bound to a protein (an apoprotein).

phytochromes   Plant growth-regulating photoreceptor proteins that absorb primarily red light and far-red light, but also absorb blue light. The holoprotein that contains the chromophore phytochromobilin.

cryptochromes   Flavoproteins implicated in many blue-light responses that have strong homology with bacterial photolyases.

phototropins   Blue-light photoreceptors that primarily regulate phototropism, chloroplast movements, and stomatal opening.

ZEITLUPE (ZTL)   A blue-light photoreceptor that regulates day-length perception (photoperiodism) and circadian rhythms.

UV RESISTANCE LOCUS 8 (UVR8)   The protein receptor that mediates various plant responses to UV-B irradiation.

de-etiolation   The rapid developmental changes associated with loss of the etiolated form due to the action of light.

Plant Photoreceptors

Pigments, such as chlorophyll and the accessory pigments of photosynthesis (see Chapter 7), are molecules that absorb visible light at specific wavelengths, and reflect or transmit the nonabsorbed wavelengths, which are perceived by us as colors. Unlike the photosynthetic pigments, photoreceptors absorb a photon of a given wavelength and use this energy as a signal that initiates a photoresponse. With the exception of UVR8 (discussed at the end of this chapter), all the known photoreceptors consist of a protein plus a light-absorbing prosthetic group (a nonprotein molecule attached to the photoreceptor protein) called a **chromophore**. The protein structures of the different photoreceptors vary, as does their sensitivity to light quantity (number of photons), light quality (wavelength dependency and associated action spectra), light intensity, and the duration of the exposure to light.

Among the photoreceptors that can promote photomorphogenesis in plants, the most important are those that absorb red and blue light. **Phytochromes** are photoreceptors that absorb red (600–750 nm) and far-red (710–850 nm) light most strongly, but they also absorb blue light (400–500 nm) and UV-A radiation (320–400 nm). As will be discussed later in the chapter, phytochromes mediate many aspects of vegetative and reproductive development, and are unique among plant photoreceptors in that they are photoreversible.

Three main classes of photoreceptors mediate the effects of UV-A/blue light: the **cryptochromes**, the **phototropins**, and the **ZEITLUPE** (**ZTL**) (German for "slow motion") family. All three of these receptor classes employ flavin molecules as chromophores. Cryptochromes, like the phytochromes, play a major role in plant photomorphogenesis, while phototropins primarily regulate phototropism, chloroplast movements, and stomatal opening. The ZEITLUPE photoreceptor plays roles in day-length perception and circadian (24-h cycling) rhythms. As in the case of hormone signaling, light signaling typically involves interactions between multiple photoreceptors and their signaling intermediates.

By convention, photoreceptors are designated in lower case (e.g., phy, cry, phot) when the holoprotein (protein plus chromophore) is described, and in upper case (PHY, CRY, PHOT) when the apoprotein (protein minus chromophore) is described. To be consistent with genetic conventions, we will use uppercase italics (*PHY, CRY, PHOT*) for the genes encoding the photoreceptor apoproteins.

Only the **UV RESISTANCE LOCUS 8** (**UVR8**) UV-B photoreceptor lacks a chromophore. A chromophore is unnecessary because high-energy UV-B photons excite the indole ring of exposed tryptophan residues in the photoreceptor protein to directly induce conformational changes and initiate photomorphogenic responses.

Photoresponses are driven by light quality or spectral properties of the energy absorbed

It can be difficult to separate specific photoresponses in plants exposed to the full solar spectrum during normal growth, because many photoreceptors may be absorbing energy at the same time. Changes in the sun's angle over the course of a day usually expose a plant to varying light intensities and wavelength mixtures. Similarly, the process of **de-etiolation**, characterized by the production of chlorophyll in dark-grown (etiolated) (see Figure 13.2) seedlings when exposed to light, results from the coaction of phytochrome absorbing red light and cryptochrome absorbing blue light from sunlight. How, then, can we functionally distinguish responses to individual photoreceptors? In many cases, a contribution from photosynthesis cannot even be excluded, since photosynthetic pigments also absorb red light and blue light.

To determine which wavelengths of light are necessary to bring about a particular plant response, photobiologists typically produce what is known as an action spectrum under carefully defined lighting conditions using precise light-emitting diodes (LEDs) and filters. Action spectra describe the wavelength specificity of a biological response within the spectral range of normal sunlight. Each photoreceptor differs in its atomic composition and arrangement and thus exhibits different absorption characteristics. In Chapter 7, the action spectrum for photosynthetic O_2 evolution shown in Figure 7.8 is a graph that plots the magnitude of a photosynthetic light response as a function of wavelength. Similarly, the action spectra of other photoresponses can then be compared with the absorption spectra of candidate photoreceptors that have been studied in isolation. The action spectra of wild-type plants can be compared to action spectra obtained from plants with mutations in genes encoding specific photoreceptors. If a particular response is missing or diminished in a particular mutant, then it is likely to be associated with the associated photoreceptor.

These methods have been used to identify photoreceptors involved in signaling pathways. For example, red light stimulates seed germination in many species, and far-red light inhibits it (shown for lettuce in **Figure 13.5**). The action spectra for these two antagonistic effects of light on Arabidopsis seed germination are shown in **Figure 13.6A**. Stimulation shows a peak in the red region (660 nm), while inhibition has a peak in the far-red region (720 nm). Under these treatments, the purified photoreceptor phytochrome exhibits a very similar **photoreversibility (Figure 13.6B)**, suggesting that phytochrome is the receptor for these responses. The close correspondence between the action and absorption spectra of phytochrome was used to confirm its identity as the photoreceptor involved in regulating seed germination, and also established that the red/far-red reversibility of seed germination was due to the photoreversibility of phytochrome itself.

Similarly, action spectra for blue light–stimulated phototropism, stomatal movements, and other key blue-light responses all exhibit a peak in the UV-A region (at 370 nm) and a peak in the blue region (400–500 nm) that has a characteristic

photoreversibility The interconversion of the Pr and Pfr forms of phytochrome.

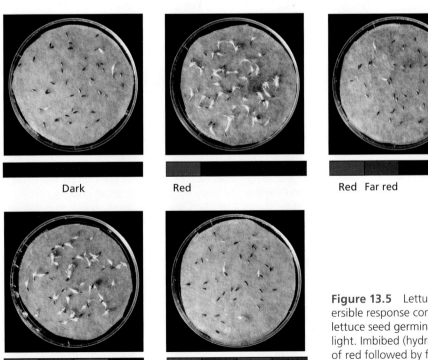

Dark

Red

Red Far red

Red Far red Red

Red Far red Red Far red

Figure 13.5 Lettuce seed germination is a typical photoreversible response controlled by phytochrome. Red light promotes lettuce seed germination, but this effect is reversed by far-red light. Imbibed (hydrated) seeds were given alternating treatments of red followed by far-red light. The effect of the light treatment depended on the last treatment given. Very few seeds germinated following the last far-red treatment. (Photos by David McIntyre.)

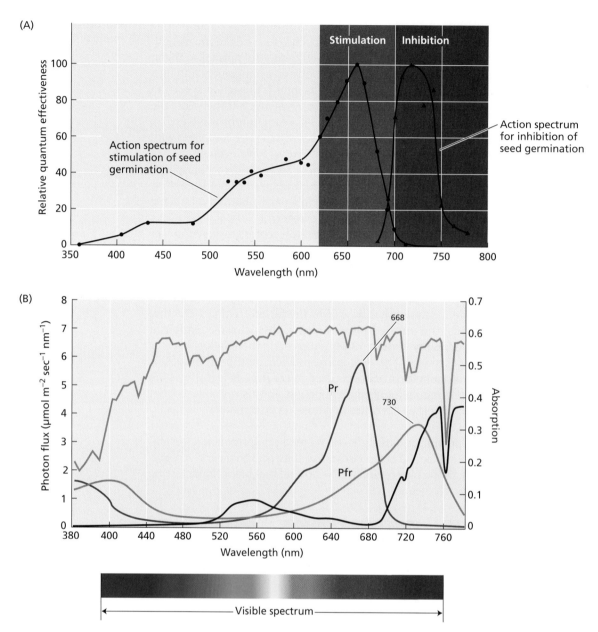

Figure 13.6 The action spectrum of phytochrome function matches its absorption spectrum. (A) Action spectra for the photoreversible stimulation and inhibition of seed germination in Arabidopsis. (B) Absorption spectra of purified oat phytochrome in the Pr (red line) and Pfr (green line) forms overlap. At the top of the canopy there is a relatively uniform distribution of visible-spectrum light (blue line), but under a dense canopy much of the red light is absorbed by plant pigments, resulting in transmittance of mostly far-red light. The black line shows the spectral properties of light that is filtered through a leaf. Thus, the relative proportions of Pr and Pfr are determined by the degree of vegetative shading in the canopy. (A after Shropshire et al. 1961; B after Kelly and Lagarias 1985, courtesy of Patrice Dubois.)

"three-finger" fine structure (**Figure 13.7A**), suggesting a common photoreceptor. The absorption spectrum for the light-sensing domain of phototropin, which contains the chromophore flavin mononucleotide (FMN), is almost identical to the action spectrum for phototropism (**Figure 13.7B**), consistent with phototropin acting as the photoreceptor for these responses.

Figure 13.7 The action spectrum of phototropism matches the absorption spectrum of the light-sensing domain of phototropin. (A) Action spectrum for blue light–stimulated phototropism in alfalfa hypocotyls. The "three-finger" pattern in the 400- to 500-nm region is characteristic of many blue-light responses. (B) The absorption spectrum of the light-sensing domain of phototropin. (A after Baskin and Iino 1987; B after Briggs and Christie 2002.)

(A)

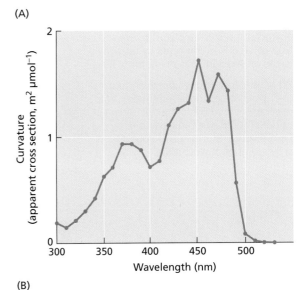

(B)

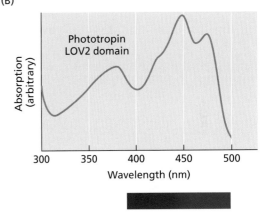

Plant responses to light can be distinguished by the amount of light required

Light responses can also be distinguished by the amount of light required to induce them. The amount of light is referred to as the **fluence**, which is defined as the total number of photons striking a unit of surface area. Total fluence = fluence rate × the length of time (duration) of irradiation. The standard units for fluence are micromoles of quanta (photons) per square meter ($\mu mol\ m^{-2}$). Some responses are sensitive not only to the total fluence but also to the **irradiance**, or fluence rate, of light. The units of irradiance are micromoles of quanta per square meter per second ($\mu mol\ m^{-2}\ s^{-1}$) (see Chapter 9).

Because photochemical responses are stimulated only when a photon is absorbed by its photoreceptor, there can be a difference between incident irradiation and absorption. For example, in photosynthesis the apparent quantum efficiency is assessed as the electron transport rate or total carbon assimilation as a function of the incident photosynthetic photon flux density (see Chapter 9). However, this measure underestimates the *actual* quantum efficiency because not all the incident photons are absorbed. This caveat is also important in assessing the dose response of the photomorphogenic responses of green plants to red or blue light, because much of the light is absorbed by chlorophyll. The same principle applies to the responses to UV radiation, since the epidermis may absorb just under 100% of the incident UV radiation. Thus, the amount of radiation required to induce a photoresponse may be quite high based on the amount of incident irradiation required, and quite low based on actual photon absorption by the photoreceptor.

Phytochromes

Phytochromes were first identified in flowering plants as the photoreceptors responsible for photomorphogenesis in response to red and far-red light. However, they are members of a gene family present in the Characaean algae and all land plants (bryophytes, ferns, fern allies, and seed plants) as well as in cyanobacteria, other bacteria, fungi, and diatoms.

Because neither red nor far-red light penetrate water to depths greater than a few meters, phytochrome would appear to be less useful as a photoreceptor for aquatic organisms. However, recent studies have shown that different algal phytochromes can sense orange, green, or even blue light, suggesting that phytochromes have the potential to be spectrally tuned during natural selection to absorb different wavelengths.

Phytochrome is the primary photoreceptor for red and far-red light

The phytochrome family of photoreceptors are red- and far-red-sensing proteins with a covalently bound bilin chromophore, termed phytochromobilin (PϕB). Purified phytochrome is a cyan-blue (midway between green and blue) or

fluence The number of photons absorbed per unit of surface area over time ($\mu mol\ m^{-2}\ s^{-1}$).

irradiance The amount of energy that falls on a flat surface of known area per unit of time. Expressed as watts per square meter ($W\ m^{-2}$). Time (seconds) is contained within the term watt: 1 W = 1 joule (J) s^{-1}, or as micromoles of quanta per square meter per second ($\mu mol\ m^{-2}\ s^{-1}$), also referred to as fluence rate.

Pr   The red light–absorbing form of phytochrome. This is the form in which phytochrome is assembled. The cyan-blue-colored Pr is converted by red light to the far-red light–absorbing form, Pfr.

Pfr   The far-red light–absorbing form of phytochrome, converted from Pr by the action of red light. The cyan-green-colored Pfr is converted back to Pr by far-red light. Pfr is the physiologically active form of phytochrome.

cyan-green protein with a molecular mass of about 125 kilodaltons (kDa). Phytochrome proteins function as dimers of either the red light–absorbing inactive form, called **Pr**, or the far-red light–absorbing active form, **Pfr**.

Many of the biological properties of phytochrome were established in the 1930s through studies of red light–induced morphogenic responses, especially seed germination. A key breakthrough in the history of phytochrome was the discovery that the effects of red light (620–700 nm) could be reversed by a subsequent irradiation with far-red light (710–850 nm). When the absorption spectra of each of the two forms of phytochrome are measured separately in a spectrophotometer designed to study photoreversible molecules, they correspond closely to the action spectra for the stimulation and inhibition of seed germination, respectively. The reversibility of the red and far-red responses ultimately led to the discovery that a single photoreversible photoreceptor, phytochrome, was responsible for both activities. It was subsequently demonstrated that the two forms of phytochrome (Pr and Pfr) could be distinguished spectrophotometrically (see Figure 13.6B).

Phytochrome can interconvert between Pr and Pfr forms

In dark-grown, or etiolated, seedlings, phytochrome is present in the red light–absorbing Pr form. This cyan-blue-colored inactive form is converted by red light to the far-red light–absorbing Pfr form.

$$\text{Pr} \; \underset{\text{Far-red light}}{\overset{\text{Red light}}{\rightleftharpoons}} \; \text{Pfr}$$

Pfr is pale cyan-green in color and is considered to be the active form of phytochrome. Upon photoconversion to Pfr, phytochrome can be rapidly degraded by the ubiquitin pathway, the rate of degradation being dependent on the type of phytochrome (discussed later in the chapter). Alternatively, Pfr can revert back to inactive Pr in darkness, but this is a relatively slow process. However, Pfr can be rapidly converted to Pr by irradiation with far-red light. A schematic diagram of the conformational change involved in phytochrome photoreversibility is shown in **Figure 13.8**.

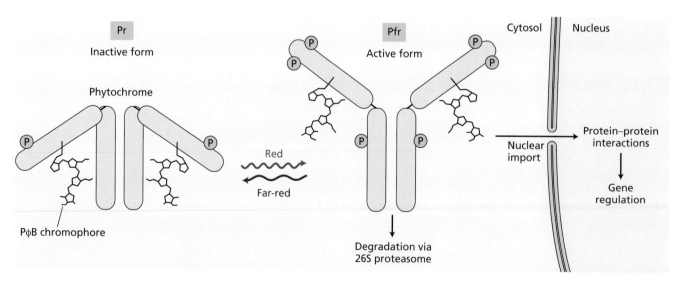

Figure 13.8   Diagram of phytochrome photoreversibility by red and far-red light. Upon perception of red light, the inactive Pr form is converted to the active Pfr form. Pfr is transported to the nucleus, where it participates in direct protein–protein interactions that result in repression or de-repression of downstream genes.

Active Pfr can revert back to inactive Pr, either rapidly following far-red irradiation or slowly upon removal of red light, or it may be sent to the 26S proteasome for degradation. PφB = the phytochrome chromophore phytochromobilin. (After art courtesy of Candace Pritchard.)

In natural settings, plants growing outdoors are exposed to a much broader spectrum of light, and the ratio of red light to far-red light varies with the conditions. Red light is abundant in direct sunlight, while far-red light is more abundant under foliage canopies in which chlorophyll has absorbed much of the incident red light (see Figure 13.6B). Plants growing beneath a canopy can use the R:FR ratio of light to regulate such processes as seed germination and shade avoidance (see Chapters 15 and 16). As we will see, phytochrome-mediated responses also play a major role in the photoperiodic control of flowering (see Chapter 17).

Phytochrome Responses

The variety of different phytochrome responses in intact plants is extensive, in terms of both the kinds of responses (**Table 13.1**) and the quantity of light needed to induce the responses. A few responses are very rapid, such as the light-induced changes in the surface potentials (the membrane potential measured extracellularly) of oat roots or coleoptiles, which occur within seconds. Such rapid changes in surface potentials result when phytochrome interacts with cytosolic factors at or near the plasma membrane and initiates ion fluxes. However, the majority of phytochrome responses occur over a longer period of time. These slower responses affect longer-term developmental events, such as growth and organ movements.

Phytochrome responses vary in lag time and escape time

Morphological responses to the photoactivation of phytochrome are often observed visually after a *lag time*—the time between stimulation and the observed response. The lag time may be as brief as a few minutes or as long as several weeks. These differences in response times result from the multiple signal transduction pathways that function downstream of phytochrome signaling as well as interactions with other developmental mechanisms. The more rapid of these responses are usually reversible movements of organelles or reversible volume changes (swelling, shrinking) in cells, but even some growth responses are remarkably fast. For instance, red-light inhibition of the stem elongation rate of light-grown pigweed (*Chenopodium album*) and Arabidopsis is observed within minutes after the proportion of Pfr to Pr in the stem is increased. However, lag

Table 13.1   Typical photoreversible responses induced by phytochrome in a variety of higher and lower plants

Group	Genus	Stage of development	Effect of red light
Angiosperms	*Lactuca* (lettuce)	Seed	Promotes germination
	Avena (oat)	Seedling (etiolated)	Promotes de-etiolation (e.g., leaf unrolling)
	Sinapis (mustard)	Seedling	Promotes formation of leaf primordia, development of primary leaves, and production of anthocyanin
	Pisum (pea)	Adult	Inhibits internode elongation
	Xanthium (cocklebur)	Adult	Inhibits flowering (photoperiodic response)
Gymnosperms	*Pinus* (pine)	Seedling	Enhances rate of chlorophyll accumulation
Pteridophytes	*Onoclea* (sensitive fern)	Young gametophyte	Promotes growth
Bryophytes	*Polytrichum* (moss)	Germling	Promotes replication of plastids
Chlorophytes	*Mougeotia* (alga)	Mature gametophyte	Promotes orientation of chloroplasts to directional dim light

times of several weeks are observed for the induction of flowering in Arabidopsis and other species.

Variety in phytochrome responses can also be seen in the phenomenon called **escape from photoreversibility**. Red light–induced events are reversible by far-red light for only a limited period of time, after which the response is said to have "escaped" from reversal control by light. This escape phenomenon can be explained by a model based on the assumption that phytochrome-controlled morphological responses are the end result of a multistep sequence of linked biochemical reactions in the responding cells. Early stages in the sequence may be fully reversible by removing Pfr, but at some point in the sequence a point of no return is reached, beyond which the reactions proceed irreversibly toward the response. The escape time therefore represents the amount of time it takes before the overall sequence of reactions becomes irreversible—essentially, the time it takes for Pfr to complete its primary action. The escape time for different phytochrome responses ranges remarkably, from less than a minute to hours.

Phytochrome responses fall into three main categories based on the amount of light required

As **Figure 13.9** shows, phytochrome responses fall into three major categories based on the amount of light they require: very low fluence responses (VLFRs), low-fluence responses (LFRs), and high-irradiance responses (HIRs). VLFRs and LFRs have a characteristic range of light fluences within which the magnitude of the response is proportional to the fluence. HIRs, by contrast, are proportional to the irradiance.

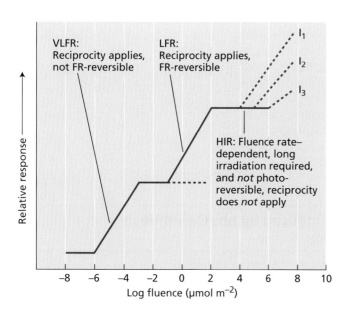

Figure 13.9   Three types of phytochrome responses, based on their sensitivities to fluence. The relative magnitudes of representative responses are plotted against increasing fluences of red light. Short light pulses are sufficient to activate very low fluence responses (VLFRs) and low-fluence responses (LFRs). The lower dashed line indicates overlap between VLFRs and LFRs. High-irradiance responses (HIRs) are mainly proportional to the irradiance. The effects of three different irradiances given continuously are illustrated ($I_1 > I_2 > I_3$). At each irradiance level the response increases with increasing fluence, indicating HIRs also respond to fluence as well as irradiance. (After Briggs et al. 1984.)

VERY LOW FLUENCE RESPONSES (VLFRs)   Some phytochrome responses can be initiated by fluences as low as 0.0001 µmol m^{-2} (a few seconds of starlight, or one-tenth of the amount of light emitted by a firefly in a single flash), and they become saturated (i.e., reach a maximum) at about 0.05 µmol m^{-2}. For example, Arabidopsis seeds can be induced to germinate with red light in the range of 0.001 to 0.1 µmol m^{-2}. In dark-grown oat (*Avena* spp.) seedlings, red light can stimulate the growth of the coleoptile and inhibit the growth of the mesocotyl (the elongated axis between the coleoptile and the root) at similarly low fluences.

VLFRs are non-photoreversible. The VLFR class of phytochrome responses, which occurs only in deep-buried seeds and seedlings, is mediated by phyA (discussed below), which is especially abundant in dark-grown tissues. VLFRs are nonreversible by far-red light because, at the extremely low light intensities involved, the amount of far-red-induced reversion of Pfr to Pr is insignificant compared with other mechanisms of Pfr degradation. Although VLFRs are non-photoreversible, the action spectra for VFLR responses (e.g., seed germination) are similar to those for LFR responses (discussed next), supporting the view that phytochrome is the photoreceptor involved in VFLRs.

LOW-FLUENCE RESPONSES (LFRs)   Another set of phytochrome responses cannot be initiated until the fluence reaches 1.0 µmol m^{-2}, and they are saturated at about 1000 µmol m^{-2}. These LFRs include processes such as the promotion of seed germination, inhibition of hypocotyl elongation, and regulation of leaf movements. As we saw in Figure 13.6A, the LFR action spectrum for Arabidopsis seed germination includes a main peak for stimulation in the red region (660 nm) and a major peak for inhibition in the far-red region (720 nm).

Both VLFRs and LFRs can be induced by brief pulses of light, provided that the total amount of light energy adds up to the required fluence. As we noted above, the total fluence is a function of two factors: the fluence rate (μmol m^{-2} s^{-1}) and the time of irradiation. Thus, a brief pulse of red light will induce a response, provided that the light is sufficiently bright; conversely, very dim light will work if the irradiation time is long enough. For both VLFRs and LFRs, the magnitude of the response (e.g., percent germination or degree of inhibition of hypocotyl elongation) exhibits **reciprocity**, or dependence on the product of the fluence rate and the time of irradiation.

HIGH-IRRADIANCE RESPONSES (HIRs)   Phytochrome responses of the third type, HIRs, require prolonged or continuous exposure to light of relatively high irradiance and do not exhibit reciprocity. The response is proportional to the irradiance until the response saturates and additional light has no further effect. The reason these responses are called high-irradiance responses rather than high-fluence responses is that they are mainly proportional to the fluence rate—the number of photons striking the plant tissue per second—rather than to the fluence—the total number of photons striking the plant in a given period of illumination. HIRs saturate at much higher fluences than LFRs—at least 100 times higher. Because neither continuous exposure to dim light nor transient exposure to bright light can induce HIRs, these responses do not show reciprocity.

Many of the LFRs also qualify as HIRs. For example, at low fluences the action spectrum for anthocyanin production in seedlings of white mustard (*Sinapis alba*) is indicative of phytochrome and shows a single peak in the red region of the spectrum. The effect is reversible with far-red light (a photochemical property unique to phytochrome), and the response shows reciprocity. However, if the dark-grown seedlings are instead exposed to high-irradiance light for several hours, the action spectrum contains peaks in the far-red and blue regions, the effect is no longer photoreversible, and the response becomes proportional to the irradiance. Thus, the same effect can be either an LFR or an HIR, depending on the history of a seedling's exposure to light.

Phytochrome A mediates responses to continuous far-red light

The different types of phytochrome responses to light dosage suggest the presence of multiple phytochrome isoforms with varying sensitivities to light. Arabidopsis contains five phytochrome isoforms (phyA–phyE). The process of Pfr degradation is conserved among all known isoforms, and phyA, which controls plant VLFRs and far-red light-induced HIRs, is rapidly degraded as Pfr by the ubiquitin pathway.

In contrast, phyB–phyE are more light-stable. Experimental evidence has shown that phyB regulates LFRs such as photoreversible seed germination. For example, wild-type Arabidopsis seeds require light for germination, and the response shows red/far-red reversibility in the low-fluence range (see Figure 13.6A). Mutants that lack phyA respond normally to red light, whereas mutants deficient in phyB are unable to respond to low-fluence red light. This experimental evidence strongly suggests that phyB mediates photoreversible seed germination.

PhyB also plays an important role in regulating plant responses to shade. Plants that are deficient in phyB often look like wild-type plants growing under dense vegetative canopies. In fact, mediating shade responses such as accelerated flowering and increased elongation growth may be one the most ecologically important roles of phytochromes. The isoforms phyC, phyD, and phyE play

reciprocity   According to the Law of Reciprocity, treating plants with a brief duration of bright light will induce the same photobiological response as treating them with a long duration of dim light.

phytochrome interacting factors (PIFs) A family of phytochrome-interacting proteins that may activate and repress gene transcription; some PIFs are targets for phytochrome-mediated degradation.

phytochrome kinase substrates (PKSs) Proteins that participate in the regulation of phytochrome via direct phosphorylation or via phosphorylation by other kinases.

unique roles in regulating other responses to red and far-red light. Responses mediated by phyD and phyE include petiole and internode elongation and the control of flowering time (see Chapter 17).

Phytochrome regulates gene expression

All phytochrome-regulated changes in plants begin with absorption of light by the photoreceptor. After light absorption, the molecular properties of phytochrome are altered, affecting the interaction of the phytochrome protein with other cellular components that ultimately bring about changes in the growth, development, or position of an organ. As noted earlier, phytochrome in the cytosol interacts with other cytosolic factors to alter plasma membrane ion fluxes. Such rapid changes in ion fluxes can cause rapid changes in turgor pressure in specialized structures called pulvini (singular pulvinus), triggering the movements of leaves and leaflets (see Figure 12.15).

However, while a significant portion of phytochrome remains in the cytosol, the majority of phytochrome signaling occurs in the nucleus, where it alters gene expression via a wide range of regulatory elements and transcription factors. Monitoring gene expression profiles over time following a shift of plants from darkness to light has led to the identification of early and late targets of *PHY* gene action. Nuclear import of phyA and phyB is highly correlated with the light quality that stimulates their activities. That is, nuclear import of phyA is activated by either red or far-red light, or low-fluence broad-spectrum light, whereas phyB import is driven by red-light exposure and is reversible by far-red light. Nuclear import of the phytochrome proteins represents a major control point in phytochrome signaling.

A family of proteins known as **phytochrome interacting factors (PIFs)** act primarily as negative regulators of various aspects of phytochrome-mediated photomorphogenesis, including seed germination, chlorophyll biosynthesis, shade avoidance, and hypocotyl elongation. Red light–induced Pfr formation initiates the degradation of PIF proteins by phosphorylation, followed by degradation via the proteasome complex (see Chapter 1). The rapid degradation of PIFs may provide a mechanism for modulating light responses that is tightly coupled to the activities of phy proteins. Phytochrome activity is further modulated by regulation of phosphorylation by a family of **phytochrome kinase substrate (PKS)** proteins.

As discussed in Chapter 12, the majority of plant signal transduction pathways involve the inactivation, degradation, or removal of repressor proteins. The phytochrome signaling pathway is consistent with this general principle. As described previously, phyA is rapidly degraded following its activation by light. Thus, protein degradation, in addition to phosphorylation, is emerging as a ubiquitous mechanism regulating many cellular processes, including light and hormone signaling, circadian rhythms, and flowering time (for examples, see Chapter 17).

Blue-Light Responses and Photoreceptors

Blue-light responses have been reported in seed plants, ferns, bryophytes, algae, fungi, and prokaryotes. In addition to phototropism, these responses include inhibition of seedling hypocotyl elongation, stimulation of chlorophyll and carotenoid synthesis, and activation of gene expression. Among motile unicellular organisms such as certain algae and bacteria, blue light mediates *phototaxis*, the movement of unicellular organisms toward or away from light. Some blue-light responses were introduced in relation to photosynthesis in Chapters 6, 7 and 9, including chloroplast movements, solar tracking, and stomatal opening. In Chapter 15 we will discuss photomorphogenesis and phototropic responses to blue light in the context of seed germination and seedling establishment.

As noted earlier, the effects of UV-A/blue light (320–500 nm) are mediated by three distinct classes of photoreceptors that use flavin molecules as chromophores: the cryptochromes, phototropins, and ZEITLUPE proteins. Cryptochromes interact with phytochromes to regulate photomorphogenesis. Phototropins (phots), by contrast, are involved in directing organ, chloroplast, and nuclear movements, solar tracking, and stomatal opening, all of which are light-dependent processes that optimize the photosynthetic efficiency of plants. The ZEITLUPE family has been shown to participate in the control of circadian clocks and flowering.

Blue-light responses have characteristic kinetics and lag times

The inhibition of stem elongation and the stimulation of stomatal opening by blue light demonstrate that blue-light responses can be relatively rapid compared with most photomorphogenic changes. Whereas typical photosynthetic responses begin the moment the chloroplast encounters photosynthetically active photons and cease as soon as the light goes off, blue-light responses exhibit a lag time of variable duration and proceed at maximum rates for several minutes after application of a light pulse. For example, blue light induces a decrease in growth rate and a transient membrane depolarization in etiolated cucumber seedlings only after a lag time of about 25 s (**Figure 13.10**).

The persistence of blue-light responses in the absence of blue light has been studied using blue-light pulses. In guard cells, blue light–induced activation of the H^+-ATPase gradually decays following a pulse of blue light, but only after several minutes have elapsed (see Figure 6.19). This persistence of the blue-light response after the pulse can be explained by a photochemical cycle in which the physiologically active form of the photoreceptor (in this case, phototropin), which has been converted from the inactive form by blue light, slowly reverts to the inactive form after the blue light is switched off. The rate of decay of the response to a blue-light pulse depends on the time course of the reversion of the active form of the photoreceptor back to the inactive form.

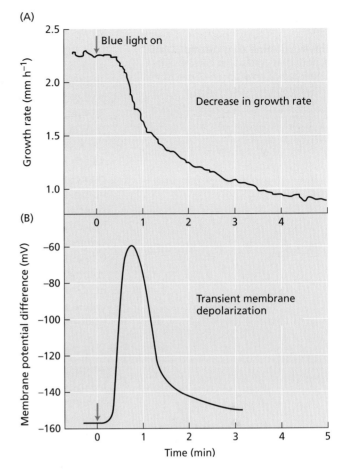

Figure 13.10 (A) Blue light–induced changes in elongation rates of etiolated cucumber hypocotyls. (B) Blue light–induced transient membrane depolarization of hypocotyl cells. (After Spalding and Cosgrove 1989.)

Cryptochromes

Cryptochromes are blue-light photoreceptors that mediate several blue-light responses, including suppression of hypocotyl elongation, promotion of cotyledon expansion, membrane depolarization, inhibition of petiole elongation, anthocyanin production, and circadian clock entrainment. The cryptochrome photoreceptor was originally identified in Arabidopsis using genetic screens for mutants whose hypocotyls were elongated when grown in white light because they lacked the light-stimulated inhibition of hypocotyl elongation described previously. The long hypocotyl phenotype of one cryptochrome mutant was shown to occur in blue, but not red, light. This indicated that the loss of function did not directly involve phytochrome. Cryptochrome proteins have been discovered in many organisms, including ferns, algae, cyanobacteria, fruit flies, mice, and humans.

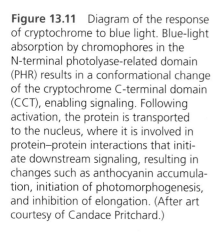

flavin adenine dinucleotide (FAD)
A riboflavin-containing cofactor that undergoes a reversible two-electron reduction to produce $FADH_2$.

Blue-light irradiation of the cryptochrome FAD chromophore causes a conformational change

Blue light–absorbing cryptochrome proteins have an N-terminal photolyase-related domain (PHR) that binds both the **flavin adenine dinucleotide (FAD)** and the pterin 5,10-methyltetrahydrofolate (MTHF). Pterins are light-absorbing pteridine derivatives often found in pigmented cells of insects, fishes, and birds. In photolyases, blue light is absorbed by the pterin, and the excitation energy is then transferred to FAD.

A similar mechanism is thought to operate in plant cryptochromes (**Figure 13.11**). Blue-light absorption alters the redox status of the bound FAD chromophore, and it is this primary event that triggers photoreceptor activation. As occurs in phytochromes and phototropins, this activation mechanism involves protein conformational changes. In the cryptochromes, light absorption by the N-terminal photolyase region appears to alter the conformation of a C-terminal extension, the cryptochrome C-terminal domain (CCT) (see Figure 13.11). This C-terminal extension is absent from photolyase enzymes but is clearly essential for cryptochrome signaling. We can therefore view plant cryptochrome as a molecular light switch whereby absorption of blue photons at the N-terminal photosensory region results in protein conformational changes at the C terminus, which, as we will see, initiates signaling by altering binding to specific partner proteins.

The nucleus is a primary site of cryptochrome action

Although the cryptochrome isoform that regulates hypocotyl elongation is stable in blue light, a second isoform that regulates cotyledon expansion and floral induction (see Chapter 17) is preferentially degraded under blue light. Both cryptochrome isoforms play key roles in regulating plant circadian rhythms. Cryptochrome homologs have also been implicated in the control of circadian clocks in flies, mice, and humans.

Cryptochromes are found in both the cytosol and nucleus. Rapid responses associated with growth, such as membrane depolarization, which occurs in 2–3 s, are associated with cytosolic cryptochromes. However, most of the cryptochrome of the cell is localized in the nucleus. Nuclear cryptochromes are involved in photomorphogenesis and act by stabilizing transcription factors and inhibiting their proteolytic degradation.

Cryptochrome interacts with phytochrome

In Arabidopsis, continuous blue or far-red light promotes flowering, and red light inhibits flowering. Far-red light acts through phyA, and the antagonistic effect of red light is produced by phyB. One might expect cryptochrome mutants to be delayed in flowering, since blue light promotes flowering. However, cryptochrome mutants flower at the same time as the wild type under either continuous

Figure 13.11 Diagram of the response of cryptochrome to blue light. Blue-light absorption by chromophores in the N-terminal photolyase-related domain (PHR) results in a conformational change of the cryptochrome C-terminal domain (CCT), enabling signaling. Following activation, the protein is transported to the nucleus, where it is involved in protein–protein interactions that initiate downstream signaling, resulting in changes such as anthocyanin accumulation, initiation of photomorphogenesis, and inhibition of elongation. (After art courtesy of Candace Pritchard.)

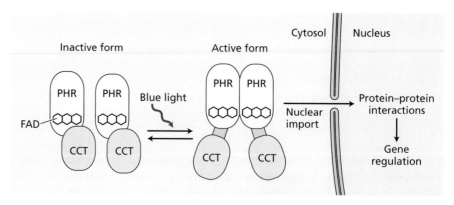

blue or continuous red light. A delay is observed only if both blue light and red light are given together, suggesting that the phytochrome and cryptochrome signaling pathways converge.

As noted previously in this chapter, several plant processes show oscillations of activity that roughly correspond to a 24-h, or circadian, cycle. This endogenous rhythm uses an oscillator that must be **entrained** (synchronized) to the daily light–dark cycles of the external environment. In experiments designed to characterize the role of photoreceptors in this process, phytochrome- and cryptochrome-deficient Arabidopsis mutants were crossed with lines carrying a reporter gene that was regulated by the circadian clock. The pace of the oscillator was slowed (i.e., the period length increased) when *phyA* mutants were grown under dim red light, whereas in *phyB* mutants, timing defects were only observed under high-irradiance red light. Similarly, cryptochrome-deficient mutants exhibited timing defects only when entrained using blue light. These studies indicate that both phytochromes and cryptochromes entrain the circadian clock in Arabidopsis.

Phototropins

Phototropins are the blue light photoreceptors that mediate seedling phototropism, chloroplast movement, guard cell opening, and some leaf movements. Angiosperms contain two phototropin genes, *PHOT1* and *PHOT2*. phot1 is the primary phototropic receptor that mediates phototropism in response to both low and high fluence rates of blue light. phot2 mediates phototropism primarily in response to high light intensities. Similar overlaps in the functions of the phot1 and phot2 photoreceptors are observed for other blue-light responses, including chloroplast movements, stomatal opening, leaf movements, and leaf expansion. Together with phototropism, these processes optimize light capture and CO_2 uptake for photosynthesis. Consequently, growth of phototropin-deficient mutants is severely compromised, particularly under low light intensities.

In contrast to cryptochromes, which are predominantly localized in the nucleus, phototropin receptors are associated with the plasma membrane, where they function as light-activated serine/threonine kinases. A schematic of blue-light induction of conformational changes in phototropins is shown in **Figure 13.12**. Phototropins contain one or two light-sensing **LIGHT-OXYGEN-VOLTAGE (LOV) domains**, each binding a chromophore flavin mononucleotide (FMN). Only one LOV domain actively participates in photoreception. When present, a second LOV domain functions in receptor dimerization. Spectroscopic studies

entrainment The synchronization of the period of biological rhythms by external controlling factors, such as light and darkness.

LIGHT-OXYGEN-VOLTAGE (LOV) domains Highly conserved protein domains that respond to light, oxygen or voltage changes to change receptor protein conformation. In phototropins, LOV domains are sites of binding of the FMN chromophore to phototropins and are thus the part of the protein that senses light.

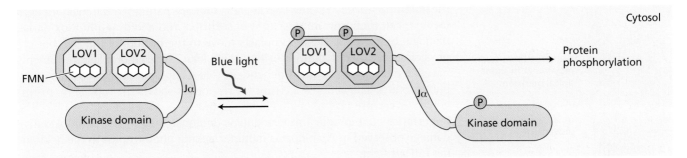

Figure 13.12 Diagram of the blue light–induced conformational change of phototropin. Phototropin proteins consist of two light-oxygen-voltage (LOV) domains that each have a bound flavin mononucleotide (FMN) chromophore, a C-terminal serine/threonine kinase domain, and a α-helical region (Jα) that joins the N-terminal and C-terminal domains. Sensing of blue light by the LOV domains results in autophosphorylation and a dramatic conformational change that exposes the kinase domain. Direct phosphorylation of target proteins via the kinase domain results in changes such as stomatal opening, phototropism, and chloroplast relocalization. (After art courtesy of Candace Pritchard.)

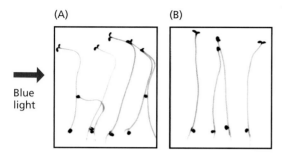

Figure 13.13 Phototropism in Arabidopsis seedlings can be used as a bioassay for phototropin activity. (A) Wild-type plants respond by bending toward the light. (B) A mutation in the LOV2 photoreceptor domain of phot1 abolishes the response, demonstrating that only the LOV2 domain is required for phototropism. (From Christie et al. 2002.)

have shown that in the dark, one FMN molecule is noncovalently bound to each LOV domain. Photoexcitation of the FMN–LOV domain "uncages" the kinase domain and leads to autophosphorylation on multiple serine residues, which is required for phot kinase activity. Blue light activation of phototropin can be reversed by a dark treatment. Dark inactivation of phototropin is mediated by a type 2A protein phosphatase that dephosphorylates the serine residues.

The ZEITLUPE family of photoreceptors also contain LOV domains with FMN cofactors, but they are F-box proteins. ZEITLUPE proteins regulate transcription factor stability via protein ubiquitination in a manner similar to that of hormone receptors (see Chapter 12).

Phototropism requires changes in auxin mobilization

Activation of phototropin kinases triggers signal transduction events that establish a variety of different responses. One of these responses is phototropism, which occurs in both mature plants and seedlings. As noted in Chapter 12, observations of this phenomenon by Charles and Francis Darwin initiated a series of experiments with grass coleoptiles that culminated in the discovery of the hormone auxin. Consistent with phototropin photoreceptor function in seedling phototropism, Arabidopsis seedlings lacking phototropins fail to bend in response to blue light (**Figure 13.13**).

Phototropic bending involves the lateral redistribution of auxin from the illuminated to the shaded side of the hypocotyl or coleoptile. As a first step in the process, light-activated phot1 directly phosphorylates an ATP-binding cassette (ABC) protein on the plasma membrane, inhibiting its activity. Because polar auxin transport (in this case, in the apical to basal direction; **Figure 13.14**) is strongly dependent on the activity of the ABC protein, phosphorylation of the latter results in the accumulation of auxin in the apical region of the hypocotyl. The next step is the lateral redistribution of auxin to the shaded side of the hypocotyl. (The physiology of phototropic bending will be discussed in more detail in Chapter 15).

Phototropins regulate chloroplast movements

Leaves can alter the intracellular distribution of their chloroplasts in response to changing light conditions. As discussed in Chapter 9, this feature is adaptive, as the redistribution of chloroplasts within the cells modulates light absorption and prevents photodamage (see Figure 9.13). Under weak illumination, chloroplasts gather near the upper and lower walls of the leaf palisade cells (accumulation), thus maximizing light absorption (**Figure 13.15**). Under strong illumination, the chloroplasts move to the lateral walls that are parallel to the incident light (avoidance), thus minimizing light absorption and avoiding photooxidative damage. In the dark, the chloroplasts move to the bottom of the cell, but the physiological function of this position is unclear. These chloroplast movements have been shown to be actin-dependent (see Chapter 1) and are regulated by phototropins. The action spectrum for responses shows the typical three-finger fine structure typical of blue light–specific responses (see Figure 13.7). The importance of chloroplast movements is demonstrated by the photooxidative damage observed in Arabidopsis mutants lacking phototropins in the field under full sunlight.

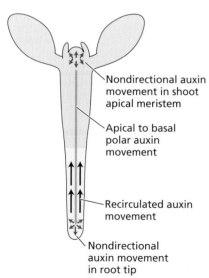

Nondirectional auxin movement in shoot apical meristem

Apical to basal polar auxin movement

Recirculated auxin movement

Nondirectional auxin movement in root tip

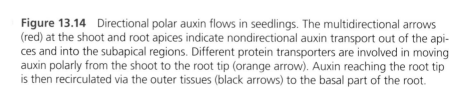

Figure 13.14 Directional polar auxin flows in seedlings. The multidirectional arrows (red) at the shoot and root apices indicate nondirectional auxin transport out of the apices and into the subapical regions. Different protein transporters are involved in moving auxin polarly from the shoot to the root tip (orange arrow). Auxin reaching the root tip is then recirculated via the outer tissues (black arrows) to the basal part of the root.

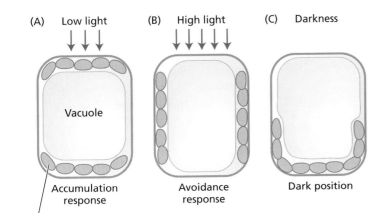

Figure 13.15 Schematic diagram of chloroplast distribution patterns in Arabidopsis leaf palisade cells in response to different light intensities. (A) Under low light conditions, chloroplasts optimize light absorption by accumulating at the upper and lower sides of palisade cells. (B) Under high light conditions, chloroplasts avoid sunlight by migrating to the side walls of palisade cells. (C) Chloroplasts move to the bottom of the cell in darkness. (After Wada 2013.)

Stomatal opening is regulated by blue light, which activates the plasma membrane H$^+$-ATPase

Stomatal photophysiology and sensory transduction in relation to water and photosynthesis were discussed in Chapters 3, 6, and 9. Blue light causes rapid stomatal opening, and can be shown to be distinct from the slower red light–induced opening brought about by photosynthesis. An action spectrum for the stomatal response to blue light under saturating background red illumination (**Figure 13.16**) is similar to that observed in phototropism (see Figure 13.7). The action spectrum for stomatal opening, typical of blue-light responses and distinctly different from the action spectrum for photosynthesis, indicates that guard cells respond specifically to blue light. The guard cell proton-pumping H$^+$-ATPase plays a central role in the regulation of stomatal movements. The activated H$^+$-ATPase transports H$^+$ across the membrane and increases the inside-negative electrical potential, driving K$^+$ uptake through the voltage-gated inward K$^+$ channels. The accumulation of K$^+$ facilitates the influx of water into the guard cells, leading to an increase in turgor pressure and stomatal opening (see Figure 6.20).

Phototropins are the primary photoreceptors for blue light–induced stomatal opening. Several key steps in the signaling pathway have been identified and are diagrammed in **Figure 13.17**. Blue light activates the H$^+$-ATPase via a multistep process. Initially, a membrane-associated, guard cell–specific protein kinase called BLUE LIGHT SIGNALING1 (BLUS1) is phosphorylated by phot1 and phot2 redundantly (that is, either phototropin is sufficient on its own for this step). The C terminus of the plasma membrane H$^+$-ATPase has an autoinhibitory domain that inhibits the activity of the enzyme. When activated by phototropin in response to blue light, BLUS1 initiates a signaling cascade that phosphorylates the C terminus of the H$^+$-ATPase, releasing the enzyme from inhibition. Further interactions with a 14-3-3 regulatory protein (named for the conditions under which the protein is biochemically purified) stabilize the H$^+$-ATPase in the active state.

Blue light also stimulates a second mechanism that contributes to increases in guard cell turgor. Activated phototropins phosphorylate cytosolic CONVERGENCE OF BLUE LIGHT AND CO$_2$ (CBC) proteins, which in turn inhibit plasma membrane anion channels. This increases cytosolic stores of Cl$^-$ that can then move into the vacuole along with malate to serve as counterions as vacuolar K$^+$ increases (see Figure 13.17).

As described above, photosynthetically active radiation (primarily red light) also slowly activates guard cell opening (see Figure 13.17). This is mainly the result of increased photosynthesis increasing the levels of cytosolic ATP. Photosynthetically active chloroplasts also produce carotenoids

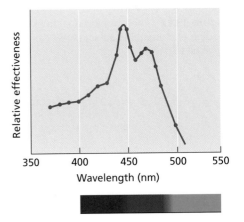

Figure 13.16 Action spectrum for blue light–stimulated stomatal opening (under a red-light background to saturate photosynthetic pigments that also absorb blue light). (After Karlsson 1986.)

1. Blue light activates phototropin photoreceptor kinases that then activate CONVERGENCE OF BLUE LIGHT AND CO_2 (CBC) kinases. Activated CBCs inhibit transport of Cl^- out of the cell, which increases Cl^- import into the vacuole.

2. Phosphorylation of a BLUE LIGHT SIGNALING 1 (BLUS1) / kinase complex activates sequential protein phosphorylation steps that activate plasma membrane H^+-ATPases by releasing inhibitory 14-3-3 proteins.

3. Activation of H^+-ATPases hyperpolarizes the plasma membrane, which increases K^+ movement into the guard cell. Vacuolar K^+ uptake is coordinated with that of Cl^- and $malate^{2-}$. These ion movements lower the water potential of the vacuole and drive water uptake, vacuolar expansion, and turgor-driven stomatal opening.

4. Photosynthetically active radiation increases production of ATP by chloroplasts, increasing cytosolic ATP levels and leading to increased rates of cytosolic ATPase activity. Photosynthetically active radiation also increases production of carotenoids, which protect membranes from photooxidative damage and are precursors for synthesis of the stomatal closure hormone abscisic acid.

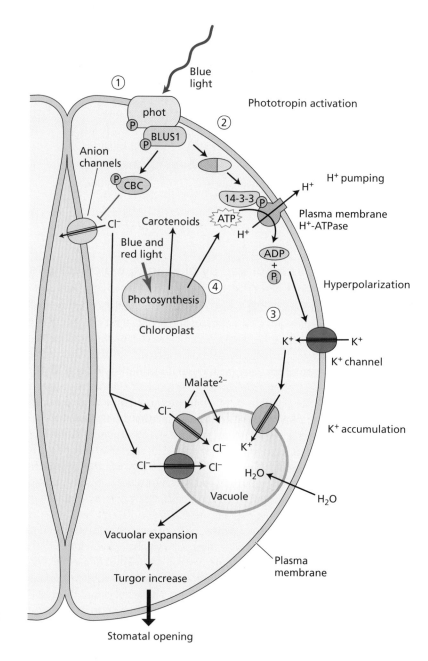

Figure 13.17   *Phototropin signal transduction leading to stomatal opening. (After Inoue and Kinoshita 2017.)*

that protect membranes and photoreceptors from photooxidative damage (see Figure 9.11). Carotenoids also are precursors in the synthesis of abscisic acid. In Chapter 19 we will discuss in detail how abscisic acid causes stomatal closure in the light.

The Coaction of Phytochrome, Cryptochrome, and Phototropins

As noted above, the stems of seedlings growing in the dark elongate very rapidly, and inhibition of stem elongation by light is a key photomorphogenic response of the seedling emerging from the soil surface (see Chapter 15). Although phytochrome is involved in this response, the blue region of the action spectrum for the decrease in elongation rate closely resembles that of phototropism (see Figure 13.7).

Experimentally, it is possible to separate a reduction in elongation rates mediated by phytochrome from a reduction mediated by a specific blue-light response. If lettuce seedlings are given low fluence rates of blue light under a strong background of yellow light, their hypocotyl elongation rate is reduced by more than 50%. The background yellow light establishes a well-defined Pr:Pfr ratio. Adding blue light at low fluence rates does not significantly change this ratio, ruling out a phytochrome effect on the reduction in elongation rate observed upon the addition of blue light. These results indicate that the elongation rate of the hypocotyl is controlled by a specific blue-light response that is independent of the phytochrome-mediated response.

A specific blue light–mediated hypocotyl response can also be distinguished from one mediated by phytochrome by their contrasting time courses. Whereas phytochrome-mediated changes in elongation rates can be detected within approximately 10–90 min, depending on the species, blue-light responses show lag times of less than 1 min. High-resolution analysis of the changes in growth rate during inhibition of hypocotyl elongation by blue light has provided valuable information about the interactions among phototropin 1 (phot1), cryptochromes (cry), and phytochrome A (phyA). After a lag of 30 s, blue light–treated, wild-type Arabidopsis seedlings show a rapid decrease in elongation rates during the first 30 min, and then they grow very slowly for several days. The rapid responses correlate with rapid membrane depolarization events. As shown in **Figure 13.18**, analysis of the same response in *phot1*, *cry1*, *cry2*, and *phyA* mutants has shown that suppression of stem elongation by blue light during seedling de-etiolation is initiated by phot1, with cry1, and to a limited extent cry2, modulating the response after 30 min. The slow growth rate of stems in blue light–treated seedlings is primarily a result of the persistent action of cry1, and this is the reason that cryptochrome mutants of Arabidopsis have longer hypocotyls than the wild type. Phytochrome A appears to play a role in at least the early stages of blue light–regulated growth, because growth inhibition does not progress normally in *phyA* mutants.

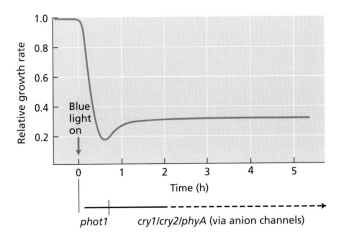

Figure 13.18 Coaction of phototropin 1 (phot1), phytochrome A (phyA), and cryptochromes (cry1/cry2) in blue-light inhibition of hypocotyl growth. Phototropin regulates initial transient inhibition, while longer-term inhibition is regulated by the combined activity of the cryptochromes and phytochrome A. (After Parks et al. 2001.)

Responses to Ultraviolet Radiation

In addition to its cytotoxic effects, UV-B radiation can elicit a wide range of photomorphogenic responses, including the inhibition of growth-promoting hormones, such as auxin and giberellins, and the enhancement of stress-induced defense hormones. The photoreceptor responsible for UV-B-induced physiological responses, UVR8, is a seven-bladed β-propeller protein, which forms functionally inactive homodimers in the absence of UV-B. Unlike phytochrome, cryptochrome, and phototropin, UVR8 lacks a prosthetic chromophore. The two identical subunits of UVR8 are linked in the dimer by a network of salt bridges formed between tryptophan residues, which serve as the primary UV-B sensors, and nearby arginine residues. The indole rings of these bridging tryptophans are excited by UV-B photons, disrupting the salt bridges (**Figure 13.19**). As a result, the monomers dissociate and become functionally active. The monomers then interact with other protein complexes to activate gene expression.

Figure 13.19 Diagram of the UV RESISTANCE LOCUS 8 (UVR8) photoreceptor response to UV-B irradiation. The inactive form of UVR8 is a dimer consisting of two monomers joined by salt bridges, with a series of tryptophan residues at the interaction surface. It is these tryptophan residues that sense UV-B, resulting in the structural changes that generate the two active monomers. Active UVR8 monomers are then transported to the nucleus where they participate in protein–protein interactions that result in downstream regulation of UV-B-responsive genes. (After art courtesy of Candace Pritchard.)

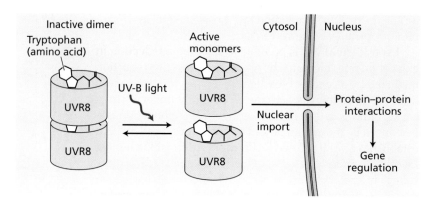

Summary

Photoreceptors, including phytochromes, cryptochromes, and phototropins, help plants regulate developmental processes over their lifetimes by sensitizing plants to incident light. Photoreceptors also initiate protective processes in response to harmful radiation.

Plant Photoreceptors

• Sunlight regulates developmental processes over the life of the plant and provides directional and nondirectional cues for growth and movement. Sunlight also contains UV and infrared radiation that can harm plant tissues (**Figures 13.1–13.4**).

• Photoreceptors undergo conformational changes after irradiation with specific light wavelengths. These conformational changes initiate downstream signaling events.

• Phytochromes (which absorb red and far-red light) and phototropins and cryptochromes (which absorb blue light and UV-A) are photoreceptors that are sensitive to light quantity, quality, and duration.

• The light-absorbing component of photoreceptors is called a chromophore. In phytochromes, cryptochromes, phototropins, and ZEITLUPE proteins, the chromophore is a small molecule cofactor. The UVR8 photoreceptor lacks a chromophore; instead an intrinsic ring of tryptophan residues functions as a chromophore.

• Action spectra and absorption spectra help researchers determine which wavelengths of light lead to specific photoresponses (**Figures 13.5–13.7**).

• Light fluence and irradiance also govern whether a photoresponse occurs.

Phytochromes

• Phytochrome is generally sensitive to red and far-red light, and exhibits the ability to interconvert between the physiologically inactive Pr form and the active Pfr form.

• Red light triggers conformational changes in both the phytochrome chromophore and protein (**Figure 13.8**).

Phytochrome Responses

• Photoresponses can vary substantially in both their lag time (the time between exposure to light and the subsequent response) and their escape time (the time before an overall sequence of reactions becomes irreversible).

• Phytochrome-initiated responses fall into one of three main categories: very low fluence responses (VLFRs), low-fluence responses (LFRs), or high-irradiance responses (HIRs) (**Figure 13.9**).

• Phytochrome A mediates responses to continuous far-red light.

• Phytochrome can rapidly change membrane potentials and ion fluxes.

• Phytochrome regulates gene expression through a wide range of regulatory elements and transcription factors.

• Phytochrome itself can be phosphorylated and dephosphorylated.

• Phytochrome-induced photomorphogenesis involves protein degradation.

Blue-Light Responses and Photoreceptors

• Blue-light responses generally exhibit a lag time after irradiation, and the effect declines gradually after the disappearance of the light signal (**Figure 13.10**).

Cryptochromes

• Activation of the flavin adenine dinucleotide (FAD) chromophore causes a conformational change in cryptochrome, enabling cryptochrome to bind to other protein partners (**Figure 13.11**).

(Continued)

Summary (*continued*)

- Cryptochrome isoforms have differential developmental effects.

Phototropins

- Similarly to cryptochromes, phototropins mediate photoresponses to blue light; phot1 and phot2 are sensitive to different and overlapping intensities of blue light.

- Phototropins are located at or near the plasma membrane and have flavin mononucleotide (FMN) chromophores that can induce conformational changes (**Figure 13.12**).

- When phototropins are activated by blue light, their kinase domain is "uncaged," causing autophosphorylation.

- Phototropins are required for seedling phototropism (**Figure 13.13**).

- Phototropins mediate chloroplast accumulation and avoidance responses to weak and strong light (**Figure 13.14**).

- Blue light, sensed by phototropins, causes activation of plasma membrane H⁺-ATPases and inhibition of plasma

membrane anion channels, and ultimately regulates stomatal opening (**Figures 13.15, 13.16**).

The Coaction of Phytochrome, Cryptochrome, and Phototropins

- Phytochrome, phototropins, and cryptochrome can all inhibit stem elongation. Phototropin inhibition of hypocotyl elongation is rapid and transient, whereas phytochrome and cryptochrome action shows a lag time and persists longer (**Figure 13.17**).

Responses to Ultraviolet Radiation

- The photoreceptor involved in responses to UV-B irradiation is UVR8 (**Figure 13.18**).

- Unlike other phytochromes, cryptochromes, and phototropins, UVR8 lacks a prosthetic chromophore.

- UVR8 interacts with other protein complexes to activate the transcription of UV-B-induced genes.

Suggested Reading

Burgie, E. S., Bussell, A. N., Walker, J. M., Dubiel, K., and Vierstra, R. D. (2014) Crystal structure of the photo-sensing module from a red/far-red light-absorbing plant phytochrome. *Proc. Natl. Acad. Sci. USA* 111: 10179–10184.

Christie, J. M., Kaiserli, E., and Sullivan, S. (2011) Light sensing at the plasma membrane. In *Plant Cell Monographs*, Vol. 19: *The Plant Plasma Membrane*, A. S. Murphy, W. Peer, and B. Schulz, eds., Springer-Verlag, Berlin, Heidelberg, pp. 423–443.

Christie, J. M., and Murphy, A. S. (2013) Shoot phototropism in higher plants: New light through old concepts. *Am. J. Bot.* 100: 35–46.

Inoue, S.-I., Takemiya, A., and Shimazaki, K.-I. (2010) Phototropin signaling and stomatal opening as a model case. *Curr. Opin. Plant Biol.* 13: 587–593.

Leivar, P., and Monte, E. (2014) PIFs: Systems integrators in plant development. *Plant Cell* 26: 56–78.

Liscum, E., Askinosie, S. K., Leuchtman, D. L., Morrow, J., Willenburg, K. T., and Coats, D. R. (2014) Phototropism: Growing towards an understanding of plant movement. *Plant Cell* 26: 38–55.

Rizzini, L., Favory, J.-J., Cloix, C., Faggionato, D., O'Hara, A., Kaiserli, E., Baumeister, R., Schäfer, E., Nagy, F., Jenkins, G. I., et al. (2011) Perception of UV-B by the *Arabidopsis* UVR8 protein. *Science* 332: 103–106.

Rockwell, R. C., Duanmu, D., Martin, S. S., Bachy, C., Price, D. C., Bhattachary, D., Worden, A. Z., and Lagarias, J. K. (2014) Eukaryotic algal phytochromes span the visible spectrum. *Proc. Natl. Acad. Sci. USA* 111: 3871–3876.

Swartz, T. E., Corchnoy, S. B., Christie, J. M., Lewis, J. W., Szundi, I., Briggs, W. R., and Bogomolni, R. A. (2001) The photocycle of a flavin-binding domain of the blue light photoreceptor phototropin. *J. Biol. Chem.* 276: 36493–36500.

Takala, H., Bjorling, A., Berntsson, O., Lehtivuori, H., Niebling, S., Hoernke, M., Kosheleva, I., Henning, R., Menzel, A., Janne, A., et al. (2014) Signal amplification and transduction in phytochrome photosensors. *Nature* 509: 245–249.

Takemiya, A., Sugiyama, N., Fujimoto, H., Tsutsumi, T., Yamauchi, S., Hiyama, A., Tadao, Y., Christie, J. M., and Shimazaki, K.-I. (2013) Phosphorylation of BLUS1 kinase by phototropins is a primary step in stomatal opening. *Nat. Commun.* 4: 2094. DOI: 10.1038/ncomms3094

Takemiya, A., Yamauchi, S., Yano, T., Ariyoshi, C., and Shimazaki, K.-I. (2013) Identification of a regulatory subunit of protein phosphatase 1, which mediates blue light signaling for stomatal opening. *Plant Cell Physiol.* 54: 24–35.

Wada, M. (2013) Chloroplast movement. *Plant Sci.* 210: 177–182.

14 Embryogenesis

Biological *development* is defined as the process by which an organism progresses from a zygote to the more complex reproductive stage. A related term, *morphogenesis*, refers to the cellular changes that are required to generate the form or shape of an organism. Development and morphogenesis are evolutionarily tied to the transition from unicellular to multicellular life, which first occurred around 600 million years ago. According to some estimates, multicellularity evolved independently at least 46 times in eukaryotic organisms, although only six major lineages (animals, fungi, brown algae, red algae, green algae, and land plants) gave rise to complex multicellular organisms. The multicellular lineage that ultimately led to land plants is believed to be derived from the characean green algae, a group of large-celled filamentous green algae. A key adaptive feature that accompanied the transition to land was the addition to the life cycle of an embryonic stage. Hence, *land plants* (mosses, liverworts, hornworts, ferns, horsetails, clubmosses, gymnosperms, and angiosperms) are also referred to as **embryophytes**.

Because multicellularity evolved independently in plants and animals, it is not surprising that they differ in their earliest stages of development, termed **embryogenesis**. However, all eukaryotes are thought to be derived from an ancient common ancestral cell, so it is also not surprising that the basic developmental processes that regulate embryogenesis are fundamentally similar in both lineages. For example, in both plants and animals development is primarily governed by *gene networks* that regulate such fundamental processes as metabolism, cell division, growth, and morphogenesis. Embryogenesis in both plants and animals also involves a division of labor among cells, with different cells assuming different metabolic

embryophyte The plant group, including all land plants, characterized by the ability of the gametophyte to contain and nurture the young sporophyte within its tissues through the earliest stages of development.

embryogenesis The formation and development of the embryo.

functions. In both plants and animals, the coordination of various cellular activities is facilitated by mechanisms of *intercellular communication*, although the specific structures involved in such intercellular signaling differ in the two lineages. Finally, in both plants and animals, *positional information*—that is, the specific location of a cell in the embryo—can exert a strong influence on the cell's developmental fate.

Plants offer intriguing developmental contrasts to animals, not only with respect to their diverse forms, but also in how those forms arise. A sequoia tree, for example, may grow for thousands of years before reaching a size big enough for an automobile to drive through its trunk. In contrast, the tiny Arabidopsis plant can complete its life cycle in little more than a month, making approximately 14 small rosette leaves before producing a short inflorescence (**Figure 14.1**). Dissimilar as they may be, both species employ growth mechanisms common to all multicellular plants. Embryogenesis in all plants arises out of patterns of cell division and expansion, rather than via cell movements as occurs in animals. Because of the highly rudimentary nature of the mature plant embryo, plant morphogenesis unfolds throughout a plant's entire life span by means of adaptive postembryonic growth processes. Animals, by contrast, typically have a more predictable pattern of development in which the basic body plan, including all the major organs, is largely determined during embryogenesis.

These differences between plants and animals can be understood partly in terms of contrasting survival strategies. Being photosynthetic, plants rely on flexible patterns of growth that allow them to adapt to fixed locations where conditions may be less than ideal, especially with respect to sunlight, and may vary over time. Animals, being heterotrophic, evolved mechanisms for mobility instead.

(A)

(B)

Figure 14.1   Two contrasting examples of plant form arising from indeterminate growth processes. (A) The Chandelier Tree in Leggett, California, a famous *Sequoia sempervirens* that has adapted to many challenges during its roughly 2400-year existence. (B) The compact form and rapid life cycle of the much smaller *Arabidopsis thaliana* have made it a useful model for understanding mechanisms that guide plant growth and development. (A © David L. Moore-CA/Alamy Stock Photo; B photo by David McIntyre.)

In this chapter we consider the essential characteristics of plant embryogenesis and the nature of the mechanisms that guide embryo formation.

Overview of Embryogenesis

The sporophytic, or diploid, phase of plant development begins with the formation of a single cell, the zygote, which results from the fusion of the two haploid gametes, egg and sperm (for a review of plant life cycles, see Chapter 1). During embryogenesis, the zygote transforms into a multicellular embryo with a characteristic rudimentary organization. In all seed plants, embryogenesis takes place within the confines of a protective structure called the *ovule*. In angiosperms ovules are enclosed within the carpels (female reproductive structures) of flowers, whereas in gymnosperms ovules are borne either on the surfaces of cone scales, as in the case of the conifers and cycads, or on short stalks, as in *Ginkgo biloba*. At maturity, the ovule turns into the seed.

The overall sequence of embryonic development follows a predictable pattern, reflecting the need for the embryo to be effectively packaged within the confines of the maternally derived tissues comprising the seed. In some small-seeded species, such as Arabidopsis and shepherd's purse (*Capsella bursa-pastoris*), the sequence of cell divisions during embryogenesis is highly predictable. Because of its predictability, embryogenesis affords some of the clearest examples of basic patterning processes in plants. At maturity, plant embryos possess the same basic tissue types present in all postembryonic seed plants: epidermis, vascular tissues, ground tissue, and meristems.

Among the initial processes required for the transformation of the zygote to a mature embryo are those involved in the establishment of cell polarity, the asymmetric distribution of organelles and macromolecules within a cell. Cell polarity helps set up the first asymmetric cell division of the zygote, which differentiates between the small apical cell that will go on to form the embryo proper, and the larger basal cell that will form the attachment to the ovule wall. The smaller cell then develops into a multicellular embryo with apical–basal polarity. This transformation involves the coordination of asymmetric cell division, cell fate specification, and signaling between adjacent cells. Different cell layers become functionally specialized to form epidermal, ground, and vascular tissues. Groups of permanently dividing cells, known as apical meristems, are established at the growing points of the shoot and root and enable the elaboration of additional tissues and organs during subsequent vegetative growth. The **shoot apical meristem (SAM)** and the **root apical meristem (RAM)** are primarily responsible for the **indeterminate growth** of plants, in contrast to the determinate growth of animals. However, although some individual plants can indeed live for thousands of years, all plants eventually senesce and die, as we will discuss in Chapter 16.

Comparative Embryology of Eudicots and Monocots

Embryogenesis provides many examples of developmental processes by which the basic architecture of the plant is established. We begin by contrasting the processes of embryogenesis in Arabidopsis, a eudicot, with those of the monocot maize (corn; *Zea mays*), a member of the grass family. Next we consider the nature of the signals that guide complex patterns of growth and differentiation in the embryo, with several lines of evidence highlighting the importance of position-dependent cues. Finally, we explore examples that illustrate how molecular and genetic approaches provide insight into the mechanisms that translate these cues into organized patterns of growth.

shoot apical meristem (SAM)
Dome-shaped region of the shoot tip composed of meristematic cells that give rise to leaves, branches, and reproductive structures.

root apical meristem (RAM)
A group of permanently dividing cells located underneath the root cap at the tip of the root that provides cells for the primary growth of the root.

indeterminate growth   The ability to keep growing and developing until the onset of senescence.

scutellum   The single cotyledon of the grass embryo, specialized for nutrient absorption from the endosperm.

coleoptile   A modified ensheathing leaf that covers and protects the young primary leaves of a grass seedling as it grows through the soil. Unilateral light perception, especially blue light, by the tip results in asymmetric growth and bending due to unequal auxin distribution in the lighted and shaded sides.

Morphological similarities and differences between eudicot and monocot embryos dictate their respective patterns of development

In general, the morphologies of eudicot and monocot embryos are similar during the establishment of the major growth axis, up to the early embryo stage. Beyond that point their developmental paths diverge significantly. For example, eudicot embryos form two cotyledons, or "seed leaves," whereas monocot embryos form only one. (The *cot* in *eudicot* and *monocot* is short for "cotyledon.") The grass family represents a specialized group of monocots in which, according to one interpretation, the single cotyledon appears to have become functionally divided into two structures, the **scutellum** and the **coleoptile**. The scutellum serves as an absorptive organ that takes up sugars from the endosperm during germination, while the coleoptile forms a tubular sheath that protects the emerging primary leaves from mechanical damage from the soil. The cell division patterns that give rise to these specialized monocot structures are more complex, and less predictable, than those required to form the two cotyledons of most eudicots. Arabidopsis and maize thus represent opposite ends of a spectrum in embryological development, from the simple to the complex.

ARABIDOPSIS   Because of the relatively small size of the Arabidopsis embryo, the patterns of cell division by which it arises are relatively simple and easily followed. Embryonic development in Arabidopsis is typical of members of the Brassicaceae (cabbage family). The five main stages of development are listed below and illustrated in **Figure 14.2**:

1. *Zygotic stage.* The first stage of the diploid life cycle commences with the fusion of the haploid egg and sperm to form the single-celled zygote. Polarized growth of this cell, followed by an asymmetric transverse (i.e., perpendicular to the cell axis) division, gives rise to a small apical cell and an elongated basal cell (see Figure 14.2A).

2. *Globular stage.* The apical cell undergoes two vertical divisions, forming a quadrant, followed by a transverse division (see Figure 14.2B–D), to generate a spherical, eight-cell (*octant*) globular embryo exhibiting *radial symmetry*. Additional cell divisions increase the number of cells in the globular embryo (see Figure 14.2D) and create the outer layer, the *protoderm*, which later becomes the epidermis.

3. *Heart stage.* Cells divide in two regions on either side of the future shoot apical meristem to form the two cotyledons, generating the embryo's *bilateral symmetry* (see Figure 14.2E and F).

4. *Torpedo stage.* Cells elongate and differentiate along the entire embryonic axis. Visible distinctions between the inner (adaxial) and outer (abaxial) surfaces of the cotyledons become apparent (see Figure 14.2G).

5. *Mature stage.* Toward the end of embryogenesis, the embryo and seed lose water and become metabolically inactive as they enter dormancy (discussed in Chapter 15) (see Figure 14.2H).

In nearly all eudicots, the first plane of cell division—which gives rise to the apical and basal cells—is transverse, as it is in Arabidopsis (see Figure 14.2A). However, in a few species the first plane of cell division is vertical (longitudinal), as in the Loranthaceae (mistletoe family), or oblique as in the Gambel oak (*Quercus gambelii*). The second division is variable, depending on the species. It may either be transverse in both the apical or basal cells, to produce a linear tetrad, or it may be vertical in the apical cell, as in Arabidopsis (see Figure 14.2B). Various schemes have been devised to categorize the different types of embryo development, typically

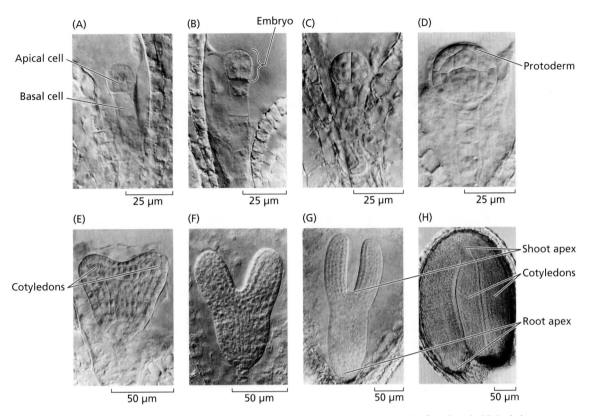

Figure 14.2 The stages of Arabidopsis embryogenesis are characterized by precise patterns of cell division. (A) One-cell embryo after the first division of the zygote, which forms the apical and basal cells. (B) Two-cell embryo. (C) Eight-cell embryo. (D) Mid-globular stage embryo, which has developed a distinct protoderm (surface layer). (E) Early heart stage embryo. (F) Late heart stage embryo. (G) Torpedo stage embryo. (H) Mature embryo. (From West and Harada 1993; photographs taken by K. Matsudaira Yee; courtesy of John Harada, © American Society of Plant Biologists, reprinted with permission.)

involving six or more classes. Unfortunately, mechanistic studies of embryogenesis in other plant species still lag behind those conducted in Arabidopsis.

MAIZE   Maize, a monocot, illustrates a more complex type of embryogenesis that is typical of members of the grass family. As in most other plants, the cell division patterns associated with embryogenesis in maize are far more variable and less well defined than those in Arabidopsis. Nevertheless, it is possible to describe embryogenesis in maize in terms of six morphologically defined developmental stages (**Figure 14.3**):

1. *Zygotic stage.* This stage begins with the fusion of the haploid egg and sperm to form the zygote (not shown in Figure 14.3). As in Arabidopsis, the zygote undergoes polarized growth followed by asymmetric transverse divisions, giving rise to a small apical cell and an elongated basal cell (see Figure 14.3A).

2. *Globular stage.* Following the creation of the apical and basal cells, a series of variable cell divisions creates a multilayered globular embryo consisting of the embryo proper and the larger multicellular suspensor (see Figure 14.3B).

3. *Transition stage.* During the early transition stage, the scutellum appears on the inner side of the embryo (relative to the future seed coat). By the late transition stage, the future shoot apical meristem is evident on the outer side of the embryo, relative to the future seed coat. (The transition stage is not shown in Figure 14.3.)

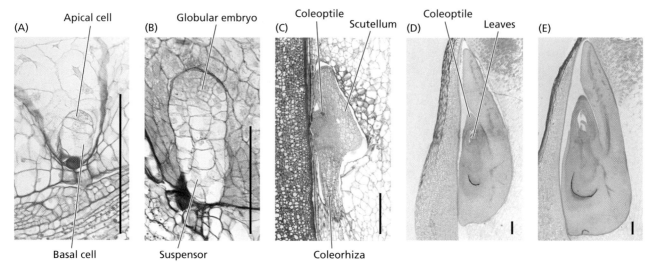

Figure 14.3   Stages of maize embryogenesis. (A) Zygotic stage. (B) Globular stage. (C) Coleoptile stage. (D) Leaf primordia stage. (E) Maturation stage. (The transition stage, between the globular and coleoptile stages, is not shown.) Longitudinal sections of maize embryos; scale bars = 200 μm. (From Sosso et al. 2012. Journal of Experimental Biology/CC BY-ND-NC 4.0.)

radicle   The embryonic root. Usually the first organ to emerge on germination.

coleorhiza   A protective sheath surrounding the embryonic radicle in members of the grass family.

polarity   Refers to the distinct ends and intermediate regions along an axis. Beginning with the single-celled zygote, the progressive development of distinctions along two axes: an apical–basal axis and a radial axis.

4. *Coleoptile stage.* This stage is marked by the formation of a distinct coleoptile, scutellum, shoot apical meristem, root apical meristem, **radicle** (embryonic root), and **coleorhiza**, a protective sheath that covers the embryonic root tip (see Figure 14.3C).

5. *Leaf primordia stage.* The shoot apical meristem initiates several leaves inside the coleoptile (see Figure 14.3D).

6. *Maturation stage.* During the final stage of embryogenesis (see Figure 14.3E), expression of maturation-related genes precedes the onset of dormancy.

Comparing embryogenesis in Arabidopsis and maize reveals several features of the process that are common to both plants. Perhaps the most fundamental of these relates to **polarity**. Beginning with the single-celled zygote, the embryo becomes progressively more polarized throughout its development along two axes: an **apical–basal axis**, which runs between the tips of the embryonic shoot and root, and a **radial axis**, perpendicular to the apical–basal axis and extending from the center of the plant outward (**Figure 14.4**). In the following section we consider how these axes are established and discuss how specific molecular processes guide their development. Much of our discussion focuses on Arabidopsis, which is not only a powerful model for molecular and genetic studies, but also displays simple and highly stereotyped cell divisions during the early stages of its embryonic development. By observing changes in this simple pattern, we can more easily recognize both physiological and genetic factors that influence embryonic development.

Apical–basal polarity is maintained in the embryo during organogenesis

Apical–basal polarity is a characteristic feature of seed plants in which tissues and organs are arrayed in a stereotypical order along an axis that extends from the shoot apical meristem to the root apical meristem. Following fertilization of the egg cell, the zygote rapidly

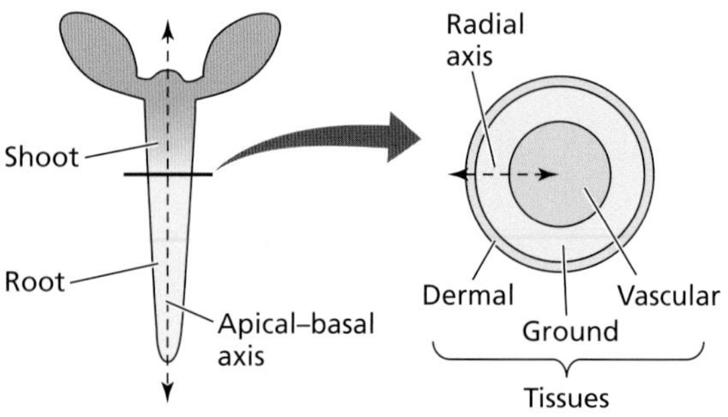

Figure 14.4   In longitudinal section (left), the apical–basal axis extends between the tips of the embryonic root and shoot. In cross section (right), the radial axis extends from the center to the surface across concentric rings of vascular, ground, and dermal tissues.

elongates along its apical–basal axis by approximately threefold and becomes polarized with respect to the distribution of its cellular contents. The apical end is densely cytoplasmic compared with the basal end, which contains a large vacuole. These differences in cytoplasmic density are passed on to the two daughter cells when the zygote divides asymmetrically to give rise to a short, cytoplasmically dense **apical cell** and a longer, vacuolated **basal cell** (**Figure 14.5**). The basal cell then further divides to form the **suspensor**, which pushes the embryo into the lumen of the developing seed, and the **hypophysis**, which contributes to the formation of the root meristem and root cap.

The apical cell continues to divide to form the body of the embryo that will mature to become the rest of the embryonic root, hypocotyl, cotyledons, and shoot meristem. After the first two longitudinal divisions and a set of transverse divisions, the eight-cell (octant) globular embryo is formed (see Figure 14.5). In the cells that make up the octant globular embryo, there is little, apart from position, to distinguish the appearance of the upper and lower tiers of cells. All eight cells then divide periclinally. During **periclinal** divisions, a new cell plate is formed parallel to the tissue surface (**Figure 14.6**). These periclinal divisions give rise to a new cell layer called the **protoderm**, which ultimately forms the epidermis. As the embryo increases in volume, cells of the protoderm undergo **anticlinal** divisions, in which the new cell plates form perpendicular to the tissue surface. Anticlinal divisions increase the number of cells in the protoderm, allowing for an expansion in circumference.

apical–basal axis   An axis that extends from the shoot apical meristem to the root apical meristem.

radial axis   An axis that extends from the center of a root or stem to its surface.

apical cell   The smaller, cytoplasm-rich cell formed by the first division of the zygote.

basal cell   The larger, vacuolated cell formed by the first division of the zygote. It gives rise to the suspensor.

suspensor   In seed plant embryogenesis, the structure that develops from the basal cell following the first division of the zygote. It supports, but is not part of, the embryo.

hypophysis   The cell located directly below the octant stage of the embryo that gives rise to the root cap and part of the root apical meristem.

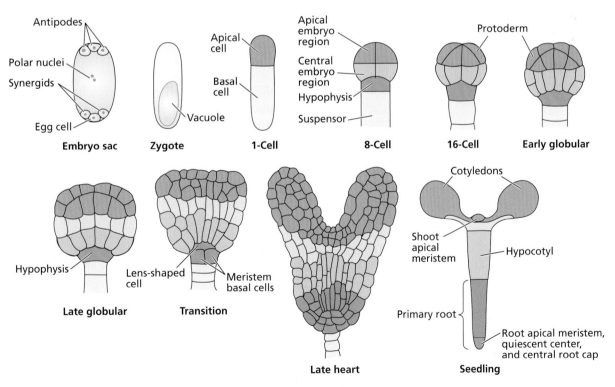

Figure 14.5   Pattern formation during Arabidopsis embryogenesis. A series of successive stages is shown to illustrate how specific cells in the young embryo contribute to specific anatomically defined features of the seedling. Clonally related groups of cells (cells that can be traced back to a common progenitor) are indicated by distinct colors. Following the asymmetric division of the zygote, the smaller, apical cell divides to form an eight-cell embryo consisting of two tiers of four cells each. The upper tier (green) gives rise to the shoot apical meristem and most of the flanking cotyledon primordia. The lower tier (beige) produces the hypocotyl and some of the cotyledons, the embryonic root, and the upper cells of the root apical meristem. The basal cell produces a single file of cells that make up the suspensor. The uppermost cell of the suspensor becomes the hypophysis (blue), which is part of the embryo. The hypophysis divides to form the quiescent center and the stem cells (initials) that form the root cap. (After Laux et al. 2004.)

Figure 14.6 Periclinal and anticlinal cell division. Periclinal divisions produce new cell walls parallel to the tissue surface, and thus contribute to the establishment of a new layer. Anticlinal divisions produce new cell walls perpendicular to the tissue surface, and thus increase the number of cells within a layer.

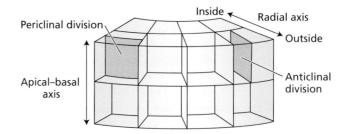

By the eight-cell (octant) stage, the apical–basal axis has become divided into three main developmental domains that give rise to different regions of the mature embryo. Later in embryogenesis two additional domains appear, for a total of five distinct developmental domains in the late heart stage embryo (see Figure 14.5):

1. The apical region (dark green), derived from the apical quartet of cells, gives rise to the upper portion of the cotyledons and the shoot apical meristem.

2. The subapical region (light green) includes the lower portion of the cotyledons and the region below the shoot apical meristem.

3. The region that gives rise to the hypocotyl (tan).

4. The region generating the basal portion of the primary root (orange), the root, and the apical regions of the root meristem.

5. The region derived from the hypophysis that gives rise to the central and basal parts of the root meristem and the root cap (blue).

The predictable patterns of cell division observed in early Arabidopsis embryogenesis suggest that a fixed sequence of cell divisions is essential to this phase of development. However, the majority of monocots and eudicots, especially those with larger embryos, have less predictable patterns of cell division. For example, in maize embryos, only the first asymmetric cell division of the zygote is predictable, leading to a small, densely cytoplasmic apical cell and a larger, vacuolated basal cell (**Figure 14.7**). Subsequent cell divisions of the maize embryo are less ordered and synchronous than those observed in Arabidopsis embryos.

Even in Arabidopsis, some variation in cell division during normal embryogenesis is often observed. Arabidopsis microtubule mutants that exhibit irregular cell division grow as short, thick, misshapen plants, consistent with the important role of cell lineages in normal development (**Figure 14.8**). However, these mutants retain the ability to form tissues and rudimentary organs in their normal spatial arrangements. Therefore, embryogenesis would seem to involve a variety of mechanisms, including those, such as positional information and cell-to-cell signaling, that do not rely solely on a fixed sequence of cell divisions.

Embryo development requires regulated communication between cells

The cells of a developing embryo can be compared to a group of people working together on a common project. To form an embryo, each cell must be able to function as an individual as well as to coordinate with others in the group, which requires communication. Instead of words, the language of cell–cell communication consists of chemical and physical signals. As described in Chapters 10 and 12, hormones are mostly small molecules that act as mobile signals during plant development. In addition, macromolecular signals, such as proteins or RNA, can move symplastically from one cell to another via plasmodesmata. As

periclinal Pertaining to the orientation of cell division such that the new cell plate forms parallel to the tissue surface.

protoderm In the plant embryo, the one-cell-thick surface layer that covers the young shoot and radicle of the embryo and gives rise to the epidermis.

anticlinal Pertaining to the orientation of cell division such that the new cell plate forms perpendicular to the tissue surface.

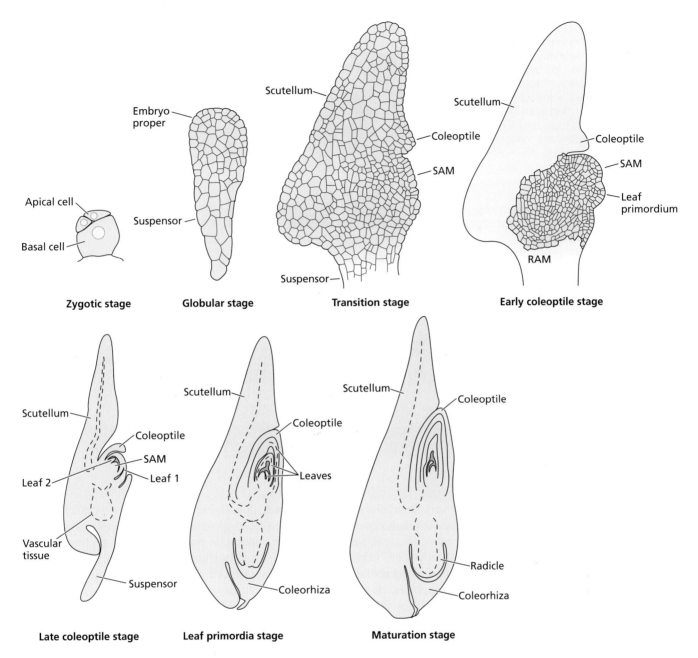

Figure 14.7 Pattern formation during maize embryogenesis. After the first division of the zygote, a series of more or less unpredictable cell divisions gives rise to the globular embryo, which includes the embryo proper at the apical end and the larger suspensor at the basal end. During the transition stage, the scutellum develops on one side of the embryo. During the early and late coleoptile stages, the shoot apical meristem (SAM) develops and the coleoptile forms. The root apical meristem (RAM) forms, surrounded by a protective coleorhiza. During the leaf primordia stage, immature leaves develop within the tubular coleoptile. During the maturation stage, metabolic changes associated with seed dormancy typically take place. Dashed lines indicate vascular tissue. (From Bommert and Werr 2001.)

discussed in Chapter 10, plasmodesmata are not simply passive conduits; they can regulate their *size exclusion limit,* that is the size and physical characteristics of the macromolecules that can pass through them.

The size exclusion limit of plasmadesmata appears to vary during embryogenesis. Studies have shown that large artificial dye molecules and fluorescently tagged proteins move more readily through plasmodesmata in the early embryo than at later stages of embryogenesis, when tissues begin to differentiate. For

Figure 14.8 Extra cell divisions do not block the establishment of basic radial pattern elements. Arabidopsis plants with mutations in the *FASS* (alternatively, *TON2*) gene are unable to form a preprophase band of microtubules in cells at any stage of cell division. (The preprophase band normally establishes where the new cell plate will form; see Figure 1.29.) Plants carrying this mutation have highly irregular cell division and expansion planes, and as a result are severely deformed. However, they continue to produce recognizable tissues and organs in their correct positions. Although the organs and tissues produced by these mutant plants are highly abnormal, a radially oriented tissue pattern is still evident. (Top) Wild-type Arabidopsis: (A) early globular stage embryo; (B) seedling seen from the top; (C) cross section of a root. (Bottom) Comparable stages of Arabidopsis homozygous for the *fass* mutation: (D) early embryogenesis; (E) mutant seedling seen from the top; (F) cross section of a mutant root, showing the random orientation of the cells but a nearly wild-type tissue order: an outer epidermal layer surrounds a multicellular cortex, which in turn surrounds the vascular cylinder. The two scale bars apply to A–C and D–F, respectively. (From Traas et al. 1995.)

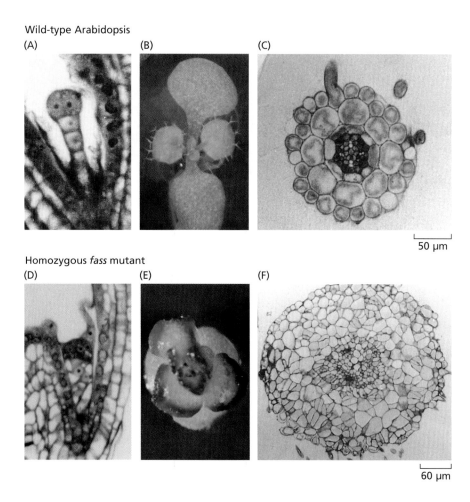

Wild-type Arabidopsis
(A) (B) (C)

50 µm

Homozygous *fass* mutant
(D) (E) (F)

60 µm

example, as shown in **Figure 14.9**, fluorescently tagged proteins migrate well beyond their site of synthesis in heart stage Arabidopsis embryos, but are more restricted at the torpedo stage. In general, as tissues differentiate, the patterns of movement of proteins, peptides, and hormones become increasingly channeled along specific pathways. This is true whether the signaling molecules are transported via the symplast or by cell-to-cell transport (discussed below). As we will see, this regulated intercellular traffic plays an essential role in a variety of developmental processes, including maintenance of the apical–basal polarity of the primary axis of the embryo.

Auxin signaling is essential for embryo development

Many mutations that cause defective embryonic development in monocots and eudicots are in genes encoding proteins involved in auxin (indole-3-acetic acid, or IAA) biosynthesis, signaling, and cellular transport. Some of these mutations impact discrete aspects of embryonic development, while others affect multiple cellular processes. Using fluorescent tags such as green fluorescent protein (GFP), researchers have been able to study exactly when and where these gene products function and how they interact. For example, apical–basal auxin concentration gradients in the embryo appear to be initially established by localized auxin synthesis, but they are gradually amplified and extended by specific transporter proteins in the plasma membrane. Auxin transport, specifically *polar* auxin transport, becomes especially important as the embryo increases in size and complexity. Before we describe the role of auxin transport in embryos in more detail, let's discuss how auxin transport mechanisms have been elucidated in seedlings.

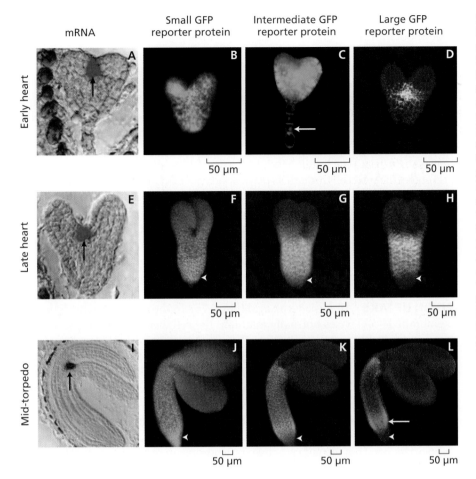

mRNA | Small GFP reporter protein | Intermediate GFP reporter protein | Large GFP reporter protein

Early heart

Late heart

Mid-torpedo

Figure 14.9 The potential for intercellular protein movement changes during development. Images show the distribution of small (B, F, J), intermediate (C, G, K), and large (D, H, L) green fluorescent protein (GFP) reporter proteins (all green) in Arabidopsis embryos of different ages (early heart, A–D; late heart, E–H; mid-torpedo, I–L). All constructs are transcribed from an *STM* promoter, which produces transcripts in relatively small regions of the embryos, as shown by in situ hybridization (A, E, I; black arrows). Small proteins appear to move readily in all stages of embryogenesis (B, F, J), but the mobility of larger proteins is lower and becomes more restricted in older embryos (C and D, G and H, K and L). White arrows indicate the nucleus in suspensor cells (C) and ectopic expression of the *STM* promoter in hypocotyls (L). White arrowheads indicate the root. (From Kim et al. 2005. © National Academy of Sciences, National Library of Medicine U.S.A.)

Polar auxin transport is mediated by localized auxin efflux carriers

Polar auxin transport is found in almost all plants, including bryophytes and ferns. Early studies of this phenomenon focused on auxin movement in apical and epidermal tissues during seedling phototropic responses (see Chapters 13 and 15). Long-distance polar auxin transport through the vascular parenchyma from sites of synthesis in apical tissues and young leaves to the root tip was subsequently shown to regulate stem elongation, apical dominance, and lateral branching (see Chapter 16). Redirection of auxin from the root apex into the root epidermis is also necessary for root gravitropic responses (see Chapter 15).

Polar auxin transport has been demonstrated by radiolabeled auxin tracer assays, mass spectroscopic analyses of auxin content in discrete tissues, and other analytical methods. More recently, scientists have used auxin reporter genes to visualize relative auxin concentrations in individual cells and tissues in intact plants. These reporter genes often consist of DR5, an artificial auxin-responsive promoter, fused to a gene whose product is easily visualized, such as β-glucuronidase (GUS), which produces a blue color when incubated with chromogenic substrates, or GFP. However, DR5-based reporter genes require gene transcription to function, which delays visualization of the response to auxin. A more rapidly responding auxin reporter, DII-Venus, is based on a fusion of yellow fluorescent protein to a portion of the AUX/IAA auxin receptor protein. Auxin causes rapid degradation of AUX/IAA (see Figure 12.19A), so DII-Venus fluorescence disappears rapidly when auxin is present.

polar auxin transport Directional auxin movement that functions in programmed development and plastic growth responses. Long-distance polar auxin transport maintains the overall polarity of the plant apical–basal axis and supplies auxin for direction into localized streams.

By longstanding convention, auxin transport from the shoot and root apices to the root–shoot transition zone of a seedling or mature plant is referred to as a *basipetal* (from the Latin for "toward the base") transport, whereas downward auxin flow into the root is referred to as *acropetal* ("toward the apex") transport. However, since this terminology can be confusing, the term *rootward* transport is also applied to all auxin flows toward the root apex, and the term *shootward* transport refers to any directional flow toward the shoot apex. Both shootward and rootward polar auxin transport are primary mechanisms for effecting programmed directional growth.

Polar transport proceeds in a cell-to-cell fashion, rather than via the symplast; that is, auxin exits a cell through the plasma membrane, diffuses across the cell wall, and enters the next cell through its plasma membrane (**Figure 14.10A**). The overall process requires metabolic energy, as evidenced by the sensitivity of polar transport to O_2 deprivation, sucrose depletion, and metabolic inhibitors. The velocity of polar auxin transport can exceed 10 mm h^{-1} in some mature plant tissues, which is faster than diffusion but much slower than phloem translocation rates (see Chapter 10). Polar transport is specific for all natural and some synthetic auxins; other weak organic acids, inactive auxin analogs, and IAA conjugates are poorly transported.

AUXIN UPTAKE   IAA is a weak acid (pK$_a$ 4.75). In the apoplast, where plasma membrane H$^+$-ATPases normally maintain the cell wall solution at pH 5 to 5.5 (see Figure 14.10A, number 2), 15 to 25% of the auxin is present in an undissociated (protonated) lipophilic form (IAAH) that diffuses passively across the plasma membrane down a concentration gradient (see Figure 14.10A, number 1). Auxin uptake is accelerated by secondary active transport of the amphipathic, anionic IAA$^-$ present in the apoplast via AUXIN1/LIKE AUXIN1 (AUX1/LAX) symporters that cotransport two protons along with the auxin anion (see Figure 14.10A, number 1). As such, auxin uptake via AUX1 results in localized membrane depolarization. This secondary active transport of auxin allows for greater auxin accumulation than does simple diffusion because anionic auxin is driven across the membrane by the proton motive force (i.e., the high proton concentration in the apoplastic solution). Although AUX1 is asymmetrically (polarly) distributed in the plasma membrane of some cells, such as the protophloem, AUX1 usually isn't responsible for the polarity of auxin transport. Rather, AUX1's primary role is to accelerate auxin uptake from the apoplast in cells destined to become auxin sinks, which helps maintain an apoplastic auxin concentration gradient from the auxin-transporting cells to the sink cells. However, the specific pathway of auxin movement is determined by other factors (see below).

The function of AUX1 has been best studied in Arabidopsis roots. In the *aux1* mutant, shootward auxin flows are completely disrupted, resulting in agravitropic root growth. Gravitropic root growth in the mutant is completely restored by expression of *AUX1* under the control of a promoter specifically expressed in the lateral root cap.

AUXIN EFFLUX   At the neutral pH of the cytosol, the anionic form of auxin, IAA$^-$, predominates (see Figure 14.10A, number 3). Transport of IAA$^-$ out of the cell is driven by the negative membrane potential inside the cell. However, because the lipid bilayer of the membrane is impermeable to the anion, auxin export out of the cell must occur via transport proteins in the plasma membrane (see Figure 14.10A, number 4). There are two main classes of transport proteins involved in auxin efflux: PINFORMED (PIN) and ABCB.

The PIN family of proteins is named after the pin-shaped inflorescences formed by the *pin1* mutant of Arabidopsis. Where PIN auxin efflux carrier proteins are polarly localized—that is, localized in the plasma membrane at only one end of a cell—auxin uptake into the cell (via diffusion and AUX1) and subsequent efflux via PIN give rise to a net polar transport (see Figure 14.10A). Different PIN family

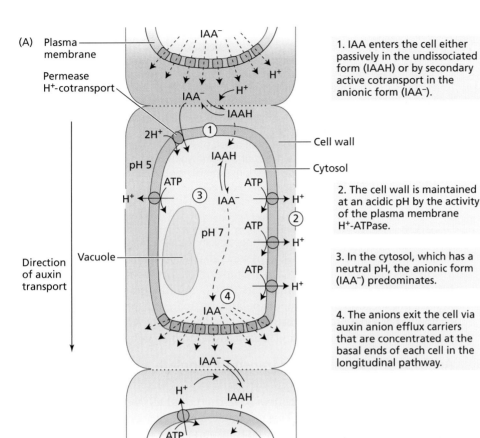

(A) Plasma membrane
Permease H⁺-cotransport

1. IAA enters the cell either passively in the undissociated form (IAAH) or by secondary active cotransport in the anionic form (IAA⁻).

Cell wall
Cytosol

2. The cell wall is maintained at an acidic pH by the activity of the plasma membrane H⁺-ATPase.

3. In the cytosol, which has a neutral pH, the anionic form (IAA⁻) predominates.

Direction of auxin transport
Vacuole

4. The anions exit the cell via auxin anion efflux carriers that are concentrated at the basal ends of each cell in the longitudinal pathway.

Figure 14.10 (A) Simplified chemiosmotic model of polar auxin transport. Shown here is a single column of elongated auxin-transporting cells. Additional mechanisms contribute to polar transport by preventing reuptake of IAA at sites of export and in adjoining cell files. (B) Model for polar auxin transport in small cells with significant back-diffusion of auxin due to a high surface-to-volume ratio. ABCB proteins are thought to maintain polar streams by preventing reuptake of auxin exported at carrier sites. In larger cells, ABCB transporters appear to exclude movement of auxin out of polar streams into adjoining cell files.

(B) 1. The plasma membrane H⁺-ATPase (purple) pumps protons into the apoplast. The acidity of apoplast affects the rate of auxin transport by altering the ratio of IAAH and IAA⁻ present in the apoplast.

2. IAAH can enter the cell via proton symporters such as AUX1 (blue) or diffusion (dashed arrows). Once inside the cytosol, IAA is an anion, and may only exit the cell via active transport.

3. ABCB proteins (red) drive active (ATP-dependent) auxin efflux. They are localized in plasma membrane subdomains that are distinct from sites of auxin uptake.

4. Synergistically enhanced active polar transport occurs when polarly localized PIN proteins (brown) associate with ABCB proteins, overcoming the effects of back-diffusion.

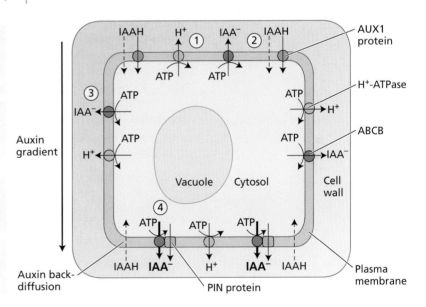

members mediate auxin efflux in each tissue, and mutants with defects in these genes exhibit phenotypes consistent with PIN's function in the different tissues. Of the PIN proteins, PIN1 is the most studied, as it is essential to virtually every aspect of polar development and organogenesis in plant shoots.

A subset of ATP-dependent transporters from the large superfamily of ATP-binding cassette (ABC) integral membrane transporters amplifies efflux and prevents reuptake of exported auxin, especially in small cells where auxin concentrations are high (**Figure 14.10B**). Defective *ABCB* (ABC "B" class) genes in Arabidopsis,

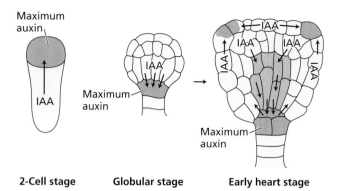

Maximum auxin

IAA

2-Cell stage

IAA

Maximum auxin

Globular stage

IAA
IAA   IAA

IAA   IAA

Maximum auxin

Early heart stage

Figure 14.11 PIN1-dependent movement of auxin (IAA) during early stages of embryogenesis. Auxin movement, as inferred from the asymmetric distribution of the PIN1 protein and the activity of a DR5 auxin-responsive reporter, is indicated by arrows. Blue areas denote cells with maximum auxin concentrations. The red region indicates the future vascular tissue that forms in response to directed auxin movement. Auxin maximums resulting from synthesis of the hormone create gradients that are then reinforced by polar orientation of PIN1. After the accumulation of auxin at the flanks of the embryo, auxin flows from the flanks in the basipetal direction, promoting vascular differentiation within the embryonic axis (red area).

maize, and sorghum result in dwarfing of varying severity and in altered gravitropism and reduced auxin efflux. In general, ABCBs are uniformly, rather than polarly, distributed in the plasma membranes of cells in shoot and root apices (see Figure 14.10B). However, when specific ABCB and PIN proteins occur together in the same location in the cell, the polarity of auxin transport is enhanced. The compound N-1-naphthylphthalamic acid (NPA) binds to ABCB auxin transport proteins and their regulators and is used as an inhibitor of auxin efflux activity.

Auxin synthesis and polar transport regulate embryonic development

DR5 and DII-Venus auxin reporters have been used to map the distribution of auxin in developing embryos. This information has been combined with fluorescent tagging of auxin biosynthesis and transport genes, to try to understand how auxin synthesis and directed transport combine to generate a patterned distribution of auxin across the developing embryo (**Figure 14.11**). In particular, fusions of fluorescent proteins to PIN1 have helped scientists visualize the auxin flows that direct organogenesis in the embryo. The position of PIN1 can be used to infer the direction of auxin flow. However, the activity of PIN1 is regulated posttranslationally by kinases, so its presence is not necessarily evidence of its activity.

Polar localization of PIN auxin efflux proteins is thought to involve three processes:

1. Initial nonpolar trafficking of PIN to the plasma membrane via the usual secretory vesicle pathway (see Chapter 1).

2. Concentration of PIN in polarized plasma membrane domains via transcytosis. In plants, **transcytosis** refers to the invagination of the plasma membrane to form vesicles at one location in the cell, followed by the movement of these vesicles to another region of the cell, where they fuse with the plasma membrane.

3. Stabilization of PIN localization via interactions with the cell wall. Genetic or pharmacological disruption of cell wall biosynthesis results in a complete loss of PIN1 polarity in Arabidopsis.

Radial patterning guides formation of tissue layers

Just as important as the apical–basal axis in the developing embryo is the radial axis, which runs perpendicular to the apical–basal axis, extending from the interior to the surface. In Arabidopsis, differentiation of tissues along the radial axis is first observed in the globular embryo (**Figure 14.12**), where periclinal divisions separate the embryo into three concentric tissue layers. The outermost cells form a one-cell-thick layer known as the protoderm, which eventually differentiates into the epidermis. Below this layer lie cells that will later become the **ground tissue**, which in turn differentiates into cortex (the ground tissue between the vascular

transcytosis Redirection of a secreted protein from one membrane domain within a cell to another, polarized domain.

ground tissue The internal tissues of the plant, other than the vascular tissues.

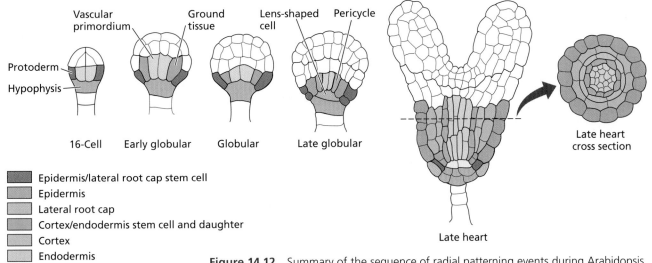

Figure 14.12 Summary of the sequence of radial patterning events during Arabidopsis embryogenesis. The five successive embryonic stages shown in longitudinal section illustrate the origin of distinct tissues, beginning with the protoderm (left) and ending with the vascular tissues (right). Note how the number of tissues increases as development proceeds. This increase is brought about by means of periclinal divisions within the ground and vascular tissues of the undifferentiated embryo. A cross-sectional view of the basal portion of the late heart stage embryo is shown at the far right (the level of the cross section is shown by the line in the longitudinal section to its left). (After Laux et al. 2004.)

system and the epidermis) and, in the root and hypocotyl, the endodermis (see Chapter 3). In the most central layer lies the **procambium**, meristematic tissue that generates the vascular tissues, including the pericycle of the root.

As we saw for apical–basal patterning of the embryo, a precise sequence of cell divisions does not appear to be essential for establishment of the radial axis. For example, closely related species of plants show significant variability in the patterns of cell divisions associated with formation of the radial axis. Furthermore, a basic radial tissue organization can still be established in mutants with disturbed patterns of cell division (see Figure 14.8), suggesting that position-dependent mechanisms of development are important. In the following sections we discuss experiments that address the nature of these mechanisms.

The protoderm differentiates into the epidermis

The protoderm forms the superficial layer of the embryo and eventually produces the epidermis, a critical tissue that mediates communication between the plant and the outside world. Protodermal cells originate early in embryogenesis and could, in theory, regulate the exchange of signals between the embryo and its surroundings. For example, studies in *Citrus* have shown the presence of a cuticle layer on the surface of the embryo from the earliest zygotic stages through maturity, suggesting that the walls of protodermal cells form a communication barrier. Some studies also suggest that the protoderm can act as a physical constraint to the growth of more internal layers.

The central vascular cylinder is elaborated by cytokinin-regulated progressive cell divisions

The vascular tissues of the stele eventually form in the most central positions of the embryo. This process involves a series of periclinal cell divisions that produce additional layers of cells along the radial axis. These layers then become patterned to particular fates by developmental programming and cell-to-cell signaling. For example, Arabidopsis mutants lacking a primary cytokinin receptor (see Chap-

procambium Primary meristematic tissue that differentiates into xylem, phloem, and cambium.

(A) Wild-type root

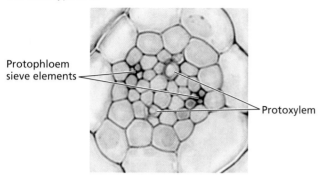

Protophloem
sieve elements

Protoxylem

(B) Mutant root lacking cytokinin receptor

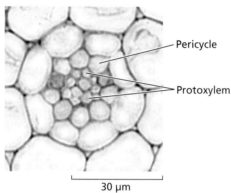

Pericycle

Protoxylem

30 μm

Figure 14.13 The cytokinin receptor is required for normal phloem development in Arabidopsis. Comparison of (A) a wild-type root and (B) a mutant root lacking the cytokinin receptor. Note that the absence of phloem elements in the mutant is accompanied by an apparent decrease in the number of cell layers. (From Mähönen et al. 2000. © 2000, Cold Spring Harbor Laboratory Press.)

ter 12) fail to undergo a critical round of cell division that normally produces precursors for xylem and phloem (**Figure 14.13**). This defect leads to development of a vascular system that contains xylem but not phloem.

Formation and Maintenance of Apical Meristems

The postembryonic development of plants shows a remarkable degree of flexibility (*plasticity*), largely because of specialized tissues called **meristems**, which are established during embryogenesis. A meristem can be broadly defined as a group of cells that retain the ability to proliferate and whose ultimate fate is not rigidly determined, but is subject to modification by external factors. This responsiveness to external cues enables the plant to best exploit the prevailing environment. In addition, the activity of meristems confers on plants the property of indeterminate growth, enabling some of them, such as bristlecone pines (*Pinus longaeva*), to live and reproduce for thousands of years. In this section we focus on the apical meristems that are formed during embryogenesis.

The root apical meristem (RAM) and the shoot apical meristem (SAM) are found at the tips of the root and shoot, respectively. Both the RAM and SAM include defined clusters of cells, termed **initials**, that are distinguished by their slow rate of division and undetermined fate. As the descendants of initials are displaced by polarized patterns of cell division, they take on various differentiated fates that contribute to the radial and longitudinal organization of the root or shoot, and to the development of lateral organs.

Auxin and cytokinin contribute to the formation and maintenance of the RAM

In addition to auxin's importance in the development of apical–basal polarity, the hormone is involved in positioning the RAM and guiding its complex behavior. The position of the quiescent center (QC), which becomes the core of the root meristem, normally coincides with an auxin concentration maximum at the base of the globular and early heart stage embryos (see Figure 14.11). When the position of the auxin maximum is shifted by chemical treatments, the position of the QC shows corresponding changes. By contrast, treatments that abolish the auxin maximum lead to loss of the QC. In both cases, the downstream effect is that the root apex becomes disordered or dysfunctional.

Similar results are observed in Arabidopsis and other species when auxin biosynthetic genes or a subset of auxin signaling components are mutated. In the embryo and very young seedlings, auxin appears to be transported from sites of production in the shoot. As seedlings mature, auxin is increasingly synthesized in the root meristem itself. Experimental evidence suggests that many of the primary transcription factors that initiate cell differentiation in the root meristem are activated by de-repression of auxin response factors. However,

meristems Localized regions of ongoing cell division that enable growth during postembryonic development.

initials Broadly defined as the cells of the root and shoot apical meristems. More specifically, a cluster of slowly dividing cells located within the meristem that gives rise to the more rapidly dividing cells of the surrounding meristem.

auxin also appears to be necessary for maintenance of a pool of undifferentiated initials in the meristem. Loss of auxin can result in premature differentiation of this pool of cells.

Root meristem development is also regulated by cytokinin. The presence of cytokinin in individual cells can be detected using a reporter gene consisting of a synthetic cytokinin-responsive two-component signaling sensor (TCS) promoter fused to a gene encoding GFP. After the hypophysis divides in early embryogenesis, *TCS::GFP* expression is observed primarily in the lens-shaped cell that becomes the QC (**Figure 14.14A and B**). At the same time, auxin reporters suggest that auxin signaling capacity is reduced in the QC at this point and elevated in the adjacent root meristem basal cell that gives rise to the columella and central root cap (**Figure 14.14E**). The meristem basal cell has high levels of two auxin-inducible proteins, ARR7 and ARR15 (**Figure 14.14C and D**); because these proteins are repressors of cytokinin responses, auxin appears to inhibit cytokinin action in this cell. Mutations of the *ARR7* and *ARR15* genes result in abnormal root phenotypes, suggesting that suppression of cytokinin signaling in the basal cell is essential for normal development.

After the formation of the RAM, differentiation of the vascular tissues, cortex, and epidermal cells takes place under the control of specific transcription factors.

SAM formation is also influenced by factors involved in auxin movement and responses

Development of the shoot apical meristem (SAM), like that of the RAM, is linked to complex patterns of intercellular auxin transport. At the two-cell stage of embryogenesis, the distribution of PIN1 is such that it leads to accumulation of auxin in the apical cell, but by the globular stage the distribution of PIN proteins is reversed, leading to a basally directed redistribution of auxin (see Figure 14.11). By the early heart stage, polar auxin transport is being directed toward the flanks of the embryo. One consequence of these auxin movements is to cause the formation of a central apical region, where the auxin concentration and auxin-dependent activities are low relative to those in flanking regions (**Figure 14.15**). This central region corresponds to the future SAM. At the same time, cotyledons develop where lateral protodermal movement of auxin away from the central intercotyledonary zone converges with upward protodermal flows along the flanks of the embryo to create two new auxin maxima. These

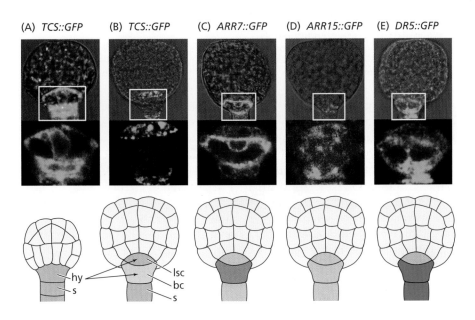

(A) *TCS::GFP* (B) *TCS::GFP* (C) *ARR7::GFP* (D) *ARR15::GFP* (E) *DR5::GFP*

Figure 14.14 Inverse correlation between cytokinin and auxin signaling in the embryo. (A) Expression of *TCS::GFP* (a reporter for cytokinin; green color) in the hypophysis at the early globular stage. (B) Decreased *TCS::GFP* expression in the basal cell lineage at the late globular stage. (C) At the same stage, expression of *ARR7::GFP* is highest in the basal cell lineage. (D) Expression pattern of *ARR15::GFP*. *ARR7* and *ARR15* encode proteins that suppress cytokinin responses. (E) Expression of *DR5::GFP* (an auxin-responsive reporter) is highest in the basal cell lineage. The boxed sections in the upper panels are magnified underneath; schematic interpretations are shown at the bottom (darker shades represent higher expression of the reporter). Abbreviations: hy, hypophysis; bc, basal cell lineage; lsc, lens-shaped cell; s, suspensor. (From Müller and Sheen 2008.)

(A)

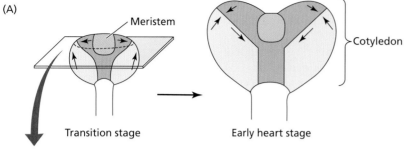

Meristem

Transition stage

Early heart stage

Cotyledon

(B)

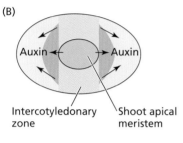

Auxin ← → Auxin

Intercotyledonary zone

Shoot apical meristem

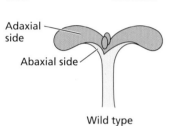

Adaxial side

Abaxial side

Wild type

Figure 14.15 Model for auxin-dependent patterning of the shoot apex. (A) The direction of auxin transport (black arrows) during transition stage and early heart stage Arabidopsis embryos. (B) Cross section (as shown in A) through the apical region of a wild-type embryo showing the region in the embryo that will develop into the shoot apical meristem, the intercotyledonary zone, and the adaxial and abaxial domains of the cotyledon. (After Jenik and Barton 2005.)

pools of auxin then feed into downward flows that converge in the hypocotyl, promoting vascular differentiation.

Mutant analyses in Arabidopsis and other species have identified a small group of transcription factors as the primary regulators of SAM formation (**Figure 14.16**). These transcription factors regulate the ordered formation of the seedling SAM from heart stage embryonic cell types. As in the RAM, this involves a balance of cellular differentiation and maintenance of a pool of initial cells to function in organogenesis and to dynamically respond to environmental conditions to expand or decrease meristem size.

Cell proliferation in the SAM is regulated by cytokinin and gibberellin

The *SHOOT MERISTEMLESS* (*STM*) gene encodes a transcription factor that regulates cell proliferation in the SAM. *stm* mutants never develop a shoot apical meristem (although the root apical meristem is normal), indicating that *STM* is required for the initiation of the SAM in embryos. Since exogenous treatments of the Arabidopsis *stm* mutant with the hormones gibberellin (GA) and cytokinin partially restore normal SAM development, GA and cytokinin seem to be involved in cell proliferation in the SAM (**Figure 14.17**). STM is a member of the KNOX family of *homeodomain* proteins, which promote synthesis of cytokinin in the SAM. This suggests that STM promotion of cytokinin synthesis helps maintain the SAM.

KNOX proteins also influence GA-mediated processes. In a variety of species, KNOX proteins suppress accumulation of GA in the SAM by stimulating transcrip-

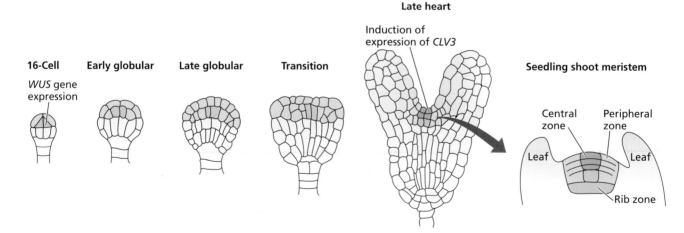

Late heart

Induction of expression of *CLV3*

16-Cell   **Early globular**   **Late globular**   **Transition**

WUS gene expression

Seedling shoot meristem

Central zone

Peripheral zone

Leaf

Leaf

Rib zone

Figure 14.16 Formation of the apical region involves a defined sequence of gene expression, including the early onset of expression of *WUS*, which encodes a transcription factor, in an internal layer (orange), which induces the expression of *CLAVATA3* (*CLV3*) in adjacent external cell layers (blue). At maturity, the SAM is composed of three distinct tissue regions: the central zone, the peripheral zone, and the rib zone (discussed in Chapter 16). (After Laux et al. 2004.)

Figure 14.17 Model for how expression of the KNOX transcription factor SHOOT MERISTEMLESS (STM) elevates cytokinin levels while repressing GA in the SAM. P4 (right) is a developing leaf, and P0 (left) is the site where the next leaf primordium will form. (After Hudson 2005.)

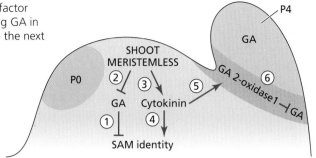

tion of *GA 20-OXIDASE1*, a gene that encodes an enzyme that degrades active GA. KNOX proteins also suppress GA activity in the meristem indirectly via cytokinin, which stimulates expression of *GA 2-OXIDASE1* at the boundaries between the meristem and the emerging leaves (see Figure 14.17). Expression of *GA 2-OXIDASE1* in this location is thought to prevent movement of active GA into the SAM from nearby developing leaves (e.g., P4 in Figure 14.17). Genetic experiments have shown that artificial activation of GA signaling in the SAM destabilizes the meristem, demonstrating that the restriction of GA levels in the SAM is likely to be a key mechanism by which *KNOX* genes contribute to meristem stability.

1. GA inhibits shoot apical meristem identity.
2. STM protein activity represses GA biosynthesis.
3. STM protein activity promotes cytokinin biosynthesis.
4. The combination of low GA and high cytokinin promotes shoot apical meristem identity.
5. Cytokinin promotes GA 2-oxidase1 activity.
6. GA 2-oxidase1 activity causes GA deactivation in the leaf primordium boundary.

Summary

The sporophyte generation of plants begins with the fertilization events that initiate embryogenesis. Regulated cell divisions produce the polar axis and bilateral symmetry of the embryo. Both mobile and positional signals function as morphogenic regulators. An extended set of these regulatory mechanisms functions in the further elaboration of plant organs during postembryonic growth. Postembryonic plants retain meristems (stem cell niches) that are sites of undifferentiated cell division to provide for plastic or adaptive growth.

Overview of Embryogenesis

- Embryogenesis and embryonic development proceed according to programmed series of cell divisions and cellular differentiations.

Comparative Embryology of Eudicots and Monocots

- The basic principals of embryonic development are similar in most plants, but specific features vary considerably between eudicots and monocots (**Figures 14.2, 14.3**).

- Among seed plants, the apical–basal polarity is established early in embryogenesis (**Figure 14.4**).

- Position-dependent mechanisms for determining cell fate guide embryogenesis (**Figure 14.5**).

- Directional cell divisions determine much of the embryonic structure (**Figure 14.6**).

- Processes other than a fixed sequence of cell divisions must guide radial pattern formation (**Figure 14.8**).

- The potential for intercellular protein movement changes during development (**Figure 14.9**).

- Auxin (indole-3-acetic acid, or IAA) may function as a mobile chemical signal during embryogenesis (**Figures 14.10, 14.11**).

- Radial patterning guides formation of tissue layers (**Figure 14.12**).

- Cytokinin signaling is required for normal phloem development in the root (**Figure 14.13**).

Formation and Maintenance of Apical Meristems

- The root and shoot apical meristems use similar strategies to enable indeterminate growth.

- The origin of different root tissues can be traced to distinct types of initial cells.

- The behavior of initials in the RAM depends on the activation of a series of transcription factors by auxin and interactions with cytokinin (**Figure 14.14**).

- The shoot apical meristem has a structure distinct from that of the root apical meristem (**Figures 14.15, 14.16**).

- Embryonic SAM formation requires the coordinated expression of specific transcription factors to establish a set of undetermined cells with potential for further proliferation.

- Expression of KNOX transcription factors promotes cytokinin production in the SAM, while limiting GA levels (**Figure 14.17**).

Suggested Reading

Aichinger, E., Kornet, N., Friedrich, T., and Laux, T. (2012) Plant stem cell niches. *Annu. Rev. Plant Biol.* 63: 615–636.

Barlow, P. W. (1994) Evolution of structural initial cells in apical meristems of plants. *J. Theor. Biol.* 169: 163–177.

Esau, K. (1965) *Plant Anatomy*, 2nd ed. Wiley, New York.

Hudson, A. (2005) Plant meristems: Mobile mediators of cell fate. *Curr. Biol.* 15: R803–R805.

Jenik, P. D., and Barton, M. K. (2005) Surge and destroy: The role of auxin in plant embryogenesis. *Development* 132: 3577–3585.

Laux, T., Wurschum, T., and Breuninger, H. (2004) Genetic regulation of embryonic pattern formation. *Plant Cell* 16 (Suppl.): S190–S202.

Maule, A. J., Benitez-Alfonso, Y., and Faulkner, C. (2011) Plasmodesmata–Membrane tunnels with attitude. *Curr. Opin. Plant Biol.* 14: 683–690.

Miyashima, S., Sebastian, J., Lee, J.-Y., and Helariutta, Y. (2013) Stem cell function during plant vascular development. *EMBO J.* 32: 178–193.

Reinhardt, D., Pesce, E. R., Stieger, P., Mandel, T., Baltensperger, K., Bennett, M., Traas, J., Friml, J., and Kuhlemeier, C. (2003) Regulation of phyllotaxis by polar auxin transport. *Nature* 426: 255–260.

Sachs, T. (1991) Cell polarity and tissue patterning in plants. *Development* 113 (Suppl. 1): 83–93.

Scheres, B. (2013) Rooting plant development. *Development* 140: 939–941.

Scheres, B., Wolkenfelt, H., Willemsen, V., Terlouw, M., Lawson, E., Dean, C., and Weisbeek, P. (1994) Embryonic origin of the *Arabidopsis* primary root and root meristem initials. *Development* 120: 2475–2487.

Sparks, E., Wachsman, G., and Benfey, P. N. (2013) Spatiotemporal signalling in plant development. *Nat. Rev. Genet.* 14: 631–644.

15 Seed Dormancy, Germination, and Seedling Establishment

"It's not dead, it's resting." –Monty Python

In Chapter 14 we discussed the process of embryogenesis that occurs in the developing seeds of angiosperms. Seeds are specialized dispersal units unique to the Spermatophyta (seed plants), which includes both the angiosperms and gymnosperms. In both angiosperms and gymnosperms, seeds develop from ovules, which contain the female gametophyte (see Chapters 1 and 17). The packaging of the embryo into a self-contained seed was one of many adaptations that freed plant reproduction from a dependence on water. The evolution of seed plants thus represents an important milestone in the adaptation of plants to dry land.

In this chapter we describe the processes of seed germination, hypocotyl and coleoptile growth (including tropisms and photomorphogenesis), and seedling establishment—by which we mean the production of the first photosynthetic leaves and a minimal root system. Between embryogenesis and germination there is typically a period of *seed maturation* that culminates in *quiescence*, a non-germinating state characterized by a reduced metabolic rate, after which the seed is released from the parent plant. The quiescent state thus ensures that germination is delayed until the seed reaches the soil, where it can receive the water and oxygen required for seedling growth. Some seeds require an additional treatment, such as light, chilling, or physical abrasion, before they can germinate, a condition known as *dormancy*. For many agricultural crops this is not a problem, because human selection for rapidly germinating seeds has resulted in the gradual loss of dormancy-inducing genes.

The seed tissues surrounding the embryo form a barrier that protects the embryo from the environment. In addition, the seed tissues provide stored foods that nourish the embryo throughout embryogenesis and early seedling development. The food reserves of seeds are stored in several types of tissues. Because germination is tightly coupled to the mobilization of these food reserves, we begin with a description of seed structure and composition. Next we consider various types of seed dormancy, which in some cases must be overcome before germination can take place. We then discuss the mobilization of stored food reserves in different types of seeds, and we describe the role of hormones in coordinating the processes of seedling growth and food mobilization.

Soon after emerging from the seed, the seedling undergoes a rapid transition from a heterotrophic to a photoautotrophic mode of nutrition. We begin our discussion of seedling establishment with an examination of photomorphogenesis—the influence of light on shoot morphology, including the role of hormones in light-regulated signaling pathways. Next we turn to root morphology and the developmental zones of the growing root tip. As sedentary organisms, plants must rely on growth responses to compete for sunlight and scavenge for water and minerals in the soil. At the cellular level, the growth rate is controlled by the mechanical properties of the cell wall, and auxin plays a central role in modulating cell wall properties. Finally, we take up the topic of phototropism and gravitropism, the processes by which plants orient themselves to light and gravity.

Seed Structure

Seeds are the dispersal units containing the mature embryos of angiosperms and gymnosperms. They are derived from the mature ovules of these two groups of plants. Although we will focus on the seeds of angiosperms because of their extraordinary diversity and importance to agriculture, it is important to appreciate the basic differences between angiosperms and gymnosperms. All seeds contain three basic structural features: an embryo, food storage tissue, and a protective outer layer of dead cells called the **seed coat** (or **testa**). In angiosperms, the food storage tissue that nourishes the growing embryo is the triploid endosperm that results from double fertilization (see Chapter 17). In some angiosperm species, the seed coat is fused to the fruit wall, or **pericarp**, which is derived from the ovary wall. Fused testa–pericarps are a feature of all cereal *caryopses*, technically making these "seeds" fruits. However, in this book we will refer to cereal caryopses as seeds. **Figure 15.1** shows a variety of familiar true seeds as well as fruits with a seedlike appearance.

Seed anatomy varies widely among different plant groups

Despite their common features, seeds exhibit a remarkable range of sizes, from the dust-sized particles of orchids, weighing 1 µg, to the massive seeds of the coco de mer (*Lodoicea maldivica*), which can reach 30 cm in length and weigh 20 kg. Despite the relative simplicity of the embryo, and the limited number of tissues surrounding it, seed anatomy exhibits considerable diversity among the different plant groups. Some representative examples of the structures of seeds of eudicots and monocots are shown in **Figure 15.2**.

Seeds can be categorized broadly as endospermic or non-endospermic, depending on the presence or absence of a well-formed triploid endosperm at maturity (see Figure 15.2). For example, beet seeds are non-endospermic because the triploid endosperm is largely used up during embryo development. Instead, the perisperm and storage cotyledons serve as the main sources of nutrients during germination. The **perisperm** is derived from the nucellus, the maternal tissue that gives rise to the ovule. Garden bean seeds (*Phaseolus vulgaris*) and

testa   The outer layer of the seed, also called the seed coat, derived from the integument of the ovule.

pericarp   The fruit wall surrounding a fruit, derived from the ovary wall.

perisperm   Storage tissue derived from the nucellus, and often consumed during embryogenesis.

Figure 15.1 Seeds and seedlike fruits. (A–D) True seeds. (A) Rapeseed (*Brassica napus*). (B) Brazil nut. (C) Coffee bean. (D) Coconut. (E–I) Seedlike fruits. (E) Maple (samara). (F) Strawberry (achene). (G) Wheat and other cereals (caryopses). (H) Oak (nut). (I) Sunflower and other composites (cypsela). Monocot caryopses and Asteraceae cypselae are routinely referred to as seeds. (A © Roman Nerud/Shutterstock.com; B © iStock.com/ sdstockphoto; C © Jiri Hera/Shutterstock.com; D © iStock.com/ kickers; E © iStock.com/SweetpeaAnna; F © Suslik1983/Shutterstock.com; G © BW Folsom/Shutterstock.com; H © Vania Zhukevych/Shutterstock.com; I © iStock.com/surabky.)

legume seeds in general are also non-endospermic, relying on their large storage cotyledons, which make up most of the bulk of the seed, for their food reserves. In contrast, castor bean (*Ricinus communis*), onion (*Allium cepa*), bread wheat (*Triticum* spp.), and maize (corn; *Zea mays*) seeds are all endospermic.

In keeping with its role as a food storage tissue, the endosperm is typically rich in starch, oils, and protein. Some endosperm tissue has thick cell walls that break down during germination, releasing a variety of sugars. The outermost layer of the endosperm in some species differentiates into a specialized secretory tissue with thickened primary walls called the *aleurone layer*, so called because it is composed of cells filled with **protein storage vacuoles**, originally called *aleurone grains*. As you will see later in the chapter, the aleurone layer plays an important role in regulating dormancy in certain eudicot seeds. In wheat seeds and those of other members of the Poaceae (grass family), secretory aleurone layers are also responsible for the mobilization of stored food reserves during germination.

The embryos of cereal grains are highly specialized and merit closer examination both because of their agricultural importance and because they have been widely used as model systems to study the hormonal regulation of food mobilization during germination. Specialized embryonic structures peculiar to

protein storage vacuoles
Specialized small vacuoles that accumulate storage proteins, typically in seeds.

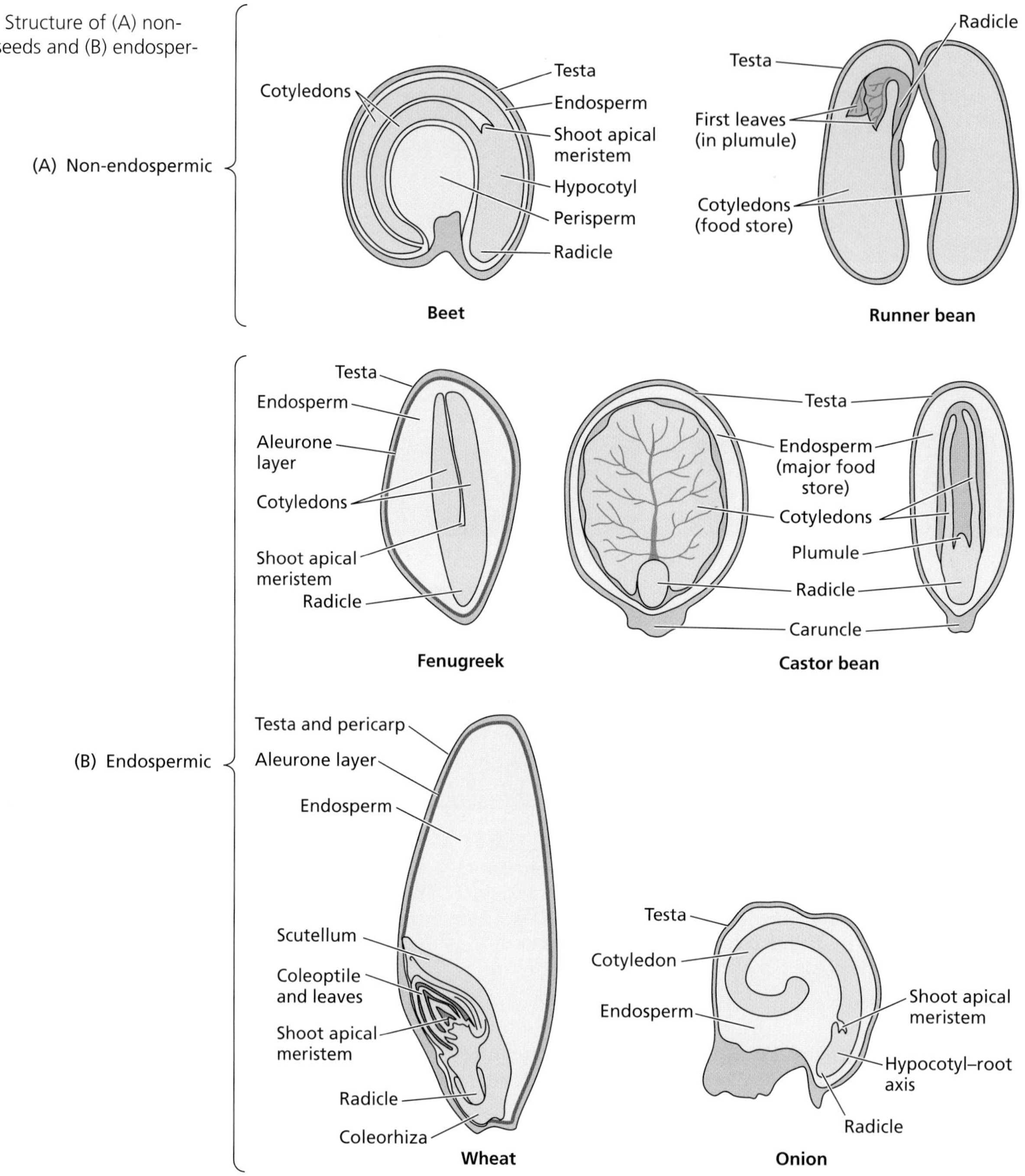

Figure 15.2 Structure of (A) non-endospermic seeds and (B) endospermic seeds.

(A) Non-endospermic

Cotyledons — Testa
— Endosperm
— Shoot apical meristem
— Hypocotyl
— Perisperm
— Radicle

Beet

Testa — Radicle
First leaves (in plumule)
Cotyledons (food store)

Runner bean

(B) Endospermic

Testa
Endosperm
Aleurone layer
Cotyledons
Shoot apical meristem
Radicle

Fenugreek

Testa
Endosperm (major food store)
Cotyledons
Plumule
Radicle
Caruncle

Castor bean

Testa and pericarp
Aleurone layer
Endosperm
Scutellum
Coleoptile and leaves
Shoot apical meristem
Radicle
Coleorhiza

Wheat

Testa
Cotyledon
Endosperm
Shoot apical meristem
Hypocotyl–root axis
Radicle

Onion

scutellum The single cotyledon of the grass embryo, specialized for nutrient absorption from the endosperm.

coleoptile A modified ensheathing leaf that covers and protects the young primary leaves of a grass seedling as it grows through the soil.

the grass family were discussed in Chapter 14 and include the following (see Figure 15.2):

- The single cotyledon has been modified by evolution to form an absorptive organ, the **scutellum**, which forms the interface between the embryo and the starchy endosperm tissue. During germination, mobilized sugars from the endosperm are absorbed by the scutellum and transported to the embryo proper.

- The basal sheath of the scutellum has been elongated to form a **coleoptile** that covers and protects the first leaves while the shoot is growing up through the soil.

- The base of the hypocotyl has been elongated to form a protective sheath around the radicle called the **coleorhiza**.

- In some members of the grass family, such as maize, the upper hypocotyl has been modified to form a **mesocotyl** (not shown in Figure 15.2). During seedling development, the growth of the mesocotyl helps raise the leaves to the soil surface, especially in the case of deeply planted seeds.

Seed Dormancy

During seed maturation, the embryo dehydrates and enters a quiescent phase. Seed germination requires rehydration and can be defined as the resumption of growth of the embryo in the mature seed. However, the process of germination encompasses all the events that take place between the start of *imbibition* (moistening) of the dry seed (discussed later in the context of seed germination) and the *emergence* of the embryo, usually starting with the radicle, from the structures that surround it. Successful completion of germination depends on the same environmental conditions as vegetative growth (see Chapter 16): Water and oxygen must be available, and the temperature must be in the *physiological range* (that is, the range that does not inhibit physiological processes). However, a viable (living) seed may not germinate even if the appropriate environmental requirements are satisfied, a phenomenon known as **seed dormancy**. Seed dormancy is an intrinsic temporal block to the initiation of germination that provides additional time for seed dispersal over greater distances. It also maximizes seedling survival by preventing germination under unfavorable conditions.

Most mature seeds typically have less than 0.1 g water g^{-1} dry weight at the time of shedding. As a consequence of dehydration, metabolism comes nearly to a halt and the seed enters a quiescent or "resting" state. In some cases the seed becomes dormant as well. Unlike **seed quiescence**, defined as the failure to germinate due to the lack of water, O_2, or proper temperature for growth, seed dormancy requires additional treatments or signals for germination to occur.

Different types of seed dormancy can be distinguished on the basis of the developmental timing of dormancy onset. Newly dispersed, mature seeds that fail to germinate under favorable conditions exhibit **primary dormancy**. Once primary dormancy has been lost, non-dormant seeds may acquire **secondary dormancy** if exposed to unfavorable conditions that inhibit germination over an extended period of time.

There are two basic types of seed dormancy mechanisms: exogenous and endogenous

Seed dormancy mechanisms have been classified in different ways. According to one scheme, primary seed dormancy can be divided into two main types, *exogenous dormancy* and *endogenous dormancy*.

Exogenous dormancy, or **coat-imposed dormancy**, refers to the inhibitory effects of the seed coat or other enclosing tissues, such as the endosperm, pericarp, or extrafloral organs, on the growth of the embryo during germination. The embryos of such seeds germinate readily in the presence of water and oxygen once the seed coat and other surrounding tissues have either been removed or damaged. There are several ways that seed coats can impose dormancy on the embryo:

- *Water impermeability.* This type of dormancy, also called hard-seededness, is common in plants found in arid and semiarid regions, especially among legumes, such as clover (*Trifolium* spp.) and alfalfa (*Medicago* spp.). The classic example is Indian lotus (*Nelumbo nucifera*) seeds, which can survive up to 1200 years because of their impermeable coats. Waxy cuticles, suberized layers, and the cell walls of palisade layers consisting

coleorhiza   A protective sheath surrounding the embryonic radicle in members of the grass family.

mesocotyl   In members of the grass family, the part of the elongating axis between the scutellum and the coleoptile.

seed dormancy   A state of arrested growth of the embryo that prevents germination even when all the necessary environmental conditions for growth, such as water, O_2, and temperature, are met.

seed quiescence   A state of suspended growth of the embryo due to a lack of water, O_2, or proper temperature for growth. Germination of quiescent seeds proceeds immediately once these conditions are met.

primary dormancy   The failure of newly dispersed, mature seeds to germinate under normal growth conditions.

secondary dormancy   Seeds that have lost their primary dormancy may become dormant again if they experience prolonged exposure to unfavorable growth conditions.

coat-imposed (or exogenous) dormancy   Seed dormancy that is caused by the seed coat and other surrounding tissues; may involve impermeability to water or oxygen, mechanical restraint, or the retention of endogenous inhibitors.

embryo (or endogenous) dormancy
Seed dormancy that is caused directly by the embryo and is not due to any influence of the seed coat or other surrounding tissues.

vivipary   The precocious germination of seeds in the fruit while still attached to the plant.

preharvest sprouting   Germination of physiologically mature wild-type seeds on the mother plant caused by wet weather.

precocious germination   Germination of viviparous mutant seeds while still attached to the mother plant.

hormone balance theory
The hypothesis that seed dormancy and germination are regulated by the balance of ABA and gibberellin.

of lignified sclereids all combine to restrict the penetration of water into the seed. This type of dormancy can be broken by mechanical or chemical scarification. In the wild, passage through the digestive tracts of animals can cause chemical scarification.

- *Interference with gas exchange.* Dormancy in some seeds can be overcome by oxygen-enriched atmospheres, suggesting that the seed coat and other surrounding tissues limit the supply of oxygen to the embryo. In wild mustard (*Sinapis arvensis*), the permeability of the seed coat to oxygen is less than that to water by a factor of 10^4. In other seeds, oxidative reactions involving phenolic compounds in the seed coat may consume large amounts of oxygen, reducing oxygen availability to the embryo.

- *Mechanical constraint.* The first visible sign of germination is typically the radicle (embryonic root) breaking through its surrounding structures, such as the endosperm, if present, and seed coat. In some cases, however, the thick-walled endosperm may be too rigid for the radicle to penetrate, as in Arabidopsis, tomato, coffee, and tobacco. For such seeds to complete germination, the endosperm cell walls must be weakened by the production of cell wall–degrading enzymes, typically where the radicle emerges.

- *Retention of inhibitors.* Dormant seeds often contain secondary metabolites, including phenolic acids, tannins, and coumarins, and repeated rinsing of such seeds with water often promotes germination. The seed coat may impose dormancy by preventing the escape of inhibitors from the seed, or the seed coat may produce inhibitors that diffuse into the embryo.

Endogenous dormancy, also called **embryo dormancy**, refers to seed dormancy that is intrinsic to the embryo and is not due to any physical or chemical influence of the seed coat or other surrounding tissues. Embryo dormancy is typically induced by abscisic acid (ABA) at the end of embryogenesis. Fully mature seeds require endogenous ABA for the regulation and maintenance of primary dormancy following imbibition of the dry seed. For example, in Arabidopsis, lettuce, barley, and tobacco seeds, the degree of seed dormancy correlates with endogenous ABA concentration in imbibed, rather than in dry, seeds. (Regulation of seed dormancy by the ABA:GA ratio is discussed later in the chapter.)

In addition to exogenous and endogenous dormancy, seeds also may fail to germinate because the embryos have not yet reached their full size or maturity. Technically, these seeds are not truly dormant, but simply require additional time for the embryo to enlarge under appropriate conditions before they can emerge from the seed. Familiar examples of seed dormancy caused by undersized embryos are celery (*Apium graveolens*) and carrot (*Daucus carota*) (**Figure 15.3**). Seeds with undifferentiated embryos are usually tiny and include the parasitic broomrapes (*Orobanche* and *Phelipanche* spp.) and orchids.

Non-dormant seeds can exhibit vivipary and precocious germination

In some estuarine species the mature seeds not only lack dormancy, but also germinate while still on the mother plant, a phenomenon known as **vivipary**. True vivipary is extremely rare in angiosperms and is largely restricted to mangroves and other plants growing in estuarine or riparian ecosystems in the tropics and subtropics. A well-known example of a viviparous species is the red mangrove (*Rhizophora mangle*) (**Figure 15.4**). Seeds of this species germinate while still inside the attached fruit and produce an elongated, dartlike propagule that can drop from the tree and root itself in the surrounding soft mud.

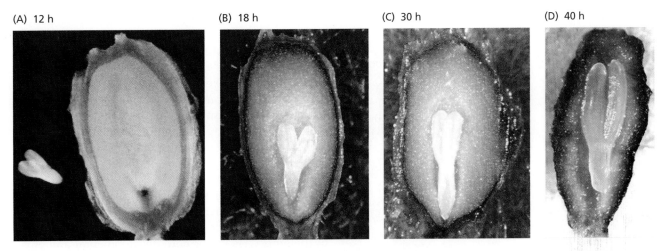

(A) 12 h   **(B)** 18 h   **(C)** 30 h   **(D)** 40 h

Figure 15.3 Growth of the undersized carrot embryo during imbibition of seeds for 12 h (A), 18 h (B), 30 h (C), and 40 h (D). The tiny embryo on the left, removed from the seed for better visibility, is embedded in a cavity in the endosperm formed by the release of cell wall–degrading enzymes. Germination begins with the emergence of the radicle from the seed 2 to 4 days after imbibition. (From Homrichhausen et al. 2003.)

The germination of physiologically mature seeds on the mother plant is known as **preharvest sprouting** and is characteristic of some grain crops when they mature in wet weather (**Figure 15.5A**). Preharvest sprouting in cereals (e.g., wheat, barley, rice, and sorghum) reduces grain quality and causes serious economic losses. In maize, *viviparous* (*vp*) mutants have been selected in which the embryos germinate directly on the cob while still attached to the parent plant, referred to as **precocious germination** (**Figure 15.5B**). Several of these mutants are ABA deficient, while one is ABA insensitive. Vivipary in the ABA-deficient mutants can be partially prevented by treatment with exogenous ABA. Vivipary in maize also requires synthesis of gibberellin (GA) early in embryogenesis as a positive signal; double mutants deficient in gibberellin and ABA do not exhibit this phenomenon. This shows that the ABA:GA ratio regulates germination, not the actual amount of ABA.

The ABA:GA ratio is the primary determinant of seed dormancy

It has long been known that ABA exerts an inhibitory effect on seed germination, whereas gibberellin exerts a positive influence. According to the **hormone balance theory**, the ratio of these two hormones serves as the primary determinant of seed dormancy and germination. The relative hormonal activities of ABA and gibberellin in the seed depend on two main factors: the amounts of each hormone present in the target tissues, and the ability of the target tissues to detect and respond to each of the hormones. Hormonal sensitivity, in turn, is determined by the hormonal signaling pathways in the target tissues.

The amounts of the two hormones are regulated by their rates of synthesis versus deactivation (see Chapter 12). The balance between the biosynthetic and deactivation pathways is regulated at the gene level by the action of transcription factors. DELLA family proteins are transcriptional repressors that, in the absence of GA, inhibit germination, in part by increasing expression of transcription factors that

Figure 15.4 Viviparous seeds of red mangrove (*Rhizophora mangle*). (Photo © Larry Larsen/Alamy.)

(A)

(B)

Figure 15.5 (A) Preharvest sprouting in a head of wheat (*Triticum aestivum*). (B) Precocious germination in an ABA-deficient *vivipary14* mutant of maize. (A from Li et al. 2009; B courtesy of Bao-Cai Tan and Don McCarty.)

promote ABA biosynthesis. As discussed below, certain environmental factors can alter the ABA:GA balance of seeds and thereby stimulate germination. Certain treatments, such as *after-ripening*, can promote germination by reducing ABA levels, while other treatments, such as chilling (or *stratification*) can promote germination by increasing GA biosynthesis (**Figure 15.6**). Gibberellin promotes germination by causing the proteolytic degradation of the DELLA repressors that normally inhibit germination. We will discuss both after-ripening and chilling in greater detail later in the chapter.

Another factor that appears to regulate seed dormancy is the relative sensitivities of the embryo to ABA and GA. According to a recent model, the hormonal sensitivity of the embryo is under both developmental and environmental control (see Figure 15.6). During the early stages of seed development, ABA sensitivity is high and gibberellin sensitivity is low, which favors dormancy over germination. Later in seed development, ABA sensitivity declines and gibberellin sensitivity increases, favoring germination. At the same time, the seed becomes progressively more sensitive to environmental cues, such as temperature and light, that can either stimulate or inhibit germination.

ABA and gibberellin are not the only hormones regulating seed dormancy. For example, both ethylene and brassinosteroids reduce the ability of ABA to inhibit germination, apparently by negatively regulating the ABA signal transduction pathway. ABA also inhibits ethylene biosynthesis, while brassinosteroids enhance it. Thus, hormonal networks are probably involved in regulating seed dormancy, as they are in regulating most developmental phenomena.

Figure 15.6 Model for ABA and gibberellin (GA) regulation of dormancy and germination in response to environmental factors. Environmental factors such as temperature affect the ABA:GA ratios and the responsiveness of the embryo to ABA and gibberellin. In dormancy, gibberellin is catabolized and ABA synthesis and signaling predominate. In the transition to germination, ABA is catabolized and gibberellin synthesis and signaling predominate. The complex interplay between ABA and gibberellin synthesis, degradation, and sensitivity in response to ambient environmental conditions can result in cycling between dormant and non-dormant states (dormancy cycling). Germination can proceed to completion when favorable environmental conditions and non-dormancy coincide. (After Finch-Savage and Leubner-Metzger 2006.)

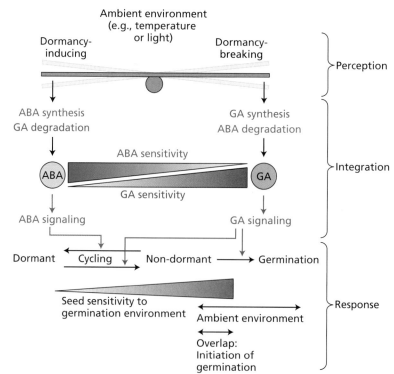

Release from Dormancy

Breaking dormancy involves a change in the metabolic state of the seed that enables the embryo to reinitiate growth. Because germination is an irreversible process that commits the seed to grow into a seedling, many species have developed sophisticated mechanisms for sensing the optimal environmental conditions for this to occur. Often there are seasonal components to the ultimate "decision" of a seed to germinate, as in the examples of secondary dormancy noted earlier in the chapter. In this section we discuss some of the environmental cues that bring about the release from dormancy. Although each external signal is discussed separately, seeds in nature must integrate their responses to multiple environmental factors, perceived either simultaneously or in succession.

Because the ABA:GA ratio plays such a decisive role in maintaining seed dormancy, environmental conditions that break dormancy ultimately are thought to activate gene networks that affect the balance between the responses to ABA and gibberellin. This hypothesis is consistent with the fact that treatment of seeds with gibberellin often can substitute for a positive environmental signal in breaking dormancy.

Light is an important signal that breaks dormancy in small seeds

Many seeds have a light requirement for germination (termed *photoblasty*), which may involve only a brief exposure, as in the case of the 'Grand Rapids' cultivar of lettuce (*Lactuca sativa*) (see Figure 13.5); an intermittent treatment (e.g., succulents of the genus *Kalanchoe*); or even a specific photoperiod involving short or long days. For example, birch (*Betula* spp.) seeds require long days (16 h) to germinate, while seeds of the conifer eastern hemlock (*Tsuga canadensis*) require short days. Phytochrome, which senses red (R) and far-red (FR) wavelengths of light (see Chapter 13), is the primary sensor for light-regulated seed germination. All light-requiring seeds exhibit coat-imposed dormancy, and removal of the outer tissues—specifically, the endosperm—allows the embryo to germinate in the absence of light. The effect that light has on the embryo is thus to enable the radicle (the embryonic root) to penetrate the endosperm, a process facilitated in some species by enzymatic weakening of the cell walls in the micropylar region, adjacent to the radicle.

Light is required by the small seeds of numerous herbaceous and grassland species, many of which remain dormant if they are buried below the depth to which light penetrates. Even when such seeds are on or near the soil surface, the amount of shading by the vegetation canopy (i.e., the R:FR light ratio the seeds receive) is likely to affect their germination. We will return to the effects of the R:FR ratio in Chapter 16 in relation to the phenomenon of shade avoidance.

Some seeds require either chilling or after-ripening to break dormancy

Many seeds require a period of low temperature (1–10°C) to germinate. In temperate-zone species, this requirement has obvious survival value, since such seeds will not germinate in the fall, but only in the following spring. Chilling seeds to break their dormancy is called **stratification**, named for the agricultural practice of overwintering dormant seeds in layered mounds of soil or moist sand. Today seeds are simply stored moist in a refrigerator. Stratification has the added benefit that it synchronizes germination, which ensures that plants will mature at the same time. **Figure 15.7A** shows the effects of chilling on apple seed germination. Intact seeds require 80 days of chilling for maximum germination, whereas isolated embryos achieve maximum germination after approximately 50 days. Thus, the presence of the surrounding tissues (seed coat and endosperm) increases the chilling requirement of the embryo by about 30 days.

stratification Technique for breaking seed dormancy by storage at cold temperatures (1–10°C) under moist conditions for several months. The term stratification is derived from the former practice of breaking dormancy by allowing seeds to overwinter in small mounds of alternating layers of seeds and soil.

(A)

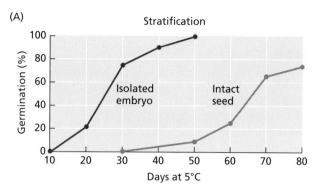

(B)

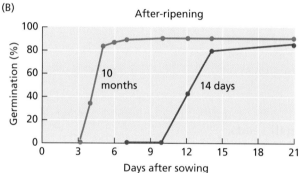

Figure 15.7 Seed dormancy can be overcome by stratification and after-ripening. (A) Release of apple seeds from dormancy by stratification, or moist chilling. Imbibed seeds were stored at 5°C and periodically removed to test the seeds or isolated embryos for germination. The germination of intact seeds was significantly delayed compared with that of isolated embryos. (B) The effect of after-ripening (dry storage at room temperature) on the germination of seeds of *Nicotiana plumbaginifolia*. After-ripening for 10 months or longer greatly accelerated germination compared with after-ripening for only 14 days. (A after Bewley 2013; B after Grappin et al. 2000.)

Some seeds may require a period of **after-ripening** before they can germinate. The duration of the after-ripening requirement may be as short as a few weeks (e.g., barley, *Hordeum vulgare*) or as long as 5 years (e.g., curly dock, *Rumex crispus*). In the field, after-ripening may occur in winter annuals in which dormancy is broken by high summer temperatures, allowing the seeds to germinate in the fall. In contrast, moist chilling during the cold winter months is effective in many summer annuals. After-ripening of horticultural and agricultural crop seeds is usually performed in special drying cupboards that maintain the appropriate temperature, aeration, and low moisture conditions.

The effect of the duration of after-ripening on the germination of seeds of *Nicotiana plumbaginifolia* is shown in **Figure 15.7B**. Seeds after-ripened for only 14 days began to germinate after about 10 days of subsequent wetting, while seeds after-ripened for 10 months began germinating after only 3 days. Seeds are considered "dry" when their water content drops below 20%. In several species, ABA decreases during after-ripening, and even a small decline may be sufficient to break dormancy. For example, in *N. plumbaginifolia* seeds, the ABA content decreases by about 40% during after-ripening. In general, after-ripening promotes a decrease in ABA concentration and sensitivity, and a rise in GA concentration and sensitivity. However, if the seeds become too dry (5% water content or less), the effectiveness of after-ripening is diminished.

Seed dormancy can be broken by various chemical compounds

Numerous chemicals, such as respiratory inhibitors, sulfhydryl compounds, oxidants, and nitrogenous compounds, have been shown to break seed dormancy in specific species. However, only a few of these chemicals occur naturally in the environment. Of these, nitrate, often in combination with light, is probably the most important. Some plants, such as hedge mustard (*Sisymbrium officinale*), have an absolute requirement for nitrate and light for seed germination. Another chemical agent that can break dormancy is nitric oxide (NO), a signaling molecule found in both plants and animals (see Chapter 18). Arabidopsis mutants unable to synthesize NO exhibit reduced germination, and the effect can be reversed by treating the seeds with exogenous NO. Another strong chemical stimulant of seed germination in many species under natural conditions is smoke, which is produced during forest fires. Smoke is likely to contain multiple germination stimulants, but one of the most active of these is **karrikinolide**, a member of the karrikin class of molecules, which structurally resemble strigolactone plant hormones (see Chapters 12 and 16).

In all three of the above examples, the chemical stimulants appear to break dormancy by the same basic mechanism: by down-regulating ABA synthesis or signaling, and up-regulating gibberellin synthesis or signaling, thus altering the ABA:GA ratio.

after-ripening Technique for breaking seed dormancy by prolonged storage at room temperature (20–25°C) under dry conditions.

karrikinolide A component of smoke that stimulates seed germination; similar in structure to strigolactones.

Seed Germination

Germination is the process that begins with water uptake by the dry seed and ends with the emergence of part of the embryo, usually the radicle, from its surrounding tissues. Strictly speaking, germination does not include seedling growth after radicle emergence. Similarly, the rapid mobilization of stored food reserves that fuels the initial growth of the seedling is considered a postgermination process.

Germination requires permissive ranges of water, temperature, and oxygen, and often light and nitrate as well. Of these, water is the most essential factor. The water content of mature, air-dried seeds is in the range of 5 to 15%, well below the threshold required for fully active metabolism. In addition, water uptake is needed to generate the turgor pressure that powers cell expansion, the basis of vegetative growth and development. As we discussed in Chapter 2, water uptake is driven by the gradient in water potential (Ψ) from the soil to the seed. For example, incubating tomato seeds at high ambient water potential ($\Psi = 0$ MPa) allows 100% germination, whereas incubation at low water potential ($\Psi = -1.0$ MPa), which nullifies the gradient in water potential, completely suppresses germination (**Figure 15.8**).

germination The events that take place between the start of imbibition of the dry seed and the emergence of part of the embryo, usually the radicle, from the structures that surround it. May also be applied to other quiescent structures, such as pollen grains or spores.

Germination and postgermination can be divided into three phases corresponding to the phases of water uptake

Under normal conditions, water uptake by the seed is triphasic (**Figure 15.9**):

- *Phase I.* The dry seed takes up water rapidly by the process of imbibition.

- *Phase II.* Water uptake by imbibition declines and metabolic processes, including transcription and translation, are reinitiated. The embryo expands, and the radicle emerges from the seed coat.

- *Phase III.* Water uptake resumes due to a decrease in Ψ as the seedling grows, and the stored food reserves of the seed are fully mobilized.

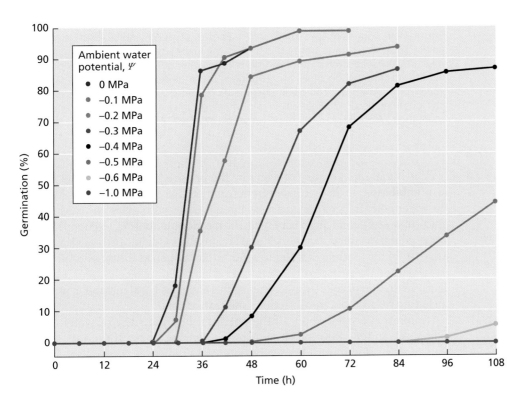

Figure 15.8 Time course of tomato seed germination at different ambient water potentials. (After G. Leubner [www.seedbiology.de] and Liptay and Schopfer 1983.)

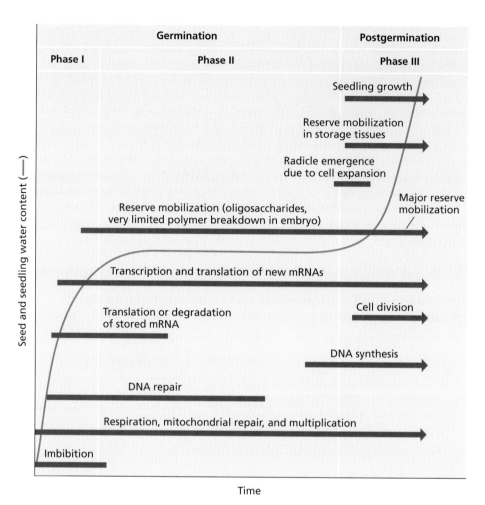

Figure 15.9 Phases of seed imbibition. In Phase I, dry seeds imbibe (take up) water rapidly. Since water flows from the higher to the lower water potential, the water uptake by the seed stops when the difference in water potential between the seed and the environment becomes zero. During Phase II the cells expand and the radicle emerges from the seed, completing germination. Metabolic activity increases and cell wall loosening occurs. In Phase III, water uptake resumes as the seedling becomes established. (After Bewley 1997 and Nonogaki et al. 2007 and 2010.)

imbibition   The initial phase of water uptake in dry seeds which is driven by the matric potential component of the water potential, that is, by the binding of water to surfaces, such as the cell wall and cellular macromolecules.

matric potential (Ψ_m)   The sum of osmotic potential (Ψ_s) + hydrostatic pressure (Ψ_p). Useful in situations (dry soils, seeds, and cell walls) where the separate measurement of Ψ_s and Ψ_p is difficult or impossible.

The initial rapid uptake of water by the dry seed during Phase I is referred to as **imbibition**, to distinguish it from water uptake during Phase III. Although the water potential gradient drives water uptake in both cases, the causes of the gradients are different. In the dry seed, the **matric potential (Ψ_m)** component of the water potential equation lowers Ψ and creates the gradient. The matric potential arises from the binding of water to solid surfaces, such as the microcapillaries of cell walls and the surfaces of proteins and other macromolecules (see Chapter 2). The rehydration of cellular macromolecules activates basal metabolic processes, including respiration, transcription, and translation.

Imbibition ceases when all the potential binding sites for water become saturated, and Ψ_m becomes less negative. During Phase II the rate of water uptake slows down until the water potential gradient is reestablished. Phase II can thus be thought of as the lag phase preceding growth, during which the solute potential (Ψ_s) of the embryo gradually becomes more negative due to the breakdown of stored food reserves and the liberation of osmotically active solutes.

The seed volume may increase as a result, rupturing the seed coat. At the same time, additional metabolic functions come online, such as the re-formation of the cytoskeleton and the activation of DNA repair mechanisms.

The emergence of the radicle through the seed coat in Phase II marks the end of the process of germination. The cell walls of the radicle are loosened and extend in response to the increase in turgor pressure that accompanies the uptake of water, which causes cell elongation. However, in many seeds the radicle must first break through the barrier imposed by the surrounding endosperm, seed coat, or pericarp before it can emerge from the seed. Radicle emergence can be a one- or two-step process. In the one-step process, either the surrounding tissues become physically weakened during imbibition, allowing the radicle to emerge unimpeded, or the radicle expands sufficiently during imbibition to rupture the surrounding tissues. In the two-step process, the surrounding tissues must first undergo metabolic weakening before the radicle can emerge from the seed.

During Phase III the rate of water uptake increases rapidly due to the onset of cell wall loosening and cell expansion, as seedling growth begins. Thus, the water potential gradient in Phase III embryos is maintained by both cell wall relaxation and solute accumulation.

Mobilization of Stored Reserves

The major food reserves of angiosperm seeds are typically stored in the cotyledons or in the endosperm. The massive mobilization of reserves that occurs after germination provides nutrients to the growing seedling until it becomes autotrophic. Carbohydrates (starches), proteins, and lipids are stored in specialized organelles within these tissues; for example, starch is stored in **amyloplasts** in the endosperm of cereals. Two enzymes responsible for initiating starch degradation are α- and β-amylase. α-Amylase hydrolyzes starch chains internally to produce oligosaccharides consisting of $\alpha(1,4)$-linked glucose residues. β-Amylase degrades these oligosaccharides from the ends to produce maltose, a disaccharide. Maltase then converts maltose to glucose.

Protein storage vacuoles are the primary source of amino acids for new protein synthesis in the seedling. In addition, protein storage vacuoles contain phytin, the K^+, Mg^{2+}, and Ca^{2+} salt of phytic acid (*myo*-inositol hexaphosphate), a major storage form of phosphate in seeds. During food mobilization, the enzyme phytase hydrolyzes phytin to release phosphate and the other ions for use by the growing seedling.

Lipids are a high-energy carbon source that is stored in oil or lipid bodies. Oil bodies from rape, mustard, cotton, flax, maize, peanut, and sesame seeds contain lipids, such as triglycerides and phospholipids, and proteins, such as oleosins (see Chapter 1). We discussed lipid catabolism during seed germination in Chapter 11.

The cereal aleurone layer is a specialized digestive tissue surrounding the starchy endosperm

Cereal seeds consist of three parts: the embryo, the endosperm, and the fused testa–pericarp (**Figure 15.10**). The embryo, which will grow into the new seedling, has a specialized absorptive organ, the scutellum. The triploid endosperm is composed of two tissues: the centrally located starchy endosperm and a peripheral **aleurone layer**. The starchy endosperm consists of thin-walled cells filled with starch grains. Living cells of the aleurone layer, which surrounds the starchy endosperm, synthesize and release α-amylase and other hydrolytic enzymes into the starchy endosperm during germination. As a consequence, the stored food reserves of the endosperm are broken down, and the solubilized sugars, amino acids, and other products are transported to the growing embryo via the scutellum.

amyloplast A starch-storing plastid found abundantly in storage tissues of shoots and roots, and in seeds. Specialized amyloplasts in the root cap and shoot also serve as gravity sensors.

aleurone layer The layer of aleurone cells surrounding and distinct from the starchy endosperm of cereal grains.

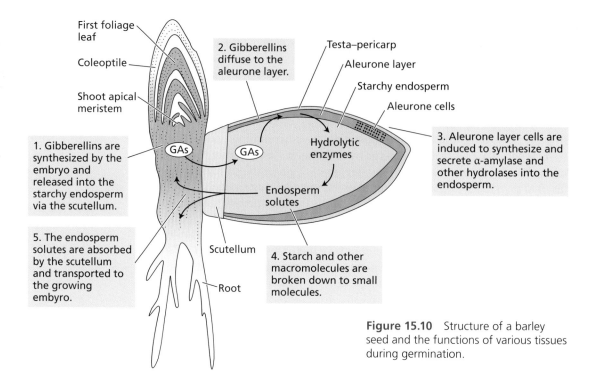

Figure 15.10 Structure of a barley seed and the functions of various tissues during germination.

As shown in Figure 15.10, gibberellins released from the embryo during germination stimulate the production and release of α-amylase by the aleurone layer of cereal seeds. Once inside the aleurone cells, GA binds to its receptor and initiates a signal transduction pathway, as described in Chapter 12. The binding of GA to its receptor initiates a response that results in increased expression of a transcriptional activator of α-amylase gene expression, leading to the production and secretion of α-amylase (**Figure 15.11**).

ABA inhibits the gibberellin-induced synthesis of hydrolytic enzymes that are essential for the breakdown of storage reserves during seedling growth. In the case of α-amylase, ABA acts by inhibiting gibberellin-dependent transcription of α-*amylase* mRNA.

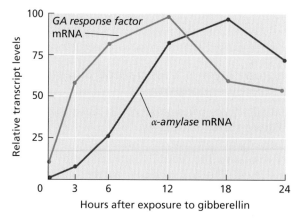

Figure 15.11 Time course for the induction of *GA response factor* and α-amylase mRNAs by GA$_3$. The production of the *GA response factor* mRNA precedes that of α-amylase mRNA by about 3 h. In the absence of gibberellin, the levels of both the *GA response factor* and α-amylase mRNAs are negligible. (After Gubler et al. 1995.)

Seedling Establishment

Seedling establishment is critical for plant survival and subsequent growth and development. This transition between germination (emergence) and growth independent of the seed is crucial, since seedlings are highly susceptible to unfavorable biotic and abiotic factors during this stage. In the field, about 10 to 55% of maize seedlings and 48 to 70% of soybean seedlings fail at this stage. Seed size is an important factor in seedling establishment because larger seeds have larger food reserves, allowing more time for seedling establishment.

Broadly defined, seedling establishment is the stage when the seedling becomes competent to photosynthesize, assimilate water and nutrients from the soil, undergo normal cellular and tissue differentiation and maturation, and respond appropriately to environmental stimuli.

The development of emerging seedlings is strongly influenced by light

A key event in seedling establishment is the emergence of the shoot from the soil into the light, which triggers profound changes in shoot development. The shoots of dark-grown seedlings are **etiolated**—that is, they have long, spindly hypocotyls, an apical hook, closed cotyledons, and nonphotosynthetic proplastids, which causes the unexpanded leaves to have a pale yellow color. In contrast, seedlings grown in nondirectional light have shorter, thicker hypocotyls, open cotyledons, and expanded leaves with photosynthetically active chloroplasts (**Figure 15.12**). Development in the dark is termed **skotomorphogenesis**, while development in the presence of light is called **photomorphogenesis**. When dark-grown seedlings are transferred to the light, photomorphogenesis takes over and the seedlings are said to be de-etiolated.

The switch between dark- and light-grown development involves genome-wide transcriptional and translational changes triggered by the perception of light by several classes of photoreceptors (see Chapter 13). Despite the complexity of the process, the transition from skotomorphogenesis to photomorphogenesis is surprisingly rapid. Within minutes of applying a single flash of light to a dark-grown bean seedling, several developmental changes occur:

- A decrease in the rate of stem elongation
- The beginning of apical hook opening
- Initiation of the synthesis of photosynthetic pigments

Light thus acts as a signal to induce a change in the shape of the seedling, from one that facilitates growth beneath the soil to one that will enable the plant to efficiently harvest light energy and convert it into the essential sugars, proteins, and lipids necessary for growth.

etiolation Effects of seedling growth in the dark, in which the hypocotyl and stem are more elongated, the cotyledons and leaves do not expand, and the chloroplasts do not mature.

skotomorphogenesis The developmental program plants follow when seeds are germinated and grown in the dark.

photomorphogenesis The influence and specific roles of light on plant development. In the seedling, light-induced changes in gene expression that support aboveground growth in the light rather than belowground growth in the dark.

(A) Light-grown maize

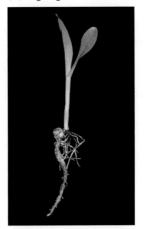

(B) Dark-grown maize

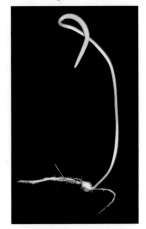

(C) Light-grown mustard

(D) Dark-grown mustard

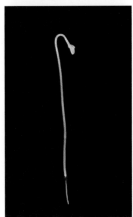

Figure 15.12 Light- and dark-grown monocot and eudicot seedlings. (A and B) Maize (corn; *Zea mays*) and (C and D) mustard (*Eruca* sp.) seedlings grown either in the light (A and C) or the dark (B and D). Symptoms of etiolation in maize, a monocot, include the absence of greening, reduction in leaf width, failure of leaves to unroll, and elongation of the coleoptile and mesocotyl. In mustard, a eudicot, etiolation symptoms include absence of greening, reduced leaf size, hypocotyl elongation, and maintenance of the apical hook. (A and B, photos courtesy of Patrice Dubois; C and D, photos by David McIntyre.)

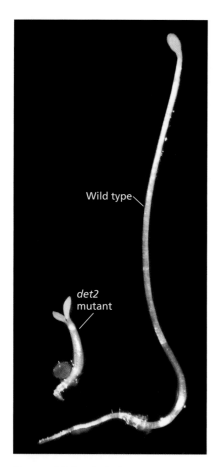

Figure 15.13 A dark-grown brassinosteroid-deficient mutant (*det2*) seedling of Arabidopsis on the left has a short, thick hypocotyl and open cotyledons. The dark-grown wild type is on the right. (Courtesy of S. Savaldi-Goldstein.)

Among the different photoreceptors that can promote photomorphogenic responses in plants, the most important are those that absorb red and blue light. Phytochrome is a protein–pigment photoreceptor that absorbs red and far-red light most strongly, but it also absorbs blue light (see Chapter 13). Cryptochromes are flavoproteins that mediate many blue-light responses involved in photomorphogenesis, including the inhibition of hypocotyl elongation, cotyledon expansion, and petiole elongation.

Gibberellins and brassinosteroids both suppress photomorphogenesis in darkness

In the dark, the level of phytochrome in the Pfr (far-red light–absorbing) form is low. Since Pfr reduces hypocotyl *sensitivity* to gibberellins, endogenous gibberellins promote hypocotyl cell elongation to a greater extent in the dark than in the light, causing the spindly appearance of dark-grown seedlings. In the light, Pr (the red light–absorbing form of phytochrome) is converted to Pfr, which causes the hypocotyl to become less sensitive to gibberellins. As a result, hypocotyl elongation is greatly reduced and the seedling appears to undergo partial photomorphogenesis. For this reason, gibberellin-deficient mutant peas grown in the dark look somewhat like light-grown seedlings, although they lack chlorophyll. Taken together, these results indicate that gibberellins suppress photomorphogenesis in the dark, and the suppression is reversed by red light.

Brassinosteroids play a parallel role in suppressing photomorphogenesis in the dark. When grown in the dark, brassinosteroid-deficient Arabidopsis mutants are short and lack the apical hooks observed in normal etiolated seedlings (**Figure 15.13**).

Hook opening is regulated by phytochrome, auxin, and ethylene

Etiolated eudicot seedlings are usually characterized by the formation of a hook located just behind the shoot apex. Hook formation and maintenance in darkness result from ethylene-induced asymmetric growth (**Figure 15.14**). The closed shape of the hook is a consequence of the more rapid elongation of the outer side of the stem compared with the inner side. When the hook is exposed to white light it opens, because the elongation rate of the inner side increases, equalizing the growth rates on both sides. This differential expansion involves auxin-induced cell elongation, which we discuss later in this chapter.

Red light induces hook opening, and far-red light reverses the effect of red, indicating that phytochrome is the photoreceptor involved in this process. A close interaction between phytochrome and

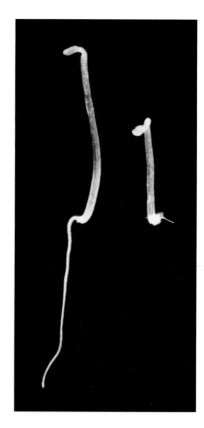

Figure 15.14 The effects of ethylene on growth in Arabidopsis seedlings. Three-day-old etiolated seedlings grown in the presence (right) or absence (left) of 10 ppm ethylene. Note the shortened hypocotyl, reduced root elongation, and exaggeration of the curvature of the apical hook that result from the presence of ethylene. (Courtesy of Joe Kieber, UNC.)

ethylene controls hook opening. As long as ethylene is produced by the hook tissue in the dark, elongation of the cells on the inner side is inhibited. Red light inhibits ethylene formation, promoting growth on the inner side, thereby causing the hook to open.

Vascular differentiation begins during seedling emergence

During embryogenesis within the seed, symplastic and apoplastic transport are sufficient to distribute water, nutrients, and signals throughout the embryo by the process of diffusion. Following germination, however, the emerging seedling requires a continuous vascular system to distribute materials quickly and efficiently throughout the plant. The vascular system of the embryo consists only of procambial strands—immature vascular tissue.

During seedling emergence, the first protoxylem and protophloem cells appear, followed by the larger metaxylem and metaphloem cells (**Figure 15.15**). Protophloem and metaphloem cells can differentiate into sieve elements, companion cells, fibers, or parenchyma cells. Protoxylem and metaxylem cells can become xylem vessels and tracheids, fibers, or parenchyma.

Growing roots have distinct zones

The development of a functional root system is just as important as shoot development for seedlings to get off to a good start. The basic features of root development can best be described by first distinguishing zones within the root with distinct cellular behaviors. Although it is not possible to define their boundaries with absolute precision, the following developmental zones provide a useful spatial framework for discussing root growth and development (**Figure 15.16**):

- The **root cap** occupies the most distal part of the root. It represents a unique set of cells that are situated below the meristematic zone and cover the apical meristem and protect it from mechanical injury as the root tip is pushed through the soil. Other functions of the root cap include the perception of gravity during gravitropism and the secretion of compounds that help the root penetrate the soil and mobilize mineral nutrients.

root cap Cells at the root apex that cover and protect the meristematic cells from mechanical injury as the root moves through the soil. Site of the perception of gravity and signaling for the gravitropic response.

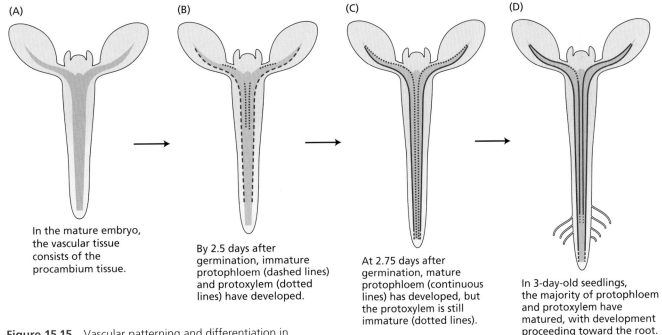

(A) In the mature embryo, the vascular tissue consists of the procambium tissue.

(B) By 2.5 days after germination, immature protophloem (dashed lines) and protoxylem (dotted lines) have developed.

(C) At 2.75 days after germination, mature protophloem (continuous lines) has developed, but the protoxylem is still immature (dotted lines).

(D) In 3-day-old seedlings, the majority of protophloem and protoxylem have matured, with development proceeding toward the root.

Figure 15.15 Vascular patterning and differentiation in Arabidopsis embryos and seedlings. (After Busse and Evert 1999.)

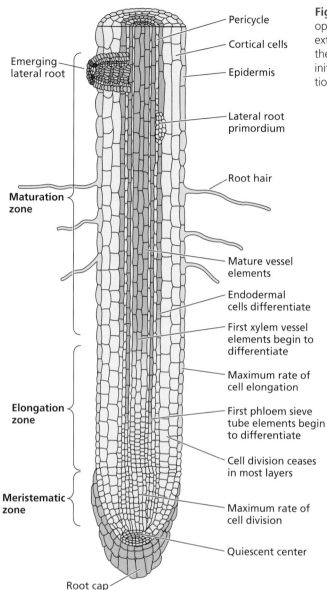

Emerging lateral root

Maturation zone

Elongation zone

Meristematic zone

Root cap

Pericycle

Cortical cells

Epidermis

Lateral root primordium

Root hair

Mature vessel elements

Endodermal cells differentiate

First xylem vessel elements begin to differentiate

Maximum rate of cell elongation

First phloem sieve tube elements begin to differentiate

Cell division ceases in most layers

Maximum rate of cell division

Quiescent center

Figure 15.16   Diagram of the primary root showing the three developmental zones. Cell division occurs in the meristematic zone, and cell extension occurs in the elongation zone. Cell differentiation occurs in the maturation zone, marked by the formation of root hairs and the initiation of lateral roots. Phloem differentiation begins in the elongation zone, and xylem differentiation begins in the maturation zone.

- The **meristematic zone** lies just under the root cap. It contains a cluster of cells that act as initials, dividing with characteristic polarities to produce cells that divide further and differentiate into the various tissues that make up the mature root. The cells surrounding these initials have small vacuoles and expand and divide rapidly. This zone is also referred to as the root apical meristem.

- The **elongation zone** is the site of rapid and extensive cell elongation. Although some cells continue to divide while they elongate within this zone, the rate of division decreases progressively to zero with increasing distance from the meristem. Phloem-conducting cells also begin to differentiate in the elongation zone.

- The **maturation zone** is the region in which cells acquire their differentiated characteristics. Cells enter the maturation zone after division and elongation have ceased, and in this region lateral organs such as lateral roots and root hairs may begin to form. Differentiation may begin much earlier, but cells do not achieve their mature state until they reach this zone. Xylem-conducting cells also begin differentiating in the maturation zone.

In Arabidopsis, these four developmental zones occupy little more than the first millimeter of the root tip. In many other species these zones extend over a longer distance, but growth is still confined to the distal regions of the root.

meristematic zone   The region of cell division at the tip of the root containing the meristem that generates the body of the root. Located between the root cap and the elongation zone.

elongation zone   The region of rapid and extensive root cell elongation in the root tip showing few, if any, cell divisions.

maturation zone   The region of the root where differentiation occurs, including the production of root hairs and functional vascular tissue.

Ethylene and other hormones regulate root hair development

Root hairs are extensions of root epidermal cells that increase the surface area of the root. Normally, only some epidermal cells produce root hairs; for example, in lettuce only those epidermal cells overlying a cortical cell junction differentiate into hair cells. However, ethylene-treated roots produce extra root hairs in abnormal locations (**Figure 15.17**). Conversely, seedlings grown in the presence of ethylene biosynthesis inhibitors (such as the silver ion, Ag^+), as well as ethylene-insensitive mutants, display a reduction in root hair formation. These observations suggest that ethylene acts as a positive regulator in the differentiation of root hairs. Jasmonic acid has also been shown to enhance root hair growth, while brassinosteroids inhibit root hair growth (see Chapter 12).

Lateral roots arise internally from the pericycle

Another way that a root system increases its surface area is by branching—by producing lateral roots. In gymnosperms and most eudicots, lateral root primordia

Figure 15.17 Promotion of root hair formation by ethylene in lettuce seedlings. Two-day-old seedlings were treated with air (left) or 10 ppm ethylene (right) for 24 h before the photo was taken. Note the profusion of root hairs on the ethylene-treated seedling. (From Abeles et al. 1992, courtesy of F. Abeles.)

are initiated in the pericycle cells adjacent to xylem cells. However, in grasses lateral root primordia form in the pericycle and endodermal cells adjacent to the phloem cells. In the majority of plants, anticlinal divisions in the pericycle cells precede periclinal divisions. These lateral root primordia cells continue cell division and cell expansion until the new lateral root emerges through the cortical and epidermal cell layers (**Figure 15.18**).

The lateral root contains all of the cell types of the primary root, and the vascular tissue of the lateral root is continuous with that of the primary root. Lateral roots initiate in the maturation zone of the primary root. Lateral roots retain the ability to form additional branches, greatly increasing the total surface area of the root system.

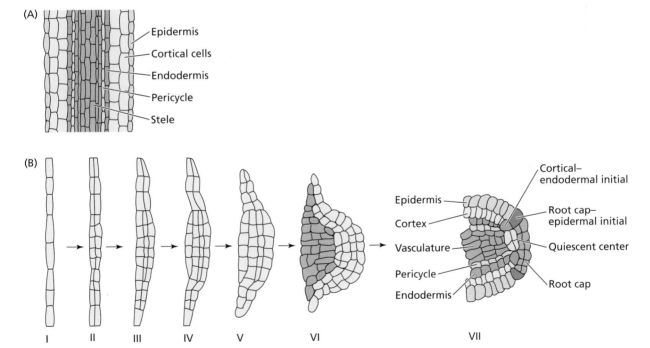

Figure 15.18 Lateral root development. (A) Longitudinal section of a root in the maturation zone. Anticlinal cell divisions in the pericycle initiate lateral root formation. (B) Stages of lateral root development. Stage I consists of a single layer of pericycle. During stage II, the pericycle cells divide periclinally to form inner and outer layers. In stages III and IV, the lateral root primordium forms a dome shape and periclinal and anticlinal divisions continue. In stage V, the cortical cells loosen so that the lateral root primordium can expand between the cells of the primary root. In stage VI, the lateral root primordium recapitulates the tissues of the primary root: epidermal, cortical, and endodermal cell layers. In stage VII, the stele differentiates, the epidermal cells separate, and the lateral root primordium emerges. (After Petricka et al. 2012.)

Cell Expansion: Mechanisms and Hormonal Controls

Growth is defined as the increase in cell number and cell size that occurs during an organism's life cycle. Growth in plants is qualitatively different from the growth of animals. In animals, cell divisions are evenly distributed throughout the entire body, and the organism reaches its full size by the end of the juvenile stage. Animal cells increase their size by synthesizing more cytoplasm, a process that depends primarily on the synthesis of proteins and other macromolecules.

Plants exhibit two types of growth: growth in height (primary growth) and growth in girth (secondary growth, see Chapter 1). Following the completion of embryogenesis, cell divisions that give rise to primary growth become restricted to the shoot and root apical meristems, which have the potential to go on dividing indefinitely. However, most of the increase in plant height is due to the process of cell expansion. In contrast to animal cells, plant cells increase their volume by water uptake into the large central vacuole (see Chapters 2 and 3). Since water uptake is energetically much cheaper than synthesizing new proteins, such a growth strategy enables plants, which earn their living as solar collectors, to compete more effectively for sunlight.

In plants, the presence of a rigid cell wall surrounding the protoplast complicates the process of cell expansion. One of the functions of the cell wall is to prevent cell lysis during water uptake. Yet, for the cell to enlarge, the cell wall's resistance must somehow be overcome without compromising its integrity. The elegant mechanism plants evolved to achieve this delicate balance of yielding without breaking is described in the following sections. In contrast, the directionality of cell enlargement (elongation versus lateral expansion) is regulated by an entirely different mechanism involving cellular microtubules. Both wall loosening and the directionality of cell expansion are strongly influenced by hormones.

The rigid primary cell wall must be loosened for cell expansion to occur

Early in their development, plant cells begin forming a primary cell wall that is both rigid and extensible. The general wall structure consists of thin layers made of long cellulose microfibrils embedded in a hydrated matrix of noncellulosic polysaccharides and a small amount of nonenzymatic proteins (see Figures 1.5 and 1.7). This structure imparts an ideal combination of flexibility and strength to the growing cell wall, which must be both extensible and strong at the same time. By dry mass, primary cell walls typically contain approximately 40% pectin, 25% cellulose, and 20% hemicellulose, with perhaps 5% protein and the remaining percentage composed of diverse other materials. However, large deviations from these values may be found among species. For example, the walls of grass coleoptiles consist of 60 to 70% hemicellulose, 20 to 25% cellulose, and only about 10% pectin.

What all primary walls have in common is that they are formed by growing cells, contain a highly hydrated matrix between the cellulose microfibrils, and have the ability to expand in surface area, at least during cell expansion. This is in contrast to secondary walls, which are packed more densely and have a structural, reinforcing role incompatible with cell wall enlargement. The primary cell wall also contains a considerable amount of water, located mostly in the matrix, which is approximately 75% water. The hydration state of the matrix is a critical determinant of the physical properties of the wall; for example, removal of water makes the wall stiffer and less extensible, and this is a factor contributing to plant growth inhibition by water deficits.

During plant cell enlargement, new wall polymers are continuously synthesized and secreted at the same time that the preexisting wall is expanding (see Figure 1.7). Wall expansion may be highly localized (as in the case of **tip growth**) or more dispersed over the wall surface (**diffuse growth**) (**Figure 15.19**). Tip

tip growth   Localized growth at the tip of a plant cell, caused by localized secretion of new wall polymers. Occurs in pollen tubes, root hairs, some sclerenchyma fibers, and cotton fibers, as well as moss protonema and fungal hyphae.

diffuse growth   A type of cell growth in plants in which expansion occurs more or less uniformly over the entire surface.

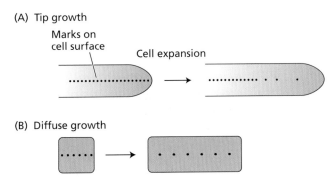

(A) Tip growth

Marks on cell surface

Cell expansion

(B) Diffuse growth

Figure 15.19 Contrasting spatial distribution of expansion during tip growth and diffuse growth. (A) Expansion of a tip-growing cell is confined to an apical dome at one end of the cell. If marks are placed on the surface of a tip-growing cell, only the marks that are initially within the apical dome grow farther apart. Root hairs and pollen tubes are examples of plant cells that exhibit tip growth. (B) If marks are placed on the surface of a diffuse-growing cell, the distance between all the marks increases as the cell grows. Most cells in multicellular plants grow by diffuse growth.

anisotropic growth Enlargement that is greater in one direction than another; for example, elongating cells in stems and roots grow more in length than in width.

growth is characteristic of pollen tubes and root hairs. Most of the other cells in the plant body exhibit diffuse growth.

Microfibril orientation influences growth directionality of cells with diffuse growth

During growth, the loosened cell wall is extended by physical forces generated from cell turgor pressure. Turgor pressure creates an outward-directed force, equal in all directions. The directionality of growth is determined in large part by the structure of the cell wall—specifically, the orientation of cellulose microfibrils.

When cells first form in the meristem, they are isodiametric; that is, they have equal diameters in all directions. If the cellulose microfibrils in the primary cell wall are randomly arranged, the cells grow isotropically (equally in all directions), expanding radially to generate a sphere (**Figure 15.20A**). In most plant cell walls, however, cellulose microfibrils are aligned in a preferential direction, resulting in **anisotropic growth** (e.g., in the stem, the cells increase in length much more than in width).

In the lateral walls of elongating cells such as the cortical and vascular cells of stems and roots, cellulose microfibrils are deposited circumferentially (transversely), at right angles to the long axis of the cell. The circumferential arrangement of cellulose microfibrils restricts growth in girth and promotes growth in length (**Figure 15.20B**).

Plant cells typically expand ten- to a thousandfold in volume before reaching maturity. In extreme cases, cells may enlarge more than ten thousandfold in volume compared with their meristematic initials (e.g., xylem vessel elements). The cell wall undergoes this massive expansion without losing its mechanical integrity and without becoming thinner. The integration of new wall polymers during cell expansion is particularly crucial for rapidly growing root hairs, pollen tubes, and other tip-growing cells, in which the region of wall deposition and surface expansion is localized to the tip of the tubelike cell.

When plant cells undergo expansion, whether by diffuse growth or tip growth, the increase in volume is due

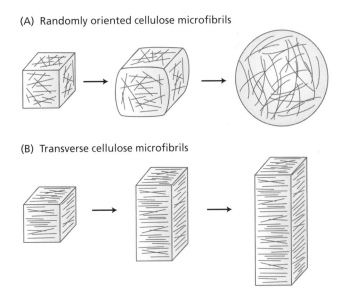

(A) Randomly oriented cellulose microfibrils

(B) Transverse cellulose microfibrils

Figure 15.20 The orientation of cellulose microfibrils determines the directionality of cell expansion. (A) Randomly oriented cellulose microfibrils. (B) Transverse cellulose microfibrils.

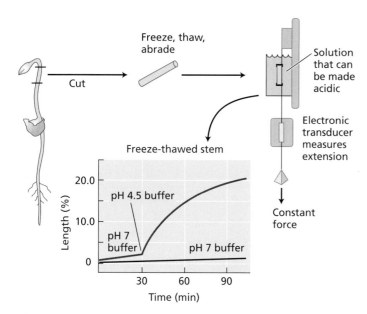

Figure 15.21 Acid-induced extension of isolated cell walls, measured in an extensometer. The wall sample from killed cells is clamped and put under tension in an extensometer, which measures the length with an electronic transducer attached to a clamp. When the solution surrounding the wall is replaced with an acidic buffer (e.g., pH 4.5), the wall extends irreversibly in a time-dependent fashion. (After Cosgrove 1997.)

primarily to water uptake. This water ends up mainly in the vacuole, which takes up an ever larger proportion of the cell volume as the cell enlarges.

Acid-induced growth and cell wall yielding are mediated by expansins

A common characteristic of growing cell walls is that they extend much faster at acidic pH than at neutral pH. This phenomenon is called **acid growth**. In living cells, acid growth is evident when growing cells are treated with acid buffers or with the drug fusicoccin, which induces acidification of the cell wall solution by activating an H⁺-ATPase in the plasma membrane. Auxin-induced growth is also associated with wall acidification (discussed below).

Acid-induced wall extension can be observed in isolated cell walls. Such an observation entails using an extensometer to place the walls under tension and to measure long-term wall extension (**Figure 15.21**). When isolated primary walls are incubated in neutral buffer (pH 7) and clamped in an extensometer, the walls extend briefly when tension is applied, but extension soon ceases. When transferred to an acidic buffer (pH 5 or less), the walls begin to extend rapidly, in some instances continuing for many hours.

Acid-induced extension is characteristic of the walls of growing cells, and it is not observed in mature (nongrowing) walls. When the walls are pretreated with heat, proteases, or other agents that denature proteins, they lose their ability to respond to acid. Such results indicate that acid growth is not due simply to the physical chemistry of the wall (e.g., a weakening of the pectin gel), but is catalyzed by one or more wall proteins.

The idea that proteins are required for acid growth was confirmed in reconstitution experiments in which heat-inactivated walls were restored to nearly full acid-growth responsiveness by the addition of proteins extracted from the walls of growing cells (**Figure 15.22**). The active components proved to be a group of proteins that were named **expansins**. Expansins catalyze the pH-dependent

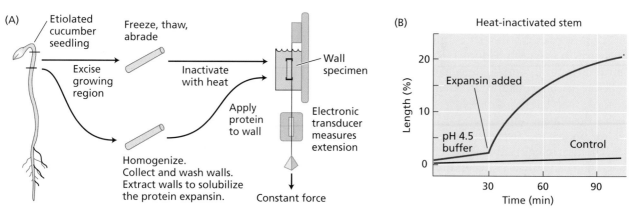

Figure 15.22 Scheme for the reconstitution of extensibility of isolated cell walls. (A) Cell walls are prepared as in Figure 15.21 and briefly heated to inactivate the endogenous acid extension response. To restore this response, proteins are extracted from growing walls and added to the solution surrounding the wall specimen. (B) Addition of proteins containing expansins restores the acid extension properties of the wall. (After Cosgrove 1997.)

yielding of cell walls. They are effective in catalytic amounts, but they do not exhibit lytic or other enzymatic activities.

Auxin promotes growth in stems and coleoptiles, while inhibiting growth in roots

Auxin synthesized in the shoot apex is transported toward the tissues below. The steady supply of auxin arriving at the subapical region of a stem or coleoptile is required for the continued elongation of these cells. Because the level of endogenous auxin in the elongation region of a normal healthy plant is nearly optimal for growth, spraying the plant with exogenous auxin causes only a modest and short-lived stimulation in growth. Such spraying may even be inhibitory in the case of dark-grown seedlings, which are more sensitive to supraoptimal auxin concentrations than light-grown plants are.

However, when the endogenous source of auxin is removed by excision of stem or coleoptile sections containing the elongation zone, the growth rate rapidly decreases to a low basal rate. Such excised sections often respond to exogenous auxin by rapidly increasing their growth rate back to the level in the intact plant (**Figure 15.23**).

Auxin control of root elongation has been more difficult to demonstrate, perhaps because auxin induces the production of ethylene, which inhibits root growth. These two hormones interact differentially in root tissue to control growth. However, even if ethylene biosynthesis is specifically blocked, low concentrations (10^{-10} to $10^{-9}\,M$) of auxin promote the growth of intact roots, whereas higher concentrations ($10^{-6}\,M$) inhibit growth. Thus, while roots may require a minimum concentration of auxin to grow, root growth is strongly inhibited by auxin concentrations that promote elongation in stems and coleoptiles.

The outer tissues of eudicot stems are the targets of auxin action

Eudicot stems are composed of many types of tissues and cells, only some of which may limit the growth rate. This point is illustrated by a simple experiment. When sections from growing regions of an etiolated eudicot stem, such as pea, are split lengthwise and incubated in buffer alone, the two halves bend outward. This result indicates that in the absence of auxin, the central tissues—including the pith, vascular tissues, and inner cortex—elongate at a faster rate than the

acid growth A characteristic of growing cell walls in which they extend more rapidly at acidic pH than at neutral pH.

expansins A class of wall-loosening proteins that accelerate wall yielding during cell elongation, typically with an optimum at acidic pH.

(A) 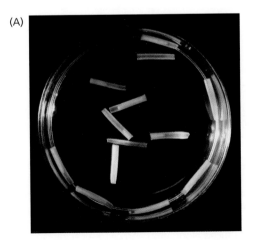 (B)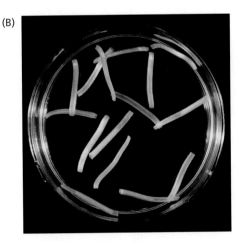

Figure 15.23 Auxin stimulates the elongation of oat coleoptile sections that have been depleted of endogenous auxin. These coleoptile sections were incubated for 18 h in either water (A) or auxin (B). The yellow material inside the translucent coleoptile is the primary leaf tissue. (Photos © M. B. Wilkins.)

outer tissues, which consist of the outer cortex and epidermis. Thus, the outer tissues must be limiting the extension rate of the stem in the absence of auxin. When similar sections are incubated in buffer plus auxin, the two halves bend inward, due to auxin-induced elongation of the outer tissues of the stem.

The minimum lag time for auxin-induced elongation is 10 minutes

When a stem or coleoptile section is excised and inserted into a sensitive growth-measuring device, the lag time for the auxin response can be monitored with great precision. For example, addition of auxin markedly stimulates the growth rates of oat (*Avena sativa*) coleoptile and soybean (*Glycine max*) hypocotyl sections after a lag period of only 10 to 12 min (**Figure 15.24A**). The maximum growth rate, which represents a five- to tenfold increase over the basal rate, is reached after 30 to 60 min of auxin treatment. As is shown in **Figure 15.24B**, a threshold concentration of auxin must be reached to initiate this response. Beyond the optimal concentration, auxin becomes inhibitory.

The stimulation of growth by auxin requires energy, and metabolic inhibitors inhibit the response within minutes. Auxin-induced growth is also sensitive to inhibitors of protein synthesis such as cycloheximide, suggesting that protein synthesis is required for the response. Inhibitors of RNA synthesis also inhibit auxin-induced growth after a slightly longer delay.

Auxin-induced proton extrusion loosens the cell wall

Auxin induces acidification of the apoplast by increasing the activity of plasma membrane H^+-ATPases. Cell wall acidification occurs 10 to 15 min after auxin exposure, consistent with the growth kinetics, as shown in **Figure 15.24C**. As discussed above, cell wall proteins called expansins mediate wall loosening at acid pH.

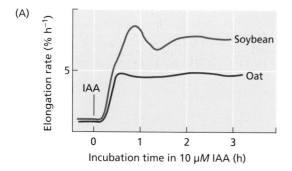

(A)

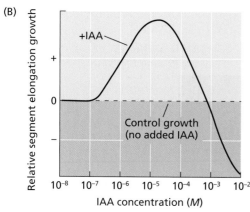

(B)

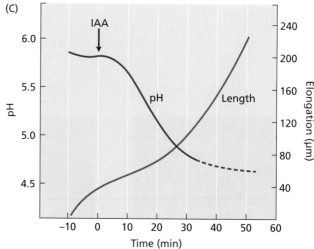

(C)

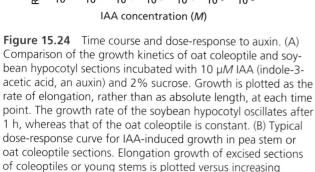

Figure 15.24 Time course and dose-response to auxin. (A) Comparison of the growth kinetics of oat coleoptile and soybean hypocotyl sections incubated with 10 µM IAA (indole-3-acetic acid, an auxin) and 2% sucrose. Growth is plotted as the rate of elongation, rather than as absolute length, at each time point. The growth rate of the soybean hypocotyl oscillates after 1 h, whereas that of the oat coleoptile is constant. (B) Typical dose-response curve for IAA-induced growth in pea stem or oat coleoptile sections. Elongation growth of excised sections of coleoptiles or young stems is plotted versus increasing concentrations of exogenous IAA. At concentrations above 10^{-5} M, IAA becomes less and less effective. Above about 10^{-4} M it becomes inhibitory, as shown by the fact that the stimulation decreases and the curve eventually falls below the dashed line, which represents growth in the absence of added IAA. (C) Kinetics of auxin-induced elongation and cell wall acidification in maize coleoptiles. The pH of the cell wall was measured with a pH microelectrode. Note the similar lag times (10–15 min) for both cell wall acidification and the increase in the rate of elongation. (A after Cleland 1995; C after Jacobs and Ray 1976.)

Figure 15.25   Ethylene affects microtubule orientation. (A) Microtubule orientation is horizontal in hypocotyls of control dark-grown transgenic Arabidopsis seedlings expressing a tubulin gene tagged with green fluorescent protein. (B) Microtubule orientation is longitudinal and diagonal in hypocotyl cells from seedlings treated with the ethylene precursor, 1-aminocyclopropane-1-carboxylic acid (ACC), which increases ethylene production. (From Le et al. 2005.)

(A) Untreated

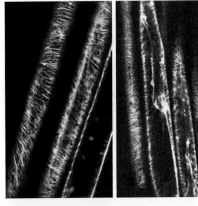

(B) ACC-treated

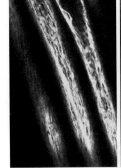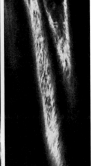

Once the cell wall is sufficiently loosened by expansin activity, turgor pressure initiates cell expansion. Other biochemical processes, such as new cell wall biosynthesis, are required to sustain cell expansion over the long term.

Ethylene affects microtubule orientation and induces lateral cell expansion

At concentrations above 0.1 μL L^{-1}, ethylene changes the growth pattern of eudicot seedlings by reducing the rate of elongation and increasing lateral expansion, leading to swelling of the hypocotyl or the epicotyl. As discussed earlier (see Figure 15.20), the directionality of plant cell expansion is determined by the orientation of the cellulose microfibrils in the cell wall. Transverse microfibrils reinforce the cell wall in the lateral direction, so that turgor pressure is channeled into cell elongation. The orientation of the microfibrils is in turn determined by the orientation of the cortical array of microtubules in the cortical (peripheral) cytoplasm. In typical elongating plant cells, the cortical microtubules are arranged transversely, giving rise to transversely arranged cellulose microfibrils.

When etiolated seedlings are treated with ethylene (see Figure 15.14), the microtubule alignment in the cells of the hypocotyl is altered. Instead of being aligned horizontally, the microtubules switch over to a diagonal or longitudinal orientation (**Figure 15.25**). This approximately 90-degree shift in microtubule orientation leads to a parallel shift in cellulose microfibril deposition. The newly deposited wall is reinforced in the longitudinal direction rather than the transverse direction, which promotes lateral expansion instead of elongation.

Tropisms: Growth in Response to Directional Stimuli

Plants respond to external stimuli by altering their growth and developmental patterns. During seedling establishment, abiotic factors such as gravity, touch, and light influence the initial growth habit of the young plant. **Tropisms** are directional growth responses in relation to environmental stimuli caused by the asymmetric growth of the plant axis (stem or root). Tropisms may be positive (growth toward the stimulus) or negative (growth away from the stimulus).

One of the first forces that emerging seedlings encounter is gravity. **Gravitropism**, growth in response to gravity, enables shoots to grow upward toward sunlight for photosynthesis, and roots to grow downward into the soil for water and nutrients. As soon as the shoot tip penetrates the soil surface, it encounters sunlight. **Phototropism** enables leafy shoots to grow toward sunlight, thus maximizing photosynthesis, while some roots grow away from sunlight. **Thigmotropism**, differential growth in response to touch, helps roots grow around obstacles, and twining vines and tendrils to wrap around other structures for support.

When dark-grown *Avena* seedlings are oriented horizontally, the coleoptiles bend upward in response to gravity. According to the **Cholodny–Went hypothesis**, auxin in a horizontally oriented coleoptile tip is transported laterally to the lower side, causing the lower side of the coleoptile to grow faster than the upper side. This general model turns out to apply to all tropism responses. Before we review some of the evidence that supports the Cholodney–Went hypothesis, let's look at two key features of long-distance auxin transport: its polarity and its gravity independence.

tropism   Oriented plant growth in response to a perceived directional stimulus from light, gravity, or touch.

gravitropism   Plant growth in response to gravity, enabling roots to grow downward into the soil and shoots to grow upward.

phototropism   The alteration of plant growth patterns in response to the direction of incident radiation, especially blue light.

thigmotropism   Plant growth in response to touch, enabling roots to grow around rocks, and shoots of climbing plants to wrap around structures for support.

Cholodny–Went hypothesis
Early mechanism proposed for tropisms involving stimulation of the bending of the plant axis by lateral transport of auxin in response to a stimulus, such as light, gravity, or touch. The original model has been supported and expanded by recent experimental evidence.

(A)

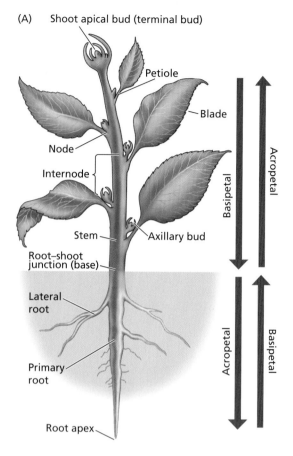

(B)

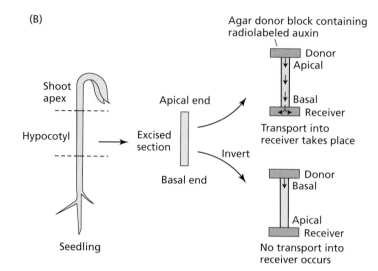

Figure 15.26 Demonstration of polar auxin transport with radiolabeled auxin. (A) Polar auxin transport is described in terms of the direction of its movement in relation to the base of the plant (the root–shoot junction). Auxin moving downward from the shoot moves *basipetally* (toward the base) until it reaches the root–shoot junction. From that point, downward movement is described as *acropetal* (toward the apex). Movement of auxin from the apex of the root toward the root–shoot junction is also described as *basipetal* (toward the base). (B) Donor–receiver agar block method for measuring polar auxin transport. A donor agar block containing radioactive auxin is placed at one end of a hypocotyl section, and a recipient agar block is placed at the other end. The amount of radioactive auxin that accumulates in the recipient block is a measure of how much auxin is transported through the hypocotyl section. The polarity of transport is apical to basal and is independent of the orientation of the plant tissue with respect to gravity.

Auxin transport is polar and gravity-independent

The cellular mechanisms underlying polar auxin transport and the use of the terms **basipetal** (toward the base), **acropetal** (toward the apex), *rootward,* and *shootward* to describe the direction of auxin flows were discussed in Chapter 14. **Figure 15.26** illustrates an experiment using radiolabeled auxin to demonstrate basipetal polar auxin transport in a seedling hypocotyl section.

A demonstration that polar auxin transport is gravity-independent is shown in **Figure 15.27**. Grape cuttings were placed in a moist chamber, which led to the formation of adventitious roots at the basal ends of the cuttings and adventitious shoots at the apical ends. When the cuttings were inverted, the polarity of root and shoot formation was preserved. Roots formed at the basal end (now pointing upward) because root differentiation was stimulated by the auxin that accumulated there due to basipetal (rootward) polar transport. Shoots tended to form at the apical end where the auxin concentration was lowest, no matter which way the cutting was oriented with respect to gravity.

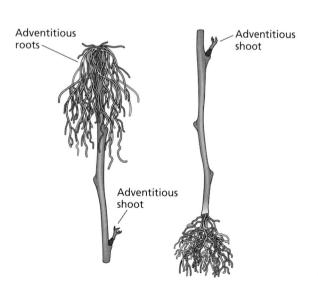

Figure 15.27 Adventitious roots grow from the basal ends of grape hardwood cuttings, and adventitious shoots grow from the apical ends, whether the cuttings are maintained in the inverted orientation (the cutting on the left) or the upright orientation (the cutting on the right). The roots always form at the basal ends because polar auxin transport is independent of gravity. (After Hartmann and Kester 1983.)

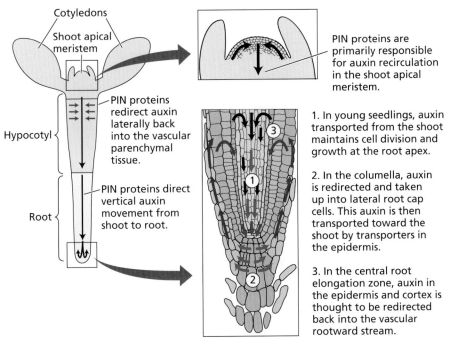

basipetal From the growing tip of a shoot or root toward the base (junction of the root and shoot).

acropetal From the base to the tip of an organ, such as a stem, root, or leaf.

PIN proteins are primarily responsible for auxin recirculation in the shoot apical meristem.

1. In young seedlings, auxin transported from the shoot maintains cell division and growth at the root apex.

2. In the columella, auxin is redirected and taken up into lateral root cap cells. This auxin is then transported toward the shoot by transporters in the epidermis.

3. In the central root elongation zone, auxin in the epidermis and cortex is thought to be redirected back into the vascular rootward stream.

Figure 15.28 The fountain model of polar auxin transport in the root. PIN transport proteins in the vascular tissue direct auxin to the root (see text for discussion). PIN proteins in the root vascular cylinder then transport auxin to the columella of the root cap. Auxin then moves into the lateral root cap cells and is redirected by PIN proteins to the epidermis. PIN proteins are also involved in redirecting auxin toward the elongation zone, after which it moves into the vascular cylinder. The term "fountain model" was suggested by the fact that the stream of auxin coming from the shoot reverses direction after reaching the root cap. (After Blilou et al. 2005).

The direction of auxin flow throughout a plant is controlled by PIN proteins. As shown in **Figure 15.28**, PIN proteins in the apical meristem are responsible for directing auxin movement, first toward the tip of the apex, and then down the stem and into the root. In the root, PIN proteins in the cells of the vascular cylinder transport auxin to the columella (central) region of the root cap. Auxin is then taken up by the lateral root cap cells by means of the AUX1 permease. PIN proteins in the cells of the lateral root cap then redirect auxin in the shootward direction via the root epidermal cells (see Figure 15.28). Upon reaching the elongation zone, auxin is transported laterally back into the vascular cylinder and returned to the root cap via PIN proteins. This recirculation of auxin from the root cap to the elongation zone and back again is referred to as the *fountain model*.

The Cholodny–Went hypothesis is supported by auxin movements and auxin responses during gravitropic growth

Early experimental studies established that the tips of coleoptiles are the site of perception for blue light-induced phototropic bending, and that the lateral movement of auxin to the shaded side was involved in the response (see Figures 12.8 and 13.3). The tips of coleoptiles are also able to sense gravity and redistribute auxin to the lower side. For example, if the excised tip of a coleoptile is placed on agar blocks and oriented horizontally, a greater amount of auxin diffuses into the agar block from the lower half of the tip than from the upper half, as demonstrated by a bioassay (**Figure 15.29**).

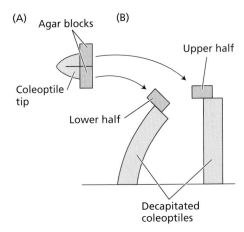

Figure 15.29 Auxin is transported to the lower side of a horizontally oriented oat coleoptile tip. (A) Auxin from the upper and lower halves of a horizontal tip is allowed to diffuse into two agar blocks. (B) The agar block from the lower half (left) induces greater curvature in a decapitated coleoptile than the agar block from the upper half (right) does.

Figure 15.30 Microsurgery experiments demonstrate that the root cap is required for redirection of auxin and subsequent differential inhibition of elongation in root gravitropic bending. (After Shaw and Wilkins 1973.)

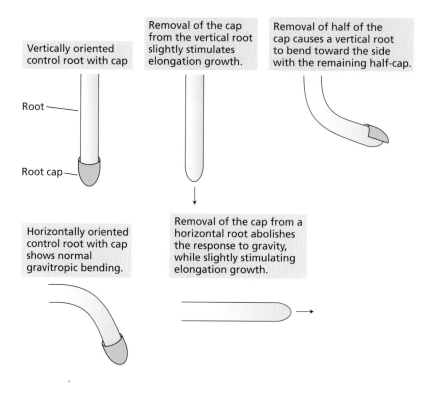

Vertically oriented control root with cap

Root

Root cap

Removal of the cap from the vertical root slightly stimulates elongation growth.

Removal of half of the cap causes a vertical root to bend toward the side with the remaining half-cap.

Horizontally oriented control root with cap shows normal gravitropic bending.

Removal of the cap from a horizontal root abolishes the response to gravity, while slightly stimulating elongation growth.

Gravitropism in roots also depends on auxin redistribution. The site of perception of gravity in roots is the root cap. When the root cap is removed from a growing root, the root no longer bends downward in response to gravity (**Figure 15.30**). In fact, the growth rate of the root actually increases slightly, suggesting that the root cap supplies an inhibitor that modulates growth in the elongation zone. Microsurgery experiments in which half of the cap was removed (see Figure 15.30) confirmed that the cap transports a root growth inhibitor, later identified as auxin, to the lower side of the root during gravitropic bending. Experiments with auxin transport inhibitors and auxin transporter mutants have shown that the shootward (basipetal) transport of auxin from the root cap to the elongation zone is required for gravitropic growth.

According to the current model for root gravitropism, shootward auxin transport in a vertically oriented root is equal on all sides. When the root is oriented horizontally, however, the signals from the cap redirect most of the auxin to the lower side, thus inhibiting the growth of that lower side. Consistent with this model, IAA rapidly accumulates on the lower side of a horizontally oriented root and concentrates in the epidermal cells of the elongation zone.

Gravity perception is triggered by the sedimentation of amyloplasts

The primary mechanism by which gravity can be detected by cells is via the motion of a falling or sedimenting intracellular body. Root cap columella cells contain large, dense amyloplasts (starch-containing plastids) called **statoliths**. These statoliths readily sediment to the bottom of the cell to align with the gravity vector (**Figure 15.31**). As we have seen, removal of the root cap from otherwise intact roots prevents root gravitropism without inhibiting growth, suggesting that the root cap columella cells function as gravity-sensing cells, or **statocytes**. Perception of the stimulus (statolith displacement by gravity) is thought to occur via membrane receptors and/or cytoskeletal interactions.

Reorientation of the root with respect to gravity causes the redistribution of PIN proteins in the root cap columella cells to the lower sides of the cells (see

statoliths Cellular inclusions such as amyloplasts that act as gravity sensors by having a high density relative to the cytosol and sedimenting to the bottom of the cell.

statocytes Specialized gravity-sensing plant cells that contain statoliths.

starch sheath A layer of cells that surrounds the vascular tissues of the shoot and coleoptile and is continuous with the root endodermis. Required for gravitropism in the shoots of eudicots.

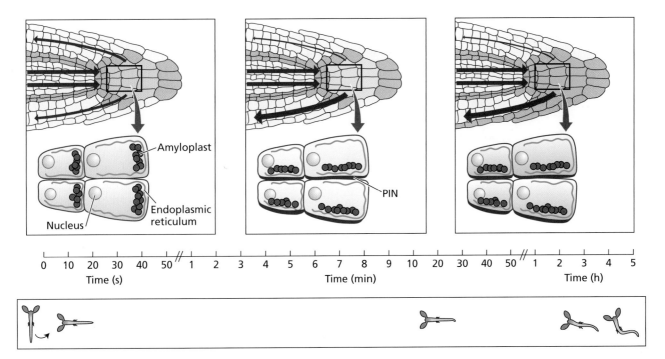

Figure 15.31 Sequence of events following gravistimulation of an Arabidopsis root. The timescale on the bottom is nonlinear. The differential growth of the shoot and roots of the seedling at different stages of the response is illustrated below the timescale. Three stages of statolith sedimentation are shown on the top. The left figure shows time zero, when the seedling is first rotated 90 degrees. The second and third stages shown are at about 6 min and 2 h after rotation. The red arrows indicate auxin flow, with thicker arrows indicating more flow. Cells with relatively high auxin concentrations are shown in orange. Columella cells of the root tip are shown in green at time zero; the color changes to blue and then to blue-green at later stages to indicate the degree of alkalinization of the cytoplasm (see Figure 15.33). The distribution of PIN proteins is diagrammed as a purple outline on the plasma membrane of the columella cells. (After Baldwin et al. 2013.)

Figure 15.31). This redistribution occurs after sedimentation of the statoliths and before the root begins to bend, consistent with a role for PIN in shunting auxin to the lower side of the root. As a result, auxin is transported out of the columella to the lower side of the root cap. From there it is transported back to the elongation zone via the epidermal cells (see Figure 15.31).

In eudicot stems and stemlike organs, the statoliths involved in gravity perception are located in the **starch sheath**, the innermost layer of cortical cells that surrounds the ring of vascular bundles of the shoot (**Figure 15.32**). The starch sheath is continuous with the endodermis of the root, but unlike in the endodermis, its cells contain amyloplasts that are redistributed when the gravity vector is changed. Genetic studies have confirmed the primary role of the starch sheath in shoot gravitropism. Arabidopsis mutants lacking amyloplasts in the starch sheath display agravitropic shoot growth; gravitropic growth of the root is unaffected in these mutants because they still have amyloplasts in the root cap.

Gravity sensing may involve pH and calcium ions (Ca²⁺) as second messengers

A variety of experiments suggest that localized changes in pH and Ca^{2+} gradients are part of the signaling that occurs during gravitropism. Changes in intracellular pH can be detected

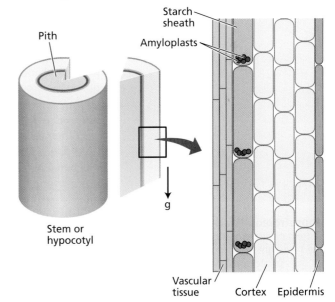

Figure 15.32 Diagram of the starch sheath located outside the ring of vascular tissue. The cutaway view shows the amyloplasts at the bottom of the cells. (After Palmieri and Kiss 2007.)

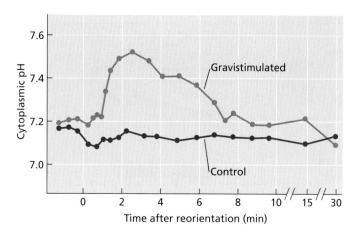

Figure 15.33 Experiments with a pH-sensitive dye suggest that pH changes in columella cells of the root cap are involved in gravitropic signal transduction. Cytoplasmic pH increases in less than 1 min after gravistimulation. (From Fasano et al. 2001.)

early in root columella cells responding to gravity (**Figure 15.33**). When pH-sensitive dyes were used to monitor both intracellular and extracellular pH in Arabidopsis roots, rapid changes were observed after roots were rotated to a horizontal position. Within 2 min of gravistimulation, the cytoplasmic pH of the columella cells of the root cap increased from 7.2 to 7.5 (see Figure 15.33), while the apoplastic pH declined from 5.5 to 4.5. These changes preceded any detectable tropic curvature by about 10 min.

The alkalinization of the cytosol combined with the acidification of the apoplast suggests that activation of the plasma membrane H+-ATPase is one of the initial events that mediates root gravity perception or signal transduction. The chemiosmotic model of polar auxin transport (see Figure 14.10) predicts that differential acidification of the apoplast and alkalinization of the cytosol would result in increased directional uptake and efflux of IAA from the affected cells.

Early physiological studies suggested that Ca^{2+} release from storage pools might be involved in root gravitropic signal transduction as well. As in the case of [³H]IAA, $^{45}Ca^{2+}$ is polarly transported to the lower half of a root cap that is stimulated by gravity. Auxin-dependent Ca^{2+} and pH signaling thus appears to regulate root gravitropic bending through the propagation of a Ca^{2+}-dependent signaling pathway.

Phototropins are the light receptors involved in phototropism

An emerging seedling is able to bend in any direction toward sunlight to optimize light absorption. This phenomenon is known as phototropism. As you saw in Chapter 13, blue light is particularly effective in inducing phototropism, and two flavoproteins, **phototropin 1** and **phototropin 2**, are the photoreceptors for phototropic bending. Phototropism results from rapid signaling events that are initiated by light-activated phototropins on the lit side of plant organs and that result in differential elongation growth. As in the case of gravitropism, the bending response to directional blue light can be explained by the Cholodny–Went model of lateral auxin redistribution.

Phototropism is mediated by the lateral redistribution of auxin

Charles and Francis Darwin provided the first clue concerning the mechanism of phototropism on coleoptiles by demonstrating that while light is perceived at the tip, bending occurs in the region below the tip. The Darwins proposed that some "influence" was transported from the tip to the growing region, thus causing the observed asymmetric growth response. This influence was later shown to be auxin.

When a shoot is growing vertically, auxin is transported polarly from the growing tip to the elongation zone. The polarity of auxin transport from shoot to root is independent of gravity (see Figure 15.27). However, auxin can also be transported laterally, and this lateral diversion of auxin lies at the heart of the Cholodny–Went model for tropisms. In gravitropic bending, auxin from the root tip that is redirected laterally to the lower side of the root inhibits cell elongation, causing the root to bend downward. In phototropic bending, the auxin for the shoot tip that is redirected laterally to the shaded side of the axis stimulates cell elongation. The resulting differential growth causes the shoot to bend toward the light (**Figure 15.34**).

Although phototropic mechanisms appear to be highly conserved across plant species, the precise sites of auxin production, light perception, and lateral transport have been difficult to define. In maize coleoptiles, auxin accumulates in the upper 1 to 2 mm of the tip. The zones of photosensing and lateral transport extend farther down, to the upper 5 mm of the tip. The response is also strongly

phototropins 1 and 2 Two flavoprotein photoreceptors that mediate the blue-light signaling pathway that induces phototropic bending in higher plants. They also mediate chloroplast movements and participate in stomatal opening in response to blue light. Phototropins are autophosphorylating protein kinases whose activity is stimulated by blue light.

dependent on the light fluence (the number of photons per unit of area). Similar zones of auxin synthesis and accumulation, light perception, and lateral transport are seen in the true shoots of all monocots and eudicots examined to date.

Acidification of the apoplast appears to play a role in phototropic growth: The apoplastic pH is more acidic on the shaded side of phototropically bending stems or coleoptiles than on the irradiated side. Decreased pH would be expected to enhance cellular elongation and amplify auxin movement from cell to cell. Both processes would be expected to contribute to bending toward light.

Shoot phototropism occurs in a series of steps

As we mentioned earlier, the events in phototropic bending occur rapidly. Although phototropins are hydrophilic proteins, they are primarily associated with the plasma membrane. In Arabidopsis, low-fluence blue light is perceived by the cells on the irradiated side of the hypocotyl and a series of signal transduction events is initiated. After approximately 3 min of unilateral blue-light irradiation, phototropin 1 undergoes autophosphorylation. Next, the activated phototropin 1 on the plasma membrane inhibits auxin efflux from the cells in the apical region of the hypocotyl. This causes an accumulation of auxin above the growing region of the hypocotyl, resulting in a rapid decrease in the rate of hypocotyl elongation (**Figure 15.35**). The mechanism of phototropin 1 inhibition of auxin efflux involves the phosphorylation of an ATP-binding cassette B-type auxin transporter (see Figure 14.10).

After the pause in elongation, the pooled auxin is laterally diverted to the shaded side of the hypocotyl via a poorly understood process. Auxin accumulation on the shaded side of the upper hypocotyl can be detected

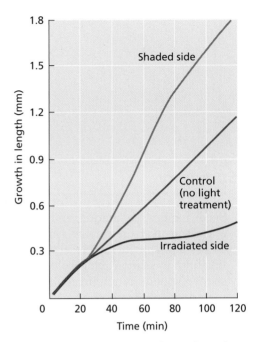

Figure 15.34 Time course of growth on the irradiated and shaded sides of a coleoptile responding to a 30-s pulse of unidirectional blue light at zero minutes. Control coleoptiles were not given a light treatment. (After Iino and Briggs 1984.)

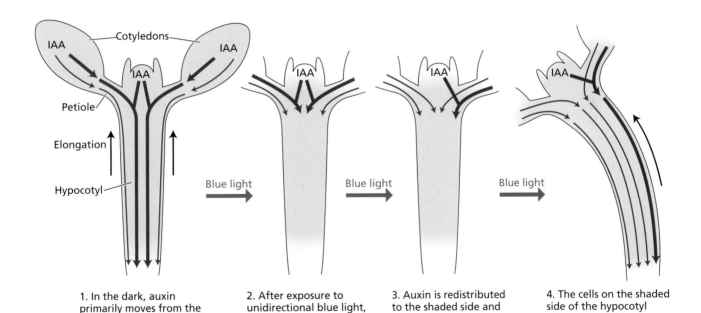

1. In the dark, auxin primarily moves from the shoot to the root through the vascular tissues in the petioles and hypocotyl, and through the epidermis.

2. After exposure to unidirectional blue light, auxin movement briefly stops at the cotyledonary node and the seedling stops growing vertically.

3. Auxin is redistributed to the shaded side and polar transport resumes.

4. The cells on the shaded side of the hypocotyl elongate, resulting in differential growth, and the seedling bends toward the light source.

Figure 15.35 Model of basipetal auxin movement (red lines) associated with phototropin 1–dependent phototropism in dark-acclimated seedlings of Arabidopsis. (After Christie et al. 2011; CC BY.)

after approximately 15 min of exposure to unilateral blue light. Diverted auxin is then transported to the hypocotyl elongation zone in the vascular tissues and epidermis. These later steps involve differential activation of PIN proteins and rapid auxin-dependent activation of H^+-ATPase activity on the shaded side of the hypocotyl. Bending toward the blue-light source can be observed after approximately 30 min.

Summary

Seeds require rehydration, and sometimes additional treatments, to germinate. During germination and establishment, food reserves maintain the seedling until it is autotrophic. Following emergence, the shoot responds to nondirectional cues from sunlight to undergo photomorphogenesis. At the same time shoots are also responding to directional signals to orient themselves with respect to sunlight (phototropism) and gravity (gravitropism). The root extends downward into the soil and forms numerous branches to provide anchorage, water, and mineral nutrients, while the shoot undergoes greening, produces photosynthetic leaves, and grows toward sunlight. Vascular tissue differentiates to facilitate the movement of water, minerals, and sugars. Hormones play central roles as signaling agents in all of the developmental pathways associated with seedling establishment.

Seed Structure

• Seeds are surrounded by a seed coat, whereas fruits are enclosed by the pericarp (**Figure 15.1**).

• Seed anatomy varies widely in the types and distributions of stored food resources and the nature of the seed coat (**Figure 15.2**).

Seed Dormancy

• Seed dormancy may be either exogenous (imposed by surrounding tissues) or endogenous (arising from the embryo itself).

• Carrot seeds exemplify seeds that require additional time to germinate because of an undersized embryo (**Figure 15.3**).

• Seeds that do not become dormant may exhibit vivipary and precocious germination (**Figures 15.4, 15.5**).

• The primary hormones regulating seed dormancy are abscisic acid and gibberellins (**Figure 15.6**).

Release from Dormancy

• Light, especially red light, breaks dormancy in many small seeds, a phenomenon mediated by phytochrome.

• Some seeds require chilling or after-ripening to break dormancy (**Figure 15.7**).

• ABA and gibberellin are not the only chemicals regulating seed dormancy. In some seeds, nitrate, nitric oxide, and chemicals in smoke can break dormancy.

Seed Germination

• Germination and postgermination take place in three phases relating to water uptake (**Figures 15.8, 15.9**).

Mobilization of Stored Reserves

• The cereal aleurone layer responds to gibberellins by secreting hydrolytic enzymes (including α-amylase) into the surrounding endosperm, making starches available to the embryo (**Figure 15.10**).

• Gibberellins secreted by the embryo also enhance the transcription of α-*amylase* mRNA, which initiates starch degradation.

• The gibberellin promotes α-amylase transcription and production (**Figure 15.11**).

• ABA inhibits α-amylase transcription.

Seedling Establishment

• Seedlings transition from skotomorphogenesis (development in the dark; i.e., underground) to photomorphogenesis (development in the presence of light) at the first instance of light (**Figure 15.12**).

• In etiolated shoots, gibberellins and brassinosteroids suppress photomorphogenesis (**Figure 15.13**).

• Phytochrome, auxin, and ethylene regulate hook opening (**Figure 15.14**).

• Vascular differentiation begins during seedling emergence (**Figure 15.15**).

(Continued)

Summary (*continued*)

- Growing root tips can be divided into three main developmental zones: the meristematic, elongation, and maturation zones (**Figure 15.16**).

- The root cap covers the root apical meristem and protects it as the root pushes through the soil.

- Root hairs are specialized epidermal cells that reach maturity in the maturation zone of the root tip (**Figure 15.16**).

- Root hair formation is regulated by ethylene (**Figure 15.17**).

- Lateral roots are initiated internally in the pericycle and emerge through cortical and epidermal cells (**Figures 15.16, 15.18**).

Cell Expansion: Mechanisms and Hormonal Controls

- Seedlings exhibit both tip growth and diffuse growth (**Figure 15.19**).

- Orientation of cellulose microfibrils regulates direction of cell expansion (**Figure 15.20**).

- Cell wall extension is induced by acidification and is mediated by cell wall proteins called expansins (**Figures 15.21, 15.22**).

- At optimal concentrations, auxin promotes stem and coleoptile growth and inhibits root growth. However, higher concentrations of auxin can inhibit stem and coleoptile growth (**Figures 15.23, 15.24**).

- Ethylene causes reorientation of microtubules and induces lateral cell expansion (**Figure 15.25**).

Tropisms: Growth in Response to Directional Stimuli

- Polarized seedling growth is directed by polar auxin streams (**Figure 15.26, Figure 15.27**).

- Most of the auxin that is redirected shootward in the apex of young seedling roots is derived from the shoot (**Figure 15.28**).

- Lateral redistribution of auxin in coleoptile tips facilitates gravitropism in coleoptiles (**Figure 15.29**).

- A horizontally oriented root redirects auxin to the lower side, inhibiting growth at the elongation zone, an activity that is mediated by the root cap (**Figure 15.30**).

- Statoliths in the columella cells of the root cap serve as gravity sensors (**Figure 15.31**).

- The statoliths that regulate gravitropism in eudicot stems and hypocotyls are located in the starch sheath (**Figure 15.32**).

- pH and calcium ions (Ca^{2+}) act as second messengers in the signaling that occurs during gravitropism (**Figure 15.33**).

- Like gravitropism, phototropism involves lateral redistribution of growth (**Figure 15.34**).

- The first step in phototropic bending in hypocotyls occurs within minutes of irradiation when phototropin 1 inhibits rootward auxin transport (**Figure 15.35**).

- Lateral auxin redirection at the shoot apex begins within 15 min, and bending begins after 30 min (**Figure 15.35**).

Suggested Reading

Baldwin, K. L., Strohm, A. K., and Masson, P. H. (2013) Gravity sensing and signal transduction in vascular plant primary roots. *Am. J. Bot.* 100: 126–142.

Bewley, J. D., Bradford, K. J., Hilhorst, H. W. M., and Nonogaki, H. (2013) *Seeds: Physiology of Development, Germination and Dormancy*, 3rd ed. Springer, New York.

Casal, J. J. (2013) Photoreceptor signaling networks in plant responses to shade. *Annu. Rev. Plant Biol.* 64: 403–427.

Finch-Savage, W. E., and Leubner-Metzger, G. (2006) Seed dormancy and the control of germination. *New Phytol.* 171: 501–523.

Graeber, K., Kakabayashi, K., Miatton, E., Leubner-Metzger, G., and Soppe, W. J. J. (2012) Molecular mechanisms of seed dormancy. *Plant Cell Environ.* 35: 1769–1786.

Lacayo, C. I., Malkin, A. J., Holman, H.-Y. N., Chen, L.,

Ding, S.-Y., Hwang, M. S., and Thelen, M. P. (2010) Imaging cell wall architecture in single *Zinnia elegans* tracheary elements. *Plant Physiol.* 154: 121–133.

Lia, Y.-C., Rena, J.-P., Cho, M.-J., Zhou, S.-M., Kim, Y.-B., Guo, H.-X., Wong, J. H., Niu, H.-B., Kim, H.-K., Morigasaki, S., et al. (2009) The level of expression of thioredoxin is linked to fundamental properties and applications of wheat seeds. *Mol. Plant* 2: 430–441.

Migliaccio, F., Tassone, P., and Fortunati, A. (2013) Circumnutation as an autonomous root movement in plants. *Am. J. Bot.* 100: 4–13.

Novo-Uzal, E., Fernández-Pérez, F., Herrero, J., Gutiérrez, J., Gómez-Ros, L. V., Ángeles Bernal, M., Díaz, J., Cuello, J., Pomar, F., and Ángeles Pedreño, M. (2013) From *Zinnia* to *Arabidopsis*: Approaching the involvement of peroxidases in lignification. *J. Exp. Bot.* 64: 3499–3518.

Palmieri, M., and Kiss, J. Z. (2007) The role of plastids in gravitropism. In *The Structure and Function of Plastids*, R. R. Wise, and J. K. Hoober, eds., Springer, Berlin, pp. 507–525.

Petricka, J. J., Winter, C. M., and Benfey, P. N. (2012) Control of *Arabidopsis* root development. *Annu. Rev. Plant Biol.* 63: 563–590.

Sawchuk, M. G., Edgar, A., and Scarpella, E. (2013) Patterning of leaf vein networks by convergent auxin transport pathways. *PLOS Genet.* 9: 1–13.

Van Norman, J. M., Xuan, W., Beeckman, T., and Benfey, P. N. (2014) Periodic root branching in *Arabidopsis* requires synthesis of an uncharacterized carotenoid derivative. *Proc. Natl. Acad. Sci USA* 111(13): E1300–E1309. DOI: 10.1073/pnas.1403016111

Van Norman, J. M., Zhang, J., Cazzonelli, C. I., Pogson, B. J., Harrison, P. J., Bugg, T. D. H., Chan, K. X., Thompson, A. J., and Benfey, P. N. (2013) To branch or not to branch: The role of pre-patterning in lateral root formation. *Development* 140: 4301–4310.

16 Vegetative Growth and Senescence

During seedling establishment, the basic polarities of the seedling axes are set up and the major tissue types are differentiated. The next developmental stage produces the mature primary plant body (see Figure 1.2). In the shoot, numerous leaves form and specialized cell types differentiate. Root system development consists largely of branch (lateral root) formation and growth. In woody perennials, the activities of the vascular and cork cambiums give rise to secondary growth (see Box 1.2). Throughout the vegetative life of the plant, the process of leaf senescence recycles organic constituents and mineral nutrients from old to newly formed leaves. Eventually, the whole plant undergoes senescence because of a combination of genetic and environmental factors. In this chapter we outline many of these developmental phenomena and the regulatory mechanisms that underlie them.

The Shoot Apical Meristem

New vegetative organs begin to develop after a seedling has become established. In the shoot, the source of this new primary growth is the **shoot apical meristem** (**SAM**). The SAM is a small, dome-shaped structure that gives rise to leaf and bud primordia on its flanks. The developing leaf primordia overlap to form a conical structure that surrounds and protects the SAM (**Figure 16.1**). The SAM plus overlapping leaf primordia are referred to as the **shoot apex** or **terminal bud**.

shoot apical meristem
Dome-shaped region of the shoot tip composed of meristematic cells that give rise to leaves, branches, and reproductive structures.

shoot apex (terminal bud)
The shoot apical meristem with its associated leaf primordia and young, developing leaves.

Figure 16.1 Structure of the shoot apex. (A) Stained, longitudinal section of the shoot apex of the bay laurel tree (*Laurus nobilis*). Note that the shoot apical meristem (the small mound in the center surrounded by two young leaf primordia) is enclosed and protected by the older leaf primordia. (B) The shoot apical meristem of a tomato plant. In this scanning electron microscope image, leaf primordia are labeled P1–P4 (from youngest to oldest); P4 has been removed to provide a better view of the apical meristem (the dome-shaped mound at the center). (A © Biology Pics/Science Source; B from Kuhlemeier and Reinhardt 2001; courtesy of D. Reinhardt.)

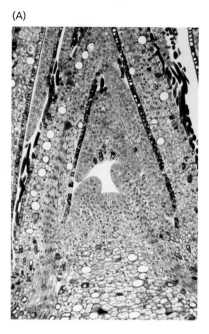

(A)

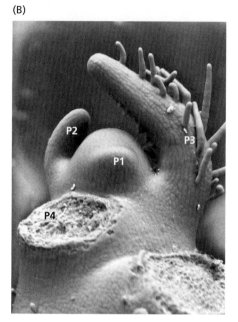

(B)

cytohistological zonation Regional and cytological differences in cell division, cell size, and vacuolization in shoot apical meristems of seed plants.

central zone (CZ) A central cluster of relatively large, highly vacuolated, slowly dividing initial cells in shoot apical meristems, comparable to the quiescent center of root meristems.

peripheral zone (PZ) A doughnut-shaped region surrounding the central zone in shoot apical meristems and consisting of small, actively dividing cells with inconspicuous vacuoles. Leaf primordia are formed in the peripheral zone.

rib zone (RZ) Meristematic cells located beneath the central zone in shoot apical meristems and that give rise to the internal tissues of the stem.

The shoot apical meristem has distinct zones and layers

As early as the mid-nineteenth century it was realized that the cells of the shoot apical meristem exhibited some degree of organization. The organization of the SAM has been interpreted according to two main theories, which are not mutually exclusive. Longitudinal sections of the SAM reveal three regions distinguishable from each other by their locations, patterns of cell division, and the derivatives they form. This type of organization, common in both gymnosperms and angiosperms, is called **cytohistological zonation (Figure 16.2A)**. The tip of the apical dome, called the **central zone (CZ)**, consists of a cluster of initial cells, comparable to the stem cells of animals, that gives rise to all the other cells of the SAM. These initial cells divide more slowly than the cells in the adjacent regions, similar to the quiescent center of roots (see Figure 15.16). A flanking region, called the **peripheral zone (PZ)**, consists of cytoplasmically dense cells that divide more frequently to produce cells that later become incorporated into lateral organs such as leaves. A centrally positioned **rib zone (RZ)** underlying the CZ (see Figure 16.2A) contains dividing cells that give rise to the internal tissues of the stem.

The second theory of SAM organization is based on the presence of different cell layers—L1, L2, and L3—with distinct patterns of cell division and cell fates **(Figure 16.2B)**. Most of the divisions are anticlinal in the L2 and L2 layers, whereas the planes of cell division are more randomly oriented in the L3 layer. During leaf development, each of the cell layers contributes to different tissues in the leaf.

Analyses of cell lineages show that cell divisions occasionally can cause cells of the SAM to be displaced by adjacent cells from one cytohistological zone to another, or from one cell layer to another. As a result of displacement, the developmental fate of the cell may change. For example, a cell that is displaced from the rib zone to the peripheral zone might be incorporated into a leaf primordium rather than the stem. Similarly, a cell that is displaced from the L2 layer to the L1 layer may become an epidermal cell rather than a mesophyll cell. This dynamic behavior indicates that the identities of cells in the SAM, including their characteristic division patterns, primarily reflect their position within the apical dome, rather than a rigidly programmed identity. This plasticity provides the opportunity for plant growth to respond to changing environmental conditions.

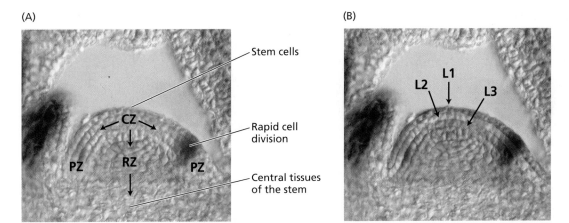

Figure 16.2 Organization of the shoot apical meristem of Arabidopsis. The organization of the shoot apical meristem can be analyzed in terms of cytohistological zones or cell layers. False color has been added to these micrographs to show the different zones and layers. (A) Cytohistological zonation. CZ, central zone (initial cells); PZ, peripheral zone (source of leaves); RZ, rib zone (source of central vascular tissues). (B) Cell layers. Most cell divisions are anticlinal in the outer, L1 and L2, layers; the planes of cell divisions are more randomly oriented in the L3 layer. The outermost (L1) layer generates the shoot epidermis; the L2 and L3 layers generate internal tissues. (From Bowman and Eshed 2000.)

For example, plants that have been repeatedly wounded produce smaller leaves, a phenomenon called the "bonsai effect." The immediate cause of the smaller leaf size is the reduction in the rate of cell division in the SAM induced by the wound hormone, jasmonic acid. Interestingly, the size of the cells is either unchanged or slightly larger than normal. The smaller number of cells in the SAM available to form the leaf primordia results in smaller, though anatomically normal, leaves. The results suggest that the developmental plasticity of the cells of the SAM may be an adaptation enabling plants to reduce their leaf size in response to environmental stresses such as herbivory (see Chapter 18).

SAMs may be continuously meristematic (indeterminate), or may cease activity (be determinate), either by differentiating into a terminal organ such as a flower, or by undergoing growth arrest or senescence. As you will see later in the chapter, the growth habits, life cycles, and senescence profiles of different plant species are intimately connected to their patterns of apical meristem determinacy. In species that reproduce only once in their lifetime, all indeterminate vegetative shoot apices become determinate floral apices, and the entire plant senesces and dies after seed dispersal. In contrast, perennials that reproduce more than once retain a population of indeterminate shoot apices as well as those apices that become reproductive and determinate.

Leaf Structure and Phyllotaxy

Morphologically, the leaf is the most variable of all the plant organs. The collective term for any type of leaf, or modified leaf, on a plant is **phyllome**. Phyllomes include the photosynthetic *foliage leaves* (what we usually mean by "leaves"), protective *bud scales*, *bracts* (leaves associated with inflorescences, or flowers), and *floral organs*. In angiosperms, the main part of the foliage leaf is expanded into a flattened structure, the **lamina** or blade. The appearance of a flat lamina in seed plants in the late Devonian was a key event in leaf evolution. A flat lamina maximizes light capture and also creates two distinct leaf domains: **adaxial** (upper surface) and **abaxial** (lower surface) (**Figure 16.3A**). Several types of leaves have evolved based on their adaxial–abaxial leaf structure.

phyllome The collective term for all the leaves of a plant, including structures that evolved from leaves, such as floral organs.

lamina The blade of a leaf.

adaxial Referring to the upper surface of a leaf.

abaxial Referring to the lower surface of a leaf.

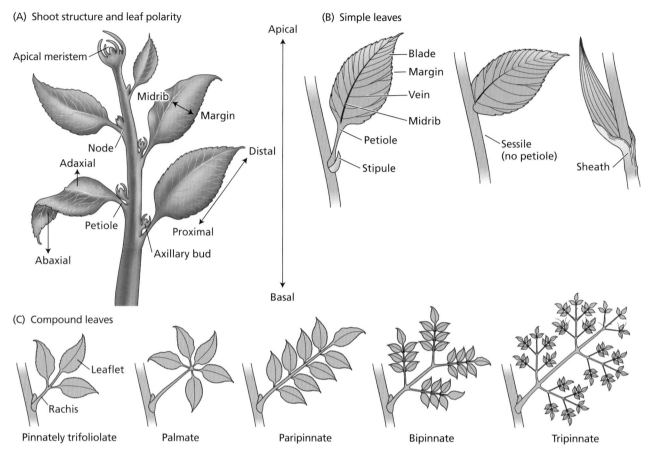

(A) Shoot structure and leaf polarity

Apical meristem

Midrib

Margin

Node

Adaxial

Petiole

Abaxial

Axillary bud

Apical

Distal

Proximal

Basal

(B) Simple leaves

Blade

Margin

Vein

Midrib

Petiole

Stipule

Sessile (no petiole)

Sheath

(C) Compound leaves

Leaflet

Rachis

Pinnately trifoliolate

Palmate

Paripinnate

Bipinnate

Tripinnate

Figure 16.3 Overview of leaf structure. (A) Shoot structure, showing three types of leaf polarity: adaxial–abaxial, distal–proximal, and midrib–margin. (B) Examples of simple leaves. Variations in lower leaf structure include the presence or absence of stipules and petioles, and leaf sheaths. (C) Examples of compound leaves.

In the majority of plants, the leaf blade is attached to the stem by a stalk called the **petiole**. However, some plants have *sessile leaves*, with the leaf blade attached directly to the stem (see Figure 16.3B). In most monocots and certain eudicots, the base of the leaf is expanded into a sheath around the stem. Many eudicots have *stipules*, small outgrowths of the leaf primordia, located on the abaxial side of the leaf base. Stipules protect the young developing foliage leaves and are sites of auxin synthesis during early leaf development.

Leaves may be **simple** or **compound** (**Figures 16.3B and C**). A simple leaf has one blade, whereas a compound leaf has two or more blades, the *leaflets*, attached to a common axis, or *rachis*. Some leaves, like the adult leaves of some *Acacia* species, lack a blade and instead have a flattened petiole simulating the blade, called the *phyllode*. In some plants the stems themselves are flattened like blades and are called *cladodes*, as in the cactus *Opuntia*.

Auxin-dependent patterning of the shoot apex begins during embryogenesis

A long-standing question in plant biology is how the characteristic arrangement of leaves on the shoot, or **phyllotaxy**, is achieved. Three basic phyllotactic patterns, termed alternate, decussate (opposite), and spiral (**Figure 16.4**), can be directly linked to the pattern of initiation of leaf primordia on the SAM. These patterns depend on several factors, including intrinsic factors that tend to produce a phyllotaxy that is characteristic of a species. However, environmental factors

petiole The leaf stalk that joins the leaf blade to the stem.

simple leaf A leaf with one blade.

compound leaf A leaf subdivided into leaflets.

phyllotaxy The arrangement of leaves on the stem.

or mutations that lead to changes in meristem size or shape can also affect phyllotaxy, suggesting that position-dependent mechanisms play important roles.

All leaves and modified leaves begin as small protuberances, called primordia, on the flanks of the SAM. The primordia then elongate, which establishes the proximal–distal leaf axis, and flatten, which defines the adaxial and abaxial leaf surfaces. As the leaf develops, the palisade parenchyma, mesophyll, and vascular tissues differentiate. Gene expression and hormonal signaling contribute to all these processes.

The sites of leaf initiation, which ultimately give rise to the phyllotaxy of the plant, correspond to localized zones of auxin accumulation, suggesting that auxin accumulation is required for leaf initiation (**Figure 16.5A**). Support for this hypothesis comes from several approaches. For example, *pin1* mutants of Arabidopsis, which are defective in polar auxin transport, fail to produce any leaf primordia on their inflorescence meristems (**Figure 16.5B**). However, application of exogenous auxin to the flank of the mutant SAM results in the production of a leaf primordium (**Figure 16.5C**). Further support for the role of auxin accumulation in leaf initiation is provided by the changes in phyllotactic patterns induced by treatment with auxin transport inhibitors.

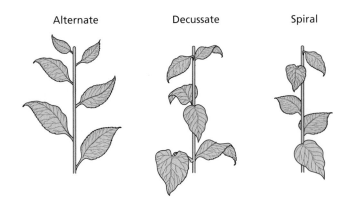

Figure 16.4 The three basic types of leaf arrangement (phyllotaxy) on the stem are alternate, decussate, and spiral. These patterns have their origin in the pattern of leaf primordia formation on the flanks of the SAM.

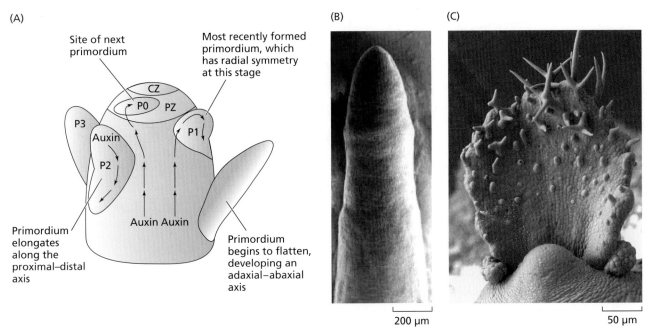

Figure 16.5 Auxin regulation of leaf initiation. (A) Sites of leaf formation are related to patterns of polar auxin transport. Patterns of auxin movement (arrows) can be inferred from asymmetric localization of PIN proteins (see Chapter 14). P0, P1, P2, and P3 refer to the ages of leaf primordia; P0 corresponds to the stage at which the leaf begins its overt development, and P1, P2, and P3 represent increasingly older leaves. Leaf primordia are initiated where auxin accumulates. Acropetal (toward the tip) movement of auxin is blocked at the boundary separating the central and peripheral zones (CZ and PZ, respectively), leading to increased auxin levels at this position and the initiation of a leaf (P0). The newly formed leaf primordium (P1) acts as an auxin sink, thus preventing initiation of new leaves directly above it. The displacement of a more mature leaf primordium (P2) away from the PZ allows acropetal auxin movements to become reestablished, thus enabling the initiation of another leaf. (B) Scanning electron micrograph of a *pin1* inflorescence meristem that fails to produce leaf primordia. (C) Leaf primordium induced on the inflorescence meristem of a *pin1* mutant by placing a microdrop of IAA (indole-3-acetic acid, an auxin) in lanolin paste on the side of the meristem. (A after Reinhardt et al. 2003; B from Vernoux et al. 2000; C from Reinhart et al. 2003.)

protoderm   In the plant embryo and in apical meristems, the one-cell-thick surface layer that covers the young shoot and radicle of the embryo and gives rise to the epidermis.

pavement cells   The predominant type of leaf epidermal cells, which secrete a waxy cuticle and serve to protect the plant from dehydration and damage from ultraviolet radiation.

trichomes   Unicellular or multicellular hairlike structures that differentiate from the epidermal cells of shoots and roots. Trichomes may be structural or glandular and function in plant responses to biotic or abiotic environmental factors.

guard cells   A pair of specialized epidermal cells that surround the stomatal pore and regulate its opening and closing.

Differentiation of Epidermal Cell Types

In addition to the palisade parenchyma and spongy mesophyll, which are specialized for photosynthesis and gas exchange, the leaf has an epidermis that also plays vital roles in leaf function. The epidermis is the outermost layer of cells on the primary plant body, including both vegetative and reproductive structures. The epidermis usually consists of a single layer of cells derived from meristem cells known as the **protoderm**. In some plants, such as members of the Moraceae and certain species of the Begoniaceae and Piperaceae, the epidermis has two to several cell layers derived from periclinal divisions of the protoderm.

There are three main types of shoot epidermal cells found in all angiosperms: pavement cells, trichomes, and guard cells. **Pavement cells** are relatively unspecialized epidermal cells that can be regarded as the default developmental fate of the protoderm. **Trichomes** are unicellular or multicellular extensions of the shoot epidermis that take on diverse forms, structures, and functions, including protection against insect and pathogen attack, reduction of water loss, and increased tolerance of abiotic stress conditions. **Guard cells** are pairs of cells that surround the **stomata**, or pores, that are present in the photosynthetic parts of the shoot. Guard cells regulate gas exchange between the leaf and the atmosphere by undergoing tightly regulated turgor changes in response to light and other factors (see Chapters 3, 6, 12, and 13). Other types of epidermal cells, such as *lithocysts*, *bulliform cells*, *silica cells*, and *cork cells* (**Figure 16.6**), play ecological roles in plant defense and drought tolerance in certain groups of plants. In the

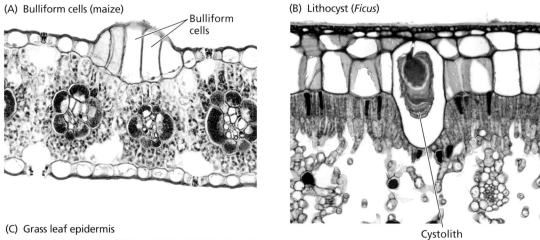

(A) Bulliform cells (maize) — Bulliform cells

(B) Lithocyst (*Ficus*) — Cystolith

(C) Grass leaf epidermis — Cork cell, Silica cell, Pavement cells, Guard cells

Figure 16.6   Examples of specialized epidermal cells. (A) Bulliform cells of maize (corn; *Zea mays*). Rolling and unrolling in grass leaves is driven by turgor changes in the bulliform cells. (B) Lithocyst cell in a *Ficus* leaf containing a cystolith, composed of calcium carbonate deposited on a cellulosic stalk attached to the upper cell wall. (C) Bread wheat (*Triticum aestivum*) leaf epidermis with pairs of silica and cork cells interspersed among the pavement cells. (A © Dr. Ken Wagner/Visuals Unlimited, Inc.; B © Jon Bertsch/Visuals Unlimited, Inc.; C © Garry DeLong/Science Source.)

next section we describe the development of stomatal guard cells as an example of epidermal cell differentiation.

A specialized epidermal lineage produces guard cells

In eudicots, only specific cells in a stomatal cell lineage differentiate into guard cells (**Figure 16.7**). In the developing protoderm (which gives rise to the leaf epidermis), a population of **meristemoid mother cells** (**MMCs**) is established. Each MMC divides asymmetrically (the so-called entry division) to give rise to two morphologically distinct daughter cells—a larger stomatal lineage ground cell (SLGC) and a smaller **meristemoid**. An SLGC can either differentiate into a pavement cell or become an MMC and found secondary or satellite lineages. The meristemoid can undergo a variable number of asymmetric amplifying divisions giving rise to as many as three SLGCs, with the meristemoid ultimately differentiating into a **guard mother cell** (**GMC**), which is recognizable because of its rounded morphology. The GMC then undergoes one symmetric division, forming a pair of guard cells surrounding the stomatal pore.

Although this lineage is called the "stomatal lineage," the ability of meristemoids and SLGCs to undergo repeated divisions means that this lineage is actually responsible for generating the majority of the epidermal cells in the leaves. Following amplifying divisions of the meristemoid, the resulting SLGCs can differentiate into pavement cells, which are the most abundant cell type in the epidermis of a mature leaf, or they can divide asymmetrically to give rise to a secondary meristemoid. The orientation of division in asymmetrically dividing SLGCs is important for enforcing the "one-cell-spacing rule," according to which

stomata (singular *stoma* or *stomate*) Microscopic pores in the leaf epidermis, each surrounded by a pair of guard cells, and in some species, also including subsidiary cells. Stomata regulate the gas exchange (water and CO_2) of leaves by controlling the dimension of a stomatal pore.

meristemoid mother cells (MMCs) The cells of the leaf protoderm that divide asymmetrically (the so-called entry division) to give rise to the meristemoid, a guard cell precursor.

meristemoid A small, triangular stomatal precursor cell that functions temporarily as an initial cell in a meristem.

guard mother cell (GMC) The cell that gives rise to a pair of guard cells to form a stomate.

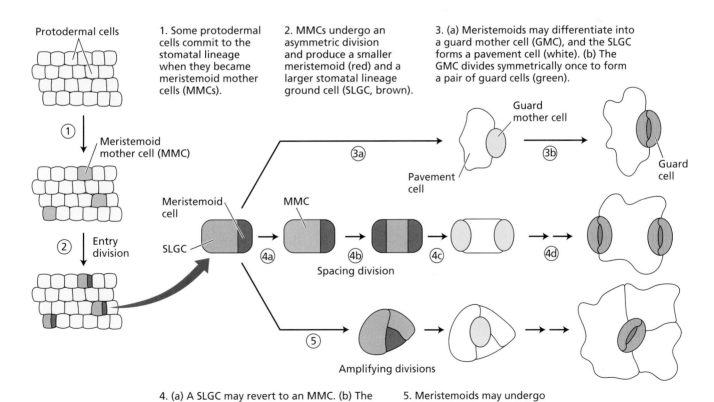

Protodermal cells

1. Some protodermal cells commit to the stomatal lineage when they became meristemoid mother cells (MMCs).

2. MMCs undergo an asymmetric division and produce a smaller meristemoid (red) and a larger stomatal lineage ground cell (SLGC, brown).

3. (a) Meristemoids may differentiate into a guard mother cell (GMC), and the SLGC forms a pavement cell (white). (b) The GMC divides symmetrically once to form a pair of guard cells (green).

Meristemoid mother cell (MMC)

Entry division

Meristemoid cell

SLGC

MMC

Spacing division

Guard mother cell

Pavement cell

Guard cell

Amplifying divisions

4. (a) A SLGC may revert to an MMC. (b) The resulting MMC may undergo an asymmetric division to create a new meristemoid. (c) The two meristemoids may differentiate into GMCs. (d) Each GMC divides to form a pair of guard cells.

5. Meristemoids may undergo additional asymmetric divisions before forming GMCs and guard cells.

Figure 16.7 Stomatal development in Arabidopsis. (After Lau and Bergmann 2012.)

(A)

(B)

Figure 16.8 Two basic patterns of leaf venation in angiosperms. (A) Reticulate venation in *Prunus serotina*, a eudicot. (B) Parallel venation in *Iris sibirica*, a monocot. (Photos by David McIntyre.)

stomata must be situated at least one cell length apart to maximize gas exchange between the leaf and the atmosphere.

Venation Patterns in Leaves

The leaf vascular system is a complex network of interconnecting veins consisting of two main conducting tissue types, xylem and phloem, as well as nonconducting cells, such as parenchyma, sclerenchyma, and fibers. The spatial organization of the leaf vascular system—its **venation pattern**—is both species- and organ-specific. Venation patterns fall into two broad categories: *reticulate venation*, found in most eudicots, and *parallel venation*, typical of many monocots (**Figure 16.8**).

Despite the diversity of leaf venation patterns, they all share a hierarchical organization. Veins are organized into distinct size classes—primary, secondary, tertiary, and so on—based on their width at the point of attachment to the parent vein (**Figure 16.9**). The smallest minor veins, called veinlets, terrminate blindly in the mesophyll. The hierarchical structure of the leaf vascular system reflects the hierarchical functions of different-sized veins, with larger diameter veins functioning in the **bulk flow** of water, minerals, sugars, and other metabolites, and smaller diameter veins functioning in **phloem loading** (see Chapter 10).

The primary leaf vein is initiated in the leaf primordium

Leaf vascular bundles arise from vascular precursor cells called the **procambium** in the emerging leaf primordia of the SAM (**Figure 16.10A**). From there the vascular bundles differentiate downward (basipetally) toward the node directly below the leaf and form a connection to the older vascular bundles that are continuous to the base of the shoot. The portion of the vascular bundle that enters the leaf is called the **leaf trace** (**Figure 16.10B**).

Auxin canalization initiates development of the leaf trace

Several lines of evidence indicate that auxin stimulates formation of vascular tissues. An example is the role of auxin in regeneration of vascular tissue after wounding (**Figure 16.11A**). Vascular regeneration is prevented by removal of the leaf and shoot above the wound but can be restored by the application of auxin to the cut petiole above the wound, suggesting that auxin from the leaf is required for vascular regeneration. As shown in **Figure 16.11B**, the files of re-

venation pattern   The pattern of veins of a leaf.

bulk flow   Translocation of water and solutes down a pressure gradient, as in the xylem or phloem.

phloem loading   The movement of photosynthetic products into the sieve elements of mature leaves.

procambium   Primary meristematic tissue that differentiates into xylem, phloem, and cambium.

leaf trace   The portion of the shoot primary vascular system that diverges into a leaf.

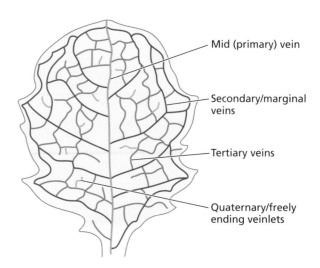

Mid (primary) vein

Secondary/marginal veins

Tertiary veins

Quaternary/freely ending veinlets

Figure 16.9 Hierarchy of venation in the mature Arabidopsis leaf based on the diameter of the veins at the site of attachment to the parent vein. (After Lucas et al. 2013.)

(A)

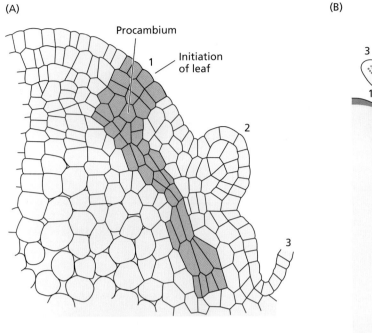

(B)

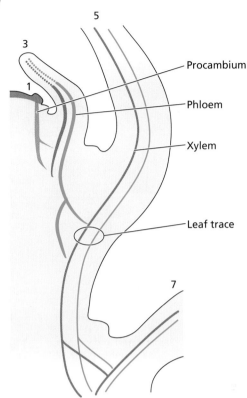

Figure 16.10 Development of the shoot vascular system. (A) Longitudinal section through the shoot tip of perennial flax (*Linum perenne*), showing the early stage in the differentiation of the leaf trace procambium at the site of a future leaf primordium. The leaf primordia and leaves are numbered, beginning with the youngest initial. (B) Early vascular development in a shoot with decussate phyllotaxy (see Figure 16.4). The dark green area at the tip indicates the SAM and two young leaf primordia; procambial strands are shown in orange. Dotted lines in the pair of leaves labeled "3" represent developing xylem and phloem. Leaf traces develop basipetally to the mature vascular tissue below to form a continuous vascular bundle. Numbers correspond to the leaf order, starting with the primordia (missing leaves are in a different plane). (A after Esau 1942; B after Esau 1953.)

generating xylem elements originate at the source of auxin at the upper cut end of the vascular bundle, and progress basipetally until they reconnect with the cut end of the vascular bundle below, matching the presumed direction of the flow of auxin. The upper end of the cut vascular bundle thus acts as the **auxin source** and the lower cut end as the **auxin sink**.

These and similar observations in other systems, such as bud grafting, have led to the hypothesis that as auxin flows through tissues it stimulates and polarizes its own transport, which gradually becomes channeled—or canalized—into files of cells leading away from auxin sources; these cell files can then differentiate to form vascular tissue.

Consistent with this idea, local auxin application (as in the wounding experiments described in Figure 16.11) induces vascular differentiation in narrow strands leading away from the application site, rather than in broad fields of cells. New vasculature usually develops toward, and unites with, preexisting vascular strands, resulting in a connected vascular network. We would therefore predict that a developing leaf trace acts as an auxin source and the existing stem vasculature as an auxin sink. Recent studies on leaf venation have supported this source–sink model, or **canalization model**, for auxin flow at the molecular level. Experimental visualization of auxin transport proteins and dynamic changes in auxin concentrations during leaf vein development suggest that both polar auxin transport and localized auxin biosynthesis are involved in vascularization during early leaf development.

auxin source A cell or tissue that exports auxin to other cells or tissues by polar transport.

auxin sink A cell or tissue that takes up auxin from a nearby auxin source. Participates in auxin canalization during vascular differentiation.

canalization model The hypothesis that as auxin flows through tissues it stimulates and polarizes its own transport, which gradually becomes channeled—or canalized—into files of cells leading away from auxin sources; these cell files can then differentiate to form vascular tissue.

(A)

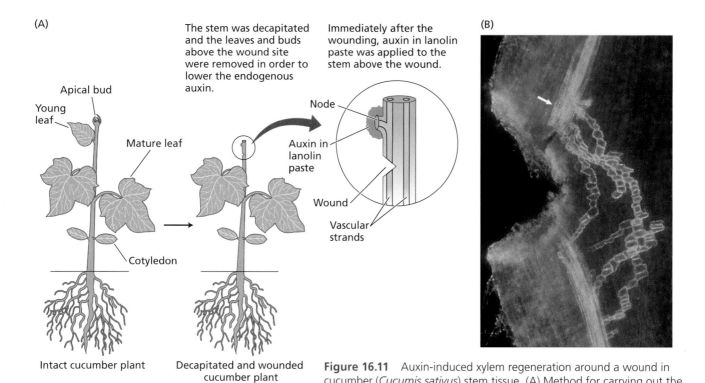

The stem was decapitated and the leaves and buds above the wound site were removed in order to lower the endogenous auxin.

Immediately after the wounding, auxin in lanolin paste was applied to the stem above the wound.

Apical bud

Young leaf

Mature leaf

Cotyledon

Node

Auxin in lanolin paste

Wound

Vascular strands

Intact cucumber plant

Decapitated and wounded cucumber plant

(B)

Figure 16.11   Auxin-induced xylem regeneration around a wound in cucumber (*Cucumis sativus*) stem tissue. (A) Method for carrying out the wound regeneration experiment. (B) Fluorescence micrograph showing regenerating vascular tissue (which is fluorescing orange) around the wound. The arrow indicates the wound site where auxin accumulates and xylem differentiation begins. (B courtesy of R. Aloni.)

SAM

Phytomer

Leaf

Node

Axillary meristem

Internode

Cotyledon

Hypocotyl

Root cap

Figure 16.12   The phytomer is the basic module of shoot organization in seed plants.

Shoot Branching and Architecture

Shoot architecture in seed plants is characterized by multiple repetitions of a basic module called the **phytomer**, which consists of an internode, a node, a leaf, and an **axillary meristem** (**Figure 16.12**). During evolution, modification of the position, size, and shape of the individual phytomer, and variations in the regulation of axillary bud outgrowth, provided the morphological basis for the remarkable diversity of shoot architecture among seed plants. Vegetative and inflorescence branches, as well as the floral primordia produced by inflorescences, are derived from axillary meristems initiated in the axils of leaves. During vegetative development, axillary meristems, like the apical meristem, initiate the formation of leaf primordia, resulting in axillary buds. These buds either become dormant or develop into lateral shoots depending on their position along the shoot axis, the developmental stage of the plant, and environmental factors. During reproductive development, axillary meristems initiate formation of inflorescence branches and flowers. Hence, the growth habit of a plant depends not only on the patterns of axillary meristem formation, but also on meristem identity and its subsequent growth characteristics.

Auxin, cytokinins, and strigolactones regulate axillary bud outgrowth

Once the axillary meristems are formed, they may enter a phase of highly restricted growth (dormancy), or they may be released to grow into axillary branches. The "grow or don't grow" decision is determined by developmental programming and environmental responses mediated by plant hormones that act as local and long-distance signals. Interactions of hormonal signaling pathways coordinate the relative growth rates of different branches and the shoot tip, which ultimately determine shoot architecture. The main hormones involved are auxin, cytokinins, and strigolactones (see Chapter 12). All three hormones are produced in varying quantities in the root and shoot, but translocation of the hormones allows them to exert effects beyond their sites of synthesis.

The role of auxin in regulating axillary bud growth is most easily demonstrated in experiments on **apical dominance**—the control exerted by the shoot tip over axillary buds and branches below. Auxin synthesized at the shoot apex is transported in a rootward polar stream (see Chapter 15). Plants with strong apical dominance are typically weakly branched and show a strong branching response to decapitation (removal of the growing or expanding leaves and shoot tip) (**Figure 16.13**). Applying auxin to the decapitated stem restores apical dominance. Plants with weak apical dominance are typically highly branched and show little, if any, response to decapitation. More than a century of experimental evidence suggests that, in plants with strong apical dominance, auxin produced in the shoot tip inhibits axillary bud outgrowth. Gardeners take advantage of this phenomenon when they "pinch back" chrysanthemums and many other plants with strong apical dominance to create dense, dome-shaped bushes of blossoms.

Strigolactones act in conjunction with auxin to regulate apical dominance. Mutants defective either in strigolactone biosynthesis or signaling show increased branching without decapitation. Although strigolactone is synthesized in both the shoot and the root, grafting studies have shown that only shoot-derived strigolactone is required for apical dominance. Direct

phytomer   A developmental unit consisting of one or more leaves, the node to which the leaves are attached, the internode below the node, and one or more axillary buds.

axillary meristem   Meristematic tissue in the axils of leaves that gives rise to axillary buds.

apical dominance   In most higher plants, the growing apical bud's inhibition of the growth of lateral buds (axillary buds).

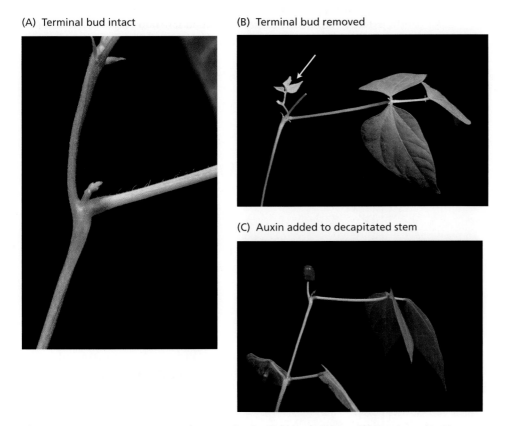

(A) Terminal bud intact

(B) Terminal bud removed

(C) Auxin added to decapitated stem

Figure 16.13   Auxin suppresses the growth of axillary buds in bean (*Phaseolus vulgaris*) plants. (A) The axillary buds are suppressed in the intact plant because of apical dominance. (B) Removal of the terminal bud releases the axillary buds from apical dominance (arrow). (C) Applying IAA in lanolin paste (contained in the gelatin capsule) to the cut surface prevents the outgrowth of the axillary buds. (Photos by David McIntyre.)

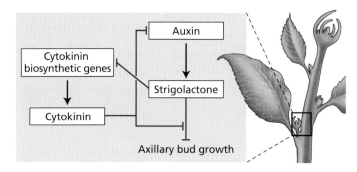

Figure 16.14 Hormonal network regulating apical dominance. Auxin from the shoot apex promotes strigolactone synthesis in the nodal region. In eudicots, strigolactone inhibits axillary bud growth and down-regulates the *cytokinin biosynthetic* genes. Because cytokinin blocks the inhibitory activity of strigolactone on axillary bud growth, reducing cytokinin activity enhances apical dominance. (After El-Showk et al. 2013.)

application of cytokinin to axillary buds, or transgenic stimulation of cytokinin production, stimulates axillary bud growth, suggesting that cytokinins are involved in breaking apical dominance.

A simplified model for the antagonistic interactions between cytokinin and strigolactone is shown in **Figure 16.14**. Auxin maintains apical dominance by stimulating strigolactone synthesis. In eudicots, strigolactone then suppresses axillary bud growth. Strigolactone also inhibits the synthesis of cytokinin by repressing expression of cytokinin biosynthetic genes, which helps suppress axillary bud growth.

The initial signal for axillary bud growth may be an increase in sucrose availability to the bud

Recent evidence indicates that sucrose itself may serve as the initial signal in controlling bud outgrowth (**Figure 16.15**). In pea plants, axillary bud growth is initiated approximately 2.5 h after decapitation. This is 24 h prior to any detectable decline in the auxin concentration in the stem adjacent to the axillary bud, which suggests that a decrease in auxin from the tip occurs much too slowly to initiate bud outgrowth.

In contrast, studies using [^{14}C]sucrose demonstrated that the concentration of leaf-derived sucrose in the stem adjacent to the bud begins to decline as early as 2 h after decapitation. This decline of sucrose in the stem is due to the uptake of sugars by the axillary bud. Thus, following decapitation, bud outgrowth is initiated *prior to* auxin depletion but *after* sucrose uptake by the axillary bud. In the intact stem, the axillary bud is starved for sucrose because the growing terminal bud is a stronger sink for sugars than the axillary bud. As a result of decapitation, the endogenous carbon supply to the axillary buds increases within the time frame sufficient to induce bud release. Apical dominance is thus regulated by

Figure 16.15 Apical dominance is regulated by sugar availability. After decapitation, sugars, which normally flow toward the shoot tip via the phloem, rapidly accumulate in axillary buds, stimulating bud outgrowth. At the same time, the loss of the apical supply of auxin results in a depletion of auxin in the stem. However, auxin depletion is relatively slow and therefore the growing buds in the upper shoot are affected before those lower on the stem. In this model, auxin is involved predominately in the later stages of branch growth. (After Mason et al. 2014.)

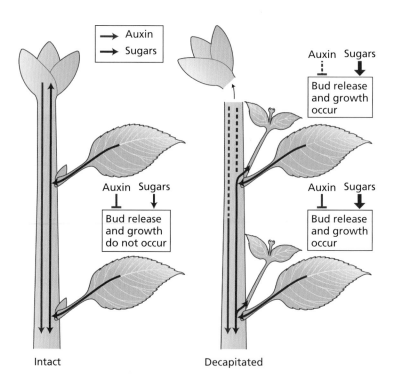

the strong sink activity of the growing tip, which limits sugar availability to the axillary buds. However, sustained bud outgrowth requires the depletion of auxin in the stem adjacent to the bud as well.

Shade Avoidance

Plants compete for light resources to maintain photosynthetic activity. Shade avoidance is the enhanced stem elongation that occurs in certain plants in response to shading by leaves. The response is specific for shade produced by green leaves, which act as filters for red and blue light, and is not induced by other types of shade. **Table 16.1** compares the total fluence rate (related to light intensity) in photons (400–800 nm) and the ratio of red (R) to far-red (FR) light in eight natural conditions and environments. Compared with daylight, there is proportionally more far-red light during sunset, under 5 mm of soil, and under the canopy of other plants (as on the floor of a forest). The canopy phenomenon results from the fact that green leaves absorb red light because of their high chlorophyll content, but are relatively transparent to far-red light.

Plants that increase stem extension in response to shading are said to exhibit a shade avoidance response. The photoreceptor that controls shade avoidance is phytochrome, which has two interconvertible forms: one is inactive and absorbs red light (Pr), and the other is active and absorbs far-red light (Pfr) (see Chapter 13). As shading increases, the R:FR ratio decreases. A higher proportion of far-red light converts more Pfr to Pr, and the ratio of Pfr to total phytochrome ($Pfr:P_{total}$) decreases. When "sun plants" (plants adapted to an open-field habitat) are grown in natural light under a system of shades so that R:FR is controlled, stem extension rates increase in response to a higher far-red content (i.e., a lower $Pfr:P_{total}$) (**Figure 16.16**). In other words, simulated canopy shading (high levels of far-red light, low $Pfr:P_{total}$) induces these plants to allocate more of their resources to growing taller. This correlation is not as strong for "shade plants," which normally grow under a leaf canopy. Shade plants exhibit less reduction in their stem extension rate than do sun plants when they are exposed to higher R:FR values. Thus, there appears to be a systematic relationship between phytochrome-controlled growth and species habitat. Such results are taken as an indication of the involvement of phytochrome in shade perception.

For a "sun plant" or "shade-avoiding plant," there is a clear adaptive value in allocating its resources toward more rapid extension growth when it is shaded by another plant. In this way it can enhance its chances of growing above the canopy and acquiring a greater share of unfiltered, photosynthetically active light. The price for increased internode elongation is usually reduced leaf area and reduced branching, but at least in the short term, this adaptation to canopy shade increases plant fitness. When the plant grows above the canopy or a

Figure 16.16 Phytochrome plays a predominant role in controlling stem elongation rate in sun plants (solid line) but not in shade plants (dashed line). (After Morgan and Smith 1979.)

Table 16.1 Ecologically important light parameters

	Fluence rate (µmol m^{-2} s^{-1})	R:FR[a]
Daylight	1900	1.19
Sunset	26.5	0.96
Moonlight	0.005	0.94
Beneath ivy canopy	17.7	0.13
Soil, at a depth of 5 mm	8.6	0.88
Lakes, at a depth of 1 m		
Black Loch	680	17.2
Loch Leven	300	3.1
Loch Borralie	1200	1.2

Source: Smith 1982, p. 493.

Note: The light intensity factor (400–800 nm) is given as the photon fluence rate, and phytochrome-active light is given as the R:FR ratio.

[a]Absolute values taken from spectroradiometer scans; the values should be taken to indicate the relationships between the various natural conditions and not as actual environmental means.

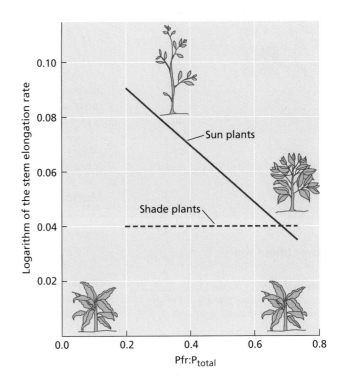

Figure 16.17   High-density planting and crop yield. Modern maize varieties are planted at high density. Traditionally, Native Americans grew maize on small hills or mounds, which were separated by more than a meter. The plants were short and often produced multiple small ears. In contrast, modern hybrids are machine-planted in dense rows with little space between them (typically 74,000–94,000 plants per hectare). Although yield per plant has not increased dramatically for many years in commercial hybrids, overall yields have continued to increase, largely because of better performance of plants at high planting density. As shown in this image from upstate New York, modern varieties of maize have upright leaves that help the plants capture sunlight energy under crowded conditions. (Courtesy of T. Brutnell.)

canopy gap occurs when a tree falls in the forest, then the plant is released from shade avoidance and competition for light.

Reducing shade avoidance responses can improve crop yields

Shade avoidance responses may be highly adaptive in a natural setting to help plants outcompete neighboring vegetation. But for many crop species, a reallocation of resources from reproductive to vegetative growth can reduce crop yield. In recent years, yield gains in crops such as maize (corn; *Zea mays*) have come largely through the breeding of new varieties with a higher tolerance to crowding (which induces shade avoidance responses) rather than through increases in basic yield per plant. As a consequence, today's maize crops can be grown at higher densities than older varieties without suffering decreases in plant yield (**Figure 16.17**).

Root System Architecture

Plant root systems are the critical link between the growing shoot and the rhizosphere, providing both vital nutrients and water to sustain growth. In addition, roots anchor and stabilize the plant, enabling the growth of aboveground vegetative and reproductive organs. Because roots function in heterogeneous and often changing soil conditions, roots must be capable of adapting to ensure a steady flow of water and nutrients to the shoot under a variety of conditions. Plants have evolved complex control mechanisms that regulate root system architecture.

Plants can modify their root system architecture to optimize water and nutrient uptake

Root system architecture refers to the geometric arrangement of individual roots within the plant's root system in the three-dimensional soil space. Root systems are composed of different root types, and plants are able to modify and control the types of roots they produce, the root angles with respect to gravity, the rates of root growth, and the degree of branching. In addition to having a primary root system, some plants can produce additional roots from shoot tissues, called **adventitious roots**, as an adaptation to a particular environment or in response to nutritional or water stress. Variations in root system architecture within and across species have been linked to resource acquisition and growth. Root system architecture varies widely among species, even among those living in the same habitat (**Figure 16.18**).

Monocots and eudicots differ in their root system architecture

Monocot and eudicot root systems are roughly similar in structure, consisting of an embryonically derived primary root (the radicle), lateral roots, and adventitious roots. However, there are significant differences between monocot and eudicot root systems. Monocot root systems are generally more fibrous and complex than the root systems of eudicots, especially in the cereals. For example, the maize seedling root system consists of a primary root that develops from the radicle, lateral roots, **seminal roots** (adventitious roots that branch from the hypocotyl of the embryo), and postembryonically derived **crown roots** (**Figure 16.19**). The primary and seminal roots are highly branched and fibrous. The crown roots, also called "prop roots," are adventitious roots derived from the

adventitious roots   Roots arising from any organ other than a root.

seminal roots   Adventitious roots that arise during embryogenesis from the stem tissue between the scutellum and the coleoptile.

crown roots   Adventitious roots that emerge from the lowermost nodes of a stem.

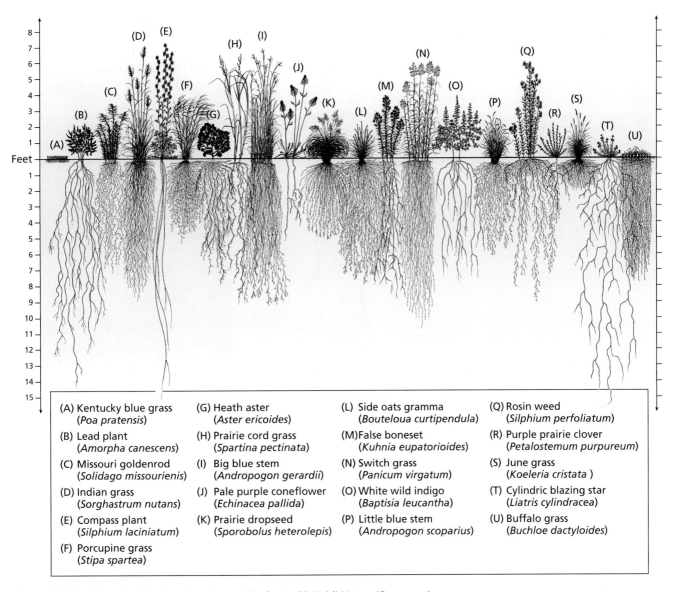

(A) Kentucky blue grass
(*Poa pratensis*)

(B) Lead plant
(*Amorpha canescens*)

(C) Missouri goldenrod
(*Solidago missourienis*)

(D) Indian grass
(*Sorghastrum nutans*)

(E) Compass plant
(*Silphium laciniatum*)

(F) Porcupine grass
(*Stipa spartea*)

(G) Heath aster
(*Aster ericoides*)

(H) Prairie cord grass
(*Spartina pectinata*)

(I) Big blue stem
(*Andropogon gerardii*)

(J) Pale purple coneflower
(*Echinacea pallida*)

(K) Prairie dropseed
(*Sporobolus heterolepis*)

(L) Side oats gramma
(*Bouteloua curtipendula*)

(M) False boneset
(*Kuhnia eupatorioides*)

(N) Switch grass
(*Panicum virgatum*)

(O) White wild indigo
(*Baptisia leucantha*)

(P) Little blue stem
(*Andropogon scoparius*)

(Q) Rosin weed
(*Silphium perfoliatum*)

(R) Purple prairie clover
(*Petalostemum purpureum*)

(S) June grass
(*Koeleria cristata*)

(T) Cylindric blazing star
(*Liatris cylindracea*)

(U) Buffalo grass
(*Buchloe dactyloides*)

Figure 16.18 Diversity of root systems in prairie plants. (© Heidi Natura/Conservation Research Institute.)

lowermost nodes of the stem. Although crown roots are relatively unimportant in seedlings (see Figure 16.19A), in contrast to the primary and seminal roots, crown roots continue to form, develop, and branch throughout vegetative growth. Thus, the crown root system makes up the vast majority of the root system in adult maize plants (see Figure 16.19B).

The root system of a young eudicot consists of the primary (or tap) root and its branch roots. As the root system matures, basal roots arise from the base of the tap root. In addition, adventitious roots can arise from subterranean stems or from the hypocotyl, and can be considered to be loosely analogous to the adventitious crown roots in the cereals. **Figure 16.20** depicts the root system of a soybean plant, a representative eudicot.

Root system architecture changes in response to phosphorus deficiencies

Phosphorus is, along with nitrogen, the most limiting mineral nutrient for crop production (see Chapters 4 and 5). Phosphorus limitation is a particular

Figure 16.19 (A) Root system of a 14-day-old maize seedling composed of primary root derived from the embryonic radicle, the seminal roots derived from the scutellar node, postembryonically formed crown roots that arise at nodes above the mesocotyl, and lateral roots. (B) Mature maize root system. (A from Hochholdinger and Tuberosa 2009; B © B W Hoffmann/AGE Fotostock.)

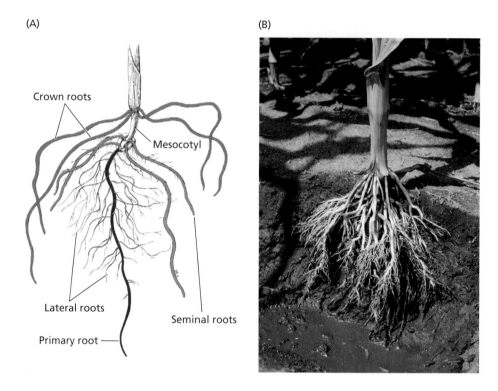

(A)

Crown roots

Mesocotyl

Lateral roots

Seminal roots

Primary root

(B)

problem in tropical soils, where the highly weathered acidic soils tend to bind phosphorus tightly, rendering much of it unavailable for acquisition by roots. Plant root systems undergo well-documented morphological alterations in response to phosphorus deficiency (**Figure 16.21**). These responses can vary somewhat from species to species, but in general they include a reduction in primary root elongation, an increase in lateral root proliferation and elongation, and an increase in the number of root hairs.

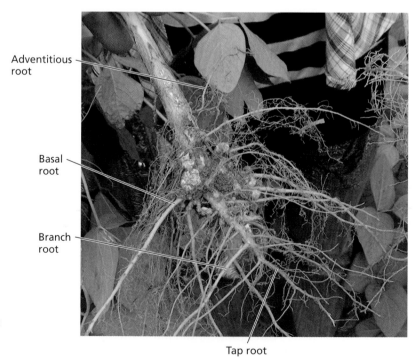

Adventitious root

Basal root

Branch root

Tap root

Figure 16.20 Soybean root system, showing the primary (tap) root, branch roots, basal roots, and adventitious roots. (Courtesy of Leon Kochian.)

Figure 16.21 In response to phosphorus deficiency, plants may alter their root architecture in ways that increase their ability to forage for the nutrient. These white lupines (*Lupinus albus*) were grown hydroponically in a nutrient solution with (+P) or without (–P) phosphate. The phosphorus-deficient plant has a shallower root system and many more lateral roots in the upper part of the system. These lateral roots in turn are covered with short, densely packed lateral roots called cluster roots; these features increase the surface area of the root system in the upper layers of the soil. Other morphological responses of plants to phosphorus deficiency include the formation of longer, denser root hairs, an increased root:shoot ratio, and formation of aerenchyma (air spaces in the cortex). (From Péret et al. 2014.)

As described in Chapter 4, soil phosphorus is bound tightly to clay particles or organic structures and is more enriched in the surface horizons (layers) of the soil. Phosphorus deficiency can trigger "topsoil foraging" by plants. For example, some bean varieties (phosphorus efficient genotypes) respond to phosphorus deficiency by producing more adventitious lateral roots that grow at a more horizontal angle so that they are shallower. There is also an overall increase in the number of lateral roots and root hairs. These bean varieties have been used extensively to breed more phosphorus efficient beans and soybeans for growth in low-phosphorus soils.

Plant Senescence

Senescence refers to the energy-dependent, autolytic (self-digesting) process that leads to the death of targeted cells. As is the case with most features of plant development, senescence is controlled by the interaction of environmental factors with genetically regulated developmental programs.

In temperate climates, the leaves of deciduous trees (trees that shed their leaves annually) turn yellow, orange, or red and fall from their branches in response to shorter day lengths and cooler temperatures, which trigger two related developmental processes: senescence and abscission. **Abscission** refers to the separation of cell layers that occurs at the bases of leaves, floral parts, and fruits, which allows them to be shed easily without damaging the plant. Before the senescing leaves are shed from the plant, however, nitrogen and other nutrients from them are transported back to the branches.

senescence An active, genetically controlled, developmental process in which cellular structures and macromolecules are broken down and translocated away from the senescing organ (typically leaves) to actively growing regions that serve as nutrient sinks. Initiated by environmental cues and regulated by hormones.

abscission The shedding of leaves, flowers, and fruits from a living plant. The process whereby specific cells in the leaf petiole (stalk) differentiate to form an abscission layer, allowing a dying or dead organ to separate from the plant.

programmed cell death (PCD)
The process whereby individual cells activate an intrinsic senescence program accompanied by a distinct set of morphological and biochemical changes similar to mammalian apoptosis.

organ senescence The developmentally regulated senescence of individual organs, such as leaves, flowers, fruits, or roots.

whole plant senescence The death of the entire plant, as opposed to the death of individual cells, tissues, or organs.

There are three types of plant senescence, based on the level of structural organization of the senescing unit: *programmed cell death, organ senescence,* and *whole plant senescence.* **Programmed cell death** (**PCD**) is a general term referring to the genetically regulated death of individual cells. During PCD, the protoplasm, and sometimes the cell wall, undergoes autolysis. In the case of the development of xylem tracheary elements and fibers, however, secondary wall layers are deposited prior to cell death. **Organ senescence**—the senescence of whole leaves, branches, flowers, or fruits—occurs at various stages of vegetative and reproductive development and typically includes abscission of the senescing organ. As previously noted, leaf senescence is strongly influenced by photoperiod and temperature. Finally, **whole plant senescence** involves the death of the entire plant. In this chapter we focus on organ senescence and whole plant senescence.

During leaf senescence, nutrients are remobilized from the source leaf to vegetative or reproductive sinks

All leaves, including those of evergreens, undergo senescence, whether in response to age-dependent factors, environmental cues, biotic stress, or abiotic stress. During developmental senescence, cells undergo genetically programmed changes in cell structure and metabolism. The earliest structural change in leaf cells is the breakdown of the chloroplast, which contains up to 70% of the leaf protein. Carbon assimilation is replaced by the breakdown and conversion of chlorophyll, proteins, and other macromolecules to exportable nutrients that can be translocated to growing vegetative organs or developing seeds and fruits. The resulting sugars, nucleosides, and amino acids are then transported via the phloem back into the main body of the plant, where they are reused for biosynthesis. Many minerals are also transported out of senescing organs back into the plant.

Since senescence redistributes nutrients to growing parts of the plant, it can serve as a survival mechanism during environmentally adverse conditions, such as drought or temperature stress (see Chapter 19). However, leaf senescence occurs even under optimal growth conditions and is therefore part of the plant's normal developmental program. As new leaves are initiated at the SAM, older leaves below may become shaded and lose the ability to function efficiently in photosynthesis, triggering senescence of the older leaves. In eudicots, senescence is usually followed by abscission, the process that enables plants to shed senescent leaves from the plants. Together, the coupled programs of leaf senescence and abscission help optimize the photosynthetic and nutrient efficiency of the plant.

The developmental age of a leaf may differ from its chronological age

Both internal and external cues influence the developmental age of leaf tissue, which may or may not correspond to the leaf's chronological age. The distinction between developmental and chronological age is nicely illustrated by a simple experiment carried out by the German plant physiologist Ernst Stahl in 1909. Stahl excised a small disc from a green leaf of the deciduous shrub mock orange (*Philadelphus grandiflora*). He then incubated the disc on a simple nutrient solution in the laboratory until the fall, by which time the leaf attached to the plant had turned yellow. The drawing in **Figure 16.22** shows the incubated leaf disc superimposed on the leaf from which it was removed, both at the end of the experiment. Although the chronological ages of the leaf and the disc were the same, the leaf was now developmentally much older than the disc tissue. The leaf that remained on the shrub

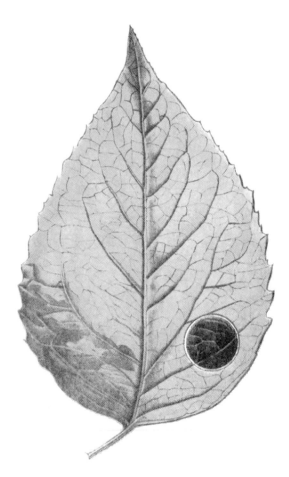

Figure 16.22 Early leaf senescence experiment showing the delayed senescence of a leaf disc cultured in the laboratory in a dilute mineral nutrient solution compared with the leaf of mock orange (*Philadelphus grandiflora*) from which the disc was excised, which remained on the bush. (From Stahl 1909.)

Figure 16.23 Sequential leaf senescence of barley stems, showing a gradient of senescence from the older leaves at the base to the younger leaves near the tip. (Courtesy of Andreas M. Fischer.)

was subjected to a variety of internal signals coming from the surrounding leaf tissue and other parts of the plant, while the disc was literally cut off from these influences. Furthermore, the leaf attached to the plant remained outdoors exposed to the changing seasons, while the disc was cultured indoors under more or less constant conditions. Shielded from both internal and external cues, the leaf disc remained at the same developmental age as at the start of the experiment, while the attached leaf became developmentally older.

Leaf senescence may be sequential, seasonal, or stress-induced

Leaf senescence under normal growth conditions is governed by the developmental age of the leaf, which is a function of hormones and other regulatory factors. Under these circumstances, there is usually a senescence gradient from the youngest leaves located near the growing tip to the oldest leaves located near the base of the shoot—a pattern known as **sequential leaf senescence** (**Figure 16.23**). In contrast, the leaves of deciduous trees in temperate climates senesce all at once in response to the shorter days and cooler temperatures of autumn, a pattern known as **seasonal leaf senescence** (**Figure 16.24**). Both sequential and seasonal leaf senescence are variations of developmental senescence, since they occur under normal growing conditions.

(A) September 8   (B) September 13   (C) September 18

(D) September 25   (E) October 3   (F) October 8

Figure 16.24 Seasonal leaf senescence in an aspen tree (*Populus tremula*). All of the leaves begin to senesce in late September and undergo abscission in early October. (From Keskitalo et al. 2005.)

sequential leaf senescence
The pattern of leaf senescence in which there is a gradient of senescence from the growing tip of the shoot to the oldest leaves at the base.

seasonal leaf senescence In temperate climates, the pattern of leaf senescence in deciduous trees in which all of the leaves undergo senescence and abscission in the autumn.

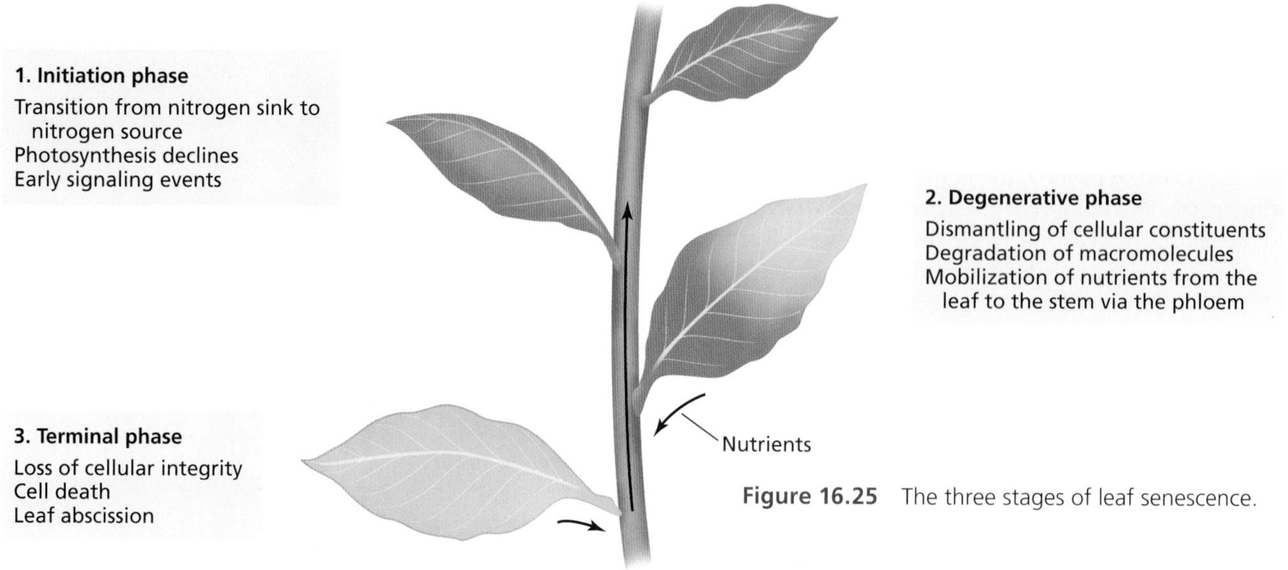

1. Initiation phase

Transition from nitrogen sink to
  nitrogen source
Photosynthesis declines
Early signaling events

2. Degenerative phase

Dismantling of cellular constituents
Degradation of macromolecules
Mobilization of nutrients from the
  leaf to the stem via the phloem

3. Terminal phase

Loss of cellular integrity
Cell death
Leaf abscission

Nutrients

Figure 16.25 The three stages of leaf senescence.

Sequential and seasonal leaf senescence can be divided into three distinct phases: initiation, degeneration, and termination (**Figure 16.25**). During the *initiation phase*, the leaf receives developmental and environmental signals that initiate a decline in photosynthesis and a transition from being a nitrogen sink to a nitrogen source. Most of the autolysis of cellular organelles and macromolecules occurs during the *degenerative phase* of leaf senescence. The solubilized mineral and organic nutrients are then remobilized via the phloem to growing sinks, such as young leaves, underground storage organs, or reproductive structures. The abscission layer forms during the degenerative phase of leaf senescence. During the *terminal phase*, autolysis is completed and cell separation takes place at the abscission layer, resulting in leaf abscission.

The earliest cellular changes during leaf senescence occur in the chloroplast

Chloroplasts contain about 70% of the total leaf protein, most of which consists of ribulose 1,5-bisphosphate carboxylase/oxygenase (Rubisco) localized in the stroma, and light-harvesting chlorophyll-binding protein II (LHCP II) associated with the thylakoid membranes (see Chapters 7 and 8). Catabolism and remobilization of chloroplast proteins thus provide the primary source of amino acids and nitrogen for sink organs, and represent the earliest changes that occur during leaf senescence (see Figure 16.25). Unlike chloroplasts, the nucleus and mitochondria, which are required for gene expression and energy production, remain intact until the later stages of senescence.

The transition from a mature, photosynthetically active leaf to a senescing leaf involves increased expression of senescence associated genes (SAGs) and decreased expression of senescence down-regulated genes. SAGs include genes regulating many processes associated with abiotic and biotic stress, including autophagy, the response to reactive oxygen species (ROS), metal ion binding, cell wall breakdown, lipid breakdown, and abscisic acid, jasmonic acid, and ethylene hormonal signaling (**Figure 16.26**).

Reactive oxygen species serve as internal signaling agents in leaf senescence

There is growing evidence that reactive oxygen species (ROS), especially H_2O_2, play important roles as signals during leaf senescence. ROS are toxic chemicals

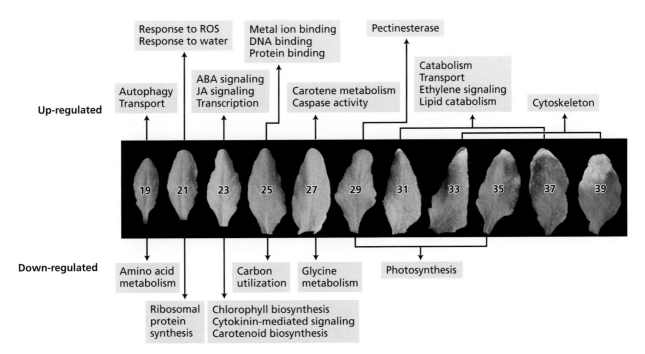

Figure 16.26 Metabolic pathways that are either up-regulated or down-regulated during leaf senescence in Arabidopsis. The seventh leaf was sampled from plants at different stages of senescence. The numbers 19 through 39 refer to the ages of the plants, expressed as the number of days after sowing, from which the seventh leaf was taken. (From Breeze et al. 2011.)

that cause oxidative damage to DNA, proteins, and membrane lipids (see Chapters 18 and 19). They are produced primarily as by-products of normal metabolic processes, such as respiration and photosynthesis, in chloroplasts, mitochondria, peroxisomes, and at the plasma membrane. Plants use ROS-scavenging systems, such as enzymes (catalase, superoxide dismutase, ascorbate peroxidase) and antioxidant molecules (e.g., ascorbate and glutathione) to protect themselves from oxidative damage. However, the plant's antioxidant concentrations decrease during leaf senescence, while ROS levels increase. In senescence, ROS act as signals that activate genetically programmed cell death events.

Plant hormones interact in the regulation of leaf senescence

Leaf senescence is an evolutionarily selected, genetically regulated process that ensures efficient remobilization of nutrients to vegetative or reproductive sink organs. No mutation, treatment, or environmental condition has yet been found that abolishes the process completely, suggesting that leaf senescence is ultimately governed either by developmental or chronological age. However, both the timing and progression of senescence are flexible, and hormones are key developmental signals that accelerate or delay the timing of leaf senescence.

The senescence-repressing role of cytokinins appears to be universal in plants and has been demonstrated in many types of studies. Although applied cytokinins do not prevent senescence completely, their effects can be dramatic, particularly when the cytokinin is sprayed directly on an intact plant. If only one leaf is treated, it remains green after other leaves of similar age have yellowed and dropped off the plant. If a small spot on a leaf is treated with cytokinin, that spot will remain green after the surrounding tissues on the same leaf begin to senesce. This "green island" effect can also be observed in leaves infected by some fungal pathogens, as well as in those hosting galls produced by insects. Such green islands have higher levels of cytokinins than surrounding leaf tissue.

Unlike young leaves, mature leaves produce little, if any, cytokinin. During senescence, the transcript abundance of genes involved in cytokinin biosynthesis declines, whereas transcripts of genes involved in degrading cytokinins, such as cytokinin oxidase, increase during senescence. Mature leaves may therefore depend on root-derived cytokinins to postpone their senescence.

Thus far, the molecular mechanism of cytokinin action in delaying leaf senescence remains unclear. According to a long-standing hypothesis, cytokinin represses leaf senescence by regulating nutrient mobilization and source–sink relations. This phenomenon can be observed when nutrients (sugars, amino acids, and so on) radiolabeled with ^{14}C or ^{3}H are fed to plants after one leaf or part of a leaf is treated with a cytokinin (**Figure 16.27**). Subsequent autoradiography of the whole plant reveals the pattern of movement and the sites at which the labeled nutrients accumulate. Experiments of this nature have demonstrated that nutrients are preferentially transported to and accumulated in the cytokinin-treated tissues, which retain the nutrient sink status associated with young, growing tissues.

Gibberellins are also senescence-repressing hormones. The abundance of active forms of GA declines in leaves as they age. For example, the senescence of excised leaf discs of *Taraxacum* and *Rumex* is delayed by treatment with GA. Auxin also plays a more limited role in delaying senescence and is associated with delays in leaf senescence and decreases in SAG expression. As described below, auxin production in leaves inhibits the initiation of leaf abscission.

Ethylene, by contrast, is regarded as a senescence-promoting hormone because ethylene treatment accelerates leaf and flower senescence, and inhibitors of ethylene synthesis and action can delay senescence. As we discuss in the next section, ethylene plays an important role in abscission as well. Abscisic acid (ABA) levels increase in senescing leaves, and exogenous application of ABA rapidly promotes leaf senescence and expression of several SAGs, which is consistent

In seedling A, the left cotyledon was sprayed with water as a control. The left cotyledon of seedling B and the right cotyledon of seedling C were each sprayed with a solution containing kinetin.

The dark stippling represents the distribution of the radioactive amino acid as revealed by autoradiography.

The results show that the cytokinin-treated cotyledon has become a nutrient sink (seedling B). However, radioactivity is retained in the cotyledon to which the amino acid was applied when the labeled cotyledon is treated with kinetin (seedling C).

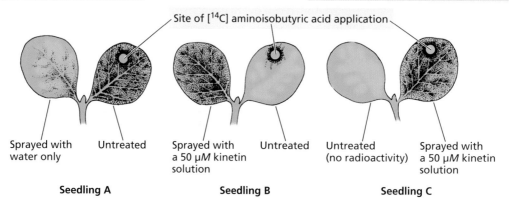

Site of [^{14}C] aminoisobutyric acid application

Sprayed with water only   Untreated
Seedling A

Sprayed with a 50 µM kinetin solution   Untreated
Seedling B

Untreated (no radioactivity)   Sprayed with a 50 µM kinetin solution
Seedling C

Figure 16.27 Effect of cytokinin (kinetin) on the movement of an amino acid in cucumber seedlings. Radioactively labeled α-aminoisobutyric acid, a non-metabolizable amino acid that cannot be used in protein synthesis, was applied as a discrete spot on the right cotyledon of each of these seedlings. After a set amount of time, the seedlings were placed on X-ray film to detect the movement of the radioactive amino acid. The black stippling indicates the distribution of radioactivity. (Drawn from data in Mothes et al. 1961.)

with ABA's effects on leaf senescence. However, like ethylene, ABA is considered an enhancer rather than a triggering factor of leaf senescence.

Leaf Abscission

The shedding of leaves, fruits, flowers, and other plant organs is termed abscission. Abscission takes place within specific layers of cells called the **abscission zone**, located near the base of the petiole (**Figure 16.28**). The abscission zone becomes morphologically and biochemically differentiated during organ development, many months before organ separation actually takes place. Often the abscission zone can be morphologically identified as one or more layers of isodiametrically flattened cells (see Figure 16.28B).

The timing of leaf abscission is regulated by the interaction of ethylene and auxin

Ethylene plays a key role in activating the events leading to cell separation within the abscission zone. The ability of ethylene gas to cause defoliation in young birch trees is shown in **Figure 16.29**. The wild-type tree on the left has lost most of its leaves; only the younger leaves at the top fail to abscise. The tree on the

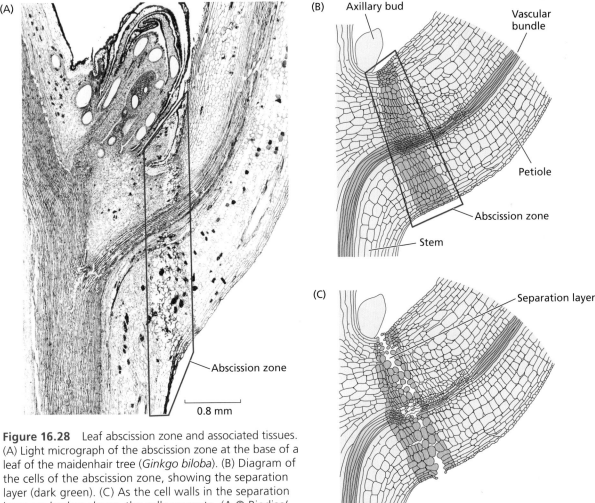

Figure 16.28   Leaf abscission zone and associated tissues. (A) Light micrograph of the abscission zone at the base of a leaf of the maidenhair tree (*Ginkgo biloba*). (B) Diagram of the cells of the abscission zone, showing the separation layer (dark green). (C) As the cell walls in the separation layer are broken down, the cells separate. (A © Biodisc/Visuals Unlimited, Inc.)

Figure 16.29 Effect of ethylene on abscission in the birch *Betula pendula*. The tree on the left is the wild type; the tree on the right has a dominant mutation than knocks out all activity of the ethylene receptor. One of the characteristics of these mutant trees is that they do not drop their leaves, as wild-type plants do, when fumigated for 3 days with 50 ppm ethylene. (From Vahala et al. 2003.)

right has been genetically transformed to be insensitive to ethylene and retains its leaves after ethylene treatment.

The process of leaf abscission can be divided into three distinct developmental phases during which the cells of the abscission zone become competent to respond to ethylene (**Figure 16.30**).

1. *Leaf maintenance phase.* Prior to the perception of any signal (internal or external) that initiates the abscission process, the leaf remains healthy and fully functional. A gradient of auxin from the leaf blade to the stem maintains the abscission zone in an insensitive state.

2. *Abscission induction phase.* A reduction or reversal in the auxin gradient from the leaf blade, normally associated with leaf senescence, causes the abscission zone to become sensitive to ethylene. Treatments that enhance leaf senescence do so by promoting abscission through interference with auxin synthesis or transport in the leaf.

3. *Abscission phase.* The sensitized cells of the abscission zone respond to low concentrations of endogenous ethylene by synthesizing and secreting cell wall–degrading enzymes and cell wall–remodeling proteins, including β-1,4-glucanase (cellulase), polygalacturonase, xyloglucan endotransglucosylase/hydrolase, and expansin, resulting in cell separation and leaf abscission.

Early in the leaf maintenance phase, auxin from the leaf prevents abscission by maintaining the cells of the abscission zone in an ethylene-insensitive state. It

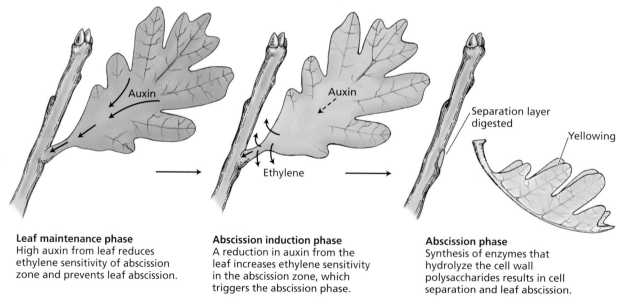

Leaf maintenance phase
High auxin from leaf reduces ethylene sensitivity of abscission zone and prevents leaf abscission.

Abscission induction phase
A reduction in auxin from the leaf increases ethylene sensitivity in the abscission zone, which triggers the abscission phase.

Abscission phase
Synthesis of enzymes that hydrolyze the cell wall polysaccharides results in cell separation and leaf abscission.

Figure 16.30 Schematic view of the roles of auxin and ethylene during leaf abscission. In the abscission induction phase, the level of auxin decreases, and the level of ethylene increases. These changes in the hormonal balance increase the sensitivity of the target cells to ethylene. (After Morgan 1984.)

has long been known that removal of the leaf blade (the site of auxin production) promotes petiole abscission. Application of exogenous auxin to petioles from which the leaf blade has been removed delays the abscission process.

In the abscission induction phase, typically associated with leaf senescence, the amount of auxin from the leaf blade decreases and the ethylene level rises. Ethylene appears to decrease the activity of auxin both by reducing its synthesis and transport and by increasing its degradation. The reduction in the concentration of free auxin increases the response of specific target cells in the abscission zone to ethylene. The abscission phase is characterized by the induction of abscission-related genes encoding specific hydrolytic and remodeling enzymes that loosen the cell walls in the abscission layer.

Whole Plant Senescence

The programmed death of individual leaves is an adaptation that benefits the plant as a whole by increasing its evolutionary fitness. However, the death of whole plants cannot easily be rationalized in evolutionary terms, even though the life spans of individual plants are, to a large extent, genetically determined.

Angiosperm life cycles may be annual, biennial, or perennial

Individual plant life spans vary from a few weeks in the case of desert ephemerals, which grow and reproduce rapidly in response to brief episodes of rain, to about 4600 years in the case of bristlecone pine. In general, **annual plants** grow, reproduce, senesce, and die in a single season. **Biennial plants** devote their first year to vegetative growth and food storage, and their second year to reproduction, senescence, and death. Because annual and biennial plants undergo whole plant senescence following fruit and seed production, both are termed **monocarpic** because they reproduce only once (**Figure 16.31**).

Perennial plants live for 3 years or longer and may be herbaceous or woody. The range of maximum life spans for perennial plants is given in **Table 16.2**. Perennial plants are usually **polycarpic**, producing fruits and seeds over multiple seasons. However, there are also examples of monocarpic perennials, such as the century plant (*Agave americana*) and Japanese timber bamboo (*Phyllostachys bambusoides*). The century plant grows vegetatively for 10 to 30 years before flowering, fruiting, and senescing, while Japanese timber bamboo can grow vegetatively for 60 to 120 years before reproduction and death. Remarkably, all clones from the same bamboo stock flower and senesce simultaneously, regardless of geographic location or climatic condition, which suggests the presence of a long-term biological clock of some kind.

Many perennial plants that form clones by asexual reproduction can proliferate into community-sized interconnected "individuals" that achieve astounding ages, such as King's lomatia (*Lomatia tasmanica*), a Tasmanian shrub in the Proteaceae family that can be over 43,000 years old. Each

annual plant A plant that completes its life cycle from seed to seed, senesces, and dies within 1 year.

biennial plant A plant that requires two growing seasons to flower and produce seed.

monocarpic Referring to plants, typically annuals, that produce fruits only once and then die.

perennial plants Plants that live for more than 2 years.

polycarpic Referring to perennial plants that produce fruit many times.

Figure 16.31 Monocarpic senescence in soybean (*Glycine max*). The entire plant on the left underwent senescence after flowering and producing fruit (pods). The plant on the right remained green and vegetative because its flowers were continually removed. (Courtesy of L. Noodén.)

Table 16.2 Longevity of various individual and clonal perennial plants

Species	Age (yr)
Individual plants	
Bristlecone pine (*Pinus longaeva*)	4600
Giant sequoia (*Sequoiadendron giganteum*)	3200
Stone pine (*Pinus cembra*)	1200
European beech (*Fagus sylvatica*)	930
Blackgum (*Nyssa sylvatica*)	679
Scots pine (*Pinus sylvestris*)	500
Chestnut oak (*Quercus montana*)	427
Red oak (*Quercus rubra*)	326
European ash (*Fraxinus excelsior*)	250
English ivy (*Hedera helix*)	200
Flowering dogwood (*Cornus florida*)	125
Bigtooth aspen (*Populus grandidentata*)	113
Scots heather (*Calluna vulgaris*)	42
Spring heath (*Erica carnea*)	21
Scandinavian thyme (*Thymus chamaedrys*)	14
Clonal plants	
King's lomatia (*Lomatia tasmanica*)	43,000+
Creosote (*Larrea tridentata*)	11,000+
Bracken (*Pteridium aquilinum*)	1400
Sheep fescue (*Festuca ovina*)	1000+
Ground pine (*Lycopodium complanatum*)	850
Reed grass (*Calamagrostis epigeios*)	400+
Wood sage (*Teucrium scorodonia*)	10

Source: Thomas 2013.

individual lomatia plant lives only about 300 years, but because it does not transfer any senescence signal to its clones, the clonal community apparently grows and proliferates indefinitely.

Nutrient or hormonal redistribution may trigger senescence in monocarpic plants

A diagnostic feature of monocarpic senescence is the ability to delay senescence well beyond the plant's normal life span by the removal of the reproductive structures. For example, repeated depodding enables soybean plants to remain vegetative for many years under favorable growing conditions (see Figure 16.31), leading to a treelike appearance. Monocarpic senescence is thought to result from the redistribution of vital nutrients via the phloem from vegetative sources to reproductive sinks. However, the critical redistributed compound that triggers monocarpic senescence is not likely to be a carbohydrate, as the carbohydrate content of leaves actually *increases* during senescence. This observation is consistent with the ability of exogenous sugars to trigger leaf senescence. Rather than carbohydrate loss, it may be alterations in source–sink relations caused by floral development that induce a global shift in the hormonal or nutrient balance of the vegetative organs. A loss of nitrogen coupled with a simultaneous accumulation of carbohydrate would cause an increase in the C:N ratio, which has been associated with autolysis in senescing leaves.

Summary

After seedling establishment, development of vegetative organs occurs primarily from the meristem tissues. Vegetative growth is controlled by developmental processes involving molecular interactions and regulatory feedback. These mechanisms create root and shoot polarity, allowing plants to produce lateral organs (e.g., leaves and branching systems), which form an overall vegetative architecture. Senescence at the cellular level, called programmed cell death, is an integral part of plant development. Senescence also occurs at the organ level, as in the case of leaf senescence, during which the leaf undergoes a genetically programmed sequence of macromolecular turnover and nutrient recycling, followed by abscission. The regulation of plant senescence differs in annuals, biennials, and perennials. In monocarpic species, senescence may be triggered by nutrient or hormonal redistribution during fruit production. Many perennial species can live for thousands of years.

The Shoot Apical Meristem

- After seedling germination, vegetative shoot organs are derived from a compact shoot apical meristem (SAM) (**Figure 16.1**).

- Organogenesis originates in discrete shoot apical meristem zones (**Figure 16.2**).

(Continued)

Summary (*continued*)

Leaf Structure and Phyllotaxy

- The development of flat laminas in seed plants was a key evolutionary event; since then, phyllome morphology has diversified dramatically (**Figure 16.3**).

- The three basic types of leaf arrangement (phyllotaxy) are alternate, decussate, and spiral (**Figure 16.4**).

- Leaf phyllotactic patterns are established at the shoot apex by localized zones of auxin accumulation resulting from polar auxin transport (**Figure 16.5**).

Differentiation of Epidermal Cell Types

- The epidermis is derived from the protoderm and has three main cell types: pavement cells, trichomes, and stomatal guard cells, as well as other cell types.

- Specialized epidermal cells reflect common and differential functions across plant species (**Figure 16.6**).

- Not only guard cells, but also the majority of leaf epidermal cells, arise from specialized meristemoid mother cells (MMCs), stomatal lineage ground cells (SLGCs), meristemoids, and guard mother cells (GMCs) (**Figure 16.7**).

Venation Patterns in Leaves

- Leaf venation patterns differ in eudicots and monocots (**Figure 16.8**).

- Leaf veins exhibit a hierarchy based on their diameter at the site of attachment to the parent vein (**Figure 16.9**).

- Leaf vascular bundles arise from the procambium and differentiate downward, forming a connection to the older vascular bundles that are continuous to the base of the shoot (**Figure 16.10**).

- The rootward transport of auxin in shoots induces the differentiation of new xylem cells after wounding (**Figure 16.11**).

- Both localized auxin biosynthesis and polar auxin transport are involved in vascularization during early leaf development.

Shoot Branching and Architecture

- Plant shoot architecture is based on a repeating unit called the phytomer, consisting of an internode, a node, a leaf, and an axillary bud (**Figure 16.12**).

- Auxin, cytokinins, and strigolactones regulate apical dominance (**Figures 16.13, 16.14**).

- Cytokinins are involved in breaking apical dominance and stimulating axillary bud growth.(**Figure 16.14**).

- Sucrose also serves as the initial signal for axillary bud growth (**Figure 16.15**).

Shade Avoidance

- Plants compete for sunlight and respond to shading from other plants by detecting decreased R:FR ratios (**Table 16.1, Figure 16.16**).

- Phytochrome is the photoreceptor that senses the R:FR ratio during shade avoidance.

- Plants react to shading by increasing elongation growth.

- Genetic modification of shade avoidance responses can increase crop yields (**Figure 16.17**).

Root System Architecture

- Species-specific root system architecture optimizes water and nutrient uptake (**Figure 16.18**).

- Monocot root systems, as exemplified by maize roots, are composed of the primary root, seminal and crown roots, and lateral roots; eudicot root systems, such as those of soybean, include the primary (tap) root and branch, basal, and adventitious roots (**Figures 16.19, 16.20**).

- Phosphorus availability can alter root system architecture (**Figure 16.21**).

Plant Senescence

- There are three types of plant senescence, based on the level of structural organization of the senescing unit: programmed cell death, organ senescence, and whole plant senescence.

- Leaf senescence involves the genetically regulated autolysis of cellular proteins, carbohydrates, and nucleic acids and the redistribution of their components back into the main body of the plant, to actively growing areas. Minerals are also transported out of senescing leaves back into the plant.

- Normal leaf senescence is regulated by internal, developmental signals as well as external, environmental signals (**Figure 16.22**).

- Leaf senescence may exhibit a sequential or seasonal pattern (**Figures 16.23, 16.24**).

- Leaf senescence can be divided into three phases: initiation, degeneration, and termination (**Figure 16.25**).

- Leaf senescence is preceded by the increased expression of senescence associated genes. (**Figure 16.26**).

- There is growing evidence that reactive oxygen species

(Continued)

Summary *(continued)*

(ROS), especially H_2O_2, can serve as internal signals to promote senescence.

- Plant hormones interact to regulate leaf senescence.

- Cytokinin may delay senescence by causing leaf cells to become sinks for nutrients (**Figure 16.27**).

Leaf Abscission

- Abscission is the shedding of leaves, fruits, flowers, or other plant organs, and is caused by the separation of cell layers within the abscission zone (**Figure 16.28**).

- High levels of auxin keep leaf tissue in an ethylene-insensitive state, but as auxin levels drop, the abscission-promoting and auxin-repressing effects of ethylene become stronger (**Figures 16.29, 16.30**).

Whole Plant Senescence

- In general, annuals and biennials reproduce only once before senescing, while perennials can reproduce multiple times before senescing.

- There is wide variation in the longevities of different plant species (**Table 16.2**).

- Nutrient or hormonal redistribution from vegetative structures to reproductive sinks may trigger whole plant senescence in monocarpic plants (**Figure 16.31**).

Suggested Reading

Bayer, I., Smith, R. S., Mandel, T., Nakayama, N., Sauer, M., Prusinkiewicz, P., and Kuhlemeier, C. (2009) Integration of transport-based models for phyllotaxis and midvein formation. *Genes Dev.* 23: 373–384.

Breeze, E., Harrison, E., McHattie, S., Hughes, L., Hickman, R., Hill, C., Kiddle, S., Kim, Y.-S., Penfold, C. A., Jenkins, D., et al. (2011) High-resolution temporal profiling of transcripts during arabidopsis leaf senescence reveals a distinct chronology of processes and regulation. *Plant Cell* 23: 873–894.

Byrne, M. E. (2012) Making leaves. *Curr. Opin. Plant Biol.* 15: 24–30.

Caño-Delgado, A., Lee, J. Y., and Demura, T. (2010) Regulatory mechanisms for specification and patterning of plant vascular tissues. *Annu. Rev. Cell Dev. Biol.* 26: 605–637.

Domagalska, M. A., and Leyser, O. (2011) Signal integration in the control of shoot branching. *Nat. Rev. Mol. Cell Biol.* 12: 211–221.

Fischer, A. M. (2012) The complex regulation of senescence. *Crit. Rev. Plant Sci.* 31: 124–147.

Heisler, M. G., Hamant, O., Krupinski, P., Uyttewaal, M., Ohno, C., Jönsson, H., Traas, J., and Meyerowitz, E. M. (2010) Alignment between PIN1 polarity and microtubule orientation in the shoot apical meristem reveals a tight coupling between morphogenesis and auxin transport. *PLOS Biol.* 8(10): e1000516. DOI:10.1371/journal.pbio.100051

Lau, S., and Bergmann, D. C. (2012) Stomatal development: A plant's perspective on cell polarity, cell fate transitions and intercellular communication. *Development* 139: 3683–3692.

Lucas, W. J., Groover, A., Lichtenberger, R., Furuta, K., Yadav, S. R., Helariutta, Y., He, X. Q., Fukuda, H., Kang, J., Brady, S. M., et al. (2013) The plant vascular system: Evolution, development and functions. *J. Integr. Plant Biol.* 55: 294–388.

Mason, M. G., Ross, J. J., Babst, B. A., Wienclaw, B. N., and Beveridge, C. A. (2014) Sugar demand, not auxin, is the initial regulator of apical dominance. *Proc. Natl. Acad. Sci. USA* 111: 6092–6097.

Noodén, L. D. (2013) Defining senescence and death in photosynthetic tissues. In *Advances in Photosynthesis and Respiration*, Vol. 36: *Plastid Development in Leaves during Growth and Senescence*, B. Biswal, K. Krupinska, and U. C. Biswal, eds., Springer, New York, pp. 283–306.

Risopatron, J. P. M., Sun, Y., and Jones, B. J. (2012) The vascular cambium: Molecular control of cellular structure. *Protoplasma* 247: 145–161.

Zhang, Z., Liao, H., and Lucas, W. J. (2014) Molecular mechanisms underlying phosphate sensing, signaling, and adaptation in plants. *J. Integr. Plant Biol.* 56: 192–220.

17 Flowering and Fruit Development

Most people look forward to the spring season and the profusion of flowers it brings. Many vacationers carefully time their travels to coincide with specific blooming seasons: *Citrus* along the Blossom Trail in southern California and tulips in the Netherlands. In Washington, DC, and throughout Japan, the cherry blossoms are received with spirited ceremonies. As spring progresses into summer, summer into fall, and fall into winter, wildflowers bloom at their appointed times. Flowering at the correct time of year is crucial for the reproductive fitness of the plant; plants that are cross-pollinated must flower in synchrony with other individuals of their species as well as with their pollinators at the time of year that is optimal for seed set.

Although the strong correlation between flowering and seasons is common knowledge, the phenomenon poses fundamental questions that we address in this chapter:

- How do plants keep track of the seasons of the year and the time of day?

- Which environmental signals influence flowering, and how are those signals perceived?

- How do different types of fruits develop?

The transition to flowering involves major changes in the pattern of morphogenesis and cell differentiation at the shoot apical meristem (SAM). Ultimately, as we will see, this process leads to the production of the floral organs—sepals, petals, stamens, and carpels.

floral evocation   The events occurring in the shoot apex that specifically commit the apical meristem to produce flowers.

phase change   The phenomenon in which the fates of the meristematic cells become altered in ways that cause them to produce new types of structures.

Floral Evocation: Integrating Environmental Cues

A particularly important developmental decision during the plant life cycle is when to flower. The process by which the SAM becomes committed to forming flowers is termed **floral evocation**. Delaying this commitment to flower increases the carbohydrate reserves that will be available for mobilization, allowing more and better-provisioned seeds to mature. Delaying flowering, however, also potentially increases the danger that the plant will be eaten, killed by abiotic stress, or outcompeted by other plants before it reproduces. Reflecting this, plants have evolved an extraordinary range of reproductive adaptations—for example, annual versus perennial life cycles.

Annual plants such as groundsel (*Senecio vulgaris*) may flower within a few weeks after germinating. But trees may grow for 20 or more years before they begin to produce flowers. Across the plant kingdom, different species flower at a wide range of ages, indicating that the age, or perhaps the size, of the plant is an internal factor controlling the switch to reproductive development.

Flowering that occurs strictly in response to internal developmental factors, independently of any particular environmental condition, is referred to as *autonomous regulation*. In species that exhibit an absolute requirement for a specific set of environmental cues in order to flower, flowering is considered to be an *obligate* or *qualitative* response. If flowering is promoted by certain environmental cues but will eventually occur in the absence of such cues, the flowering response is *facultative* or *quantitative*. A species with a facultative flowering response, such as Arabidopsis, relies on both environmental and autonomous signals to promote reproductive growth.

Photoperiodism and vernalization are two of the most important mechanisms underlying seasonal responses. Photoperiodism is a response to the length of day or night; vernalization is the promotion of flowering by prolonged cold temperature. Other signals, such as light quality, ambient temperature, and abiotic stress, are also important external cues for plant development.

The evolution of both internal (autonomous) and external (environment-sensing) control systems enables plants to precisely regulate flowering so that it occurs at the optimal time for reproductive success. For example, in many populations of a particular species, flowering is synchronized, which favors crossbreeding. Flowering in response to environmental cues also helps ensure that seeds are produced under favorable conditions, particularly with respect to water and temperature. However, this makes plants especially vulnerable to rapid climate change, such as global warming, which can alter the regulatory networks that govern floral timing. Several studies have shown that many plant species are now flowering several days to weeks earlier than they did in the nineteenth century.

The Shoot Apex and Phase Changes

All multicellular organisms pass through a series of more or less defined developmental stages, each with its characteristic features. In humans, infancy, childhood, adolescence, and adulthood represent four general stages of development, with puberty as the dividing line between the nonreproductive and the reproductive phases. Similarly, plants pass through distinct developmental phases. The timing of these transitions often depends on environmental conditions, allowing plants to adapt to a changing environment. This is possible because plants continuously produce new organs from the SAM.

The transitions between different phases are tightly regulated developmentally, since the plant must integrate information from the environment as well as autonomous signals to maximize its reproductive fitness. The following sections describe the major pathways that control these decisions.

Plant development has three phases

Postembryonic development in plants can be divided into three phases:

1. The juvenile phase
2. The adult vegetative phase
3. The adult reproductive phase

The transition from one phase to another is called **phase change**.

The primary distinction between the juvenile and the adult vegetative phases is that the latter has the ability to form reproductive structures: flowers in angiosperms, cones in gymnosperms. However, flowering, which represents the expression of the reproductive competence of the adult phase, often depends on specific environmental and developmental signals. Thus, the absence of flowering itself is not a reliable indicator of juvenility.

The transition from juvenile to adult is frequently accompanied by changes in vegetative characteristics, such as leaf morphology, phyllotaxy (the arrangement of leaves on the stem), thorniness, rooting capacity, and leaf retention in deciduous plants such as English ivy (*Hedera helix*) (**Figure 17.1**). Such changes are most evident in woody perennials, but they are apparent in many herbaceous species as well. Unlike the abrupt transition from the adult vegetative phase to the reproductive phase, the transition from juvenile to adult vegetative is usually gradual, involving intermediate forms.

Juvenile tissues are produced first and are located at the base of the shoot

The time sequence of the three developmental phases results in a spatial gradient of juvenility along the shoot axis. Because growth in height is restricted to the apical meristem, the juvenile tissues and organs, which form first, are located at the base of the shoot. In rapidly flowering herbaceous species, the juvenile phase may last only a few days, and few juvenile structures are produced. In contrast, woody species have a more prolonged juvenile phase, in some cases lasting 30 to 40 years (**Table 17.1**). In these cases the juvenile structures can account for a significant portion of the mature plant.

Once the meristem has switched to the adult phase, only adult vegetative structures are produced, culminating in flowering. The adult and reproductive phases are therefore located in the upper and peripheral regions of the shoot.

Attainment of a sufficiently large size appears to be more important than the plant's chronological age in determining the transition to the adult phase. Conditions that retard growth, such as mineral deficiencies, low light, water stress, defoliation, and low temperature, tend to prolong the juvenile phase or even cause reversion to juvenility of adult shoots. In contrast, conditions that promote vigorous growth accelerate the transition to the adult phase. When growth is accelerated, exposure to the correct flower-inducing treatment can result in flowering.

Ovate adult leaves

Fruit

Lobed juvenile leaves

Figure 17.1 Juvenile and adult forms of English ivy (*Hedera helix*). The juvenile form has lobed palmate leaves arranged alternately, a climbing growth habit, and no flowers. The adult form (projecting out to the right) has entire ovate leaves arranged in spirals, an upright growth habit, and flowers that develop into fruits. (Courtesy of L. Rignanese.)

Table 17.1   Length of juvenile period in some woody plants

Species	Length of juvenile period
Rose (*Rosa* [hybrid tea])	20–30 days
Grape (*Vitis* spp.)	1 year
Apple (*Malus* spp.)	4–8 years
Citrus spp.	5–8 years
English ivy (*Hedera helix*)	5–10 years
Redwood (*Sequoia sempervirens*)	5–15 years
Sycamore maple (*Acer pseudoplatanus*)	15–20 years
English oak (*Quercus robur*)	25–30 years
European beech (*Fagus sylvatica*)	30–40 years

Source: Clark 1983.

Although plant size seems to be the most important factor, it is not always clear which specific component associated with size is critical. In some *Nicotiana* species, it appears that plants must produce a certain number of leaves to transmit a sufficient amount of the floral stimulus to the apex.

Once the adult phase has been attained, it is relatively stable and is maintained during vegetative propagation or grafting. For example, cuttings taken from the basal region of mature plants of English ivy develop into juvenile plants, while those taken from the tip develop into adult plants. When scions were taken from the base of a flowering silver birch (*Betula verrucosa*) and grafted onto seedling rootstocks, there were no flowers on the grafts for the first 2 years. In contrast, grafts taken from the top of the mature tree flowered freely.

The term *juvenility* has different meanings for herbaceous and woody species. Whereas juvenile herbaceous meristems flower readily when grafted onto flowering adult plants, juvenile woody meristems generally do not. Juvenile woody meristems are thus said to lack the competence to flower.

Phase changes can be influenced by nutrients, gibberellins, and epigenetic regulation

The transition at the shoot apex from the juvenile to the adult phase can be affected by transmissible factors from the rest of the plant. In many plants, exposure to low-light conditions prolongs juvenility or causes reversion to juvenility. A major consequence of a low-light regime is a reduction in the supply of carbohydrates to the apex; thus, carbohydrate supply, especially sucrose, may play a role in the transition between juvenility and maturity. Carbohydrate supply as a source of energy and raw material can affect the size of the apex. For example, in the florist's chrysanthemum (*Chrysanthemum morifolium*), flower primordia are not initiated until a minimum apex size has been reached.

Gibberellin signaling may be involved in these responses in some plants. In pines and some other conifers, gibberellins accumulate under conditions that promote cone production (e.g., root removal, water stress, and nitrogen starvation), and application of gibberellins is used to stimulate production of reproductive structures in juvenile trees.

In Arabidopsis and other herbaceous species, juvenile status is maintained by the epigenetic repression of genes that are associated with transition to the adult phase. A common mechanism for temporary repression of a large number of genes is the production of short, noncoding RNA sequences called microRNAs (miRNAs). These miRNAs contain sequences that are complementary to regions of mRNAs that are transcribed from target genes. Cellular mechanisms that recognize double-stranded RNAs degrade the target mRNA, thus preventing the synthesis of the protein. In Arabidopsis, two of these miRNAs, numbered 155 and 156, are produced during juvenility. At the onset of the transition to the adult phase, increased methylation of the histones that organize chromatin in the nucleus (see Chapter 1) results in decreased expression of miR155/156, and derepression of their adult/reproductive phase target genes.

Photoperiodism: Monitoring Day Length

Photoperiodism is the ability of an organism to detect day length to ensure that events occur at the appropriate time of year, thus allowing for a seasonal response. Photoperiodic phenomena are found in both animals and plants. In the animal kingdom, day length controls such seasonal activities as hibernation, development of summer and winter coats, and reproductive activity. Plant responses controlled by day length are numerous; they include the initiation of flowering, asexual reproduction, the formation of storage organs, and the onset

photoperiodism A biological response to the length and timing of day and night, making it possible for an event to occur at a particular time of year.

(A)

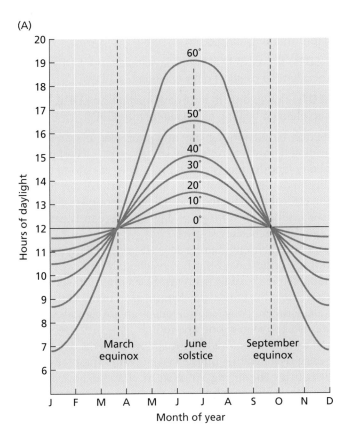

(B)

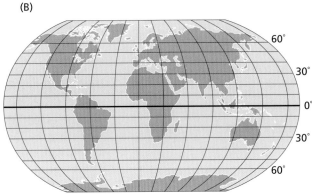

Figure 17.2 (A) Effect of latitude on day length at different times of the year in the Northern Hemisphere. Day length was measured on day 20 of each month. (B) Global map showing longitudes and latitudes.

of dormancy. In a natural environment, light and dark periods change seasonally according to latitude (**Figure 17.2**), and plants must have mechanisms to adjust to these changes to ensure survival and reproduction.

Plants can be classified according to their photoperiodic responses

Numerous plant species flower during the long days of summer, and for many years plant physiologists believed that the correlation between long days and flowering was a consequence of the accumulation of photosynthetic products synthesized during long days.

This hypothesis was shown to be incorrect by the work of Wightman Garner and Henry Allard, conducted in the 1920s at the U.S. Department of Agriculture laboratories in Beltsville, Maryland. Garner and Allard found that a mutant variety of tobacco, 'Maryland Mammoth', grew profusely to about 5 m in height, but failed to flower in the prevailing conditions of summer (**Figure 17.3**). However, the plants flowered in the greenhouse during the winter under natural light conditions.

These results ultimately led Garner and Allard to test the effect of artificially shortened days by covering plants grown during the long days of summer with a light-tight tent from late in the afternoon until the following morning. These artificial short days also caused the plants to flower. Garner and Allard concluded that day length, rather than the accumulation of photosynthate, was the determining factor in flowering. They were able to confirm their hypothesis in many different species and conditions. This work laid the foundations for the extensive subsequent research on photoperiodic responses.

Although many other aspects of plants' development may also be affected by day length, flowering is the response that has been studied the most.

Figure 17.3 'Maryland Mammoth' mutant of tobacco (right) compared with wild-type tobacco (left). Both plants were grown during summer in the greenhouse. (University of Wisconsin graduate students used for scale.) (Courtesy of R. Amasino.)

short-day plant (SDP)   A plant that flowers only in short days (qualitative SDP) or whose flowering is accelerated by short days (quantitative SDP).

long-day plant (LDP)   A plant that flowers only in long days (qualitative LDP) or whose flowering is accelerated by long days (quantitative LDP).

critical day length   The minimum length of the day required for flowering of a long-day plant; the maximum length of day that will allow short-day plants to flower. However, studies have shown that it is the length of the night, not the length of the day, that is important.

long–short-day plant (LSDP) A plant that flowers in response to a shift from long days to short days.

short–long-day plant (SLDP) A plant that flowers only after a sequence of short days followed by long days.

Flowering species tend to fall into one of two main photoperiodic response categories: short-day plants and long-day plants.

- **Short-day plants** (**SDPs**) flower only in short days (*qualitative* SDPs), or their flowering is accelerated by short days (*quantitative* SDPs).

- **Long-day plants** (**LDPs**) flower only in long days (*qualitative* LDPs), or their flowering is accelerated by long days (*quantitative* LDPs).

The essential distinction between long-day and short-day plants is that flowering in LDPs is promoted only when the day length *exceeds* a certain duration, called the **critical day length**, in every 24-h cycle, whereas promotion of flowering in SDPs requires a day length that is *less than* the critical day length. The absolute value of the critical day length varies widely among species, and only when flowering is examined for a range of day lengths can the correct photoperiodic classification be established (**Figure 17.4**).

LDPs can effectively measure the lengthening days of spring or early summer and delay flowering until the critical day length is reached. Many varieties of bread wheat (*Triticum aestivum*) behave in this way. SDPs often flower in the fall when the days shorten below the critical day length, as in many varieties of *Chrysanthemum morifolium*. However, day length alone is an ambiguous signal, because it cannot distinguish between spring and fall.

Plants exhibit several adaptations for avoiding the ambiguity of the day-length signal. One is the presence of a juvenile phase that prevents the plant from responding to day length during the spring. Another mechanism for avoiding the ambiguity of day length is the coupling of a temperature requirement to a photoperiodic response. Certain plant species, such as winter wheat, a variety of bread wheat, do not respond to photoperiod until after a cold period (vernalization or overwintering) has occurred. (We discuss vernalization later in this chapter.)

Other plants avoid seasonal ambiguity by distinguishing between *shortening* and *lengthening* days. Such "dual–day length plants" fall into two categories:

- **Long–short-day plants** (**LSDPs**) flower only after a sequence of long days followed by short days. LSDPs, such as *Bryophyllum*, *Kalanchoe*, and night-blooming jasmine (*Cestrum nocturnum*), flower in the late summer and fall, when the days are shortening.

- **Short–long-day plants** (**SLDPs**) flower only after a sequence of short days followed by long days. SLDPs, such as white clover (*Trifolium repens*), Canterbury bells (*Campanula medium*), and echeveria (*Echeveria harmsii*), flower in the early spring in response to lengthening days.

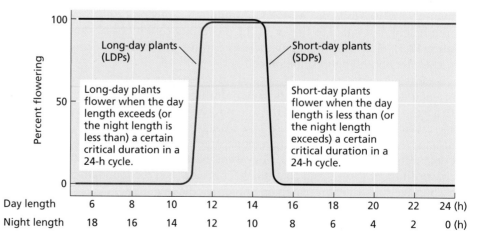

Figure 17.4 Photoperiodic response in long- and short-day plants. The critical day length varies among species. In this example, both the SDPs and the LDPs would flower in photoperiods between 12 and 14 h long.

Finally, species that flower under any photoperiodic condition are referred to as day-neutral plants. **Day-neutral plants** (**DNPs**) are insensitive to day length. Flowering in DNPs is typically under autonomous regulation—that is, internal developmental control. Some day-neutral species, such as maize (*Zea mays*), evolved near the equator where the day length is virtually constant throughout the year (see Figure 17.2). Many desert annuals, such as desert paintbrush (*Castilleja chromosa*) and desert sand verbena (*Abronia villosa*), evolved to germinate, grow, and flower quickly whenever sufficient water is available. These are also DNPs.

Photoperiodism is one of many plant processes controlled by a circadian rhythm

Organisms are normally subjected to daily cycles of light and darkness, and both plants and animals often exhibit rhythmic behavior in association with these changes. Examples of such rhythms include leaf and petal movements (day and night positions), stomatal opening and closing, growth and sporulation patterns in fungi (e.g., *Pilobolus* and *Neurospora*), time of day of pupal emergence (the fruit fly *Drosophila*), and activity cycles in rodents, as well as daily changes in the rates of metabolic processes such as photosynthesis and respiration.

When organisms are transferred from daily light–dark cycles to continuous darkness or continuous light, many of these rhythms continue to be expressed, at least for several days. Under such uniform conditions the period of the rhythm is close to 24 h, and consequently the term **circadian rhythm** (from the Latin *circa*, "about," and *diem*, "day") is applied. Because they continue under constant light or darkness, these circadian rhythms cannot be direct responses to the presence or absence of light, but must be based on an internal pacemaker mechanism, often called an *endogenous oscillator*. A single oscillator mechanism can be tied to multiple downstream processes at different times. Endogenous oscillators are thought to be regulated by the interactions of four sets of genes expressed in the dawn, morning, afternoon, and evening hours. Light may augment the amplitude of the oscillation by activating the morning and evening genes (**Figure 17.5**).

The endogenous oscillator is coupled to a variety of physiological processes, such as leaf movement or photosynthesis, and it maintains the rhythm. For this reason the endogenous oscillator can be considered the clock mechanism, and the physiological functions that are being regulated, such as leaf movements or photosynthesis, are sometimes referred to as the hands of the clock.

Circadian rhythms exhibit characteristic features

Circadian rhythms arise from cyclic phenomena that can be depicted as wave forms and are defined by three parameters:

1. **Period** is the time between comparable points in the repeating cycle. Typically the period is measured as the time between consecutive maxima (peaks) or minima (troughs) (**Figure 17.6A**).

2. **Phase**[1] is any point in the cycle that is recognizable by its relationship to the rest of the cycle. The most obvious phase points are the peak and trough positions.

3. **Amplitude** is usually considered to be the distance between peak and trough. The amplitude of a biological rhythm can often vary while the period remains unchanged (as, for example, in **Figure 17.6B**).

In constant light or darkness, rhythms depart from an exact 24-h period. The rhythms then drift in relation to solar time, either gaining or losing time

day-neutral plant (DNP) A plant whose flowering is not regulated by day length.

circadian rhythm A physiological process that oscillates endogenously with a cycle of approximately 24 hours.

period In cyclic (rhythmic) phenomena, the time between comparable points in the repeating cycle, such as peaks or troughs.

phase In cyclic (rhythmic) phenomena, any point in the cycle recognizable by its relationship to the rest of the cycle, for example, the maximum and minimum positions.

amplitude In a biological rhythm, the distance between peak and trough; it can often vary while the period remains unchanged.

---

[1]The term phase in this context should not be confused with the term phase change in meristem development discussed earlier in this chapter.

(A)
Mutual inhibition of Dawn and Afternoon genes. This motif is known as a "toggle switch." When one of the pair is high, the other is low. This motif alone does not oscillate.

(B)
The switch is located within a four-member ring, in which each set of genes turns off the set expressed before. In steady state, the ring alone does not oscillate either: Diagonally opposite pairs are on or off. This outcome is prevented by (A).

(C)
The Evening genes turn off all genes except Dawn genes in the early night, "clearing the decks" before Dawn gene expression starts again in the mid- to late night.

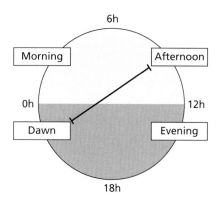

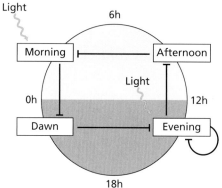

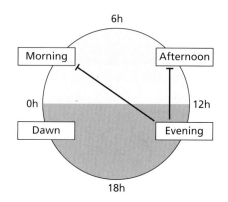

(D) The daily cycle, step by step

1. Dawn gene expression starts in the mid- to late night, continuing repression of Afternoon genes (A).

2. Light activates Morning and Evening gene expression. Evening genes are repressed by Dawn genes (B), so only Morning genes are expressed.

3. Morning genes turn off Dawn genes (B), allowing expression of Afternoon genes (A) and Evening genes (B). Ongoing light activation of Evening genes also helps.

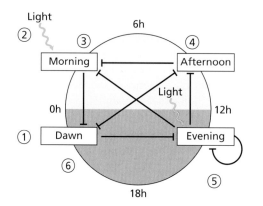

4. Afternoon genes turn off Morning genes (B) and also keep Dawn genes off (A).

5. Evening genes turn off all genes except Dawn genes (C).

6. Dawn genes are expressed again in the late night.

Figure 17.5   Model for the endogenous circadian oscillator. The circles represent a 24-h cycle, marked at 6-h intervals (0 h, 6 h, 12 h, and 18 h); daylight is from 0 h to 12 h. (Based on Purcell et al. 2010 and Andrew Millar, pers. comm.)

entrainment   The synchronization of the period of biological rhythms by external controlling factors, such as light and darkness.

Zeitgebers   Environmental signals such as light-to-dark or dark-to-light transitions that synchronize the endogenous oscillator to a 24-h periodicity.

free-running   Designation of the biological rhythm that is characteristic for a particular organism when environmental signals are removed, as in total darkness.

depending on whether the endogenous period is shorter or longer than 24 h. Under natural conditions, the endogenous oscillator is **entrained** (synchronized) to a true 24-h period by environmental signals, the most important of which are the light-to-dark transition at dusk and the dark-to-light transition at dawn (**Figure 17.6C**).

Such environmental signals are termed **Zeitgebers** (German for "time givers"). When such signals are removed—for example, by transfer to continuous darkness—the rhythm is said to be **free-running**, and it reverts to the circadian period that is characteristic of the particular organism. Although the rhythms are generated internally, they normally require an environmental signal, such as exposure to light or a change in temperature, to synchronize their expression. In addition, many rhythms damp out (i.e., the amplitude decreases) when the organism is subjected to a constant environment for several cycles. When this occurs, a Zeitgeber, such as a transfer from light to dark or a change in temperature, is required to restart the rhythm (see Figure 17.6B). Note that the clock itself need not damp out; only the coupling between the molecular clock (endogenous oscillator) and the physiological function is affected.

(A)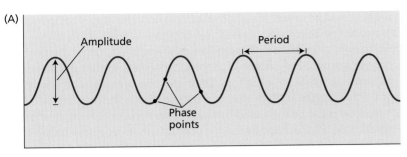

A typical circadian rhythm. The period is the time between comparable points in the repeating cycle; the phase is any point in the repeating cycle recognizable by its relationship with the rest of the cycle; the amplitude is the distance between peak and trough.

(B)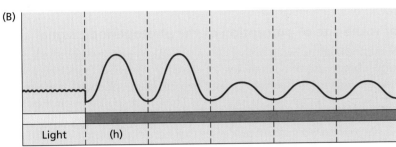

Suspension of a circadian rhythm in continuous bright light and the release or restarting of the rhythm following transfer to darkness.

(C)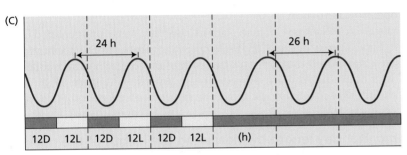

A circadian rhythm entrained to a 24-h light–dark (L–D) cycle and its reversion to the free-running period (26 h in this example) following transfer to continuous darkness.

(D)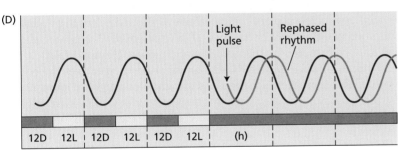

Typical phase-shifting response to a light pulse given shortly after transfer to darkness. The rhythm is rephased (delayed) without its period being changed.

Figure 17.6 Some characteristics of circadian rhythms.

The circadian clock would be of no value to the organism if it could not keep accurate time under the fluctuating temperatures experienced in natural conditions. Indeed, temperature has little or no effect on the period of the free-running rhythm. The feature that enables the clock to keep time at different temperatures is called **temperature compensation**.

Circadian rhythms adjust to different day–night cycles

How do circadian rhythms remain constant when the daily durations of light and darkness change with the seasons? Investigators typically test the response of the endogenous oscillator by placing the organism in continuous darkness and examining the response to a short pulse of light (usually less than 1 h) given at different time points in the free-running rhythm. If a light pulse is given during the first few hours of the original night period, the phase of the rhythm is delayed; the organism interprets the light pulse as the end of the previous day (**Figure 17.6D**). As would be expected, a light pulse given toward the end of the

temperature compensation A characteristic of circadian rhythms, which can maintain their circadian period over a broad range of temperatures within the physiological range.

Figure 17.7 Demonstration by grafting of a leaf-generated floral stimulus in the SDP *Perilla crispa*. (Left) Grafting an induced leaf from a plant grown under short days onto a noninduced shoot causes the axillary shoots to produce flowers. The donor leaf has been trimmed to facilitate grafting, and the upper leaves have been removed from the stock to promote phloem translocation from the scion to the receptor shoots. (Right) Grafting a noninduced leaf from a plant grown under long days results in the formation of vegetative branches only. (Courtesy of J. A. D. Zeevaart.)

photoperiodic induction
The photoperiod-regulated processes that occur in leaves resulting in the transmission of a floral stimulus to the shoot apex.

night break   An interruption of the dark period with a short exposure to light that makes the entire dark period ineffective.

original night period advances the rhythm phase; now the organism interprets the light pulse as the beginning of the following day.

The biochemical mechanism that enables a light signal to cause these phase shifts is not yet known, but photoreceptor studies (see Chapter 13) have improved our understanding of how light regulates the process. The low levels and specific wavelengths of light that can induce phase shifting indicate that the light response must be mediated by specific photoreceptors rather than by photosynthetic rate. Phytochromes are the primary photoreceptors influencing circadian rhythms, but cryptochromes also participate in blue-light entrainment of the clock in plants, as they do in insects and mammals.

The leaf is the site of perception of the photoperiodic signal

The photoperiodic stimulus in both LDPs and SDPs is perceived by the leaves. For example, treatment of a single leaf of the SDP *Xanthium* (cocklebur) with short photoperiods is sufficient to cause the formation of flowers, even when the rest of the plant is exposed to long days. Thus, in response to photoperiod the leaf transmits a signal that regulates the transition to flowering at the shoot apex. The photoperiod-regulated processes that occur in the leaves resulting in the transmission of a floral stimulus to the shoot apex are referred to collectively as **photoperiodic induction**.

Photoperiodic induction can take place in a leaf that has been separated from the plant. For example, in the SDP *Perilla crispa* (a member of the mint family), an excised leaf exposed to short days can cause flowering when subsequently grafted to a noninduced plant maintained in long days (**Figure 17.7**). This result indicates that photoperiodic induction depends on events that take place exclusively in the leaf.

Plants monitor day length by measuring the length of the night

Under natural conditions, day and night lengths configure a 24-h cycle of light and darkness. In principle, a plant could perceive a critical day length by measuring the duration of either light or darkness. Flowering of SDPs has been shown to be primarily determined by the duration of darkness (**Figure 17.8A**). It was possible to induce flowering in SDPs with light periods longer than the critical value, provided that these were followed by sufficiently long nights (**Figure 17.8B**). Similarly, SDPs did not flower when short days were followed by short nights.

More detailed experiments demonstrated that the mechanism of photoperiodic timekeeping in SDPs is based on the duration of darkness. For example, flowering occurred only when the dark period exceeded 8.5 h in cocklebur (*Xanthium strumarium*) or 10 h in soybean (*Glycine max*). The duration of darkness was also shown to be important in LDPs (see Figure 17.8B). These plants were found to flower in short days, provided that the accompanying night length was also short; however, a regime of long days followed by long nights was ineffective.

Night breaks can cancel the effect of the dark period

A feature that underscores the importance of the dark period is that it can be made ineffective by interruption with a short exposure to light, called a **night break** (see Figure 17.8A). In contrast, interrupting a long day with a brief dark period does not cancel the effect of the long day (see Figure 17.8B). Night-break treatments of only a few minutes are effective in *preventing* flowering in many SDPs, including *Xanthium* and *Pharbitis*, but much longer exposures are often required to *promote* flowering in LDPs. Consistent with the involvement of a circadian rhythm, the effect of a night break varies greatly according to the time when it is given. For both LDPs and SDPs, a night break was found to be most effective when given near the middle of a dark period of 16 h (**Figure 17.9**).

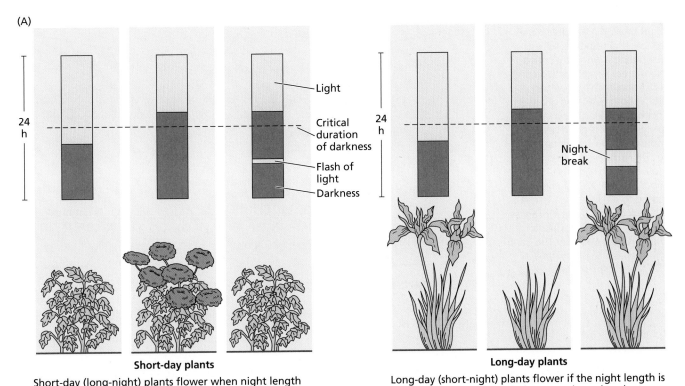

Short-day plants

Short-day (long-night) plants flower when night length exceeds a critical dark period. Interruption of the dark period by a brief light treatment (a night break) prevents flowering.

Long-day plants

Long-day (short-night) plants flower if the night length is shorter than a critical period. In some long-day plants, shortening the night with a night break induces flowering.

The discovery of the night-break effect, and its time dependence, had several important consequences. It established the central role of the dark period and provided a valuable probe for studying photoperiodic timekeeping. Because only small amounts of light are needed, it became possible to study the action and identity of the photoreceptor without the interfering effects of photosynthesis and other nonphotoperiodic phenomena. This discovery has also led to the development of commercial methods for regulating the time of flowering in horticultural species, such as *Kalanchoe*, chrysanthemum, and poinsettia (*Euphorbia pulcherrima*).

Photoperiodic timekeeping during the night depends on a circadian clock

The decisive effect of night length on flowering indicates that measuring the passage of time in darkness is central to photoperiodic timekeeping. Photoperiodic timekeeping appears to depend on an endogenous circadian oscillator of the type discussed earlier (see Figure 17.5). Measurements of the effect on flowering of interrupting an extended night period with a short light treatment (*night break*) can be used to investigate the role of circadian rhythms in photoperiodic timekeeping. For example, when soybean plants, which are SDPs, are transferred from an 8-h light period to a 64-h dark period, the flowering response to 4-h night breaks given at different times during the dark period shows a circadian rhythm (**Figure 17.10**). Soybean plants, which are SDPs, would normally flower in response to 8-h day lengths. *Inhibition* of flowering by a night break (see Figure 17.8A) therefore corresponds to a period of light

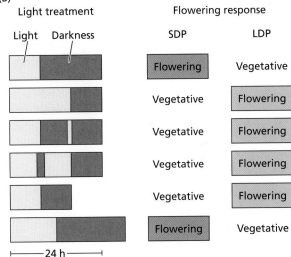

Figure 17.8 Photoperiodic regulation of flowering. (A) Effects on SDPs and LDPs. (B) Effects of the duration of the dark period on flowering. Treating SDPs and LDPs with different photoperiods clearly shows that the critical variable for flowering is actually the length of the dark period rather than the day length.

Figure 17.9 The time at which a night break is given determines the flowering response. When given during a long dark period, a night break promotes flowering in LDPs and inhibits flowering in SDPs. In both cases, the greatest effect on flowering occurs when the night break is given near the middle of the 16-h dark period. The LDP *Fuchsia* was given a 1-h exposure to red light in a 16-h dark period. The SDP *Xanthium* was exposed to red light for 1 min in a 16-h dark period. (Data for *Fuchsia* from Vince-Prue 1975; data for *Xanthium* from Salisbury 1963 and Papenfuss and Salisbury 1967.)

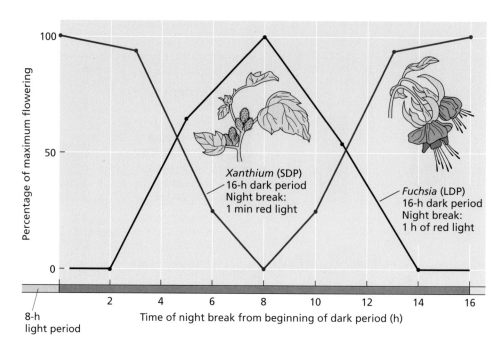

sensitivity. As shown in Figure 17.10, during a 64-h night period, the peaks of light inhibition of flowering occur at 24-h intervals, indicating that the phases of light sensitivity and insensitivity show circadian periodicity even during continuous darkness.

A coincidence model links oscillating light sensitivity and photoperiodism

How does an oscillation with a 24-h period measure a critical duration of darkness of, say, 8 to 9 h, as in the SDP *Xanthium*? In 1936 Erwin Bünning proposed that the control of flowering by photoperiodism is achieved by an oscillation of phases with different sensitivities to light. This proposal has evolved into the **coincidence model**, in which the endogenous oscillator controls the timing of light-sensitive and light-insensitive phases.

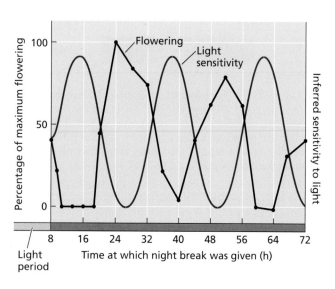

Figure 17.10 Rhythmic flowering in response to night breaks. In this experiment, the SDP soybean (*Glycine max*) received cycles of an 8-h light period followed by a 64-h dark period. A 4-h night break (light treatment) was given at various times during the long inductive dark period. The flowering response, plotted as the percentage of the maximum, was then plotted for each night break given. Note, for example, that a night break given at 26 h resulted in maximum flowering, while no flowering was obtained when the night break was given at 40 h. Since in SDPs light acts as an inhibitor of flowering during the long night period, we can infer that 26 h corresponds to a minimum for light sensitivity, while 40 h corresponds to a maximum for light sensitivity. The flowering data can thus be used to infer the periodicity of the plants' sensitivity to the effect of the night break over time (the red curve), which exhibits a circadian rhythm. These data support a model in which flowering in SDPs is induced only when dawn (or a night break) occurs after the completion of the light-sensitive phase. Since in LDPs light stimulates flowering during the dark period, the light break must coincide with the light-sensitive phase for flowering to occur in LDPs. (Data from Coulter and Hamner 1964.)

The ability of light either to promote or to inhibit flowering depends on the phase in which the light is given. When a light signal is administered during the light-sensitive phase of the rhythm, the effect is either to *promote* flowering in LDPs or to *prevent* flowering in SDPs. For example, flowering in SDPs is induced only when exposure to light during a night break or at dawn occurs after completion of the light-sensitive phase of the rhythm.

If a similar experiment is performed with an LDP, flowering is induced only when the night break occurs *during* the light-sensitive phase of the rhythm. In other words, *flowering in both SDPs and LDPs is induced when the light exposure is coincident with the appropriate phase of the rhythm.* This continued oscillation of sensitive and insensitive phases in the absence of dawn and dusk light signals is characteristic of a variety of processes controlled by the circadian oscillator.

Molecular evidence from monocot and eudicot species supports the function of a coincidence model in the control of floral induction in both SDPs and LDPs. The mechanisms are highly conserved and involve the expression in the leaf of a conserved floral induction gene coding for a transcriptional activator that serves as a master switch. The expression of this master switch gene is controlled by the circadian clock, with the peak of gene expression occurring about 12–16 h after dawn (**Figure 17.11**). In LDPs such as Arabidopsis, the master switch protein acts as an inducer of flowering, while in SDPs such as rice it acts as an inhibitor. Because the master switch mRNA reaches its peak around 16 h after dawn, it will

coincidence model A model for flowering in photoperiodic plants in which the circadian oscillator controls the timing of light-sensitive and light-insensitive phases during the 24-h cycle.

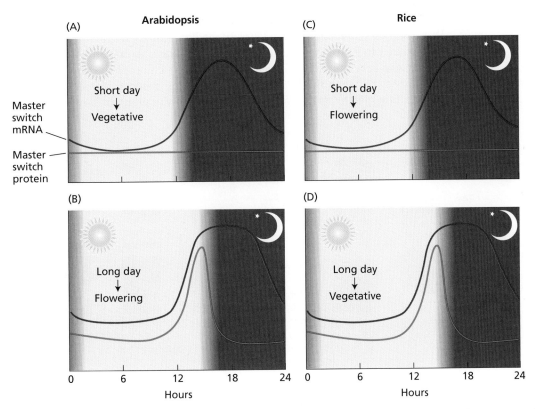

Figure 17.11 The coincidence model of photoperiodic induction in LDP Arabidopsis (A and B) and SDP rice (C and D). (A) In Arabidopsis under short days, there is little overlap between the expression of the master switch gene and daylight, and the plant remains vegetative. (B) Under long days, the peak of expression of the master switch gene (at hours 12 through 16) overlaps with daylight (sensed by phytochrome), allowing the protein to accumulate. (C) In rice under short days, the lack of coincidence between the master switch gene mRNA expression and daylight prevents the accumulation of the protein, which in this case is an inhibitor of flowering. In the absence of the inhibitor, the plants flower. (D) Under long days (sensed by phytochrome), the peak of the master switch gene expression overlaps with the day, allowing the accumulation of the repressor protein. As a result, the plant remains vegetative. (After Hayama and Coupland 2004.)

Pr The red light–absorbing form of phytochrome. This is the form in which phytochrome is assembled. The cyan-blue colored Pr is converted by red light to the far-red light–absorbing form, Pfr.

Pfr The far-red light–absorbing form of phytochrome converted from Pr by the action of red light. The cyan-green colored Pfr is converted back to Pr by far-red light. Pfr is the physiologically active form of phytochrome.

vernalization In some species, the cold temperature requirement for flowering. The term is derived from the Latin word for "spring."

be coincident with light exposure only under long days. Therefore, under short-day conditions, when the master switch protein is *not* coincident with light, the master switch protein does not accumulate and LDPs will remain vegetative (see Figure 17.11A). In contrast, SDPs will flower under short-day conditions because the master switch protein acts as an inhibitor (see Figure 17.11C). Under long day conditions, when the master switch mRNA *is* coincident with light, the master switch protein accumulates and LDPs will flower (see Figure 17.11B), whereas SDPs will remain vegetative (see Figure 17.11D).

As we discuss next, for both LDPs and SDPs phytochrome is the primary photoreceptor (see Chapter 13) that mediates the interactions between the master switch proteins and light under long day conditions.

Phytochrome is the primary photoreceptor in photoperiodism

Night-break experiments are well suited for studying the nature of the photoreceptors that perceive light signals during the photoperiodic response. The inhibition of flowering in SDPs by night breaks was one of the first physiological processes shown to be under the control of phytochrome (**Figure 17.12**).

In many SDPs, a night break becomes effective only when the supplied dose of light is sufficient to saturate the photoconversion of **Pr** (phytochrome that absorbs red light) to **Pfr** (phytochrome that absorbs far-red light) (see Chapter 13). A subsequent exposure to far-red light, which photoconverts the pigment back to the physiologically inactive Pr form, restores the flowering response.

Action spectra for the inhibition and restoration of the flowering response in SDPs are shown in **Figure 17.13**. A peak at 660 nm, the absorption maximum of Pr, is obtained when dark-grown *Pharbitis* seedlings are used to avoid interference from chlorophyll. In contrast, the spectra for *Xanthium* provide an example of the response in green plants, in which the presence of chlorophyll can cause some discrepancy between the action spectrum and the absorption spectrum of Pr. These action spectra plus the red/far-red reversibility of the night-break responses

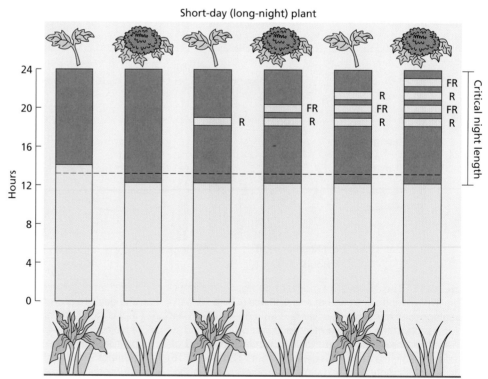

Figure 17.12 Phytochrome control of flowering by red (R) and far-red (FR) light. A flash of red light during the dark period induces flowering in an LDP, and the effect is reversed by a flash of far-red light. This response indicates the involvement of phytochrome. In SDPs, a flash of red light prevents flowering, and the effect is reversed by a flash of far-red light.

Figure 17.13 Action spectra for the control of flowering by night breaks implicate phytochrome. Flowering in SDPs grown under inductive short-day conditions is inhibited by a short light treatment (night break) during the long night. In the SDP *Xanthium strumarium*, red-light night breaks of 620 to 640 nm are the most effective wavelengths. Reversal of the red-light effect is maximal at 725 nm. In the dark-grown SDP *Pharbitis nil*, which is devoid of chlorophyll and its interference with light absorption, night breaks of 660 nm are the most effective. This 660-nm maximum coincides with the absorption maximum of phytochrome. The data was normalized relative to the dark control at each wavelength. (Data for *Xanthium* from Hendricks and Siegelman 1967; data for *Pharbitis* from Saji et al. 1983.)

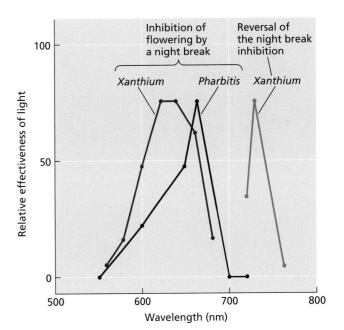

confirm the role of phytochrome as the photoreceptor that is involved in photoperiod measurement in SDPs. Confirming the function of phytochrome, flowering is promoted by red-light night breaks, and a subsequent exposure to far-red light prevents this response (see Figure 17.13).

A circadian rhythm in the promotion of flowering by far-red light has been observed in the LDPs darnel ryegrass (*Lolium temulentum*), barley (*Hordeum vulgare*), and Arabidopsis. The response is proportional to the irradiance and duration of far-red light and is therefore a high-irradiance response (HIR; see Chapter 13). As in other HIRs, phyA is the phytochrome that mediates the response to far-red light.

Vernalization: Promoting Flowering with Cold

Vernalization is the process whereby repression of flowering is alleviated by a cold treatment given to a hydrated seed (i.e., a seed that has imbibed water) or to a growing plant (dry seeds do not respond to the cold treatment because vernalization is an active metabolic process). Without the cold treatment, plants that require vernalization show delayed flowering or remain vegetative, and they are not competent to respond to floral signals such as inductive photoperiods. This requirement is important for production of overwintering cereal crops such as barley (**Figure 17.14**) and in overwintering annual eudicots that grow as rosettes with no elongation of the stem until longer days signal the end of the frost period.

Plants differ considerably in the age at which they become sensitive to vernalization. Winter annuals, such as the winter forms of cereals (which are sown in the fall and flower in the following summer), respond to low temperature very early in their life cycle. In fact, many winter annuals can be vernalized before germination (i.e., radicle emergence from the seed) if the seeds have imbibed water and become metabolically active. Other plants, including most biennials

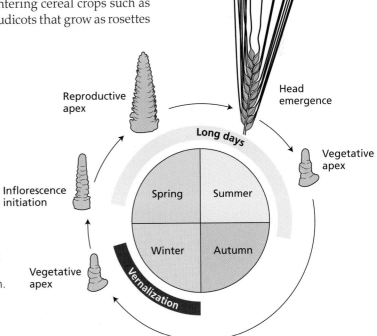

Figure 17.14 Influence of seasonal cues on shoot apex development in the temperate cereals. Varieties that require vernalization are sown in late summer or autumn. The shoot apex develops vegetatively until winter, when vernalization occurs. This promotes inflorescence initiation, which occurs as temperatures increase in spring. (After Trevaskis et al. 2007.)

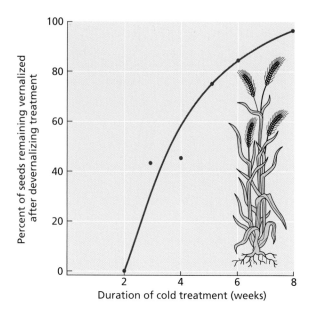

Figure 17.15 Duration of exposure to low temperature increases the stability of the vernalization effect. The longer that winter rye (*Secale cereale*) is exposed to a cold treatment, the greater the number of plants that remain vernalized when the cold treatment is followed by a devernalizing treatment. In this experiment, seeds of rye that had imbibed water were exposed to 5°C for different lengths of time, then immediately given a devernalizing treatment of 3 days at 35°C. (Data from Purvis and Gregory 1952.)

(which grow as rosettes during the first season after sowing and flower in the following summer), must reach a minimum size before they become sensitive to low temperature for vernalization.

The effective temperature range for vernalization is from just below freezing to about 10°C, with a broad optimum usually between about 1 and 7°C. The effect of cold increases with the duration of the cold treatment until the response is saturated. The response usually requires several weeks of exposure to low temperature, but the precise duration varies widely with species and variety. High temperature and other stresses can "devernalize" overwintering annuals, but the longer the exposure to low temperature, the more permanent the vernalization effect (**Figure 17.15**).

Vernalization appears to take place primarily in the SAM. Localized cooling causes flowering when only the shoot apex is chilled, and this effect appears to be largely independent of the temperature experienced by the rest of the plant. This effect is most easily visualized in winter cereals, where experimental temperature treatments of isolated stem apices are readily accomplished.

In developmental terms, vernalization results in the acquisition of competence of the meristem to undergo the floral transition. Yet as discussed earlier in the chapter, competence to flower does not guarantee that flowering will occur. A vernalization requirement is often linked to a requirement for a particular photoperiod. The most common combination is a requirement for cold treatment *followed* by a requirement for long days—a combination that leads to flowering in early summer at high latitudes.

Long-distance Signaling Involved in Flowering

Although floral evocation occurs at the apical meristems of shoots, in photoperiodic plants inductive photoperiods are sensed by the leaves. This suggests that a long-range signal must be transmitted from the leaves to the apex, which has been shown experimentally through extensive grafting experiments in many different plant species. The transmitted floral stimulus originating in leaves acts in combination with other factors and is sometimes referred to as **florigen**.

In some plant species, noninduced receptor plants can be stimulated to flower by having a leaf or shoot from a photoperiodically induced donor plant grafted to them. For example, in the SDP *Perilla crispa*, grafting a leaf from a plant grown under inductive short days onto a plant grown under noninductive long days causes the latter to flower (see Figure 17.7). The movement of the floral stimulus from a donor leaf to the rest of the plant requires establishment of vascular continuity across the graft union and correlates closely with the translocation of ^{14}C-labeled assimilates from the donor. Disruptions of the phloem also prevent flowering, and experimental measurements and modeling show that floral induction correlates with phloem translocation rates.

The floral stimulus seems to be the same in plants with different photoperiodic requirements. Thus, grafting an induced shoot from the LDP *Nicotiana sylvestris*, grown under long days, onto the SDP 'Maryland Mammoth' tobacco caused the

florigen   A systemically mobile transcription factor synthesized by leaves and translocated to the shoot apical meristem via the phloem to stimulate flowering.

Table 17.2 Transmission of the flowering signal occurs through a graft junction

Donor plants maintained under flower-inducing conditions	Photoperiod type[a,b]	Vegetative receptor plant induced to flower	Photoperiod type[a,b]
Helianthus annus	DNP in LD	*H. tuberosus*	SDP in LD
Nicotiana tabacum 'Delcrest'	DNP in SD	*N. sylvestris*	LDP in SD
Nicotiana sylvestris	LDP in LD	*N. tabacum* 'Maryland Mammoth'	SDP in LD
Nicotiana tabacum 'Maryland Mammoth'	SDP in SD	*N. sylvestris*	LDP in SD

Note: The successful transfer of a flowering induction signal by grafting between plants of different photoperiodic response groups shows the existence of a transmissible floral hormone that is effective.

[a]LDPs = long-day plants; SDPs = short-day plants; DNPs = day-neutral plants.

[b]LD, long days; SD, short days.

latter to flower under noninductive (long-day) conditions. The leaves of DNPs have also been shown to produce a graft-transmissible floral stimulus (**Table 17.2**). For example, grafting a single leaf of a day-neutral variety of soybean, 'Agate', onto the short-day variety, 'Biloxi', caused flowering in 'Biloxi' even when the latter was maintained in noninductive long days. Similarly, a shoot from a day-neutral variety of tobacco (*Nicotiana tabacum*, cv. Trapezond) grafted onto the LDP *Nicotiana sylvestris* induced the latter to flower under noninductive short days.

In Arabidopsis and other species, the transmissible flowering stimulus has been shown to be a small globular protein called FLOWERING LOCUS T (FT), which is expressed in companion cells and interacts with SAM transcription factors to induce flowering. The gene that acts as a light-inducible master switch in leaves, whose transcription is regulated by the circadian clock (see Figure 17.11), regulates the movement of FT into the phloem, where it is translocated along the source–sink gradient to the shoot apex. Once in the floral meristem, the FT protein enters the nuclei of the cells and forms an active complex with the transcription factor FLOWERING LOCUS D (FD). The FT–FD complex then activates various floral identity genes (discussed later) that convert the vegetative meristem into a floral meristem. This process is diagrammed in **Figure 17.16**.

Gibberellins and ethylene can induce flowering

As described previously, gibberellins promote transition to reproductive phases in many plants. Exogenous gibberellin can evoke flowering when applied either to rosette LDPs such as Arabidopsis, or to dual–day length plants such as *Bryophyllum*, when grown under short days. Gibberellin can also evoke flowering in cold-requiring plants that have not been vernalized, and promote cone formation in juvenile plants of several gymnosperm families. Thus, in some plants exogenous gibberellins can bypass the endogenous trigger of age in autonomous flowering, as well as the primary environmental signals of day length and temperature.

Gibberellin metabolism in the plant is strongly affected by day length. For example, in the LDP spinach (*Spinacia oleracea*), the levels of gibberellins are relatively low in short days, and the plants maintain a rosette form. After the plants are transferred to long days, there is a fivefold increase in the level of physiologically active gibberellin, GA_1, which causes the marked stem elongation that accompanies flowering.

In addition to gibberellins, other growth hormones can either inhibit or promote flowering. One commercially important example is the promotion of flowering in pineapple (*Ananas comosus*) by ethylene and ethylene-releasing compounds—a response that appears to be restricted to members of the pineapple family (Bromeliaceae).

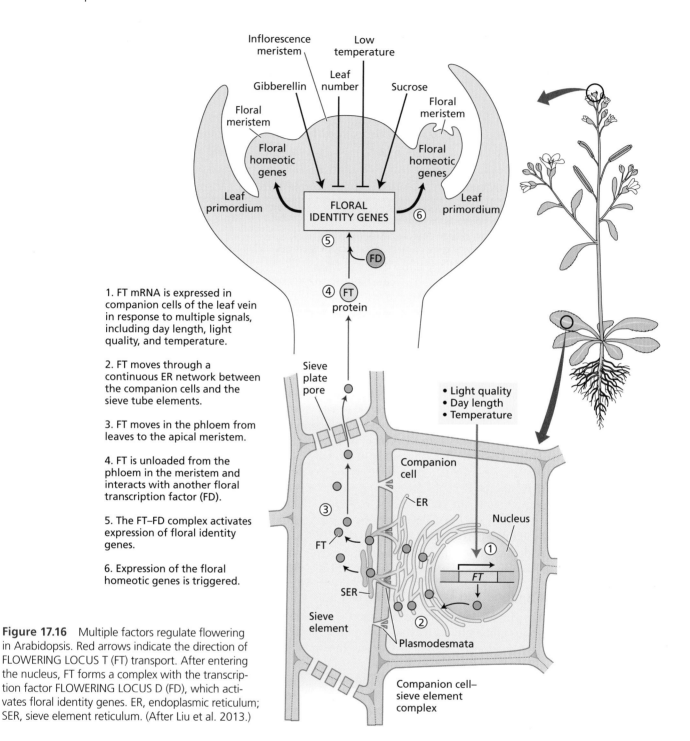

1. FT mRNA is expressed in companion cells of the leaf vein in response to multiple signals, including day length, light quality, and temperature.

2. FT moves through a continuous ER network between the companion cells and the sieve tube elements.

3. FT moves in the phloem from leaves to the apical meristem.

4. FT is unloaded from the phloem in the meristem and interacts with another floral transcription factor (FD).

5. The FT–FD complex activates expression of floral identity genes.

6. Expression of the floral homeotic genes is triggered.

Figure 17.16 Multiple factors regulate flowering in Arabidopsis. Red arrows indicate the direction of FLOWERING LOCUS T (FT) transport. After entering the nucleus, FT forms a complex with the transcription factor FLOWERING LOCUS D (FD), which activates floral identity genes. ER, endoplasmic reticulum; SER, sieve element reticulum. (After Liu et al. 2013.)

Floral Meristems and Floral Organ Development

Once flowering has been evoked, the business of building flowers begins. The shapes of flowers are extremely diverse, reflecting adaptations to protect developing gametophytes, attract pollinators, promote self-pollination or cross-pollination as appropriate, and produce and disperse fruits and seeds. Despite this diversity, genetic and molecular studies have now identified a network of genes that control floral morphogenesis in flowers as different as those of Arabidopsis and snapdragon (*Antirrhinum majus*). Variations on this regulatory network now seem to account for floral morphogenesis in other species as well.

In this section we focus on floral development in Arabidopsis, which has been studied extensively. First we outline the basic morphological changes that occur during the transition from the vegetative to the reproductive phase. Next we consider the arrangement of the floral organs in four whorls on the meristem, and the types of genes that govern the normal pattern of floral development.

The SAM in Arabidopsis changes with development

Floral meristems can usually be distinguished from vegetative meristems by their larger size. In the vegetative meristem, the cells of the central zone complete their division cycles slowly. The transition from vegetative to reproductive development is marked by an increase in the frequency of cell divisions within the central zone of the SAM. The increase in the size of the meristem is largely a result of the increased division rate of these central cells.

During the vegetative phase of growth, SAMs give rise to leaf primordia on the flanks of the shoot apex (**Figure 17.17A**). When reproductive development is initiated, the vegetative meristems are transformed either directly into a floral meristem, or indirectly into a primary inflorescence meristem, depending on the species. In rosette plants such as Arabidopsis, the **primary inflorescence meristem** produces an elongated inflorescence axis bearing two types of lateral organs: caulescent (stem-borne) leaves and flowers. The axillary buds of the caulescent leaves develop into **secondary inflorescence meristems**, and their activity repeats the pattern of development of the primary inflorescence meristem. The Arabidopsis inflorescence meristem has the potential to grow indefinitely and thus exhibits *indeterminate* growth. Flowers arise from **floral meristems** that form on the flanks of the inflorescence meristem (**Figure 17.17B**). In contrast to the inflorescence meristem, the floral meristem is determinate.

The four different types of floral organs are initiated as separate whorls

Floral meristems initiate four different types of floral organs: sepals, petals, stamens, and carpels. These sets of organs are initiated in concentric rings, called **whorls**, around the flanks of the meristem (**Figure 17.18**). The initiation of the innermost organs, the carpels, consumes all of the meristematic cells in the apical dome, and only the floral organ primordia (localized regions of cell division) are present as the floral bud develops. In Arabidopsis, the whorls are arranged as follows:

- The first (outermost) whorl consists of four sepals, which are green at maturity.

- The second whorl is composed of four petals, which are white at maturity.

- The third whorl contains six stamens (the male reproductive structures), two of which are shorter than the other four.

- The fourth (innermost) whorl is a single complex organ, the gynoecium or pistil (the female reproductive structure), which is composed of an ovary with two fused carpels, each containing numerous ovules, and a short style capped with a stigma (see Figure 17.27A).

primary inflorescence meristem The meristem that produces stem-bearing flowers; it is formed from the shoot apical meristem.

secondary inflorescence meristem The inflorescence meristem that develops from the axillary buds of stem-borne leaves of the primary inflorescence.

floral meristem The meristem that forms floral (reproductive) organs: sepals, petals, stamens, and carpels. May form directly from a vegetative meristem or indirectly via an inflorescence meristem.

whorl Pertaining to the concentric pattern of a set of organs that are initiated around the flanks of the meristem.

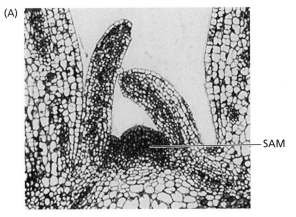

(A)

SAM

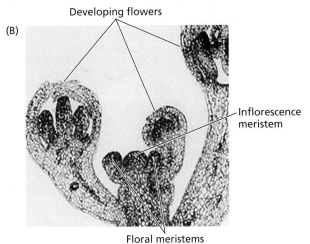

Developing flowers

(B)

Inflorescence meristem

Floral meristems

Figure 17.17 Longitudinal sections through a vegetative (A) and a reproductive (B) shoot apical region of Arabidopsis. (Courtesy of V. Grbic and M. Nelson.)

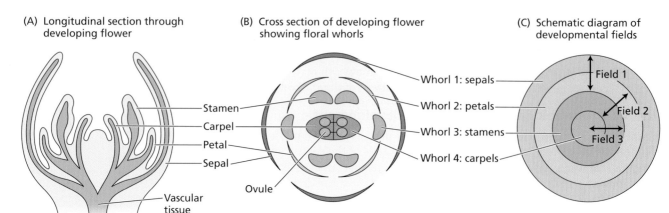

(A) Longitudinal section through developing flower

Stamen
Carpel
Petal
Sepal
Vascular tissue

(B) Cross section of developing flower showing floral whorls

Whorl 1: sepals
Whorl 2: petals
Whorl 3: stamens
Whorl 4: carpels
Ovule

(C) Schematic diagram of developmental fields

Field 1
Field 2
Field 3

Figure 17.18 Floral organs are initiated sequentially by the floral meristem of Arabidopsis. (A and B) The floral organs are produced as successive whorls (concentric circles), starting with the sepals and progressing inward. (C) According to the combinatorial model, the functions of each whorl are determined by three overlapping developmental fields. These fields correspond to the expression patterns of specific floral organ identity genes. (After Bewley et al. 2000.)

Two major categories of genes regulate floral development

Studies of mutations have enabled identification of two key categories of genes that regulate floral development: floral meristem identity genes and floral organ identity genes.

1. **Floral meristem identity genes** encode transcription factors that are necessary for the initial induction of floral organ identity genes. In general, these are the positive regulators of floral organ identity in the developing floral meristem.

2. **Floral organ identity genes** directly control floral organ identity. The proteins encoded by these genes are transcription factors that interact with other protein cofactors to control the expression of downstream genes whose products are involved in the formation or function of floral organs. Mutations of these genes often result in dramatic changes in flower structures, with sepals, petals, stamens, and carpels formed in the wrong whorl. Many of these mutants are used extensively in flower breeding to introduce flowers with unique and striking forms. Developmental mutations of this type, which cause one organ to be replaced by another, are called *homeotic mutations*, and the associated genes are called *homeotic genes*.

While certain genes fit neatly within these two general categories, it is important to keep in mind that floral development involves complex, nonlinear gene networks. In these networks, individual genes often play multiple roles.

The ABC model partially explains the determination of floral organ identity

Floral organ identity genes fall into three classes—A, B, and C—defining three different kinds of activities encoded by three distinct types of genes (**Figure 17.19**):

- Class A activity controls organ identity in the first and second whorls. Loss of Class A activity results in the formation of carpels instead of sepals in the first whorl, and of stamens instead of petals in the second whorl.

- Class B activity controls organ determination in the second and third whorls. Loss of Class B activity results in the formation of sepals instead of petals in the second whorl, and of carpels instead of stamens in the third whorl.

floral meristem identity genes
Two classes of genes, one required for the conversion of a shoot apical meristem into a floral meristem, the other that maintains the identity of an inflorescence meristem (as opposed to a floral meristem).

floral organ identity genes   Three classes of genes that control the specific locations of floral organs in the flower.

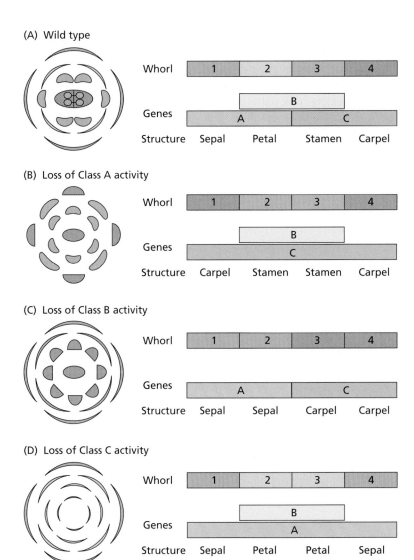

(A) Wild type

(B) Loss of Class A activity

(C) Loss of Class B activity

(D) Loss of Class C activity

Figure 17.19   Interpretation of the phenotypes of floral homeotic mutants based on the ABC model. (A) All three activity classes are functional in the wild type. (B) Loss of Class A activity results in the spread of Class C activity throughout the meristem. (C) Loss of Class B activity results in the expression of only Class A and C activities. (D) Loss of Class C activity results in expansion of Class A activity throughout the floral meristem. (After Bewley et al. 2000.)

- Class C activity controls events in the third and fourth whorls. Loss of Class C activity results in the formation of petals instead of stamens in the third whorl. Moreover, in the absence of Class C activity, the fourth whorl (normally a carpel) is replaced by a *new flower*. As a result, the fourth whorl of a Class C gene mutant flower is occupied by sepals. The floral meristem is no longer determinate. Flowers continue to form *within* flowers, and the pattern of organs (from outside to inside) is: sepal, petal, petal; sepal, petal, petal; and so on.

The **ABC model** accounts for many observations in two distantly related eudicot species (snapdragon and Arabidopsis), and provides a way of understanding how relatively few key regulators can combinatorially provide a complex outcome. The ABC model postulates that organ identity in each whorl is determined by a unique combination of the three organ identity gene activities:

- Class A activity alone specifies sepals.
- Class A and B activities are required for the formation of petals.
- Class B and C activities form stamens.
- Class C activity alone specifies carpels.

ABC model   A proposal for the way in which floral homeotic genes control organ formation in flowers. According to the model, organ identity in each whorl is determined by a unique combination of the three organ identity gene activities.

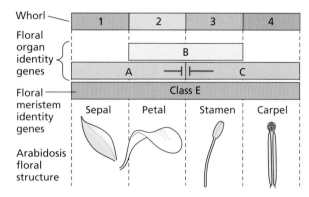

Figure 17.20 The ABCE model of floral development. (From Krizek and Fletcher 2005.)

The model further proposes that Class A and C activities mutually repress each other; that is, both A- and C-Class genes exclude each other from their expression domains, in addition to their function in determining organ identity.

Although the patterns of organ formation in wild-type flowers and most of the mutants are predicted and explained by this model, not all observations can be accounted for by the ABC genes alone. For example, expression of the ABC genes throughout the plant does not transform vegetative leaves into floral organs. Thus, the ABC genes, while necessary, are not sufficient to impose floral organ identity onto a leaf developmental program.

A fourth class of genes, the Class E genes, have been shown to be required for the other three floral homeotic genes to be active. E-Class genes can therefore be described as floral meristem identity genes. The augmented ABC model is referred to as the ABCE model (**Figure 17.20**). The ABCE model was formulated based on genetic experiments in Arabidopsis and snapdragon. Flowers from different species have evolved diverse structures by modifying the regulatory networks described by the ABCE model.

Pollen Development

As described in Chapter 1, the male and female gametophtyes are the true sexual structures in angiosperms. The male gametophyte is formed in the stamen of the flower. Typically, the stamen is made up of a delicate filament attached to an anther composed of four microsporangia arranged in opposite pairs (**Figure 17.21**). Each pair of microsporangia is separated from the oth-

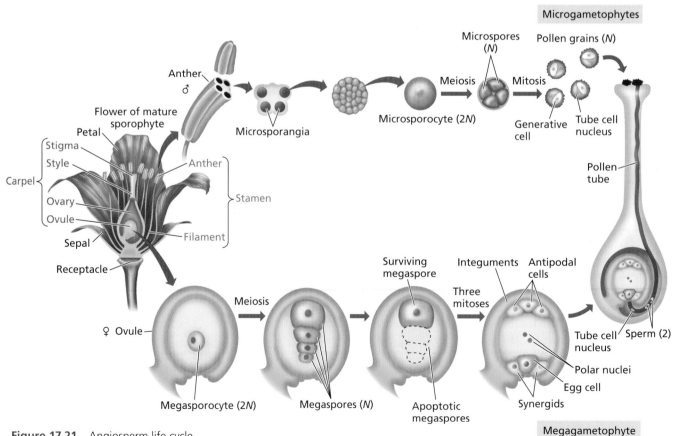

Figure 17.21 Angiosperm life cycle.

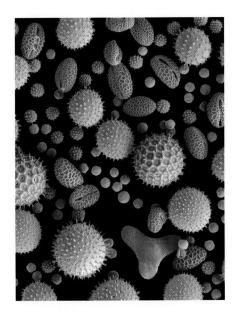

Figure 17.22 Pollen grain structure. Scanning electron micro-scope image of pollen grains from different species exhibiting distinct ornamentation. (© Scientifica/RMF/Visuals Unlimited, Inc.)

er by a central region of sterile tissue surrounding a vascular bundle. Development of the male gametophyte, or pollen grain, begins with microsporogenesis in which cells in the anther divide to form diploid pollen mother cells (microsporocytes) that subsequently undergo meiosis to produce haploid microspores (see Figure 17.21). The microspores further divide and differentiate to form pollen grains with highly sculpted cell walls, which play important ecological roles in transferring the pollen from flower to flower (**Figure 17.22**).

Pollen grains contain two sperm cells containing DNA surrounded by a vegetative cell that later forms the pollen tube (see Figure 17.24), which functions in sperm delivery during fertilization.

Female Gametophyte Development in the Ovule

Ovule primordia arise in a specialized ovary tissue called the *placenta*. The female gametophyte, or *embryo sac*, which develops within the ovule, is more complex and more diverse than the development of the male gametophyte within the anther. According to one classification scheme, there are more than 15 different patterns of embryo sac development in angiosperms. The most common pattern was first described in the genus *Polygonum* ("knotweed") and is therefore called the *Polygonum* type of embryo sac. In angiosperms, *ovules* are located within the *ovary* of the *gynoecium*, the collective term for the carpels (see Figure 17.21). Upon fertilization of the female gamete, or egg, by a sperm cell, embryogenesis is initiated and the ovule develops into a seed. Simultaneously, the ovary enlarges and becomes a fruit. We will discuss fertilization and the development of fruits later in this chapter.

Functional megaspores undergo a series of free nuclear mitotic divisions followed by cellularization

Female gametogenesis begins with the formation of the megaspore mother cell (megasporocyte), which, in *Polygonum*-type embryo sacs, undergoes meiosis to form four haploid megaspores (see Figure 17.21). Three of the megaspores subsequently undergo programmed cell death, leaving only one functional megaspore. Functional megaspores then undergo three rounds of free nuclear mitotic divisions (mitoses without cytokinesis) to produce a *syncytium*—a multinucleate cell formed by nuclear divisions. The result is an eight-nucleate, immature embryo sac. Four of the nuclei then migrate to one pole, and the other four migrate to the opposite pole. Three of the nuclei at each pole undergo cellularization, while the remaining two nuclei, called polar nuclei, migrate toward the central region of the embryo sac, which also contains a large vacuole. The cytoplasm and the two polar nuclei develop their own plasma membrane and cell wall, giving rise to a large binucleate cell. The fully cellularized embryo sac represents the mature female gametophyte or embryo sac. At maturity, the *Polygonum*-type embryo sac consists of seven cells and eight nuclei.

egg cell   The female gamete.

synergid cells   Two cells adjacent to the egg cell of the embryo sac, one of which is penetrated by the pollen tube upon entry into the ovule.

micropyle   The small opening at the distal end of the ovule, through which the pollen tube passes prior to fertilization.

central cell   The cell in the embryo sac that fuses with the second sperm cell, giving rise to the primary endosperm cell.

cross-pollination   Pollination of a flower by pollen from the flower of a different plant.

outcrossing   The mating of two plants with different genotypes by cross-pollination.

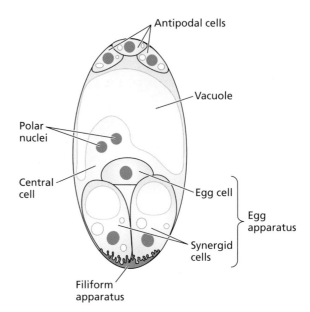

Figure 17.23   Diagram of the egg apparatus and filiform apparatus of the *Polygonum*-type embryo sac.

The **egg cell** (the female gamete that combines with a sperm cell to form the zygote) and the two **synergid cells** are located at one end of the embro sac adjacent to the **micropyle**, a small pore in the ovule that allows pollen tube entry during pollination (**Figure 17.23**; see also Figure 17.25). The synergid cells are involved in the final stages of pollen tube attraction, the discharge of pollen tube contents into the embryo sac, and gamete fusion.

The large binucleate cell in the middle of the embryo sac, composed of two **polar nuclei**, is called the **central cell**. Although its developmental fate is quite different from that of the egg, the central cell is also regarded as a gamete because it fuses with one of the sperm cells during a process that is unique to angiosperms called *double fertilization*.

Pollination and Double Fertilization in Flowering Plants

Pollination in angiosperms is the process of transferring pollen grains from the anther of the stamen, the male organ of the flower, to the stigma of the pistil, the female organ of the flower. In some species, such as Arabidopsis and rice, reproduction typically occurs via self-pollination, or selfing—that is, the pollen and the stigma belong to the same individual sporophyte. In other species, **cross-pollination**, or **outcrossing**, is the norm—the male parent and the female parent are separate sporophytic individuals. Many species can reproduce by either self- or cross-pollination; other species have various mechanisms for promoting cross-pollination, and may even be incapable of reproducing by self-pollination.

In the case of cross-pollination, pollen might travel great distances before landing on a suitable stigma. Produced in excess, pollen grains are dispersed by wind, insects, birds, and mammals, which carry the nonmotile male gametes of angiosperms much farther than the motile sperm of lower plants could ever swim.

Successful pollination depends on several factors, including ambient temperature, timing, and the receptivity of the stigma of a compatible flower. Many pollen grains can tolerate desiccation and high temperatures during their journey to the stigma. However, some pollen grains, such as those of tomato, are damaged by heat. Understanding how some pollen grains tolerate periods of high temperature will help ensure our food supply as the global climate changes.

Two sperm cells are delivered to the female gametophyte by the pollen tube

The female gametes are protected from the environment by the ovary tissues. Consequently, to reach an unfertilized egg, sperm cells must be delivered by a pollen tube that grows from the stigma to the ovule. Successful delivery of the two sperm cells to the two female gametes (egg and central cell) depends on extensive interactions and communication between the pollen tube, the pistil, and the female gametophyte.

Pollination begins with adhesion and hydration of a pollen grain on a compatible flower

Angiosperm reproduction is highly selective. Female tissues are able to discriminate among diverse pollen grains, accepting those from the appropriate species and rejecting others from unrelated species. When pollen lands on a compatible stigma, the grains physically adhere to the stigma papillar cells, probably due to biophysical and chemical interactions between pollen proteins and lipids and stigma surface proteins. Pollen grains adhere poorly to stigmas of plants of other families.

The initial step in pollen tube development is the hydration of the pollen grain. During hydration, the pollen grain becomes physiologically activated. Calcium ion influx into the vegetative cell triggers reorganization of the cytoskeleton and causes the cell to become physiologically and ultrastructurally polarized. The source of the Ca^{2+} may be either the cytoplasm or the cell wall of the papillar cell.

Pollen tubes grow by tip growth

Following germination on the stigma surface, the pollen tube begins to grow downward through the style by *tip growth* (**Figure 17.24**; see also Figure 15.19). The pollen tube growth rate of angiosperms ranges from approximately 10 μm per h to more than 20,000 μm (2 mm) per h, about 100 times faster than the growth rate of gymnosperm pollen tubes. Some pollen tubes can reach up to 40 cm in length, as in the case of maize (corn; *Zea mays*) pollen tubes growing through the threadlike styles ("silk") to reach the ovaries.

Growing pollen tubes restrict the cytoplasm, the two sperm cells, and the vegetative nucleus to the growing apical region by forming large vacuoles and callose partitions that seal off the rear portion of the tube (see Figure 17.24). At the apical end of the pollen tube is a region packed with small secretory vesicles delivering wall materials and new membranes to the growing tip.

For successful fertilization to occur, the pollen tube must find its way to the micropyle of an ovule. In fact, there is often competition among pollen tubes to arrive at the micropyle first and thus be the one to fertilize the egg. The surrounding maternal tissues may even influence the outcome of this "race"—a type of female mate selection.

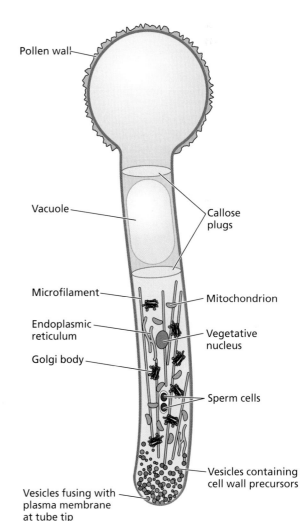

Figure 17.24 Pollen tubes elongate by tip growth. The cytoplasm is concentrated in the growing region of the tube by large vacuoles and callose partitions. (After Jones et al. 2013.)

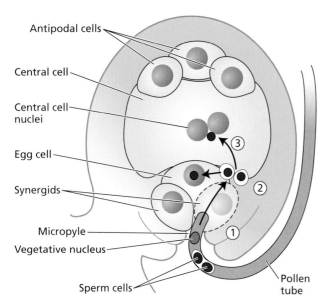

Antipodal cells

Central cell

Central cell nuclei

Egg cell

Synergids

Micropyle

Vegetative nucleus

Sperm cells

Pollen tube

1. The pollen tube bursts and discharges. Sperm cells are delivered rapidly from the pollen tube into the female gametophyte. The receptive synergid cell is likely to break down just after the start of pollen tube discharge.

2. Two sperm cells remain at the boundary region between the egg cell and the central cell for several minutes.

3. One sperm cell fuses with the egg cell and the other fuses with the central cell, and their nuclei move toward the target gamete nuclei.

Figure 17.25 Sperm cell behavior during double fertilization in Arabidopsis can be divided into three stages.

primary endosperm cell Following fusion of the two polar nuclei with the sperm nucleus to produce a triploid nucleus, the central cell becomes the primary endosperm cell, which divides mitotically to give rise to the endosperm.

fruit In angiosperms, one or more mature ovaries containing seeds and sometimes adjacent attached parts.

dehiscence The spontaneous opening of a mature anther or fruit, releasing its contents.

indehiscent Lacking spontaneous opening of a mature anther or fruit.

berry A simple, fleshy fruit produced from a single ovary and consisting of an outer pigmented exocarp, a fleshy, juicy mesocarp, and a membranous inner endocarp.

drupe A structure similar to a berry, but with a hardened, shell-like endocarp (pit or stone) that contains a seed.

pome A fruit, such as an apple, composed of one or more carpels surrounded by accessory tissue derived from the receptacle.

Double fertilization results in the formation of the zygote and the primary endosperm cell

When the pollen tube senses chemical attractants secreted by the synergids, the tube grows through the micropyle, penetrates the embryo sac, and enters one of the synergid cells. Once inside the synergid, the pollen tube stops growing and the tip bursts, releasing the two sperm cells. During double fertilization in the *Polygonum*-type embryo sac, one sperm cell fuses with the egg to produce the zygote, and the other fuses with the binucleate central cell (which includes the two polar nuclei) to produce the triploid **primary endosperm cell**, which divides mitotically to give rise to the nutritive endosperm of the seed (**Figure 17.25**). Because different types of embryo sacs contain different numbers of polar nuclei, the ploidy level of the endosperm ranges from $2N$ in *Oenothera* to $15N$ in *Peperomia*. After double fertilization takes place the process of embryogenesis begins, as described in Chapter 14.

Fruit Development and Ripening

True fruits are found only in the flowering plants. In fact, fruits are a defining feature of the angiosperms, since *angio* means "vessel" or "container" in Greek and *sperm* means "seed." Diverse fruit types are represented in early Cretaceous fossils, including nuts and fleshy drupes and berries. Fruits are typically derived from a mature ovary containing seeds, but they can also include a variety of other tissues. For instance, the fleshy part of the strawberry is actually the receptacle, while the true fruits are the dry achenes embedded in this tissue.

Fruits are seed-dispersal units and they can be grouped according to several features. Based on their composition and moisture content, they can be either dry or fleshy. If the fruit splits to release its seeds, it is termed **dehiscent**. The fleshy fruits that we are most familiar with are **indehiscent** and occur in a variety of forms. Tomatoes, bananas, and grapes are defined botanically as **berries**, in which the seeds are embedded in a fleshy mass, while peaches, plums, apricots, and almonds are classified as **drupes**, in which the seed is enclosed in a stony endocarp (the "pit"). Apples and pears are **pome** fruits, in which the edible tissue is derived from accessory structures such as floral parts or the receptacle. Fruits can also be defined as *simple*, with a mature single or compound ovary, as in hazelnut, Arabidopsis, and tomato. Alternatively, they may be *aggregate*, where flowers have multiple carpels that are not joined together, as in strawberry and raspberry. Finally, they may be *multiple*, where the fruit is formed from a cluster of flowers and each flower produces a fruit segment, as in pineapple. Some examples of fleshy and dry fruit types are illustrated in **Figure 17.26**.

The developmental switch that turns a pistil into a growing fruit depends on fertilization of the ovules. In most angiosperms, the gynoecium senesces and dies if unfertilized.

Arabidopsis and tomato are model systems for the study of fruit development

Members of the Brassicaceae, including Arabidopsis, have been widely used as model systems for the study of dry, dehiscent fruits. The gynoecium in Arabidopsis arises from the fusion of two carpels, referred to collectively as the pistil,

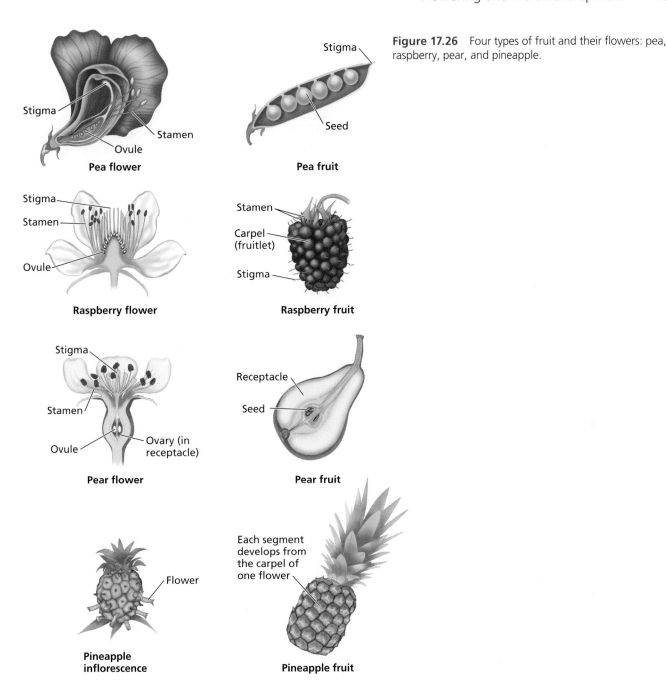

Figure 17.26 Four types of fruit and their flowers: pea, raspberry, pear, and pineapple.

Pea flower

Stigma
Stamen
Ovule

Pea fruit

Stigma
Seed

Raspberry flower

Stigma
Stamen
Ovule

Raspberry fruit

Stamen
Carpel (fruitlet)
Stigma

Pear flower

Stigma
Stamen
Ovule
Ovary (in receptacle)

Pear fruit

Receptacle
Seed

Pineapple inflorescence

Flower

Pineapple fruit

Each segment develops from the carpel of one flower

and forms in the center of the flower. In Arabidopsis and many other members of Brassicaceae, several fruit tissues develop, including the carpel walls or pericarp (known also as the valves) and other internal structures (**Figure 17.27**). The valve margins differentiate into zones that will be involved in the dehiscence (splitting) of the mature fruit. Dry fruits have relatively few cell layers in the carpel walls (see Figure 17.27D), and some of these may be lignified, especially in areas associated with fruit dehiscence.

Much of what we know about the development of fleshy, indehiscent fruits has come from work on tomato (*Solanum lycopersicum*), a member of the nightshade family (Solanaceae) (**Figure 17.28A**). In tomato, as in Arabidopsis, the fruit is derived from the fusion of carpels. The carpel walls are called the pericarp (equivalent to the valves in Arabidopsis), and the seeds are attached to the placenta. Unlike Arabidopsis fruits, tomato fruits are indehiscent and the carpels remain completely

(A)

Stigma

Style

Valve

Valve margin

Replum

(B)

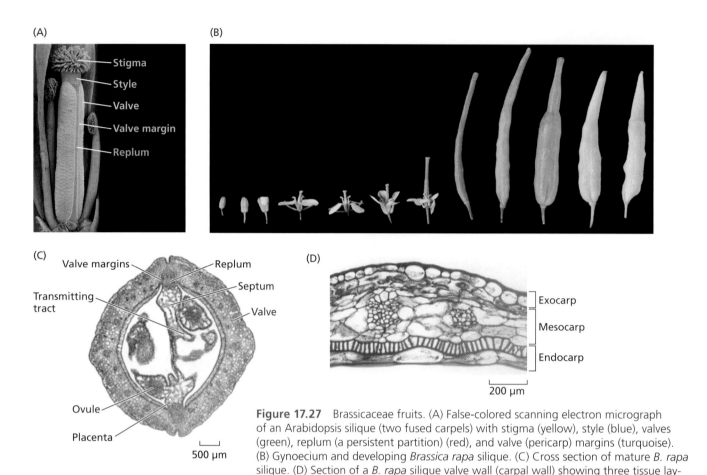

(C)

Valve margins — — Replum

— Septum

Transmitting tract

— Valve

Ovule

Placenta

500 µm

(D)

Exocarp

Mesocarp

Endocarp

200 µm

Figure 17.27 Brassicaceae fruits. (A) False-colored scanning electron micrograph of an Arabidopsis silique (two fused carpels) with stigma (yellow), style (blue), valves (green), replum (a persistent partition) (red), and valve (pericarp) margins (turquoise). (B) Gynoecium and developing *Brassica rapa* silique. (C) Cross section of mature *B. rapa* silique. (D) Section of a *B. rapa* silique valve wall (carpal wall) showing three tissue layers. (A, C, and D from Seymour et al. 2013; B courtesy of Lars Østergaard.)

fused. In fleshy fruits, cell division is usually followed by massive cell expansion (**Figure 17.28B**). In some varieties of tomato, for example, pericarp cell diameters may reach 0.5 mm. About 30 genetic loci have been shown to control fruit size in tomato.

Fleshy fruits undergo ripening

Ripening in fleshy fruits refers to the changes that make them attractive (to humans and other animals) and ready to eat. Such changes typically include color development (see Figure 17.28A), softening, starch hydrolysis, sugar accumulation, production of aroma compounds, and the disappearance of organic acids and phenolic compounds, including tannins. Dry fruits do not undergo a true ripening process, but many of the same families of genes that control dehiscence in dry fruits seem to have been recruited to new functions in ripening in fleshy fruits. Because of the importance of fruits in agriculture and their health benefits, the vast majority of studies on fruit ripening have focused on edible fruits. Tomato is the established model for studying fruit ripening, as it has proved highly amenable to biochemical, molecular, and genetic studies on the mechanism of ripening.

Ripening involves changes in the color of fruit

Fruits ripen from green to a range of colors, including red, orange, yellow, purple, and blue. The pigments involved not only affect the visual appeal of the fruit, but also taste and aroma, and are known to have health benefits for humans. Fruits typically contain a mixture of pigments, including the green chlorophylls; yellow, orange, and red carotenoids; red, blue, and violet anthocyanins; and yellow flavo-

ripening The process that causes fruits to become more palatable, including softening, increasing sweetness, loss of acidity, and changes in coloration.

(A)

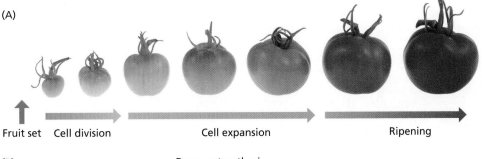

Fruit set   Cell division   Cell expansion   Ripening

(B)   Days post-anthesis

2   4   8   24

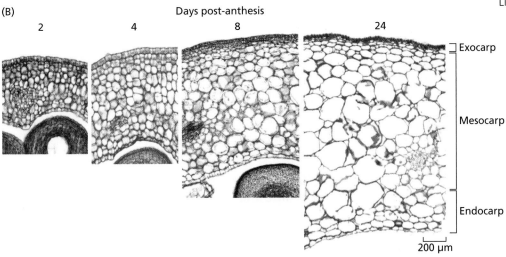

Exocarp

Mesocarp

Endocarp

200 μm

Figure 17.28 Tomato fruit growth. (A) Photographs of developmental stages in a miniature tomato. (B) Light microscope images of a cross section of tomato pericarp at 2, 4, 8, and 24 days after flower opening (anthesis). (A © bro-zova/istock; B from Seymour et al. 2013, after Pabón-Mora and Litt 2011.)

noids. The loss of the green pigment at the onset of ripening is caused by the degradation of chlorophyll and the conversion of chloroplasts to chromoplasts, which act as the site for the accumulation of carotenoids (see Chapter 1, Figure 1.21).

Carotenoids are responsible for the red color of tomato fruits. During ripening in tomato, the concentration of carotenoids increases between 10- and 14-fold, mainly due to the accumulation of the deep red pigment lycopene. Fruit ripening involves the active biosynthesis of carotenoids, the chemical precursors for which are synthesized in the plastids. The first committed step is the formation of the colorless molecule phytoene by the enzyme phytoene synthase. In tomato, phytoene is then converted to the red pigment lycopene through a series of further reactions. Experiments with transgenic tomatoes have demonstrated that silencing the gene for phytoene synthase using molecular methods prevents the formation of lycopene (**Figure 17.29**).

Anthocyanins are the pigments responsible for the blue and purple color in some berries (**Figure 17.30**). Anthocyanins are made through the phenylpropanoid pathway; that is, they are derived from the amino acid phenylalanine. Phenylpropanoids constitute some of the most important sets of secondary metabolites in plants. They contribute not only to the characteristic color and flavor of fruits, but also to unfavorable traits, such as browning of fruit tissues via enzymatic oxidation of phenolic compounds by polyphenol oxidases.

Fruit softening involves the coordinated action of many cell wall–degrading enzymes

Softening of fruit involves changes to the fruit's cell walls. In most fleshy fruits, the cell walls consist of a semirigid composite of cellulose microfibrils—thought to be tethered by a xyloglucan network—which

Figure 17.29 Phytoene synthase plays a role in lycopene production in tomato pericarp. The tomato on the left is a wild-type, red ripe fruit. The tomato on the right has reduced levels of phytoene synthase gene expression and therefore fails to accumulate the red pigment lycopene. (From Fray and Grierson 1993, courtesy of R. G. Fray.)

Figure 17.30 Blueberries accumulate more than a dozen different anthocyanins during ripening, including malvidin-, delphinidin-, petunidin-, cyanidin-, and peonidin-glycosides, which give them a deep purple color. (Photo by David McIntyre.)

is embedded in a gel-like pectin matrix. In tomato, more than 50 genes related to cell wall structure show changes in expression during ripening, indicating a highly complex set of events connected with cell wall remodeling during the ripening process.

Experiments in transgenic plants have demonstrated that no single cell wall–degrading enzyme can account for all aspects of softening in tomato or other fruits. It appears that texture changes result from the synergistic action of a range of cell wall–degrading enzymes and that suites of texture-related genes give different fruits their unique melting, crisp, or mealy textures. However, even in tomato, the precise contribution of each type of enzyme to fruit texture is still poorly understood. Changes in the cuticle of the fruit that affect water loss also affect texture and shelf life.

Taste and flavor reflect changes in acids, sugars, and aroma compounds

Fruits have evolved to act as vehicles for the dispersal of seeds, and most fleshy fruits that are consumed by humans undergo changes that make them especially palatable to eat when they are ripe. These chemical changes include alterations in sugars and acids and the release of aroma compounds. In many fruits, starch is converted at the onset of ripening to glucose and fructose, and citric and malic acids are also abundant. However, although sugars and acids are vital for taste, volatile chemicals are what really determine the unique flavor of fruits such as tomato.

Flavor volatiles arise from a wide range of compounds. Some of the most detailed studies have been undertaken in tomato. They show that of the 400 or so volatiles produced by tomato, only a small number have a positive effect on flavor. The most important flavor volatiles in tomato are derived from the catabolism of fatty acids such as linoleic acid (hexanal) and linolenic acid (*cis*-3-hexenal, *cis*-3-hexenol, *trans*-2-hexenal) via lipoxygenase activity.

Volatile production is intimately connected with the ripening process, but the regulation of these events is not well understood. It is probably controlled by some of the diverse transcription factors that show altered expression during ripening.

The causal link between ethylene and ripening was demonstrated in transgenic and mutant tomatoes

Ethylene has long been recognized as the hormone that can accelerate the ripening of many edible fruits. It is possible to silence the gene encoding ACC synthase, which catalyzes the rate-limiting step in ethylene biosynthesis (see Figure 12.14D), using antisense RNA. The transgenic fruits fail to ripen normally, and exogenous ethylene restores normal ripening (**Figure 17.31**).

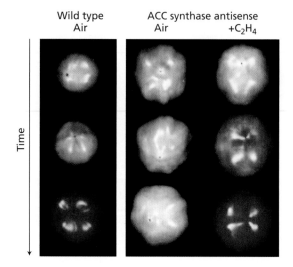

Wild type
Air

ACC synthase antisense
Air   +C$_2$H$_4$

Time

Figure 17.31 Molecular silencing of ACC synthase. Fruit expressing an antisense gene for ACC synthase, which inhibits expression of the wild-type gene, together with controls (wild type). Note that in air the ACC synthase-deficient fruit did not ripen but did senesce after 70 days (yellow fruit at bottom); ripening could be restored by adding external ethylene (C$_2$H$_4$). (From Oeller et al. 1991; reprinted in Grierson 2013.)

Further demonstration of the requirement for ethylene in fruit ripening came from analysis of the *Never-ripe* mutation in tomato. As the name implies, this mutation completely blocks the ripening of tomato fruit. Molecular analysis has revealed that the *Never-ripe* phenotype is caused by a mutation in an ethylene receptor that renders the receptor unable to bind ethylene. These results, together with the demonstration that inhibiting ethylene biosynthesis blocks ripening, provided unequivocal proof of the role of ethylene in fruit ripening.

Climacteric and non-climacteric fruits differ in their ethylene responses

Fleshy fruits have traditionally been placed in two groups as defined by the presence or absence of a characteristic respiratory rise called a **climacteric** at the onset of ripening. Climacteric fruits show this respiratory increase, and also a spike of ethylene production immediately before, or coincident with, the respiratory rise (**Figure 17.32**). Apple, banana, avocado, and tomato are examples of climacteric fruits. In contrast, fruits such as citrus fruits and grape do not exhibit such large changes in respiration and ethylene production and are termed **non-climacteric** fruits.

In plants with climacteric fruits, two systems of ethylene production operate, depending on the stage of development:

- In System 1, which acts in immature climacteric fruit, ethylene inhibits its own biosynthesis by negative feedback.

- In System 2, which occurs in mature climacteric fruit and in senescing petals in some species, ethylene stimulates its own biosynthesis—that is, it is autocatalytic.

The positive feedback loop for ethylene biosynthesis in System 2 ensures that the entire fruit ripens evenly once ripening has commenced.

When mature climacteric fruits are treated with ethylene, the onset of the climacteric and the biochemical changes associated with ripening are accelerated. In contrast, when immature climacteric fruits are treated with ethylene, the respiration rate increases gradually as a function of the ethylene concentration, but the treatment does not trigger production of endogenous ethylene or induce ripening. Ethylene treatment of non-climacteric fruits, such as citrus, strawberry, and grape, does not cause an increase in respiration and does not accelerate ripening. However, some non-climacteric fruits do respond to ethylene; for example, exogenous ethylene causes citrus fruits to de-green, although the effect is limited to the outer cell layers of the peel and is therefore not considered true ripening.

climacteric   Marked rise in respiration at the onset of ripening that occurs in all fruits that ripen in response to ethylene, and in the senescence process of detached leaves and flowers.

non-climacteric   Referring to a type of fruit that does not undergo a climacteric, or respiratory burst, during ripening.

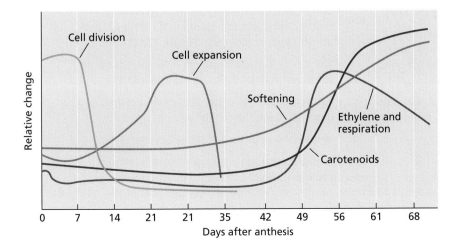

Figure 17.32 Growth and development of tomato fruits in relation to the effects of ethylene and ripening. Climacteric fruits show a characteristic increase in respiration and ethylene production which signals the onset of ripening. (After Giovanni 2004.)

Summary

Formation of the floral organs (sepals, petals, stamens, and carpels) occurs at the SAM and is linked to both internal (autonomous) and external (environmental) signals. A network of genes that control floral morphogenesis has been identified in several species. Fruit development has evolved differentially, but involves expression of many common genes as well.

Floral Evocation: Integrating Environmental Cues

- Internal (autonomous) and external (environment-sensing) control systems enable plants to precisely regulate and time flowering for reproductive success.

- Two of the most important seasonal responses that affect floral development are photoperiodism (response to changes in day length) and vernalization (response to prolonged cold).

- Synchronized flowering favors crossbreeding and helps ensure seed production under favorable conditions.

The Shoot Apex and Phase Changes

- Postembryonic development in plants can be divided into three phases with different physiological and morphological features: juvenile, adult vegetative, and reproductive (**Figure 17.1**).

- The transition from one phase to another is called phase change.

- Juvenile tissues are produced first and are located at the base of the shoot.

- Gibberellins play a role in regulating phase changes in some species.

Photoperiodism: Monitoring Day Length

- Plants can detect seasonal changes in day length that occur at latitudes away from the equator (**Figure 17.2**).

- Flowering in LDPs requires that the day length exceed a certain duration, called the critical day length. Flowering in SDPs requires a day length that is less than the critical day length (**Figure 17.4**).

- Circadian rhythms are based on an endogenous oscillator, involving the interactions of Dawn, Morning, Afternoon, and Evening genes (**Figure 17.5**).

- Circadian rhythms are defined by three parameters: period, phase, and amplitude (**Figure 17.6**).

- Temperature compensation prevents temperature changes from affecting the period of the circadian clock.

- Phytochromes and cryptochromes entrain the circadian clock

- Leaves are the sites of perception of the photoperiodic stimulus in both LDPs and SDPs.

- Leaf grafting experiments have shown that a photoperiodically induced leaf can induce an uninduced plant to flower (**Figure 17.7**).

- Plants monitor day length by measuring the length of the night; flowering in both SDPs and LDPs is determined primarily by the duration of darkness (**Figure 17.8**).

- For both LDPs and SDPs, the dark period can be made ineffective by interruption with a short exposure to light (a night break) (**Figure 17.9**).

- The flowering response to night breaks shows a circadian rhythm, supporting the clock hypothesis (**Figure 17.10**).

- In the coincidence model, flowering is induced in both SDPs and LDPs when light exposure is coincident with the appropriate phase of the oscillator.

- A transcriptional activator serves as a master switch controlling the expression of floral stimulus genes (**Figure 17.11**).

- Flowering in LDPs is promoted when the inductive light treatment coincides with a peak in light sensitivity, which follows a circadian rhythm (**Figure 17.11**).

- Flowering in SDPs is inhibited when the inductive light treatment coincides with a peak in light sensitivity, which follows a circadian rhythm (**Figure 17.11**).

- The effects of red and far-red night breaks implicate phytochrome in the control of flowering in SDPs and LDPs (**Figures 17.12, 17.13**).

Vernalization: Promoting Flowering with Cold

- In vernalization-requiring plants, a cold treatment is required for plants to respond to floral signals such as inductive photoperiods (**Figure 17.14**).

- The longer the cold treatment, the more stable the vernalized state of the plant is (**Figure 17.15**).

- For vernalization to occur, active metabolism is required during the cold treatment.

- Vernalization takes place primarily in the SAM.

- High temperatures can cause devernalization in vernalized plants.

(Continued)

Summary (*continued*)

Long-distance Signaling Involved in Flowering

- In photoperiodic plants, a long-range signal is transmitted via the phloem from the leaves to the apex, permitting floral evocation.

- FT is a small globular protein that exhibits the properties that would be expected of florigen.

- FT moves via the phloem from the leaves to the SAM under inductive photoperiods. In the meristem, FT forms a complex with the transcription factor FD to activate floral identity genes (**Figure 17.16**).

Floral Meristems and Floral Organ Development

- Floral meristems may develop directly from vegetative meristems or indirectly from inflorescence meristems (**Figure 17.17**).

- The four different types of floral organs are initiated sequentially in separate, concentric whorls (**Figure 17.18**).

- Formation of floral meristems requires active floral meristem identity genes, while development of the floral organs requires active floral organ identity genes.

- Mutations in floral organ identity genes alter the types of floral organs produced in each of the whorls (**Figure 17.19**).

- The ABC model states that organ identity in each whorl is determined by the combined activity of three floral organ identity genes (**Figure 17.19**).

- According to the ABCE model, a fourth gene is required for the three floral organ identity genes to be active (**Figure 17.20**).

Pollen Development

- Plants undergo both a diploid and a haploid generation in order to make gametes and reproduce (**Figure 17.21**).

- The diverse forms of pollen grains reflect adaptations that facilitate pollination (**Figure 17.22**).

Female Gametophyte Development in the Ovule

- Eggs are formed in the female gametophyte (embryo sac) first by megaspore formation and then by the development of the female gametophyte (**Figure 17.23**).

- Most angiosperms exhibit *Polygonum*-type embryo sac development, in which meiosis of a diploid megaspore mother cell produces four haploid megaspores, only one of which goes on to form the female gametophyte (embryo sac).

- Embryo sac development begins with three mitotic divisions without cytokinesis, followed by cellularization.

Pollination and Fertilization in Flowering Plants

- Pollen tubes grow by tip growth (**Figure 17.24**).

- Once the pollen tube has reached the ovule, two sperm cells are released into the embryo sac (**Figure 17.25**).

- During double fertilization, one sperm cell fertilizes the egg and the other fuses with the binucleate central cell, forming the zygote and primary endosperm cell, respectively.

- The primary endosperm cell divides mitotically to form triploid endosperm tissue, the main nutritive tissue of angiosperm seeds.

Fruit Development and Ripening

- Fruits are seed-dispersal units that arise from the pistil and contain the seed(s) (**Figures 17.26, 17.27**).

- Fleshy fruits undergo ripening, which involves color changes, highly coordinated fruit softening, and other changes (**Figures 17.28–17.30**).

- Acids, sugars, and volatiles determine the flavors of ripe and unripe fleshy fruits.

- Ethylene accelerates ripening, particularly in climacteric fruits (**Figures 17.31, 17.32**).

Suggested Reading

Amasino, R. (2010) Seasonal and developmental timing of flowering. *Plant J.* 61: 1001–1013. DOI: 10.1111/j.1365-313X.2010.04148.x

Burg, S. P., and Burg, E. A. (1965) Ethylene action and ripening of fruits. *Science* 148: 1190–1196.

Dresselhaus, T., and Franklin-Tong, N. (2013) Male-female crosstalk during pollen germination, tube growth and guidance, and double fertilization. *Mol. Plant* 6: 1018–1036.

Huijser, P., and Schmid, M. (2011) The control of developmental phase transitions in plants. *Development* 138: 4117–4129. DOI:10.1242/dev.063511

Klee, H. J., and Giovannoni, J. J. (2011) Genetics and control of tomato fruit ripening and quality attributes. *Annu.*

Rev. Genet. 45: 41–59. DOI: 10.1146/annurev-genet-110410-132507

Krizek, B. A., and Fletcher, J. C. (2005) Molecular mechanisms of flower development: An armchair guide. *Nat. Rev. Genet.* 6: 688–698.

Li, J., and Berger, F. (2012) Endosperm: Food for humankind and fodder for scientific discoveries. *New Phytol.* 195: 290–305.

Liu, L., Zhu, Y., Shen, L., and Yu, H. (2013) Emerging insights into florigen transport. *Curr. Opin. Plant Biol.* 16: 607–613.

Seymour, G. B., Østergaard, L., Chapman, N. H., Knapp, S., and Martin, C. (2013) Fruit development and ripening.

Annu. Rev. Plant Biol. 64: 219–241. DOI: 10.1146/annurev-arplant-050312-120057

Song, Y. H., Ito, S., and Imaizumi, T. (2013) Flowering time regulation: Photoperiod- and temperature-sensing in leaves. *Trends Plant Sci.* 18: 575–583.

Twell, D. (2010) Male gametophyte development. In *Plant Developmental Biology—Biotechnological Perspectives*, Vol. 1, E. C. Pua and M. R. Davey, eds., Springer-Verlag, Berlin, pp. 225–244.

Yang, W.-C., Shi, D.-Q., and Chen, Y.-H. (2010) Female gametophyte development in flowering plants. *Annu. Rev. Plant Biol.* 61: 89–108.

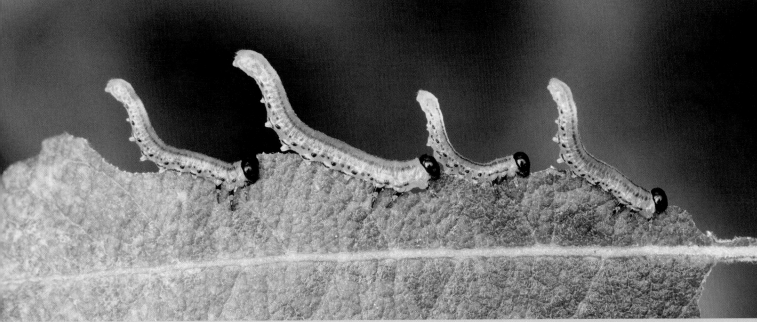

© Wendy Kreeftenberg/Minden Pictures

18 Biotic Interactions

In natural habitats, plants live in complex, diverse environments in which they interact with a wide variety of organisms (**Figure 18.1**). Some interactions are clearly beneficial, if not essential, to both the plant and the other organism. Such mutually beneficial biotic interactions are termed **mutualisms**. Examples of mutualism include plant–pollinator interactions, the symbiotic relationship between nitrogen-fixing bacteria (rhizobia) and leguminous plants, mycorrhizal associations between roots and fungi, and the fungal endophytes of leaves. Other types of biotic interactions, including **herbivory**, infection by **microbial pathogens** or **parasites**, and **allelopathy** (chemical warfare between plants), are detrimental. In response to the latter, plants have evolved intricate defense mechanisms to protect themselves against harmful organisms, and harmful organisms have evolved counter mechanisms to defeat these defenses. Such tit-for-tat evolutionary processes are examples of **coevolution**, which accounts for the complex interactions between plants and other organisms we see today.

However, it would be an oversimplification to characterize all organisms that interact with plants as either beneficial or harmful. For example, grazing of flowers by mammals decreases fitness in some plant species, but in others it can lead to an increase in the number of flowering stalks, thus increasing fitness. There are also organisms that benefit from their interaction with the plant without causing any harmful effects. Such neutral interactions (from the plant's perspective) are termed **commensalism**. Commensal organisms can become beneficial if they protect the plant from a second, harmful organism. For example, nonpathogenic soil rhizobacteria and fungi, which cause no harm to the plant, may nevertheless stimulate the plant's innate immune system (discussed later in the chapter)

mutualism A symbiotic relationship in which both organisms benefit.

herbivory Consumption of plants or parts of plants as a food source.

microbial pathogens Bacterial or fungal organisms that cause disease in a host plant.

parasite An organism that lives on or in an organism of another species, known as the host, from the body of which the parasite obtains nutriment.

allelopathy Release by plants of substances into the environment that have harmful effects on neighboring plants.

coevolution Linked adaptations of two or more organisms.

commensalism A relationship between two organisms in which one organism benefits without negatively affecting the other.

Figure 18.1 Virtually every part of
the plant is adapted to coexist with
the organisms in its immediate environment. (After van Dam 2009.)

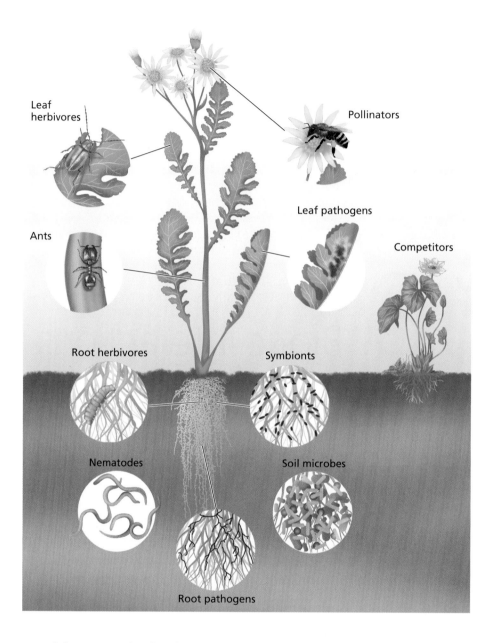

Leaf
herbivores

Pollinators

Ants

Leaf pathogens

Competitors

Root herbivores

Symbionts

Nematodes

Soil microbes

Root pathogens

and thus protect the plant from pathogenic microorganisms. Even plant–pathogen
interactions can be either highly negative or relatively neutral. While obligate
pathogens must cause a disease in order to complete their life cycles, facultative
pathogens are able to complete their life cycles with little or no host interactions.

The first line of defense against potentially harmful organisms is the plant
surface. The cuticle (a waxy outer layer), the periderm, and other mechanical
barriers help block bacterial, fungal, and insect entry. The second line of defense
typically involves biochemical mechanisms that may be either constitutive or
inducible. **Constitutive defenses** are always present, whereas **inducible defenses**
are triggered in response to attack. Unlike constitutive defenses, inducible defenses require specific detection systems and signal transduction pathways that
can sense the presence of an herbivore or pathogen and alter gene expression
and metabolism accordingly.

We begin our discussion of biotic interactions with examples of beneficial
associations between plants and microorganisms. Next we consider various
types of harmful interactions between plants, herbivores, and pathogens and the

constitutive defenses   Plant defenses
that are always immediately available or
operational; that is, defenses that are not
induced.

inducible defenses   Defense responses
that exist at low levels until a biotic or abiotic stress is encountered.

defensive mechanisms used by plants against them. We then survey the wide range of constitutive and inducible defenses that plants deploy to fend off insect herbivores. We also note the important roles that volatile organic compounds play in repelling herbivores, attracting insect predators, and acting as distress signals between different parts of plants and between neighboring plants.

Beneficial Interactions between Plants and Microorganisms

Land plants are colonized by a wide variety of beneficial microorganisms, including endophytic and mycorrhizal fungi (see Chapter 4), bacteria in the form of biofilms on the surfaces of roots and leaves, endophytic bacteria, and nitrogen-fixing bacteria contained in the root or stem nodules (see Chapter 5). It is likely that symbiotic associations between marine algae and fungi predate the emergence of the first bryophyte-like land plants about 475 to 450 million years ago (MYA). The first lichens, which are obligate associations between a fungus and either a cyanobacterium or green alga, appear in the fossil record about 400 MYA—around the time that the first mycorrhizal associations with early vascular land plants made their appearance. This suggests that the invasion of land by green plants may, as in the case of algae, have been aided by symbiotic associations with fungi.

In Chapter 4 we discussed the role of mycorrhizae in supplying phosphorus to plant roots. Both ecto- and arbuscular mycorrhizae involve complex exchange structures that enhance the capacity of plant roots to extract phosphorus from the soil (see Figures 4.13 and 4.14). These mycorrhizal structures require extensive signaling between the fungi and plant to ensure that intimate cell-to-cell exchange surfaces are elaborated without provoking defense responses. Recent advances in microscopy and transcriptomic sequencing have provided insights suggesting a great deal of similarity between the processes of mycorrhizal associations and rhizobial nodulation.

In legumes, the same core symbiotic genes are required both for legume nodulation and for arbuscular mycorrhiza formation, suggesting that the interaction between legumes and nitrogen-fixing bacteria (rhizobia) evolved from the more ancient interaction between plants and mycorrhizal fungi. The key symbiotic signals produced by rhizobia are lipochitin oligosaccharides called Nod factors. Similarly, the arbuscular mycorrhizal fungus *Rhizophagus intraradices* releases lipochitin oligosaccharides called Myc factors that stimulate the formation of mycorrhizas in a wide variety of plants.

Nitrogen-fixing symbioses and arbuscular mycorrhizae may also involve related protein receptors. *Parasponia andersonii* is a nonlegume tropical tree that forms nitrogen-fixing symbioses with rhizobia. Even though this plant is only distantly related to the legumes, it has a Nod factor receptor that is required for both mycorrhiza formation *and* rhizobium-induced root nodulation. The Nod factor receptors are also closely related to two receptors identified in the non-nodulating angiosperms Arabidopsis and rice (*Oryza sativa*). These receptors are required for the defense-related perception of chitin oligomers, compounds that are structurally related to Nod factors and Myc factors. This suggests that during evolution, a plant receptor involved in defense signaling has been recruited to activate genes involved in symbiotic associations.

Other types of rhizobacteria can increase nutrient availability, stimulate root branching, and protect against pathogens

Plant roots provide a nutrient-rich habitat for the proliferation of soil bacteria that thrive on root exudates and lysates, which can represent as much as 40% of the total carbon fixed during photosynthesis. Population densities of bacteria in the rhizosphere can be up to 100-fold higher than in bulk soil, and up to

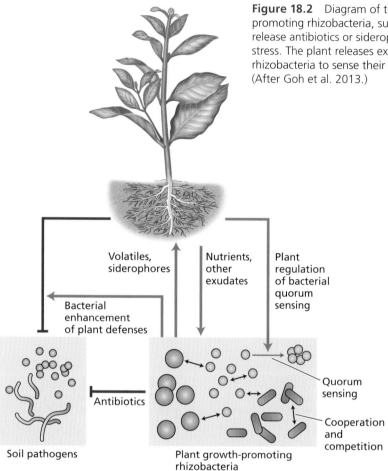

Figure 18.2 Diagram of the interactions between plants and plant growth-promoting rhizobacteria, such as *Pseudomonas aeruginosa*, which is thought to release antibiotics or siderophores into the soil that alleviate plant abiotic or biotic stress. The plant releases exudates from roots to alter the mechanisms used by rhizobacteria to sense their own optimal population density (quorum sensing). (After Goh et al. 2013.)

Volatiles, siderophores

Nutrients, other exudates

Plant regulation of bacterial quorum sensing

Bacterial enhancement of plant defenses

Antibiotics

Quorum sensing

Cooperation and competition

Soil pathogens

Plant growth-promoting rhizobacteria

15% of the root surface may be covered by microcolonies of a variety of bacterial strains. While these bacteria use the nutrients that are released from the host plant, they also secrete metabolites into the rhizosphere.

A loosely defined group of plant growth-promoting rhizobacteria provides several beneficial services to growing plants (**Figure 18.2**). For example, volatiles produced by the bacterium *Bacillus subtilis* enhance proton release by roots in iron-deficient growth media, thereby facilitating increased iron uptake. The resulting increase in iron content of the plants treated with *B. subtilis* volatiles is correlated with higher chlorophyll content, higher photosynthetic efficiency, and increased size. *Bacillus subtilis* volatiles also alter root architecture by changing root length and lateral root density.

Rhizobacteria can also control the buildup of harmful soil organisms, as in the case of the suppression of the pathogenic fungus *Gaeumannomyces graminis* by a *Pseudomonas* species that synthesizes an antifungal compound. Rhizobial microbes can also provide cross-protection against pathogenic organisms by activating systemic resistance mechanisms (discussed later in the chapter). In addition, several studies have suggested that *Pseudomonas aeruginosa* can alleviate the symptoms of biotic and abiotic stress by the release of antibiotics or iron-scavenging siderophores (see Figures 5.16B and 18.2).

Harmful Interactions of Pathogens and Herbivores with Plants

Microorganisms that cause infectious diseases in plants include fungi, oomycetes, bacteria, and viruses. The majority of fungal pathogens belong to the Ascomycetes, which produce their spores inside a saclike *ascus*, and Basidiomycetes, which produce spores outside club-shaped cells called *basidia*. Oomycetes are funguslike organisms that include some of the most destructive plant pathogens in history, including the genus *Phytophthora*, cause of the disastrous potato late blight of the Great Irish Famine (1845–1849). Plant pathogenic bacteria, such as *Xylella fastidiosa* and *Erwinia amylovora*, also cause many serious diseases of trees, fruits, and vegetables.

In addition to microbial pathogens, about half of the nearly 2 million insect species feed on plants. In over 350 million years of plant–insect coevolution, insects have developed diverse feeding behaviors, while plants have evolved mechanisms to defend themselves against insect herbivory, including mechanical barriers, constitutive chemical defenses, and direct and indirect inducible defenses. These defense mechanisms have apparently been quite effective, since most plant species are resistant to most insect species. Indeed, about 90% of insect herbivores are restricted to a single family of plants or a few closely related plant species, while only 10% are generalists. This suggests that the vast majority of plant–herbivore interactions have involved coevolution.

Mechanical barriers provide a first line of defense against insect pests and pathogens

Mechanical barriers, including surface structures, mineral crystals, and thigmonastic (touch-induced) leaf movements, often provide a first line of defense against pests and pathogens for many plant species.

The most common surface structures are thorns, spines, prickles, and trichomes (**Figure 18.3**). **Thorns** are modified branches, as in citrus and acacia; **spines** are modified leaves, as in cacti; and **prickles** are derived mainly from the epidermis and thus can be easily snapped off the stem, as in roses. These structures possess sharp, pointed tips that physically protect plants from larger herbivores, such as mammals, although they are less effective against smaller, insect herbivores, which can easily penetrate such defenses and reach the edible parts of the shoot. **Trichomes**, or hairs, provide a much more effective defense against insect pests, based on their physical and chemical deterrence mechanisms. Trichomes occur in a variety of forms, either as simple hairs or as glandular trichomes. Glandular trichomes store species-specific secondary metabolites (discussed in the next section) such as terpenoids and phenolics in a pocket formed between the cell wall and the cuticle. These pockets burst and release their contents upon contact, and the strong smell and bitter taste of these compounds repel insect herbivores.

thorns   Sharp plant structures that physically deter herbivores and are derived from branches.

spines   Sharp, stiff plant surface structures, thought to be modified leaves, that physically deter herbivores and may assist in water conservation.

prickles   Pointed plant structures that physically deter herbivory and are derived from epidermal cells.

trichomes   Unicellular or multicellular hairlike structures that differentiate from the epidermal cells of shoots and roots. Trichomes may be structural or glandular and function in plant responses to biotic or abiotic environmental factors.

(A)

(B)

(C)

(D)

Figure 18.3   Examples of mechanical barriers developed by plants. (A) Thorns on a lemon tree (*Citrus* sp.) are modified branches, as can be seen by their position in the axil of a leaf. (B) Spines, which are characteristic of cacti (*Opuntia* spp.) in the New World, are modified leaves. (C) Prickles can be found on the stem and petioles of roses (*Rosa* spp.) and are formed by the epidermis. (D) Trichomes on shoot and leaves of tomato (*Solanum lycopersicum*) are also derived from epidermal cells. (Photos © J. Engelberth.)

Figure 18.4 Trichomes of nettles (*Urtica dioica*) have a multicellular base with a single stinging cell protruding. The cell wall of this single cell is reinforced by silicates and breaks easily upon contact, releasing a "cocktail" of secondary metabolites that can cause severe irritation of the skin of animals.

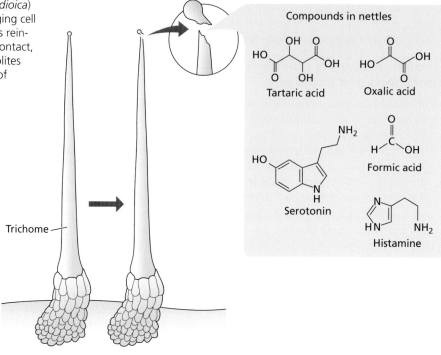

The leaves of the stinging nettle (*Urtica dioica*) possess highly specialized "stinging hairs" that form an effective physical and chemical barrier against larger herbivores. The cell walls of these hollow, needlelike trichomes are reinforced with glass (silicates) and filled with a nasty "cocktail" of histamine, oxalic acid, tartaric acid, formic acid, and serotonin (**Figure 18.4**), which can cause severe irritation and inflammation. Prior to contact, the tip of the stinging hair is covered by a tiny glass bulb that readily snaps off when touched by an herbivore (or an unlucky human who happens to brush against it), leaving an extremely sharp point at the end. The pressure of contact pushes the needlelike trichome down onto the spongy tissue at the base, which acts like the plunger of a syringe to inject the cocktail into the skin.

Besides serving as barriers to insect herbivory, trichomes—when bent or damaged—may also act as herbivore sensors by sending electrical or chemical signals to surrounding cells. Such signals can trigger the formation of inducible defense compounds in the leaf mesophyll.

A different type of mechanical obstacle to herbivory is provided by mineral crystals that are present in many plant species. For example, silica crystals, called **phytoliths**, form in the epidermal cell walls, and sometimes in the vacuoles, of the Poaceae. Phytoliths add toughness to the cell walls and make the leaves of grasses difficult for insect herbivores to chew. The cell walls of the horsetail *Equisetum hyemale* contain so much silica that Native Americans and Mexicans used the stems to scour cooking pots.

Calcium oxalate crystals are present in the vacuoles of many species. Some calcium oxalate crystals form bunches of needlelike structures called **raphides (Figure 18.5)**, which can be harmful to larger herbivores. More than 200 families of plants contain these crystals, including species in the genera

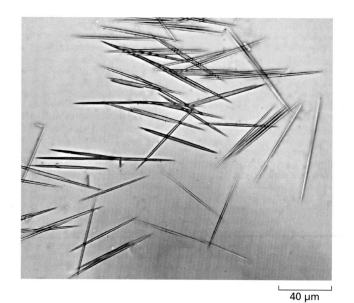

40 μm

Figure 18.5 Calcium oxalate crystals (raphides) from the leaf of an agave (*Agave weberi*). These raphides are tightly packed in specialized cells called idioblasts and are released upon damage. Note the size and pointed ends of these structures. (Courtesy of Agong1/Wikipedia.)

Vitis, Agave, and *Medicago.* Raphides have extremely sharp tips that can penetrate the soft tissue in the throat and esophagus of an herbivore. The raphide-rich *Dieffenbachia,* a popular tropical houseplant, is called "dumb cane" because chewing the leaves reputedly renders the victim speechless due to inflammation. Besides causing mechanical damage, raphides may also allow other toxic compounds produced by the plant to penetrate through the wounds they produce. Even prismatic calcium oxalate crystals have abrasive effects on the mouthparts of insect herbivores, especially the mandibles, thus serving as a mechanical deterrent for insects, mollusks, and other herbivores.

A very different means of avoiding herbivory is employed by the sensitive plant (*Mimosa* spp.). *Mimosa* leaves are composite leaves consisting of many individual leaflets that are connected to the midrib by a jointlike structure called the pulvinus. This pulvinus acts as a turgor-driven hinge, which allows each pair of leaflets to fold in response to various stimuli, including touch, damage, heat, diurnal cycles (in what is called *nyctinasty,* or sleep movements), and in response to water stress. If an insect herbivore attempts to nibble a *Mimosa* leaflet, the damaged leaflet immediately folds, and the response soon spreads to the other undamaged leaflets of the leaf. If the stress signal is strong enough, the entire leaf collapses downward due to the action of another pulvinus located at the base of the petiole. Such rapid movements of leaflets and leaves may deter feeding insects and grazing herbivores by startling them (**Figure 18.6**).

Specialized plant metabolites can deter insect herbivores and pathogen infection

Chemical defense mechanisms comprise a second line of defense against plant pests and pathogens. Plants produce a wide array of chemicals that can be categorized into primary metabolites and secondary metabolites. Primary metabolites are those compounds that all plants produce and that are directly involved in growth and development. This includes sugars, amino acids, fatty acids and lipids, and nucleotides, as well as larger units that are made of them, such as proteins, polysaccharides, membranes, and nucleic acids. In contrast, **secondary metabolites** are often species-specific and are derived from the terpenoid, fatty acid β-oxidation, phenylpropanoid, or amino acid modification pathways that also produce plant hormones and secondary cell wall components.

phytoliths Discrete cells that accumulate silica in leaves or roots.

raphides Needles of calcium oxalate or carbonate that function in plant defense.

secondary metabolites Plant compounds that have no direct role in plant growth and development but that function as defenses against herbivores and infection by microbial pathogens, as attractants for pollinators and seed-dispersing animals, and as agents of plant–plant competition.

(A)

(B)

Figure 18.6 Leaves of the sensitive plant (*Mimosa* spp.) respond rapidly to touch by folding in their individual leaflets within seconds. This rapid movement may deter potential insect herbivores. (A) Untouched (control) leaves. (B) Leaves 5 sec after touch. (Photos © J. Engelberth.)

Additional ring formed from a C₅ unit from the terpene pathway

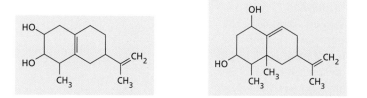

Medicarpin (from alfalfa)   Glyceollin I (from soybean)

Isoflavonoids from the Leguminosae (the pea family)

Rishitin (from potato and tomato)   Capsidiol (from pepper and tobacco)

Sesquiterpene defense compounds from the Solanaceae (the potato family)

Figure 18.7 Structures of some phytoalexins found in two different plant families.

Phytoalexins are a chemically diverse group of secondary metabolites with strong antimicrobial activities that often accumulate around infection sites. Phytoalexin production appears to be a common mechanism of resistance to pathogenic microbes in a wide range of plants. However, different plant families employ different types of phytoalexins. For example, in leguminous plants, such as alfalfa and soybean, isoflavonoids are common phytoalexins, whereas in solanaceous plants, such as potato, tobacco, and tomato, various sesquiterpenes are produced as phytoalexins (**Figure 18.7**). Phytoalexins can be constitutively produced but are often synthesized rapidly after microbial attack. The point of control is usually the expression of genes encoding enzymes for phytoalexin biosynthesis.

Plants store constitutive toxic compounds in specialized structures

Plants can synthesize a broad range of secondary metabolites that have negative effects on the growth and development of other organisms and that can therefore be regarded as toxic. Classic examples of plants that are toxic to humans are hemlock (*Cicuta* spp.) and foxglove (*Digitalis* spp.) (**Figure 18.8**). The metabolites that cause symptoms in humans are well known and demonstrate the potential of secondary metabolites as defensive agents against mammalian herbivores. In some cases these compounds have proven useful for medical research or

(A) *Cicuta* sp.

Cicutoxin

Figure 18.8 Constitutive chemical defenses are effective against many different herbivores, including insects and mammals. Hemlock (*Cicuta* sp.) produces cicutoxin, a diacetylene that prolongs the repolarization of neuronal action potentials. The active principle in foxglove (*Digitalis* sp.) is digitoxin, a cardiac glycoside that inhibits ATPase activity and can increase myocardial contraction. (Photos © J. Engelberth.)

(B) *Digitalis* sp.

Digitoxin

therapies. For example, the diacetylene cicutoxin from hemlock is used in neurological research, as it prolongs the repolarization phase of neuronal action potentials, presumably through blockage of voltage-dependent K^+ channels. The active principle in foxglove, digitoxin, is a **cardenolide**, one of two groups of steroidal cardiac glycosides produced by plants. **Cardiac glycosides** are drugs used to treat congestive heart failure and cardiac arrhythmia. Digitoxin inhibits the Na^+/K^+ ATPase pump in the plasma membranes of heart cells, leading to increased myocardial contraction.

Constitutively produced secondary metabolites that accumulate in cells potentially could have toxic effects on the plant itself. To avoid toxicity, these compounds must be safely stored in leak-proof cellular compartments, and they must also be relatively isolated from susceptible tissues in the event of leakage due to cellular damage. Plants therefore tend to accumulate toxic secondary metabolites in storage organelles such as vacuoles, or in specialized anatomical structures such as *resin ducts, laticifers* (latex-producing cells), or glandular trichomes. Following an attack by herbivores or pathogens, the toxins are released and become active at the site of damage, without adversely affecting vital growing areas. **Conifer resin ducts**, which are found in the cortex and phloem, contain a mixture of diverse terpenoids, including bicyclic monoterpenes such as α-pinene and β-pinene, monocyclic terpenes such as limonene and terpinolene, and tricyclic sesquiterpenes, including longifolene, caryophyllene, and δ-cadinene, as well as resin acids, which are released immediately upon damage by herbivores (**Figure 18.9**). Once released, they can either be toxic to an attacking insect herbivore or act as an adhesive that may glue together the mouthparts of an herbivore. In extreme cases the resin may even engulf the whole insect or pathogen, effectively killing it.

Most of the resin ducts in coniferous plants are considered constitutive defenses, but they can also be induced upon herbivore damage. The formation of these adventitious resin ducts, sometimes referred to as *traumatic resin ducts*,

phytoalexins A chemically diverse group of secondary metabolites with strong antimicrobial activity that are synthesized following infection and that accumulate at the site of infection.

cardenolides Steroidal glycosides that taste bitter and are extremely toxic to higher animals through their action on Na^+/K^+-activated ATPases. Extracted from foxglove (*Digitalis*) for treatment of human heart disorders.

cardiac glycosides Glycosylated organic plant defense compounds–similar to oleandrin from oleander bushes–that are poisonous to animals and inhibit Na^+/K^+ channels to cause contractions in cardiac muscles.

conifer resin ducts Ducts or channels in conifer leaves and woody tissue that conduct terpenoid defense compounds. They may be constitutive, or their formation may be induced by wounding or defense responses.

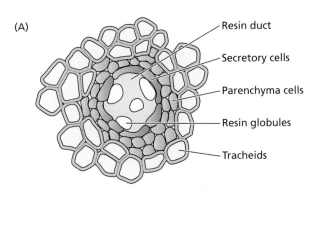

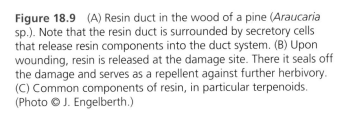

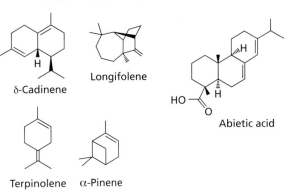

Figure 18.9 (A) Resin duct in the wood of a pine (*Araucaria* sp.). Note that the resin duct is surrounded by secretory cells that release resin components into the duct system. (B) Upon wounding, resin is released at the damage site. There it seals off the damage and serves as a repellent against further herbivory. (C) Common components of resin, in particular terpenoids. (Photo © J. Engelberth.)

laticifers In many plants, an elongated, often interconnected network of separately differentiated cells that contain latex (hence the term laticifer), rubber, and other secondary metabolites.

latex A complex, often milky solution exuded from cut surfaces of some plants that represents the cytoplasm of laticifers and may contain defensive substances.

as well as the biosynthesis of their resin, are regulated by the hormone methyl jasmonate, a derivative of jasmonic acid (discussed later in the chapter).

Laticifers are composed of cells that produce a milky fluid of emulsified components that coagulate upon exposure to air. This liquid is also often referred to as **latex**. Compared with resins, latex is usually much more complex and may also contain proteins and sugars besides toxic or repelling secondary metabolites. Laticifers may consist of a series of fused cells (articulated laticifers) or one long syncytial cell (non-articulated laticifers) (**Figure 18.10**). Most notable among the latex-producing plants is the rubber plant (*Hevea brasiliensis*), which has been grown commercially as a source of natural rubber. Upon wounding, this plant releases huge amounts of latex, which is collected and later converted into rubber. This rubber consists of a polymer of isoprene (*cis*-1,4-polyisoprene) and can have a molecular weight of up to 1 million Da. Under natural conditions, the rubber released by wounded trees defends the plant against herbivores and pathogens, either by repelling or engulfing them.

Another commercially important plant that produces latex is the opium poppy (*Papaver somniferum*). The latex of this plant contains a high concentration of opiates, in particular morphine and codeine. When consumed, these compounds bind to opiate receptors in the nervous systems of herbivores and exert analgesic effects that deter feeding.

The milkweed (*Asclepias curassavica*) and related genera, such as oleander (*Nerium oleander*), produce latex that contains significant amounts of cardenolides, which are present at high concentrations in laticifers. The activity of these poisonous steroids is similar to that of digitoxin (see above) and at high concentrations may lead to heart arrest. Cardenolides also activate nerve centers in the vertebrate brain that induce vomiting. Generalist insect herbivores respond to these compounds either by being repelled or by suffering spasms that lead to death. In contrast, the specialist caterpillars of the monarch butterfly

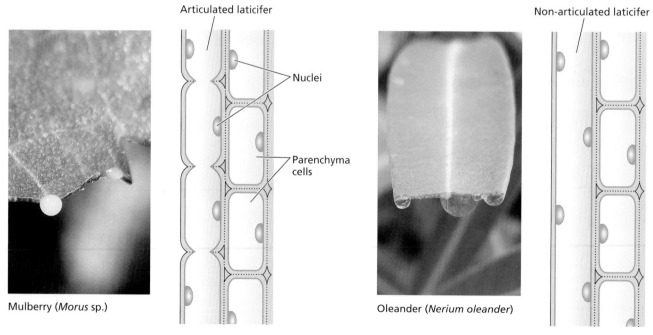

Mulberry (*Morus* sp.)

Oleander (*Nerium oleander*)

Figure 18.10 Laticifers are made of individual cells and can occur either as articulated systems (single cells connected by a small tube) or non-articulated systems (one large syncytial cell). Latex in the laticifers is released upon damage and often contains cardiac glycosides which fend off herbivores. While mulberry (*Morus* sp.) produces a milky latex in its articulated laticifers, oleander (*Nerium oleander*) releases a clear latex from non-articulated laticifers. (Photos © J. Engelberth.)

(A) Monarch butterfly
     (*Danaus plexippus*)

(B) Milkweed bug
     (*Oncopeltus fasciatus*)

(C) Milkweed aphid
     (*Aphis nerii*)

Figure 18.11   While most herbivores are very susceptible to the toxic metabolites in the latex of milkweed and oleander plants, some insect herbivores incorporate these compounds into their bodies and show this to potential predators by presenting bright colors. Shown here are three specialist insect herbivores that feed on these latex-producing plants: the caterpillar of the monarch butterfly (A), the milkweed bug (B), and the milkweed aphid (C). Of these, the milkweed bug and the milkweed aphid will use oleander as a food source if milkweed plants are not available. (Photos © J. Engelberth.)

(*Danaus plexippus*) are insensitive to the toxins. They feed on milkweed leaves and retain the cardenolides. As a result, most insectivorous birds quickly learn to avoid eating monarch caterpillars and adult monarch butterflies. The bright and distinctive coloration of the caterpillars and butterflies serves to warn off birds. The large milkweed bug (*Oncopeltus fasciatus*) and the milkweed aphid (*Aphis nerii*) can also incorporate cardenolides into their bodies and become toxic themselves (**Figure 18.11**). Although all of these insects feed preferentially on milkweeds, the milkweed aphid and the large milkweed bug can also feed on oleander, which produces oleandrin as its major cardenolide.

In a further embellishment to the milkweed saga, the parasitic fly *Zenillia adamsoni* can obtain its cardenolide secondhand from the monarch caterpillar. When the female fly is ready to lay eggs, she seeks out a monarch caterpillar and deposits her eggs on its surface. Upon hatching, the fly larvae bore into the caterpillar and consume it from within. Besides using the caterpillar for nourishment, the fly larvae are able to store the caterpillar's toxic cardenolide and retain it throughout their adult stage.

Plants often store defensive chemicals as nontoxic water-soluble sugar conjugates in specialized vacuoles

A common mechanism for storing toxic secondary metabolites is to conjugate them with a sugar, which inactivates them and makes them more water soluble. As described above, most cardenolides and other related and toxic steroids are highly abundant as glycosides in the latex, and also in other compartments of the plant cell such as the vacuole. To become active, the glycosidic linkages often need to be hydrolyzed. Uncontrolled activation is prevented by the spatial separation of the activating hydrolases and their respective toxic substrates.

A good example of this spatial separation is found in the order Brassicales. Members of the Brassicales produce glucosinolates—sulfur-containing organic compounds derived from glucose and an amino acid—as their major defensive secondary metabolites. The hydrolyzing enzyme, myrosinase (a thioglucosidase),

Figure 18.12 Hydrolysis of glycosides into active defense compounds. (A) Glucosinolates are hydrolyzed to form mustard-smelling volatiles. (B) Enzyme-catalyzed hydrolysis of cyanogenic glycosides releases hydrogen cyanide. R and R' represent various alkyl or aryl substituents.

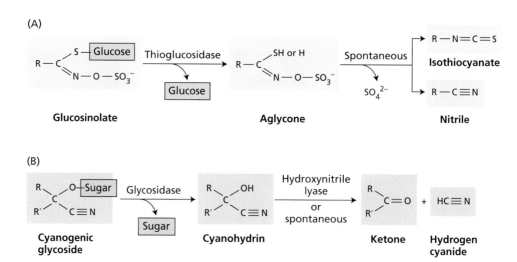

is stored in different cells than its substrates. While myrosinase-containing cells are mostly glucosinolate-free, the so-called sulfur-rich cells (or S-cells) contain high concentrations of glucosinolates. When the tissue is damaged, the released myrosinase and glucosinolates mix, resulting in the irreversible production of an unstable aglycone, which then rearranges into a variety of biologically active compounds, mostly nitriles and isothiocyanates (**Figure 18.12A**). These two-component "mustard oil bombs," in particular the isothiocyanates, are very effective against most generalist insect herbivores. The flavors of mustard, wasabi, radish, Brussel sprouts, and other related vegetables are due to isothiocyanates.

Cyanogenic glycosides represent a similar, but more toxic, class of N-containing secondary metabolites. Upon tissue damage, these glycosides break down and release hydrogen cyanide (**Figure 18.12B**). Cyanide inhibits cytochrome *c* oxidase in the mitochondria, which blocks the electron transport chain. As a consequence, electron transport and ATP synthesis come to a halt and the cell ultimately dies. Several plant species of economic and nutritional importance, including sorghum (*Sorghum bicolor*) and cassava (*Manihot esculenta*), produce different types of cyanogenic glycosides. The major cyanogenic glycoside in sorghum is dhurrin, which is derived from tyrosine. Dhurrin is stored as a glycoside, but when consumed by herbivores, the glycoside is quickly hydrolyzed to sugar and an aglycone, which is very unstable and releases cyanide.

Cassava accumulates linamarin and lotaustralin as its major cyanogenic glycosides. Cassava roots are a major source of nutrition in tropical regions, but they must be carefully prepared to avoid cyanide toxicity. The continuous consumption of improperly processed cassava, even at low endogenous concentrations of cyanogenic glycosides, may lead to paralysis as well as liver and kidney damage.

There are many more plants that produce constitutive secondary metabolites and store them in specific cells or compartments, where they can be released upon damage by herbivores and pathogens. Despite the presence of defensive chemicals, humans often prize these plants, or parts of them, for their medicinal properties or culinary flavors. No spice rack would be complete without the dried leaves of basil, sage, thyme, rosemary, and oregano, even though the sole reason for the high contents of secondary metabolites in these plants is to protect them from damage by herbivores and pathogens.

The indeterminate nature of plant vegetative growth (see Chapter 16) means that there will always be an age gradient from the younger leaves positioned near the apical bud to the mature leaves lower down on the stem. Most plant defense mechanisms are not uniformly distributed across this age gradient, but are continuously adjusted by developmental and environmental cues.

Defense compounds tend to be concentrated in younger rather than older or senescing leaves. Younger leaves also exhibit higher levels of inducible defense compound production.

Inducible Defense Responses to Insect Herbivores

While constitutive chemical defenses provide plants with basic protection against many pests and pathogens, and are common among plants in nature, there are disadvantages to this type of defense strategy. First, constitutive defenses are costly to the plant. The production of secondary metabolites requires a significant energy investment derived from primary metabolism, which is then unavailable for use in growth and reproduction. This trade-off is most obvious in agricultural crops, in which yield is increased, in part, by reducing the capacity of the plant to defend itself. Second, pests and pathogens can adapt to the plant's constitutive chemical defenses, as you saw in the case of the monarch caterpillar and milkweed. Certain species of insect herbivores and microbial pathogens have evolved physiological mechanisms to detoxify otherwise lethal secondary metabolites, and may even use these compounds to defend themselves against their own predators or parasites. Accordingly, most plants have evolved inducible defense systems in addition to whatever constitutive defenses they may have. Inducible defense systems enable plants to respond more flexibly to the full panoply of threats presented by pests and pathogens.

Based on their feeding behaviors, three major types of insect herbivores can be distinguished:

1. *Phloem feeders*, such as aphids and whiteflies, cause little damage to the epidermis and mesophyll cells. Phloem feeders insert their narrow *stylet*, which is an elongated mouthpart, into the phloem sieve tubes of leaves and stems.

2. *Cell-content feeders*, such as mites and thrips, are piercing-and-sucking insects that cause an intermediate amount of physical damage to plant cells.

3. *Chewing insects*, such as caterpillars (the larvae of moths and butterflies), grasshoppers, and beetles, cause the most significant damage to plants.

In the discussion that follows, we are mostly concerned with damage caused by chewing insects.

Plants can recognize specific components of insect saliva

To mount an effective inducible defense against pests or pathogens, the host plant must be able to distinguish between mechanical causes, such as wind or hail, and an actual biotic attack. Most plant responses to insect herbivores involve both a wound response and the recognition of certain compounds abundant in the insect's saliva. These compounds belong to a broad group of chemicals called **elicitors**, which trigger defense responses to a wide variety of herbivores and pathogens. Although repeated mechanical wounding can induce responses similar to those caused by insect herbivory in some plants, certain molecules in insect saliva can serve as enhancers of this stimulus. In addition, insect-derived elicitors can trigger signaling pathways *systemically*—that is, throughout the plant—thereby initiating defense responses that can minimize further damage in distal regions of the plant.

The first elicitors identified in insect saliva were *fatty acid–amino acid conjugates* (or *fatty acid amides*) in the oral secretions of larvae of the beet armyworm (*Spodoptera exigua*). These compounds have been shown to elicit a response closely resembling the response to chewing insects, as opposed to the response to wounding alone. The biosynthesis of these conjugates depends on the plant as

elicitors Specific pathogen molecules or cell wall fragments that bind to plant proteins and thereby act as signals for activation of plant defense against a pathogen.

the source of the fatty acids linolenic acid and linoleic acid. After the insect ingests plant tissue containing these fatty acids, an enzyme in the gut conjugates the plant-derived fatty acid to an insect-derived amino acid, typically glutamine. In some caterpillars the resulting conjugate of linolenic acid and glutamine is further processed by the introduction of a hydroxyl group at position 17 of linolenic acid. This compound was named **volicitin** for its potential to induce volatile secondary metabolites in maize (corn; *Zea mays*) plants. Since the discovery of volicitin, a variety of fatty acid amides have been identified not only in lepidopteran species (moths and butterflies) but also in crickets and fruit flies, and most of them were found to exhibit elicitor-like activities when applied to plants.

Another class of insect-derived elicitors has been isolated and characterized from the oral secretions of a grasshopper (*Schistocerca americana*). These elicitors are named *caeliferins* for their association with the grasshopper suborder Caelifera. Application of one of these fatty acid–based compounds to Arabidopsis leaves results in a transient spike of ethylene production and significantly higher jasmonic acid accumulation compared with mechanical wounding alone. However, the biological activity of caeliferins appears to be very species-specific. Unlike the fatty acid amides, the fatty acid–based caeliferins are not derived from the plant.

Phloem feeders activate defense signaling pathways similar to those activated by pathogen infections

Although phloem feeders, such as aphids, cause little mechanical damage to plants, they are nonetheless serious agricultural pests that can significantly reduce crop yields. Plants in nature have evolved mechanisms to recognize and defend against phloem feeders. In contrast to chewing and piercing-and-sucking insects, which inflict severe tissue damage resulting in the activation of the jasmonic acid signaling pathway, phloem feeders activate the **salicylic acid** signaling pathway, which is usually associated with pathogen infections. Because the defense response to phloem feeders involves receptor–ligand complexes that are closely related to those involved in the response to pathogens, we will describe the signaling mechanisms of this class of herbivores later in the chapter when we discuss microbial infections.

Jasmonic acid activates defense responses against insect herbivores

When plants recognize elicitors from insect saliva, a complex signal transduction network is activated. An increase in the cytosolic Ca^{2+} concentration ($[Ca^{2+}]_{cyt}$) is an early signal (see Chapter 12) that mediates insect elicitor–induced responses. The duration and extent of these responses are further regulated by the *octadecanoid pathway*, which leads to the production of the hormone jasmonic acid (JA). JA concentrations rise steeply in response to insect herbivore damage and trigger the production of many proteins involved in plant defenses. Direct demonstration of the role of JA in insect resistance has come from research with JA-deficient mutant lines of Arabidopsis, tomato, and maize. Such mutants are easily killed by insect pests that normally cannot damage wild-type plants. Application of exogenous JA restores resistance nearly to the levels of the wild-type plant. JA induces the transcription of multiple genes that encode key enzymes in all major pathways for secondary metabolites.

The structure and biosynthesis of JA have intrigued plant biologists because of the parallels to oxylipins, such as the prostaglandins, which are central to inflammatory responses and other physiological processes in mammals. Two organelles participate in jasmonate biosynthesis, chloroplasts and peroxisomes. In the chloroplast, an intermediate derived from linolenic acid is cyclized and then transported to the peroxisome, where enzymes of the β-oxidation pathway complete the conversion to JA (**Figure 18.13**).

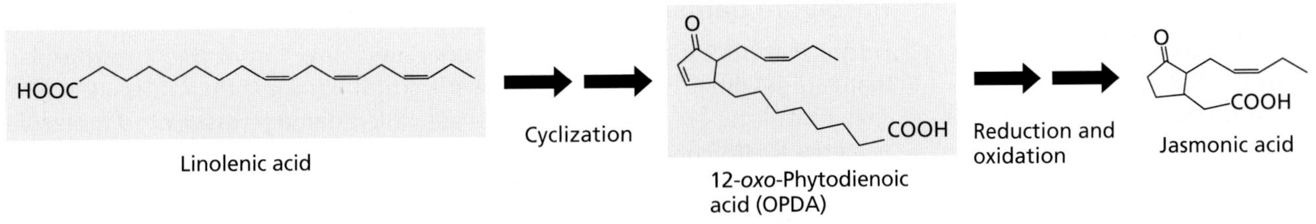

Figure 18.13 Simplified biosynthetic pathway for jasmonic acid.

Hormonal interactions contribute to plant–insect herbivore interactions

Several other signaling agents—including ethylene, salicylic acid, and methyl salicylate—are often induced by insect herbivory. In particular, ethylene appears to play an important role in this context. When applied alone to plants, ethylene has little effect on defense-related gene activation. However, when applied together with JA it seems to enhance JA responses. Similarly, when plants are treated with elicitors such as fatty acid amides (which by themselves do not induce the production of significant amounts of ethylene) in combination with ethylene, defense responses are significantly increased. Results such as these suggest that a concerted action of these signaling compounds is required for the full activation of induced defense responses. Multifactorial control allows plants to integrate numerous environmental signals in modulating the defense response.

JA initiates the production of defense proteins that inhibit herbivore digestion

Besides activating pathways for the production of toxic or repelling secondary metabolites, JA also initiates the biosynthesis of defense proteins. Most of these proteins interfere with the herbivore digestive system. For example, some legumes synthesize **α-amylase inhibitors**, which block the action of the starch-digesting enzyme α-amylase. Other plant species produce **lectins**, defensive proteins that bind to carbohydrates or carbohydrate-containing proteins. After ingestion by an herbivore, lectins bind to the epithelial cells lining the digestive tract and interfere with nutrient absorption.

A more direct attack on the insect herbivore's digestive system is performed by some plants through the production of a specific cysteine protease, which disrupts the peritrophic membrane that protects the gut epithelium of many insects. While none of these genes are essential for the vegetative growth of the plant, they have probably evolved from normal "housekeeping" genes during the coevolution of plants and their insect herbivores.

The best-known antidigestive proteins in plants are the proteinase inhibitors. Found in legumes, tomato, and other plants, these substances block the action of herbivore proteolytic enzymes. After entering the herbivore's digestive tract, they hinder protein digestion by binding tightly and specifically to the active site of protein-hydrolyzing enzymes such as trypsin and chymotrypsin. Insects that feed on plants containing proteinase inhibitors suffer reduced rates of growth and development that can be offset by supplemental amino acids in their diet.

The defensive role of proteinase inhibitors has been confirmed by experiments with transgenic tobacco. Plants that had been transformed to accumulate increased amounts of proteinase inhibitors suffered less damage from insect herbivores than did untransformed control plants. As with glucosinolates, some insect herbivores have become adapted to plant proteinase inhibitors by production of digestive proteinases resistant to inhibition.

α-amylase inhibitors Substances synthesized by some legumes that interfere with herbivore digestion by blocking the action of the starch-digesting enzyme α-amylase.

lectins Defensive plant proteins that bind to carbohydrates; or carbohydrate-containing proteins inhibiting their digestion by a herbivore.

damage associated molecular patterns (DAMPs) Molecules originating from nonpathogenic sources that can initiate immune responses.

Herbivore damage induces systemic defenses

During herbivore attack, mechanical damage releases lytic enzymes from the plant that can potentially compromise the structural barriers of plant tissues. Some of the resulting breakdown products are called **damage associated molecular patterns** (**DAMPs**), as they function as elicitors of damage response pathways. As we will discuss later in the chapter, DAMPs are recognized by pattern recognition receptors (PRRs) located on the cell surface. DAMPs usually appear in the apoplast and can induce protection against a broad range of organisms, a response known as *innate immunity*. For example, oligogalacturonides released from the cell wall can act as endogenous elicitors.

In tomato, insect feeding leads to the rapid accumulation of proteinase inhibitors throughout the plant, even in undamaged areas far from the initial feeding site. The systemic production of proteinase inhibitors in young tomato plants is triggered by a complex sequence of events (**Figure 18.14**).

Although peptide signals such as systemin were originally thought to be restricted to solanaceous plants, it has become clear in recent years that other plants also produce peptides as signaling molecules in response to insect herbivory.

Long-distance electrical signaling occurs in response to insect herbivory

In response to herbivory, jasmonic acid accumulates within minutes—both locally, at the site of herbivore damage, and distally, in undamaged tissues of the same leaf and in other leaves. Although plants lack nervous systems, several lines of evidence are consistent with a role for electrical signaling in defense responses that occur some distance from the site of herbivore damage. For example, the feeding of Egyptian cotton leafworm (*Spodoptera littoralis*) on bean leaves induces a wave of plasma membrane depolarizations that spreads to undamaged areas of the leaf. In addition, ionophore-induced plasma membrane depolarization in tomato cells results in the expression of jasmonate-regulated genes.

Measurements of membrane potentials at the surface of Arabidopsis leaves in response to feeding by the larvae of Egyptian cotton leafworm have now confirmed the role of electrical signaling in the spread of the jasmonate defense response to undamaged leaves. During feeding, electrical signals induced near the site of attack subsequently spread to neighboring leaves at a maximum speed of 9 cm per min (**Figure 18.15**). Since the relay of the electrical signal is most efficient for leaves directly above or below the wounded leaf, the vascular system is a good candidate for the transmission of the electrical signals to other leaves. At all sites that receive the electrical signals, jasmonate-mediated gene expression is turned on and initiates defense-response gene expression. A family of glutamate receptor-like proteins has been shown to function in this electrical signaling, but its relationship with other types of long-distance defense signaling is unclear.

Figure 18.14 Proposed systemin signaling pathway for the rapid induction of protease inhibitor biosynthesis in a wounded tomato plant. Wounded leaves (bottom left of plant) synthesize prosystemin in phloem parenchyma cells, and the prosystemin is proteolytically processed to systemin. Systemin is released from phloem parenchyma cells and binds to receptors in the plasma membrane of adjacent companion cells. This binding activates a signaling cascade involving phospholipase A2 (PLA2) and mitogen-activated protein (MAP) kinases, which results in the biosynthesis of jasmonic acid (JA). JA is then transported via sieve elements, possibly in a conjugated form (JA–*X*), to unwounded leaves. There, JA initiates a signaling pathway in target mesophyll cells, resulting in the expression of genes that encode protease inhibitors. Plasmodesmata facilitate the spread of the signal at various steps in the pathway. ▶

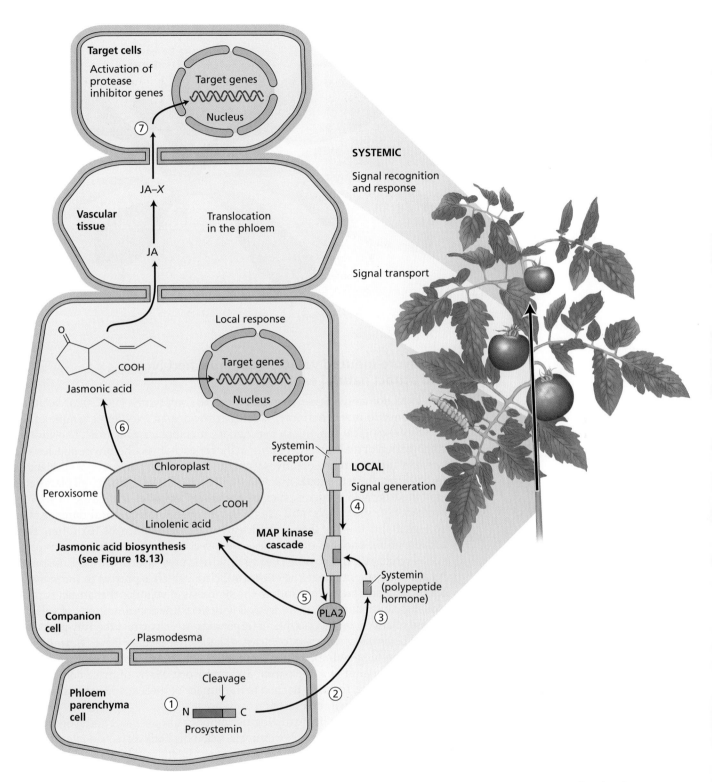

Target cells

Activation of protease inhibitor genes

Target genes

Nucleus

⑦

SYSTEMIC

Signal recognition and response

JA–*X*

Vascular tissue

Translocation in the phloem

JA

Signal transport

Local response

Target genes

Nucleus

Jasmonic acid

COOH

⑥

Systemin receptor

LOCAL

Signal generation

④

Peroxisome

Chloroplast

COOH

Linolenic acid

Jasmonic acid biosynthesis (see Figure 18.13)

MAP kinase cascade

⑤

Systemin (polypeptide hormone)

③

PLA2

Companion cell

Plasmodesma

Phloem parenchyma cell

Cleavage

① N ▬▬ C

Prosystemin

②

1. Wounded tomato leaves synthesize prosystemin, a 200–amino acid precursor protein.

2. Prosystemin is proteolytically processed to produce the short (18 amino acids) polypeptide DAMP called systemin.

3. Systemin is released from damaged cells into the apoplast.

4. In adjacent intact tissue (phloem parenchyma), systemin binds to a pattern recognition receptor on the plasma membrane.

5. The activated systemin receptor becomes phosphorylated and activates a phospholipase A2 (PLA2).

6. The activated PLA2 generates the signal that initiates JA biosynthesis.

7. JA is then transported through the phloem, possibly in a conjugated form, to a target cell in unwounded leaves, initiating a signaling pathway.

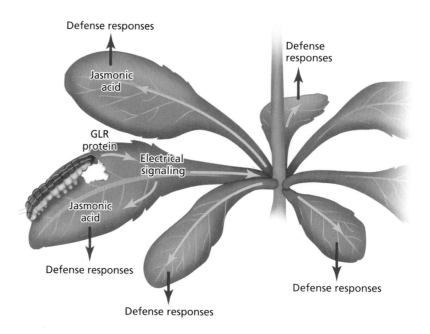

Figure 18.15 Model for the electrical signaling response of Arabidopsis to herbivore attack. Injury to the leaf caused by herbivory appears to activate glutamate receptor-like (GLR) proteins in the vascular system to stimulate jasmonic acid (JA) production both locally and in other leaves (yellow arrows). JA production then initiates defense responses that discourage further herbivory (red arrows). (After Christmann and Grill 2013.)

Herbivore-induced volatiles can repel herbivores and attract natural enemies

The induction and release of volatile organic compounds (VOCs), or volatiles, in response to insect herbivore damage provides an excellent example of the complex ecological functions of secondary metabolites in nature. The emitted combination of molecules is often specific for each insect herbivore species and typically includes representatives from the three major pathways of specialized metabolism: the terpenoids, alkaloids, and phenolics. In addition, all plants also emit lipid-derived products, such as **green-leaf volatiles** (a mixture of six-carbon aldehydes, alcohols, and esters) in response to mechanical damage. The ecological functions of these volatiles are manifold (**Figure 18.16**). Often, they attract natural enemies of the attacking insect herbivores—predators or parasites—which use the volatile cues to find their prey or host for their offspring. As noted earlier, in maize the elicitor volicitin, which is present in the saliva of beet armyworm larvae, can induce the synthesis of volatiles that attract parasitoids. Maize seedlings that are treated with very low concentrations of volicitin release relatively large amounts of terpenoids, which attract the tiny parasitoid wasp *Microplitis croceipes*. In contrast, volatiles released by leaves during moth oviposition (egg laying) can act as repellents to other female moths, thereby preventing further egg deposition and herbivory. Many of these compounds, although volatile, remain attached to the surface of the leaf and serve as feeding deterrents because of their taste.

Plants have the ability to distinguish among various insect herbivore species and to respond differentially. For example, following herbivory, *Nicotiana attenuata*, a wild tobacco that grows in the deserts of the Great Basin in the western United States, typically produces higher amounts of nicotine, which poisons the insect central nervous system. However, when wild tobacco plants are attacked by nicotine-tolerant caterpillars, the plants show no increase in nicotine. Instead, they release volatile terpenes that attract insect predators of the caterpillars. Clearly, wild tobacco and other plants must have ways of determining what type of insect herbivore is damaging their foliage. Herbivores might signal their presence by the type of damage they inflict or the distinctive chemical compounds they release in their oral secretions.

green-leaf volatiles A mixture of lipid-derived six-carbon aldehydes, alcohols, and esters released by plants in response to mechanical damage.

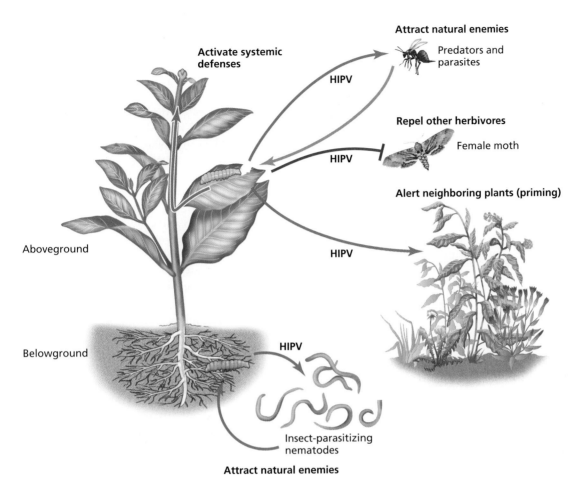

Figure 18.16 Ecological functions of insect herbivore–induced plant volatiles (HIPVs). Many plants release a specific bouquet of volatile organic compounds when attacked by insect herbivores. These volatiles can consist of compounds from all major pathways for specialized metabolites, including terpenoids (mono- and sesquiterpenes), alkaloids (indole), and phenylpropanes (methyl salicylate), as well as green-leaf volatiles. These volatiles can act as cues for natural enemies of the insect herbivore, for example parasitic wasps. Belowground parts of plants can also release volatiles when attacked by herbivores. It has been shown that these volatiles attract insect-parasitizing nematodes, which then attack the herbivore. Volatiles may also serve as a repellant for female moths, thereby avoiding further egg deposition. Most recently, volatiles have been found to act as a systemic defense signal in highly sectorial plants with interrupted vascular connections, and also between plants over short distances. There, these volatile signals prepare the receiving plant against impending herbivory by priming (preparing) defense responses, resulting in a stronger and faster response when the receiving plant is actually attacked.

Herbivore-induced volatiles can serve as long-distance signals within and between plants

The role of herbivore-induced plant volatiles is not limited to the mediation of ecological interactions between plants and insects. Volatiles have been shown to act as inducers of herbivore resistance between different branches of sagebrush (*Artemisia tridentata*). It was found that airflow was essential for the induction of the induced resistance. Sagebrush, like other desert plants, is highly *sectorial*, meaning the vascular system of the plant is not well integrated by interconnections. Although many plants are capable of responding systematically to herbivores by means of chemical signals that move internally through vascular interconnections, sagebrush and many other desert plants are unable to do so. Instead, volatiles are used to overcome these constraints and provide systemic signaling.

A similar effect of volatiles was observed in lima bean (*Phaseolus lunatus*), which uses extrafloral nectaries located at the base of leaf blades to attract

extrafloral nectar   Nectar produced outside the flower and not involved in pollination events.

predacious and parasitoid arthropods to protect them against various types of herbivores. When leaf beetles attack lima bean, volatiles, in particular green-leaf volatiles, are released immediately from the damage site and signal other parts of the same plant to activate their defenses, including the production of **extrafloral nectar**.

Certain volatiles emitted by infested plants can also serve as signals for neighboring plants to initiate expression of defense-related genes (see Figure 18.16). In addition to several terpenoids and methyl jasmonate, green-leaf volatiles act as potent signals in this process. Green-leaf volatiles such as 3-hexenal, 3-hexenol, 3-hexenyl acetatal, and traumatin are produced from linolenic acid and are the major components of the familiar scent of freshly cut grasses. These volatiles serve as potent signals in inter- and intraplant signaling. For example, when maize plants were exposed to green-leaf volatiles, JA and JA-related gene expression was rapidly induced. More important, exposure to green-leaf volatiles primed maize plant defenses to respond more strongly to subsequent attacks by insect herbivores. Green-leaf volatiles have been shown to prime or sensitize the defensive mechanisms of a variety of other plant species, including Arabidopsis, poplar (*Populus tremula*), and blueberry (*Vaccinium* spp.). In addition, they activate the production of phytoalexins and other antimicrobial compounds.

Insects have evolved mechanisms to defeat plant defenses

In spite of all the chemical mechanisms plants have evolved to protect themselves, herbivorous insects have evolved mechanisms for circumventing or overcoming these plant defenses by the process of *reciprocal evolutionary change* between plant and insect, a type of coevolution. These adaptations, like plant defense responses, can be either constitutive or induced. Constitutive adaptations are more widely distributed among specialist herbivorous insects, which can feed on only a few plant species, whereas induced adaptations are more likely to be found among insects that are dietary generalists. Although it is not always obvious, in most natural environments plant–insect interactions have led to a standoff in which each can develop and survive under suboptimal conditions.

Plant Defenses against Pathogens

Despite the absence of an immune system comparable to that of animals, plants are surprisingly resistant to diseases caused by the fungi, bacteria, viruses, and nematodes that are ever present in the environment. In this section we examine the diverse array of mechanisms that plants have evolved to resist infection locally, including microbe-associated molecular pattern (MAMP)-triggered immunity, effector-triggered immunity, the production of antimicrobial agents, and a type of programmed cell death called the hypersensitive response. We also discuss a type of systemic plant immunity called *systemic acquired resistance (SAR)*.

Microbial pathogens have evolved various strategies to invade host plants

Throughout their lives, plants are continuously exposed to a diverse array of pathogens. Successful pathogens have evolved various mechanisms to invade their host plant and cause disease (**Figure 18.17**). Some penetrate the cuticle and cell wall directly by secreting lytic enzymes, which digest these mechanical barriers. Others enter the plant through natural openings such as stomata, hydathodes, and lenti-

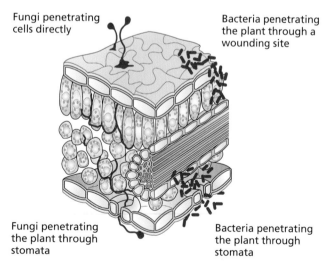

Fungi penetrating cells directly

Bacteria penetrating the plant through a wounding site

Fungi penetrating the plant through stomata

Bacteria penetrating the plant through stomata

Figure 18.17   Plant pathogens such as bacteria and fungi have developed various methods for invading plants. Some fungi have mechanisms that allow them to directly penetrate the cuticle and cell wall of the plant. Other fungi, and also pathogenic bacteria, enter through natural openings such as stomata or through existing wounds caused by herbivores.

cels. A third category invades the plant through wound sites, for example those caused by insect herbivores. Additionally, many viruses, as well as other types of pathogens, are transferred by insect herbivores, which serve as vectors, and invade the plant from the insect feeding site. Phloem feeders such as whiteflies and aphids deposit these pathogens directly into the vascular system, from which they can easily spread throughout the plant.

Once inside the plant, pathogens generally employ one of three main attack strategies to use the host plant as a substrate for their own proliferation. **Necrotrophic pathogens** attack their host by secreting cell wall–degrading enzymes or toxins, which eventually kill the affected plant cells, leading to extensive tissue maceration (softening of the tissues after death by autolysis). This dead tissue is then colonized by the pathogens and serves as a food source. A different strategy is used by **biotrophic pathogens**; after infection, most of the plant tissue remains alive and only minimal cell damage can be observed, as the pathogens feed on substrates provided by their host. **Hemibiotrophic pathogens** are characterized by an initial biotrophic stage, in which the host cells are kept alive. This phase is followed by a necrotrophic stage, in which the pathogens can cause extensive tissue damage.

Although these invasion and infection strategies are individually successful, plant disease epidemics are rare in natural ecosystems. This is because plants have evolved effective defenses against this diverse array of pathogens.

Pathogens produce effector molecules that aid in the colonization of their plant host cells

Plant pathogens can produce a wide array of effectors that support their ability to successfully colonize their host and gain nutritional benefits. **Effectors** are molecules that change the plant's structure, metabolism, or hormonal regulation to the advantage of the pathogen. They can be divided into three major classes: *enzymes, toxins,* and *growth regulators*. Because invasion of a suitable host is often the most difficult step for a pathogen, many pathogens produce enzymes that can degrade the plant cuticle and cell wall. Among those enzymes are cutinases, cellulases, xylanases, pectinases, and polygalacturonases. These enzymes have the ability to compromise the integrity of the cuticle as well as the primary and secondary cell walls.

Pathogens can also produce a wide array of toxins that act by targeting specific proteins of the plant (**Figure 18.18**). For example, the **HC-toxin** from the fungus *Cochliobolus carbonum*, which causes northern leaf blight disease, inhibits specific histone deacetylases in maize. In general, decreased deacetylation of histones, which are essential for the organization of chromatin, tends to increase the expression of associated genes (see Chapter 1). However, it is not yet known whether this is how HC-toxin causes disease in maize.

Fusicoccin (see Figure 18.18) is a toxin from the fungus *Fusicoccum amygdali*. **Fusicoccin** constitutively activates the

necrotrophic pathogens Pathogens that attack their host plant first by secreting cell wall–degrading enzymes or toxins, which will lead to massive tissue laceration and plant death.

biotrophic pathogens Pathogens that leave infected tissue alive and only minimally damaged while the pathogen continues to feed on host resources.

hemibiotrophic pathogens Pathogens that show an initial biotrophic stage, which is followed by a necrotrophic stage in which the pathogen causes extensive tissue damage.

effector A molecule that binds a protein to change its activity. Bacterial effectors are secreted by pathogens to act on proteins within a host cell.

HC-toxin A cell permeant cyclic tetrapeptide produced by the maize pathogen *Cochliobolus carbonum* that inhibits histone deacetylases.

fusicoccin A fungal toxin that induces acidification of plant cell walls by activating H^+-ATPases in the plasma membrane. Fusicoccin stimulates rapid acid growth in stem and coleoptile sections. It also stimulates stomatal opening by stimulating proton pumping at the guard cell plasma membrane.

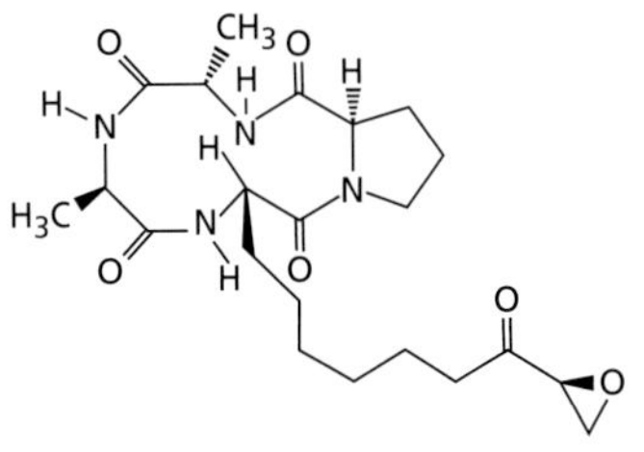

HC-toxin

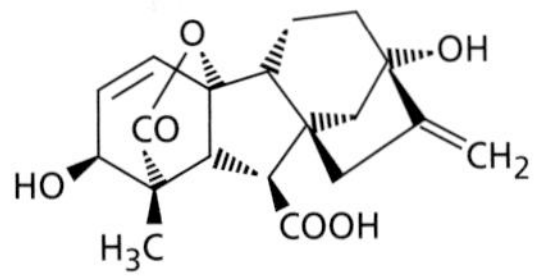

Gibberellic acid GA$_3$

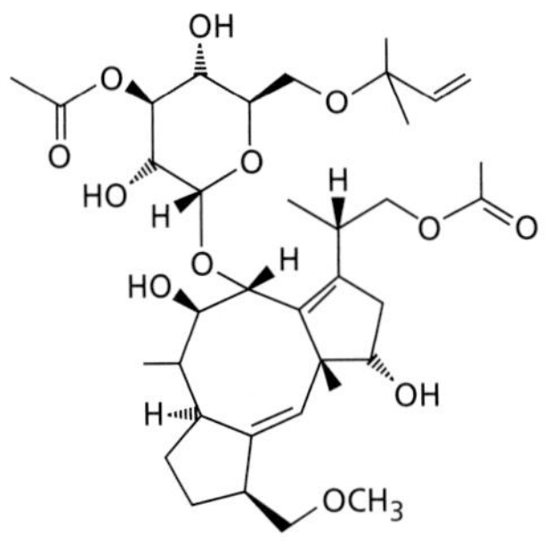

Fusicoccin

Figure 18.18 Effector molecules produced by pathogens help invade plants. Some pathogens produce specific effector molecules that significantly alter the physiology of the plant. The HC-toxin, a cyclic peptide, acts on the enzyme histone deacetylase in the nucleus, and may compromise the expression of genes involved in defense. Fusicoccin binds to plant plasma membrane H^+-ATPases (see Chapter 6), in particular those in stomata, and activates them irreversibly. Gibberellins, produced by the fungus *Gibberella fujikuroi*, accelerate growth, resulting in bigger plants when compared with uninfested plants. The gibberellins produced by the fungi are identical to those produced endogenously by the plant.

pattern recognition receptors (PRRs)
Innate immune system proteins that are associated with microbe-associated molecular patterns (MAMPs) and damage-associated molecular patterns (DAMPs).

microbe-associated molecular patterns (MAMPs) Microbially-produced molecules that are recognized by host cells.

injectisome A name for a specialized secretion system appendage of some pathogenic bacteria.

resistance (R) proteins Proteins that function in plant defense against fungi, bacteria, and nematodes by binding to specific pathogen molecules, elicitors.

effector-triggered immunity Immune responses that are mediated by a class of intracellular resistance (R) proteins.

plant plasma membrane H^+-ATPase by first binding to a specific protein of the 14-3-3 group of regulators. This complex then binds to the C-terminal region of the H^+-ATPase and activates it irreversibly, leading to cell wall overacidification and plasma membrane hyperpolarization. These effects of fusicoccin are of particular importance for stomatal guard cells (see Chapter 6). Fusicoccin-induced plasma membrane hyperpolarization in guard cells causes massive K^+ uptake and permanent stomatal opening, which leads to wilting and ultimately death of the plant. It is not yet clear if and how the pathogen benefits from the excessive wilting of its host.

Some pathogens produce effector molecules that significantly interfere with the hormonal balance of the plant host. The fungus *Gibberella fujikuroi*, which causes infected rice shoots to grow much faster relative to uninfected plants, produces gibberellic acid (GA_3) and other gibberellins. Gibberellins are thus responsible for the "foolish seedling disease" of rice. It is thought that fungal spores released from the taller, infected plants are more likely to spread to surrounding plants because of their height advantage. It was subsequently demonstrated that gibberellins are naturally occurring plant hormones (see Chapter 12).

The effectors of some pathogenic bacteria, such as *Xanthomonas*, are proteins that target the plant cell nucleus and cause marked changes in gene expression. These so-called transcription activator-like (TAL) effectors bind to the host plant DNA and activate the expression of genes beneficial to the pathogen's growth and dissemination.

Pathogen infection can give rise to molecular "danger signals" that are perceived by cell surface pattern recognition receptors (PRRs)

To distinguish between "self" and "nonself" during pathogen infection, plants possess **pattern recognition receptors (PRRs)** that perceive signals called **microbe-associated molecular patterns (MAMPs)**, which are conserved among a specific class of microorganisms (such as chitin for fungi, flagella for bacteria) but are absent in the host (**Figure 18.19**). As we mentioned earlier, molecular alarm signals can also arise from the plant itself, either from damage caused by microbes or as the result of damage inflicted by chewing insects. Such plant-derived signals are collectively referred to as damage-associated molecular patterns (DAMPs).

Perception of MAMPs or DAMPs by cell surface PRRs initiates a localized basal defense response, which inhibits the growth and activity of nonadapted pathogens or pests. For example, control over stomatal aperture, a common site of pathogen invasion, serves as the first line of defense against pathogen invasion. When an Arabidopsis leaf is exposed to bacteria on the leaf surface, or to a MAMP derived from bacterial flagellar proteins, the stomatal apertures decrease, thereby retarding pathogen invasion.

R proteins provide resistance to individual pathogens by recognizing strain-specific effectors

Well-adapted microbial pathogens are able to subvert MAMP-triggered immunity by introducing a wide variety of effectors directly into the cytoplasm of the host cell. For example, some pathogenic gram-negative bacteria have evolved a syringe-shaped structure called the **injectisome** that spans the inner and outer bacterial membranes and includes a needlelike extracellular projection. Fungi and oomycetes have evolved other methods of transporting effectors directly into plant cells. Once inside the cells, these effectors can no longer be detected by the membrane-bound PRRs, and without a backup system the plant would be defenseless against the attack.

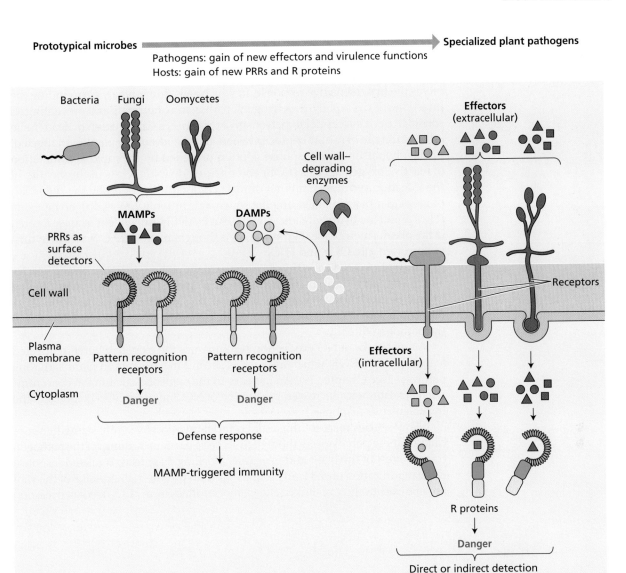

Figure 18.19 Plants have evolved defense responses to a variety of danger signals of biotic origin. These danger signals include microbe-associated molecular patterns (MAMPs), damage-associated molecular patterns (DAMPs), and effectors. Extracellular MAMPs produced by microbes, and DAMPs released by microbial enzymes, bind to pattern recognition receptors (PPRs) on the cell surface. As plants coevolved with pathogens, the pathogens acquired effectors as virulence factors, and plants evolved new PRRs to perceive extracellular effectors, and new resistance (R) proteins to perceive intracellular effectors. When MAMPs, DAMPs, and effectors bind to their PRRs and R proteins, two types of defense responses are induced: MAMP-triggered immunity and effector-triggered immunity. (After Boller and Felix 2009.)

This microbial innovation places plants under strong evolutionary pressure. For example, the bacterial toxin coronatine reverses the stimulatory effects on stomatal closing of MAMPs derived from bacterial flagellar proteins. In return, plants evolved a second line of defense based on a class of specialized **resistance (R) proteins** that recognize these intracellular effectors and trigger defense responses to render them harmless (see Figure 18.19). As a result, plants possess a second type of immunity called **effector-triggered immunity**, mediated by a set of highly specific intracellular receptors.

The hypersensitive response is a common defense against pathogens

A common physiological phenotype associated with effector-triggered immunity is the hypersensitive response, in which cells immediately surrounding the infection site die rapidly, depriving the pathogen of nutrients and preventing its spread. After a successful hypersensitive response, a small region of dead tissue is left at the site of the attempted invasion, but the rest of the plant is unaffected.

The hypersensitive response is often preceded by the rapid accumulation of reactive oxygen species (ROS) and nitric oxide (NO). Cells in the vicinity of the infection synthesize a burst of toxic compounds formed by the reduction of molecular oxygen, including the superoxide anion ($O_2^{\bullet}$), hydrogen peroxide (H_2O_2), and the hydroxyl radical (OH•). An NADPH-dependent oxidase located at the plasma membrane (**Figure 18.20**) is thought to produce $O_2^{\bullet}$, which in turn is converted into OH• and H_2O_2.

The hydroxyl radical is the strongest oxidant of these reactive oxygen species and can initiate radical chain reactions with a range of organic molecules, leading to lipid peroxidation, enzyme inactivation, and nucleic acid degradation. ROS may contribute to cell death as part of the hypersensitive response or act to kill the pathogen directly.

A rapid spike of NO production accompanies the oxidative burst in infected leaves. NO is a ROS, which also act as second messengers in plant signaling pathways (see Chapter 19). An increase in the cytosolic calcium ion concentration correlates with increased production of NO and other ROS involved in the hypersensitive response.

Many species react to fungal or bacterial invasion by synthesizing lignin or callose. These polymers are thought to serve as barriers, walling off the pathogen from the rest of the plant and physically blocking its spread. A related response is the modification of cell wall proteins. Certain proline-rich proteins of the wall become oxidatively cross-linked after pathogen attack in an H_2O_2-mediated reaction

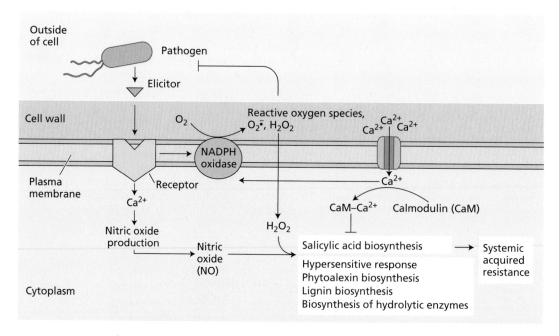

Figure 18.20   Many types of antipathogen defenses are induced by infection. Fragments of pathogen molecules called elicitors initiate a complex signaling pathway leading to the activation of defense responses. A burst of oxidation activity and nitric oxide production stimulates the hypersensitive response and other defense mechanisms.

(see Figure 18.20). This process strengthens the walls of the cells in the vicinity of the infection site, thereby increasing their resistance to microbial digestion.

Another defense response to infection is the formation of hydrolytic enzymes that attack the cell wall of the pathogen. An assortment of glucanases, chitinases, and other hydrolases are induced by fungal invasion. Chitin, a polymer of *N*-acetylglucosamine residues, is a principal component of fungal cell walls. Increased production of chitinases is often observed after fungal infection.

A single encounter with a pathogen may increase resistance to future attacks

In addition to triggering defense responses locally, microbial pathogens also induce the production of signals such as salicylic acid, methyl salicylate, and other compounds that lead to systemic expression of the antimicrobial **pathogenesis-related (PR) genes**. PR genes comprise a small multigene family that encodes low-molecular-weight proteins (6–43 kDa) composed of a diverse group of hydrolytic enzymes, wall-modifying enzymes, antifungal agents, and components of signaling pathways. PR proteins are localized either in vacuoles or in the apoplast and are most abundant in leaves, where they are presumed to confer protection against secondary infections. This phenomenon of local pathogen challenge enhancing resistance to secondary infection, called **systemic acquired resistance (SAR)**, normally develops over a period of several days. SAR appears to result from increased amounts of certain defense compounds that we have already mentioned, including chitinases and other hydrolytic enzymes.

Measurements of the rate of SAR transmission from the site of attack to the rest of the plant indicate that movement is too rapid (3 cm per h) for simple diffusion and further supports the hypothesis that the mobile signal must be transported through the vascular system. Most of the evidence points to the phloem as the primary pathway of translocation of the SAR signal.

Although the mechanism of SAR induction is still unknown, one of the endogenous signals is salicylic acid. The concentration of this benzoic acid derivative rises dramatically in the zone of infection after initial attack, and it is thought to establish SAR in other parts of the plant. However, grafting experiments in tobacco showed that infected, salicylic acid–deficient rootstocks could nevertheless trigger SAR in wild-type scions. These results indicate that salicylic acid is neither the initial trigger at the infection site nor the mobile signal that induces SAR throughout the plant. Other experiments point to methyl salicylate as the mobile signal that induces SAR. Although methyl salicylate is volatile, it appears to be transported via the vascular system in tobacco.

Plant Defenses against Other Organisms

While herbivorous insects and pathogenic microorganisms represent the greatest threat to plants, other organisms, including nematodes and parasitic plants, can also cause significant damage. However, relatively little is known about the factors that regulate the interactions of nematodes and parasitic plants with their respective hosts. There is, however, emerging evidence that secondary metabolites play an important role in this process.

Some plant parasitic nematodes form specific associations through the formation of distinct feeding structures

Nematodes, or roundworms, are water and soil inhabitants that often outnumber all other animals in their respective environments. Many nematodes exist as parasites relying on other living organisms, including plants, to complete their life cycle. Nematodes can cause severe losses of agricultural crops and ornamental plants. Plant parasitic nematodes can infect all parts of a plant, from roots to leaves, and

pathogenesis-related (PR) genes
Genes that encode small proteins that function either as antimicrobials or in initiating systemic defense responses.

systemic acquired resistance (SAR)
The increased resistance throughout a plant to a range of pathogens following infection by a pathogen at one site.

may even live in the bark of forest trees. Nematodes feed through a hollow stylet that can easily penetrate plant cell walls. In the soil, nematodes can move from plant to plant, thereby causing extensive damage. Arguably the best studied among the plant parasitic nematodes are the cyst nematodes and those causing gall formation on infected roots, the so-called root knot nematodes. Both are endoparasites that depend on a living plant as host to complete their life cycles, and are therefore categorized as biotrophs. The life cycles of parasitic nematodes begin when dormant eggs recognize specific compounds secreted by the plant root (**Figure 18.21**). Once hatched, the juvenile nematodes swim to the root and penetrate it. There they migrate to the vascular tissue where they begin feeding on the cells of the vascular system.

At the permanent feeding site, usually in the root cortex, a cyst nematode larva pierces a cell with its stylet and injects saliva. As a result, the cell walls break down and neighboring cells are incorporated into a **syncytium** (see Figure 18.21A). The syncytium is a large, metabolically active feeding site that becomes multinucleate as neighboring plant cells are incorporated into it by cell wall dissolution and cell fusion. The syncytium continues to spread centripetally toward the vascular tissue, incorporating pericycle cells and xylem parenchyma. The outer walls of the syncytium adjacent to the conducting elements form protuberances that function in transfer of nutrients.

The cyst nematode, after establishing itself in such a feeding structure, grows and undergoes three molting stages while becoming a vermiform (wormlike) adult. At maturity, the female produces eggs internally, swells, and protrudes from the root surface. The mature male nematodes are released from the root into the soil and are attracted by pheromones to protruding females on the root surface. After fertilization, the female dies, forming a cyst containing the fertilized eggs.

(A) Cyst nematodes

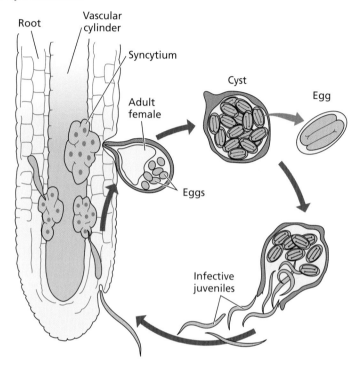

(B) Root knot nematodes

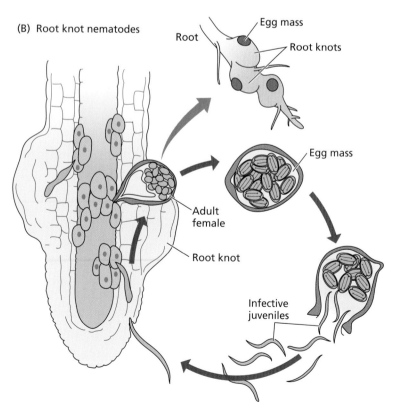

Figure 18.21 Nematodes can cause significant damage to plants. Most plant pathogenic nematodes attack the roots of plants. Free-living juvenile nematodes are attracted to secretions by the roots. After penetration, the nematode starts feeding on cells in the vasculature. (A) Cyst nematodes cause the formation of a specific feeding structure (syncytium) in the vasculature but do not cause other morphological changes. After fertilization the female cyst nematode dies, thereby forming a cyst containing the fertilized eggs, from which a new generation of infective juveniles hatch. (B) Infection by root knot nematodes causes the formation of giant cells, resulting in the typical root knots. Upon maturation, the female nematode releases an egg mass from which new infective juveniles hatch and cause further infestations of plants.

Roots infected by root knot nematodes form large cells, resulting in the establishment of the characteristic knot or gall, which also remains in close contact with the vasculature and provides the nematode with nutrients (see Figure 18.21B).

Plants compete with other plants by secreting allelopathic secondary metabolites into the soil

Plants release compounds (root exudates) into their environment that change soil chemistry, thus increasing nutrient uptake or protecting against metal toxicity. Plants also secrete chemical signals that are essential for mediating interactions between plant roots and nonpathogenic soil bacteria, including nitrogen-fixing bacterial symbionts. However, microbes are not the only organisms that are influenced by secondary metabolites released by plant roots. Some of these chemicals are also involved in direct communication between plants. Plants release secondary metabolites to the soil to inhibit the roots of other plants, a phenomenon known as allelopathy.

Interest in allelopathy has increased in recent years because of the problem of invasive species that outcompete native species and take over natural habitats. A devastating example is the spotted knapweed (*Centaurea maculosa*), an invasive exotic weed introduced to North America that releases phytotoxic secondary metabolites into the soil. Spotted knapweed, a member of the aster family (Asteraceae), is native to Europe, where it is not a dominant or problematic species. However, in the northwestern United States it has become one of the worst invasive weeds, infesting over 1.8 million ha (~4.4 million acres) in Montana alone. Spotted knapweed often colonizes disturbed areas in North America, but it also invades rangelands, pastures, and prairies, where it displaces native species and establishes dense monocultures.

The phytotoxic secondary metabolites that spotted knapweed roots release into the soil have been identified as a racemic mixture of (±)-catechin (hereafter catechin) (**Figure 18.22**). The mechanism by which catechin acts as a phytotoxin has been elucidated. In susceptible species such as Arabidopsis, catechin triggers a wave of ROS initiated at the root meristem, which leads to a Ca^{2+} signaling cascade that triggers genome-wide changes in gene expression. In Arabidopsis, catechin doubled the expression of about 1000 genes within 1 h of treatment. By 12 h many of these same genes were repressed, which may reflect the onset of cell death. Laboratory experiments examining the effects of catechin on plant germination and growth showed that native North American grassland species vary considerably in their sensitivity to catechin. Resistant species may produce root exudates that detoxify this allelochemical.

Some plants are biotrophic pathogens of other plants

While most plants are autotrophic, some plants have evolved into parasites that rely on other plants to provide essential nutrients for their own growth and development. Parasitic plants can be divided into two main groups depending on the degree of parasitism. **Hemiparasitic plants** retain the ability to perform at least some photosynthesis, while **holoparasitic plants** are completely parasitic

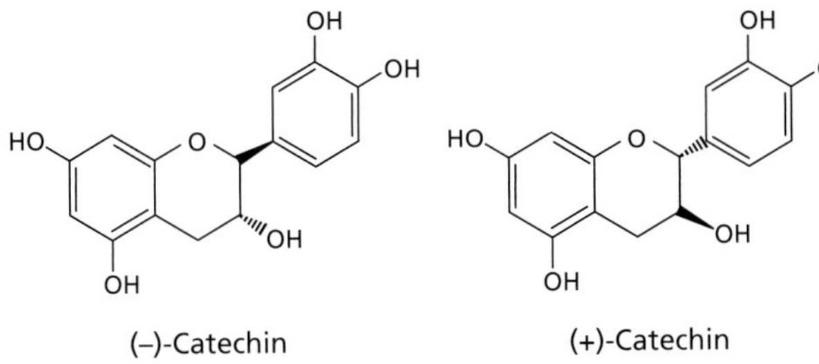

syncytium A multinucleate cell that can result from multiple cell fusions of uninuclear cells, usually in response to viral infection.

hemiparasitic plants Photosynthetic plants that are also parasites.

holoparasitic plants Nonphotosynthetic plants that are obligate parasites.

(–)-Catechin   (+)-Catechin

Figure 18.22 Phytotoxic allelopathic compounds produced by spotted knapweed (*Centaurea maculosa*).

Figure 18.23 Parasitic plants. (A) Mistletoe (*Viscum* sp.) on a mesquite tree (*Prosopis* sp.). (B) Clearly visible is the green stem of the mistletoe growing through the bark of the host plant. (C) Dodder (*Cuscuta* sp.) growing on a patch of sand verbena (*Abronia umbellata*) on dunes at the Pacific coast in California. (D) Close-up showing the high density of infestation of dodder on its host plant. (Photos © J. Engelberth.)

on their host plants and have lost the ability to carry out photosynthesis. For example, mistletoe (genus *Viscum*), which has green leaves and is able to perform photosynthesis, is a hemiparasite (**Figure 18.23A and B**). In contrast, dodder (genus *Cuscuta*), which has lost the ability to photosynthesize and depends entirely on the host for sugars, is a holoparasite (**Figure 18.23C and D**).

Parasitic plants have developed a specialized structure, the **haustorium**, which is a modified root (**Figure 18.24**). After establishing contact with its host plant, the haustorium penetrates the epidermis or bark and then the parenchyma to grow into the vascular tissue and absorb nutrients from the host. To reach the host plant, seeds of parasitic plants are either directly deposited by birds or are more randomly distributed by wind or other means. After germination, the seedlings must rely for a time on their seeds for their food supply, until they can find a suitable host. Recent research has shown that low amounts of species-specific plant volatiles may serve as cues for dodder seedlings and direct their growth toward the host. Alternatively, in the case of root parasites, such as *Striga*, compounds secreted by the host root guide the growth of the seedling roots toward the host. Upon contacting the host root, the *Striga* seedling root develops into a haustorium. The haustorium then penetrates the host root and grows directly into the vascular system through the pits of xylem vessels, where it absorbs the necessary nutrients through tubelike protoplasmic structures not covered by a cell wall.

The mechanisms of these interactions between parasitic plants and their hosts have been studied mostly at the morphological level, and little is known about the signaling mechanisms involved. It is clear that metabolites that are secreted or emitted as volatiles by the host plant provide important cues for the

haustorium The hyphal tip of a fungus or root tip of a parasitic plant that penetrates host plant tissue.

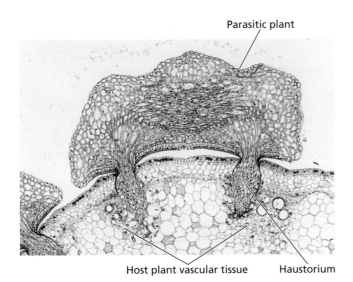

Parasitic plant

Host plant vascular tissue   Haustorium

Figure 18.24 Micrograph showing the haustorium of dodder penetrating the tissues of its host plant. (Photo © Biodisc/Visuals Unlimited, Inc.)

parasite. However, other factors, such as light, may also play a significant role in this process. There is also little known about the defense mechanisms of the host plant. It is likely that common defense signaling pathways, including jasmonic acid, salicylic acid, and ethylene, play an important role in the defense against parasitic plants, but much more research is needed.

Summary

Plants have evolved many strategies to cope with the threats by pests and pathogens. Strategies include sophisticated detection mechanisms and the production of toxic and repelling secondary metabolites. While some of these responses are constitutive, others are inducible. Overall, these strategies have led to a standoff in the coevolutionary race between plants and their pests.

Beneficial Interactions between Plants and Microorganisms

• Plants grow in an environment that includes positive, neutral, and negative interactions with other living organisms (**Figure 18.1**).

• Mycorrhizal interactions can form at the root surface (ectomycorrhizae) or within root cells (arbuscular mycorrhizae) plant roots to enhance phosphorus assimilation from soils.

• Rhizobacteria can release metabolites that assist plant growth by increasing nutrient availability and pathogen protection (**Figure 18.2**).

Harmful Interactions of Pathogens and Herbivores with Plants

• Mechanical barriers that provide a first line of defense against pests and pathogens include thorns, spines, prickles,

trichomes, and raphides. (**Figures 18.3–18.5**).

• Touch-sensitive *Mimosa* plants deter herbivores by folding their leaflets when grazed (**Figure 18.6**).

• Specialized metabolites such as phytoalexins act as antimicrobial and antiherbivory agents (**Figure 18.7**).

• Specialized metabolites that serve defensive functions are stored in adapted structures that release their contents only upon damage (**Figures 18.8–18.11**).

• Some secondary metabolites are stored in specialized vacuoles as water-soluble sugar conjugates that are spatially separated from their activating enzymes (**Figure 18.12**).

Inducible Defense Responses to Insect Herbivores

• Rather than producing defensive secondary metabolites continuously, plants can save energy by producing defensive compounds only when induced by mechanical damage or specific components of insect saliva (elicitors).

• The concentration of jasmonic acid (JA) increases rapidly in response to insect damage and induces transcription of genes involved in plant defense (**Figure 18.13**).

(Continued)

Summary (continued)

- Herbivore damage can induce systemic defenses by causing the synthesis of polypeptide signals. For example, systemin is released to the apoplast and binds to receptors in undamaged tissues, activating JA synthesis there (**Figure 18.14**).

- In addition to producing polypeptide signals, plants can also project electrical signals to initiate defense responses in as-yet undamaged tissues (**Figure 18.15**).

- Plants may release volatile compounds to attract natural enemies of herbivores, or to signal neighboring plants to initiate defense mechanisms (**Figure 18.16**).

Plant Defenses against Pathogens

- Pathogens can invade plants through cell walls by secreting lytic enzymes, through natural openings such as stomata and lenticels, and through wounds. Insect herbivores may also be pathogen vectors (**Figure 18.17**).

- Pathogens generally use one of three attack strategies: necrotrophism, biotrophism, or hemibiotrophism.

- Pathogens often produce effector molecules (enzymes, toxins, or growth regulators) that aid in initial infection (**Figure 18.18**).

- All plants have pattern recognition receptors (PRRs) that set off defense responses when activated by evolutionarily conserved microbe-associated molecular patterns (MAMPs) (e.g., flagella, chitin) (**Figure 18.19**).

- Another antipathogen defense is the hypersensitive response, in which cells surrounding the infected site die

rapidly, thereby limiting the spread of infection. The hypersensitive response is often preceded by rapid production of ROS and NO, which may kill the pathogen directly or aid in cell death (**Figure 18.20**).

- A plant that survives local pathogen infection often develops increased resistance to subsequent attack, a phenomenon called systemic acquired resistance (SAR).

Plant Defenses against Other Organisms

- Nematodes (roundworms) are parasites that can move between hosts and that induce formation of feeding structures and galls from vascular plant tissues (**Figure 18.21**).

- Some plants produce allelopathic secondary metabolites that enable them to outcompete nearby plant species (**Figure 18.22**).

- Some plants are parasitic on other plants. Parasitic plants can be divided into two main groups (hemiparasites and holoparasites) depending on their ability to perform some photosynthesis (**Figure 18.23**).

- Parasitic plants use a specialized structure, the haustorium, to penetrate their host, grow into the vasculature, and absorb nutrients (**Figure 18.24**).

- Some parasitic plants detect their host by the specific volatile profile that is constitutively released.

Suggested Reading

Belkhadir, Y., Yang, L., Hetzel, J., Dangl, J. L., and Chory, J. (2014) The growth-defense pivot: Crisis management in plants mediated by LRR-RK surface receptors. *Trends Biochem Sci.* 39: 447–456. DOI: 10.1016/j.tibs.2014.06.006

Elzinga, D. A., and Jander, G. (2013) The role of protein effectors in plant-aphid interactions. *Curr. Opin. Plant Biol.* 16: 451–456. DOI: 10.1016/j.pbi.2013.06.018

Gleadow, R. M., and Møller, B. L. (2014) Cyanogenic glycosides: Synthesis, physiology, and phenotypic plasticity. *Annu. Rev. Plant Biol.* 65: 155–185. DOI: 10.1146/annurev-arplant-050213-040027

Holeski, L. M., Jander, G., and Agrawal, A. A. (2012) Transgenerational defense induction and epigenetic inheritance in plants. *Trends Ecol. Evol.* 27: 618–626. DOI: 10.1016/j.tree.2012.07.011

Jung, S. C., Martinez-Medina, A., Lopez-Raez, J. A., and Pozo, M. J. (2012) Mycorrhiza-induced resistance and priming of plant defenses. *J. Chem. Ecol.* 38: 651–664. DOI: 10.1007/s10886-012-0134-6

Kachroo, A., and Robin, G. P. (2013) Systemic signaling during plant defense. *Curr. Opin. Plant Biol.* 16: 527–533. DOI: 10.1016/j.pbi.2013.06.019

Kandoth, P. K., and Mitchum, M. G. (2013) War of the worms: How plants fight underground attacks. *Curr. Opin. Plant Biol.* 16: 457–463. DOI: 10.1016/j.pbi.2013.07.001

Kazan, K., and Lyons, R. (2014) Intervention of phytohormone pathways by pathogen effectors. *Plant Cell* 26: 2285–2309.

Romeis, T., and Herde, M. (2014) From local to global: CDPKs in systemic defense signaling upon microbial and herbivore attack. *Curr. Opin. Plant Biol.* 20: 1–10. DOI: 10.1016/j.pbi.2014.03.002

Yan, S., and Dong, X. (2014) Perception of the plant immune signal salicylic acid. *Curr. Opin. Plant Biol.* 20: 64–68. DOI: 10.1016/j.pbi.2014.04.006

19 Abiotic Stress

Plants grow and reproduce in potentially stressful environments containing a multitude of abiotic (nonliving) chemical and physical factors, which vary both with time and geographic location. The primary abiotic environmental parameters that affect plant growth are light (intensity, quality, and duration), water (soil availability and humidity), carbon dioxide, oxygen, soil nutrient content and availability, temperature, and toxins (i.e., heavy metals and salinity). Fluctuations of these abiotic factors outside their normal ranges usually have negative biochemical and physiological consequences for plants. Being sessile, plants are unable to avoid abiotic stress simply by moving to a more favorable environment. Instead, plants have evolved the ability to compensate for stressful conditions by altering physiological and developmental processes to maintain growth and reproduction.

In this chapter we provide an integrated view of how plants adapt and respond to abiotic stresses in the environment. Like all living organisms, plants are complex biological systems comprising thousands of different genes, proteins, regulatory molecules, signaling agents, and chemical compounds that form hundreds of interlinked pathways and networks. Under normal growing conditions, the different biochemical pathways and signaling networks must act in a coordinated manner to balance environmental inputs with the plant's genetic imperative to grow and reproduce. When exposed to unfavorable environmental conditions, this complex interactive system adjusts *homeostatically* to minimize the negative impacts of stress and maintain metabolic equilibrium (**Figure 19.1**).

We begin by distinguishing between adaptation and acclimation in relation to abiotic stress. Next, we describe the various abiotic factors in the environment that can negatively affect plant growth and development.

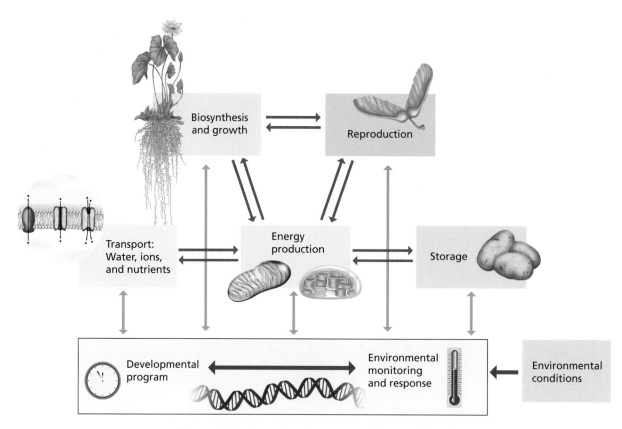

Figure 19.1 Interactions between environmental conditions and plant development, growth, energy production, and ion and nutrient balance and storage. The balance between these processes is controlled by the plant genome (lower blue box), which encodes sensors and signal transduction pathways that monitor and adjust for environmental parameters. Based on the different environmental stress signals, the plant genome can thus direct the flow of energy between the different processes (brown arrows) to establish a new homeostatic state matched to the specific stress conditions.

In the remainder of the chapter, we discuss plant stress-sensing mechanisms and the processes that transform sensory signals into physiological responses. Finally, we describe the specific metabolic, physiological, and anatomical changes that result from these signaling pathways and that enable plants to adapt or acclimate to abiotic stress.

Defining Plant Stress

The ideal growth conditions for a given plant can be defined as the conditions that allow the plant to achieve its maximum growth and reproductive potential as measured by plant weight, height, and seed number, which are all components of the *total biomass* of the plant. **Stress** can be defined as any environmental condition, biotic or abiotic, that prevents the plant from achieving its full genetic potential. For example, a decrease in light intensity would cause a reduction in photosynthetic activity with a concomitant decrease in the energy supply to the plant. Under these conditions, the plant could compensate either by slowing down biosynthesis, thus reducing its growth rate, or by drawing on its stored food reserves in the form of starch (see Figure 19.1).

As anyone who has ever forgotten to water their garden can attest, drought stress can cause severe wilting (**Figure 19.2A and B**). A decrease in water availability can also have a deleterious effect on growth. One way that plants

stress Disadvantageous influences exerted on a plant by external abiotic or biotic factor(s), such as herbivory, infection, heat, water, or anoxia. Measured in relation to plant survival, crop yield, biomass accumulation, or CO_2 uptake.

(A)

Control

(B)

Severe drought

(C)

Control   Moderate   Severe
           drought   drought

Figure 19.2 Comparisons of control (non-drought) and drought-stressed squash (*Cucurbita pepo*) and rice (*Oryza sativa*) plants. (A) Well-watered squash plant. (B) Drought-stressed squash plant. (C) Control, moderately drought-stressed, and severely drought-stressed rice plants. (A and B © iStock.com/PlazacCameraman; C courtesy of Eduardo Blumwald.)

compensate for a decrease in water potential is by closing their stomata, which reduces water loss by transpiration. However, stomatal closure also decreases CO_2 uptake by the leaf, thereby reducing photosynthesis and suppressing growth. An example of the effects of drought on the growth of rice plants is shown in **Figure 19.2C**. Rice is able to tolerate moderate drought without any measurable effect on growth, but severe drought strongly inhibits vegetative growth.

Physiological adjustment to abiotic stress involves trade-offs between vegetative and reproductive development

How do changes in environmental conditions affect reproduction? Under optimal growing conditions, the competition for resources among the different plant organs or developmental phases is minimal. The transition to reproductive growth occurs only after the vegetative adult phase completes its genetically determined developmental program (see Chapter 17). Under stress conditions, however, the vegetative growth program may terminate prematurely, and an annual plant may go immediately to the reproductive phase. In this case, the plant undergoes a transition to flowering, fertilization, and seed set before the plant has reached its full size, resulting in a smaller plant. With fewer leaves to provide photosynthate, plants growing under suboptimal conditions may also produce fewer and smaller seeds.

The particular developmental pathway that a plant uses to maximize its reproductive potential under abiotic stress depends to a large extent on the plant's life cycle. For example, *annual plants* complete their life cycle in a single season. It is thus advantageous for annual plants to adjust their metabolism and developmental programs so as to produce the maximum number of viable seeds under whatever environmental conditions are encountered during the season.

By contrast, *perennial plants*, which have multiple seasons in which to produce seeds, tend to adjust their metabolism and developmental programs to ensure the optimal storage of food resources that will enable the plants to survive to the next season, even at the expense of seed production in the current season.

Acclimation versus Adaptation

Individual plants respond to changes in the environment by directly altering their physiology or morphology to enhance survival and reproduction. Such responses require no new genetic modifications. If the response of the individual plant improves with repeated exposure to the environmental stress, then the response is termed **acclimation**. Acclimation represents a nonpermanent change in the physiology or morphology of the individual that can be reversed if the prevailing environmental conditions change. Epigenetic mechanisms that alter the expression of genes without changing the genetic code of an organism can extend the duration of acclimation responses and make them heritable.

One example of acclimation, from gardening, is a process known as *hardening off*. To speed up the growth of plants, gardeners often start by growing them indoors in pots under optimal growth conditions. The gardeners then move the plants outdoors for part of the day over a period long enough to acclimate, or "harden," the plants to outdoor weather before moving them outdoors permanently.

Another example of acclimation is the response of salt-sensitive plants, termed *glycophytic* plants, to salinity. Although glycophytic plants are not genetically adapted to growth in saline environments, when exposed to elevated salinity they can activate several stress responses that allow the plants to cope with the physiological perturbations imposed by elevated salinity in their environment. Salt exposure stimulates enhanced efflux of Na^+ from cells to reduce salinity-induced toxicity. This response is not sufficient to protect plants in hypersaline conditions, but it can afford short-term protection when salt concentrations temporarily increase in drying soils.

In contrast to acclimation, which is a result of transient changes in an individual plant, morphological or physiological changes in a population that have become genetically fixed over many generations by natural selection are referred to as **adaptations**. A remarkable example of adaptation to an extreme abiotic environment is the growth of plants in serpentine soils. Serpentine soils are characterized by low moisture, low concentrations of macronutrients, and elevated levels of nickel, cobalt, and chromium ions. These conditions would result in severe stress conditions for most plants. However, it is not unusual to find populations of plants that have become genetically adapted to serpentine soils growing not far from closely related nonadapted plants growing on "normal" soils. Simple transplant experiments have shown that only the adapted populations can grow and reproduce on the serpentine soil, and genetic crosses reveal the stable genetic basis of this adaptation.

The evolution of adaptive mechanisms in plants to a particular set of environmental conditions generally involves processes that allow *avoidance* of the potentially damaging effects of these conditions. For example, populations of the weed Yorkshire fog grass (*Holcus lanatus*) that are adapted to growth on arsenic-contaminated mine sites in southwestern England contain a specific genetic modification that reduces the uptake of arsenate via the phosphate uptake system, allowing the plants to avoid arsenic toxicity and thrive on contaminated mine sites. In contrast, populations growing on uncontaminated soils are less likely to contain this genetic modification.

Both adaptation and acclimation can contribute to plants' overall tolerance of extremes in their abiotic environment. In the example above, genetic adaptation

acclimation   The increase in plant stress tolerance due to exposure to prior stress. May involve changes in gene expression.

adaptation   An inherited level of stress resistance acquired by a process of selection over many generations.

in the arsenic-tolerant Yorkshire fog grass population only *reduces* arsenate uptake—it does not stop it. To mitigate the toxic effects of the arsenate that does accumulate, the adapted plants use the same biochemical mechanism that nonadapted plants use to respond to the toxic effects of arsenate accumulation in tissues. This mechanism involves the biosynthesis of low-molecular-weight, metal-binding molecules called *phytochelatins*, which can bind heavy metals or arsenic to reduce cellular toxicity. Thus, the ability of Yorkshire fog grass to thrive on arsenic-contaminated mine waste depends on both a specific genetic adaptation found in the tolerant population (arsenate exclusion) and on acclimation, which is common to all plants that respond to arsenic by producing phytochelatins.

Environmental Stressors

In this section we briefly describe the ways in which various environmental stresses can disrupt plant metabolism. As with every biological system, plant survival and growth depend on complex networks of coupled anabolic and catabolic pathways that direct the flow of energy and resources within and between cells. Uncoupling of the pathways by environmental stressors can disrupt these networks. For example, metabolic enzymes can, and often do, have different temperature optima. An increase or decrease in ambient temperature can inhibit a subset of enzymes without affecting other enzymes in the same or connected pathways. Such functional uncoupling of metabolic pathways could result in the accumulation of intermediate compounds that could be converted to toxic by-products.

Environmental stress can also disrupt compartmentation of metabolic processes that isolates them from other cellular components. The same temperature extremes that can inhibit enzyme activity can also affect membrane fluidity: High temperature increases fluidity, and low temperature decreases fluidity. Changes in membrane fluidity can disrupt the coupling between different protein complexes in chloroplast or mitochondrial membranes, resulting in the uncontrolled transfer of electrons to oxygen and the formation of reactive oxygen species (ROS) that can cause cellular damage.

As shown in **Table 19.1**, individual abiotic stresses have primary and secondary effects on plant growth. Some of these effects overlap. This often makes diagnosis of plant stress more difficult. In general, the greater the combination of stress factors, the more likely it is that a plant will succumb over a given interval.

Water deficit decreases turgor pressure, increases ion toxicity, and inhibits photosynthesis

As in most other organisms, water makes up the largest proportion of the cellular volume in plants and is the most limiting resource. About 97% of water taken up by plants is lost to the atmosphere (mostly by transpiration). About 2% is used for volume increase or cell expansion, and 1% for metabolic processes, predominantly photosynthesis (see Chapters 2 and 3). Water deficit (insufficient water availability) occurs in most natural and agricultural habitats and is caused mainly by intermittent to continuous periods without precipitation. *Drought* is the meteorological term for a period of insufficient precipitation that results in plant water deficit. However, this definition is somewhat misleading, since a crop can absorb water from the soil under conditions without rainfall, depending on the soil's water-holding capacity and the depth of the water table.

Water deficit can affect plants differently during vegetative versus reproductive growth. When plant cells experience water deficit, cell dehydration occurs. Cell dehydration adversely affects many basic physiological processes. For example, during water deficit the water potential (Ψ) of the apoplast becomes more negative than that of the symplast, causing reductions in pressure

Table 19.1 **Physiological and biochemical perturbations in plants caused by fluctuations in the abiotic environment**

Environmental factor	Primary effects	Secondary effects
Water deficit	Water potential (Ψ) reduction	Reduced cell/leaf expansion
	Cell dehydration	Reduced cellular and metabolic activities
	Hydraulic resistance	Stomatal closure
		Photosynthetic inhibition
		Leaf abscission
		Altered carbon partitioning
		Cavitation
		Membrane and protein destabilization
		ROS production
		Ion cytotoxicity
		Cell death
Salinity	Water potential (Ψ) reduction	Same as for water deficit (see above)
	Cell dehydration	
	Ion cytotoxicity	
Flooding and soil compaction	Hypoxia	Reduced respiration
	Anoxia	Fermentative metabolism
		Inadequate ATP production
		Production of toxins by anaerobic microbes
		ROS production
		Stomatal closure
High temperature	Membrane and protein destabilization	Photosynthetic and respiratory inhibition
		ROS production
		Cell death
Chilling	Membrane destabilization	Membrane dysfunction
Freezing	Water potential (Ψ) reduction	Same as for water deficit (see above)
	Cell dehydration	Physical destruction
	Symplastic ice crystal formation	
Trace element toxicity	Disturbed cofactor binding to proteins and DNA	Disruption of metabolism
	ROS production	
High light intensity	Photoinhibition	Inhibition of PSII repair
	ROS production	Reduced CO_2 fixation

potential (turgor) (Ψ_p) and volume. A secondary effect of cell dehydration is that ions become more concentrated and may become cytotoxic. Water deficit also induces the accumulation of abscisic acid (ABA), which promotes stomatal closure, reducing gas exchange and inhibiting photosynthesis (**Figure 19.3**). Dehydration can also disrupt chloroplast function, causing the photosystems to become uncoupled. When this occurs, free electrons produced by the reaction centers are not transferred to NADP+, leading to the generation of ROS. Excess ROS damage DNA, inhibit protein synthesis, oxidize photosynthetic pigments, and cause the peroxidation of membrane lipids.

Salinity stress has both osmotic and cytotoxic effects

Excess soil salinity brought about by a combination of over-irrigation and poor soil drainage affects large regions of the world's land mass and has a severe impact on agriculture (**Figure 19.4**). It is estimated that 20% of all irrigated land is currently affected by salinity stress. Salinity stress has two components: nonspecific **osmotic stress** that causes water deficits, and specific ion effects resulting from the accumulation of toxic ions, which interfere with nutrient uptake and cause cytotoxicity. Salt-tolerant plants genetically adapted to salinity are termed **halophytes** (from the Greek word *halo*, "salty"), while less salt-tolerant plants that are not adapted to salinity are termed **glycophytes** (from the Greek word *glyco*, "sweet"). Under nonsaline conditions, the cytosol of higher plant cells contains about 100 mM K$^+$ and less than 10 mM Na$^+$, an ionic environment in which enzymes are optimally functional. In saline environments, cytosolic Na$^+$ may increase to more than 100 mM, and these ions become cytotoxic. High cellular concentrations of Na$^+$ and K$^+$ can cause protein denaturation and membrane destabilization by reducing the hydration of these macromolecules. However, Na$^+$ is a more potent denaturant than K$^+$. At high concentrations, apoplastic Na$^+$ also competes for sites on transport proteins that are necessary for high-affinity uptake of K$^+$, an essential macronutrient (see Chapter 4).

The effects of high salinity in plants occur through a two-phase process: a fast response to the high osmotic pressure at the root–soil interface and a slower response caused by the accumulation of Na$^+$ (and Cl$^-$) in the leaves. In the osmotic phase there is a reduction in shoot growth, with reduced leaf expansion and inhibition of lateral bud formation. The second phase starts with the accumulation of toxic amounts of Na$^+$ in the leaves, which inhibits photosynthesis and biosynthetic processes. Although in most species Na$^+$ reaches toxic concentrations before Cl$^-$, some plant species, such as citrus, grapevine, and soybean, are highly sensitive to excess Cl$^-$.

Temperature stress affects a broad spectrum of physiological processes

Temperature stress disrupts plant metabolism because of its differential effect on protein stability and enzymatic reactions, causing the uncoupling of different reactions and the accumulation of toxic intermediates and ROS. As noted ear-

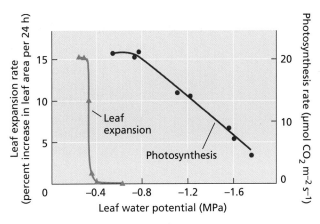

Figure 19.3 Effects of water stress on photosynthesis and leaf expansion of sunflower (*Helianthus annuus*). In this species, leaf expansion is completely inhibited under mild stress levels that barely affect photosynthetic rates. (After Boyer 1970.)

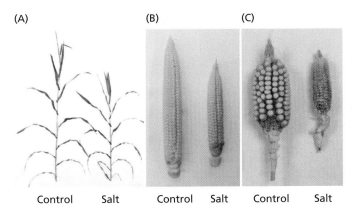

Figure 19.4 Effects of salt stress on maize production. (A) Stalk height. (B) Ear size. (C) Kernel number. (From Henry et al. 2015.)

osmotic stress Stress imposed on cells or whole plants when the osmotic potential of external solutions is more negative than that of the solution inside the plant.

halophytes Plants that are native to saline soils and complete their life cycles in that environment.

glycophytes Plants that are not able to resist salts to the same degree as halophytes. Show growth inhibition, leaf discoloration, and loss of dry weight at soil salt concentrations above a threshold.

lier, extreme temperatures (high or low) affect membrane fluidity, uncoupling multiprotein complexes and disrupting electron flow, energetic reactions, and ion homeostasis and regulation. Heat and cold stress can also destabilize and melt, or overstabilize and harden, respectively, RNA and DNA secondary structures, causing the disruption of transcription, translation, or RNA processing and turnover. In addition, temperature stress can block protein degradation, causing the buildup of protein aggregates. Such protein clumps disrupt normal cellular functions by interfering with the function of the cytoskeleton and associated organelles.

Plants subjected to freezing temperatures must contend with the formation of ice crystals, either inside or outside the cells. Intracellular ice crystal formation nearly always proves lethal to the cell. However, the water in the apoplast is relatively dilute and therefore has a higher freezing point than that of the more concentrated symplast. As a result, ice crystals tend to form in the apoplast and in the xylem tracheids and vessels, along which the ice can quickly propagate. The formation of ice crystals lowers the apoplastic water potential (Ψ), which becomes more negative than that of the symplast. Unfrozen water within the cell moves down this gradient out of the cell toward the ice crystals in the intercellular spaces. As water leaves the cell, the plasma membrane contracts and pulls away from the cell wall. During this process the plasma membrane, rigidified by the low temperature, may become damaged. The colder the temperatures, the more water travels down the gradient toward the frozen water. For example, at −10°C the symplast loses about 90% of its osmotically active water to the apoplast. In this respect, freezing stress has much in common with drought stress. As with drought stress, cells that are already dehydrated, like those in seeds and pollen, are less likely to undergo further dehydration and damage by extracellular ice crystal formation.

Flooding results in anaerobic stress to the root

When a field is flooded, the O_2 levels at the root surface decrease dramatically because most of the air in the soil is displaced by water, and the O_2 concentration of water is significantly lower than that of air: The atmosphere contains about 20% O_2 or 200,000 ppm, compared with less than 10 ppm dissolved O_2 in flooded soil. Under these conditions, respiration in roots is suppressed and fermentation is enhanced (see Chapter 11). This metabolic shift can cause energy depletion, acidification of the cytosol, and toxicity from ethanol accumulation. As a consequence of energy depletion, many processes such as protein synthesis are suppressed. Anaerobic stress can cause cell death within hours or days, depending on the degree of genetic adaptation of the species.

Even if the O_2-deprived plant is returned to normal O_2 levels, the recovery process itself can pose a hazard. While the roots are under anaerobic stress, the absence of O_2 prevents the formation of ROS. But if the O_2 level in the soil is rapidly increased, ROS formation increases sufficiently to cause oxidative damage to the root cells. This damage is similar to that observed in plants exposed to excess atmospheric ozone associated with industrial and automotive emissions.

Light stress can occur when shade-adapted or shade-acclimated plants are subjected to full sunlight

Light stress can occur when excess high-intensity light absorbed by the plant overwhelms the capacity of the photosynthetic machinery to convert this light into sugars, as in the case of a shade-adapted or shade-acclimated plant suddenly subjected to full sunlight. In response to shade, most land plants either add more light-harvesting chlorophyll units (LHCII) to PSII, augmenting antenna size, or increase the number of PSII reaction centers relative to PSI to enhance light capture and energy transfer (see Chapters 7 and 9). If the shade-adapted or shade-acclimated plants are suddenly subjected to full sunlight, the excess light energy absorbed by the enlarged antenna complexes and transferred to the reaction centers can overwhelm the carbon reactions' ability to convert the energy into sugars. Instead, the electrons feeding into the reaction centers are diverted to atmospheric oxygen, generating ROS, which can, in turn, cause cellular damage. Similar damage can be observed when intense sunlight strikes conifer needles in arctic and subarctic regions while temperatures are still subzero and carbon fixation reactions cannot keep up with light reactions.

Light stress is also associated with increased UV radiation. Excess UV light disrupts photosynthesis, damages DNA, and induces the formation of ROS. UV stress has been shown to suppress plant growth and reduce agronomic yields, especially in temperate-climate crops that are grown at higher elevations in the tropics during the winter season.

Heavy metal ions can both mimic essential mineral nutrients and generate ROS

The uptake of heavy metal ions such as cadmium (Cd) ions can disrupt normal cellular processes such as photosynthesis, mineral nutrient utilization, and enzymatic functions. One of the reasons that heavy metal ions are so toxic is that they can mimic essential metal ions (e.g., Ca^{2+} and Mg^{2+}), take their place in essential reactions, and disrupt these reactions. Cd^{2+}, for example, can replace Mg^{2+} in chlorophyll or Ca^{2+} in the Ca^{2+}-signaling protein calmodulin, disrupting both photosynthesis and signal transduction. The mimicking of essential elements can also explain the uptake of Cd^{2+} and other heavy metal ions into cells via channels that evolved to transport essential nutrient ions. Heavy metals can also bind to and inhibit different enzymes, as well as directly interact with oxygen to form ROS.

Although aluminium (Al) ions are a major component of most soils, they become phytotoxic when enriched in tropical acidic soil solutions, which results in stunted root growth. Similarly, the metalloid (i.e., intermediate between metal and nonmetal) arsenic is present in many shale-derived soils, and arsenate is produced by natural weathering. Arsenate inhibits metabolic processes and disrupts DNA when it displaces phosphate in biological reactions. Even essential nutrients such as copper and iron ions are toxic and cause stress when soil concentrations are excessive.

Combinations of abiotic stresses can induce unique signaling and metabolic pathways

In the field, plants are often subjected to a combination of different abiotic stresses simultaneously. Drought and heat stress are examples of two different abiotic stresses that almost always occur together in the environment, with devastating results. Between 1980 and 2004 in the United States, the cost of crop damage due to drought plus heat was six times greater than the cost due to drought alone (**Figure 19.5A**).

The physiological acclimation of plants to a combination of different abiotic stresses is different from the acclimation of plants to different abiotic stresses applied individually. **Figure 19.5B** shows the effects of heat and drought, applied separately, on four physiological parameters of Arabidopsis: photosynthesis, respiration, stomatal conductance, and leaf temperature. The physiological profiles under the two stresses applied individually were quite different. Heat alone caused an elevation of leaf temperature and a large increase in stomatal conductance. Drought, however, was more inhibitory to photosynthesis and stomatal opening. The main effect of the combination of drought plus heat was a significant elevation in leaf temperature that could be deadly to the plant.

The combination of heat plus drought also induced different patterns of gene expression and metabolite biosynthesis than either stress alone. As shown in

(A)

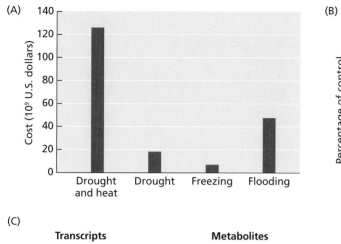

(B)

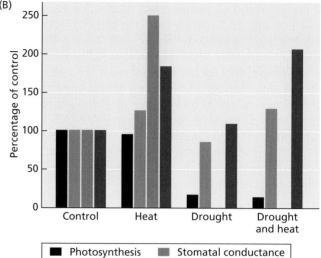

(C)

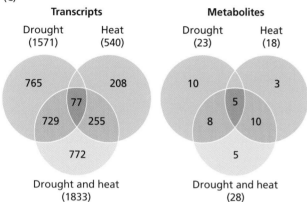

Figure 19.5 Effect of combined abiotic stresses on plant productivity, physiology, and molecular responses. (A) Losses to U.S. agriculture resulting from a combination of drought and heat stress were much higher than losses caused by drought, freezing, or flooding alone between 1980 and 2004. (B) The effect of combined drought and heat on plant physiology. Note the complete closure of plant stomata, which results in a higher leaf temperature. (C) Venn diagrams showing the effect of combined drought and heat on the transcriptome (left) and metabolome (right) of plants. (After Mittler 2006.)

Figure 19.5C, drought plus heat caused the accumulation of 772 unique transcripts (left, yellow) and 5 unique metabolites (right, yellow), demonstrating that the acclimation of plants to the combination of drought and heat is different in many aspects from the acclimation of plants to drought or heat stress applied individually. Differences in physiological parameters, transcript accumulation, and metabolites could be a result of conflicting physiological responses to the two stresses. For example, during heat stress, plants *increase* their stomatal conductance, which cools their leaves by transpiration. However, if heat stress occurs simultaneously with drought, the stomata are closed, causing leaf temperature to be 20 to 25 degrees Celsius higher.

Salinity or heavy metal stress could pose a similar problem when combined with heat stress, because enhanced transpiration could result in enhanced uptake of salt or heavy metals. In contrast, some stress combinations could have beneficial effects on plants compared with the individual stresses applied separately. For example, drought, which causes stomatal closure, could potentially enhance tolerance to ozone. Among the several stress combinations that could have a deleterious effect on crop productivity are drought and heat, salinity and heat, cold and high light, nutrient deficiency and drought, and nutrient deficiency and salinity. Interactions that could have a beneficial impact include drought and ozone, ozone and UV, and high CO_2 combined with drought, ozone, or high light.

Perhaps the most studied stress interactions are those of different abiotic stresses with biotic stresses, such as pests or pathogens. In most cases, prolonged exposure to abiotic stress conditions, such as drought or salinity, results in the weakening of plant defenses and enhanced susceptibility to pests or pathogens.

Sequential exposure to different abiotic stresses sometimes confers cross-protection

Several studies have reported that the application of a particular abiotic stress condition can enhance the tolerance of plants to a subsequent exposure to a different type of abiotic stress. This phenomenon is called **cross-protection**. It occurs because many stresses result in the accumulation of the same general stress-response proteins and metabolites—for example, ROS-scavenging enzymes, molecular chaperones, and osmoprotectants—and they persist in plants for some time even after the stress conditions have subsided. The application of a second stress to the same plants that experienced the initial stress may therefore have a decreased effect because the plants are already primed and ready to deal with several different aspects of the new stress condition.

cross-protection   A plant response to one environmental stress that confers resistance to another stress.

Plants use a variety of mechanisms to sense abiotic stress

As discussed above, environmental stress disrupts or alters many physiological processes in the plant by affecting lipid fluidity, protein or RNA stability, ion transport, the coupling of reactions, or other cellular functions. Interruptions of cellular processes can disrupt electron transport chains or electrophilic reactions to produce ROS that, in turn, produce cellular damage. Any of these primary disruptions could be signaling the plant that a change in environmental conditions has occurred and that it's time to respond by altering existing pathways or by activating stress-response pathways. Stress can also activate receptors and downstream second messenger or phosphorylation signaling to activate physiological responses (see Chapter 12). Nonspecific stress activation of receptors can sometimes add to damage, especially in a plant species that is not adapted to a particular form of stress. Stress-sensing mechanisms can act individually or in combination to activate downstream signal transduction pathways.

Initial stress-sensing mechanisms trigger downstream responses involving multiple signal transduction pathways. As diagrammed in **Figure 19.6**, these pathways involve receptors, calcium ions, protein kinases, protein phosphatases, ROS signaling, transcriptional regulators, and plant hormones. The signals that emerge from these pathways, in turn, activate or suppress various networks that may either allow growth and reproduction to continue under stress conditions, or enable the plant to survive the stress until more favorable conditions return.

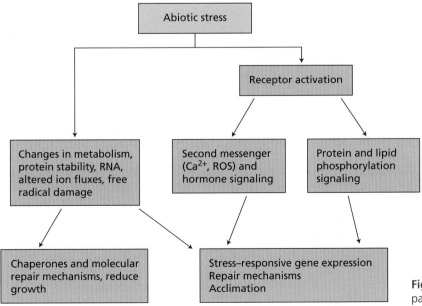

Figure 19.6   Signal transduction and acclimation pathways activated by abiotic stress in plants.

phenotypic plasticity Physiological or developmental responses of a plant to its environment that do not involve genetic changes.

Physiological Mechanisms That Protect Plants against Abiotic Stress

Thus far in the chapter we have discussed the various types of abiotic stress and how they are perceived by plants. In this section we discuss the fruits of the labors of all these genetic networks—the metabolic, physiological, and anatomical changes that are produced to counter the effects of abiotic stress. Terrestrial plants first emerged onto land ~450 million years ago. Thus, they have had ample time to evolve mechanisms to cope with various types of abiotic stress. These mechanisms include the abilities to accumulate protective metabolites and proteins and to regulate growth, morphogenesis, photosynthesis, membrane transport, stomatal apertures, and resource allocation. The effects of these and other changes are to retain cellular homeostasis so that the plant's life cycle can be completed under the new environmental regime. Below we discuss some of the major physiological mechanisms of acclimation.

Plants can alter their morphology in response to abiotic stress

In response to abiotic stress, plants can activate developmental programs that alter their phenotype, a phenomenon known as **phenotypic plasticity**. Phenotypic plasticity can result in adaptive anatomical changes that enable plants to avoid some of the harmful effects of abiotic stress.

An important example of phenotypic plasticity is the ability to alter leaf shape. As biological solar collectors, leaves must be exposed to sunlight and air, which makes them vulnerable to environmental extremes. Plants have thus evolved the ability to modify leaf morphology in ways that enable them to avoid or mitigate the effects of abiotic extremes. Such mechanisms are regulated by multiple factors and include changes in leaf area, leaf orientation, leaf rolling, trichomes, and waxy cuticles, as outlined below.

LEAF AREA Large, flat leaves provide optimal surfaces for the production of photosynthate, but they can be detrimental to crop growth and survival under stressful conditions because they provide a large surface area for evaporation of water, which can lead to quick depletion of soil water or excessive, damaging absorption of solar energy. Plants can reduce their leaf area by reducing leaf cell division and expansion, by altering leaf shapes, and by initiating senescence and abscission of leaves.

LEAF ORIENTATION As described in Chapter 9, the leaves of some plants may orient themselves parallel or perpendicular to the sun's rays to reduce damage. Other factors that can alter the interception of radiation include wilting and leaf rolling. Wilting changes the angle of the leaf, and leaf rolling minimizes the profile of tissue exposed to the sun.

TRICHOMES Many leaves and stems have hairlike epidermal cells known as trichomes or hairs (see Chapter 18). Trichomes can either be ephemeral or persist throughout the life of the organ. Some persisting trichomes remain alive, while others undergo programmed cell death, leaving only their cell walls. Densely packed trichomes on a leaf surface keep leaves cooler by reflecting radiation and reducing evaporation by creating a thicker unstirred surface layer. Leaves of some plants have a silvery white appearance because the densely packed trichomes reflect a large amount of light. However, pubescent leaves are a disadvantage in the cooler spring months because the trichomes also reflect the visible light needed for photosynthesis.

CUTICLE The cuticle is a multilayered structure of waxes and related hydrocarbons deposited on the outer cell walls of the leaf epidermis. The cuticle, like trichomes, can reflect light, thereby reducing heat load. The cuticle appears to also restrict the diffusion of water and gases, as well as the entrance of pathogens. A developmental response to water deficit in some plants is the production of a thicker cuticle, which decreases apoplastic evaporation.

Metabolic shifts enable plants to cope with a variety of abiotic stresses

Changes in the environment may stimulate shifts in metabolic pathways that decrease the effect of stress on plant metabolism. Facultative crassulacean acid metabolism (see Chapter 9) is an adaptation that is used to counter intermittent drought and osmotic stress in many plants. A mechanism common to flooded roots, particularly those of plants routinely exposed to intermittent flooding, is the fermentation of pyruvate to lactate (lactic acid) through the action of lactate dehydrogenase (see Chapter 11). Production of lactate lowers the intracellular pH; this activates pyruvate decarboxylase, which in turn leads to a switch from lactate to ethanol production. The net yield of ATP in fermentation is only 2 moles of ATP per mole of hexose sugar catabolized (compared with 30 moles of ATP per mole of hexose respired in aerobic respiration) and is therefore inadequate to support normal root growth. However, it is sufficient to keep root cells alive until normal oxygen levels are restored.

Heat shock proteins maintain protein integrity under stress conditions

Protein structure is sensitive to disruption by changes in temperature, pH, or ionic strength associated with different types of abiotic stress. **Molecular chaperone proteins** physically interact with other proteins to facilitate protein folding, reduce misfolding, stabilize tertiary structure, and prevent aggregation or mediate disaggregation. A unique set of chaperones, called **heat shock proteins** (**HSPs**), are synthesized in response to a variety of environmental stresses. HSPs are named according to their approximate molecular mass measured in kilodaltons (kDa). Cells that synthesize HSPs in response to heat stress show improved thermal tolerance and can tolerate subsequent exposures to higher temperatures that would otherwise be lethal. HSPs are induced by widely different environmental conditions, including water deficit, wounding, low temperature, and salinity. In this way, cells that have previously experienced one stress may gain cross-protection against another stress. There are several different classes of HSPs, including HSP70s that bind and release misfolded proteins, HSP60s that produce huge barrel-like complexes that are used as chambers for protein folding, HSP101s that mediate the disaggregation of protein aggregates, and small HSPs (sHSPs) that bind and stabilize different complexes and membranes (**Figure 19.7**).

Numerous other proteins have been identified that act in a similar fashion to stabilize proteins and membranes during dehydration, temperature extremes, and ion imbalance. These include the Late Embryogenesis Abundant (LEA)/ dehydrin protein family. LEA proteins accumulate in response to dehydration during the latter stages of seed maturation. Most LEA proteins belong to a more widespread group of proteins called *hydrophilins*. Hydrophilins have a strong attraction for water, fold into α helices upon drying, and have the ability to reduce aggregation of dehydration-sensitive proteins, a property termed *molecular shielding*. **Dehydrins** accumulate in plant tissues in response to a variety of abiotic stresses, including salinity, dehydration, cold, and freezing stress. Dehydrins, like LEA proteins, are highly hydrophilic and are intrinsically disordered proteins. Their ability to serve as molecular shields and as cryoprotectants has been attributed to their flexibility and minimal secondary structure. Since both LEAs and dehydrins are often induced by ABA, they are sometimes referred to as *responsive to ABA proteins*.

molecular chaperone proteins
Proteins that maintain or restore the active three-dimensional structures of other macromolecules.

heat shock proteins (HSPs)
A specific set of proteins that are induced by a rapid rise in temperature, and by other factors that lead to protein denaturation. Most act as molecular chaperones.

dehydrins Hydrophilic plant proteins that accumulate in response to drought stress and cold temperatures.

Figure 19.7 Molecular chaperone network in cells. Nascent proteins requiring the assistance of molecular chaperones to reach a proper conformation are associated with HSP70 chaperones (top). Native proteins that undergo denaturation during stress (right) associate with HSP70 (top right) and HSP60 (bottom right) chaperones. If aggregates are formed (left middle), they are disaggregated by HSP101 and HSP70 (left). Additional stress-related chaperones such as HSP31, HSP33, and sHSPs can also associate with denatured proteins during stress. (After Baneyx and Mujacic 2004.)

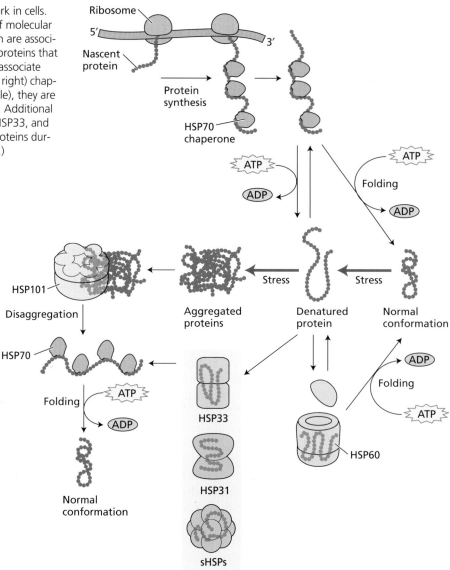

Membrane lipid composition can adjust to changes in temperature and other abiotic stresses

As temperatures drop, plant membranes may go through a phase transition from a flexible liquid-crystalline structure to a solid gel structure. The phase transition temperature varies depending on the lipid composition of the membranes. Chilling-resistant plants tend to have membranes with more unsaturated fatty acids that increase their fluidity, whereas chilling-sensitive plants have a high percentage of saturated fatty acid chains that tend to solidify at low temperatures. In general, saturated fatty acids that have no double bonds solidify at higher temperatures than do lipids that contain polyunsaturated fatty acids, because the latter have kinks in their hydrocarbon chains and do not pack as closely as saturated fatty acids (see Chapter 1).

During acclimation to cold temperatures, the activities of **desaturase enzymes** increase and the proportion of unsaturated lipids rises. Differences in fatty acid composition from mitochondria derived from chilling-sensitive and chilling-resistant species are shown in **Table 19.2**. The resultant changes in lipid composition affect the function of receptors, ion channels, transport proteins,

desaturase enzymes Enzymes that remove hydrogens in a carbon chain to create a double bond between carbons.

Table 19.2 **Fatty acid composition of mitochondria isolated from chilling-resistant and chilling-sensitive species**

| | Percent weight of total fatty acid content | | | | | |
| | Chilling-resistant species | | | Chilling-sensitive species | | |
Major fatty acids[a]	Cauliflower bud	Turnip root	Pea shoot	Bean shoot	Sweet potato	Maize shoot
Palmitic (16:0)	21.3	19.0	17.8	24.0	24.9	28.3
Stearic (18:0)	1.9	1.1	2.9	2.2	2.6	1.6
Oleic (18:1)	7.0	12.2	3.1	3.8	0.6	4.6
Linoleic (18:2)	16.1	20.6	61.9	43.6	50.8	54.6
Linolenic (18:3)	49.4	44.9	13.2	24.3	10.6	6.8
Ratio of unsaturated to saturated fatty acids	3.2	3.9	3.8	2.8	1.7	2.1

Source: After Lyons et al. 1964.

[a]Shown in parentheses are the number of carbon atoms in the fatty acid chain and the number of double bonds.

and other cellular machinery associated with membranes. ABA signaling often mediates perception of these changes and downstream expression of genes associated with stress responses.

Chloroplast genes respond to high-intensity light by sending stress signals to the nucleus

We usually think of the nucleus as the master organelle of the cell, controlling the activities of the other organelles by regulating nuclear gene expression. However, retrograde, or reverse, signaling from the chloroplast (and from mitochondria) to the nucleus also appears to mediate abiotic stress perception. Many abiotic stress conditions affect chloroplasts, either directly or indirectly, and can potentially generate signals that can influence nuclear gene expression and acclimation responses. Light stress, for example, can cause over-reduction of the electron transport chain, enhanced accumulation of ROS, and altered redox potential.

During the acclimation to light stress, the levels of light-harvesting complex II (LHCII) decline due to the down-regulation of the *LHCB* gene, which encodes the apoprotein of the LHCII complex (see Chapter 7). Since *LHCB* is a nuclear gene, the chloroplast appears to send an unidentified stress signal to the nucleus that activates a transcriptional suppressor of *LHCB* gene expression.

A self-propagating wave of ROS mediates systemic acquired acclimation

As in systemic acquired resistance (SAR) during biotic stress (see Chapter 18), abiotic stress applied to one part of the plant can generate signals that can be transported to the rest of the plant, initiating acclimation even in parts of the plant that have not been subjected to the stress. This process is called **systemic acquired acclimation** (**SAA**). Rapid SAA responses to different abiotic stress conditions, including heat, cold, salinity, and high light intensity, have been shown to be propagated by a wave of ROS production, which travels at a rate of approximately 8.4 cm min^{-1} and is dependent on the presence of a specific NADPH oxidase, **respiratory burst oxidase homolog D** (**RBOHD**), which is located on the plasma membrane (**Figure 19.8**). The rapid rates of abiotic stress–systemic signals detected with luciferase imaging in these experiments suggest

systemic acquired acclimation (SAA) A system whereby exposure of one part of a plant to an abiotic stress generates signals that can initiate acclimation to the stress in other, unexposed parts of the plant.

respiratory burst oxidase homolog D (RBOHD) An enzyme that generates superoxide using NADPH as an electron donor.

(A)

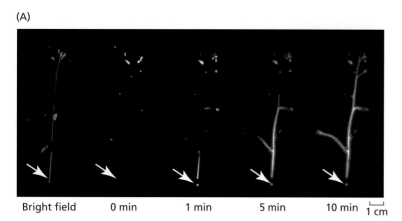

Bright field | 0 min | 1 min | 5 min | 10 min   1 cm

Figure 19.8 Rapid systemic signaling in response to the physical sensing of a wound. (A) Time-lapse imaging of a rapid systemic signal initiated by a wound (arrow) using a luciferase reporter fused to the promoter of the ROS-responsive *ZAT12* gene. Light is emitted from tissues where luciferase is expressed. (B) Schematic model of the ROS wave that is required to mediate rapid systemic signaling in response to abiotic stress. The ROS wave is generated by an active, self-propagating wave of ROS that starts at the initial tissue subjected to stress and spreads to the entire plant. Each cell along the path of the signal activates its RBOHD proteins (NADPH oxidase) and generates ROS. When the signal reaches its systemic targets, it activates acclimation mechanisms in those parts of the plant. The ROS wave is accompanied by a Ca^{2+} wave and electrical signals. (From Mittler et al. 2011.)

(B)

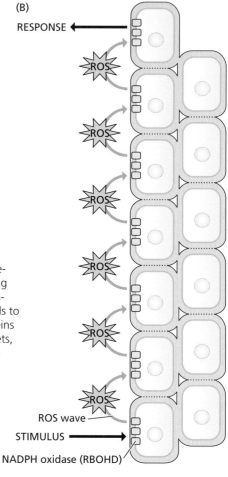

RESPONSE

ROS wave

STIMULUS

NADPH oxidase (RBOHD)

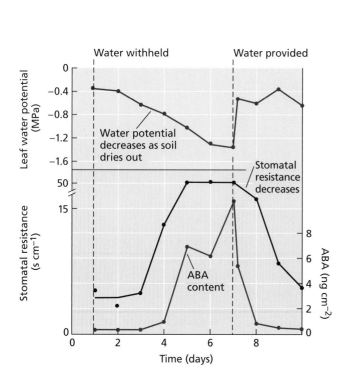

Figure 19.9 Changes in leaf water potential, stomatal resistance (the inverse of stomatal conductance), and ABA content in maize in response to water stress. As the soil dries out, the water potential of the leaf decreases, and the ABA content and stomatal resistance increase. Rewatering reverses the process. (After Beardsell and Cohen 1975.)

that many of the responses to abiotic stresses may occur at a rate much faster than previously thought.

Abscisic acid and cytokinins are stress-response hormones that regulate drought responses

Plant hormones mediate a wide range of acclimation responses and are essential for the ability of plants to adapt to abiotic stresses. ABA biosynthesis is among the most rapid responses of plants to abiotic stress. As shown in **Figure 19.9**, ABA concentrations in leaves can increase up to 20-fold under drought conditions—the most dramatic change in concentration reported for any hormone in response to an environmental signal. Redistribution or biosynthesis of ABA is very effective in causing stomatal closure, and ABA accumulation in stressed leaves plays an important role in the reduction of water loss by transpiration under water-stress conditions (see Figure 19.9). Increases in humidity reduce ABA concentrations by increasing ABA breakdown, thereby permitting stomata to reopen. ABA-biosynthesis or -response mutants are unable to close their stomata under drought conditions and wilt rapidly.

Within the limits set by the plant's genetic potential, a shoot tends to grow until water uptake by the roots becomes limiting to further growth; conversely, roots tend to grow until their demand for photosynthate from the shoot exceeds the supply. This functional balance is shifted if the water supply decreases. When water to the shoot becomes limiting, leaf expansion is reduced before photosynthetic activity is affected. Inhibition of leaf expansion reduces the consumption of carbon and energy, and a greater proportion of the plant's assimilates can be allocated to the root system, where they can support further root growth. This root growth is sensitive to the water status of the soil microenvironment; the root apices in dry soil lose turgor, while roots in the soil zones that remain moist continue to grow.

ABA plays an important role in regulating the root:shoot ratio during water stress. As shown in **Figure 19.10**, under water-stress conditions the root-to-shoot biomass ratio increases, allowing the roots to grow at the expense of the leaves. However, ABA-deficient mutants are unable to change their root:shoot ratio in response to water stress. Thus, ABA is required for the change in the root:shoot ratio to occur.

Another plant hormone that plays a key role in acclimation to various abiotic stresses is cytokinin. Cytokinin and ABA have antagonistic effects on stomatal opening, transpiration, and photosynthesis. Drought results in decreased cytokinin levels and increased ABA levels. Although ABA is normally required for stomatal closure, preventing excessive water loss, drought stress conditions can also inhibit photosynthesis and cause premature leaf senescence. Cytokinins appear to be able to ameliorate the effects of drought. As shown in **Figure 19.11**, transgenic plants that overexpress the gene encoding the enzyme isopentenyl transferase (the enzyme that catalyzes the rate-limiting step in cytokinin synthesis) exhibit enhanced drought tolerance compared with wild-type plants. Thus, cytokinins are able to protect biochemical processes associated with photosynthesis and delay senescence during drought stress. Auxin, brassinosteroids, and gibberellins have also been shown to interact with ABA signaling in acclimation responses.

Plants adjust osmotically to drying soil by accumulating solutes

Water can only move through the soil–plant–atmosphere continuum if water potential decreases along that path (see Chapters 2 and 3). Recall from Chapter 2

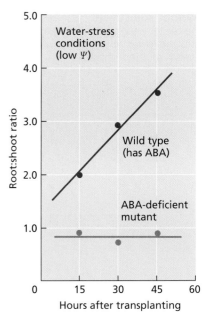

Figure 19.10   Under water-stress conditions (low Ψ, defined slightly differently for shoot and root), the ratio of root growth to shoot growth is much higher when ABA is present (i.e., in the wild type) than when it is absent (in the mutant). (After Saab et al. 1990.)

(A) Wild type

(B) Transgenic tobacco plants synthesizing cytokinin during drought stress

Figure 19.11   Effects of drought on wild-type and transgenic tobacco plants expressing isopentenyl transferase (a key enzyme in the production of cytokinin) under the control of a maturation and stress-induced promoter. Shown are (A) wild-type and (B) transgenic plants after 15 days of drought followed by 7 days of rewatering. (Courtesy of E. Blumwald.)

osmotic adjustment The ability of the cell to accumulate compatible solutes and lower water potential during periods of osmotic stress.

that $\Psi = \Psi_s + \Psi_p$, where Ψ = water potential, Ψ_s = osmotic potential, and Ψ_p = pressure potential (turgor). When the water potential of the rhizosphere (the microenvironment surrounding the root) decreases due to water deficiency or salinity, plants can continue to absorb water only as long as Ψ is lower (more negative) than it is in the soil water. **Osmotic adjustment** is the capacity of plant cells to accumulate solutes and use them to lower Ψ during periods of osmotic stress. The adjustment involves a net increase in solute content per cell that is independent of the volume changes that result from loss of water. The decrease in Ψ_s is typically limited to about 0.2 to 0.8 MPa, except in plants adapted to extremely dry conditions.

There are two main ways by which osmotic adjustment can take place, one involving the vacuole and the other the cytosol. A plant may take up ions from the soil, or transport ions from other plant organs to the root, so that the solute concentration of the root cells increases. For example, increased uptake and accumulation of K^+ will lead to decreases in Ψ_s, due to the effect of the K^+ on the osmotic pressure within the cell. This response is common in plants growing in saline soils, where ions such as K^+ and Ca^{2+} are readily available to the plant. The uptake of K^+ and other cations must be electrically balanced either by the uptake of inorganic anions, such as Cl^-, or by the production and vacuolar accumulation of organic acids such as malate or citrate.

There is a potential problem, however, when ions are used to decrease Ψ_s. Some ions, such as Na^+ and Cl^-, are essential to plant growth in low concentrations but in higher concentrations can have a detrimental effect on cellular metabolism. Other ions, such as K^+, are required in larger quantities but at high concentrations can still have a detrimental effect on the plant, mostly through disruption of plasma membranes. The accumulation of ions during osmotic adjustment is predominantly restricted to the vacuoles, where the ions are kept out of contact with cytosolic enzymes or organelles. For example, many plants use vacuolar compartmentation of Na^+ and Cl^- in conjunction with export of Na^+ at the plasma membrane to facilitate osmotic adjustment (**Figure 19.12**). Plants that are adapted to saline environments (halophytes) often use aggressive vacuolar sequestration of Na^+ as a detoxification mechanism.

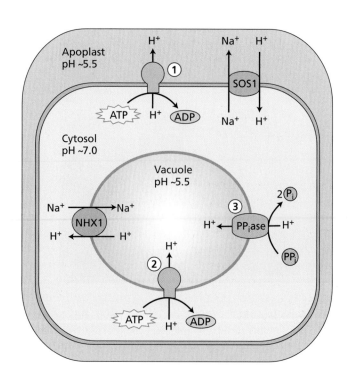

Figure 19.12 Primary and secondary active transport mechanisms function in salt stress responses. The plasma membrane–localized H^+-pumping ATPase (P-type ATPase) (1) and the tonoplast-localized H^+-pumping ATPase (V-type ATPase) (2), and pyrophosphatase (PP$_i$ase) (3) are primary active transport systems that energize the plasma membrane and the tonoplast, respectively. By coupling the energy released by hydrolysis of ATP or pyrophosphate, these proton pumps are able to transport H^+ across the plasma membrane and the tonoplast against an electrochemical gradient. The H^+–Na^+ antiporters SOS1 and NHX1 are secondary active transport systems that couple the transport of Na^+ against its electrochemical gradient with that of H^+ down its electrochemical gradient. SOS1 transports Na^+ out of the cell, whereas NHX1 transports Na^+ into the vacuole (see also Chapter 6).

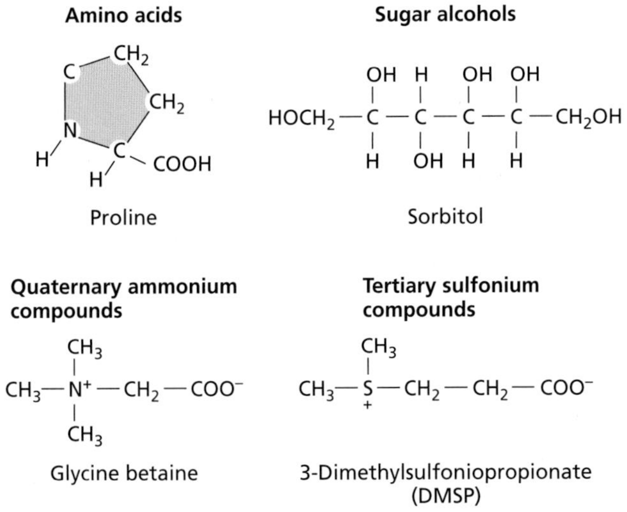

Figure 19.13 Four groups of molecules frequently serve as compatible solutes: amino acids, sugar alcohols, quaternary ammonium compounds, and tertiary sulfonium compounds. Note that these compounds are small and have no net charge.

When the ion concentration in the vacuole increases, other solutes must accumulate in the cytosol to maintain water potential equilibrium between the two compartments. These solutes are called compatible solutes (or compatible osmolytes). **Compatible solutes** are organic compounds that are osmotically active in the cell but do not destabilize membranes or interfere with enzyme function at high concentrations, as ions do. Plant cells can tolerate high concentrations of these compounds without detrimental effects on metabolism. Common compatible solutes include amino acids such as proline, sugar alcohols such as sorbitol, and quaternary ammonium compounds such as glycine betaine (**Figure 19.13**). Some of these solutes, such as proline, also seem to have an osmoprotectant function whereby they protect plants from toxic by-products produced during periods of water shortage, and provide a source of carbon and nitrogen to the cell when conditions return to normal. Each plant family tends to use one or two compatible solutes in preference to others. Because the synthesis of compatible solutes is an active metabolic process, energy is required. The amount of carbon used for the synthesis of these organic solutes can be rather large, and for this reason the synthesis of these compounds tends to reduce crop yield.

Epigenetic mechanisms and small RNAs provide additional protection against stress

Thus far we have discussed responses to abiotic stress in terms of signaling cascades and altered gene expression—acclimation processes that can be reversed when more favorable conditions arise. Recently attention has focused on epigenetic changes, which potentially can provide long-term adaptation to abiotic stress. Because some chromatin modifications are heritable, stress-induced epigenetic changes might even have evolutionary implications. Chromatin immunoprecipitation of DNA cross-linked to modified histones, coupled with modern sequencing technologies, has opened the door to genome-wide analyses of changes in the **epigenome**. Stable or heritable DNA methylation and histone modifications have now been linked with specific abiotic stresses (**Figure 19.14**).

The role of epigenetic regulation of flowering time has been studied in Arabidopsis in relation to genes known to be involved in abiotic stress. Mutations in some of the genes involved in epigenetic processes during stress cause changes in flowering times.

compatible solutes Organic compounds that are accumulated in the cytosol during osmotic adjustment. Compatible solutes do not inhibit cytosolic enzymes, unlike high concentrations of ions. Examples of compatible solutes include proline, sorbitol, mannitol, and glycine betaine.

epigenome Heritable chemical modifications to DNA and chromatin, including DNA methylation, histone methylation and acetylation.

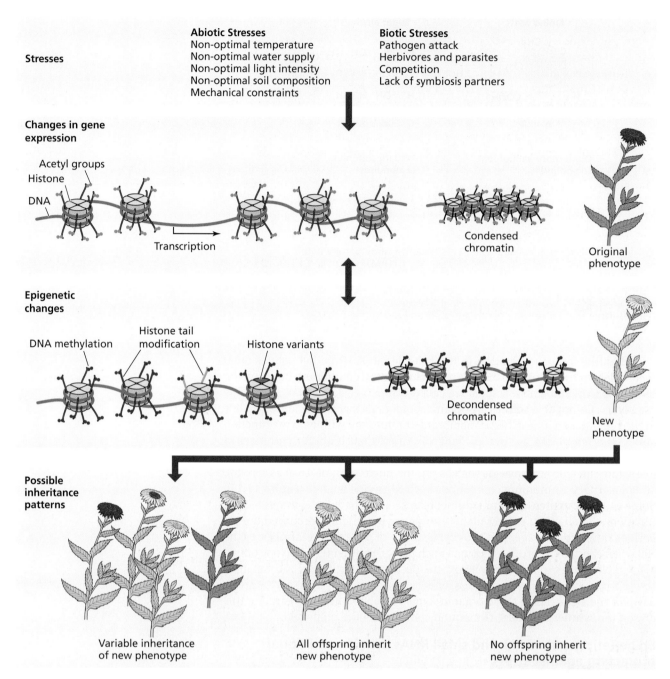

Stresses

Abiotic Stresses
Non-optimal temperature
Non-optimal water supply
Non-optimal light intensity
Non-optimal soil composition
Mechanical constraints

Biotic Stresses
Pathogen attack
Herbivores and parasites
Competition
Lack of symbiosis partners

Changes in gene expression

Acetyl groups
Histone
DNA

Transcription

Condensed chromatin

Original phenotype

Epigenetic changes

DNA methylation
Histone tail modification
Histone variants

Decondensed chromatin

New phenotype

Possible inheritance patterns

Variable inheritance of new phenotype

All offspring inherit new phenotype

No offspring inherit new phenotype

Figure 19.14 Stress-induced changes in gene expression can be mediated by protein/lipid/nucleic acid modification, second messengers, or hormones (e.g., ABA, salicylic acid, jasmonic acid, and ethylene). Transcription changes or stress factors can affect chromatin via DNA methylation, histone tail modifications, histone variant replacements, or nucleosome loss and chromatin decondensation. These changes are reversible and can modify plant metabolism or morphology under stress conditions. Usually the new phenotypes are not transmitted to progeny, however, chromatin-associated changes have the potential to be heritable and could result in uniform maintenance of new features. (After Gutzat and Mittelsten-Scheid 2012.)

Submerged organs develop aerenchyma tissue in response to hypoxia

In most wetland plants, such as rice, and in plants that acclimate well to wet conditions, the stem and roots develop longitudinally interconnected, gas-filled channels that provide a low-resistance pathway for the movement of O_2 and other gases. The gases (air) enter through stomata, or through lenticels (porous regions of the periderm that allow gas exchange) on woody stems and roots, and travel by diffusion or by convection driven by small pressure gradients. In many plants adapted to wetland growth, root cells are separated by prominent, gas-filled spaces that form a tissue called aerenchyma. These cells develop in the roots of wetland plants independently of environmental stimuli. In some non-wetland monocots and eudicots, however, O_2 deficiency induces the formation of aerenchyma in the stem base and newly developing roots.

An example of induced aerenchyma occurs in maize (corn; *Zea mays*) (**Figure 19.15**). Hypoxia stimulates the activity of ACC synthase and ACC oxidase in the root tips of maize and causes ACC and ethylene to be produced faster (see Chapter 12). Ethylene triggers programmed cell death and disintegration of cells and cell walls in the root cortex. The spaces formerly occupied by these cells provide gas-filled voids that facilitate movement of O_2. Ethylene-triggered cell death is highly selective; only some cells have the potential to initiate the developmental program that creates the aerenchyma.

When aerenchyma formation is induced, a rise in cytosolic Ca^{2+} concentration is thought to be part of the ethylene signal transduction pathway leading to cell death. Signals that elevate cytosolic Ca^{2+} concentration can promote cell death in the absence of hypoxia. Conversely, signals that lower cytosolic Ca^{2+} concentration block cell death in hypoxic roots that would normally form aerenchyma.

Some tissues can tolerate anaerobic conditions in flooded soils for an extended period (weeks or months) before developing aerenchyma. These include the embryo and coleoptile of rice (*Oryza sativa*) and rice grass (*Echinochloa crus-galli* var. *oryzicola*) and the rhizomes (underground horizontal stems) of giant bulrush (*Schoenoplectus lacustris*), salt marsh bulrush (*Scirpus maritimus*), and narrow-leafed cattail (*Typha angustifolia*). These rhizomes can survive for several months and expand their leaves under anaerobic conditions.

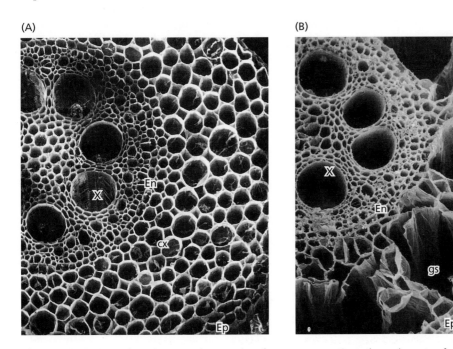

(A)   (B)

Figure 19.15   Scanning electron micrographs of transverse sections through roots of maize, showing changes in structure with O_2 supply. (150×) (A) Control root, supplied with air, with intact cortical cells. (B) O_2-deficient root growing in a nonaerated nutrient solution. Note the prominent gas-filled spaces (gs) in the cortex (cx), formed by degeneration of cells. The stele (all cells interior to the endodermis, En) and the epidermis (Ep) remain intact. X, xylem. (Courtesy of J. L. Basq and M. C. Drew.)

ROS scavenging Detoxification of reactive oxygen species via interactions with proteins and electron acceptor molecules.

In nature, these rhizomes overwinter in anaerobic mud at the edges of lakes. In spring, once the leaves have expanded above the mud or water surface, O_2 diffuses down through aerenchyma into the rhizome. Metabolism then switches from an anaerobic (fermentative) to an aerobic mode, and roots begin to grow using the available O_2. Likewise, during germination of paddy (wetland) rice and of rice grass, the coleoptile breaks through the water surface and becomes a diffusion pathway for O_2 into submerged parts of the plant, including the roots. Although rice is a wetland species, its roots are as intolerant of anoxia as maize roots are. As the root extends into O_2-deficient soil, the continuous formation of aerenchyma just behind the tip allows O_2 movement within the root to supply the apical zone.

In roots of rice and other typical wetland plants, structural barriers composed of suberized and lignified cell walls prevent O_2 diffusion outward to the soil. The O_2 thus retained supplies the apical meristem and allows growth to extend 50 cm or more into anaerobic soil. In contrast, roots of nonwetland species, such as maize, leak O_2. Internal O_2 becomes insufficient for aerobic respiration in the root apex of these nonwetland plants, and this lack of O_2 severely limits the depth to which such roots can extend into anaerobic soil.

Antioxidants and ROS-scavenging pathways protect cells from oxidative stress

Reactive oxygen species accumulate in cells during many different types of environmental stresses and are detoxified by specialized enzymes and antioxidants, a process referred to as **ROS scavenging**. Biological antioxidants are small organic compounds or small peptides that can accept electrons from ROS, such as superoxide or H_2O_2, and neutralize them. Common antioxidants in plants (**Figure 19.16**) include the water-soluble ascorbate (vitamin C) and reduced tripeptide glutathione (GSH in reduced form, GSSG in oxidized form), flavonoids (soluble or lipophilic depending on glycosylation), and the lipid-soluble α-tocopherol (vitamin E) and β-carotene (vitamin A). To maintain an adequate supply of these compounds in the reduced state, cells rely on various reductases, such as glutathione reductase, dehydroascorbate reductase, and monodehydroascorbate reductase, which use the reducing power of NADH or NADPH produced by respiration or photosynthesis.

Some ROS may react spontaneously with cellular antioxidants, and some are unstable and decay before they cause cellular damage. However, plants have

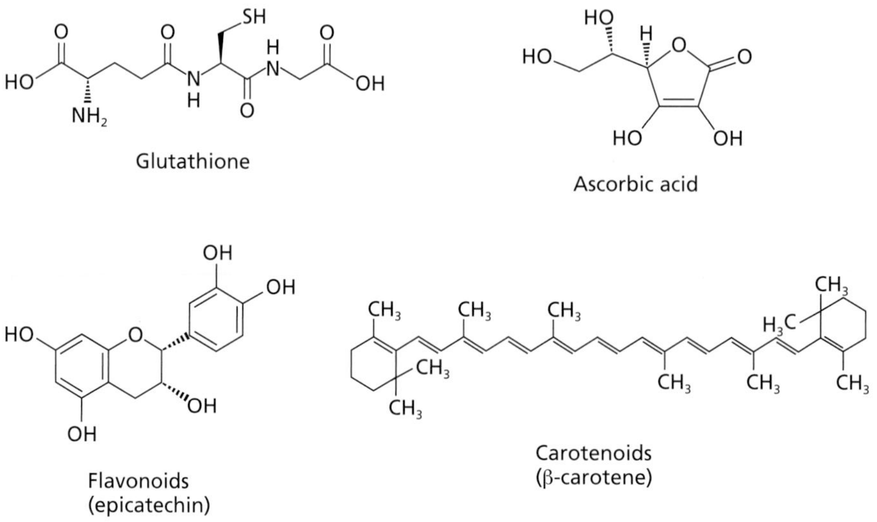

Glutathione

Ascorbic acid

Flavonoids
(epicatechin)

Carotenoids
(β-carotene)

Figure 19.16 Antioxidant ROS scavengers.

evolved several different **antioxidative enzymes** that dramatically increase the efficiency of these processes. For example, **superoxide dismutase** is an enzyme that simultaneously oxidizes and reduces the superoxide anion to produce hydrogen peroxide and oxygen according to the reaction: $2\,O_2^{\bullet-} + 2\,H^+ \rightarrow O_2 + H_2O_2$. Variants of superoxide dismutase are found in chloroplasts, peroxisomes, mitochondria, and the cytosol and apoplast. Different forms of **ascorbate peroxidase** are present in the same cellular compartments as superoxide dismutase. Ascorbate peroxidase catalyzes the destruction of hydrogen peroxide using ascorbic acid as a reducing agent in the following reaction: $2\,L\text{-ascorbate} + H_2O_2 + 2\,H^+ \rightarrow 2\,\text{monodehydroascorbate} + 2\,H_2O$. Another example is **catalase**, which catalyzes the detoxification of hydrogen peroxide into water and oxygen in peroxisomes according to the reaction: $2\,H_2O_2 \rightarrow 2\,H_2O + O_2$.

Exclusion and internal tolerance mechanisms allow plants to cope with toxic metal and metalloid ions

Two basic mechanisms are employed by plants to tolerate the presence of high concentrations of toxic ions in the environment, including aluminum, arsenic, cadmium, copper, nickel, selenium, and zinc ions: exclusion and internal tolerance. Exclusion refers to the ability to block the uptake of toxic ions into the cell, thereby preventing the concentrations of these ions from reaching a toxic threshold level. Exclusion mechanisms often involve activation of plasma membrane ion transport proteins. **Internal tolerance** typically involves biochemical adaptations that enable the plant to tolerate, compartmentalize, or chelate elevated concentrations of potentially toxic ions. Typical detoxifying chelators are organic acids such as citric acid and specialized small peptides known as phytochelatins (**Figure 19.17**).

Chelation is the binding of an ion with at least two ligating atoms within a chelating molecule. Chelating molecules can have different atoms available for ligation, such as sulfur (S), nitrogen (N), or oxygen (O), and these different atoms have different affinities for the ions they chelate. The chelating molecule renders the ion less chemically active by wrapping itself around the ion it binds to form a complex, thereby reducing the potential toxicity of the ion. The complex is then usually translocated to other parts of the plant, or stored away from the cytoplasm (typically in the vacuole).

Phytochelatins are low-molecular-weight thiols consisting of the amino acids glutamate, cysteine, and glycine, with the general form of $(\gamma\text{-Glu-Cys})_n\text{Gly}$ (see Figure

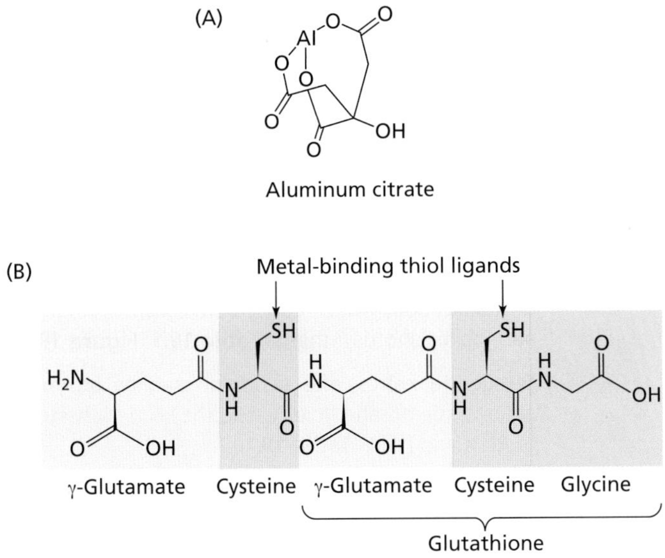

antioxidative enzymes Proteins that detoxify reactive oxygen species.

superoxide dismutase An enzyme that converts superoxide radicals to hydrogen peroxide.

ascorbate peroxidase An enzyme that converts peroxide and ascorbate to dehydroascorbate and water.

catalase An enzyme that breaks hydrogen peroxide down into water. When it is abundant in peroxisomes it may form crystalline arrays.

internal tolerance Tolerance mechanisms that function in the symplast (as opposed to exclusion mechanisms).

(A)

Aluminum citrate

(B)

Metal-binding thiol ligands

γ-Glutamate   Cysteine   γ-Glutamate   Cysteine   Glycine

Glutathione

Figure 19.17 Molecular structure of metal chelators. (A) Citrate, shown coordinated with aluminum. (B) Phytochelatin, which uses the sulfur in cysteine to bind metal ions such as cadmium, zinc, and arsenic.

hyperaccumulation   Accumulation of metals in a healthy plant to levels much higher than those found in the soil and that are generally toxic to non-accumulators.

supercool   The condition in which cellular water remains liquid because of its solute content, even at temperatures several degrees below its theoretical freezing point.

antifreeze proteins   Proteins that lower the temperature at which aqueous solutions freeze. When induced by cold temperatures, these plant proteins bind to the surfaces of ice crystals to prevent or slow further crystal growth, thereby limiting or preventing freeze damage. Some antifreeze proteins are similar to pathogenesis-related proteins.

19.17B). Phytochelatins are synthesized by the enzyme phytochelatin synthase. The thiol groups act as ligands for ions of trace elements such as cadmium and arsenic. Once formed, the phytochelatin–metal complex is transported into the vacuole for storage. Synthesis of phytochelatins has been shown to be necessary for resistance to cadmium and arsenic. In addition to chelation, active transport of metal ions into the vacuole and out of the cell also contributes to internal tolerance to metals.

Hyperaccumulation is an extreme example of internal tolerance to toxic ions. Hyperaccumulating plants can tolerate foliar concentrations of various trace elements, such as arsenic, cadmium, nickel, zinc, and selenium, of up to 1% of their shoot dry weight (10 mg per gram dry weight). Hyperaccumulation is a relatively rare plant adaptation to potentially toxic ions that requires heritable genetic changes that enhance the expression of the ion transporters involved in the uptake and vacuolar compartmentation of these ions.

Plants use cryoprotectant molecules and antifreeze proteins to prevent ice crystal formation

During rapid freezing, the protoplast, including the vacuole, may **supercool**; that is, the cellular water may remain liquid even at temperatures several degrees below its theoretical freezing point. Supercooling is common in many species of the hardwood forests of southeastern Canada and the eastern United States. Cells can supercool only to about $-40°C$, the temperature at which ice forms spontaneously. Spontaneous ice formation sets the *low-temperature limit* at which many alpine and subarctic species that undergo deep supercooling can survive.

Several specialized plant proteins, termed **antifreeze proteins**, limit the growth of ice crystals through a mechanism independent of lowering of the freezing point of water. Synthesis of these antifreeze proteins is induced by cold temperatures. The proteins bind to the surfaces of ice crystals to prevent or slow further crystal growth. Sugars, polysaccharides, osmoprotectant solutes, dehydrins, and other cold-induced proteins also have cryoprotective effects.

Summary

Plants sense potentially harmful changes in the abiotic environment and respond to them by means of dedicated stress-response pathways. These pathways involve gene networks, regulatory proteins, and signaling intermediates, as well as proteins, enzymes, and molecules that act to protect cells from the deleterious effects of abiotic stress. Together these antistress mechanisms enable plants to acclimate or adapt to stresses such as drought, heat, cold, and salinity, and their possible combinations.

Defining Plant Stress

- Stress can be defined as any condition that prevents the plant from achieving its maximum growth and reproductive potential (**Figures 19.1, 19.2**).

Acclimation versus Adaptation

- Acclimation is the process whereby individual plants respond to periodic changes in the environment by directly altering their morphology or physiology. The physiological changes associated with acclimation require no new genetic modifications, and many are reversible.

- Adaptation is characterized by genetic changes in an entire population that have been fixed by natural selection over many generations.

Environmental Stressors

- Water deficit leads to decreased turgor pressure, increased ion toxicity, reduced growth, and inhibition of photosynthesis (**Table 19.1, Figure 19.3**).

- Salinity stress causes protein denaturation and membrane destabilization, which reduce aboveground plant growth and inhibit photosynthesis (**Table 19.1, Figure 19.4**).

- Temperature stress affects protein stability, enzymatic reactions, membrane fluidity, and the secondary structures of RNA and DNA (**Table 19.1**).

(Continued)

Summary *(continued)*

- Freezing stress, like drought stress, causes cell dehydration (**Table 19.1**).

- Flooded soil is depleted of oxygen, leading to hypoxia and anaerobic stress to the root (**Table 19.1**).

- Light stress occurs when plants receive more sunlight than they can use photosynthetically (**Table 19.1**).

- Heavy metal ions can replace essential metal ions and disrupt key reactions (**Table 19.1**).

- Combinations of abiotic stresses can have effects on plant physiology and productivity that are different than those of individual stresses (**Figure 19.5**).

- Plants can gain cross-protection when sequentially exposed to different abiotic stresses.

- Plants use physical, biophysical, metabolic, biochemical, and epigenetic mechanisms to detect stresses and activate response pathways (**Figure 19.6**).

Physiological Mechanisms That Protect Plants against Abiotic Stress

- Plants can alter their leaf morphology to avoid or mitigate abiotic stress.

- Molecular chaperone proteins protect sensitive proteins and membranes during abiotic stress (**Figure 19.7**).

- Prolonged exposure to extreme temperatures can alter the composition of membrane lipids, thereby allowing plants to maintain membrane fluidity (**Table 19.2**).

- Chloroplast and mitochondria can send distress signals to the nucleus.

- A self-propagating wave of ROS production alerts as-yet unstressed parts of the plant of the need for a response (**Figure 19.8**).

- Hormones act separately and together to regulate abiotic stress responses.

- Stomatal closure is prompted by ABA-induced changes in guard cell turgor (**Figure 19.9**).

- Plants can alter their root-to-shoot biomass ratio to avoid or mitigate abiotic stress (**Figure 19.10**).

- Cytokinin functions in drought stress responses (**Figure 19.11**).

- Plants lower root osmotic potential to continue to absorb water in drying soil.

- Plant roots reduce salt toxicity by transporting Na^+ into the vacuole or outside the cell (**Figure 19.12**) and substitute compatible solutes (**Figure 19.13**).

- Epigenetic changes result in gene expression that is compatible with long-term stress conditions (**Figure 19.14**).

- Aerenchyma tissue allows O_2 to diffuse down to submerged organs (**Figure 19.15**).

- ROS can be detoxified through antioxidants (**Figure 19.16**).

- Heavy metal and metalloid ion toxicity is reduced by exclusion and/or internal detoxification, often involving chelators (**Figure 19.17**).

- Plants generate antifreeze proteins to prevent ice crystal formation.

Suggested Reading

Ahuja, I., de Vos, R. C., Bones, A. M., and Hall, R. D. (2010) Plant molecular stress responses face climate change. *Trends Plant Sci.* 15: 664–674.

Atkinson, N. J., and Urwin, P. E. (2012) The interaction of plant biotic and abiotic stresses: From genes to the field. *J. Exp. Bot.* 63: 3523–3543.

Chinnusamy, V., and Zhu, J. K. (2009) Epigenetic regulation of stress responses in plants. *Curr. Opin. Plant Biol.* 12: 133–139.

Lobell, D. B., Schlenker, W., and Costa-Roberts, J. (2011) Climate trends and global crop production since 1980. *Science* 333: 616–620.

Mittler, R. (2002) Oxidative stress, antioxidants and stress tolerance. *Trends Plant Sci.* 7: 405–410.

Mittler, R., and Blumwald, E. (2010) Genetic engineering for modern agriculture: Challenges and perspectives. *Annu. Rev. Plant Biol.* 61: 443–462.

Mittler, R., Vanderauwera, S., Gollery, M., and Van Breusegem, F. (2004) Reactive oxygen gene network of plants. *Trends Plant Sci.* 9: 490–498.

Peleg, Z., Apse, M. P., and Blumwald, E. (2011) Engineering salinity and water stress tolerance in crop plants: Getting closer to the field. *Adv. Bot. Res.* 57: 405–443.

Peleg, Z., and Blumwald, E. (2011) Hormone homeostasis and abiotic stress tolerance in crop plants. *Curr. Opin. Plant Biol.* 14: 1–6.

Glossary

A

abaxial Referring to the lower surface of a leaf.

ABC model A proposal for the way in which floral homeotic genes control organ formation in flowers. According to the model, organ identity in each whorl is determined by a unique combination of the three organ identity gene activities.

abscisic acid (ABA) A plant hormone that functions in regulation of seed dormancy as well as in stress responses such as stomatal closure during water deficit and responses to cold and heat. ABA is derived from carotenoid precursors.

abscission The shedding of leaves, flowers, and fruits from a living plant. The process whereby specific cells in the leaf petiole (stalk) differentiate to form an abscission layer, allowing a dying or dead organ to separate from the plant.

abscission zone The region that contains the abscission layer and is located near the base of the petiole of leaves.

absorption spectrum A graphic representation of the amount of light energy absorbed by a substance plotted against the wavelength of the light.

ACC oxidase The enzyme that catalyzes the conversion of 1-aminocyclopropane-1-carboxylic acid (ACC) to ethylene, the last step in ethylene biosynthesis.

ACC synthase The enzyme that catalyzes the synthesis of 1-aminocyclopropane-1-carboxylic acid (ACC) from S-adenosylmethionine.

accessory pigments Light-absorbing molecules in photosynthetic organisms that work with chlorophyll a in the absorption of light used for photosynthesis. They include carotenoids, other chlorophylls, and phycobiliproteins.

acclimation The increase in plant stress tolerance due to exposure to prior stress. May involve changes in gene expression. Contrast with adaptation.

acid growth A characteristic of growing cell walls in which they extend more rapidly at acidic pH than at neutral pH.

acropetal From the base to the tip of an organ, such as a stem, root, or leaf.

actin A major ATP-binding cytoskeletal protein. The monomeric, globular form of actin is called G-actin; the polymerized form in microfilaments is F-actin.

actinorhizal Pertaining to several woody plant species, such as alder trees, in which symbiosis occurs with soil bacteria of the nitrogen-fixing genus *Frankia*.

action potential A transient event in which the membrane potential difference rapidly decreases (depolarizes) and abruptly increases (hyperpolarizes). Action potentials, which are triggered by the opening of ion channels, can be self-propagating along linear files of cells, especially in the vascular systems of plants.

action spectrum A graphic representation of the magnitude of a biological response to light as a function of wavelength.

active transport The use of energy to move a solute across a membrane against a concentration gradient, a potential gradient, or both (electrochemical potential gradient). Uphill transport.

adaptation An inherited level of stress resistance acquired by a process of selection over many generations. Contrast with acclimation.

adaxial Referring to the upper surface of a leaf.

adenosine triphosphate (ATP) The major carrier of chemical energy in the cell, which by hydrolysis is converted to adenosine diphosphate (ADP) or adenosine monophosphate (AMP) with release of energy.

adequate zone In plant tissue, the range of concentrations of a mineral nutrient beyond which further addition of the nutrient no longer increases growth or yield.

adhesion The attraction of water to a solid phase such as a cell wall or glass surface, due primarily to the formation of hydrogen bonds.

adventitious roots Roots arising from any organ other than a root.

aerenchyma An anatomical feature of roots found in hypoxic conditions, showing large, gas-filled intercellular spaces in the root cortex.

aerobic respiration The complete oxidation of carbon compounds to CO_2 and H_2O, using oxygen as the final electron acceptor. Energy is released and conserved as ATP.

aeroponics The technique by which plants are grown with their roots suspended in air while being sprayed continuously with a nutrient solution.

after-ripening Technique for breaking seed dormancy by prolonged storage at room temperature (20–25°C) under dry conditions.

aleurone layer The layer of aleurone cells surrounding and distinct from the starchy endosperm of cereal grains.

allelopathy Release by plants of substances into the environment that have harmful effects on neighboring plants.

allocation The regulated diversion of photosynthate into storage, utilization, and/or transport.

alternation of generations The presence of two genetically distinct multicellular stages, one haploid and one diploid, in the plant life cycle. The haploid gametophyte generation begins with meiosis, while the diploid sporophyte generation begins with the fusion of sperm and egg.

alternative oxidase An enzyme in the mitochondrial electron transport chain that reduces oxygen to water and oxidizes ubiquinol (ubihydroquinone).

amplitude In a biological rhythm, the distance between peak and trough; it can often vary while the period remains unchanged.

α-amylase inhibitors   Substances synthesized by some legumes that interfere with herbivore digestion by blocking the action of the starch-digesting enzyme α-amylase.

amyloplast   A starch-storing plastid found abundantly in storage tissues of shoots and roots, and in seeds. Specialized amyloplasts in the root cap and shoot also serve as gravity sensors.

anaphase   The stage of mitosis during which the two chromatids of each replicated chromosome are separated and move toward opposite poles.

anchored proteins   Proteins that are bound to the membrane surface via lipid molecules, to which they are covalently attached.

angiosperms   The flowering plants. With their innovative reproductive organ, the flower, they are a more advanced type of seed plant and dominate the landscape. Distinguished from gymnosperms by the presence of a carpel that encloses the seeds.

anisotropic growth   Enlargement that is greater in one direction than another; for example, elongating cells in stems and roots grow more in length than in width.

annual plant   A plant that completes its life cycle from seed to seed, senesces, and dies within 1 year.

annual rings   Alternating rings of spring and summer wood (formed by secondary growth of the xylem) seen in cross-sections of stems of woody species.

antenna complex   A group of pigment molecules that cooperate to absorb light energy and transfer it to a reaction center complex.

anticlinal   Pertaining to the orientation of cell division such that the new cell plate forms perpendicular to the tissue surface.

antifreeze proteins   Proteins that lower the temperature at which aqueous solutions freeze. When induced by cold temperatures, these plant proteins bind to the surfaces of ice crystals to prevent or slow further crystal growth, thereby limiting or preventing freeze damage. Some antifreeze proteins are similar to pathogenesis-related proteins.

antioxidative enzymes   Proteins that detoxify reactive oxygen species.

antiport   A type of secondary active transport in which the passive (downhill) movement of protons or other ions drives the active (uphill) transport of a solute in the opposite direction.

apical cell   The smaller, cytoplasm-rich cell formed by the first division of the zygote.

apical dominance   In most higher plants, the growing apical bud's inhibition of the growth of lateral buds (axillary buds).

apical meristems   Localized regions made up of undifferentiated cells undergoing cell division without differentiation at the tips of shoots and roots.

apical–basal axis   An axis that extends from the shoot apical meristem to the root apical meristem.

apoplast   The mostly continuous system of cell walls, intercellular air spaces, and xylem vessels in a plant.

apoplastic transport   Movement of molecules through the cell wall continuum that is called the apoplast. Molecules may move through the linked cell walls of adjacent cells, and in that way move throughout the plant without crossing a plasma membrane.

aquaporins   Integral membrane proteins that form channels across a membrane, many of which are selective for water (hence the name). Such channels facilitate water movement across a membrane.

arbuscular mycorrhizae   Symbioses between fungi in the phylum Glomeromycota and the roots of a broad range of angiosperms, gymnosperms, ferns, and liverworts. The hyphae of arbuscular mycorrhizae penetrate the cortical cells of the root.

arbuscules   Branched structures formed by mycorrhizal fungi inside cortical cells of the host plant root; the sites of nutrient transfer between the fungus and the plant.

ascorbate peroxidase   An enzyme that converts peroxide and ascorbate to dehydroascorbate and water.

asparagine synthetase (AS)   An enzyme that transfers nitrogen as an amino group from glutamine to aspartate, forming asparagine.

aspartate aminotransferase (Asp-AT)   An aminotransferase that transfers the amino group from glutamate to the carboxyl atom of oxaloacetate to form aspartate.

ATP synthase   The multi-subunit protein complex that synthesizes ATP from ADP and phosphate (P_i). F_0F_1 and CF_0-CF_1 types are present in mitochondria and chloroplasts, respectively. Also called ATPase.

auxins   Major plant hormones involved in numerous developmental processes, including cell elongation, organogenesis, apical-basal polarity, differentiation, apical dominance, and tropic responses. The most abundant chemical form is indole-3-acetic acid.

auxin sink   A cell or tissue that takes up auxin from a nearby auxin source. Participates in auxin canalization during vascular differentiation.

auxin source   A cell or tissue that exports auxin to other cells or tissues by polar transport.

axillary buds   Secondary meristems that are formed in the axils of leaves.

axillary meristem   Meristematic tissue in the axils of leaves that gives rise to axillary buds.

B

bacteriochlorophylls   Light-absorbing pigments active in photosynthesis in anoxygenic photosynthetic organisms.

bacteroids   Endosymbiotic bacteria that have differentiated into a nondividing nitrogen-fixing state.

bark   Collective term for all the tissues outside the cambium of a woody stem or root, and composed of phloem and periderm.

basal cell   The larger, vacuolated cell formed by the first division of the zygote. It gives rise to the suspensor.

basipetal   From the growing tip of a shoot or root toward the base (junction of the root and shoot).

berry   A simple, fleshy fruit produced from a single ovary and consisting of an outer pigmented exocarp, a fleshy, juicy mesocarp, and a membranous inner endocarp.

β-oxidation   Oxidation of fatty acids into fatty acyl-CoA, and the sequential breakdown of the fatty acids into acetyl-CoA units. NADH is also produced.

biennial plant   A plant that requires two growing seasons to flower and produce seed.

biosphere   The parts of the surface and atmosphere of Earth that support life as well as the organisms living there.

biotrophic pathogens   Pathogens that leave infected tissue alive and only minimally damaged while the pathogen continues to feed on host resources.

bleaching The loss of chlorophyll's characteristic absorbance due to its conversion into another structural state, often by oxidation.

boundary layer resistance (r_b) The resistance to the diffusion of water vapor, CO_2, and heat due to the layer of unstirred air next to the leaf surface. A component of diffusional resistance.

Bowen ratio The ratio of sensible heat loss to evaporative heat loss, the two most important processes in the regulation of leaf temperature.

brassinolide A plant steroid hormone with growth-promoting activity, first isolated from *Brassica napus* pollen. One of a group of plant hormones with similar structures and activities, called brassinosteroids.

brassinosteroids (BRs) A group of plant steroid hormones that play important roles in many developmental processes, including cell division and cell elongation in stems and roots, photomorphogenesis, reproductive development, leaf senescence, and stress responses.

bryophyte *See* nonvascular plants.

bulk flow Translocation of water and solutes down a pressure gradient, as in the xylem or phloem.

bundle sheath One or more layers of closely packed cells surrounding the small veins of leaves and the primary vascular bundles of stems.

C

C$_4$ photosynthesis Photosynthetic carbon metabolism in which the initial fixation of CO_2 is catalyzed by phospho-*enol*pyruvate carboxylase (not by Rubisco as in C$_3$ photosynthesis), producing a four-carbon compound (oxaloacetate). The fixed carbon is subsequently released and refixed by the Calvin–Benson cycle.

callose A β-1,3-glucan synthesized in the plasma membrane and deposited between the plasma membrane and the cell wall. Synthesized by sieve elements in response to damage, stress, or as part of a normal developmental process.

calmodulin A Ca^{2+}-binding protein that regulates many cellular processes in a Ca^{2+}-dependent manner.

Calvin–Benson cycle The biochemical pathway for the reduction of CO_2 to carbohydrate. The cycle involves three phases: the carboxylation of ribulose 1,5-bisphosphate with atmospheric CO_2, catalyzed by Rubisco; the reduction of the formed 3-phosphoglycerate to triose phosphates; and the regeneration of ribulose 1,5-bisphosphate through the concerted action of ten enzymatic reactions.

CAM *See* Crassulacean acid metabolism.

cambium Layer of meristematic cells between the xylem and phloem that produces cells of these tissues and results in the lateral (secondary) growth of the stem or root.

canalization model The hypothesis that as auxin flows through tissues it stimulates and polarizes its own transport, which gradually becomes channeled—or canalized—into files of cells leading away from auxin sources; these cell files can then differentiate to form vascular tissue.

capillarity The movement of water for small distances up a glass capillary tube or within the cell wall, due to water's cohesion, adhesion, and surface tension.

carbon fixation reactions The synthetic reactions occurring in the stroma of the chloroplast that use the high-energy compounds ATP and NADPH for the incorporation of CO_2 into carbon compounds.

cardenolides Steroidal glycosides that taste bitter and are extremely toxic to higher animals through their action on Na^+/K^+-activated ATPases. Extracted from foxglove (*Digitalis*) for treatment of human heart disorders.

cardiac glycosides Glycosylated organic plant defense compounds—similar to oleandrin from oleander bushes—that are poisonous to animals and inhibit Na^+/K^+ channels to cause contractions in cardiac muscles.

carotenoids Linear polyenes arranged as a planar zigzag chain, with conjugated double bonds. These orange pigments serve both as antenna pigments and photoprotective agents.

carriers Membrane transport proteins that bind to a solute, undergo conformational change, and release the solute on the other side of the membrane.

Casparian strip A band in the cell walls of the endodermis that is impregnated with lignin. Prevents apoplastic movement of water and solutes into the stele.

catalase An enzyme that breaks hydrogen peroxide down into water. When it is abundant in peroxisomes it may form crystalline arrays.

cation exchange The replacement of mineral cations adsorbed to the surface of soil particles by other cations.

cavitation The collapse of tension in a column of water resulting from the formation and expansion of tiny gas bubbles.

cell autonomous response A response to an environmental stimulus or genetic mutation that is localized to a particular cell.

cell plate A wall-like structure that separates newly divided cells. Formed by the phragmoplast and later becomes the cell wall.

cell wall The rigid cell surface structure external to the plasma membrane that supports, binds, and protects the cell. Composed of cellulose and other polysaccharides and proteins.

cellulose A linear chain of (1,4)-linked β-D-glucose. The repeating unit is cellobiose.

cellulose synthase Enzyme that catalyzes the synthesis of individual (1,4)-linked β-D-glucans that make up cellulose microfibrils.

central cell The cell in the embryo sac that fuses with the second sperm cell, giving rise to the primary endosperm cell.

central zone (CZ) A central cluster of relatively large, highly vacuolated, slowly dividing initial cells in shoot apical meristems, comparable to the quiescent center of root meristems.

centromere The constricted region on the mitotic chromosome where the kinetochore forms and to which spindle fibers attach.

CF$_0$CF$_1$-ATP synthase See F$_o$F$_1$-ATP synthase.

channels Transmembrane proteins that function as selective pores for passive transport of ions or water across the membrane.

checkpoint One of several key regulatory points in the cell cycle. One checkpoint, late in G$_1$, determines if the cell is committed to the initiation of DNA synthesis.

chelator A carbon compound (e.g., malic acid or citric acid) that can form a soluble noncovalent complex with certain cations, thereby facilitating their uptake.

chemical fertilizers   Fertilizers that provide nutrients in inorganic forms.

chemical potential   The free energy associated with a substance that is available to perform work.

chemiosmotic hypothesis   The mechanism whereby the electrochemical gradient of protons established across a membrane by an electron transport process is used to drive energy-requiring ATP synthesis. It operates in mitochondria and chloroplasts.

chilling (or stratification)   Technique for breaking seed dormancy by storage at cold temperatures (1–10°C) under moist conditions for several months. The term stratification is derived from the former practice of breaking dormancy by allowing seeds to overwinter in small mounds of alternating layers of seeds and soil.

chlorophyll *a/b* antenna proteins   Chlorophyll-containing proteins associated with one or the other of the two photosystems in eukaryotic organisms. Also known as light-harvesting complex proteins (LHC proteins).

chlorophylls   A group of light-absorbing green pigments active in photosynthesis.

chloroplast envelope   The double-membrane system surrounding the chloroplast.

chloroplast   The organelle that is the site of photosynthesis in photosynthetic eukaryotic organisms.

chlorosis   The yellowing of plant leaves such as occurs as a result of mineral deficiency. The leaves affected and the parts of the leaves that yellow can be diagnostic for the type of deficiency.

Cholodny–Went hypothesis   Early mechanism proposed for tropisms involving stimulation of the bending of the plant axis by lateral transport of auxin in response to a stimulus, such as light, gravity, or touch. The original model has been supported and expanded by recent experimental evidence.

chromatin   The DNA–protein complex found in the interphase nucleus. Condensation of chromatin occurs during cell division to form the rod-shaped mitotic and meiotic chromosomes.

chromophore   A light-absorbing pigment molecule that is usually bound to a protein (an apoprotein).

chromoplasts   Plastids that contain high concentrations of carotenoid pigments, rather than chlorophyll.

chromosome   A threadlike or rod-shaped structure of DNA and protein found in the nucleus of most living cells, encoding genetic information in the form of genes.

chronic photoinhibition   Photoinhibition of photosynthentic activity in which both quantum efficiency and the maximum rate of photosynthesis are decreased. Occurs under high levels of excess light.

circadian rhythm   A physiological process that oscillates endogenously with a cycle of approximately 24 hours.

cisternae   (singular *cisterna*) A network of flattened saccules and tubules that compose the endoplasmic reticulum.

climacteric   Marked rise in respiration at the onset of ripening that occurs in all fruits that ripen in response to ethylene, and in the senescence process of detached leaves and flowers.

CO_2 compensation point   The CO_2 concentration at which the rate of respiration balances the photosynthetic rate.

coat-imposed (or exogenous) dormancy   Seed dormancy that is caused by the seed coat and other surrounding tissues; may involve impermeability to water or oxygen, mechanical restraint, or the retention of endogenous inhibitors.

coevolution   Linked adaptations of two or more organisms.

cohesion   The mutual attraction between water molecules due to extensive hydrogen bonding.

coincidence model   A model for flowering in photoperiodic plants in which the circadian oscillator controls the timing of light-sensitive and light-insensitive phases during the 24-h cycle.

coleoptile   A modified ensheathing leaf that covers and protects the young primary leaves of a grass seedling as it grows through the soil. Unilateral light perception, especially blue light, by the tip results in asymmetric growth and bending due to unequal auxin distribution in the lighted and shaded sides.

coleorhiza   A protective sheath surrounding the embryonic radicle in members of the grass family.

collection phloem   Sieve elements of sources.

collenchyma   A specialized parenchyma with irregularly thickened, pectin-rich, primary cell walls.

commensalism   A relationship between two organisms in which one organism benefits without negatively affecting the other.

companion cells   In angiosperms, metabolically active cells that are connected to their sieve element by large, branched plasmodesmata and that take over many of the metabolic activities of the sieve element. In source leaves, they function in the transport of photosynthate into the sieve elements.

compatible solutes   Organic compounds that are accumulated in the cytosol during osmotic adjustment. Compatible solutes do not inhibit cytosolic enzymes, unlike high concentrations of ions. Examples of compatible solutes include proline, sorbitol, mannitol, and glycine betaine.

compound leaf   A leaf subdivided into leaflets.

conifer resin ducts   Ducts or channels in conifer leaves and woody tissue that conduct terpenoid defense compounds. They may be constitutive, or their formation may be induced by wounding or defense responses.

conifers   Cone-bearing trees.

constitutive defenses   Plant defenses that are always immediately available or operational; that is, defenses that are not induced.

contact angle   A quantitative measure of the degree to which a water molecule is attracted to a solid phase versus to itself.

cork cambium   A layer of lateral meristem that develops within mature cells of the cortex and the secondary phloem. Produces the secondary protective layer, the periderm. Also called phellogen.

cortex   Ground tissue in the region of the primary stem or root located between the vascular tissue and the epidermis, mainly consisting of parenchyma.

cortical ER   The network of endoplasmic reticulum that lies just under the plasma membrane and is associated with the plasma membrane at specific contact points.

crassulacean acid metabolism (CAM)   A biochemical process for concentrating CO_2 at the carboxylation site of Rubisco. Found in the family Crassulaceae (*Crassula, Kalanchoe, Sedum*) and numerous other families of angiosperms. In CAM, CO_2 uptake and initial fixation take place at night, and decarboxylation and reduction of the internally released CO_2 occur during the day.

cristae Folds in the inner mitochondrial membrane that project into the mitochondrial matrix.

critical concentration (of a nutrient) In plant tissue, the minimum concentration of a mineral nutrient that is correlated with maximum growth or yield.

critical day length The minimum length of the day required for flowering of a long-day plant; the maximum length of day that will allow short-day plants to flower. However, studies have shown that it is the length of the night, not the length of the day, that is important.

cross-pollination Pollination of a flower by pollen from the flower of a different plant.

cross-protection A plant response to one environmental stress that confers resistance to another stress.

cross-regulation The interaction of two or more signaling pathways.

crown roots Adventitious roots that emerge from the lowermost nodes of a stem.

cryptochromes Flavoproteins implicated in many blue-light responses that have strong homology with bacterial photolyases.

cyclic electron flow In photosystem I, the flow of electrons from the electron acceptors through the cytochrome b_6f complex and back to P700, coupled to proton pumping into the lumen. This electron flow energizes ATP synthesis but does not oxidize water or reduce $NADP^+$.

cyclin-dependent kinases (CDKs) Protein kinases that regulate the transitions from G_1 to S, from S to G_2, and from G_2 to mitosis, during the cell cycle.

cyclins Regulatory proteins associated with cyclin-dependent kinases that play a crucial role in regulating the cell cycle.

cytochrome b_6f complex A large multi-subunit protein complex containing two b-type hemes, one c-type heme (cytochrome f), and a Rieske iron–sulfur protein. A relatively immobile complex distributed equally between the grana and the stroma regions of the thylakoid membranes.

cytochrome c A peripheral, mobile component of the mitochondrial electron transport chain that oxidizes complex III and reduces complex IV.

cytochrome f A subunit in the cytochrome b_6f complex that plays a role in electron transport between photosystems I and II.

cytochrome P450 monooxygenase (CYP) A generic term for a large number of related, but distinct, mixed-function oxidative enzymes localized on the endoplasmic reticulum. CYPs participate in a variety of oxidative processes, including steps in the biosynthesis of gibberellins and brassinosteroids.

cytohistological zonation Regional cytological differences in cell division in the shoot apical meristems of seed plants.

cytokinesis In plant cells, following nuclear division, the separation of daughter nuclei by the formation of new cell wall.

cytokinins A class of plant hormones that function in cell division and differentiation, as well as axillary bud growth, and leaf senescence. Cytokinins are adenine derivatives, the most common form of which is zeatin.

cytoplasm The cellular matter enclosed by the plasma membrane exclusive of the nucleus.

cytoplasmic streaming The coordinated movement of particles and organelles through the cytosol.

cytoskeleton Composed of polarized microfilaments of actin or microtubules of tubulin, the cytoskeleton helps control the organization and polarity of organelles and cells during growth.

cytosol The aqueous phase of the cytoplasm containing dissolved solutes but excluding supramolecular structures, such as ribosomes and components of the cytoskeleton.

D

damage associated molecular patterns (DAMPs) Molecules originating from nonpathogenic sources that can initiate immune responses.

day-neutral plant (DNP) A plant whose flowering is not regulated by day length.

de-etiolation The rapid developmental changes associated with loss of the etiolated form due to the action of light. *See* photomorphogenesis.

deficiency zone In plant tissue, the range of concentrations of a mineral nutrient below the critical concentration, the highest concentration where reduced plant growth or yield is observed

dehiscence The spontaneous opening of a mature anther or fruit, releasing its contents.

dehydrins Hydrophilic plant proteins that accumulate in response to drought stress and cold temperatures.

dermal tissue The tissue system that covers the outside of the plant body; the epidermis or periderm.

desaturase enzymes Enzymes that remove hydrogens in a carbon chain to create a double bond between carbons.

difference in water vapor concentration (Δc_{wv}) The difference between the water vapor concentration of the air spaces inside the leaf and that of the air outside the leaf.

diffuse growth A type of cell growth in plants in which expansion occurs more or less uniformly over the entire surface. Contrast with tip growth.

diffusion coefficient (D_s) The proportionality constant that measures how easily a specific substance s moves through a particular medium. The diffusion coefficient is a characteristic of the substance and depends on both the medium and the temperature.

diffusion potential The potential (voltage) difference that develops across a semipermeable membrane as a result of the differential permeability of solutes with opposite charges (for example K^+ and Cl^-).

diffusion The movement of substances due to random thermal agitation from regions of high free energy to regions of low free energy (e.g., from high to low concentration).

diffusional resistance (r) The restriction posed by the boundary layer and the stomata to the free diffusion of gases from and into the leaf.

dioecious Refers to plants in which male and female flowers are found on different individuals, such as spinach (*Spinacia*) and hemp (*Cannabis sativa*). Contrast with monoecious.

directed organelle movement Movement of an organelle in a particular direction, which can be driven by the interaction with molecular motors associated with the cytoskeleton.

double fertilization A unique feature of all angiosperms whereby, along with the fusion of a sperm with the egg to create a diploid zygote, a second male gamete fuses with the polar nuclei in the embryo sac to generate the endosperm tissue (with a triploid or higher number of chromosomes).

drupe   A structure similar to a berry, but with a hardened, shell-like endocarp (pit or stone) that contains a seed.

dynamic photoinhibition   Photo-inhibition of photosynthesis in which quantum efficiency decreases but the maximum photosynthetic rate remains unchanged. Occurs under moderate, not high, excess light.

E

ectomycorrhizae   Symbioses in which the fungus typically forms a thick sheath, or mantle, around the roots. The root cells themselves are not penetrated by the fungal hyphae, and instead are surrounded by a network of hyphae called the Hartig net.

effector   A molecule that binds a protein to change its activity. Bacterial effectors are secreted by pathogens to act on proteins within a host cell.

effector-triggered immunity   Immune responses that are mediated by a class of intracellular R proteins.

egg cell   The female gamete.

electrochemical potential   The chemical potential of an electrically charged solute.

electrochemical proton gradient   The sum of the electrical charge gradient and the pH gradient across the membrane, resulting from a concentration gradient of protons.

electrogenic transport   Active ion transport involving the net movement of charge across a membrane.

electron spin resonance (ESR) spectroscopy   A magnetic-resonance technique that detects unpaired electrons in molecules. Instrumental measurements that identify intermediate electron carriers in the photosynthetic or respiratory electron transport system.

electronegative   Having the capacity to attract electrons and thus producing a slightly negative electric charge.

electroneutral transport   Active ion transport that involves no net movement of charge across a membrane.

elicitors   Specific pathogen molecules or cell wall fragments that bind to plant proteins and thereby act as signals for

activation of plant defense against a pathogen.

elongation zone   The region of rapid and extensive root cell elongation showing few, if any, cell divisions.

embryo (or endogenous) dormancy   Seed dormancy that is caused directly by the embryo and is not due to any influence of the seed coat or other surrounding tissues.

embryogenesis   The formation and development of the embryo.

embryophyte   The plant group, including all land plants, characterized by the ability of the gametophyte to contain and nurture the young sporophyte within its tissues through the earliest stages of development.

endodermis   A specialized layer of cells surrounding the vascular tissue in roots and some stems.

endoplasmic reticulum (ER)   A continuous membrane system within the cytoplasm of eukaryotic cells that serves multiple functions, including the synthesis, modification, and intracellular transport of proteins.

endoreduplication   Cycles of nuclear DNA replication without mitosis resulting in polyploidization.

energy transfer   In the light reactions of photosynthesis, the direct transfer of energy from an excited molecule, such as β-carotene, to another molecule, such as chlorophyll. Energy transfer can also take place between chemically identical molecules, as in chlorophyll-to-chlorophyll transfer.

enhancement effect   The synergistic (higher) effect of red and far-red light on the rate of photosynthesis, as compared with the sum of the rates when the two different wavelengths are delivered separately.

entrainment   The synchronization of the period of biological rhythms by external controlling factors, such as light and darkness.

epidermis   The outermost layer of plant cells, typically one cell thick.

epigenome   Heritable chemical modifications to DNA and chromatin, including DNA methylation, histone methylation and acetylation.

escape from photoreversibility   The physiological state in which far-red light will not act to reverse events induced by red light interactions with phytochrome.

essential element   A chemical element that is an intrinsic component of the structure or metabolism of a plant. When the element is in limited supply, a plant suffers abnormal growth, development, or reproduction.

ethylene   A gaseous plant hormone involved in fruit ripening, abscission, and the growth of etiolated seedlings. The chemical formula of ethylene is C_2H_4.

etiolation   Effects of seedling growth in the dark, in which the hypocotyl and stem are more elongated, the cotyledons and leaves do not expand, and the chloroplasts do not mature.

etioplast   Photosynthetically inactive form of chloroplast found in etiolated seedlings.

euchromatin   The dispersed, transcriptionally active form of chromatin. Compare with heterochromatin.

expansins   A class of wall-loosening proteins that accelerate wall yielding during cell elongation, typically with an optimum at acidic pH.

export   The movement of photosynthate in sieve elements away from the source tissue.

extracellular space   In plants, the space continuum outside the plasma membrane made up of interconnecting cell walls through which water and mineral nutrients readily diffuse.

extrafloral nectar   Nectar produced outside the flower and not involved in pollination events.

F

F_1   The ATP-binding, matrix-facing part of the F_oF_1-ATP synthase.

F_o   The integral membrane part of the F_oF_1-ATP synthase.

F_oF_1-ATP synthase   A multi-subunit protein complex associated with the inner mitochondrial membrane that couples the passage of protons across the membrane to the synthesis of ATP from ADP and phosphate. The subscript "o" in F_o refers to the binding of the inhibitor oligomycin. Similar to CF_0CF_1-ATP synthase in photophosphorylation, to which oligomycin does not bind and inhibit (hence the subscript is "0").

facilitated diffusion   Passive transport across a membrane using a carrier.

Fe–S centers   Prosthetic groups consisting of inorganic iron and sulfur that are abundant in proteins in respiratory and photosynthetic electron transport.

fermentation   The metabolism of pyruvate in the absence of oxygen, leading to the oxidation of the NADH generated in glycolysis to NAD^+. Allows glycolytic ATP production to function in the absence of oxygen.

ferredoxin (Fd)   A small, water-soluble iron–sulfur protein involved in electron transport in photosystem I.

ferredoxin–NADP$^+$ reductase (FNR)   A membrane-associated flavoprotein that receives electrons from photosystem I and reduces $NADP^+$ to NADPH.

ferredoxin–thioredoxin system   Three chloroplast proteins (ferredoxin, ferredoxin–thioredoxin reductase, thioredoxin). The three proteins use reducing power from the photosynthetic electron transport chain to reduce protein disulfide bonds in a cascade of thiol–disulfide exchanges. As a result, light controls the activity of several enzymes of the Calvin–Benson cycle.

ferritin   A protein that functions in cellular iron storage in multiple compartments, including plastids and mitochondria.

fertilization   The formation of a diploid ($2N$) zygote from the cellular and nuclear fusion of two haploid ($1N$) gametes, the egg and the sperm. In angiosperms, fertilization also involves fusion of a second sperm nucleus with the haploid nuclei (usually two) of the central cell to form the endosperm (usually triploid).

FeS$_R$ An iron- and sulfur-containing subunit of the cytochrome b_6f complex, involved in electron and proton transfer. See also Rieske iron–sulfur protein.

FeS$_X$, FeS$_A$, FeS$_B$ Membrane-bound iron–sulfur proteins that transfer electrons between photosystem I and ferredoxin.

fiber   An elongated, tapered sclerenchyma cell that provides support in vascular plants.

flavin adenine dinucleotide (FAD)   A riboflavin-containing cofactor that undergoes a reversible two-electron reduction to produce $FADH_2$.

flavin mononucleotide (FMN)   A riboflavin-containing cofactor that undergoes a reversible one- or two-electron reduction to produce FMNH or $FMNH_2$. FAD and FMN occur as cofactors in flavoproteins.

floral evocation   The events occurring in the shoot apex that specifically commit the apical meristem to produce flowers.

floral meristem identity genes   Two classes of genes, one required for the conversion of a shoot apical meristem into a floral meristem, the other that maintains the identity of an inflorescence meristem (as opposed to a floral meristem).

floral meristem   The meristem that forms floral (reproductive) organs: sepals, petals, stamens, and carpels. May form directly from a vegetative meristem or indirectly via an inflorescence meristem.

floral organ identity genes   Three classes of genes that control the specific locations of floral organs in the flower. *See also* ABC model.

florigen   A systemically-mobile transcription factor synthesized by leaves and translocated to the shoot apical meristem via the phloem to stimulate flowering.

fluence   The number of photons absorbed per unit of surface area over time ($\mu mol\ m^{-2}\ s^{-1}$).

fluid-mosaic model   The common molecular lipid–protein structure for all biological membranes. A double layer (bilayer) of polar lipids (phospholipids or, in chloroplasts, glycosylglycerides) has a hydrophobic, fluid-like interior. Sterols and sphingolipids embedded in the bilayer create ordered domains within the fluid layers. Membrane proteins are embedded in the bilayer and may move laterally due to its fluid-like properties, or may be less mobile if associated with ordered domains.

fluorescence   Following light absorption, the emission of light at a slightly longer wavelength (lower energy) than the wavelength of the absorbed light.

fluorescence resonance energy transfer (FRET)   The physical mechanism by which excitation energy is conveyed from a molecule that absorbs light to an adjacent molecule.

flux density (J_s)   The rate of transport of a substance s across a unit of area per unit of time. J_s may have units of moles per square meter per second ($mol\ m^{-2}\ s^{-1}$).

foliar application   The application, and subsequent absorption, of mineral nutrients or other chemical compounds to leaves as sprays.

forisomes   P-proteins that rapidly and reversibly disperse and block the sieve tube. Occur only in certain legumes.

free-running   Designation of the biological rhythm that is characteristic for a particular organism when environmental signals are removed, as in total darkness. See Zeitgeber.

frequency (ν)   A unit of measurement that characterizes waves, in particular light energy. The number of wave crests that pass an observer in a given time.

fruit   In angiosperms, one or more mature ovaries containing seeds and sometimes adjacent attached parts.

fusicoccin   A fungal toxin that induces acidification of plant cell walls by activating H^+-ATPases in the plasma membrane. Fusicoccin stimulates rapid acid growth in stem and coleoptile sections. It also stimulates stomatal opening by stimulating proton pumping at the guard cell plasma membrane.

G

G$_1$ The phase of the cell cycle preceding the synthesis of DNA.

G$_2$ The phase of the cell cycle following the synthesis of DNA.

GABA shunt   A pathway supplementing the tricarboxylic acid cycle with the ability to form and degrade GABA.

gamete   A haploid ($1N$) reproductive cell.

gametophyte   The haploid ($1N$) multicellular structure that produces haploid gametes by mitosis and differentiation. Contrast with sporophyte.

gate   A structural domain of the channel protein that opens or closes the channel in response to external signals such as voltage changes, hormone binding, or light.

germination   The events that take place between the start of imbibition of the dry seed and the emergence of part of the embryo, usually the radicle, from the structures that surround it. May also be applied to other quiescent structures, such as pollen grains or spores.

gibberellins (GAs)   A large group of chemically related plant hormones synthesized by a branch of the terpenoid pathway and associated with the promotion of stem growth (especially in dwarf and rosette plants), seed germination, and many other functions.

Gibbs free energy   The energy that is available to do work; in biological systems, the work of synthesis, transport, and movement.

glucan   A polysaccharide made from glucose units.

gluconeogenesis   The synthesis of carbohydrates through the reversal of glycolysis.

glucose 6-phosphate dehydrogenase A cytosolic and plastidic enzyme that catalyzes the initial reaction of the oxidative pentose phosphate pathway.

glutamate dehydrogenase (GDH) An enzyme that catalyzes a reversible reaction that synthesizes or deaminates glutamate as part of the nitrogen assimilation process.

glutamate synthase (GOGAT)   An enzyme that transfers the amide group of glutamine to 2-oxoglutarate, yielding two molecules of glutamate. Also known as glutamine:2-oxoglutarate aminotransferase.

glutamine synthetase (GS)   An enzyme that catalyzes the condensation of ammonium and glutamate to form glutamine. The reaction is critical for the assimilation of ammonium into essential amino acids. Two forms of GS exist—one in the cytosol and one in chloroplasts.

glyceroglycolipids   Glycerolipids in which sugars form the polar head group. Glyceroglycolipids are the most abundant glycerolipids in chloroplast membranes.

glycerophospholipids   Polar glycerolipids in which the hydrophobic portion consists of two 16-carbon or 18-carbon fatty acid chains esterified to positions 1 and 2 of a glycerol backbone. The phosphate-containing polar head group is attached to position 3 of the glycerol.

glycolysis   A series of reactions in which a sugar is oxidized to produce two molecules of pyruvate. A small amount of ATP and NADH is produced.

glycophytes   Plants that are not able to resist salts to the same degree as halophytes. Show growth inhibition, leaf discoloration, and loss of dry weight at soil salt concentrations above a threshold. Contrast with halophytes.

glyoxylate cycle   The sequence of reactions that convert two molecules of acetyl-CoA to succinate in the glyoxysome.

glyoxysome   An organelle found in the oil-rich storage tissues of seeds in which fatty acids are oxidized. A type of microbody.

Goldman equation   An equation that predicts the diffusion potential across a membrane, as a function of the concentrations and permeabilities of all ions (e.g., K^+, Na^+, and Cl^-) that permeate the membrane.

grana lamellae   Stacked thylakoid membranes within the chloroplast. Each stack is called a granum.

granum   (plural *grana*) In the chloroplast, a stack of thylakoids.

gravitational potential (Ψ_g)   The part of the water potential caused by gravity. It is only of a significant size when considering water transport into trees and drainage in soils.

gravitropic response   The growth initiated by the root cap's perception of gravity and the signal that directs the roots to grow downward.

gravitropism   Plant growth in response to gravity, enabling roots to grow downward into the soil and shoots to grow upward.

green-leaf volatiles   A mixture of lipid-derived six-carbon aldehydes, alcohols, and esters released by plants in response to mechanical damage.

greenhouse effect   The warming of Earth's climate, caused by the trapping of long-wavelength radiation by CO_2 and other gases in the atmosphere. The term is derived from the heating of a greenhouse that results from the penetration of long-wavelength radiation through the glass roof, the conversion of the long-wave radiation to heat, and the blocking of the heat escape by the glass roof.

ground tissue   The internal tissues of the plant, other than the vascular tissues.

growth respiration   The respiration that provides the energy needed for converting sugars into the building blocks that make up new tissue. Contrast with maintenance respiration.

guard cells   A pair of specialized epidermal cells that surround the stomatal pore and regulate its opening and closing.

guard mother cell (GMC)   The cell that gives rise to a pair of guard cells to form a stomate.

guttation   An exudation of liquid from the leaves due to root pressure.

gymnosperms   An early group of seed plants. Distinguished from angiosperms by having seeds borne unprotected (naked) in cones.

H

H⁺-pyrophosphatase   An electrogenic pump that moves protons into the vacuole, energized by the hydrolysis of pyrophosphate.

halophytes   Plants that are native to saline soils and complete their life cycles in that environment. Contrast with glycophytes.

Hartig net   A network of fungal hyphae that surround, but do not penetrate, the cortical cells of roots in ectomycorrhizal symbioses.

haustorium   The hyphal tip of a fungus or root tip of a parasitic plant that penetrates host plant tissue.

HC-toxin   A cell permeant cyclic tetrapeptide produced by the maize pathogen *Cochliobolus carbonum* that inhibits histone deacetylases.

heat shock proteins (HSPs)   A specific set of proteins that are induced by a rapid rise in temperature, and by other factors that lead to protein denaturation. Most act as molecular chaperones.

heliotropism   The movements of leaves toward or away from the sun.

hemibiotrophic pathogens   Pathogens that show an initial biotrophic stage, which is followed by a necrotrophic stage in which the pathogen causes extensive tissue damage.

hemicelluloses   Heterogeneous group of polysaccharides that bind to the surface of cellulose, linking cellulose microfibrils together into a network.

hemiparasitic plants   Photosynthetic plants that are also parasites.

herbivory   Consumption of plants or parts of plants as a food source.

heterochromatin   Chromatin that is densely packed and transcriptionally modified or suppressed. Compare with euchromatin.

hexose phosphates   Six-carbon sugars with phosphate groups attached.

histones   A family of proteins that interact with DNA and around which DNA is wound to form a nucleosome.

Hoagland solution   A type of nutrient solution for plant growth, originally formulated by Dennis R. Hoagland.

holoparasitic plants   Nonphotosynthetic plants that are obligate parasites.

hormone balance theory   The hypothesis that seed dormancy and germination are regulated by the balance of ABA and gibberellin.

hydraulic conductivity A measure of how readily water can move across a membrane; it is expressed in terms of volume of water per unit of area of membrane per unit of time per unit of driving force (i.e., $m^3 \ m^{-2} \ s^{-1} \ MPa^{-1}$).

hydrogen bonds Weak chemical bonds formed between a hydrogen atom and an oxygen or nitrogen atom.

hydrostatic pressure Pressure generated by compression of water into a confined space. Measured in units called pascals (Pa) or, more conveniently, megapascals (MPa).

hyperaccumulation Accumulation of metals in a healthy plant to levels much higher than those found in the soil and that are generally toxic to non-accumulators.

hyphal coils Coiled structures formed by mycorrhizal fungi within the root cortical cells of its host plant; the sites of nutrient transfer between the fungus and the plant. Also called arbuscules.

hypophysis The cell located directly below the octant stage of the embryo that gives rise to the root cap and part of the root apical meristem.

I

imbibition The initial phase of water uptake in dry seeds which is driven by the matric potential component of the water potential, that is, by the binding of water to surfaces, such as the cell wall and cellular macromolecules.

import The movement of photosynthate in sieve elements into sink organs.

indehiscent Lacking spontaneous opening of a mature anther or fruit.

indeterminate growth The ability to keep growing and developing until the onset of senescence.

inducible defenses Defense responses that exist at low levels until a biotic or abiotic stress is encountered.

induction period The period of time (time lag) between the perception of a signal and the activation of the response. In the Calvin–Benson cycle, the time elapsed between the onset of illumination and the full activation of the cycle.

infection thread An internal tubular extension of the plasma membrane of root hairs through which rhizobia enter root cortical cells.

initials Broadly defined as the cells of the root and shoot apical meristems. More specifically, a cluster of slowly dividing cells located within the meristem that gives rise to the more rapidly dividing cells of the surrounding meristem.

injectisome A name for a specialized secretion system appendage of some pathogenic bacteria.

inner mitochondrial membrane The inner of the two mitochondrial membranes, containing the electron transport chain, the F_oF_1-ATP synthase, and numerous transporters.

integral membrane proteins Proteins that are embedded in the lipid bilayer of a membrane via at least one transmembrane domain.

intercellular air space resistance The resistance or hindrance that slows down the diffusion of CO_2 inside a leaf, from the substomatal cavity to the walls of the mesophyll cells.

interface light scattering The randomization of the direction of photon movement within plant tissues due to the reflecting and refracting of light from the many air–water interfaces. Greatly increases the probability of photon absorption within a leaf.

intermediary cell A type of companion cell with numerous plasmodesmatal connections to surrounding cells, particularly to the bundle sheath cells.

intermembrane space The fluid-filled space between the two mitochondrial membranes or between the two chloroplast envelope membranes.

internal tolerance Tolerance mechanisms that function in the symplast (as opposed to exclusion mechanisms).

internode The portion of a stem between nodes.

interphase Collectively the G_1, S, and G_2 phases of the cell cycle.

inwardly rectifying Refers to ion channels that open only at potentials more negative than the prevailing Nernst potential for a cation, or more positive than the prevailing Nernst potential for an anion, and thus mediate inward current.

irradiance The amount of energy that falls on a flat surface of known area per unit of time. Expressed as watts per square meter (W m^{-2}). Time (seconds) is contained within the term watt: 1 W = 1 joule (J) s^{-1}, or as micromoles of quanta

per square meter per second (μmol m^{-2} s^{-1}), also referred to as fluence rate.

J

jasmonic acid A plant hormone that functions in plant defenses to biotic and abiotic stresses as well as some other aspects of development. Jasmonic acid is derived from the octadecanoid pathway. Can function as a volatile signal when methylated (methyl jasmonate) and is also active in some plants when conjugated to amino acids.

K

karrikinolide A component of smoke that stimulates seed germination; similar in structure to strigolactones.

kinases Enzymes that have the capacity to transfer phosphate groups from ATP to other molecules.

kinetochore The site of spindle fiber attachment to the chromosome in anaphase.

Kranz anatomy (German *kranz*, "wreath.") The wreathlike arrangement of mesophyll cells around a layer of bundle sheath cells. The two concentric layers of photosynthetic tissue surround the vascular bundle. This anatomical feature is typical of leaves of many C_4 plants.

L

lamina The blade of a leaf.

latent heat of vaporization The energy needed to separate molecules from the liquid phase and move them into the gas phase at constant temperature.

lateral roots Arise from the pericycle in mature regions of the root through establishment of secondary meristems that grow out through the cortex and epidermis, establishing a new growth axis.

latex A complex, often milky solution exuded from cut surfaces of some plants that represents the cytoplasm of laticifers and may contain defensive substances.

laticifers In many plants, an elongated, often interconnected network of separately differentiated cells that contain latex (hence the term laticifer), rubber, and other secondary metabolites.

leaf blade The broad, expanded area of the leaf; also called the lamina.

leaf stomatal resistance (r_s) The resistance to CO_2 diffusion imposed by the stomatal pores.

leaf trace   The portion of the shoot primary vascular system that diverges into a leaf.

leaves   The main lateral appendages radiating out from stems and branches. Green leaves are usually the major photosynthetic organs of the plant.

lectins   Defensive plant proteins that bind to carbohydrates; or carbohydrate-containing proteins inhibiting their digestion by a herbivore.

leghemoglobin   An oxygen-binding heme protein produced by legumes during nitrogen-fixing symbioses with rhizobia. Found in the cytoplasm of infected nodule cells, it facilitates the diffusion of oxygen to the respiring symbiotic bacteria.

leucoplasts   Nonpigmented plastids, the most important of which is the amyloplast.

light channeling   In photosynthetic cells, the propagation of some of the incident light through the central vacuole of the palisade cells and through the air spaces between the cells.

light compensation point   The amount of light reaching a photosynthesizing leaf at which photosynthetic CO_2 uptake exactly balances respiratory CO_2 release.

light energy   The energy associated with photons.

light-harvesting complex II (LHCII)   The most abundant antenna protein complex, associated primarily with photosystem II.

LIGHT-OXYGEN-VOLTAGE (LOV) domains   Highly conserved protein domains that respond to light, oxygen or voltage changes to change receptor protein conformation. In phototropins, LOV domains are sites of binding of the FMN chromophore to phototropins and are thus the part of the protein that senses light.

lignin   Highly branched phenolic polymer made up of phenylpropanoid alcohols that is deposited in secondary cell walls.

long-day plant (LDP)   A plant that flowers only in long days (qualitative LDP) or whose flowering is accelerated by long days (quantitative LDP).

long-distance transport   Translocation through the phloem to the sink.

long–short-day plant (LSDP)   A plant that flowers in response to a shift from long days to short days.

lowest excited state   The excited state with the lowest energy attained when a chlorophyll molecule in a higher energy state gives up some of its energy to the surroundings as heat.

lytic vacuoles   Analogous to lysosomes in animal cells, plant lytic vacuoles contain hydrolytic enzymes that break down cellular macromolecules during senescence.

M

M phase   The phase of the cell cycle in which the replicated chromosomes condense, move to opposite poles, and come to reside in the nuclei of two identical cells.

maintenance respiration   The respiration needed to support the function and turnover of existing tissue. Contrast with growth respiration.

malic enzyme   An enzyme that catalyzes the oxidation of malate to pyruvate, permitting plant mitochondria to oxidize malate or citrate to CO_2 without involving pyruvate generated by glycolysis.

MAP (mitogen-activated protein) kinase cascade   A series of protein kinases that transmit an activation signal from a cell surface receptor to DNA in the nucleus.

margo   A porous and relatively flexible region of the pit membranes in tracheids of conifer xylem, surrounding a central thickening, the torus.

mass transfer rate   The quantity of material passing through a given cross section of phloem or sieve elements per unit of time.

matric potential (Ψ_m)   The sum of osmotic potential (Ψ_s) + hydrostatic pressure (Ψ_p). Useful in situations (dry soils, seeds, and cell walls) where the separate measurement of Ψ_s and Ψ_p is difficult or impossible to obtain.

matrix polysaccharides   Polysaccharides comprising the matrix of plant cell walls. In primary cell walls they consist of pectins, hemicelluloses, and proteins.

matrix   The aqueous, gel-like phase of a mitochondrion that occupies the internal space into which the cristae extend. Contains the DNA, ribosomes, and soluble enzymes required for the tricarboxylic acid cycle, oxidative phosphorylation, and other metabolic reactions.

maturation zone   The region of the root where differentiation occurs, including the production of root hairs and functional vascular tissue.

maximum quantum yield   The ratio between photosynthetic product and the number of photons absorbed by a photosynthetic tissue. In a graphic plot of photon flux and photosynthetic rate, the maximum quantum yield is given by the slope of the linear portion of the curve.

megaspore   The haploid (1N) spore that develops into the female gametophyte.

meiosis   The "reduction division" whereby two successive cell divisions produce four haploid (1N) cells from one diploid (2N) cell. In plants with alternation of generations, spores are produced by meiosis. In animals, which don't have alternation of generations, gametes are produced by meiosis.

membrane permeability   The extent to which a membrane permits or restricts the movement of a substance.

meristematic zone   The region at the tip of the root containing the meristem that generates the body of the root. Located between the root cap and the elongation zone.

meristemoid   A small, triangular stomatal precursor cell that functions temporarily as an initial cell in a meristem.

meristemoid mother cells (MMCs)   The cells of the leaf protoderm that divide asymmetrically (the so-called entry division) to give rise to the meristemoid, a guard cell precursor.

meristems   Localized regions of ongoing cell division that enable growth during postembryonic development.

mesocotyl   In members of the grass family, the part of the elongating axis between the scutellum and the coleoptile.

mesophyll   Leaf tissue found between the upper and lower epidermal layers.

mesophyll resistance   The resistance to CO_2 diffusion imposed by the liquid phase inside leaves. The liquid phase includes diffusion from the intercellular leaf spaces to the carboxylation sites in the chloroplast.

metabolic redundancy   A common feature of plant metabolism in which different pathways serve a similar function. They can therefore replace each other without apparent loss in function.

metaphase   A stage of mitosis during which the nuclear envelope breaks down and the condensed chromosomes align in the middle of the cell.

microbe-associated molecular patterns (MAMPs)   Microbially-produced molecules that are recognized by host cells.

microbial pathogens   Bacterial or fungal organisms that cause disease in a host plant.

microbodies   A class of spherical organelles surrounded by a single membrane and specialized for one of several metabolic functions.

microfibril   The major fibrillar component of the cell wall composed of layers of cellulose molecules packed tightly together by extensive hydrogen bonding.

microfilament   A component of the cell cytoskeleton made of actin; it is involved in organelle motility within cells.

micropyle   The small opening at the distal end of the ovule, through which the pollen tube passes prior to fertilization.

microspores   The haploid (1N) cell that develops into the pollen tube or male gametophyte.

microtubule   A component of the cell cytoskeleton and the mitotic spindle, and a player in the orientation of cellulose microfibrils in the cell wall. Made of tubulin.

middle lamella   A thin layer of pectin-rich material at the junction where the primary walls of neighboring cells come into contact.

mineral nutrition   The study of how plants obtain and use mineral nutrients.

mineralization   The process of breaking down organic compounds, usually by soil microorganisms, and thereby releasing mineral nutrients in forms that can be assimilated by plants.

mitochondrion   (plural *mitochondria*) The organelle that is the site for most reactions in the respiratory process in eukaryotes.

mitosis   The ordered cellular process by which replicated chromosomes are distributed to daughter cells formed by cytokinesis.

mitotic spindle   The mitotic structure involved in chromosome movement. Polymerized from α- and β-tubulin monomers formed by the disassembly of the preprophase band in early metaphase.

molecular chaperone proteins   Proteins that maintain or restore the active three-dimensional structures of other macromolecules.

monocarpic   Referring to plants, typically annuals, that produce fruits only once and then die.

monoecious   Refers to plants in which male and female flowers are found on the same individuals, such as cucumber (*Cucumis sativus*) and maize (corn; *Zea mays*). Contrast with dioecious.

movement proteins   Nonstructural proteins encoded by a virus's genome that facilitate movement of that virus through the symplast.

mutualism   A symbiotic relationship in which both organisms benefit.

mycorrhiza   (plural *mycorrhizae*) The symbiotic (mutualistic) association of certain fungi and plant roots. Facilitates the uptake of mineral nutrients by roots.

mycorrhizal fungi   Fungi that can form mycorrhizal symbioses with plants.

N

NAD(P)H dehydrogenases   A collective term for membrane-bound enzymes that oxidize NADH or NADPH, or both, and reduce quinone. Several are present in the electron transport chain of mitochondria; for example, the proton-pumping complex I, but also simpler non-proton-pumping enzymes.

NADH dehydrogenase (complex I)   A multi-subunit protein complex in the mitochondrial electron transport chain that catalyzes oxidation of NADH and reduction of ubiquinone linked to the pumping of protons from the matrix to the intermembrane space.

necrotic spots   Small spots of dead leaf tissue.

necrotrophic pathogens   Pathogens that attack their host plant first by secreting cell wall–degrading enzymes or toxins, which will lead to massive tissue laceration and plant death.

Nernst potential   The electrical potential described by the Nernst equation.

night break   An interruption of the dark period with a short exposure to light that makes the entire dark period ineffective.

nitrate reductase   An enzyme located in the cytosol that reduces nitrate (NO_3^-) to nitrite (NO_2^-). It catalyzes the first step by which nitrate absorbed by roots is assimilated into organic form.

nitrogen fixation   The natural or industrial processes by which atmospheric nitrogen N_2 is converted to ammonia (NH_3) or nitrate (NO_3^-).

nitrogenase enzyme complex   The two-component protein complex that catalyzes the biological nitrogen fixation reaction in which ammonia is produced from molecular nitrogen.

Nod factors   Lipochitin oligosaccharide signal molecules active in regulating gene expression during nitrogen-fixing nodule formation. All Nod factors have a chitin β-1,4-linked *N*-acetyl-D-glucosamine backbone (varying in length from three to six sugar units) and a fatty acid chain on the C-2 position of the nonreducing sugar.

nodal roots   In monocots, adventitious roots that form after the emergence of primary roots.

node   Position on the stem where leaves are attached.

nodulation (*nod*) genes   Rhizobial genes, the products of which participate in nodule formation.

nodules   Specialized organs of a plant host that contain symbiotic nitrogen-fixing bacteria.

nodulin genes   Plant genes specific to nodule formation.

non-climacteric   Referring to a type of fruit that does not undergo a climacteric, or respiratory burst, during ripening.

non-seed plants   Plant families that do not produce a seed.

non–cell autonomous response   A cellular response to an environmental stimulus or genetic mutation that is induced by other cells.

nonvascular plants   Plants that do not have vascular tissues, such as xylem and phloem.

nuclear envelope   The double membrane surrounding the nucleus.

nuclear genome   The entire complement of DNA found in the nucleus.

nuclear pores   Sites where the two membranes of the nuclear envelope join, forming a partial opening between the interior of nucleus and the cytosol.

nucleolar organizer region (NOR)   Site in the nucleolus where chromosomal regions that code for ribosomal RNA are clustered and are transcribed.

nucleolus (plural *nucleoli*) A densely granular region in the nucleus that is the site of ribosome synthesis.

nucleoplasm The soluble matrix of the nucleus in which the chromosomes and nucleolus are suspended.

nucleosome A structure consisting of eight histone proteins around which DNA is coiled.

nucleus (plural *nuclei*) The organelle that contains the genetic information primarily responsible for regulating cellular metabolism, growth, and differentiation.

nutrient assimilation The incorporation of mineral nutrients into carbon compounds such as pigments, enzyme cofactors, lipids, nucleic acids, or amino acids.

nutrient depletion zone The region surrounding the root surface showing diminished nutrient concentrations due to uptake into the roots and slow replacement by diffusion.

nutrient film growth system A form of hydroponic culture in which the plant roots lie on the surface of a trough, and the nutrient solution flows over the roots in a thin layer along the trough.

nutrient solution A solution containing only inorganic salts that supports the growth of plants in sunlight without soil or organic matter.

nyctinasty Sleep movements of leaves. Leaves extend horizontally to face the light during the day and fold together vertically at night.

O

oil bodies Organelles that accumulate and store triacylglycerols. They are bounded by a single phospholipid leaflet ("half–unit membrane" or "phospholipid monolayer") derived from the endoplasmic reticulum. Also known as oleosomes or spherosomes.

ordinary companion cell A type of companion cell with a variable number of plasmodesmata connecting it to neighboring cells other than its associated sieve element.

organ senescence The developmentally regulated senescence of individual organs, such as leaves, flowers, fruits, or roots.

organic fertilizer A fertilizer that contains nutrient elements derived from

natural sources without any synthetic additions.

osmolarity A unit of concentration expressed as moles of total dissolved solutes per liter of solution (mol L^{-1}). In biology, the solvent is usually water.

osmosis The net movement of water across a selectively permeable membrane toward the region of more negative water potential, Ψ (lower concentration of water).

osmotic adjustment The ability of the cell to accumulate compatible solutes and lower water potential during periods of osmotic stress.

osmotic potential (Ψ_s) The effect of dissolved solutes on water potential. Also called solute potential.

osmotic stress Stress imposed on cells or whole plants when the osmotic potential of external solutions is more negative than that of the solution inside the plant.

outcrossing The mating of two plants with different genotypes by cross-pollination.

outer mitochondrial membrane The outer of the two mitochondrial membranes, which appears to be freely permeable to all small molecules.

outwardly rectifying Refers to ion channels that open only at potentials more positive than the prevailing Nernst potential for a cation, or more negative than the prevailing Nernst potential for an anion, and thus mediate outward current.

oxidative pentose phosphate pathway A cytosolic and plastidic pathway that oxidizes glucose and produces NADPH and several sugar phosphates.

oxidative phosphorylation The transfer of electrons to oxygen in the mitochondrial electron transport chain that is coupled to ATP synthesis from ADP and phosphate by ATP synthase.

P

P-protein bodies Discrete spheroidal, spindle-shaped, or twisted and coiled structures of P-proteins present in the cytosol of immature sieve tube elements of the phloem. Generally disperse into tubular or fibrillar forms during cell maturation.

P-proteins Phloem proteins that act to seal damaged sieve elements by plugging the sieve element pores. Abundant in the

sieve elements of most angiosperms, but absent from gymnosperms.

P680 The chlorophyll of the photosystem II reaction center that absorbs maximally at 680 nm in its neutral state.

P700 The chlorophyll of the photosystem I reaction center that absorbs maximally at 700 nm in its neutral state. The P stands for pigment.

palisade cells Below the leaf upper epidermis, the top one to three layers of pillar-shaped photosynthetic cells.

parasite An organism that lives on or in an organism of another species, known as the host, from the body of which the parasite obtains nutriment.

parenchyma Metabolically active ground tissue consisting of thin-walled cells.

partitioning The differential distribution of photosynthate to multiple sinks within the plant.

passive transport Diffusion across a membrane. The spontaneous movement of a solute across a membrane in the direction of a gradient of (electro) chemical potential (from higher to lower potential). Downhill transport.

pathogenesis-related (PR) genes Genes that encode small proteins that function either as antimicrobials or in initiating systemic defense responses.

pattern recognition receptors (PRRs) Innate immune system proteins that are associated with microbe-associated molecular patterns (MAMPs) and damage-associated molecular patterns (DAMPs).

pavement cells The predominant type of leaf epidermal cells, which secrete a waxy cuticle and serve to protect the plant from dehydration and damage from ultraviolet radiation.

pectins A heterogeneous group of complex cell wall polysaccharides that form a gel in which the cellulose–hemicellulose network is embedded.

PEP carboxylase A cytosolic enzyme that forms oxaloacetate by the carboxylation of phospho*enol*pyruvate.

perennial plants Plants that live for more than 2 years.

perforation plate The perforated end wall of a vessel element in the xylem.

pericarp The fruit wall surrounding a fruit, derived from the ovary wall.

periclinal Pertaining to the orientation of cell division such that the new cell plate forms parallel to the tissue surface.

pericycle Meristematic cells forming the outermost layer of the vascular cylinder in the stem or root, interior to the endodermis.

periderm Tissue produced by the cork cambium that contributes to the outer bark of stems and roots during secondary growth of woody plants, replacing the epidermis. Also forms over wounds and abscission layers after the shedding of plant parts.

period In cyclic (rhythmic) phenomena, the time between comparable points in the repeating cycle, such as peaks or troughs.

peripheral membrane proteins Proteins that are bound to the membrane surface by noncovalent bonds, such as ionic bonds or hydrogen bonds.

peripheral zone (PZ) A doughnut-shaped region surrounding the central zone in shoot apical meristems and consisting of small, actively dividing cells with inconspicuous vacuoles. Leaf primordia are formed in the peripheral zone.

perisperm Storage tissue derived from the nucellus, and often consumed during embryogenesis.

peroxisome Organelle in which organic substrates are oxidized by O_2. These reactions generate H_2O_2 that is broken down to water by the peroxisomal enzyme catalase.

petiole The leaf stalk that joins the leaf blade to the stem.

Pfr The far-red light–absorbing form of phytochrome converted from Pr by the action of red light. The cyan-green colored Pfr is converted back to Pr by far-red light. Pfr is the physiologically active form of phytochrome.

phase change The phenomenon in which the fates of the meristematic cells become altered in ways that cause them to produce new types of structures.

phase In cyclic (rhythmic) phenomena, any point in the cycle recognizable by its relationship to the rest of the cycle, for example, the maximum and minimum positions.

phellogen *See* cork cambium.

phenotypic plasticity Physiological or developmental responses of a plant

to its environment that do not involve genetic changes.

pheophytin A chlorophyll in which the central magnesium atom has been replaced by two hydrogen atoms.

phloem The tissue that transports the products of photosynthesis from mature leaves (or storage organs) to areas of growth and storage, including the roots.

phloem loading The movement of photosynthetic products into the sieve elements of mature leaves. *See also* phloem unloading.

phloem unloading The movement of photosynthates from the sieve elements to neighboring cells that store or metabolize them or pass them on to other sink cells via short-distance transport. *See also* phloem loading.

phosphatidylinositol bisphosphate (PIP$_2$) A phosphorylated derivative of phosphatidylinositol.

photochemistry The very rapid chemical reactions in which light energy absorbed by a molecule causes a chemical reaction to occur.

photoinhibition The inhibition of photosynthesis by excess light.

photomorphogenesis The influence and specific roles of light on plant development. In the seedling, light-induced changes in gene expression that support aboveground growth in the light rather than belowground growth in the dark.

photon A discrete physical unit of radiant energy.

photon flux density (PFD) The amount of energy striking a leaf per unit of time, expressed as moles of quanta per square meter per second (mol m^{-2} s^{-1}). Also referred to as fluence rate.

photonasty Plant movements in response to nondirectional light.

photoperiodic induction The photoperiod-regulated processes that occur in leaves resulting in the transmission of a floral stimulus to the shoot apex.

photoperiodism A biological response to the length and timing of day and night, making it possible for an event to occur at a particular time of year.

photophosphorylation The formation of ATP from ADP and inorganic phosphate (P_i), catalyzed by the CF_0F_1-ATP synthase and using light energy stored in the proton gradient across the thylakoid membrane.

photoreceptors Proteins that sense the presence of light and initiate a response via a signaling pathway.

photorespiration Uptake of atmospheric O_2 with a concomitant release of CO_2 by illuminated leaves. Molecular oxygen serves as substrate for Rubisco, producing 2-phosphoglycolate that enters the photorespiratory cycle (the oxidative photosynthetic carbon cycle). The activity of the cycle recovers some of the carbon found in 2-phosphoglycolate, but some is lost to the atmosphere.

photoreversibility The interconversion of the Pr and Pfr forms of phytochrome.

photosynthate A carbon-containing product of photosynthesis.

photosystem I (PSI) A system of photoreactions that absorbs maximally far-red light (700 nm), oxidizes plastocyanin, and reduces ferredoxin.

photosystem II (PSII) A system of photoreactions that absorbs maximally red light (680 nm), oxidizes water, and reduces plastoquinone. Operates very poorly under far-red light.

phototropins 1 and 2 Two flavoprotein photoreceptors that mediate the blue-light signaling pathway that induces phototropic bending in higher plants. They also mediate chloroplast movements and participate in stomatal opening in response to blue light. Phototropins are autophosphorylating protein kinases whose activity is stimulated by blue light.

phototropins Blue-light photoreceptors that primarily regulate phototropism, chloroplast movements, and stomatal opening.

phototropism The alteration of plant growth patterns in response to the direction of incident radiation, especially blue light.

phragmoplast An assembly of microtubules, membranes, and vesicles that forms during late anaphase or early telophase and precedes fusion of vesicles to form the cell plate.

phyllome The collective term for all the leaves of a plant, including structures that evolved from leaves, such as floral organs.

phyllotaxy The arrangement of leaves on the stem.

phytoalexins A chemically diverse group of secondary metabolites with strong antimicrobial activity that are synthesized following infection and that accumulate at the site of infection.

phytochrome interacting factors (PIFs) A family of phytochrome-interacting proteins that may activate and repress gene transcription; some PIFs are targets for phytochrome-mediated degradation.

phytochrome kinase substrates (PKSs) Proteins that participate in the regulation of phytochrome via direct phosphorylation or via phosphorylation by other kinases.

phytochromes Plant growth-regulating photoreceptor proteins that absorb primarily red light and far-red light, but also absorb blue light. The holoprotein that contains the chromophore phytochromobilin.

phytoliths Discrete cells that accumulate silica in leaves or roots.

phytomer A developmental unit consisting of one or more leaves, the node to which the leaves are attached, the internode below the node, and one or more axillary buds.

pit A microscopic region where the secondary wall of a tracheary element is absent and the primary wall is thin and porous

pit membrane The porous layer in the xylem between pit pairs, consisting of two thinned primary walls and a middle lamella.

pit pair Two pits occurring opposite one another in the walls of adjacent tracheids or vessel elements. Pit pairs constitute a low-resistance path for water movement between the conducting cells of the xylem.

pith The ground tissue in the center of the stem or root.

plant tissue analysis In the context of mineral nutrition, the analysis of the concentrations of mineral nutrients in a plant sample.

plants All the families of plants inclusive of the non-seed, nonvascular plants.

plasma membrane (plasmalemma) A bilayer of polar lipids (phospholipids or glycosylglycerides) and embedded proteins that together form a selectively permeable boundary around a cell.

plasma membrane H⁺-ATPase An ATPase that pumps H^+ across the plasma membrane energized by ATP hydrolysis.

plasmodesmata (singular *plasmodesma*) Microscopic membrane-lined channel connecting adjacent cells through the cell wall and filled with cytoplasm and a central rod derived from the ER called the desmotubule.

plasticity The ability to adjust morphologically, physiologically, and biochemically in response to changes in the environment.

plastids Cellular organelles found in eukaryotes, bounded by a double membrane, and sometimes containing extensive membrane systems. They perform many different functions: photosynthesis, starch storage, pigment storage, and energy transformations.

plastocyanin (PC) A small (10.5 kDa), water-soluble, copper-containing protein that transfers electrons between the cytochrome b_6f complex and P700. This protein is found in the lumenal space.

plastohydroquinone (PQH₂) The fully reduced form of plastoquinone.

polar auxin transport Directional auxin movement that functions in programmed development and plastic growth responses. Long-distance polar auxin transport maintains the overall polarity of the plant apical–basal axis and supplies auxin for direction into localized streams.

polar glycerolipids The main structural lipids in membranes, in which the hydrophobic portion consists of two 16-carbon or 18-carbon fatty acid chains esterified to positions 1 and 2 of a glycerol.

polarity (1) Property of some molecules, such as water, in which differences in the electronegativity of some atoms result in a partial negative charge at one end of the molecule and a partial positive charge at the other end. (2) Refers to the distinct ends and intermediate regions along an axis. Beginning with the single-celled zygote, the progressive development of distinctions along two axes: an apical–basal axis and a radial axis.

pollen Small structures (microspores) produced by anthers of seed plants. Contain haploid male nuclei that will fertilize the egg in the ovule.

polycarpic Referring to perennial plants that produce fruit many times.

polymer-trapping model A model that explains the specific accumulation of sugars in the sieve elements of symplastically loading species.

pome A fruit, such as an apple, composed of one or more carpels surrounded by accessory tissue derived from the receptacle.

Pr The red light–absorbing form of phytochrome. This is the form in which phytochrome is assembled. The cyanblue colored Pr is converted by red light to the far-red light–absorbing form, Pfr.

precocious germination Germination of viviparous mutant seeds while still attached to the mother plant.

preharvest sprouting Germination of physiologically mature wild-type seeds on the mother plant caused by wet weather.

preprophase band A circular array of microtubules and microfilaments formed in the cortical cytoplasm just prior to cell division that encircles the nucleus and predicts the plane of cytokinesis following mitosis.

preprophase In mitosis, the stage just before prophase during which the G_2 microtubules are completely reorganized into a preprophase band.

pressure potential (Ψₚ) The hydrostatic pressure of a solution in excess of ambient atmospheric pressure.

pressure-flow model A widely accepted model of phloem translocation in angiosperms. It states that transport in the sieve elements is driven by a pressure gradient between source and sink. The pressure gradient is osmotically generated and results from the loading at the source and unloading at the sink.

prickles Pointed plant structures that physically deter herbivory and are derived from epidermal cells.

primary active transport The direct coupling of a metabolic energy source such as ATP hydrolysis, oxidation-reduction reaction, or light absorption to active transport by a carrier protein.

primary cell walls The thin (less than 1 μm), unspecialized cell walls that are characteristic of young, growing cells.

primary cross-regulation Involves distinct signaling pathways regulating a shared transduction component in a positive or a negative manner.

primary dormancy The failure of newly dispersed, mature seeds to germinate under normal growth conditions.

primary growth The phase of plant development that gives rise to new organs and to the basic plant form.

primary inflorescence meristem
The meristem that produces stem-bearing flowers; it is formed from the shoot apical meristem.

primary plant axis   The longitudinal axis of the plant defined by the positions of the shoot and root apical meristems.

primary plant body   The part of the plant directly derived from the shoot and root apical meristems and primary meristems.

primary root   In monocots, a root generated directly by growth of the embryonic root or radicle.

procambium   Primary meristematic tissue that differentiates into xylem, phloem, and cambium.

programmed cell death (PCD)   A process whereby individual cells activate an intrinsic senescence program accompanied by a distinct set of morphological and biochemical changes similar to mammalian apoptosis.

prolamellar bodies   Elaborate semicrystalline lattices of membrane tubules that develop in plastids that have not been exposed to light (etioplasts).

prophase   The first stage of mitosis (and meiosis) prior to disassembly of the nuclear envelope, during which the chromatin condenses to form distinct chromosomes.

proplastid   Type of immature, undeveloped plastid found in meristematic tissue.

26S proteasome   A large proteolytic complex that degrades intracellular proteins marked for destruction by the attachment of one or more copies of the small protein, ubiquitin.

protein bodies   Protein storage organelles enclosed by a single membrane; found mainly in seed tissues.

protein storage vacuoles   Specialized small vacuoles that accumulate storage proteins, typically in seeds.

protoderm   In the plant embryo and in apical meristems, the one-cell-thick surface layer that covers the young shoot and radicle of the embryo and gives rise to the epidermis.

protofilament   A column of polymerized tubulin monomers (α- and β-tubulin heterodimers) or a chain of polymerized actin subunits.

proton motive force (PMF)   The energetic effect of the electrochemical H^+ gradient across a membrane, expressed in units of electrical potential.

pulvinus   (plural *pulvini*) A turgor-driven organ found at the junction between the blade and the petiole of the leaf that provides the mechanical force for leaf movements.

pumps   Membrane proteins that carry out primary active transport across a biological membrane. Most pumps transport ions, such as H^+ or Ca^{2+}.

pyruvate dehydrogenase   An enzyme in the mitochondrial matrix that decarboxylates pyruvate, producing NADH (from NAD^+), CO_2, and acetic acid in the form of acetyl-CoA (acetic acid bound to coenzyme A).

Q

Q cycle   A mechanism for oxidation of plastohydroquinone (reduced plastoquinone, also called plastoquinol) in chloroplasts and of ubihydroquinone (reduced ubiquinone, also called ubiquinol) in mitochondria.

quantum   (plural *quanta*) A discrete packet of energy contained in a photon.

quantum yield (ϕ)   The ratio of the yield of a particular product of a photochemical process to the total number of quanta absorbed.

quiescent center   The central region of the root meristem where cells divide more slowly than surrounding cells, or do not divide at all. Serves as a reservoir of meristematic cells for tissue regeneration in case of wounding.

R

radial axis   An axis that extends from the center of a root or stem to its outer surface.

radicle   The embryonic root. Usually the first organ to emerge on germination.

raphides   Needles of calcium oxalate or carbonate that function in plant defense.

rays   Tissues of various height and width that radiate through the secondary xylem and phloem, and are formed from ray initials in the vascular cambium.

reaction center complex   A group of electron transfer proteins that receive energy from the antenna complex and convert it into chemical energy using oxidation–reduction reactions.

reactive oxygen species (ROS)
These include the superoxide anion ($O_2^{\bullet-}$), hydrogen peroxide (H_2O_2), the hydroxyl radical (HO•), and singlet oxygen. They are generated in several cell compartments and can act as signals or cause damage to cellular components.

reciprocity   According to the Law of Reciprocity, treating plants with a brief duration of bright light will induce the same photobiological response as treating them with a long duration of dim light.

release phloem   Sieve elements of sinks where sugars and other photosynthetic products are unloaded into sink tissues.

resistance (R) proteins   Proteins that function in plant defense against fungi, bacteria, and nematodes by binding to specific pathogen molecules, elicitors.

respiratory burst oxidase homolog D (RBOHD)   An enzyme that generates superoxide using NADPH as an electron donor.

respiratory quotient (RQ)   The ratio of CO_2 evolution to O_2 consumption.

response regulator   One component of the two-component regulatory systems that are composed of a histidine kinase sensor protein and a response regulator protein. Response regulators have a *receiver domain*, which is phosphorylated by the sensor protein, and an *output domain*, which carries out the response.

rhizobia   A collective term for the genera of soil bacteria that form symbiotic (mutualistic) relationships with members of the plant family Fabaceae (Leguminosae).

rhizosphere   The immediate microenvironment surrounding the root.

rib zone (RZ)   Meristematic cells located beneath the central zone in shoot apical meristems and that give rise to the internal tissues of the stem.

ribosome   The site of cellular protein synthesis; consists of RNA and protein.

ribulose 5-phosphate   In the pentose phosphate pathway, the initial five-carbon product of the oxidation of glucose 6-phosphate; in subsequent reactions, it is converted into sugars containing three to seven carbon atoms.

ribulose-1,5-bisphosphate carboxylase/oxygenase   *See* Rubisco.

Rieske iron–sulfur protein A protein subunit in the cytochrome b_6f complex, in which two iron atoms are bridged by two sulfur atoms, with two histidine and two cysteine ligands.

ripening The process that causes fruits to become more palatable, including softening, increasing sweetness, loss of acidity, and changes in coloration.

root apical meristem A group of permanently dividing cells located underneath the root cap at the tip of the root that provides cells for the primary growth of the root.

root cap Cells at the root apex that cover and protect the meristematic cells from mechanical injury as the root moves through the soil. Site for the perception of gravity and signaling for the gravitropic response in roots.

root hairs Microscopic extensions of root epidermal cells that greatly increase the surface area of the root for absorption.

root pressure A positive hydrostatic pressure in the xylem of roots that typically occurs at night in the absence of transpiration.

root The organ, usually underground, that serves to anchor the plant in the soil, and to absorb water and mineral ions, and conduct them to the shoot. In contrast to shoots, roots lack buds, leaves, or nodes.

ROS scavenging Detoxification of reactive oxygen species via interactions with proteins and electron acceptor molecules.

rough ER The endoplasmic reticulum to which ribosomes are attached. Rough ER synthesizes proteins that are transported by vesicles either to internal organelles or the plasma membrane.

Rubisco The acronym for the chloroplast enzyme *ribulose 1,5-bisphosphate carboxylase/oxygenase*. In a carboxylase reaction, Rubisco uses atmospheric CO_2 and ribulose 1,5-bisphosphate to form two molecules of 3-phosphoglycerate. It also functions as an oxygenase that adds O_2 to ribulose 1,5-bisphosphate to yield 3-phosphoglycerate and 2-phosphoglycolate. The competition between CO_2 and O_2 for ribulose 1,5-bisphosphate limits net CO_2 fixation.

S

S phase The phase in the cell cycle during which DNA is replicated; it follows G_1 and precedes G_2.

salicylic acid A benzoic acid derivative believed to be an endogenous signal for systemic acquired resistance.

salt stress The adverse effects of excess minerals on plants.

salt-tolerant plants Plants that can survive or even thrive in high-salt soils. *See also* halophytes.

sclerenchyma Plant tissue composed of cells (sclereids and fibers), often dead at maturity, with thick, lignified secondary cell walls.

scutellum The single cotyledon of the grass embryo, specialized for nutrient absorption from the endosperm.

seasonal leaf senescence In temperate climates, the pattern of leaf senescence in deciduous trees in which all of the leaves undergo senescence and abscission in the autumn.

second messenger A small intracellular molecule (e.g., cyclic AMP, cyclic GMP, calcium, IP_3, or diacylglycerol) whose concentration increases or decreases in response to the activation of a receptor by an external signal, such as hormones or light. It then diffuses intracellularly to the target enzymes or intracellular receptor to produce and amplify the physiological response.

secondary active transport Active transport that uses energy stored in the proton motive force or other ion gradient and operates by symport or antiport.

secondary cell wall Cell wall synthesized by nongrowing cells. Often multilayered and containing lignin, it differs in composition and structure from the primary wall.

secondary cross-regulation Regulation by the output of one signal pathway of the abundance or perception of a second signal.

secondary dormancy Seeds that have lost their primary dormancy may become dormant again if they experience prolonged exposure to unfavorable growth conditions.

secondary growth Growth in diameter that occurs after primary growth (stem and root elongation) is complete. It involves the vascular cambium (producing the secondary xylem and phloem) and the cork cambium (producing the periderm).

secondary inflorescence meristem The inflorescence meristem that develops from the axillary buds of stem-borne leaves of the primary inflorescence.

secondary metabolites Plant compounds that have no direct role in plant growth and development but that function as defenses against herbivores and infection by microbial pathogens, as attractants for pollinators and seed-dispersing animals, and as agents of plant–plant competition.

secondary phloem Phloem produced by the vascular cambium.

secondary xylem Xylem produced by the vascular cambium.

seed dormancy A state of arrested growth of the embryo that prevents germination even when all the necessary environmental conditions for growth, such as water, O_2, and temperature, are met.

seed plants Plants in which the embryo is protected and nourished within a seed. The gymnosperms and angiosperms.

seed quiescence A state of suspended growth of the embryo due to a lack of water, O_2, or proper temperature for growth. Germination of quiescent seeds proceeds immediately once these conditions are met.

selective permeability The membrane property that allows diffusion of some molecules across the membrane to a different extent than other molecules.

seminal roots Adventitious roots that arise during embryogenesis from the stem tissue between the scutellum and the coleoptile.

senescence An active, genetically controlled, developmental process in which cellular structures and macromolecules are broken down and translocated away from the senescing organ (typically leaves) to actively growing regions that serve as nutrient sinks. Initiated by environmental cues and regulated by hormones.

sensor proteins Bacterial receptor proteins that perceive external or internal signals as part of two-component regulatory systems. They consist of two domains, an *input domain*, which receives the environmental signal, and a *transmitter domain*, which transmits the signal to the response regulator.

sequential leaf senescence The pattern of leaf senescence in which there is a gradient of senescence from the growing tip of the shoot to the oldest leaves at the base.

shoot apex (terminal bud)   The shoot apical meristem with its associated leaf primordia and young, developing leaves.

shoot apical meristem   Dome-shaped region of the shoot tip composed of meristematic cells that give rise to leaves, branches, and reproductive structures.

shoots   The organ, usually aboveground, that includes the stem, leaves, buds, and reproductive structures. Function in photosynthesis and reproduction.

short-day plant (SDP)   A plant that flowers only in short days (i.e., shorter than a critical day length) or whose flowering is accelerated by short days (quantitative SDP).

short-distance transport   Transport over a distance of only two or three cell diameters. Precedes phloem loading, when sugars move from the mesophyll to the vicinity of the smallest veins of the source leaf, and follows phloem unloading, when sugars move from the veins to the sink cells.

short–long-day plant (SLDP)   A plant that flowers only after a sequence of short days followed by long days.

sieve area   A depression in the cell wall of a sieve tube element that contains a field of plasmodesmata.

sieve cells   The relatively unspecialized sieve elements of gymnosperms.

sieve effect   The penetration of photosynthetically active light through several layers of cells due to the gaps between chloroplasts permitting the passage of light.

sieve elements   Cells of the phloem that conduct sugars and other organic materials throughout the plant. Refers to both sieve tube elements (angiosperms) and sieve cells (gymnosperms).

sieve plates   Sieve areas found in angiosperm sieve tube elements; they have larger pores (sieve plate pores) than other sieve areas and are generally found in end walls of sieve tube elements.

sieve tube elements   The highly differentiated sieve elements typical of the angiosperms.

sieve tube   Tube formed by the joining together of individual sieve tube elements at their end walls.

signal transduction pathway   A sequence of biochemical processes by which an extracellular signal (typically light or a hormone) interacts with a receptor, causing a change in the level of a second messenger and ultimately a change in cell function.

simple leaf   A leaf with one blade.

sink   Any organ that imports photosynthate, including nonphotosynthetic organs and organs that do not produce enough photosynthetic products to support their own growth or storage needs, such as roots, tubers, developing fruits, and immature leaves. Contrast with source.

sink activity   The rate of uptake of photosynthate per unit of weight of sink tissue.

sink size   The total weight of the sink.

sink strength   The ability of a sink organ to mobilize assimilates toward itself. Depends on two factors: sink size and sink activity.

size exclusion limit (SEL)   The restriction on the size of molecules that can be transported via the symplast. It is imposed by the width of the cytoplasmic sleeve that surrounds the desmotubule in the center of the plasmodesma.

skotomorphogenesis   The developmental program plants follow when seeds are germinated and grown in the dark.

smooth ER   The endoplasmic reticulum lacking attached ribosomes and usually consisting of tubules. Functions in lipid synthesis.

soil analysis   The chemical determination of the nutrient content in a soil sample, typically from the root zone.

soil hydraulic conductivity   A measure of the ease with which water moves through a soil.

solar tracking   The movement of leaf blades throughout the day so that the planar surface of the blade remains perpendicular to the sun's rays.

solution culture   A technique for growing plants with their roots immersed in nutrient solution without soil. Also called hydroponics.

source   Any organ that is capable of exporting photosynthetic products in excess of its own needs, such as a mature leaf or a storage organ. Contrast with sink.

specific heat   Ratio of the heat capacity of a substance to the heat capacity of a reference substance, usually water. Heat capacity is the amount of heat needed to change the temperature of a unit of mass by 1 degree Celsius. The heat capacity of water is 1 calorie (4.184 Joule) per gram per degree Celsius.

spines   Sharp, stiff plant surface structures, thought to be modified leaves, that physically deter herbivores and may assist in water conservation.

spongy mesophyll   Mesophyll cells of very irregular shape located below the palisade cells and surrounded by large air spaces. Functions in photosynthesis and gas exchange.

spores   Reproductive cells formed in plants by meiosis in the sporophyte generation. They give rise by mitotic divisions to the gametophyte generation.

sporophyte   The diploid ($2N$) multicellular structure that produces haploid spores by meiosis. Contrast with gametophyte.

starch   A polyglucan consisting of long chains of 1,4-linked glucose molecules and branch points where 1,6-linkages are used. Starch is the carbohydrate storage form in most plants.

starch sheath   A layer of cells that surrounds the vascular tissues of the shoot and coleoptile and is continuous with the root endodermis. Required for gravitropism in the shoots of eudicots.

statocytes   Specialized gravity-sensing plant cells that contain statoliths.

statoliths   Cellular inclusions such as amyloplasts that act as gravity sensors by having a high density relative to the cytosol and sedimenting to the bottom of the cell.

stele   In the root, the tissues located interior to the endodermis. The stele contains the vascular elements of the root: the phloem and the xylem.

stem   The typically above ground primary axis of the shoot that bears leaves and buds. May also occur underground in the form of rhizomes, corms, and tubers.

stomata   (singular *stoma* or *stomate*) Microscopic pores in the leaf epidermis, each surrounded by a pair of guard cells, and in some species, also including subsidiary cells. Stomata regulate the gas exchange (water and CO_2) of leaves by controlling the dimension of a stomatal pore.

stomatal complex The guard cells, subsidiary cells, and stomatal pore, which together regulate leaf transpiration.

stomatal resistance A measurement of the limitation to the free diffusion of gases from and into the leaf posed by the stomatal pores. The inverse of stomatal conductance.

stress Disadvantageous influences exerted on a plant by external abiotic or biotic factor(s), such as herbivory, infection, heat, water, or anoxia. Measured in relation to plant survival, crop yield, biomass accumulation, or CO_2 uptake.

strigolactones Carotenoid-derived plant hormones that inhibit shoot branching. They also play roles in the soil by stimulating the growth of arbuscular mycorrhizae and the germination of parasitic plant seeds, such as those of *Striga* (the source of their name).

stroma lamellae Unstacked thylakoid membranes within the chloroplast.

stroma The fluid component surrounding the thylakoid membranes of a chloroplast.

subsidiary cells Specialized epidermal cells that flank the guard cells and work with the guard cells in the control of stomatal apertures.

substrate-level phosphorylation A process that involves the direct transfer of a phosphate group from a substrate molecule to ADP to form ATP.

sucrose A disaccharide consisting of a glucose and a fructose molecule linked via an ether bond between C-1 on the glucosyl subunit and C-2 on the fructosyl unit. Sucrose is the carbohydrate transport form (e.g., in the phloem between source and sink).

sunflecks Patches of sunlight that pass through openings in a forest canopy to the forest floor. A major source of incident radiation for plants growing under the forest canopy.

supercool The condition in which cellular water remains liquid because of its solute content, even at temperatures several degrees below its theoretical freezing point.

superoxide dismutase An enzyme that converts superoxide radicals to hydrogen peroxide.

surface tension A force exerted by water molecules at the air–water interface, resulting from the *cohesion* and *adhesion* properties of water molecules.

This force minimizes the surface area of the air–water interface.

suspensor In seed plant embryogenesis, the structure that develops from the basal cell following the first division of the zygote. It supports, but is not part of, the embryo.

symbiosis The close association of two organisms in a relationship that may or may not be mutually beneficial. Often applied to beneficial (mutualistic) relationships.

symplast The continuous system of cell protoplasts interconnected by plasmodesmata.

symplastic transport The intercellular transport of water and solutes through plasmodesmata.

symport A type of secondary active transport in which two substances are moved in the same direction across a membrane.

syncytium A multinucleate cell that can result from multiple cell fusions of uninuclear cells, usually in response to viral infection.

synergid cells Two cells adjacent to the egg cell of the embryo sac, one of which is penetrated by the pollen tube upon entry into the ovule.

systemic acquired acclimation (SAA) A system whereby exposure of one part of a plant to an abiotic stress generates signals that can initiate acclimation to that stress in other, unexposed parts of the plant.

systemic acquired resistance (SAR) The increased resistance throughout a plant to a range of pathogens following infection by a pathogen at one site.

systems biology An approach to examining complex living processes that employs mathematical and computational models to simulate nonlinear biological networks and to better predict their operation.

T

taproot In eudicots, the main single root axis from which lateral roots develop.

telomeres Regions of repetitive DNA that form the chromosome ends and protect them from degradation.

telophase Prior to cytokinesis, the final stage of mitosis (or meiosis) during which the chromatin decondenses, the

nuclear envelope re-forms, and the cell plate extends.

temperature coefficient (Q_{10}) The increase in the rate of a process (e.g., respiration) for every 10°C increase in temperature.

temperature compensation A characteristic of circadian rhythms, which can maintain their circadian periodicity over a broad range of temperatures within the physiological range.

tensile strength The ability to resist a pulling force. Water has a high tensile strength.

tension Negative hydrostatic pressure.

tertiary cross-regulation Regulation that involves the outputs of two distinct signaling pathways exerting influences on one another.

testa The outer layer of the seed, also called the seed coat, derived from the integument of the ovule.

thigmotropism Plant growth in response to touch, enabling roots to grow around rocks, and shoots of climbing plants to wrap around structures for support.

thorns Sharp plant structures that physically deter herbivores and are derived from branches.

thylakoid reactions The chemical reactions of photosynthesis that occur in the specialized internal membranes of the chloroplast (called thylakoids). Include photosynthetic electron transport and ATP synthesis.

thylakoids The specialized, internal chlorophyll-containing membranes of the chloroplast where light absorption and the chemical reactions of photosynthesis take place.

tip growth Localized growth at the tip of a plant cell, caused by localized secretion of new wall polymers. Occurs in pollen tubes, root hairs, some sclerenchyma fibers, and cotton fibers, as well as moss protonema and fungal hyphae. Contrast with diffuse growth.

tonoplast The vacuolar membrane.

torus A central thickening found in the pit membranes of tracheids in the xylem of most gymnosperms.

toxic zone In plant tissue, the range of concentrations of a mineral nutrient in excess of the adequate zone and where growth or yield declines.

tracheary elements   Water-transporting cells of the xylem.

tracheids   Spindle-shaped, water-conducting cells with tapered ends and pitted walls without perforations found in the xylem of both angiosperms and gymnosperms.

tracheophyte   *See* vascular plant.

transcription   The process by which the base sequence information in DNA is copied into an RNA molecule.

transcytosis   Redirection of a secreted protein from one membrane domain within a cell to another, polarized domain.

transfer cell   A type of companion cell similar to an ordinary companion cell, but with fingerlike projections of the cell wall that greatly increase the surface area of the plasma membrane and increase the capacity for solute transport across the membrane from the apoplast.

translation   The process whereby a specific protein is synthesized on ribosomes according to the sequence information encoded by the mRNA.

translocation   The movement of photosynthate from sources to sinks in the phloem.

translocons   Pores in the rough endoplasmic reticulum that enable proteins synthesized on ribosomes to enter the ER lumen.

transpiration ratio   The ratio of water loss to photosynthetic carbon gain. Measures the effectiveness of plants in moderating water loss while allowing sufficient CO_2 uptake for photosynthesis.

transpiration   The evaporation of water from the surface of leaves and stems.

transport   Molecular or ionic movement from one location to another; may involve crossing a diffusion barrier such as one or more membranes.

transport proteins   Transmembrane proteins that are involved in the movement of molecules or ions from one side of a membrane to the other side.

treadmilling   During interphase, a process by which microfilaments or microtubules in the cortical cytoplasm appear to migrate around the cell periphery due to addition of G-actin or tubulin heterodimers, respectively, to the plus end at the same rate as their removal from the minus end.

triacylglycerols   Three fatty acyl groups in ester linkage to the three hydroxyl groups of glycerol. Fats and oils.

tricarboxylic acid (TCA) cycle   A cycle of reactions catalyzed by enzymes localized in the mitochondrial matrix that leads to the oxidation of pyruvate to CO_2. ATP and NADH are generated in the process. Also called the citric acid cycle.

trichomes   Unicellular or multicellular hairlike structures that differentiate from the epidermal cells of shoots and roots. Trichomes may be structural or glandular and function in plant responses to biotic or abiotic environmental factors.

triglycerides   Three fatty acyl groups in ester linkage to three hydroxyl groups of glycerol. Fats and oils.

triose phosphate   A three-carbon sugar phosphate.

tropism   Oriented plant growth in response to a perceived directional stimulus from light, gravity, or touch.

tubulin   A family of cytoskeletal GTP-binding proteins with three members, α-, β-, and γ-tubulin. α-tubulin forms heterodimers with β-tubulin, which polymerize to form microtubules.

turgor pressure   Force per unit of area in a liquid. In a plant cell, turgor pressure pushes the plasma membrane against the rigid cell wall and provides a force for cell expansion.

turnover   The balance between the rate of synthesis and the rate of degradation, usually applied to protein or RNA. An increase in turnover typically refers to an increase in degradation.

two-component regulatory systems   Signaling pathways common in prokaryotes. They typically involve a membrane-bound histidine kinase sensor protein that senses environmental signals and a response regulator protein that mediates the response. Although rare in eukaryotes, systems resembling bacterial two-component systems are involved in both ethylene and cytokinin signaling.

U

ubiquinone   A mobile electron carrier of the mitochondrial electron transport chain. Chemically and functionally similar to plastoquinone in the photosynthetic electron transport chain.

ubiquitin   A small polypeptide that is covalently attached to proteins, and that serves as a recognition site for a large proteolytic complex, the proteasome.

ubiquitin–proteasome pathway   A mechanism for the specific degradation of cellular proteins involving two discrete steps: the polyubiquitination of proteins via the E3 ubiquitin ligase and the degradation of the tagged protein by the 26S proteasome.

uncoupler   A chemical compound that increases the proton permeability of membranes and thus uncouples the formation of the proton gradient from ATP synthesis.

uncoupling   A process by which coupled reactions are separated in such a way that the free energy released by one reaction is not available to drive the other reaction.

uncoupling protein   A protein that increases the proton permeability of the inner mitochondrial membrane and thereby decreases energy conservation.

UV RESISTANCE LOCUS 8 (UVR8)   The protein receptor that mediates various plant responses to UV-B irradiation.

V

vacuolar H⁺-ATPase (V-ATPase)   A large, multi-subunit enzyme complex, related to F_oF_1-ATPases, present in endomembranes (tonoplast, Golgi). Acidifies the vacuole and provides the proton motive force for the secondary transport of a variety of solutes into the lumen. V-ATPases also function in the regulation of intracellular protein trafficking.

vacuolar sap   The fluid contents of a vacuole, which may include water, inorganic ions, sugars, organic acids, and pigments.

vascular cambium   A lateral meristem consisting of fusiform and ray stem cells, giving rise to secondary xylem and phloem elements, as well as ray parenchyma.

vascular plant   A plant that has xylem and phloem.

vascular tissues   Plant tissues specialized for the transport of water (xylem) and photosynthetic products (phloem).

venation pattern   The pattern of veins of a leaf.

vernalization   In some species, the cold temperature requirement for flowering. The term is derived from the Latin word for "spring."

vessel A stack of two or more vessel elements in the xylem.

vessel elements Nonliving water-conducting cells with perforated end walls found only in angiosperms and a small group of gymnosperms.

vivipary The precocious germination of seeds in the fruit while still attached to the plant.

volicitin A volatile compound produced by beet armyworm (*Spodoptera exigua*) when feeding on host grasses that attracts the generalist parasitoid wasp *Cotesia marginiventris*.

water potential (Ψ) A measure of the free energy associated with water per unit of volume ($J\ m^{-3}$). These units are equivalent to pressure units such as pascals. Ψ is a function of the solute potential, the pressure potential, and the gravitational potential: $\Psi = \Psi_s + \Psi_p + \Psi_g$. The term Ψ_g is often ignored because it is negligible for heights under 5 m.

W

wavelength (λ) A unit of measurement for characterizing light energy. The distance between successive wave crests. In the visible spectrum, it corresponds to a color.

whole plant senescence The death of the entire plant, as opposed to the death of individual cells, tissues, or organs.

whorl Pertaining to the concentric pattern of a set of organs that are initiated around the flanks of the meristem.

wilting Loss of rigidity, leading to a flaccid state, due to turgor pressure falling to zero.

wound callose Callose deposited in the sieve pores of damaged sieve elements that seals them off from surrounding intact tissue. As sieve elements recover, the callose disappears from pores.

X

xylem loading The process whereby ions exit the symplast and enter the conducting cells of the xylem.

xylem The vascular tissue that transports water and ions from the root to the other parts of the plant.

xyloglucan A hemicellulose with a backbone of 1→4-linked β-D-glucose residues and short side chains that contain xylose, galactose, and sometimes fucose.

Z

Zeitgebers Environmental signals such as light-to-dark or dark-to-light transitions that synchronize the endogenous oscillator to a 24-h periodicity.

ZEITLUPE A blue-light photoreceptor that regulates day-length perception (photoperiodism) and circadian rhythms.

Illustration Credits

CHAPTER 1

Figure 1.5A Zhang, T., Mahgsoudy-Louyeh, S., Tittmann, B., and Cosgrove, D. J. (2013) Visualization of the nanoscale pattern of recently-deposited cellulose microfibrils and matrix materials in never-dried primary walls of the onion epidermis. *Cellulose* 21: 853–862. DOI: 10.1007/s10570-013-9996-1. **Figure 1.5B-D** Matthews, J. F., Skopec, C. E., Mason, P. E., Zuccato, P., Torget, R. W., Sugiyama, J., Himmel, M. E., and Brady, J. W. (2006) Computer simulation studies of microcrystalline cellulose Ib. *Carbohydr. Res.* 341: 138–152. **Figure 1.7** Cosgrove, D. J. (2005) Growth of the plant cell wall. *Nat. Rev. Mol. Cell Biol.* 6: 850–861. **Figure 1.8B** Robinson-Beers, K., and Evert, R. F. (1991) Fine structure of plasmodesmata in mature leaves of sugar cane. *Planta* 184: 307–318. **Figure 1.8C** Bell, K., and Oparka, K. (2011) Imaging plasmodesmata. *Protoplasma* 248: 9–25. **Figure 1.10B-F** Esau, K. (1977) *Anatomy of Seed Plants*, 2nd edition. Wiley, New York. **Figure 1.12B** Buchanan, B. B., Gruissem, W., and Jones, R. L., eds. (2000) *Biochemistry and Molecular Biology of Plants*. American Society of Plant Biologists, Rockville, MD. **Figure 1.13B** Fiserova, J., Kiseleva, E., and Goldberg, M. W. (2009) Nuclear envelope and nuclear pore complex structure and organization in tobacco BY-2 cells. *Plant J.* 59: 243–255. **Figure 1.14** Alberts, B., Johnson, A., Lewis, J., Raff, M., Roberts, K., and Walter, P. (2007) *Molecular Biology of the Cell*. 5th ed. Garland Science, New York. **Figures 1.17A-C, 1.21, & 1.22** Gunning, B. E. S., and Steer, M. W. (1996) *Plant Cell Biology: Structure and Function of Plant Cells*. Jones and Bartlett, Boston. **Figure 1.24A,B** Lodish, H., Berk, A.,Kaiser, C. A.,Krieger, M., Scott, M. P., Bretscher, A., Ploegh, H., and Matsudaira, P. 2007. *Molecular Cell Biology*, 6th edition. W. H. Freeman and Co., New York. **Figure 1.30** Seguí-Simarro, J. M., Austin, J. R., White, E. A., and Staehelin, L. A. (2004) Electron tomographic analysis of somatic cell plate formation in meristematic cells of Arabidopsis preserved by high-pressure freezing. *Plant Cell* 16: 836–856.

CHAPTER 2

Figure 2.1 Day, W., Legg, B. J., French, B. K., Johnston, A. E., Lawlor, D. W., and Jeffers, W. de C. (1978) A drought experiment using mobile shelters: The effect of drought on barley yield, water use and nutrient uptake. *J. Agric. Sci.* 91: 599–623; Innes, P., and Blackwell, R. D. (1981) The effect of drought on the water use and yield of two spring wheat genotypes. *J. Agric. Sci.* 96: 603–610; Jones, H. G. (1992) *Plants and Microclimate*, 2nd ed., Cambridge University Press, Cambridge. **Figure 2.2** Schuur, E. A. G. (2003) Productivity and global climate revisited: The sensitivity of tropical forest growth to precipitation. *Ecology* 84: 1165–1170. **Figure 2.8** Nobel, P. S. (1991) *Physicochemical and Environmental Plant Physiology*. Academic Press, San Diego, CA. **Figure 2.11** Hsiao, T. C., and Xu, L. K. (2000) Sensitivity of growth of roots versus leaves to water stress: Biophysical analysis and relation to water transport. *J. Exp. Bot.* 51: 1595–1616. **Figure 2.15** Hsiao, T. C. (1973) Plant responses to water stress. *Annu. Rev. Plant Physiol.* 24: 519–70.

CHAPTER 3

Table 3.1 Nobel, P. S. (1999) *Physicochemical and Environmental Plant Physiology*, 2nd ed. Academic Press, San Diego, CA. **Figure 3.3A** Kramer, P. J. (1983) *Water Relations of Plants*. Academic Press, San Diego, CA. **Figure 3.7A** Zimmermann, M. H. (1983) *Xylem Structure and the Ascent of Sap*. Springer, Berlin. **Figure 3.8** Gunning, B. E. S., and Steer, M. W. (1996) *Plant Cell Biology: Structure and Function*. Jones and Bartlett, Boston. **Figure 3.10** Sperry, J. S. (2000) Hydraulic constraints on plant gas exchange. *Agric. For. Meteorol.* 104: 13–23. **Figure 3.12** Bange, G. G. J. (1953) On the quantitative explanation of stomatal transpiration. *Acta Bot. Neerl.* 2: 255–296. **Figure 3.13A** Zeiger, E., and Hepler, P. K. (1976) Production of guard cell protoplasts from onion and tobacco. *Plant Physiol.* 58: 492–498. **Figure 3.14A** Palevitz, B. A. (1981) The structure and development of guard cells. In *Stomatal Physiology*, P. G. Jarvis and T. A. Mansfield, eds., Cambridge University Press, Cambridge, pp. 1–23. **Figure 3.14B** Sack, F. D. (1987) The development and structure of stomata. In *Stomatal Function*, E. Zeiger, G. Farquhar, and I. Cowan, eds., Stanford University Press, Stanford, CA, pp. 59–90. **Figure 3.14C** Meidner, H., and Mansfield, D. (1968) *Physiology of Stomata*. McGraw-Hill, London. **Figure 3.16** Srivastava, A., and Zeiger, E. (1995) Guard cell zeaxanthin tracks photosynthetically active radiation and stomatal apertures in *Vicia faba* leaves. *Plant Cell Environ.* 18: 813–817. **Figure 3.17** Schwartz, A., and Zeiger, E. (1984) Metabolic energy for stomatal opening. Roles of photophosphorylation and oxidative phosphorylation. *Planta* 161: 129–136.

CHAPTER 4

Table 4.1 Epstein, E. (1999) Silicon. *Annu. Rev. Plant Physiol. Plant Mol. Biol.* 50: 641–664; Epstein, E. (1972) *Mineral Nutrition of Plants: Principles and Perspectives*. John Wiley and Sons, New York. **Table 4.2** Evans, H. J., and Sorger, G. J. (1966) Role of mineral elements with emphasis on the univalent cations. *Annu. Rev. Plant Physiol.* 17: 47–76; Mengel, K., and Kirkby, E. A. (2001) *Principles of Plant Nutrition*, 5th ed. Kluwer Academic Publishers, Dordrecht, Netherlands. **Table 4.3** Epstein, E., and Bloom, A. J. (2005) Mineral Nutrition of Plants: Principles and Perspectives, 2nd ed. Sinauer Associates, Sunderland, MA. **Table 4.5** Brady, N. C. (1974) *The Nature and Properties of Soils*, 8th ed. Macmillan, New York. **Figure 4.2** Epstein, E., and Bloom, A. J. (2005) *Mineral Nutrition of Plants: Principles and Perspectives*, 2nd ed. Sinauer Associates, Sunderland, MA. **Figure 4.3** Sievers, R. E., and Bailar, J. C., Jr. (1962) Some metal chelates of ethylenediaminetetraacetic acid, diethylenetriaminepentaacetic acid, and triethylenetriaminehexaacetic acid. *Inorg. Chem.* 1: 174–182. **Figure 4.5** Lucas, R. E., and Davis, J. F. (1961) Relationships between pH values of organic soils and availabilities of 12 plant nutrients. *Soil Sci.* 92: 177–182. **Figures 4.7 & 4.8** Weaver, J. E. (1926) *Root Development of Field Crops.* McGraw-Hill, New York. **Figure 4.10** Mengel, K., and Kirkby, E. A. (2001) *Principles of Plant Nutrition,* 5th ed. Kluwer Academic Publishers, Dordrecht, Netherlands. **Figure 4.11** Bloom, A. J., Jackson, L. E., and Smart, D. R. (1993) Root growth as a function of ammonium and nitrate in the root zone. *Plant Cell Environ.* 16: 199–206. **Figure 4.12** Giovannetti, M., Avio, L., Fortuna, P., Pellegrino, E., Sbrana, C., and Strani, P. (2006) At the root of the wood wide web. *Plant Signal. Behav.* 1: 1–5. **Figure 4.14** Rovira, A. D., Bowen, C. D., and Foster, R. C. (1983) The significance of rhizosphere microflora and mycorrhizas in plant nutrition. In *Encyclopedia of Plant Physiology,* New Series, Vol. 15B: *Inorganic Plant Nutrition,* A. Läuchli and R. L. Bieleski, eds., Springer, Berlin, pp. 61–93.

CHAPTER 5

Figure 5.2 Bloom, A. J. (1997) Nitrogen as a limiting factor: Crop acquisition of ammonium and nitrate. In *Ecology in Agriculture,* L. E. Jackson, ed., Academic Press, San Diego, CA, pp. 145–172. **Figure 5.4** Kleinhofs, A., Warner, R. L., and Melzer, J. M. (1989) Genetics and molecular biology of higher plant nitrate reductases. In *Recent Advances in Phytochemistry,* Vol. 23: *Plant Nitrogen Metabolism,* J.E. Poulton, J.T. Romeo and E. Conn, eds., Plenum, New York, pp. 117–155. **Figure 5.6** Pate, J. S. (1973) Uptake, assimilation and transport of nitrogen compounds by plants. *Soil Biol. Biochem.* 5: 109–119. **Figure 5.11** Stokkermans, T. J. W., Ikeshita, S., Cohn, J., Carlson, R. W., Stacey, G., Ogawa, T., and Peters, N. K. (1995) Structural requirements of synthetic and natural product lipo-chitin oligosaccharides for induction of nodule primordia on *Glycine soja. Plant Physiol.* 108: 1587–1595. **Figure 5.14** Dixon, R. O. D., and Wheeler, C. T. (1986) *Nitrogen Fixation in Plants.* Chapman and Hall, New York; Buchanan, B., Gruissem, W., and Jones, R., eds. (2000) *Biochemistry and Molecular Biology of Plants.* American Society of Plant Physiologists, Rockville, MD. **Figure 5.16** Guerinot, M. L., and Yi, Y. (1994) Iron: Nutritious, noxious,

and not readily available. *Plant Physiol.* 104: 815–820.

CHAPTER 6

Table 6.1 Higinbotham, N., Etherton, B., and Foster, R. J. (1967) Mineral ion contents and cell transmembrane electropotentials of pea and oat seedling tissue. *Plant Physiol.* 42: 37–46. **Figure 6.4** Higinbotham, N., Graves, J. S., and Davis, R. F. (1970) Evidence for an electrogenic ion transport pump in cells of higher plants. *J. Membr. Biol.* 3: 210–222. **Figure 6.7A** Leng, Q., Mercier, R. W., Hua, B. G., Fromm, H., and Berkowitz, G. A. (2002) Electrophysiological analysis of cloned cyclic nucleotide-gated ion channels. *Plant Physiol.* 128: 400–410. **Figure 6.7B** Buchanan, B. B., Gruissem, W., and Jones, R. L., eds. (2000) *Biochemistry and Molecular Biology of Plants.* American Society of Plant Physiologists, Rockville, MD **Figure 6.12** Lin, W., Schmitt, M. R., Hitz, W. D., and Giaquinta, R. T. (1984) Sugar transport into protoplasts isolated from developing soybean cotyledons. *Plant Physiol.* 75: 936–940. **Figure 6.14A** Lebaudy, A., Véry, A., and Sentenac, H. (2007) K^+ channel activity in plants: Genes, regulations and functions. *FEBS Lett.* 581: 2357–2366. **Figure 6.14B– D** Very, A. A., and Sentenac, H. (2002) Cation channels in the *Arabidopsis* plasma membrane. *Trends Plant Sci.* 7: 168–175. **Figure 6.16** Palmgren, M. G. (2001) Plant plasma membrane H^+-ATPases: Powerhouses for nutrient uptake. *Annu. Rev. Plant Physiol. Plant Mol. Biol.* 52: 817–845. **Figure 6.17** Kluge, C., Lahr, J., Hanitzsch, M., Bolte, S., Golldack, D., and Dietz, K. J. (2003) New insight into the structure and regulation of the plant vacuolar H^+-ATPase. *J. Bioenerg. Biomembr.* 35: 377–388. **Figure 6.18A, right** Zeiger, E., and Hepler, P. K. (1977) Light and stomatal function: Blue light stimulates swelling of guard cell protoplasts. *Science* 196: 887–889. **Figure 6.18B** Amodeo, G., Srivastava, A., and Zeiger, E. (1992) Vanadate inhibits blue light–stimulated swelling of *Vicia* guard cell protoplasts. *Plant Physiol.* 100: 1567–1570. **Figure 6.19B** Serrano, E. E., Zeiger, E., and Hagiwara, S. (1988) Red light stimulates an electrogenic proton pump in *Vicia* guard cell protoplasts. *Proc. Natl. Acad. Sci. USA* 85: 436–440. **Figure 6.19C** Assmann, S. M., Simoncini, L., and Schroeder, J. I. (1985) Blue light activates electrogenic ion pumping in guard cell protoplasts of *Vicia faba. Nature* 318: 285–287. **Figure 6.20** Inoue, S., and Kinoshita, T.

(2017) Blue light regulation of stomatal opening and the plasma membrane H^+-ATPase. *Plant Physiol.* 174: 531-538.

CHAPTER 7

Figure 7.15 Becker, W. M. (1986) *The World of the Cell.* Benjamin/ Cummings, Menlo Park, CA. **Figure 7.16A** Allen, J. F., and Forsberg, J. (2001) Molecular recognition in thylakoid structure and function. *Trends Plant Sci.* 6: 317–326. **Figure 7.16B** Nelson, N., and Ben-Shem, A. (2004) The complex architecture of oxygenic photosynthesis. *Nat. Rev. Mol. Cell Biol.* 5: 971–982. **Figure 7.18** Barros, T., and Kühlbrandt, W. (2009) Crystallisation, structure and function of plant light-harvesting Complex II. *Biochim. Biophys. Acta* 1787: 753–772. **Figures 7.19 &7.21** Blankenship, R. E., and Prince, R. C. (1985) Excited-state redox potentials and the Z scheme of photosynthesis. *Trends Biochem. Sci.* 10: 382–383. **Figure 7.22A,B** Ferreira, K. N., Iverson, T. M., Maghlaoui, K., Barber, J., and Iwata, S. (2004) Architecture of the photosynthetic oxygen-evolving center. *Science* 303: 1831–1838. **Figure 7.22C** Umena, Y., Kawakami, K., Shen, J.-R., and Kamiya, N. (2011) Crystal structure of oxygen-evolving photosystem II at a resolution of 1.9 Å. *Nature* 473: 55–60. **Figure 7.24** Kurisu, G., Zhang, H. M., Smith, J. L., and Cramer, W. A. (2003) Structure of cytochrome b6f complex of oxygenic photosynthesis: tuning the cavity. *Science* 302: 1009–1014. **Figure 7.26A** Buchanan, B. B., Gruissem, W., and Jones, R. L., eds. (2000) *Biochemistry and Molecular Biology of Plants.* American Society of Plant Physiologists, Rockville, MD. **Figure 7.26B** Nelson, N., and Ben-Shem, A. (2004) The complex architecture of oxygenic photosynthesis. *Nat. Rev. Mol. Cell Biol.* 5: 971–982. **Figure 7.28** Jagendorf, A. T. (1967) Acid-based transitions and phosphorylation by chloroplasts. *Fed. Proc. Am. Soc. Exp. Biol.* 26: 1361–1369.

CHAPTER 8

Figure 8.9C Edwards, G. E., Franceschi, V. R., and Voznesenskaya, E. V. (2004) Single-cell C4 photosynthesis versus the dual-cell (Kranz) paradigm. *Annu. Rev. Plant Biol.* 55: 173–196.

CHAPTER 9

Figure 9.3 Smith, H. (1986). The perception of light quality. In *Photomorphogenesis in Plants,* R. E. Kendrick and G. H. M. Kronenberg, eds., Nijhoff, Dordrecht, Netherlands, pp. 187–217. **Figure 9.4** Smith, H.

(1994) Sensing the light environment: The functions of the phytochrome family. In *Photomorphogenesis in Plants*, 2nd ed., R. E. Kendrick and G. H. M. Kronenberg, eds., Nijhoff, Dordrecht, Netherlands, pp. 377–416. **Figure 9.5** Vogelmann, T. C., and Björn, L. O. (1983) Response to directional light by leaves of a sun-tracking lupine (*Lupinus succulentus*). *Physiol. Plant.* 59: 533–538. **Figure 9.7** Harvey, G. W. (1979) Photosynthetic performance of isolated leaf cells from sun and shade plants. *Carnegie Inst. Wash. Yearb.* 79: 161–164. **Figure 9.8** Björkman, O. (1981) Responses to different quantum flux densities. In *Encyclopedia of Plant Physiology*, New Series, Vol. 12A, O. L. Lange, P. S. Nobel, C. B. Osmond, and H. Zeigler, eds., Springer, Berlin, pp. 57–107. **Figure 9.9** Jarvis, P. G., and Leverenz, J. W. (1983) Productivity of temperate, deciduous and evergreen forests. In *Encyclopedia of Plant Physiology*, New Series, Vol. 12D, O. L. Lange, P. S. Nobel, C. B. Osmond, and H. Ziegler, eds., Springer, Berlin, pp. 233–280. **Figure 9.10** Osmond, C. B. (1994) What is photoinhibition? Some insights from comparisons of shade and sun plants. In *Photoinhibition of Photosynthesis: From Molecular Mechanisms to the Field*, N. R. Baker and J. R. Bowyer, eds., BIOS Scientific, Oxford, pp. 1–24. **Figure 9.12** Adams, W.W. & Demmig-Adams, B. (1992) Operation of the xanthophyll cycle in higher plants in response to diurnal changes in incident sunlight. *Planta* 186: 390–398. **Figure 9.13** Tlalka, M., and Fricker, M. (1999) The role of calcium in blue-light-dependent chloroplast movement in *Lemna trisulca* L. *Plant J.* 20: 461–473. **Figure 9.14** Demming-Adams, B., and Adams, W. (2000) Harvesting sunlight safely. *Nature* 403: 371–372. **Figure 9.16** Björkman, O., Mooney, H.A., Ehleringer, J. (1975) Photosynthetic responses of plants from habitats with contrasting thermal environments: comparison of photosynthetic characteristics of intact plants. *Carnegie Inst. Washington Yearb.* 74: 743–748. **Figure 9.17** Ehleringer, J. and Björkman, O. (1977) Quantum yields for CO_2 uptake in C3 and C4 plants. *Plant Physiol.* 59: 86–90. **Figure 9.18** Ehleringer, J. R. (1978) Implications of quantum yield differences on the distributions of C3 and C4 grasses. *Oecologia* 31: 255–267. **Figure 9.19A** Barnola, J. M., Raynaud, D., Lorius, C., and Barkov, N.I. (2003) Historical CO_2 record from the Vostok ice core. In *Trends: A Compendium of Data on Global Change*, T. A. Boden, D. P. Kaiser, R. J. Sepanski, and F. W. Stoss, eds., Carbon Dioxide Information Analysis Center, Oak Ridge National Laboratory, U.S. Dept. of Energy, Oak Ridge, TN, pp. 7–10. **Figure 9.19B** Etheridge, D. M., Steele, L. P., Langenfelds, R. L., Francey, R. J., Barnola, J.-M., and Morgan, V. I. (1998) Historical CO_2 records from the Law Dome DE08, DE08-2, and DSS ice cores. In *Trends: A Compendium of Data on Global Change*. Carbon Dioxide Information Analysis Center, Oak Ridge National Laboratory, U.S. Department of Energy, Oak Ridge, Tenn., U.S.A. **Figure 9.19C** Keeling, C. D., and Whorf, T. P. (1994) Atmospheric CO_2 records from sites in the SIO air sampling network. In *Trends '93: A Compendium of Data on Global Change*, T. A. Boden, D. P. Kaiser, R. J. Sepanski, and F. W. Stoss, eds., Carbon Dioxide Information Center, Oak Ridge National Laboratory, Oak Ridge, TN, pp. 16–26; updated using data from Dr. Pieter Tans, NOAA/ESRL (www.esrl.noaa.gov/gmd/ccgg/trends/) and Dr. Ralph Keeling, Scripps Institution of Oceanography (scrippsco2.ucsd.edu/). **Figure 9.21** Berry, J. A., and Downton, J. S. (1982) Environmental regulation of photosynthesis. In *Photosynthesis: Development, Carbon Metabolism and Plant Productivity*, Vol. 2, Govindjee, ed., Academic Press, New York, pp. 263–343. **Figure 9.22** Ehleringer, J. R., Cerling, T. E., and Helliker, B. R. (1997) C_4 photosynthesis, atmospheric CO_2, and climate. *Oecologia* 112: 285–299. **Figure 9.23** Jeon, M.-W., Ali, M. B., Hahn, E.-J., and Paek, K.-Y. (2006) Photosynthetic pigments, morphology and leaf gas exchange during ex vitro acclimatization of micropropagated CAM *Doritaenopsis* plantlets under relative humidity and air temperature. *Environ. Exp. Bot.* 55: 183–194. **Figure 9.24C** Long, S. P., Ainsworth, E. A., Leakey, A. D., Nosberger, J., and Ort, D. R. (2006) Food for thought: Lower-than-expected crop stimulation with rising CO_2 concentrations. *Science* 312: 1918–1921.

CHAPTER 10

Table 10.2 Hall, S. M., and Baker, D. A. (1972) The chemical composition of *Ricinus* phloem exudate. *Planta* 106: 131–140. **Figure 10.2A** Joy, K. W. (1964) Translocation in sugar beet. I. Assimilation of $^{14}CO_2$ and distribution of materials from leaves. *J. Exp. Bot.* 15: 485–494. **Figure 10.6** Warmbrodt, R. D. (1985) Studies on the root of *Hordeum vulgare* L.—Ultrastructure of the seminal root with special reference to the phloem. *Am. J. Bot.* 72: 414–432. **Figure 10.7A,B** Evert, R. F. (1982) Sieve-tube structure in relation to function. *Bioscience* 32: 789–795. **Figure 10.7C,D** Truernit, E., Bauby, H., Dubreucq, B., Grandjean, O., Runions, J., Barthelemy, J., and Palauqui, J.-C. (2008) High-resolution whole-mount imaging of three-dimensional tissue organization and gene expression enables the study of phloem development and structure in Arabidopsis. *Plant Cell* 20: 1494–1503. **Figure 10.9** Nobel, P. S. (2005) *Physicochemical and Environmental Plant Physiology*, 3rd ed., Academic Press, San Diego, CA. **Figure 10.10** Geiger, D. R., and Sovonick, S. A. (1975) Effects of temperature, anoxia and other metabolic inhibitors on translocation. In *Transport in Plants, 1: Phloem Transport* (*Encyclopedia of Plant Physiology*, New Series, Vol. 1), M. H. Zimmerman and J. A. Milburn, eds., Springer, New York, pp. 256–286. **Figure 10.12** Fondy, B. R. (1975) Sugar selectivity of phloem loading in *Beta vulgaris, vulgaris* L. and *Fraxinus americanus, americana* L. Ph.D. diss., University of Dayton, Dayton, OH. **Figure 10.14** van Bel, A. J. E. (1992) Different phloem-loading machineries correlated with the climate. *Acta Bot. Neerl.* 41: 121–141. **Figure 10.15** Turgeon, R., and Webb, J. A. (1973) Leaf development and phloem transport in *Cucurbita pepo*: Transition from import to export. *Planta* 113: 179–191. **Figure 10.16** Turgeon, R. (2006) Phloem loading: How leaves gain their independence. *Bioscience* 56: 15–24. **Figure 10.17** Schneidereit, A., Imlau, A., and Sauer, N. (2008) Conserved cis-regulatory elements for DNA-binding-with-one-finger and homeo-domain-leucine-zipper transcription factors regulate companion cell-specific expression of the *Arabidopsis thaliana* SUCROSE TRANSPORTER 2 gene. *Planta* 228: 651–662. **Figure 10.19** Preiss, J. (1982) Regulation of the biosynthesis and degradation of starch. *Annu. Rev. Plant Physiol.* 33: 431–454.

CHAPTER 11

Table 11.2 Brand, M. D. (1994) The stoichiometry of proton pumping and ATP synthesis in mitochondria. *Biochem.* (Lond) 16: 20–24. **Figure 11.5A** Perkins, G., Renken, C., Martone, M. E., Young, S. J., Ellisman, M., and Frey, T. (1997) Electtron tomography of neuronal mitochondria: Three-dimensional structure and organization of cristae and membrane contacts. *J. Struct. Biol.* 119: 260–

272.   **Figure 11.5B** Gunning, B. E.
S., and Steer, M. W. (1996) *Plant Cell
Biology: Structure and Function*. Jones and
Bartlett, Boston.

CHAPTER 12

Figure 12.3 Santer, A., and Estelle, M.
(2009) Recent advances and emerging
trends in plant hormone signaling.
Nature 459: 1071–1078.   **Figure
12.10A** Riou-Khamlichi, C., Huntley,
R., Jacqmard, A., and Murray, J.
A. (1999) Cytokinin activation of
Arabidopsis cell division through
a D-type cyclin. *Science* 283: 1541–
1544.   **Figure 12.10B** Aloni, R.,
Wolf, A., Feigenbaum, P., Avni, A.,
and Klee, H. J. (1998) The Never
ripe mutant provides evidence that
tumor-induced ethylene controls
the morphogenesis of *Agrobacterium
tumefaciens*-induced crown galls
in tomato stems. *Plant Physiol.* 117:
841–849.   **Figure 12.13** Hartwig, T. et
al. (2011) Operation of the xanthophyll
cycle in higher plants in response to
diurnal changes in incident sunlight.
Proc. Natl. Acad. Sci. USA 108: 19814–
19819.   **Figure 12.16B** Escalante-
Pérez, M., Krola, E., Stangea, A.,
Geigera, D., Al-Rasheidb, K. A. S.,
Hausec, B., Neherd, E., and Hedrich, R.
(2011) A special pair of phytohormones
controls excitability, slow closure, and
external stomach formation in the Venus
flytrap. *Proc. Natl. Acad. Sci. USA* 108:
15492–15497.

CHAPTER 13

Figure 13.6A Shropshire, W., Jr., Klein,
W. H., and Elstad, V. B. (1961) Action
spectra of photomorphogenic induction
and photoinactivation of germination in
Arabidopsis thaliana. *Plant Cell Physiol.* 2:
63–69.   **Figure 13.6B** Kelly, J. M., and
Lagarias, J. C. (1985) Photochemistry
of 124-kilodalton *Avena* phytochrome
under constant illumination in vitro.
Biochemistry 24: 6003–6010.   **Figure
13.7A** Baskin, T. I., and Iino, M.
(1987) An action spectrum in the
blue and ultraviolet for phototropism
in alfalfa. *Photochem. Photobiol.* 46:
127–136.   **Figure 13.7B** Briggs,
W. R. and Christie, J. M. (2002)
Phototropins 1 and 2: versatile plant
blue-light receptors. *Trends Plant Sci.* 7:
204–210.   **Figure 13.9** Briggs, W. R.,
Mandoli, D. F., Shinkle, J. R., Kaufman,
L. S., Watson, J. C., and Thompson, W.
F. (1984) Phytochrome regulation of
plant development at the whole plant,
physiological, and molecular levels. In
*Sensory Perception and Transduction in
Aneural Organisms*, G. Colombetti, F.
Lenci, and P.-S. Song, eds., Plenum,

New York, pp. 265–280.   **Figure
13.10** Spalding, E. P., and Cosgrove, D.
J. (1989) Large membrane depolarization
precedes rapid blue-light induced
growth inhibition in cucumber. *Planta*
178: 407–410.   **Figure 13.13** Christie,
J. M., Swartz, T. E., Bogomolni, R. A.
and Briggs, W. R. (2002) Phototropin
LOV domains exhibit distinct roles in
regulating photoreceptor function. *Plant
J.* 32: 205–219.   **Figure 13.15** Wada, M.
(2013) Chloroplast movement. *Plant Sci.*
210: 177–182.   **Figure 13.16** Karlsson,
P. E. (1986) Blue light regulation
of stomata in wheat seedlings. II.
Action spectrum and search for
action dichroism. *Physiol. Plant.* 66:
207–210.   **Figure 13.17** Inoue, S., and
Kinoshita, T. (2017) Blue light regulation
of stomatal opening and the plasma
membrane H^+-ATPase. *Plant Physiol.*
174: 531–538.   **Figure 13.18** Parks,
B. M., Folta, K. M., and Spalding, E. P.
(2001) Photocontrol of stem growth.
Curr. Opin. Plant Biol. 4: 436–440.

CHAPTER 14

Figure 14.2 West, M. A. L., and
Harada, J. J. (1993) Embryogenesis in
higher plants: An overview. *Plant Cell*
5: 1361–1369.   **Figure 14.3** Sosso, D.,
Canut, M., Gendrot, G., et al. (2012)
PPR8522 encodes a chloroplast-targeted
pentatricopeptide repeat protein
necessary for maize embryogenesis
and vegetative development. *J. Exp. Bot.*
63: 5843–5857.   **Figure 14.5** Laux,
T., Würschum, T., and Breuninger, H.
(2004) Genetic regulation of embryonic
pattern formation. *Plant Cell* 16: S190–
S202.   **Figure 14.7** Bommert, P.,
and Werr, W. (2001) Gene expression
patterns in the maize caryopsis:
clues to decisions in embryo and
endosperm development. *Gene* 271:
131–142.   **Figure 14.8** Traas, J., Bellini,
C., Nacry, P., Kronenberger, J. Bouchez,
D., and Caboche, M. (1995) Normal
differentiation patterns in plants lacking
microtubular preprophase bands. *Nature*
375: 676–677.   **Figure 14.9** Kim, I.,
Kobayashi, K., Cho, E., and Zambryski,
P. C. (2005) Subdomains for transport
via plasmodesmata corresponding to
the apical-basal axis are established
during Arabidopsis embryogenesis.
Proc. Natl. Acad. Sci. USA 102:
11945–11950.   **Figure 14.12** Laux,
T., Würschum, T., and Breuninger, H.
(2004) Genetic regulation of embryonic
pattern formation. *Plant Cell* 16: S190–
S202.   **Figure 14.13** Mähönen, A. P.,
Bonke, M., Kauppinen, L., Riikonen,
M., Benfey, P. N., and Helariutta,
Y. (2000) A novel two-component
hybrid molecule regulates vascular

morphogenesis of the Arabidopsis root.
Genes Dev. 14: 2938–2943.   **Figure
14.14** Müller, B., and Sheen, J. (2008)
Cytokinin and auxin interaction in
root stem-cell specification during
early embryogenesis. *Nature* 453:
1094–1097.   **Figure 14.15** Jenik, P.
D., and Barton, M. K. (2005) Surge
and destroy: The role of auxin in plant
embryogenesis. *Development* 132:
3577–3585.   **Figure 14.16** Laux, T.,
Würschum, T., and Breuninger, H.
(2004) Genetic regulation of embryonic
pattern formation. *Plant Cell* 16: S190–
S202.   **Figure 14.17** Hudson, A. (2005)
Plant meristems: Mobile mediators of
cell fate. *Curr. Biol.* 15: R803–805.

CHAPTER 15

Figure 15.3 Homrichhausen, T.
M., Hewitt, J. R., and Nonogaki, H.
(2003) Endo-β-mannanase activity
is associated with the completion
of embryogenesis in imbibed carrot
(*Daucus carota* L.) seeds. *Seed Sci. Res.*
13: 219–227.   **Figure 15.5A** Li, Y.-C.,
Rena, J.-P., Cho, M.-J., Zhou, S.-M.,
Kim, Y.-B., Guo, H.-X., Wong, J.-H.,
Niu, H.-B., Kim, H.-K., Morigasaki, S.,
et al. (2009) The level of expression of
thioredoxin is linked to fundamental
properties and applications of wheat
seeds. *Mol. Plant* 2: 430–441.   **Figure
15.6** Finch-Savage, W. E. and Leubner-
Metzger, G. (2006) Seed dormancy and
the control of germination. *New Phytol.*
171: 501–523.   **Figure 15.7A** Bewley,
J. D., Bradford, K. J., Hilhorst, H. W. M.,
and Nonogaki, H. (2013) Dormancy
and the control of germination. In *Seeds:
Physiology of Development, Germination
and Dormancy*, 3rd Edition. Springer,
New York.   **Figure 15.7B** Grappin,
P., Bouinot, D., Sotta, B., Migniac, E.,
and Julien, M. (2000) Control of seed
dormancy in *Nicotiana plumbaginifolia*:
post-imbibition abscisic acid synthesis
imposes dormancy maintenance. *Planta*
210: 279–285.   **Figure 15.8** Liptay,
A., and Schopfer, P. (1983) Effect of
water stress, seed coat restraint, and
abscisic acid upon different germination
capabilities of two tomato lines at
low temperature. *Plant Physiol.* 73:
935–938.   **Figure 15.9** Bewley,
J. D. (1997) Seed germination and
dormancy. *Plant Cell* 9: 1055–1066;
Nonogaki, H., Chen, F., and Bradford,
K. J. (2007) Mechanisms and genes
involved in germination sensu stricto.
In *Annual Plant Reviews*, vol. 27: *Seed
Development, Dormancy and Germination*,
K. J. Bradford and H. Nonogaki, eds.
Blackwell Publishing Ltd, Oxford;
Nonogaki, H., Bassel, G. W., and
Bewley, J. D. (2010) Germination—

Still a mystery. *Plant Sci.* 179: 574–581. **Figure 15.11** Gubler, F., Kalla, R., Roberts, J. K., and Jacobsen, J. V. (1995) Gibberellin-regulated expression of a myb gene in barley aleurone cells: Evidence of Myb transactivation of a high-pI alpha-amylase gene promoter. *Plant Cell* 7: 1879–1891. **Figure 15.15** Busse, J. S., and Evert, R. F. (1999) Pattern of differentiation of the first vascular elements in the embryo and seedling of *Arabidopsis thaliana*. *Int. J. Plant Sci.* 160: 1–13. **Figure 15.17** Abeles, F. B., Morgan, P. W., and Saltveit, M. E., Jr. (1992) *Ethylene in Plant Biology*, 2nd ed. Academic Press, San Diego, CA. **Figure 15.18** Petricka, J. J., Winter, C. M., and Benfey, P. N. (2012) Control of Arabidopsis root development. *Annu. Rev. Plant Biol.* 63: 563–590. **Figures 15.21 & 15.22** Cosgrove, D. J. (1997) Assembly and enlargement of the primary cell wall in plants. *Annu. Rev. Cell Dev. Biol.* 13: 171–201. **Figure 15.24A** Cleland, R. E. (1995) Auxin and cell elongation. In *Plant Hormones and Their Role in Plant Growth and Development*, 2nd ed., P. J. Davies, ed., Kluwer, Dordrecht, Netherlands, pp. 214–227. **Figure 15.24C** Jacobs, M., and Ray, P. M. (1976) Rapid auxin-induced decrease in free space pH and its relationship to auxin-induced growth in maize and pea. *Plant Physiol.* 58: 203–209. **Figure 15.25** Le, J., Vandenbussche, F., De Cnodder, T., Van Der Straeten, D., and Verbelen, J.-P. (2005) Cell elongation and microtubule behavior in the Arabidopsis hypocotyl: Responses to ethylene and auxin. *J. Plant Growth Regul.* 24: 166–178. **Figure 15.27** Hartmann, H. T., and Kester, D. E. (1983) *Plant Propagation: Principles and Practices*, 4th ed. Prentice-Hall, Inc., NJ. **Figure 15.28** Blilou, I., Xu, J., Wildwater, M., Willemsen, V., Paponov, I., Friml, J., Heidstra, R., Aida, M., Palme, K., and Scheres, B. (2005) The PIN auxin efflux facilitator network controls growth and patterning in Arabidopsis roots. *Nature* 433: 39–44. **Figure 15.30** Shaw, S., and Wilkins, M. B. (1973) The source and lateral transport of growth inhibitors in geotropically stimulated roots of *Zea mays* and *Pisum sativum. Planta* 109: 11–26. **Figure 15.31** Baldwin, K. L., Strohm, A. K., and Masson, P. H. (2013) Gravity sensing and signal transduction in vascular plant primary roots. *Am. J. Bot.* 100: 126–142. **Figure 15.32** Palmieri M., and Kiss J. Z. (2007) The role of plastids in gravitropism. In *The Structure and Function of Plastids. Advances in Photosynthesis and Respiration*, vol. 23, R. R. Wise and J.

K. Hoober, eds. Springer, Dordrecht, pp. 507–525. **Figure 15.33** Fasano, J. M., Swanson, S. J., Blancaflor, E. B., Dowd, P. E., Kao, T. H., and Gilroy, S. (2001) Changes in root cap pH are required for the gravity response of the Arabidopsis root. *Plant Cell* 13: 907–921. **Figure 15.34** Iino, M., and Briggs, W. R. (1984) Growth distribution during first positive phototropic curvature of maize coleoptiles. *Plant Cell Environ.* 7: 97–104. **Figure 15.35** Christie, J. M., Yang, H., Richter, G. L., Sullivan, S., Thomson, C. E., Lin, J., Titapiwatanakun, B., Ennis, M., Kaiserli, E., Lee, O. R., et al. (2011) phot1 inhibition of ABCB19 primes lateral auxin fluxes in the shoot apex required for phototropism. *PLoS Biol.* 9: e1001076. DOI: 10.1371/journal.pbio.1001076.

CHAPTER 16

Table 16.1 Smith, H. (1982) Light quality photoperception and plant strategy. *Annu. Rev. Plant Physiol.* 33: 481–518. **Table 16.2** Thomas, H. (2013) Senescence, ageing and death of the whole plant. *New Phytol.* 197: 696–711. DOI: 10.1111/nph.12047. **Figure 16.1B** Kuhlemeier, C., and Reinhardt, D. (2001) Auxin and phyllotaxis. *Trends Plant Sci.* 6: 187–189. **Figure 16.2** Bowman, J. L., and Eshed, Y. (2000) Formation and maintenance of the shoot apical meristem. *Trends Plant Sci.* 5: 110–115. **Figure 16.5A** Reinhardt, D., Pesce, E. R., Stieger, P., Mandel, T., Baltensperger, K., Bennett, M., Traas, J., Friml, J., and Kuhlemeier, C. (2003) Regulation of phyllotaxis by polar auxin transport. *Nature* 426: 255–260 **Figure 16.5B** Vernoux, T., Kronenberger, J., Grandjean, O., Laufs, P., and Traas, J. (2000) PIN-FORMED 1 regulates cell fate at the periphery of the shoot apical meristem. *Development* 127: 5157–5165. **Figure 16.5C** Reinhardt, D., Pesce, E. R., Stieger, P., Mandel, T., Baltensperger, K., Bennett, M., Traas, J., Friml, J., and Kuhlemeier, C. (2003) Regulation of phyllotaxis by polar auxin transport. *Nature* 426: 255–260 **Figure 16.7** Lau, S., and Bergmann, D. C. (2012) Stomatal development: A plant's perspective on cell polarity, cell fate transitions and intercellular communication. *Development* 139: 3683–3692. **Figure 16.9** Lucas, W. J., Groover, A., Lichtenberger, R., Furuta, K., Yadav, S.-R., Helariutta, Y., He, X.-Q., Fukuda, H., Kang, J., Brady, S. M., et al. (2013) The plant vascular system: Evolution, development and functions. *J. Integr. Plant Biol.* 55: 294–388. **Figure 16.10A** Esau, K. (1942). Vascular

differentiation in the vegetative shoot of *Linum*. I. The procambium. *Am. J. Bot.* 29: 738–747. **Figure 16.10B** Esau, K. (1953) *Plant Anatomy*. Wiley, New York. **Figure 16.14** El-Showk, S., Ruonala, R., Helariutta, Y. (2013) Crossing paths: Cytokinin signalling and crosstalk. *Development* 140: 1373–1383. **Figure 16.15** Mason, M. G., Ross, J. J., Babst, B. A., Wienclaw, B. N., and Beveridge, C. A. (2014) Sugar demand, not auxin, is the initial regulator of apical dominance. *Proc. Natl. Acad. Sci. USA* 111: 6092–6097. **Figure 16.16** Morgan, D. C., and Smith, H. (1979) A systematic relationship between phytochrome-controlled development and species habitat, for plants grown in simulated natural irradiation. *Planta* 145: 253–258. **Figure 16.19A** Hochholdinger, F., and Tuberosa, R. (2009) Genetic and genomic dissection of maize root development and architecture. *Curr. Opin. Plant Biol.* 12: 172–177. **Figure 16.21** Péret, B., Desnos, T., Jost, R., Kanno, S., Berkowitz, O., and Nussaume, L. (2014) Root architecture responses: in search of phosphate. *Plant Physiol.* 166: 1713–1723. **Figure 16.22** Stahl, E. (1909) *Zur biologie des chlorophylls: Laubfarbe und himmelslicht, vergilbung und etiolement*. G. Fisher Verlag, Jena, Germany. **Figure 16.24** Keskitalo, J., Bergquist, G., Gardestrom, P., and Jansson, S. (2005) A cellular timetable of autumn senescence. *Plant Physiol.* 139: 1635–1648. **Figure 16.26** Breeze, E., Harrison, E., McHattie, S., Hughes, L., Hickman, R., Hill, C., Kiddle, S., Kim, Y.-S., Penfold, C. A., Jenkins, D., et al. (2011) High-resolution temporal profiling of transcripts during Arabidopsis leaf senescence reveals a distinct chronology of processes and regulation. *Plant Cell* 23: 873–894. **Figure 16.27** Mothes, K., Engelbrecht, L., and Schütte, H. (1961) Über die akkumulation von alpha-aminoisobuttersäure in blattgewebe unter dem einfluss von kinetin. *Physiol. Plant.* 14: 72–75. **Figure 16.29** Vahala, J., Ruonala, R., Keinänen, M., Tuominen, H., and Kangasjärvi, J. (2003) Ethylene insensitivity modulates ozone-induced cell death in birch (*Betula pendula*). *Plant Physiol.* 132: 185–195. **Figure 16.30** Morgan, P. W. (1984) Is ethylene the natural regulator of abscission? In *Ethylene: Biochemical, Physiological and Applied Aspects*, Y. Fuchs and E. Chalutz, eds., Martinus Nijhoff, The Hague, Netherlands, pp. 231–240.

CHAPTER 17

Table 17.1 Clark, J. R. (1983) Age-related changes in trees. *J. Arboriculture* 9: 201–205. **Figure 17.5** Purcell, O., Savery, N. J., Grierson, C. S., and di Bernardo, M. (2010) A comparative analysis of synthetic genetic oscillators. *J. R. Soc. Interface* 7: 1503-1524. **Figure 17.9** Vince-Prue, D. (1975) *Photoperiodism in Plants.* McGraw-Hill, London; Salisbury, F. B. (1963) Biological timing and hormone synthesis in flowering of *Xanthium*. *Planta* 49: 518–524; Papenfuss, H. D., and Salisbury, F. B. (1967) Aspects of clock resetting in flowering of *Xanthium*. *Plant Physiol.* 42: 1562–1568. **Figure 17.10** Coulter, M. W., and Hamner, K. C. (1964) Photoperiodic flowering response of Biloxi soybean in 72 hour cycles. *Plant Physiol.* 39: 848–856. **Figure 17.11** Hayama, R., and Coupland, G. (2004) The molecular basis of diversity in the photoperiodic flowering responses of Arabidopsis and rice. *Plant Physiol.* 135: 677–684. **Figure 17.13** Hendricks, S. B., and Siegelman, H. W. (1967) Phytochrome and photoperiodism in plants. *Comp. Biochem.* 27: 211–235; Saji, H., Vince-Prue, D., and Furuya, M. (1983) Studies on the photoreceptors for the promotion and inhibition of flowering in dark-grown seedlings of *Pharbitis nil* choisy. *Plant Cell Physiol.* 67: 1183–1189. **Figure 17.14** Trevaskis, B., Hemming, M. N., Dennis, E. S., and Peacock, W. J. (2007) The molecular basis of vernalization-induced flowering in cereals. *Trends Plant Sci.* 12: 352-357. **Figure 17.15** Purvis, O. N., and Gregory, F. G. (1952) Studies in vernalization of cereals. XII. The reversibility by high temperature of the vernalized condition in Petkus winter rye. *Ann. Bot.* 1: 569–592. **Figure 17.16** Liu, L., Zhu, Y., Shen, L., and Yu, H. (2013) Emerging insights into florigen transport. *Curr. Opin. Plant Biol.* 16: 607–613. **Figures 17.18 & 17.19** Bewley, J. D., Hempel, F. D., McCormick, S., and Zambryski, P. (2000) Reproductive Development. In: *Biochemistry and Molecular Biology of Plants*, B. B. Buchanan, W. Gruissem, and R. L. Jones, eds., American Society of Plant Biologists, Rockville, MD, pp. 988–1034. **Figure 17.20** Krizek, B. A., and Fletcher, J. C. (2005) Molecular mechanisms of flower development: An armchair guide. *Nat. Rev. Genet.* 6: 688–698. **Figure 17.24** Jones, R. L., Ougham, H., Thomas H., Waaland, S. (2013) *The Molecular Life of Plants*. Wiley-Blackwell. **Figure 17.27A,C,D** Seymour, G. B., Østergaard, L., Chapman, N. H., Knapp, S., and Martin, C. (2013) Fruit development and ripening. *Annu. Rev. Plant Biol.* 64: 219–241. **Figure 17.28B** Seymour, G. B., Østergaard, L., Chapman, N. H., Knapp, S., and Martin, C. (2013) Fruit development and ripening. *Annu. Rev. Plant Biol.* 64: 219–241; Pabón-Mora, N., and Litt, A. (2011) Comparative anatomical and developmental analysis of dry and fleshy fruits of Solanaceae. *Am. J. Bot.* 98: 1415–1436. **Figure 17.29** Fray, R. F., and Grierson, D. (1993) Identification and genetic analysis of normal and mutant phytoene synthase genes of tomato by sequencing, complementation and co-suppression. *Plant Mol. Biol.* 22: 589–602. **Figure 17.31** Grierson, D. (2013) Ethylene and the control of fruit ripening. In *The Molecular Biology and Biochemistry of Fruit Ripening*, G. B. Seymour, G. A. Tucker, M. Poole, and J. J. Giovannoni, eds., Wiley-Blackwell, Oxford, UK, p. 216; Oeller, P. W., Lu, M. W., Taylor, L. P., Pike, D. A., and Theologis, A. (1991) Reversible inhibition of tomato fruit senescence by anti-sense RNA. *Science* 254: 437–439. **Figure 17.32 Figure 17.32** Giovannoni, J. J. (2004) Genetic regulation of fruit development and ripening. *Plant Cell* 16 (suppl 1): S170–S180.

CHAPTER 18

Figure 18.1 van Dam, N. M. (2009) How plants cope with biotic interactions. *Plant Biol.* 11: 1–5. **Figure 18.2** Goh, C.-H., Veliz Vallejos, D. F., Nicotra, A. B., and Mathesius, U. (2013) The impact of beneficial plant-associated microbes on plant phenotypic plasticity. *J. Chem. Ecol.* 39: 826–839. **Figure 18.15** Christmann, A., and Grill, E. (2013) Electric defence. *Nature* 500: 404–405. **Figure 18.19** Boller, T., and Felix, G. (2009) A Renaissance of elicitors: Perception of microbe-associate molecular patterns and danger signals by pattern-recognition receptors. *Annu. Rev. Plant Biol.* 60: 379–406.

CHAPTER 19

Table 19.2 Lyons, J. M., Wheaton, T. A., and Pratt, H. K. (1964) Relationship between the physical nature of mitochondrial membranes and chilling sensitivity in plants. *Plant Physiol.* 39: 262–268. **Figure 19.3** Boyer, J. S. (1970) Leaf enlargement and metabolic rates in corn, soybean, and sunflower at various leaf water potentials. *Plant Physiol.* 46: 233–235. **Figure 19.4** Henry, C., Bledsoe, S. W., Griffiths, C. A., Kollman, A., Paul, M. J., Sakr, S., and Lagrimini, L. M. (2015) Differential role for trehalose metabolism in salt-stressed maize. *Plant Physiol.* 169: 1072–1089. **Figure 19.5** Mittler, R. (2006) Abiotic stress, the field environment and stress combination. *Trends Plant Sci.* 11: 15–19. **Figure 19.7** Baneyx, F., and Mujacic, M. (2004) Recombinant protein folding and misfolding in *Escherichia coli*. *Nat. Biotechnol.* 22: 1399–1408. **Figure 19.8** Mittler, R., Vanderauwera, S., Suzuki, N., Miller, G., Tognetti, V. B., Vandepoele, K., Gollery, G., Shulaev, V., and Van Breusegem, F. (2011) ROS signaling: The new wave? *Trends Plant Sci.* 16: 300–309. **Figure 19.9** Beardsell, M. F., and Cohen, D. (1975) Relationships between leaf water status, abscisic acid levels, and stomatal resistance in maize and sorghum. *Plant Physiol.* 56: 207–212. **Figure 19.10** Saab, I. N., Sharp, R. E., Pritchard, J., and Voetberg, G. S. (1990) Increased endogenous abscisic acid maintains primary root growth and inhibits shoot growth of maize seedlings at low water potentials. *Plant Physiol.* 93: 1329–1336. **Figure 19.14** Gutzat, R., and Mittelsten-Scheid, O. (2012) Epigenetic responses to stress: Triple defense? *Curr. Opin. Plant. Biol.* 15: 568–573.

Index

Page numbers in *italic* type indicate that the information will be found in an illustration.

About the Book

Editor: Andrew D. Sinauer
Project Editor: Laura Green
Production Manager: Christopher Small
Book Design: Ann Chiara
Cover Design: Beth Roberge Friedrichs
Photo Researcher: Mark Siddall
Copyeditor: Liz Pierson
Indexer: Grant Hackett
Illustrator: Elizabeth Morales
Book and Cover Manufacturer: LSC Communications